Outstanding Contributions to Logic

Volume 11

More information about this series at http://www.springer.com/series/10033

Can Başkent · Lawrence S. Moss
Ramaswamy Ramanujam
Editors

Rohit Parikh on Logic, Language and Society

Springer

Editors
Can Başkent
Department of Computer Science
University of Bath
Bath
UK

Ramaswamy Ramanujam
Department of Theoretical Computer Science
The Institute of Mathematical Sciences
Chennai, Tamil Nadu
India

Lawrence S. Moss
Department of Mathematics
Indiana University Bloomington
Bloomington, IN
USA

ISSN 2211-2758 ISSN 2211-2766 (electronic)
Outstanding Contributions to Logic
ISBN 978-3-319-47842-5 ISBN 978-3-319-47843-2 (eBook)
DOI 10.1007/978-3-319-47843-2

Library of Congress Control Number: 2016954021

Printed on acid-free paper

This Springer imprint is published by Springer Nature
The registered company is Springer International Publishing AG
The registered company address is: Gewerbestrasse 11, 6330 Cham, Switzerland

Cover and in-page photographs of Rohit Parikh: Lauren Fleishman

Editors' Preface

This volume is part of Springer's book series *Outstanding Contributions to Logic*. Without doubt the contributions made by Rohit Jivanlal Parikh to logic have been deep and outstanding. Rohit is a leader in the realm of ideas, offering concepts and definitions that enrich the field and lead to new research directions.

Rohit has contributed to a variety of areas in logic, computer science and game theory: in mathematical logic they have been in recursive function theory, proof theory and non-standard analysis; in computer science, in the areas of modal, temporal and dynamic logics of programs and semantics of programs, as well as logics of knowledge; in artificial intelligence in the area of belief revision; in game theory in the formal analysis of social procedures; in all this there is a strong undercurrent of philosophy as well. With such a wide-ranging set of contributions and ever-expanding intellectual interests, we have no possibility of summarizing his work or even take the rider for an extensive tour in Parikh-land. What we do is more in the spirit of the tour brochure, listing attractions 'not to be missed'.

Finally, this volume attempts to underline the academic trajectory of a brilliant scholar whose work opened up various new avenues in research. We will briefly discuss the milestones of Parikh's scholarly work, hoping to give a sense of how he has developed his recent research program *social software* and the formal and philosophical toolkit that lies behind it. We believe that such an impressive trajectory can only help motivate the young researchers and scholars.

As our mentor, colleague, co-author, and always as our teacher, Rohit taught us a lot. This volume is a simple but honest thank you to him from all of us.

A Brisk Tour of Parikh-Land

Parikh's (1966a) paper is a classic: it introduces the notion of *semi-linearity* and proves that commutative maps of context-free languages are semi-linear sets. The paper is a revised version of a 1961 MIT report, and it was published at the invitation of Donald Knuth. Today Parikh maps and Parikh vectors are such common usage in formal language theory that many papers use lower case as in 'parikh maps'. The paper also establishes the existence of inherently ambiguous context-free languages, which again led to a slew of results informal language theory.

Parikh (1966b) is an early paper of Parikh on *well-orderings*, a theme that would recur in his work over the next couple of decades. This paper shows that there are linear orders which are pseudo-well-ordered in that they have no recursive descending sequences, but exponentiating them yields orders that have primitive recursive descending sequences. A decade later, the paper (Parikh and de Jongh 1977) sets up a very interesting connection between hierarchies defined by closure operations and well-partial-orderings. A decade later, Parikh (1986) makes connections between this topic and two very different ones: formulas in logics of knowledge and automata theory. This paper shows precisely what levels of knowledge are achievable by a collection of communicating processes.

Parikh (1971) may well have launched the area of *bounded arithmetic*. 0 is a *small* number and whenever n is small, surely $n + 1$ is small. But a number like 10^{20} is of course not small. The paper then suggests that the length of proof establishing smallness of a number should be relevant (rather than unconstrained induction), and we can have an arithmetic that is conservative for proofs of low complexity. The paper also shows that there are formulae whose proofs are long, but the proof that they are provable could be short. Today, the study of systems of bounded arithmetic and their connection to complexity theory is a fertile area of research.

Parikh (1982, 1994) on *vagueness* are what we might call a typical Parikh phenomenon, addressing a problem of intense philosophical interest and offering conceptual guidance to those addressing it mathematically. Even more than the results, what the paper offers is clarity on concepts, and has influenced a generation of research on approximate and inexact reasoning.

The paper (Parikh 1978) and its companion (Kozen and Parikh 1981) provide an essential element that is now founded in the toolkit of every graduate student in logics of computation today. Following Kozen and Parikh, one proves a completeness theorem for the propositional dynamic logic of regular programs and combines this with a small model property leading to a decision procedure. This is perhaps the most authoritative handling of Kleene iteration in logic yet. Parikh went on to made extensive mathematical contributions to the study of dynamic and process logics.

Parikh (1983) is another landmark, a brilliant logicisation of two-person games that offered technical questions that remain open to this day. If logic can be given its meaning entirely in game theoretic terms, it is reasonable to ask what reasoning about existence of strategies in such games may mean. Game logics are a fertile area of study today.

Parikh (1984) connects logics of knowledge and non-monotonic reasoning in essential ways that would influence thinking in research on dynamic epistemic logics almost two decades later. In a series of papers over the next decade, Parikh has enriched the theory of knowledge in its interaction with communication and action, exploring intrinsic logical questions.

Moss and Parikh (1992) opens up another area, the study of topology via logic. The use of epistemic logic in this context enriches and expands our understanding of topological spaces and this study flourishes to date.

Parikh (1995) formally launches Parikh's idea of *social software*, one that has kept him creatively occupied for the last two decades, enriching interaction between logic and game theory. Social software is the analysis of social procedures by exploring their logical underpinnings. For instance, notions like social obligation, why politicians lie during election campaigns, can be studied logically.

Parikh (1999) is another landmark, this time offering a new technique, that of language splitting, in the extensively studied area of belief revision. The paper shows how we can incorporate a formal notion of relevance, which allows one's information to be split uniquely into a number of disjoint subject areas. This idea has been subsequently been extended fruitfully by Makinson and others, leading to many new directions of research.

Through all this, Parikh's contributions to fundamental logic (e.g. Parikh and Väänänen 2005) and mathematics (Parikh and Nathanson 2007) continue. His recent contributions to philosophical thought (e.g. Parikh 2013; Ginés and Parikh 2015) in the arena of logic, games, language and computation, raise a number of conceptual issues and offer new approaches which will guide research for a long time to come.

This Volume

The contributors to this volume honor Rohit Parikh and his *oeuvre* in many ways. This is not a collection of articles on one theme, or even directly connected to specific works of Parikh, but inspired and influenced by Parikh in some way, adding structures to Parikh-land, enriching it. Our goal in what follows is to say a few brief words about each of the contributions.

Juliet Floyd illustrates the transition in Parikh's interest from formal languages to natural languages, and how Parikh approached Wittgenstein's philosophy of language. In fact, the article describes, as Floyd put it why "Wittgenstein owes Parikh a big 'Thank You'."

Prashant Parikh notes that "Rohit Parikh may have been the first person to study how communication with vague expressions can be fruitful even when the speaker and addressee interpret them differently". Prashant's paper continues this line of work, introducing models from cognitive psychology to analyze vague communication.

Robert van Rooij's paper employs non-classical logic in an analysis of a well-known epistemic paradox, Fitch's Paradox. The topic is attractive for both philosophers and epistemic logicians alike, and van Rooij's contribution hopefully will initiate more interaction between these relatively separate research communities.

If we wish to take seriously Parkih's idea of social software, *money* would seem to be a central social and computational resource. Jan van Eijck and Philip Elsas orchestrate a thought-provoking Socratic dialogue on the function of money, its organizing role in society and its underlying logical principles.

Dominik Klein and Eric Pacuit focus on another important aspect of social software, and one of Parikh's recent interests: voting and political campaigns. Klein and Pacuit discuss a qualitative analysis of voters' changing opinions during a political campaign. Such analyses will be crucial to advance Parikh's program of analysis of social procedures.

Can Başkent discusses the role of classical logic in the social sofware enterprise, and offers an interesting extension of social software to paraconsistent logic, arguing that non-classical logics provide the theme with a broader domain of applications. The paper follows the footsteps of Parikh (2002) and takes a logical pluralist stand on the subject.

Epistemic logic is a central arena of Parikh's work, and much of his work is on delineating notions of knowledge, syntactically and semantically. Knowing a proposition is distinct from the sentential knowledge of the proposition. While economists in general prefer to work directly with propositions at the model level, logicians prefer to work with compositional syntax. This distinction runs through Parikh's work (Parikh 2005) and the paper by Joseph Halpern addresses the issue. He gives strong arguments why syntax can help make finer distinctions and describe notions in a model-independent manner.

Another contribution to this volume on knowledge is Johan van Benthem's exploration of epistemology. The paper is foundational, a far-reaching exploration of a number of themes weaving epistemology, dynamic logic, information, modality, and action. Many of these themes run through Parikh's oeuvre as well. (It should be noted here that Parikh was an early enthusiast for Jan Plaza's work, and that work itself is a major forerunner of dynamic epistemic logic.)

A central lesson of Parikh's work on knowledge is that communication and knowledge ought to be studied together. In their paper on *Gossip protocols*, Maduka Attamah, Hans van Ditmarsch, Davide Grossi and Wiebe van der Hoek consider agents who exchange private information about others over a telephone network. They study how many message exchanges are needed for a particular state of knowledge to be achieved.

Analyzing puzzles through logic and games is an endearing component of Parikh's style, and Sandu and Velica's paper in this volume pays tribute to it. Hintikka and Sandu (1989) introduced independence-friendly (IF) logic in order to express patterns of dependencies among quantifiers which go beyond what can be expressed in first-order logic. The paper by Gabriel Sandu and Silviu Velica offers a formulation of the Monty Hall puzzle in IF logic via a game-based modeling of the problem. In the process, they endow IF logic with a probabilistic semantics.

Amy Greenwald, Jiacui Li and Eric Sodomka establish a formally appealing connection between a specific presentation of games (called *extensive normal form games*) and a process of randomized decisions. Uniting two different approaches to decision making provides us with a broader understanding of game theoretical decision processes.

Similarly, Jouko Väänänen discusses an iteration of a logical framework which arises from a situation of limited information. The solution is given within the context of dependence logic with team semantics.

Melvin Fitting's paper in this volume is a contribution to an important issue in modal logic, the relation between *intensions* and *extensions*. His paper proposes an appealing formal treatment of *predicate abstraction*. This paper not only discusses examples, it presents a formal proof system which is sure to be of independent interest.

Konstantinos Georgatos' paper on *epistemic conditionals* addresses belief revision, an important topic in artificial intelligence and related areas of philosophical logic. Yet another connection to Parikh's work comes in this paper's use of the logic of *subset spaces*, a topic founded by Parikh and Moss.

Dexter Kozen's contribution to this volume is a contribution to *Kleene algebra*, an area pioneered by Kozen himself. Kleene algebras may be thought of as algebraic structures, which generalize and illuminate the algebra of regular expressions. Thus, it harks back to Parikh's interest in formal language theory and dynamic logic.

Vaughan Pratt's paper shares with Kozen's a decidedly algebraic flavor. In fact, Pratt's paper takes up several topics relating category theory and syllogistic logic. Although at first this seems an unlikely match, Pratt shows how algebraic perspectives can illuminate the technical sides of logical systems which we thought we knew well.

Noson Yanofsky discusses a deeply foundational issue in theoretical computer science: algorithms. Parikh's work on the logic of programs is carried to a more abstract level. The relation between programs and algorithms is discussed using a novel category theoretic approach.

All Aboard

We hope to have presented you with a brochure-view of Parikh-land and then given an "introductory video" on the sights and sounds that you will experience when reading the book. We now invite you to board the bus. The drivers are grateful to all the authors for their contributions, and to the Series Editor Sven Ove Hansson for giving us the opportunity for this wonderful ride.

Bath, UK — Can Başkent
Bloomington, USA — Lawrence S. Moss
Chennai, India — Ramaswamy Ramanujam

References

Ginés, A. S., & Parikh, R. (2015). A pragmatistic approach to propositional knowledge based on the successful behavior of belief. In *Logic and Its Applications—6th Indian Conference, ICLA 2015, Mumbai, India, January 8–10, 2015. Proceedings*, pp. 146–157.

Hintikka, J., & Sandu, G. (1989). Information independence as a semantical phenomenon. In Fenstad, J. E., Frolov, I. T., & Hilpinen, R. (Eds.), *Logic, Methodology and Philosophy of Science VIII* (pp. 571–589). Elsevier.

Kozen, D., & Parikh, R. (1981). An elementary proof of the completness of PDL. *Theoretical Computer Science*, *14*, 113–118.

Moss, L. S., & Parikh, R. (1992). Topological reasoning and the logic of knowledge. In *Proceedings of the 4th conference on Theoretical Aspects of Reasoning about Knowledge, Monterey, CA, March 1992*, pp. 95–105.

Parikh, R. (1966a). On context-free languages. *Journal of the ACM*, *13*(4), 570–581.

Parikh, R. (1966b). Some generalisations of the notion of well ordering. *Zeit. Math. Logik und Grund. Math.*, *12*, 333–340.

Parikh, R. (1971). Existence and feasibility in arithmetic. *Journal of Symbolic Logic*, *36*(3), 494–508.

Parikh, R. (1978). The completeness of propositional dynamic logic. In *Mathematical Foundations of Computer Science 1978, Proceedings, 7th Symposium, Zakopane, Poland, September 4–8, 1978*, pp. 403–415.

Parikh, R. (1982). The problem of vague predicates. In Cohen, R. S. & Wartofsky, M. W. (Eds.), *Logic, Language and Method, volume 31 of Boston Studies in the Philosophy of Science*. D. Reidel Publishing.

Parikh, R. (1983). Propositional game logic. In *24th Annual symposium on Foundations of Computer Science, Tucson, Arizona, USA, 7–9 November 1983*, pp. 195–200.

Parikh, R. (1984). Logics of knowledge, games and dynamic logic. In *Foundations of Software Technology and Theoretical Computer Science, Fourth Conference, Bangalore, India, December 13–15, 1984, Proceedings*, pp. 202–222.

Parikh, R. (1986). Levels of knowledge in distributed computing. In *Proceedings of the First Annual IEEE Symposium on Logic in Computer Science* (pp. 314–321). LICS 1986, IEEE Computer Society Press.

Parikh, R. (1994). Vagueness and utility: The semantics of common nouns. *Linguistics and Philosophy*, *17*(6), 521–535.

Parikh, R. (1995). Language as social software. In *International Congress on Logic, Methodology and Philosophy of Science*, p. 417.

Parikh, R. (1999). Beliefs, belief revision and splitting languages. In Moss, L. S., Ginzburg, J. & de Rijke, M. (Eds.), *Logic, Language and Computation* (pp. 266–278). CSLI.

Parikh, R. (2002). Social software. *Synthese*, *132*(3), 187–211.

Parikh, R. (2005). Logical omniscience and common knowledge: What do we know and what do we know? In van der Meyden, R. (Ed.), *Proceedings of the 10th conference on Theoretical aspects of rationality and knowledge* (pp. 62–77). TARK 2005.

Parikh, R. (2013). On kripke's puzzle about time and thought. In *Logic and Its Applications, 5th Indian Conference, ICLA 2013, Chennai, India, January 10–12, 2013*. Proceedings, pp. 121–126.

Parikh, R., & de Jongh, D. (1977). Well partial orderings and hieararchies. *Proc. Kon. Ned. Akad. Scie, Series A*, *80*, 195–207.

Parikh, R., & Nathanson, M. (2007). Density of natural numbers and levy group. *The Journal of Number Theory*, *124*, 151–158.

Parikh, R., & Väänänen, J. A. (2005). Finite information logic. *Annals of Pure and Applied Logic*, *134*(1), 83–93.

Contents

Chapter 1
Parikh and Wittgenstein

Juliet Floyd

Abstract A survey of Parikh's philosophical appropriations of Wittgensteinian themes, placed into historical context against the backdrop of Turing's famous paper, "On computable numbers, with an application to the *Entscheidungsproblem*" (Turing in Proc Lond Math Soc 2(42): 230–265, 1936/1937) and its connections with Wittgenstein and the foundations of mathematics. Characterizing Parikh's contributions to the interaction between logic and philosophy at its foundations, we argue that his work gives the lie to recent presentations of Wittgenstein's so-called metaphilosophy (e.g., Horwich in Wittgenstein's metaphilosophy. Oxford University Press, Oxford, 2012) as a kind of "dead end" quietism. From early work on the idea of a feasibility in arithmetic (Parikh in J Symb Log 36(3):494–508, 1971) and vagueness (Parikh in Logic, language and method. Reidel, Boston, pp 241–261, 1983) to his more recent program in social software (Parikh in Advances in modal logic, vol 2. CSLI Publications, Stanford, pp 381–400, 2001a), Parikh's work encompasses and touches upon many foundational issues in epistemology, philosophy of logic, philosophy of language, and value theory. But it expresses a unified philosophical point of view. In his most recent work, questions about public and private languages, opportunity spaces, strategic voting, non-monotonic inference and knowledge in literature provide a remarkable series of suggestions about how to present issues of fundamental importance in theoretical computer science as serious *philosophical* issues.

Keywords Social software · Wittgenstein · Turing · Parikh · Common knowledge

1.1 Introduction

The influence of Wittgenstein's later philosophy on Rohit Parikh's work has been thoroughgoing, quite distinctive, and partly explains the creativity and breadth of

J. Floyd (✉)
Department of Philosophy, Boston University,
745 Commonwealth Avenue, Boston, MA, USA
e-mail: jfloyd@bu.edu

C. Başkent et al. (eds.), *Rohit Parikh on Logic, Language and Society*,
Outstanding Contributions to Logic 11, DOI 10.1007/978-3-319-47843-2_1

Parikh's contributions. This essay surveys the long and brilliant career of his appropriations of Wittgensteinian themes, using this to show why it is that his biggest notion, that of *social software*, should be attended to by mathematicians, philosophers, and computer scientists.

Of course there are many Wittgensteins, just as there are many Peirces, many Kants, many Platos. The best advice here is to think *with* a philosopher, rather than primarily *about* that philosopher; to make as good sense of as much of the thought as you can, given its context; and to see how the best parts might be projected forward in your own thinking. This is what Parikh has done, aided and abetted by four twentieth century philosophers from whom he has learned a great deal: W.V. Quine, Burton Dreben, Nelson Goodman, and Hilary Putnam.[1] For Parikh, in a very distinctive way, Wittgenstein isn't the subject matter, he's the way through.

So in what follows I shall not debate interpretations, so much as try to say why I think Parikh's Wittgenstein is worthwhile. The important point is that he is putting Wittgensteinian insights to *use*. If philosophy lives and dies by its applications, then that is a very important thing. In fact I believe Wittgenstein owes Parikh a big "Thank You", no matter if the biographical Wittgenstein would have balked at the appropriation and development of his ideas under the name "Wittgenstein". The real, historical Wittgenstein, obsessed with controlling his out-of-control vanity, worried that all his writing and lecturing might do would be to sow the seeds of a certain jargon, and he feared and hated the humiliation of anyone writing about *him*.[2] But he did still dare to hope that someone else would draw out and apply his thinking on the foundations of mathematics.[3] That hope has not been in vain.

From Parikh's earliest work on the idea of "feasibility" in arithmetic (Parikh 1971), through his work on vagueness (Parikh 1983, 1994, 1996b) and on social software (Parikh 2001a, 2002), he has touched on many foundational issues in epistemology, philosophy of logic and mathematics, philosophy of language, and value theory. But his work embodies and develops a unified philosophical point of view, particularly about what "foundations" really are (and are not). Although in a broad sense the point of view he has developed belongs in the "finitist" tradition, it offers a surprising number of undogmatic series of new twists, reaching horizontally, and not just vertically, and thereby going to the heart of what we *mean* by a "foundation". In general, Parikh has suggested that we learn how to replace the sort of familiar objections to finitism, and to Wittgenstein, namely,

> But isn't the statement, concerning a particular system of logistic, that "*there is no contradiction with fewer than* 10^{100} *lines*" true or false? This is not just a question of our "form of life"![4]

---

[1] Strictly speaking, Parikh studied officially only with Quine and Dreben, for he left Harvard in 1961 (cf. Parikh 1996a, p. 89). We might also add to the list of philosophers who have influenced him Peirce and Ramsey. If we did, that might well explain the attraction Parikh's work has had for the present author, who was similarly influenced and educated, at least in part, two generations later on.

[2] Monk and Wittgenstein (1990), Chap. 24.

[3] Monk and Wittgenstein (1990), Chap. 22.

[4] Putnam (1979), p. 511, n. 4.

with questions such as the following—and here I distill the subject matter of many of Parikh's papers, without quoting precisely from him:

> Can we rigorously formalize aspects of our "form of life"—
> including the idea of a "feasible" formalization? In what sense of "works" is it that we may hope to find a philosophy of mathematics (and of knowledge) that *does* "work"? What is knowledge really like?

Parikh thus serves as a counterexample to Paul Horwich's recent claim in *The New York Times* that Wittgenstein's thought today is ignored by "all a small and ignored clique of hard-core supporters" (Horwich 2013).[5] In fact, given the interdisciplinary breadth of Parikh's influence, and his lack of dogmatism, Parikh shows that Wittgenstein left us something much more important than a *philosophie du jour* or a method of exposing nonsense or a great name to add to a pantheon; rather, he left us open problems and a *way of thinking* that is constructive, pointing a way forward.[6] Like much philosophy, Parikh's lies in a long and developing, if ever-contested tradition. He has specified a vision of a certain kind. Even if the vision is different from others, and may not even *look* to some like a vision, it is one.

In his most recent work, Parikh is surveying and analyzing questions about public and private languages, the limits of formalization, opportunity spaces, elections and social spaces of coordination, knowledge's effects on obligations, truthfulness and sincerity, and knowledge in literature—thereby drawing together game theory, logic, and computer science into a remarkable series of (rigorously presented) bouquets. Thereby he often shows, quite critically, how certain ideas that people think go together, ideas that appear to give us an ultimate foundation (say, of *homo economicus*, for example, or our notions of *truth* and *belief* and *meaning*) do not, but are better seen as "preliminary" analyses (Parikh 2007, Answer 3), embedded in collections of other structures and systems, in which the characterization of a total collection is not the point, so much as the study of interactions among them in ordinary life, which includes the slippages, bendings, and breakings that go on.

Parikh's mathematical facility—as well as his conviviality and generosity as a teacher and colleague—have led to a constant stream of interesting and fruitful putting-togethers and rearrangements of heretofore disparate areas of logic and the theory of "rationality": proof theory and bounded arithmetic (Parikh 1971, 1998), temporal logic and social levels (Parikh 2001a, 2003), Bayesian probability theory and defeasible inference (Arló-Costa and Parikh 2005), epistemic and dynamic epistemic logic (Parikh 2006, 2008), modal, deontic, and finite information logics (Parikh and Väänänen 2005; Parikh 2006; Pacuit et al. 2006; Baskent 2012), belief revision

---

[5]In a recent YouTube debate with Timothy Williamson http://www.youtube.com/\penalty\exhyphenpenaltywatch?v=IpOmFTRcwUM), Horwich appealed to problems about logical omniscience to draw out criticisms of taking quantified modal logic without several grains of salt. It is unfortunate that he didn't take the next step, which would have been to point out the positive results and work that is going on as a result of those criticisms, rooted in Wittgenstein himself. However, this would have contradicted his own reading of Wittgenstein as having an essentially negative "metaphilosophy".

[6]Cf. Collins (2013).

theory, relevance and topology (Chopra and Parikh 2000; Chopra et al. 2001; Parikh 2001b; Dabrowski et al. 1996), electoral and political theory (Pacuit and Rohit 2005; Dean and Parikh 2011), and even literature and life (Parikh 2009b, forthcoming). Juxtaposing these traditions of research with one another by asking philosophical questions yields interesting accounts of tensions and presuppositions among them.

Yet Parikh's papers, while mathematically creative and rigorous, also provide a remarkable series of suggestions about how to present issues of fundamental importance in computer science *as* serious philosophical and mathematical issues: perhaps most important, *ways of thinking* that may be accepted, changed, confronted, or turned, allowing us to rethink and perhaps change our lives.

It is from this, above all, that emerges the characteristic Wittgensteinian touch within Parikh's work: he is able to see philosophical problems where others see nothing but "rags and dust" (*Philosophical Investigations* (Wittgenstein 2009) (hereafter "PI") §52). There are only a few of his papers in which Parikh takes Wittgenstein up as an explicit topic (cf. Parikh 1995; Parikh and Renero, forthcoming), and even in these he is not doing straight exposition, but making a series of points: drawing connections, working out analogies, testing the limits of other analogies.

Why are the analogies so important? Because analogies always lie at the basis of any rigorization, and one can fail to see something as philosophical when it truly is so. One can also mistake a problem in one's analogical thinking for a metaphysical literalism, eclipsing the root of the matter. So the frequent and informed allusions and quotations Parikh often adduces to Wittgenstein's remarks should be taken seriously, as part of what he is doing, and not merely decorative or literary asides. They make a series of points that economists, philosophers, game theorists and theoretical computer scientists should listen to.

First of all, as Parikh has himself pointed out, logic in the broad sense of critical reflection on rationality and discourse is an activity, not a particular standpoint or theory, and in this sense belongs to many traditions.[7] (As Quine and Wittgenstein might have said, each sounding a rather different note, logic in the broadest sense is part and parcel of having any view at all, it is not just one optional point of view.) Yet secondly, and more specifically, Parikh's work stands in close proximity to a certain quite particular philosophical tradition of reflection on foundations, stemming not only from the classical works of Gödel and Turing in the 1930s, but also from the earliest reception of Wittgenstein's later writings on mathematics. I shall be emphasizing this stage-setting in what follows, in order to make the case that there is something philosophical at stake in every paper Parikh writes.

Wittgenstein's reception began in public space, as a piece of "social software", in 1956, when the first edition of *Remarks on the Foundations of Mathematics* (Wittgenstein 1978, hereafter "RFM") first appeared. This book was heavily and posthumously reconstructed by Wittgenstein's literary executors from manuscripts never intended for publication, aided and abetted in their editorial task by Kreisel, an important interlocutor of Wittgenstein's (and Parikh's). (Afterwards Kreisel reviewed the book, calling it "the surprisingly insignificant product of a sparkling mind", perhaps to cover

[7] See the discussions of Indian sagas in Parikh's (2009b).

himself.[8]) Dummett (1959) saw numerous mistakes and errors in RFM, especially in the remarks on Gödel.[9] (These remarks, suffering from a variety of weaknesses, had been eliminated by Wittgenstein himself from all drafts of the PI that appeared after 1937.)

Bernays's somewhat less critical review of RFM (Bernays 1959), though it still accused Wittgenstein of embracing nihilism and irrationality, did note that one might see emerging from the scattered remarks a point of view, which he called "anthropological". Hao Wang, partly on the basis of conversations with Gödel, wrote several papers on this perspective, and associated it with his own interest in automated theorem proving.[10] Nevertheless, at least initially, Wang was proud of never mentioning Wittgenstein at all (cf. e.g., Wang 1961): the *brouhaha* of enthusiasm for the famed philosopher during this early period of the 1950 s was oppressive, and like Hilary Putnam and John Rawls, Wang resisted labels, certainly not wanting to become known as a "Wittgensteinian" (cf. my Floyd 2011). Parikh, entering the philosophical scene later on, has felt it important, instead, to develop Wittgenstein's ideas quite explicitly, and this under the rubric of "common sense" (Parikh 1971, p. 494).

Since then, Dummett and others developed the point of view of "strict finitism" (Dummett 1975; Wright 1980, 1993), and Kripke wrote his famed book on Wittgenstein's rule-following remarks (Kripke 1982). These made somewhat more kosher to the wider philosophical public at least some of Wittgenstein's writings on mathematics. But outside of these writers, there has not been a more serious appropriator of this reading of Wittgenstein's ideas about logic, foundations, and mathematics than Parikh. Actually Parikh has done far more than any of these authors to spread and develop the "anthropologism" associated with Wittgenstein. For one thing, he is far more sympathetic to the literal usefulness of Wittgenstein's ways of thinking. For another, he has absorbed the anti-dogmatic side of Wittgenstein more deeply. For a third, his facility as a proof theorist developed very early, in a very focussed way, building on Gödel's pioneering work on speed up theorems, where tradeoffs between how we represent the length of a proof, a "step", or a symbol are placed front and center as part of mathematics (Parikh 1986). For a fourth, he was immersed at the early stages in developing proof and complexity theory as part and parcel of theoretical computer science—a turn anticipated by Wang, but only dimly.

Wang (1996) took the interaction between Gödel and Wittgenstein to have been one of the most important in twentieth century philosophy, but he sided with Gödel and Bernays in regarding "anthropologism" as a limited and partial, though possibly entertainable, point of view. He also attempted, following Gödel, to develop a notion of "intuition" of concepts. Parikh, by contrast, has no use for "intuition" in any

[8]See Parikh's reminiscence of Kreisel in his (Parikh 1996a).

[9]For a survey of reactions and a response, see Floyd (2001).

[10]Wang (1996), p. 214; Wang (1961). As we are now beginning to learn (Floyd and Kanamori 2016), Gödel was thoroughly immersed in Russell's philosophy in 1942–1943, and thoroughly involved in combatting what he took, rightly, to have been the constructivistic effects of Wittgenstein on the second edition of *Principia*. Gödel himself wrote of an "anthropological" sense of truth in his notebooks, and one may conjecture that although the later Wittgenstein's writings were not known to him then, he discussed this idea with Bernays.

classical philosophical sense, but works with common sense examples in a strategic way, developing "anthropologism" to see just how far it can go. He has shown that it can go very far. Not of course by contradicting the classical results of Gödel, without which Parikh's earliest foundational work in proof theory would have been impossible. But working against this whole approach when it is conceived of as the *only* foundational way. Parikh is a kind of conceptual and ethical pluralist in his very bones, and in this sense a true student, both of Quine and Putnam, if not also the Buddha. He is also a realist, in Cora Diamond's a-metaphysical sense of someone who is *realistic* about what logic is and can do (Diamond 1991).

Gödel's completeness theorem gave us the non-standard models of arithmetic Parikh would later exploit as engines of semantic development. It also laid into neat conceptual space the classical consequence relation, on which so many later results (Gödel's incompleteness theorems included) turned. Classical first order logic tells how logic would be, if things were simple.[11] Parikh urges us to develop logic in the face of the fact that things are not so simple. Better and more deeply put, Parikh's work emphasizes that what counts as "simplicity" is relative to who and where we are, and whom we are with, and how we incur obligations and draw consequences differently in the face of particular acts of speaking. This is explicitly a theme, not only in Plato, but also in the later Wittgenstein.[12]

Tarski's analysis of truth for formalized languages not only served the development of model and set theory, it also allowed philosophers to skirt, through the schematic approach to structures, head-on confrontation with the distinction between the infinite and the finite in their syntax for treating the quantifiers. For Tarski's schematic, metatheoretic analysis of truth in formalized languages is neutral with respect to the size of the domain, as well it should be in connection with the formal aspect of truth in general, as it serves us in the development of a general notion of definability.

But, as Parikh shows, analogous versions of the old problems keep on rearing their heads throughout logic. In a sense, even to claim that

> From two integers k, l one passes immediately to k^l; this process leads in a few steps to numbers which are far larger than any occurring in experience, e.g., $67^{(257^{729})}$.

is in the end "an application of the general method of analogy, consisting in extending to inaccessible numbers the relations which we can concretely verify for accessible numbers" as Bernays had said (Parikh 1971, p. 494). Parikh showed (Parikh 1971) that the analogy could be cashed in differently, that the intuitive notions of "feasible" and "reasonable length" are rigorizable, and the complexity of proofs in that sense may be made mathematically rigorous. As Buss has said, this work (Parikh 1971, 1973) was "seminal and influential and led to large research areas which are still active and

---

[11] I owe this way of putting the matter to Colin McClarty, in conversation.

[12] Cf. PI §§48ff, which allude to Plato's *Theaetetus* discussion of whether knowledge can be reduced to perception. For an explicit connection with Turing, see Parikh and Renero (forthcoming). As Floyd and Kanamori (2016) and Floyd (forthcoming) show, Gödel himself discussed the *Theaetetus* in his Max Phil notebooks, 1942–1943, following Russell (1940, pp. 116ff).

fruitful 25 years later" (Buss 1999, p. 43).[13] It also showed, philosophically speaking, that the *conventional* answer to the question

> Does the Bernays number $67^{(257^{729})}$ actually belong to every set containing 0 and closed under the successor function?

—viz., "Yes"—may still be accepted. Nevertheless,

> ...we have seen that there is a very large element of phantasy in conventional mathematics which one may accept if one finds it pleasant, but which one could equally sensibly (perhaps more sensibly) reject (Parikh 1971, p. 507).

Going further, into the foundations of logic itself, we are not—individually or collectively—logically omniscient, either about the logical consequences of our thoughts or one another's points of view. We can of course study how things would be if we were, and develop notions of *proposition* using either Frege's structured thought approach, or the Hintikka-Lewis-Stalnaker idea of propositions as sets of possible worlds. But, like Wittgenstein, Parikh's suggestion is to complicate the logic so as to keep it, and the notion of "correctness", applicable, and show how conceptually relative are classical idealizations.[14] This sheds light on what knowledge and logic and truth are, and not merely on how they might be, if things were simple.

1.2 Surveyability (*Übersichtlichkeit*)

In looking at Wittgenstein's RFM, Parikh was struck by how many questions bearing on fundamentals of theoretical computer science were already being explored in Wittgenstein's writings from the 1930s (Parikh 1995, p. 92):

> I would like to make the case that people in AI who are actually unaware of much of what Wittgenstein says on this issue are in fact actually carrying out some of what might have been his program, namely that if you look at various activities that are going on right now in Artificial Intelligence and Logics oriented towards it, then you'll find that very many of these developed theories can be seen as expansions of relatively off-hand remarks that Wittgenstein makes.
>
> A rather interesting historical fact is that when the RFM first came out in the early 1950s, Complexity Theory as we know it now did not exist. The book was criticized very sharply by Kreisel (1958). It was criticized also by Dummett (1959), though somewhat more mildly. Now, whether it was because of this or simply because the time was not right, RFM was not taken as seriously as the PI was.
>
> If knowing Complexity Theory one goes back and reads the RFM, then what are impressive are the things that he says which could not make sense in the 50's but which make perfect sense now.

[13] For example, Sazonov (1995).

[14] Parikh (1971) acknowledges earlier work on the notion of "feasible" by Esenine (1970), but states (p. 494, n. 1) that Parikh "preferred the more conservative approach of using 'standard' methods for [his] technical proofs".

There are some historical reasons for this, and they confirm Parikh's sense of what is going on. They are important to appreciate philosophically. For the history shows that Parikh's questions were always part and parcel of the classical tradition in logic and foundations, as well as the foundations of computer science and artificial intelligence, even before they emerged as separate departments in universities, or separate ventures, and even though philosophers tended to carve them off from what was called "philosophy" of "foundations of mathematics" as the twentieth century progressed. The issues may have been forgotten or overlooked while logicians worked out the fundamental classical notions of *truth*, *definability*, and *logical consequence*. But they were always there, entangled with that classical project.

The notion of *Übersichtlichkeit* as a developed philosophical idea in connection with mathematical proof derives from Wittgenstein's writings of 1939–1940, and, significantly, precisely those manuscripts that were written in the wake of his conversations with Alan Turing in 1937 and 1939 (cf. Floyd 2001, 2012, forthcoming). Turing personally sent off only five offprints of his famous paper "On Computable Numbers, with an application to the *Entscheidungsproblem*" (Turing 1936/1937) in the first round, and one was to Wittgenstein, who was then trying to put together the PI.[15] Wittgenstein's interest in the machine as *symbolizing its own action*—famously explored in PI §§193–194— probably derives, in part, from his pondering the relation of Turing's results to Gödel's. What is clear is that after discussions with Turing the notion of *Übersichtlichkeit* springs to the fore of his thinking about the nature of logic and proof in mathematics, and is brought to bear in his criticisms of the idea that *Principia Mathematica* can serve as a "foundation" of arithmetic in logic. Wittgenstein also became interested in the context-bound status of the distinction between the notions of *proof*, *calculation*, and *experiment* after discussions with Turing.[16]

In RFM Wittgenstein appears to offer a version of the Poincaré objection to logicism: because formal *Principia* proofs and terms are not always surveyable, we need to use mathematical induction in setting out the very formulae of the system, thereby helping ourselves to arithmetic and counting already at the outset, in trying to "take in" a formal proof.[17] In working through a proof, for example, we may need to index and count the variables, draw lines in the margins connecting parts with one another, and so on. But if these pieces of "software" are necessary for us to take in the proof, then they are no less working parts of the proof than the formalism, and then

[15]Turing to his mother February 11, 1947, from Princeton [AMT/K/1/54, at the King's College archives (http://www.turingarchive.org/viewer/?id=414\&title=54)].

[16]Contrary to what many have assumed, Wittgenstein did not read Turing's famous paper on the "Turing Test" for intelligence (Turing 1950), at least until the very end of his life, when there is no evidence of its having influenced him. This is shown by his letter to Malcolm of January 12, 1950, where he tells Malcolm he has not read it, but states, "I imagine it is no leg pull" (Malcolm and Wittgenstein 2001, pp. 129–130, also in Wittgenstein 2005). Wittgenstein did discuss the question, "Can a machine think?" as early as 1926 with Schlick. This was because Russell asked the question already in his (Russell 1921) Lecture XIII, "On Truth and Falsehood", arguing that while there is no difficulty in attributing to a machine "correctness" of response, the issue of truth is more complex.

[17]Parikh (1971, p. 502) already interprets the objection in terms of what is called there an "anthropomorphic system" of bounded arithmetic, in which the notion of "feasibility" is rigorized.

there is no true "reduction" of arithmetic to logic. Instead we are using mathematics to make things "surveyable" or *übersichtlich*, just as Hilbert would have said we must.

The difficulty with this objection, of course, is not to have it bring in irrelevancies of psychology into discussions of the foundations of logic and mathematics.[18] Frege and Russell aimed to extrude psychology altogether from their respective conceptions of justification, and Wittgenstein followed them here. Justifications do not reduce to strikings, experiences, acts of acquaintance, or seemings: we cannot experience in that way the "because" (PI § 176). Nevertheless, if mathematics itself is necessary for us to take logic in, then it is arguably a matter of logic, and not merely psychology, that "a proof must be surveyable" (*übersichtlich, übersehbar, überblickbar*) (RFM III §§1,21–22,39,55; IV §41).

The word "*Übersicht*" is difficult to translate; some readers like the term "perspicuous" or "open to view", or "surview", but this can imply that one has achieved a kind of mountain-top angle on things, and is able to take in all details in a glance, and understand the place of all of them within a larger whole. This is not Wittgenstein's meaning, and as has been argued by Mühlhölzer (2006), a better translation would be "surveyable". When Wittgenstein writes that one of his methods in philosophy is to present us with *übersichtliche Darstellungen*, or "surveyable representations" of grammar, he means that he is providing snapshots of portions of human linguistic activity designed to let us work through them usefully, to survey them in that sense. What matters is, first of all, that the proof can be easily reproduced or copied, "in the manner of a picture": a proof must be communicable, easily reidentifiable, recognizable, as *this* proof and not another, it must not be an experiment each time it is ventured, but rather it must serve as a kind of *calculation* in the sense that it entangles us with what *has* to come out, if the procedure is correctly followed (cf. RFM I §§1,22,39,55). Thus what also matters, second, is that a proof is a proof without something (some thing) behind it: it shows *itself* as a proof, no one general foundational theory of it being needed.

Let us generalize this idea of *Übersichtlichkeit*. The saying "I only know my way about like a rat in a maze" voices an idea Parikh has elaborated for many years: the rat may well know the right cues about when to turn left or right, but lack an *overview* or *map* of the situation—anyway, a map that she can use to see how to read other maps, using maps as members of a category. Deferring a fuller discussion of animal consciousness as such to §4 below, let us consider ourselves, as human beings with our widely shared capabilities. As a human with faculties and experience enough, I can easily memorize one or two station changes on the Boston MBTA transportation line. But holding in mind the whole, so as to make creative or novel choices, requires a model or structure that I am capable of projecting in a variety of ways. Fortunately, officials provide posters of the surview in all stations, and make it available on the web, and the map is comprehensive enough to do the job. In this regard, more than its *accuracy*, we should emphasize the map's *usefulness* in being the sort of representation that can be taken in just a few seconds, followed and discussed by

[18] On this see Goldfarb (1985) and Stenlund (1996).

most of us (including those who may not speak very fluent English), easily posted on the wall of a train car, and used. In a sense, its *truthfulness* is its loyalty, loyalty being the originally root meaning of the word "true", at least in English.

Parikh substitutes "truthfulness" for "truth" in many of his writings, and this is no accident. He is suggesting that truth is a matter of degree, insofar, at least, as we know and act on it. Peirce, for one, seems to have thought of a true belief, not necessarily as an ideal limit point, i.e., a fixed ideal belief that scientists would agree about in the ultimate long run, but, rather, a tenacious and loyal one that will not let us down.[19] Generally speaking, the MBTA map does not let us down and we expect it not to. It works, and in this sense serves as an *Übersicht*, i.e., as a guide to action and communication, not merely a piece of information. Or, perhaps better, insofar as it codifies things to be known, it is *sensitive* to understandings and perceived and expected needs among us.[20] Could this "usefulness" break down? Of course. New lines may be built, and it becomes outdated; the map could be systematically misprojected by a group, Martians could land who couldn't understand it. But it is not the job of a piece of social software to take into account all possibilities. Nor even to justify us sufficient unto the day we act. Instead, the map opens up a space of possible actions: it is analogous to an operating system, within which we make choices.

Parikh is not suggesting that we reduce "knowing that" to "knowing how" in any facile way. More deeply, he is after what the knowing of "knowing that" really amounts to, if it is logically structured. The MBTA map is a representation, surely, a piece of information, perhaps a paradigmatic one. We may not wish to honor it with the title of "depiction", if by depiction we mean something like a portrait, *in* which we may see, not only the Mona Lisa, but Lisa herself. Or perhaps it is a "depiction" in a very malleable sense of a "symbol", but then it may not be considered a very good one, except in a school of graphic design.

The important point here is that there is a broader point of view, both on a depictive portrait and on the MBTA map: both are also, as Parikh has called them since his originating works (Parikh 2001a, 2002), pieces of *social software*. The map of the MBTA helps us bear in mind more easily a great deal of action-points, as individuals. The portrait of Lisa tells us something, not only about her, but about the artist, the time and manner in which she lived, and her person. That is why it can serve as a touchstone for each of us, gazing upon it: it is a gesture of Leonardo's—which is

[19]Read this way, as Cheryl Misak does (Misak 2000), Peirce does not fall into the errors attributed to him by Quine at the end of Chap. 1 of *Word and Object*, where Quine complains about the unclarity of "ideal truth in the long run" as a *definition* of truth (Quine 1960, §6). Parikh's work with San Ginés (manuscript) explores the relevance of a pragmatic approach to belief, augmented by a 3-dimensional approach analyzing the "success" of a belief into its relativity to the (first person) thinker, the situation of a co-participant (second person) as their joint activities unfold over time, and the (third person) unfolding of circumstances themselves over time.

[20]On context sensitivity and understandings, see Travis (2005, 2006, 2008).

why it matters whether he painted two Mona Lisa portraits, and not just one.[21] The social aspect opens up for us humans the possibility of all kinds of collective action: communication, teaching, learning, and, generally, the enlargement of spaces of opportunities for knowing, believing, and acting—things we do, after all, in concert *with* one another, not merely *to* one another.

It is in these "social software" respects that the rat's powers to articulate, advise its fellows, generalize, and create new opportunities for all of the above is quite limited in comparison with our own. This is shown everywhere in our daily lives. Shopping for an apple in a grocery store—to choose a canonical example from Wittgenstein's PI §2 that has been discussed in detail by Parikh, emphasizing the importance of parameters and protocols (Parikh 2009a)—or, to take some other examples, recounting a story of why someone in the office is angry, or "liking" on the web, or voting, or calling 911: all of these activities are embedded within and by means of social forms of infrastructure that we mostly do not bring to mind as forms of social software when we act within them. Of course we do not: their function is, after all, to *offload* the problem of actually reaching decisions to something else, just like the MBTA map. That offloading can then itself be represented as a piece of social software, and that representation in turn again regarded as part of what we do, hence part of social software as well.

The idea of taking this notion of *Übersichtlichkeit*, or "surveyability", in at the foundational level is this. The notion is nearly as comprehensive as that of *computability* à la Turing, if not more so, since we don't calculate with all of our concepts, but only with special ones, whereas concepts as such are meant for sharing and use in a social setting in which we act, communicate, articulate, express, influence, grow and jointly engage with one another.

Parikh has recently asked, somewhat tongue in cheek, whether there might not be a Church's Thesis for social algorithms (Parikh, forthcoming). That would place what Turing accomplished into a new light—without, I suppose, contradicting his analysis. The point would be, as Parikh says (Parikh 1995, pp. 89–90), a fourfold, Wittgensteinian one:

- **Mathematics as an Applied Science:** The truth of Mathematics is that it fits into our lives.
- **The Importance of the Social:** Language as also Mathematics and even thinking—certainly we think of the last as a private process—are in fact, according to [Wittgenstein], social activities.
- **Locality or Context Dependence:** We do not have general notions like knowledge, truth, number but context dependent ones.
- **Flexibility:** A formal system does not fully determine our behaviour nor how we use it.

[21] Cf. http://www.pbs.org/wnet/secrets/mona-lisa-mystery-full-episode/1821/ and http://www.openculture.com/2013/12/first-mona-lisa.html.

1.3 Turing Machines: From Language Games to Social Software

Let us return to Turing, to whom we owe the idea, if not in some respects the fact, of the stored program computer, *via* what Church would call his notion of a "Turing Machine". Wittgenstein and social software—hence philosophy—may be seen, in retrospect, to have been entangled with Turing's model, just as Parikh has been arguing.

Many philosophers since the early 1960s have thought that logical analysis gave us great insight into the individual: states of mind, cognitive processes and their computational modeling of cognitive states, and so on. Computational Functionalism—invented in part by Putnam, but later rejected by him—dominated the approach to cognitive content for a very long time. But "Can Machines Think?" was imagined by Turing in his (Turing 1936/1937, §8) explicitly as nothing more (and nothing less) than a *comparison* between a human computer—an individual—and a machine: as what Wittgenstein would have called a *language-game*, i.e., ultimately, an *analogy*. Turing's proofs do not depend upon their serving as descriptions of what actually goes on in our minds, ultimately, when we calculate, any more than Wittgenstein's imagined language-games are intended to describe everything that happens when we use language. They could not, after all, if he was to resolve the *Entscheidungsproblem*, for no mathematical theorem can turn on a thesis in the philosophy of mind.

Turing signaled this in §1 of his famed (Turing 1936/1937) by stating that "we may *compare* a man in the process of computing a real number to a machine which is only capable of a finite number of conditions" (my italics). A language game is not exactly a *description* of what we do, it is instead a *comparison*, as Wittgenstein says (PI §130):

> Our clear and simple language-games are not preliminary studies for a future regimentation of language, as it were, first approximations, ignoring friction and air resistance. Rather, the language-games stand there as *objects of comparison* which, through similarities and dissimilarities, are meant to throw light on relations of our language.

In other words, the idea of a "language game" is intended by Wittgenstein to speak to questions concerning our very idea of what logic is, rather than specifically to matters of psychology. As Parikh has written (Parikh 2009a), Wittgenstein is using language games to stress the importance of social software, protocols, and partial renditions of games in logic.

Turing begins §1 of his (Turing 1936/1937) this way:

> We have said that the computable [real] numbers are those whose decimals are calculable by finite means. This requires rather more explicit definition. No real attempt will be made to justify the definitions given until we reach §9. For the present I shall only say that the justification lies in the fact that the human memory is necessarily limited.

There has been a lot of discussion in the literature of this remark about human memory. Many have assumed that Turing *must* have been committed to a very particular theory of human mentality, and used this theory of mind to achieve his result.[22] In particular, the thought is that he was reducing consciousness to merely bodily or physical processes, limited in space and time, and construing a human thinker as really nothing more than a machine. It is true that throughout his paper Turing speaks of the finite number of "states of mind" of a human computer, and the ability of a human to take in only a small number of figures "at a glance" (Turing 1936/1937, §9). It sounds as if Turing is making a series of epistemological points, directly continuing the Hilbert finitistic or formalistic tradition, or perhaps broaching a behavioristic theory of cognition and/or perception.[23] This invited the idea of a "language of thought" operating automatically inside the head, as if philosophy of mind and/or our perceptual ability to cognize were central or foundational to his model. This in turn lead to the rather irrelevant criticism that in his modeling of the classical consequence relation by way of an infinite tape Turing was analyzing thought in a way useful only for Martians or cognitive scientists.

However, Turing himself was more careful. In fact this opening remark constitutes no theory of mind at all, in the sense that a traditional metaphysics is intended to offer. The response to Hilbert is a response, but also a reorienting of the whole subject back to *us*, to scrutinize the nature of the conditions involved in resolving the *Entscheidungsproblem*. What Turing does, right at the beginning, is to simply reiterate the *point* of effectiveness in the context of mathematics, and therefore, in human life. He is calling attention to what is "right before our eyes".

Turing could not have proved a *mathematical* result based on any theory of mind, however powerful: mathematical theorems about mathematics cannot be based on theories of minds, but only on mathematics. Instead, in the spirit of Wittgenstein's idea of routines that are "plain to view", I suggest we take Turing to be remarking on what is, in the end, a rather obvious point. He is making the whole idea of an "effective calculation", hence the whole idea of a "formal system", *plain*.

The general idea of a Turing Machine boils down the idea of a "step-by-step" routine to its simplest, most intuitive elements, those derived from what we human beings *do*. The tape is unbounded ("infinite") in length: one can always add on to each routine another, then another, and so on (Turing points out that the linearization into a 2-dimensional tape is only one possibility, used for his purposes). Moreover the "paper", or "tape" contains "squares" to be marked with "symbols", and it does not matter *which* particular symbols they are: algorithms can be communicated by means of a whole variety of methods, languages, diagrams, pictures, and so on. The (human) computer is held to (a) "see" only a bounded collection of symbols at each single step, "at a glance"—shades of *Übersichtlichkeit*—and (b) only to have written down, at any specific point in the process, a finite number of symbols that it can move to locally.[24] As Turing says explicitly (Turing 1936/1937, §9, III),

---

[22]See Gödel (1972). For criticisms of Gödel's assumptions see Webb (1990), Sieg (2007).

[23]Cf. Shanker (1998).

[24]Wilfried Sieg (2008) has "axiomatized" Turing's model with these "boundedness" and "locality" conditions, showing that nothing in what Turing did refutes Hilbert's approach. One might however say that Turing encompasses that approach as but one among others in mathematics and philosophy.

> It is always possible for the computer to break off from his work, to go away and forget all about it, and later to come back and go on with it. If he does this he must leave a note of instructions (written in some standard form) explaining how the work is to be continued. This note is the counterpart of the "state of mind".

This "counterpart", is a shareable command: part of the general *interface* environment. It forms part of our most ordinary sense of a routine that offloads, to save effort, for human beings—as individuals, and within groups. There is no "there" there, if we ask "*Where* is the interface?"

The most comprehensive perspective we have of what Turing modeled is, then, that of social software. In Turing (1936/1937) Turing showed, step by step, how his machines could carry out any effective calculation that the Gödel and Church systems could carry out— and, by imaginative extension, any one we might dream up, thereby fixing a very widely-applicable parameter for titrating the notion of a "step" in a computation. That this parameter is provably robust, impervious to the vagaries of particular formal languages or local conventions of symbolism, struck Gödel as nothing short of a "miracle", once and for all determining the precise generality of his incompleteness results.[25] Turing, he said, had gotten to "the right perspective" mathematically.[26] Gödel objected, however, to what he took to be a prejudicial assumption of Turing's about our mental lives: that our experiences are discrete or discretizeable.

However, philosophically what needs to be stressed is that a Turing machine, as a construct, is double-faced: from one point of view, it is nothing but just another formal system. But from another point of view, it tells us what a formal system *is*, by showing us what it is *for*. To resolve the question what a formal system (or "effective calculation") is, it did no good to write down another formal system (or perform another computation). It also would have done no good to have furthered a theory of mind, if Turing had done so. Instead, one had to *do* something, to make the question clear.

Turing of course also showed that since any Turing machine's recipe of directions, configurations, and symbols can *itself* be coded by numbers and thereby worked on by another, *there is a universal Turing machine* that can carry out the routine of any and every machine. It is this, as Martin Davis has argued, that leads to our sense of the "ubiquity" of computable processes in our world.[27] There is no diagonal "escape" from the class of effective computations by means of anything effectively computable, for the universal computer can always incorporate any effectively computable process into itself.

As Davis has also emphasized, Turing's model shows us that a distinction that may be—and was—naturally drawn by early computer designers between software,

---

(Footnote 24 continued)
Wittgenstein himself makes very few remarks on the axiomatic method, although there is nothing in his approach, *per se*, that refutes its importance for tagging assumptions in deductive reasoning contexts, and in fact the axiomatic method may be said to fit nicely with the view of mathematical sentences as norms or commands (cf. Friederich 2011).

[25] See Gödel (1946), p. 1.

[26] Gödel to Wang, in Wang (1974), p. 85; Copeland (2004), pp. 45, 48, (Copeland 2012), Chap. 2.

[27] Davis, forthcoming.

hardware, and data is not rigid, but rather contextual and shifting.[28] This is the essential insight behind the idea of the stored program computer: a single device, as we all know nowadays, can perform multiple tasks, as we open up different routines, or programs, and it can work on its own program and activities as well, for example joining in the activities with other machines and users.[29]

One might add something else, drawing out a philosophical implication that is also a supposition: the distinction between what *I* and *we* do and what the *machine* does is fluid in life, within the interface. This suggests that what it *is* to perform a calculation (correctly or incorrectly) is itself an occasion-sensitive matter; an agent-relative matter sometimes, a matter of social purpose at other times, a matter of particular collectivities within multi-agent systems at others. Our standards of correctness are norms, like commands, and their standards of fulfillment are brought together with language in ways that involve much plasticity, as well as structure.

In the initial philosophical reception of Turing's ideas, however, the community, the environment, the social setting, and the culture, were shoved offstage. This was a sign of the times: philosophers wanted to bring psychology back into the fold of their thinking. Even Paul Grice, whose notion of "conversational implicature" has been an important stimulus for Parikh's work (Parikh 2002, 2003), remained in thrall to a certain picture of correctness based on the notion of "intention" conceived of as a psychological state. And in this Grice continues to be followed by many philosophers of a naturalistic bent, as well as researchers in Artificial Intelligence who believe that we are Turing Machines all the way down.

In his later essays, however, Turing himself was quite clear that a social context would be required for the development of his ideas. His 1948 technical report "Intelligent Machinery: A Report by A.M. Turing" contains the observation that "an isolated [human] does not develop any intellectual power" (Turing 1948/1992), and he emphasized the need for creativity and intuition, as well as rule-following. In fact, as has been argued (Sterrett 2012, forthcoming), his vision of computing machinery presupposed that frequent communication and contact among human beings would be crucial for developing human cognitive abilities that are integrated with machines.

Only with the development of the web's architecture, the ubiquity of analysis of our communications and the crowd-sourcing of intellectual projects has the obviousness of the point risen to philosophical consciousness clearly. Early on, logic and philosophy were too much in thrall to the idea that, as Russell put it in his William James Lectures of 1940 (Russell 1940) (Chap. XIII), "'correct' cannot be used in defining 'true', since 'correct' is a social concept, but 'true' is not".

For Parikh, however, it was always and only about the *system*, about the interplay between truth and correctness, the individual and society, and never about the individual's state of mind or truth as such independent of the opportunity spaces we inhabit.[30] Somehow—perhaps because of his upbringing in India, perhaps because

[28] See Davis (2000, 2001, forthcoming).

[29] See Copeland (2012).

[30] Lately Parikh has absorbed the importance of Amartya Sen's thinking about justice as a notion rooted in capabilities and opportunities, rather than abstract principles of opportunity.

of study with Quine, Goodman and Putnam and others in the philosophy department at Harvard—Parikh made it a point to pursue culture and society as, so to speak, foundational in the development of logic. He never fell for functionalism, but instead kept Turing's analysis of computability in mind as a designed device, an analogy, as Wittgenstein would have always suggested that it was, and as Turing himself said. Already in his early work on bounded arithmetic (Parikh 1971), Parikh was turning away from the classical consequence relation and placing epistemic limitations on it, working with notions such as "feasibility" in arithmetic and in vagueness (Parikh 1983, 1994). He took Turing, quite correctly, to have come up with something that itself was only an "analogy" in analyzing the notion of "effective calculability".

There are two final historical sides about Wittgenstein and Turing that are relevant.

1. Action and Turing's Argument. The importance to Turing's analysis of what it is that we *do* with rules is very clear when we examine the particular argument by means of which Turing resolved Hilbert's *Entscheidungsproblem* in his (Turing 1936/1937), showing that there can be no general way of effectively determining whether or not a given sentence of a language does or does not follow from another.[31] He did so, *not* by producing a contradiction, as is usual in presentations of the proof nowadays *via* the Halting argument, but instead by showing how, if one assumes that there is a decision procedure of this general kind, one would then be committed to the constructability of a *tautological* machine, one that could be made to do something like turning up a card in a game that says "Do whatever it is that you are doing". This would be an empty command that *cannot* be followed, though it can be defined. There can be no such machine, and so no such general procedure.[32]

The argument is a kind of "diagonal" argument, but one that works by reduction to tautology, rather than to contradiction. It bears an analogy to the truth teller paradox ("This sentence is true"), but with an important difference: the focus on action and correctness, the possibility of being able to carry out or *follow* the command, is clearer. What Turing wrote is that the more straightforward approach, via a general use of negation in a diagonal argument, may "leave the reader with a feeling that 'there must be something wrong'" (§8). That person might have been, e.g., an intuitionist—it may even have been the historical Wittgenstein. Be that as it may, in Turing's situation, he did not want to get into the problem of negation, whether or not the "law of contradiction" is a universal law. So instead he artfully dodges foundational controversy, something Bernays suggested he flag explicitly in the revisions to (Turing 1936/1937) that he got Turing to publish right away.[33] He defines

[31]Or, "what is the same thing" (Turing 1936/1937, §11), how to effectively determine in general whether when one adjoins a new axiom in the series of theories Gödel (1931) showed us how to generate, one ends up with a consistent theory.

[32]The negation of this rule, namely, "Do whatever you are *not* doing" could also be defined; yet in the context of Turing's argument in his (Turing, 1936/1937), this would equally well be a command that could not really be *followed*, for the problem would remain of how to specify what it is that you *are* doing at that particular step, such that you are not to do it.

[33]See Turing (1937), discussed in Floyd (2012). Bernays urged, in a suggested revision to Turing's association of machines with real numbers, that Turing make explicit that a Brouwerian fan construction could be used.

a "rule" that cannot be followed, communicated in the service of action, or worked with. What he is showing is that the idea of *that* goes outside the game of *effective calculability*, as we understand it humanly in the Hilbertian context.

As I have shown (Floyd 2012), Wittgenstein knew Turing's 1936/1937 tautological diagonal argument quite well. Well enough that in 1947, recalling that argument apparently by heart, he remarked in a notebook (Wittgenstein et al. 1980, §1096):

> Turing's 'Machines'. These machines are humans who calculate. And one might express what he says also in the form of games. And the interesting games would be such as brought one via certain rules to nonsensical instructions. I am thinking of games like the "racing game". One has received the order "Go on in the same way" when this makes no sense, say because one has got into a circle. For that order makes sense only in certain positions.[34]

2. Turing derived his notion of types from Wittgenstein. In his (Turing 1944) paper "The Reform of Mathematical Notation and Phraseology" Turing explicitly acknowledged Wittgenstein's Cambridge 1939 lectures on the foundations of mathematics [LFM], stating that they inspired his approach to data types. This approach is earmarked by its attention to what Turing calls the language of "the mathematician in the street", ordinary language, which is conceived of as given to us *already* typed (cf. Floyd 2013b for an introduction to Turing's paper). The point of Turing's paper was to take stock of the fact that mathematical logic is "an alarming mouthful" for the average mathematician, so that it would be desirable to design logic in such a way that none of the classical formalisms would even be necessary to learn. The method proposed was to go through textbooks of ordinary mathematics, to see what words mathematicians actually use, and build logic from there.

So Parikh was also on firm historico-philosophical ground in stating (Parikh 2002, p. 187) that Wittgenstein "introduced" the notion of a *data type* with his notion of a language-game.

1.4 Animal Cognition

The "form of life" that is human forms its life ubiquitously with social software. The reason Parikh's notion is philosophical is that it runs this wide and this deep. Even deeper still, however: when one adopts a Wittgensteinian "snapshot" perspective, thinking methodologically by means of "language-games"—partial, truncated descriptions of a portion of human linguistic activity—then comparisons with animal behavior come to seem, for certain games, as important as anything else. In §§2–5 of PI, elaborating on his language-game 2 ("block", "pillar", "slab", "beam") Wittgenstein uses the terms "*rufen*" (to call) and "*abrichten*" (to train), both of which apply to animals as naturally as to humans. To say—as we ordinarily and naturally do—that

[34] Wittgenstein's remarks date from 1947. We do not know precisely which "racing game" Wittgenstein had in mind, but at Blackpool and other amusement parks there were mechanical games with this name played in pinball parlors, and there were board games as well.

"the cock calls [*ruft*] the hens by crowing" already involves us in "a comparison with our own language" (§493). As Wittgenstein writes (§494),

> I want to say: it is *above all* the apparatus of our ordinary language, of our word-language, that we call "language"; and then other things by analogy or comparability with it.

Social software—and much of Parikh's work in proof theory—similarly works by way of analogies, carefully chosen to suggest novel ways of re-designing our logic. A certain degree of cooperation and communication and signaling of feeling, hope, disappointment and happiness are surely part of what we hold in common with animals—perhaps especially dogs and horses, more than cats or rats or fish (Parikh, born a Jain and now a Buddhist, is an admirably thoroughgoing and inspiring vegetarian). It is not necessary for a "belief" to be linguistically articulated, as Russell and Ramsey and Quine all held: watching the behavior of an animal, placing a bet, observing, are enough for belief-attribution to get going, and Parikh has modeled these forms of belief (Parikh 2008). Of course, one may say with justification that *specific* directives, of an articulated propositional sort, carrying forth to far-flung, nearly unlimited regions of discourse about discourse and about cooperation and its role in discourse and hence in philosophies: these are not part of what fish or rats or dogs have yet evolved to. They lack social software of these kinds. They lack logic and mathematics. We lack knowing our ways about with them in certain contexts in which we feel more at home with humans, as they have lacks with us. Yet we do feel for and with them, and there are times when they may even out-do us in poignancy and appropriateness of response, anticipation, and instinct. Social software, used as a model, tells us about belief, including animal belief, because what beliefs are depends in part on *our* resources in representing them.

This is the spirit in which Wittgenstein wrote in PI Part II (PPF) that (xi, §327) "if a lion could talk, we couldn't understand him" (used as an epigraph to Parikh and Ramanujam's 2003), or, perhaps better put (Part II (PPF) i §1):

> One can imagine an animal angry, fearful, sad, joyful, startled. But hopeful? And why not?
>
> A dog believes his master is at the door. But can he also believe that his master will come the day after tomorrow? —And *what* can he not do here? — How do I do it? —What answer am I supposed to give to this?
>
> Can only those hope who can talk? Only those who have mastered the use of a language. That is to say, the manifestations of hope are modifications of this complicated form of life. (If a concept points to a characteristic of human handwriting, it has no application to beings that do not write.)

The point here is not to deny that a dog can expect or hope for the return of a human master (or be disappointed, angry, jealous, and so on), or to say, with Davidson (1982), that animals do not have beliefs at all. Instead, it is to insist that *our* concept(s) of hope are embedded in highly ramified and articulated forms of interlocking systems of parametrization.

This helps to clarify some of the ways in which Davidson may be thought of as having been correct about animals, and some of the ways he may be considered to have been wrong. We, unlike the dog, can hope for someone's arrival the day after tomorrow, can therefore *be* hoping for someone's arrival day the after tomorrow or the next day, can hope for all kinds of situations we can write down in sentences; can be wrong (and disappointed and frustrated) in all kinds of ways a dog cannot be. What we have in addition to (beside? over?) the dog is connected with our evolved powers of language, including, for example, the fact that we can and do write to one another to set appointments, express hopes, fears, expectations, and so on. We have what one might call a "knowledge" based semantics of messages, as Parikh and Ramanujam have argued (Parikh and Ramanujam 2003).

This implies that the reach of our power of grasping negation, in the classical sense, is an important benchmark.[35] While the distinction (and relation(s)) between that power and our powers of direct perception comprise disputed philosophical territory, in which we may coin new terms for half- or part-way representations (cf. Millikan (2006) on "Popperian" animals, Travis (2013) on "Pyrrhonian attitudes"), the important point all these powers comprise, in Parikh's view, elements of social software. Animals surely share some of this with us. Not all of it. This is not a principled argument about rights, or good and bad. It is instead an ordinary observation, if we let what is right before us in ordinary life shine through as something important, awaiting modeling.

We, unlike the dog, face a question whether and to what extent propositionhood (propositional structuring) is *real*. If, with Quine, one rejects the idea of an ether-like absolute framework of propositions, or works without the full the idea of propositions as sets of possible worlds, there remains the question of *how* we are to see these notions at work in our talk and action. Our resistance to making full propositional attitude attributions to animals ought to get us to ask an important question: since we too are animals, isn't there a large degree of imagined "phantasy" (Parikh 1971, p. 507) surrounding the very idea of a *proposition* as it has been articulated from Frege onwards? If that notion is precise, then are what we do and say vague? Is vagueness the wrong way to talk about our notion of something specific that is said on a particular occasion? What are the tradeoffs in different approaches to meaning and belief?

1.5 Rationality

It is an implication of Parikh's analogy between what we do and software systems that propositional utterances, true or false, are not the only sorts of moves we make in a language-game—another Wittgensteinian point (cf. PI §23). In fact Parikh wants to work, as Wittgenstein did not, with questions about *why* people act in the ways that

---

[35] Wittgenstein registers this Fregean, Russellian point in many places, perhaps most interestingly in RFM I Appendix III §4, in his earliest remarks on Gödel's incompleteness result, when he mentions our making gestures to stop a dog acting in a certain way. Here he is not so much attempting to refute Gödel, as to understand how the result bears on our notion of truth (cf. my Floyd 1995). It is no accident that the language-games at the opening of PI lack negation: part of Wittgenstein's point is to register the importance of negation to our notion(s) of thought.

they do, rather than simply model knowledge and justification. This, the province of a Ramseyian approach to belief, entangles Parikh with incentive-representations, and how it is that obligations emerge.

Parikh's notion of social software is not something exhausted by the ordinary analysis of rationality as self-interest that has traditionally taken place in game theory. On the contrary: it is game theory's usefulness in modeling certain aspects of market and other transactional human behavior that needs explaining, from his point of view. Game theory is, to repeat, only a "preliminary" step, so far as he is concerned.

First of all, game theory is able to represent only a portion of who we are, as humans. Second, game theory was long plagued by too narrow a Hobbesian focus on individual self-interest, and too underdeveloped a sense of language as a kind of stipulative convention, rather than treating language itself, and what we do with and in it, as part and parcel of the structure within which incentives and normative elements actually arise. Finally, game theorists—as Kahneman, Tversky and Aumann and others have shown—for a long time retained the eggshells of game theory's birth from classical modern logic: it did not systematically confront problems of how deontic obligations are actually generated on the basis of beliefs; it customarily assumed infinitary common knowledge; and in general it suppressed the significance of limited and partial knowledge in real life. There are so many other games we play, so much else we in fact do, that may be subjected to rigorous structural representation, and thereby the generation of designed *Übersichtlichkeit* or surveyability in logic.

Many philosophers interested in the notion of *convention* have been influenced by Lewis (1969), who elaborated our interactions with one another as "score-keeping" devices at work in what Lewis, following Wittgenstein, called "language-games" (Lewis 1979). But Lewis, interested in the *causal* unfolding of language's evolution, was not stressing the social and logical structure of logical structuring itself, even if he did bring out numerous ways in which presuppositions affect reasoning. Instead, he was, like the game theorist, thinking of the evolution as a process of coordination, as in the rowing of a boat by two persons, each commanding her own oar separately. He was a realist as a metaphysician. But in a real sense, he remained in the Hobbesian fold, with possible world semantics used as a way to make sense of truth *via* a general metaphysics of propositions. From this he drew many philosophical implications; among them, for example, that (on pain of triviality) probability conditionals are uninterpretable as *propositions* (Lewis 1976). In the end, like Quine, though from a different perspective, Lewis ended by regarding our notion of knowledge as "elusive", perhaps as elusive as an indexical like "big": a matter of degree.[36]

Lewis's approach has evoked responses from other philosophers, (such as Gilbert 1989, 2014), who defend the important notion of a "social" or "institutional" fact, something real held in common among us. Searle (2010) elaborated this idea more recently by means of an extension of Austin's speech act theory, and a realistic theory of social ontology. And yet: this dialectic between a conception of individuals coordinating through self-interest and a competing conception of individuals sharing common goods and common knowledge, while it unfolds an ontological controversy,

---

[36]See Quine (1984), p. 295 and Lewis's (1996).

requires proof theory and logic for its resolution: the subjection of the structuring of possible systems to detailed mathematical investigation.

Part of Parikh's alternative approach has been to rigorize and design non-monotonic theories of belief revision that are sensitive to such philosophical questions. "Near" monotonicity may well do for a large number of cases (Johnson and Parikh 2008). In these systems knowing is all right if it usually works, or works for a time, and modulations and corrections can be seen to update generalizations as we go. While philosophers of language and linguists are still puzzling over generics ("The tiger sleeps during the day"), computer scientists know that there is no simple treatment of them, no simple "generics" operator that will serve. Parikh's transposition of probabilistic conditionals out of the setting of the infinite (possible worlds semantics) and into the settings of topology and probability allows for a rigorous approach to the idea of a generalization that "usually" works. Usually, certain generics don't let us down. And if we are content with beliefs that are nearly enough right ("Birds fly", "Humans require contact with loved ones to thrive", "Lydia detests pigs"), we can get completeness of a sort, with respect to "feasible" knowledge.

A large part of Parikh's contribution to these fields of philosophy comes from his having brought an implementable and rigorous epistemology, logically and mathematically articulated, to bear on these disputes. It is indeed knowing that one can know that others know such and such that permeates our sense of our own actions' directedness at specific outcomes.

As Austin noted in "Other Minds" (Austin 1946), saying "I know" involves *recognized* procedures and expectations, just as promising does. Thus epistemology must be built for human beings, or for animals, and not for omniscient ideal beings or Gods. The implementation of such "recognition" by mathematical and computational means, the structuring of parametrization, is crucial to our world of connected scientific infrastructure. But the constraints of what we can *do* and *use* should be taken into account at the fundamental level, when the mathematics and logic are built. This is Parikh's picture of logic and games.

Social software is therefore an important alternative to readings of Wittgenstein that take him for a sceptic about how it is that in general we follow rules—Kripke (1982) being the most articulate defender of this sort of interpretation. In the very first article Parikh published on the notion of social software (Parikh 2001a), he showed, step by step, how the programmer's point of view could accommodate Quine's indeterminacy of translation (Quine 1960), Kripke's "no fact of meaning" reading of Wittgenstein (Kripke 1982), and Searle's Chinese Room argument (Searle 1984) in a quite ordinary and rigorous way using Dynamic Logic. A compiler may compile correctly or incorrectly, but only relative to the whole situation that takes into account an operating language (with decisions made by the programmer about how to handle its syntax and conventions), and, in the end, the purpose of the program that is written—its standards of correctness by *our* lights—hence in light of social software. Typically, Parikh argued, observers of linguistic interactions who see the behavior of their fellows are in the same boat as an outsider attempting to scrutinize the operational language alone, wondering what it means to say that a compilation is correct or incorrect—unless and until they can "take in" the point of the situation, the "compilations" going on by those around them, and the purpose of the exercise.

In cases where it is difficult to say whether one is following the same rule as another, the user end—that is, the whole social context in which we act—decides, a point to which we shall return in §6 below.

Parikh's analogy between the activity of a programmer and Wittgenstein's idea of a language-game is not one whose point is to show that humans are nothing but machines. Instead, it serves to question whether humans are in a necessarily superior boat, with respect to meaning, than observers of an operating language who are not yet in a position to judge whether or not a program has compiled correctly. The metaphor is this: social software is the operating system on which we write our lives, and it includes data types at the start, although the protocols for updating alternatives in these types depend upon numerous factors, including the particular perspectives and interests of the participants, and are very complex. That this is an extended *analogy* is no bar at all to having it play the role of a serious philosophical, logico-mathematical, foundational device. In fact Parikh's point, like Wittgenstein's in PI, would be to suggest that philosophical points made by means of reflection on what we are inclined to say about language-games often consist of no more and no less than the use of designed "objects of comparison" (§130) to throw light on the "relations" (*Verhältnisse*) our language (including what we *do* with it) bears to itself and to other things.[37]

Parikh's relation to Frege's notion of a *proposition* or *thought* is thus more complicated than that suggested, also on the basis of a reading of Wittgenstein, by Travis. Travis has insisted that Wittgenstein is capable of fully accommodating the Fregean notion of a possible thing to be said in a language: a thought or proposition. The twist he takes Wittgenstein to have placed on Frege is an insistence that, on any given occasion of use, there remains the question of *which* among the range of possible things to be said that are there to be said *is* being said. This notion of individuation can only happen if the speaker and audience share a sense of the *point* of the conversation, the situation, the occasion of use. Travis calls this Wittgensteinian supplement to Frege Wittgenstein's insistence that there must be "understandings" among us (Travis 2006). This implies that there can, if we vary the situation and actors and/or point of the conversation enough, be misunderstandings. For the very same form of words may bear several different "understandings" depending upon these (and other) factors. On this view, the primary force of what Travis calls the "occasion sensitivity" of propositions (and property and belief and meaning talk) is to individuate truth values on specific occasions of saying (cf. Putnam 2002).

But Parikh's analogy of social software brings in more than occasion sensitivity and the need for "understandings". It is intended to complicate the logic itself, the semantics, the way and manner—structurally speaking—in which understandings may be said to shape our sense of the point of a given exchange.

---

[37] I translate "*Verhältnisse*" as "relations": whereas Anscombe translated it "facts [of our language]", Hacker and Schulte suggest "features", i.e., "internal" grammatical relations. I think Wittgenstein has in mind here, by contrast, the idea of treating language itself, and we users of language, as part and parcel of the world, and in turn the world's samples of meaning as part of "language".

Travis's "brill-bream" example will serve (Travis 1995, 2008, p. 106). In this example Jones asks in the fish-shop, "Could I have brill?", meaning her "brill" to mean *brill*. The default, *ceteris paribus*, is that she asked for brill. But of course there are exceptional cases: she may think that "brill" refers to that piece of bream, having glimpsed the bream in the shop-window and taken it for just what was recommended to her by a friend as delicious. Then it is not so clear how we are to interpret exchanges in the shop, i.e., exchanges in which the language-game is played. If, confronting the piece of fish (of what we call "bream"), she says "No, I don't think that looks good at all", which of her beliefs are we to say was revised (that bream or that brill is delicious)? If instead she says, "Thanks, I'll take what you call 'bream'" after the shopkeeper corrects her, explaining that this is not brill, but rather bream, for brill is horrid tasting, whereas this bream is delicious—then what are we to say was really revised? The thought that prompted her to enter the shop (that brill or bream is delicious)? Her understanding of linguistic meaning? Or of certain generalities that hold?

Travis's point is that in order to secure a clear sense of something she thinks (truly or falsely), understandings must be in place that give those concepts weight. Parikh's question—not denying this—is: How is it that we actually spell things out in practice? Do we need to insist, for every speech act, that there must *be* a precise proposition expressed, one that fits into a total semantic space of possibility?

There are worries to be expressed here about truth, and also about beliefs and desires. One might reject the sense of notions Parikh has often formalized, "almost true" or "taken by default to be true" (Arló-Costa and Parikh 2005). If one works with such notions, then one cannot say—as some philosophers have—that truth (in the classical semantic sense) is a *constitutive* element of belief, but at best a regulative norm (shades here of Kant). And there will be stress placed also, therefore—given Parikh's approach—on certain ideas about incentives, rationality, and desire. The analogy between truth- and satisfaction-conditions will become more plastic, and thereby also the notions of intention and agency.

However, one gain, at least in relation to certain variously contested interpretations of Wittgenstein, is that the concept of social software provides an alternative to the "social practices" explanation of meaning common to a number of different Wittgensteinian schools.[38] On the "practices explain our ability to follow rules" view, the notions of "social" and "practice" are taken for granted, and issues of hermeneutical, interpretive stance are taken to be fundamental to all cases of rule-following. There are variants of the reading: some regard the appeal to practices as explanatory, others believe it simply evinces the irreducibility of the notion as part and parcel of our understanding of what it is to follow a rule. The beneficial side of this hermeneutical tradition is its emphasis on cultural and conceptual relativity, on drawing in reflection on the contingencies and alternative possibilities for construing our intellectual and cultural practices, including whatever it is that gets called "objective" (if anything does). The less beneficial side of the hermeneutical and social constructivist point of view is loss of a hold on truth and knowledge, as opposed to "correctness", hence

[38] Most prominent here are Winch (2008) and Williams (1999, 2009).

the temptation to embrace a deflationary, anti-theory, end of philosophy reading of Wittgenstein (such as Horwich's). Parikh, however, is not a "deflationist": he has not lost hold of the notion of truth as the logician works with it, but instead shown how it unfolds, and how knowledge and belief may be structured in relation to it.

What Parikh has shown is that there may be a number of very different levels of social software at work, among them of course the agent's point of view, but also the points of view of other agents and, especially, the emergence of the idea of a *collective* space of opportunity among them. This was perhaps less clear in some of his early papers, such as his papers on vagueness (Parikh 1983, 1994), where his motivation for rejecting supervaluational models, appealing to actual experiments he ran, were based on an (at that time) fairly undifferentiated notion of "utility". Then Parikh's argument was more Quinean and pragmatic. He expressed the point of view that anyone who believes in structured propositions that transcend our practices is bound to come up with a useless (from the utility point of view) framework for analyzing such phenomena. But this left untouched a deeper point: if "knowledge" and "utility" are taken as primitive notions, in relation to how we actually do operate with them in real life, then *how* is this to be done?

Since Parikh was also a student of Putnam's (hence aware of Putnam's arguments for "semantic externalism" (Putnam 1975), in which it is made clear that we rely, not primarily or solely upon meanings in the head, but rather upon a "division of linguistic labor"), he had a way to move forward with this question. He took a pragmatic, experimental approach to vagueness and color terms, dripping one yellow drop at a time into a red can, and querying himself and others as to the point at which orange appeared. (Those who have recently argued that philosophers face a dichotomous dilemma of method between "experimental" and "armchair" philosophy haven't been reading Parikh.) Parikh (1994) rejected fuzzy logic as a solution, and any kind of multi-valued logic. Instead, he said, at the cost of making the logic a little more complicated, we could generate arguments that we can "take in", in the way we take in the MBTA map—and still retain a working sense of bivalence, true and false, if we want it. And this Parikh did want to do, just as he did in Parikh (1971). For, like Wittgenstein and Putnam, he has been thinking through the classical notion of truth as it figured in figures such as Frege and Russell, rather than rejecting the notion altogether in favor of a global constructivism, verificationism, or intuitionism about meaning—at the same time acknowledging the viability of alternative, non-classical points of view, in virtue of the fundamental circularities and residues that must be faced, philosophically, in foundational work at the fundamental level.

The key is that social software of this kind is connected to what we *do*, but also what we can *take in*. In Parikh's recent (2013) the division of linguistic labor is adduced to disarm Kripke's puzzle about time and thought: Parikh reduces the paradox to a version of a simpler problem: Can I enter an empty elevator? Yes, if all I ask is that the elevator be empty prior to my entering it. But if I try to take into account all times at once, and approach this top-down, I will end up demanding that the elevator be empty after I have entered it—"and then I am going to be frustrated" (p. 124).

1.6 Attunements [*Übereinstimmungen*]

In contemporary epistemology, "Gettier" problems, so-called for a famous paper (cf. Gettier 1963), are stock-in-trade. These examples derive, at least, from Russell, who provided several important counterexamples to the ancient attempt, broached (but rejected) in Plato's *Theaetetus*, to analyze knowledge as "justified true belief" (the "JTB" analysis). In considering the fundamentals of epistemology Russell wrote (1912, Chap. XIII):

> At first sight we might imagine that knowledge could be defined as 'true belief'. When what we believe is true, it might be supposed that we had achieved a knowledge of what we believe. But this would not accord with the way in which the word is commonly used. To take a very trivial instance: If a man believes that the late Prime Minister's name began with a B, he believes what is true, since the late Prime Minister was Sir Henry Campbell Bannerman. But if he believes that Mr. Balfour was the late Prime Minister, he will still believe that the late Prime Minister's last name began with a B, yet this belief, though true, would not be thought to constitute knowledge. If a newspaper, by an intelligent anticipation, announces the result of a battle before any telegram giving the result has been received, it may by good fortune announce what afterwards turns out to be the right result, and it may produce belief in some of its less experienced readers. But in spite of the truth of their belief, they cannot be said to have knowledge. Thus it is clear that a true belief is not knowledge when it is deduced from a false belief.
>
> In like manner, a true belief cannot be called knowledge when it is deduced by a fallacious process of reasoning ...

At one level, the problem concerns the fact that we seem in such cases to want to say that someone does not know a proposition even though they meet the JTB conditions: it looks as if there are numerous Gettier counterexamples to the JTB theory. More interestingly, though, are the features of our concepts of knowledge and justification that emerge through reflection on such examples. What is at issue may not only be justification and belief, but the character of their "ordinary" environments, including the "normality" of the participants to the occasion. Is the environment, unbeknownst to the believer, somehow unusual? Perhaps one is driving in "fake barn country", but happens to spot a real barn by accident, getting to the right belief. Perhaps, in the fog, one dimly glimpses a sheep-shaped rock, announces "there's a sheep down there!", and lo, there is one behind it. We say one doesn't know, but that emerges from the abnormalities of the environment: non-accidental accidents, so to speak. And yet: are not such "abnormalities" often very normal?

One feature of the environment is, of course, the audience and the social context in which the knowledge claim takes place. Is the audience complicit in the justificatory process? Perhaps the newspaper was sent a passenger pigeon message saying the opposite of what it came to print, but the clerk misread it, leading to the happy "prediction". If all are satisfied in the end, and the newspaper's circulation increases, are we still to be concerned about not having fulfilled the JTB conditions? This raises the question of how many possibilities are to be taken into account in analyzing the notion of knowledge.[39]

From the point of view of social software, however, there is an even deeper point: *every* claim to have analyzed or understood a knowledge claim is itself at issue in taking an articulation for granted. The terms used in the knowledge representation, the principles formulated, the very structure of how the environment is conceived (as dynamic, evolving, or fixed), all matter. Philosophy at its foundations must be, in the end, *first-order*. That is to say, "explanations come to an end somewhere" (PI §1), and in particular, they must be intersubjectively shareable, in the social light of things. There may be no particular place at which they all end, but it is part and parcel of our notion of justification that some group can implement an explanation, can use it. A justification must be *communicable*.

As Parikh and Renero (forthcoming) explain, the point runs deep, affecting even the notion of a Turing machine. In Turing's model, symbols need to be, not merely taken in at a glance, but parsed and grasped and understood as things to be *used*. So it is not just sequences of symbols and their length that matter, but also the individuation of the symbols themselves, and that within the social (perhaps even the mathematical) context in which the program is carried out or applied.

In Plato's *Theaetetus* the JTB analysis is examined at length, but in the end rejected by Socrates as not improving upon the initial, purported analysis of knowledge as perception. This does not imply that the notion of perception is itself clear, and it is to be hoped that Parikh might attack the logic of perception and at some future point (cf. Putnam's 1999 for an important step in this direction). At some point, however, something must be taken for granted. This does not mean that point cannot itself be called into question in turn. But not everything always can be called into question at the same time. And there is always a point at which analysis stops.

This, I would say, is a very important, and very traditional piece of the notion of *rigor* itself (cf. Floyd 2013a). To be fully rigorous, one must admit that at some point we must rest with something taken for granted. Moreover, the endpoint sought after from Plato to Dedekind to Frege to Turing is something we ought to strive to produce as commonsensical, and shared, naïve and everyday. This implies a *variety* of different forms of rigorization.

This brings us to another fundamental notion of Wittgenstein's later philosophy that Parikh has been developing (wittingly or not) with the notion of social software: the notion of an attunement [*Übereinstimmung*].

Let us begin from the thought that in everyday life, our commerce and interchange with one another, in a very everyday sense, ubiquitously involve understandings and attunements, what Wittgenstein calls "*Übereinstimmungen*" among us. The first thing to see is that this is a phenomenon that is to be treated as a given. The second thing to stress is that it is very difficult to see how it may be given.

---

[39]As Feltz (2008) and others argue, there is evidence of cross-cultural variation and bias documentable in speakers' responses to Gettier and other problems in philosophy. Another "fact" of social software, suggesting that the rise of big data sets in philosophy would be helped by attention to Parikh's point of view.

The German word *Übereinstimmung*, as Stanley Cavell emphasized in *The Claim of Reason* (1979), may be translated as "harmony" or "attunement"; examples would be choral voicings (in German, "*Stimme*" means voice, as well as, through the verb, "agree": "*Das stimmt!*", "That's right!". A good translation for the general notion in Wittgenstein would be "attunement" or "accord", as in pitches or tones or devices like clocks or metronomes.

Human beings are capable, in groups, of being mutually voiced with respect to their attunements (Cavell 1979, p. 32). How far it is necessary for a person to feel herself to have a voice in a discussion will depend upon the case: some situations are cooperative, some are not, and some speakers are situated differently from others—sometimes hiding certain specific pieces of knowledge, or using infrequently used pass-phrases in unusual ways, or holding back on information for some other reason. Yet of course "attunements" may either be in place, or disrupted on any given occasion. They may be displayed (or thwarted) in a performance—say, of Schubert's quintet in C major, to take an instrumental piece—to a greater or lesser extent, and in one or more dimensions: emotional, expressive, with respect to the specific notes played, the tuning of the instruments, musicality, allusions to other performances and traditions, contrasts or similarities in emphases among specific instruments, players, and so on. Attunement is thus a matter of degree, and is multifarious, even if it can be pointed out as occurring on a given occasion. As in choral singing, there may be attunement among different voices, and a performance may consists of many smaller ones. There may be a "polyphonic" quality of orchestration or timed dissonance—just as Wittgenstein orchestrates throughout the PI.[40] All this implies that in order to see "agreement" or "accord" for what it is, one needs to find the right level of aspects, parameters, and context to draw out the protocols, purposes, interests, and attunements, and this may not be a trivial exercise.

Of course it may be easier to see attunement in action in certain cases rather than in others. As the opening of PI reminds us, the purchase of five apples in a store may display certain specific attunements, though these may be limited and partial with respect to many of the attunements we would ordinarily say we share with our local greengrocer. In the presence of the world wide web, the embedding of such a relatively simple transaction into wider and wider contexts of analysis may well also be going on. Then it is not so clear what the transaction, or the utterances of the parties to the transaction, or we discussing them, are "*really* talking about". Obviously there is a knowledge component in such cases, but whose knowledge is at issue matters, and just as obviously there are incentive-components, cultural components, components of desire, of value, of expertise, social standing, and the shared understandings and expectations of the participants. The multiplicity of these differing components suggests that without at least some of these attunements among us, meaning and significance would be drained from our lives. And yet it may well

[40]See Stern (2005) and Soulez (2005) on reading the PI as a "polyphonic" text of different voices: the interpretation is important for indicating the rationale for Wittgenstein's later style of writing as internal to what he has to say about our notion of logic.

never be possible to survey all the attunements that accompany our actions—even very "simple" ones.

Attunements may or may not be "agreements", if the latter are understood on the model of explicit renderings of recognized rules that are correctly applied, as in legal contexts or stipulations of parameters for specific trains of reasoning. Attunements are also not reducible to evolved convention or coordination as these are frequently understood in much philosophy, as in Lewis's (1969) example of two people rowing a boat, each with an oar, or a society's driving on the right or the left hand side of the road. This is because several different pieces of social software may be being implemented at once. "Attunements" should also not be reduced wholly to rationality in the sense of the classical game theorist, as what *counts* as rational on a given occasion may need to take "attunements" or the lack thereof into account. In fact, sometimes the enunciation of an explicit agreement, e.g., that one will pursue classical game-theoretic rationality, may even disrupt an "attunement" and change the game. On other occasions an implicit "attunement" may generate an "agreement", or the expected obligation that would ordinarily accompany one.

The most fundamental point here is that the Wittgensteinian notion of "attunement" is broader than that of "agreement", including it, but also relevant "disagreement" that exhibits a meaningful pattern of accord. In the famous example in PI of the wayward pupil who follows the series +2 up to 1000, and then begins to write 1000,1004,1008,1012 (PI §§185ff), a question arises whether the teacher could possibly have meant the order "+2" in such a "fully" general way in enunciating that it is determined at each step by the teacher's intention what we are to call an *Übereinstimmung*, a step "in accordance with" the intended order. The pupil who adds four, as we would say, may not be doing precisely what the teacher wanted. Nevertheless, as we see s/he has still managed to evince a certain degree of "accord" in error. Then, notes Wittgenstein, there might still be enough common ground that the teacher could try to teach the student her way "as an offshoot or variant" of the pupil's, thus transforming a command into an invitation, and an absence of attunement into an understanding (PI §143).

But not all attunements may necessarily involve understandings that are explicit: they may only be generated in the course of a discussion. The famed "paradox" of PI §201 is that no course of action could be determined by a rule, because every course of action can be "brought into accord with" (*in Übereinstimmung zu bringen*) the rule, so that there would be neither contradiction nor "accord" (*Übereinstimmung*). But since there is such accord, there cannot be an intrinsic paradox. The words "accord" ("*Übereinstimmung*") and rule are, writes Wittgenstein (§224), cousins, they are taught together: we might say that the notion of "correctness" is erected by their means.

This gives Wittgenstein the means to reply to the interlocutor who objects that he has collapsed the notions of truth and falsity into that of *Übereinstimmung* (§241), thus losing sight of truth and knowledge:

> "So you are saying that human agreement decides what is true and what is false?"

Wittgenstein answers:

> —What is true or false is what human beings *say*; and it is in their *language* that human beings agree (*stimmen die Menschen überein*). This is agreement (*Übereinstimmung*) not in opinions, but rather in form of life.

He adds (§242) that

> ...communication by means of language requires, not only agreement (*Übereinstimmung*) in definitions, but (odd as this may sound) agreement in judgments. This seems to abolish logic, but does not do so.

Attunement in Wittgenstein's sense will depend upon our sharing, roughly speaking, a sense of what is outrageous, what is sensible, what is reasonable, and what is what. This does not eliminate the possibility of error, it illuminates it. Again (§345), what has to be accepted are *forms of life*. And as always in Wittgenstein—early, middle and late—form is the *possibility* of structure, not a description of facts that actually are. So in addition to our *actual* behavior, there is the question—one for logic, not for psychology alone—of what is possible. We illuminate what is possible for us by varying the sorts of language-games we imagine modeling from the bottom (everyday life) up, rather than taking a top-down approach and beginning with a total space of possibility ahead of time.

Parikh's work, however, goes beyond any easy metaphor of "bottom up" or "top down"—or even "casting a sideways glance" at our impressions of things (cf. PI §274). For he is *implementing* Wittgenstein's notion of surveyability. In the end, this means that there is no fixed stopping point or ground for logical analysis, no ultimate simples or protocols outside of what we (are willing to) do and use. Instead, each analysis ends where it ends, and may be used accordingly. Each analysis erects its own terms, protocols, schematic understandings, etc. from somewhere. But, as I have emphasized, there is no "there" there, no final point where the metaphors are cashed in once and for all possible times and persons and occasions and places.

Nothing could be more purely philosophical, or more ancient. Somewhere, at some point, whatever the symbols, vocabularies, and logical principles we choose to use, we must take it that our interlocutors can "see" what we are getting at. We appeal to, and for, not merely perception, but perceived attunements. We do not always find them. This does not reduce knowledge to perception (or intuition). Instead, we must work with comparisons, and discuss. This is the import of PI §144, which alludes to the ancient practice in mathematics of *exhibiting* a proof:

> What do I mean when I say "the pupil's ability to learn *may* come to an end here"? Do I report this from my own experience? Of course not. (Even if I have had such experience.) Then what am I doing with that remark? After all, I'd like you to say: "Yes, it's true, one could imagine that too, that might happen too!"—But was I trying to draw someone's attention to the fact that he is able to imagine that? —I wanted to put that picture before him, and his *acceptance* of the picture consists in his now being inclined to regard a given case differently: that is, to compare it with *this* sequence of pictures. I have changed his *way of looking at things*. (Indian mathematicians: "Look at this!")

Since we have numerous subroutines and elements of logic to work with, we can get on with discussion, understanding that our hold on our notions of truth, knowledge, and the project of logic itself is a complicated affair, one involving, ultimately, culture, history, value, and philosophy. Our preferences, acknowledgments and attunements, which are culturally and historically contingent, require, ultimately, our involvement and responsibility in discussion and disputation with others in our community and tradition.

For Parikh, at their most general, Wittgensteinian attunements are, perhaps among other things, *implementable procedures* that are *recognized* not merely in the sense of the reidentification of particulars, but—just as importantly and not separably—in the sense in which an authority or administration *recognizes* the dominion of an individual or community over a place, idea, or role. (In German one has *erkennen* and *anerkennen* for this, in French *savoir* and *reconnaître*.) There are institutional factors, a system, and there are individuals, and all of these are evolving together. The point of Parikh's notion of social software is to insist that recognition in this "performative" sense is entangled with recognition in the sense of predication or re-identification of a particular, and *vice versa*, and this throughout all of logic and mathematics.

Acknowledgments The author is grateful to Akihiro Kanamori, Juliette Kennedy and two anonymous referees for comments that led to a much improved final version of this paper.

References

Arló-Costa, H., & Parikh, R. (2005). Conditional probability and defeasible inference. *Journal of Philosophical Logic*, *34*, 97–119.

Austin, J. L. (1946). Other minds. *Proceedings of the Aristotelian Society, Supplementary Volume, XX* (1946), 148–187. Reprinted in Austin, J. L. (1976). In J. O. Urmson & G. J. Warnock (Eds.), *Philosophical papers* (pp. 76–116). Oxford University Press.

Başkent, C., Loohuis, O. L., & Rohit, P. (2012). On knowledge and obligation. *Episteme*, *9*(2), 171–188.

Bernays, P. (1959). Betrachtungen zu Ludwig Wittgensteins *Bemerkungen über die Grundlagen der Mathematik*/Comments on Ludwig Wittgenstein's *Remarks on the Foundations of Mathematics.Ratio*, *2*(1), 1–22.

Buss, S. R. (1999). Bounded arithmetic, proof complexity and two papers of Parikh. *Annals of Pure and Applied Logic*, *96*(13), 43–55.

Cavell, S. (1979). *The claim of reason: Wittgenstein, skepticism, morality, and tragedy*. Oxford, New York: Clarendon Press; Oxford University Press.

Chopra, S., Georgatos, K., & Parikh, R. (2001). Relevance sensitive non-monotonic inference on belief sequences. *Journal of Applied Non-Classical Logics*, *11*(1–2), 131–150.

Chopra, S., & Parikh, R. (2000). Relevance sensitive belief structures. *Annals of Mathematics and Artificial Intelligence*, *28*, 259–285.

Collins, A. (2013). Review of Paul Horwich. Wittgenstein's Metaphilosophy. *Notre Dame Philosophical Reviews*. http://ndpr.nd.edu/news/41347-wittgenstein-s-metaphilosophy/.

Copeland, B. J. (2012). *Turing: Pioneer of the information age*. Oxford: Oxford University Press.

Copeland, J. (2004). *The essential turing*. Oxford: Clarendon Press.

Dabrowski, A., Moss, L. S., & Parikh, R. (1996). Topological reasoning and the logic of knowledge. *Annals of Pure and Applied Logic*, *78*(1–3), 73–110.

Davidson, D. (1982). Rational animals. *Dialectica, 36*(4), 317–328. Reprinted in Davidson, D. (2001). *Subjective, intersubjective, and objective* (pp. 95–106). Oxford: Oxford University Press.
Davis, M. (2000). *The Universal computer: The road from Leibniz to Turing* (1st ed.). New York: Norton.
Davis, M. (2001). *Engines of logic: Mathematicians and the origin of the computer*. New York: Norton.
Davis, M. (forthcoming). Universality is ubiquitous. In: J. Floyd & A. Bokulich (Eds.), *Philosophical explorations of the legacy of Alan Turing: Turing 100*, Boston Studies in the Philosophy and History of Science. Berlin: Springer.
Dean, W., & Parikh, R. (2011). The logic of campaigning. In M. Banerjee & A. Seth (Eds.), *Logic and its applications, Lecture Notes in Computer Science, Volume 6521, 4th Indian Conference, ICLA 2011, Delhi, India, January 5–11, 2011* (pp. 38–49). Berlin: Springer.
Diamond, C. (1991). *The realistic spirit: Wittgenstein, philosophy and the mind*. Cambridge, MA: MIT Press.
Dummett, M. (1959). Wittgenstein's philosophy of mathematics. *The Philosophical Review, 68*(3), 324–348.
Dummett, M. (1975). Wang's paradox. *Synthese, 30*(3), 301–324.
Esenine, V. A. (1970). The ultra-intuitionistic criticism and the antitraditional program for foundations of mathematics. In A. Kino, J. Myhill, & R. E. Vesley (Eds.), *Intuitionism and proof theory, Proceedings of the conference at Buffalo, New York, 1968* (pp. 3–46). Amsterdam: North-Holland.
Feltz, A. (2008). Problems with the appeal to intuition in epistemology. *Philosophical Explorations, 11*(2), 131–141.
Floyd, J. (1995). On saying what you really want to say: Wittgenstein, Gödel and the trisection of the angle. In J. Hintikka (Ed.), *From Dedekind to Gödel: The foundations of mathematics in the early twentieth century series: Vol. 251 of the Synthese library* (pp. 373–426). Boston: Kluwer Academic.
Floyd, J. (2001). Prose versus proof: Wittgenstein on gödel, tarski and truth. *Philosophia Mathematica, 3*(9), 901–928.
Floyd, J. (2011). Wang and Wittgenstein. In C. Parsons & M. Link (Eds.), *Hao Wang, logician and philosopher, Texts in Philosophy* (pp. 145–194). London: College Publications.
Floyd, J. (2012). Wittgenstein's diagonal argument: A variation on Cantor and Turing. In P. Dybjer, S. Lindström, E. Palmgren, & G. Sundholm (Eds.), *Epistemology versus ontology: Essays on the philosophy and foundations of mathematics in honour of per Martin-Löf* (Vol. 27, pp. 25–44). Dordrecht: Springer.
Floyd, J. (2013a). The varieties of rigorous experience. In M. Beaney (Ed.), *The Oxford handbook of the history of early analytic philosophy* (pp. 1003–1042). New York: Oxford University Press.
Floyd, J. (2013b). Turing, Wittgenstein and types: Philosophical aspects of Turing's "The reform of mathematical notation and phraseology" (1944–5). In B. S. Cooper & J. van Leeuwen (Eds.), *Alan Turing: His work and impact* (pp. 250–253). Amsterdam: Elsevier Science.
Floyd, J. (forthcoming). Turing on 'Common Sense': Cambridge resonances. In J. Floyd & A. Boklich (Eds.), (forthcoming).
Floyd, J., & Bokulich, A. (Eds.). (2016). *Philosophical explorations of the legacy of alan turing: Turing 100*. Springer: Boston Studies in the Philosophy of Sciences Series.
Floyd, J., & Kanamori, A. (2016). Gödel vis-à-vis Russell: Logic and set theory to philosophy. In G. Crocco, & E.-M. Engelen (Eds.), *Kurt Gödel: Philosopher-Scientist* (pp. 243–326). Presses Universitaires de France.
Friederich, S. (2011). Motivating Wittgenstein's perspective on mathematical sentences as norms. *Philosophia Mathematica, 19*(1), 1–19.
Gettier, E. L. (1963). Is justified true belief knowledge? *Analysis, 23*(6), 121–123.
Gilbert, M. (1989). *On social facts*. London: Routledge.
Gilbert, M. (2014). *Joint commitment: How we make the social world*. Oxford: Oxford University Press.

Gödel, K. (1931). On formally undecidable propositions of Principia mathematica and related Systems I [1931]. In S. Feferman, et al. (Eds.), *Kurt Gödel collected works: Vol. I, Publications 1929–1936* (pp. 145–195). New York: Oxford University Press.
Gödel, K. (1946). Remarks before the princeton bicentennial conference on problems in mathematics [1946]. In S. Feferman, et al. (Eds.), *Kurt Gödel collected works: Vol. II, Publications, 1938–1974* (pp. 144–153). New York: Oxford University Press.
Gödel, K. (1972). Some remarks on the undecidability results. In S. Feferman, et al. (Eds.), *Kurt Gödel collected works: Vol. II, Publications 1938–1974* (pp. 305–306). Oxford: Oxford University Press.
Goldfarb, W. (1985). Poincaré against the logicists. In W. Aspray & P. Kitcher (Eds.), *History and philosophy of modern mathematics, Minnesota Studies in the Philosophy of Science* (Vol. 11, pp. 61–81). Minneapolis, MN: University of Minnesota Press.
Horwich, P. (2012). *Wittgenstein's metaphilosophy*. Oxford: Oxford University Press.
Horwich, P. (2013). *Was Wittgenstein right? The New York Times*, Opinionator, March 3.
Johnson, M. P., & Parikh, R. (2008). Probabilistic conditionals are almost monotonic. *Review of Symbolic Logic*, *1*(1), 73–80.
Kreisel, G. (1958). Wittgenstein's *remarks on the foundations of mathematics*. *British Journal for the Philosophy of Science*, *9*, 135–158.
Kripke, S. A. (1982). *Wittgenstein on rules and private language: An elementary exposition*. Cambridge, MA: Harvard University Press.
Lewis, D. K. (1969). *Convention: A philosophical study*. Cambridge, MA: Harvard University Press.
Lewis, D. K. (1976). Probabilities of conditionals and conditional probabilities. *Philosophical Review*, *85*, 297–315.
Lewis, D. K. (1979). Scorekeeping in a language game. *Journal of Philosophical Logic*, *8*(1), 339–359.
Lewis, D. K. (1996). Elusive knowledge. *Australasian Journal of Philosophy*, *74*(4), 549–567.
Malcolm, N., & Wittgenstein, L. (2001). *A memoir*. Oxford: Clarendon Press.
Millikan, R. G. (2006). Styles of rationality. In M. Nudds & S. Hurley (Eds.), *Rational animals?* (pp. 117–126). Oxford: Oxford University Press.
Misak, C. (2000). *Truth, politics, morality: Pragmatism and deliberation*. New York: Routledge.
Monk, R., & Wittgenstein, L. (1990). *The duty of genius*. New York: Free Press/Jonathan Cape.
Mühlhölzer, F. (2006). "A mathematical proof must be surveyable": What Wittgenstein meant by this and what it implies. *Grazer Philosophische Studien*, *71*, 57–86.
Pacuit, E., & Parikh, R. (2005). Safe votes, sincere votes, and strategizing. Presented at the 16th International conference on game theory, Stony Brook.
Pacuit, E., Parikh, R., & Cogan, E. (2006). The logic of knowledge based obligation. *Synthese*, *149*(2), 311–341.
Parikh, R. (1971). Existence and feasibility in arithmetic. *The Journal of Symbolic Logic*, *36*(3), 494–508.
Parikh, R. (1973). Some results on the lengths of proofs. *Transactions of the American Mathematical Society*, *177*, 29–36.
Parikh, R. (1983). The problem of vague predicates. In R. Cohen & M. W. Wartofsky (Eds.), *Logic, language and method* (pp. 241–261). Boston: Reidel.
Parikh, R. (1986). Introductory Note to Gödel (1936a), "On the Length of Proofs". In S. Feferman (Ed.), *Kurt Gödel collected works: Vol. I, Publications, 1929–1936* (pp. 394–395). Oxford: Oxford University Press.
Parikh, R. (1994). Vagueness and utility: The semantics of common nouns. *Linguistics and Philosophy*, *17*(6), 521–535.
Parikh, R. (1995). How Far can we formalize language games? In E. K. DePauli-Scimanovich & F. Stadler (Eds.), *The foundational debate* (pp. 89–100). Amsterdam: Kluwer Academic.
Parikh, R. (1996a). Some reminiscences of Kreisel. In P. Odifreddi (Ed.), *Kreiseliana. About and around Georg Kreisel* (pp. 89–96). A. K. Peters.
Parikh, R. (1996b). Vague predicates and language games. *Theoria*, *11*(3), 97–107.

Parikh, R. (1998). Length and structure of proofs. *Synthese, 114*(1), 41–48.
Parikh, R. (2001a). Language as social software. In J. Floyd & S. Shieh (Eds.), *Future pasts: The analytic tradition in twentieth century philosophy* (pp. 339–350). New York: Oxford University Press.
Parikh, R. (2001b). Propositions, propositional attitudes and belief revision. In M. Zakharyaschev, K. Segerberg, M. de Rijke, & H. Wansing (Eds.), *Advances in modal logic* (Vol. 2, pp. 381–400). Stanford, CA: CSLI Publications.
Parikh, R. (2002). Social software. *Synthese, 132*(3), 187–211.
Parikh, R. (2003). Levels of knowledge, games, and group action. *Logic and the Foundations of the Theory of Games and Decisions, 57*(3), 267–281.
Parikh, R. (2006). Modal logic and possible worlds. In H. Lagerlund, S. Lindstrm, & R. Sliwinski (Eds.), *Modality matters: Twenty-five essays in honour of Krister Segerberg*, Uppsala Philosophical Studies *53 (pp. 53–339).*
Parikh, R. (2007). Game theory: 5 questions. In V. F. Hendricks & P. G. Hansen (Eds.), *Epistemology: 5 Questions* (pp. 145–153). New York: Automatic Press/VIP.
Parikh, R. (2008). Sentences, belief and logical omniscience, or what does deduction tell us? *Review of Symbolic Logic, 1*(4), 459–476.
Parikh, R. (2009a). From language games to social software. Manuscript. April 16, 2009.
Parikh, R. (2009b). Knowledge, games and tales from the east. In R. Ramanujan & S. Sarukkai (Eds.), *ICLA 2009, LNAI 5378* (pp. 65–76). Berlin: Springer.
Parikh, R. (2013). On Kripke's puzzle about time and thought. In K. Lodaya (Ed.), *Logic and its applications* (pp. 121–126). Berlin: Springer.
Parikh, R. (forthcoming). Is there a church-turing thesis for social algorithms? In J. Floyd & A. Bokulich (Eds.), (forthcoming).
Parikh, R., & Ramanujam, R. (2003). A knowledge based semantics of messages. *Journal of Logic, Language and Information, 12*(4), 453–467.
Parikh, R., & Renero, A. (forthcoming). Justified True Belief: Plato, Gettier and Turing. In J. Floyd & A. Bokulich (Eds.), (forthcoming).
Parikh, R., & Väänänen, J. (2005). Finite information logic. *Annals of Pure and Applied Logic, 134*(1), 83–93.
Putnam, H. (1975). The Meaning of "Meaning". In K. Gunderson (Ed.), *Language, mind and knowledge*. Minnesota Studies in the Philosophy of Science, (Vol. 7, pp. 131–193). Minneapolis: University of Minnesota Press. Reprinted in *Mind, language and reality: Philosophical papers* (Vol. 2, pp. 215–271). New York: Cambridge University Press (1975).
Putnam, H. (1979). Philosophy of mathematics: Why nothing works. In H. Putnam, & J. Conant, (Eds.), *Words and life* (pp. 499–512). Cambridge, MA: Harvard University Press. Originally published in *Current research in philosophy of science: Proceedings of the P.S.A. critical research problems conference*, P.D. Asquith & Henry Kyburg, Jr. (Eds.). East Lansing: Philosophy of Science Association (1979).
Putnam, H. (1999). *The threefold cord: Mind, body, and world*. New York: Columbia University Press.
Putnam, H. (2002). Travis on meaning, thought and the ways the world is. *The Philosophical Quarterly, 52*(206), 96–106.
Quine, W. V. (1984). Relativism and absolutism. *The Monist, 67*(3), 293–296.
Quine, W. V. (1960). *Word and object*. Cambridge, MA: Harvard University Press.
Russell, B. (1912). *The problems of philosophy, Home university library of modern knowledge, 40*. London: Oxford University Press.
Russell, B. (1921). *The analysis of mind*. London, New York: G. Allen and Unwin Ltd.; The Macmillan company.
Russell, B. (1940). *An inquiry into meaning and truth*. New York: W. W. Norton & Company Inc.
Sazonov, V. Y. (1995). On feasible numbers. In D. Leivant (Ed.), *Logic and computational complexity international workshop LCC '94 Indianapolis, IN, USA, October 1316, 1994 Selected Papers Lecture Notes in Computer Science* (Vol. 960, pp. 30–51). New York: Springer.

Searle, J. R. (1984). *Minds, brains and science*. Cambridge, MA: Harvard University Press.
Searle, J. R. (2010). *Making the social world: The structure of human civilization*. Oxford: Oxford University Press.
Shanker, S. G. (1998). *Wittgenstein's remarks on the foundations of AI*. New York: Routledge.
Sieg, W. (2007). On mind and Turing's machines. *Natural Computing*, *6*, 187–205.
Sieg, W. (2008). Church without dogma: Axioms for computability. In B. S. Cooper, B. Löwe, & A. Sorbi (Eds.), *New computational paradigms: Changing conceptions of what is computable* (pp. 139–152). New York: Springer.
Soulez, A. (2005). A case of early Wittgensteinian dialogism: Stances on the impossibility of "red and green in the same place". In A. Pichler & S. Säätelä, (Eds.), *Wittgenstein: The Philosopher and his works* (pp. 313–325). Working Papers from the Wittgenstein Archives at the University of Bergen. University of Bergen.
Stenlund, S. (1996). Poincaré and the Limits of Formal Logic. In J.-L. Greffe, G. Heinzmann, & K. Lotenz (Eds.), *Henri Poincaré, Science and Philosophy, International Congress, Nancy, France 1994* (pp. 467–479). Berlin/Paris: Akademie Verlag/Albert Blanchard.
Stern, D. (2005). How many Wittgensteins? In A. Pichler, & S. Säätelä (Eds.), *Wittgenstein: The philosopher and his works* (pp. 164–188). Working Papers from the Wittgenstein Archives at the University of Bergen. University of Bergen.
Sterrett, S. G. (2012). Bringing up Turing's 'Child-Machine'. In S. B. Cooper, A. Dawar, & B. Löwe, (Eds.), *How the world computes: Turing centenary conference and 8th conference on computability in Europe, CiE 2012, Cambridge, UK, June 18–23, 2012, proceedings* (pp. 703–713).
Sterrett, S. G., (forthcoming). Alan Turing on the integration of machine and human intellectual power. In A. Bokulich & J. Floyd (Eds.), (forthcoming).
Travis, C. (1995). Order out of messes, review of Akeel Bilgrami, *Belief and Meaning*. *Mind*, *104*, 133–144.
Travis, C. (2005). A sense of occasion. *The Philosophical Quarterly*, *55*(219), 286–314.
Travis, C. (2006). *Thought's footing*. New York: Oxford University Press.
Travis, C. (2008). *Occasion-sensitivity: Selected essays*. Oxford: Oxford University Press.
Travis, C. (2013). *Perception: Essays after Frege*. Oxford: Oxford University Press.
Turing, A. M. (1937). Correspondence with Paul Bernays. ETH-Bibliothek, Archive, Zürich, Hs 947: 276. *Hs*, *975*, 4616–4617.
Turing, A. M. (1948/1992). Intelligent machinery. In D. C. Ince (Ed.), *Collected works of A.M. Turing: Mechanical intelligence*. Amsterdam: North-Holland, pp. 107–128.
Turing, A. M. (1936/1937). On computable numbers, with an application to the Entscheidungsproblem. *Proceedings of the London Mathematical Society*, *2*(42), 230–265.
Turing, A. M. (1944–1945). The reform of mathematical notation and phraseology. In R. O. Gandy & C. E. M. Yates (Eds.), *The collected works of A.M. Turing: Mathematical logic* (pp. 211–222). Amsterdam: North-Holland/Elsevier Science.
Turing, A. M. (1950). Computing machinery and intelligence. *Mind*, *59*, 433–460.
Wang, H. (1961). Process and existence in mathematics. In Y. Bar-Hillel, E. I. J. Poznanski, M. O. Rabin, & A. Robinson (Eds.), *Essays on the foundations of mathematics, dedicated to Prof. A. A. Fraenkel on his 70th anniversary* (pp. 328–351). Jerusalem: Magnes Press, the Hebrew University of Jerusalem.
Wang, H. (1974). *From mathematics to philosophy. International library of philosophy and scientific method*. New York: Humanities Press.
Wang, H. (1996). *A logical journey: From Gödel to philosophy*. Cambridge, MA: The MIT Press.
Webb, J. (1990). Introductory Note to Gödel 1972a. In S. Feferman, et al. (Eds.), *Kurt Gödel collected works: Vol. II, Publications, 1938–1974* (pp. 292–304). New York: Oxford University Press.
Williams, M. (1999). *Wittgenstein, mind, and meaning: Toward a social conception of mind*. London: Routledge.

Williams, M. (2009). *Blind obedience: The structure and content of Wittgenstein's later philosophy*. London: Routledge.
Winch, P. (2008). *The idea of a social science and its relation to philosophy*. London: Routledge classics, Routledge.
Wittgenstein, L. (1978). *Remarks on the foundations of mathematics* (revised ed.). Cambridge, MA: M. I. T. Press.
Wittgenstein, L. (2005). *Ludwig Wittgenstein: Gesamtbriefwechsel/Complete Correspondence. Electronic Edition*. Oxford University Press/Bergen University Wittgenstein Archives.
Wittgenstein, L. (2009). *Philosophische Untersuchungen = Philosophical investigations*, rev. 4th ed. (trans: G. E. M. Anscombe, P.M.S. Hacker, & J. Schulte). Chichester: Wiley-Blackwell.
Wittgenstein, L., Anscombe, G. E. M., & von Wright, G. H. (1980). *Remarks on the philosophy of psychology* (Vol. 1). Oxford: Blackwell.
Wright, C. (1980). *Wittgenstein on the foundations of mathematics*. Cambridge: Harvard University Press.
Wright, C. (1993). Strict finitism. *Realism, meaning and truth* (2nd ed., pp. 107–175). Oxford: Blackwell.

Chapter 2
Vagueness, Communication, and the Sorites Paradox

Prashant Parikh

Abstract In this paper, I model first-order and higher-order vagueness, look at certain aspects of vague communication, and offer an intuitively appealing resolution of the sorites paradox.

Keywords Vagueness · Cognitive models · Theory of communication · Sorites' paradox

2.1 Introduction

Practically every word in a natural language is vague. Despite a large philosophical literature, the attention this fact has received within semantics is surprisingly scant, perhaps because quite new ideas and tools are required to accommodate it.

Parikh (1994) may have been the first person to study how communication with vague expressions can be fruitful even when the speaker and addressee interpret them differently. His paper contributed three main things. First, it showed that, within limits, vagueness is not by itself an obstacle to effective communication, a pervasive but relatively unnoticed problem. Second, it challenged the orthodoxy that communication involves the transmission of a single *objective* proposition from one agent to another. Last, it made *agency* central in such considerations in a simple and appealing way. These Wittgensteinian—and, arguably, ancient Jaina[1]—insights helped move the conversation about vagueness away from purely metaphysical and logical issues towards questions of use.

In this paper, I attend to three problems he left open in his analysis. I start by modeling vagueness, that is, characterizing clear cases and borderline cases of a

[1]See Mohanty (2000, p. 4).

I would like to thank James Hampton, Gregory Murphy, Daniel Osherson, Rohit Parikh, and Stephen Schiffer for helpful conversations.

P. Parikh (✉)
New York, NY, USA
e-mail: pparikh1@gmail.com

C. Başkent et al. (eds.), *Rohit Parikh on Logic, Language and Society*,
Outstanding Contributions to Logic 11, DOI 10.1007/978-3-319-47843-2_2

vague concept. Then I ask: *how* do speakers and addressees decide to use or interpret a vague expression in the first place? Borderline cases and, indeed, borders that shift with context pose additional difficulties in understanding natural language generation and interpretation. Finally, I tackle the *sorites paradox*. It is perhaps this puzzle that has occupied most writers on vagueness, especially philosophers and logicians. Parikh (1983) has also discussed variants of this paradox.

I approach this task by adapting models from cognitive psychology as it appears that psychologists understand this domain better than philosophers or linguists owing to their emphasis on experimental data. I also apply these adapted models in new philosophical and linguistic ways. Once I describe vagueness, it becomes easy to incorporate it in a larger theory of communication and meaning. With the right models, the sorites paradox also yields to a natural and intuitive resolution.

2.2 Vagueness

I first clarify some basic terminology. Concepts are taken to be mental representations of collections of things and categories the collections themselves. Concepts correspond to properties or attributes or features, all terms referring to abstract entities. So an agent's vague concept *bald* corresponds to a vague property of baldness. Since each agent will have a slightly different concept *bald*, the corresponding vague property can be thought of in two ways: as a kind of social and abstract average of these individual representations[2] and as a kind of abstract individual counterpart to the concept, one for each agent. Both kinds of property are important, the average kind and the individual kind. Vague words such as BALD have vague concepts conventionally attached to them that serve as their conventional meanings.

It has been taken for granted from classical times until Wittgenstein (1953/1968, sections 66 and 67, pp. 31–32) questioned it that most concepts have clear definitions, that is, non-circular necessary and sufficient conditions. This implies every object is or is not a member of the corresponding category. As discussed by Smith and Medin (1981), this classical view and its variants are untenable primarily because most concepts are vague and have borderline cases.[3] Here is Murphy's (2004, p. 21) account of *why* vagueness is ubiquitous:

> The Necessity of Category Fuzziness
>
> The existence of unclear examples can be understood in part as arising from the great variation of things in the world combined with the limitations on our concepts. We do not wish to have a concept for every single object—such concepts would be of little use and would require enormous memory space. Instead, we want to have a relatively small number of concepts that

[2] In general, properties are individuated from *reality* by agents and so are social but abstract constructs that nevertheless have a certain objectivity. For example, the number 5 can be thought of as being abstracted from collections of five objects just as the property of being blue can be thought of as being discriminated from blue objects.

[3] The other important theoretical reason is that they are unable to account for typicality effects. And much of the experimental evidence disconfirms them.

> are still informative enough to be useful (Rosch 1978). The ideal situation would probably be one in which these concepts did pick out objects in a classical-like way. Unfortunately, the world is not arranged so as to conform to our needs.
>
> The gradation of properties in the world means that our smallish number of categories will never map perfectly onto all objects: the distinction between members and nonmembers will always be difficult to draw or will even be arbitrary in some cases. If the world existed as much more distinct clumps of objects, then perhaps our concepts could be formed as the classical view says they are. But if the world consists of shadings and gradations and of a rich mixture of different kinds of properties, then a limited number of concepts would almost have to be fuzzy.

As described by Murphy (2004, Chap. 3), there are three new views that have emerged: the exemplar approach, the prototype approach, and the knowledge approach. The first uses the information provided by each encounter with an exemplar for a category separately; the second works with a *summary representation* of a category derived from the experience of exemplars; and the third integrates concepts with the broader knowledge schemata in which they must reside via plausible reasoning.

Because concepts are used in very diverse tasks, none of these approaches is able to account for all the empirical data. Indeed, as Murphy (2004, Chap. 13) concludes, some amalgam of the three will probably be required for a "Big Theory of Concepts" as they also appear to be somewhat complementary in their explanatory adequacy. In other words, each approach focuses on a different source of information, and any one or more of these sources may be summoned for a particular task based on their suitability.[4]

I will adapt the first two approaches for my purposes. The third knowledge approach is, in a sense, not really an independent stand-alone approach but one that operates by combining knowledge effects with one or both of the other approaches. I construct the simplest models required for the tasks at hand and do not aim at more comprehensive versions.

Basic Setup

Both the exemplar and prototype approaches rely on the idea that each exemplar of a concept has multiple properties that take on particular values. For example, the concept *bald* may involve features such as the number of hairs on the scalp, the number of completely hairless patches on the scalp, the fraction of the scalp that is hairless, and so on;[5] each exemplar will instantiate these attributes with particular numbers. In other words, each concept is associated with an n-dimensional attribute space and each exemplar can be represented by a point in this space. Some dimensions may be continuous and some may be discrete but this should not affect the basic idea.

Let b_i with $i = 1, 2, \ldots, N$ be clear exemplars of *bald* that an agent $\mathcal{A}$ has encountered in his experience. Likewise, let $b'_{i'}$ with $i' = 1, 2, \ldots, N'$ be clear exemplars of

---

[4] I apologize to the reader for this extremely abstract summary of a very large and fascinating field.

[5] Such a listing of features is an idealization as they are not entirely independent of one another. It is not clear, however, whether agents actually operate with completely independent attributes. Presumably, this depends on what they know and this is one way in which knowledge effects enter.

not bald. A negative category such as the latter is a little unusual in that it contains not just persons with a full head of hair but also random items such as clocks and cars. There is no problem with this as all potential exemplars are assessed relative to the relevant attributes which come from the corresponding positive category. Thus, only persons with relatively full heads of hair will qualify as exemplars and items such as clocks and cars will be discarded as junk. A different way to think about negative categories is that in any particular situation where its exemplars are accessed there will always be a default reference category that will automatically limit the possibilities to the relevant types of individuals. In the case of *not bald*, the possibilities will be limited to persons; in the case of *not tall* to men or women or basketball players depending on the situation; and in the case of *not chair*, the default category might be items of furniture.

Let x_{ij} be the value of the jth attribute of b_i and, similarly, let $x'_{i'j}$ be the value of the jth attribute of $b'_{i'}$. That is, $b_i = (x_{ij})$ and $b'_{i'} = (x'_{i'j})$, the right-hand side of both equalities being vectors with $j = 1, 2, \ldots, n$.

Now suppose $\mathcal{A}$ has to judge whether the candidate a is bald or not bald or borderline bald in some situation u. Then a will also be a point in the same space with value x_{aj} in the jth dimension. That is, $a = (x_{aj})$.

The basic idea underlying both approaches is to see how "far" a is from all the exemplars taken separately or from an "average"exemplar (i.e. the prototype) and, based on this, to see how similar a is to the other members in the category. This computation allows $\mathcal{A}$ to decide where a stands with respect to *bald*.

The Exemplar Model

This model has its roots in Medin and Schaffer (1978), Nosofsky (1992), and Nosofsky and Palmeri (1997). I build upon the description in Murphy (2004, pp. 65–71). Schiffer (2010) informally mentions the possibility of using weighted distance in the context of vagueness.

In order to get at the psychological distance between a and b_i, we need to first note the following. For certain attributes such as the number of hairs on an individual's scalp, if a's value x_{a1} is less than b_i's value x_{i1} then the psychological difference between these values along this dimension is not $|x_{a1} - x_{i1}|$ but 0 because b_i is an *exemplar* and a has, so to speak, met the *bar* set by b_i. Likewise, if the attribute is the number of completely hairless patches on the individual's scalp, if $x_{a2} > x_{i2}$ then again the difference is 0 by the same reasoning. There may, of course, be attributes where only an exact equality $x_{aj} = x_{ij}$ results in a zero difference.[6] Which of these cases obtains depends on the particular concept and the attribute in question. So we can define a psychological difference function $\delta(x_{aj}, x_{ij})$ which is either 0 or $|x_{aj} - x_{ij}|$ based on the nature of the concept and attribute being considered.[7]

[6] For example, the category *blue* is such because overshooting the relevant color frequency in either direction counts as a non-zero difference. With such attributes, only an exact equality results in a zero difference.

[7] There is some empirical warrant for such a result as reported in, for example, Hampton et al. (2005).

Now define the weighted psychological distance between a and b_i as follows:

$$d_u(a, b_i) = \sqrt{\sum_{j=1}^{n} w_j(u)\, \delta(x_{aj}, x_{ij})^2}$$

Here, $w_j(u)$ are weights issuing from the situation u. The psychological distance that $\mathcal{A}$ perceives between a candidate and an exemplar thus varies with the situation he is in. This variation implies that certain attributes and therefore certain exemplars will play a more or less important role in $\mathcal{A}$'s judgment.

This distance function is *not* a metric in the technical sense as it is not symmetric: $d_u(a, b_i)$ may not equal $d_u(b_i, a)$ because the underlying psychological difference function δ is not symmetric. Also, many different forms for it can be used; I have restricted myself to the commonest Euclidean variety.

Correspondingly, the weighted psychological distance between a and $b'_{i'}$ will be:

$$d_u(a, b'_{i'}) = \sqrt{\sum_{j=1}^{n} w'_j(u)\, \delta(x_{aj}, x'_{i'j})^2}$$

Shepard (1987) has shown that behavioral similarity between items is an exponentially decreasing function of their psychological distance.

$$s_u(a, b_i) = e^{-c(u)d_u(a,b_i)}$$

where $c(u) > 0$ is a situation-based parameter. Again, a larger or smaller $c(u)$ will determine the relative importance of items that are near and items that are far.

Analogously:

$$s_u(a, b'_{i'}) = e^{-c'(u)d_u(a,b'_{i'})}$$

Finally, define the psychological probability that a is bald rather than not bald for the agent $\mathcal{A}$ as follows:

$$P(bald \mid a; u) = \frac{\sum_{i=1}^{N} s_u(a, b_i)}{\sum_{i=1}^{N} s_u(a, b_i) + \sum_{i'=1}^{N'} s_u(a, b'_{i'})}$$

Then:

$$P(not\ \ bald \mid a; u) = \frac{\sum_{i'=1}^{N'} s_u(a, b'_{i'})}{\sum_{i=1}^{N} s_u(a, b_i) + \sum_{i'=1}^{N'} s_u(a, b'_{i'})}$$

Note that $P(bald \mid a; u) + P(not\ \ bald \mid a; u) = 1$. These psychological probabilities are measured with respect to the agent $\mathcal{A}$ as they are based on his exemplars. So they are agent-relative probabilities. But they are not "subjective" probabilities in the usual sense of being $\mathcal{A}$'s beliefs. That is, an agent's beliefs are not related in the way psychological probabilities are to his exemplars. This means the latter are not open to the standard charges against epistemic probabilistic accounts that are based on an agent's beliefs.

The probabilities depend on the situation u which is a parameter, not a conditioning random variable. If we wish, we can introduce weights $v_i(u)$, $v'_{i'}(u)$ for each pair of similarities $s_u(a, b_i)$, $s_u(a, b'_{i'})$ but I will not. They would allow us to weight different exemplars directly.

I will return to these identifications after describing the prototype model.

The Prototype Model

This model has its roots in Rosch and Mervis (1975) but the account below is based on a certain natural construal of a summary representation of a category.

The only difference between the exemplar model and the prototype model is that the latter does not compute the psychological distance between the candidate and each exemplar separately as above but first averages the values of all the exemplars and then computes the distance.

So we first define the average values as follows:

$$\overline{x_j} = \frac{\sum_{i=1}^{N} w_i(u) x_{ij}}{N}$$

$$\overline{x'_j} = \frac{\sum_{i'=1}^{N'} w'_i(u) x'_{i'j}}{N'}$$

This tells us that the prototypes for *bald* and *not bald* are just $\overline{b} = (\overline{x_j})$ and $\overline{b'} = (\overline{x'_j})$. The weights $w_i(u)$, $w'_i(u)$ are different from the earlier weights above described in the exemplar model, and are also indexed with respect to i and not j as before, but I have used the same letters to avoid using an excessive number of symbols. These weights play an important role because with many categories such as *bald* it is not ordinary averages that count as prototypes but extreme examples such as a completely hairless person. Such extreme exemplars can be selected as the relevant prototypical average by adjusting the weights suitably. Alternatively, they can be selected as the minimum or maximum of the relevant attribute values. Generalized means are a family of functions for aggregating sets of numbers and we can draw upon any of these based on the nature of the concept and its attributes.[8]

We can use the same idea for the psychological difference as before except that it is measured with respect to the average values. So $\delta(x_{aj}, \overline{x_j})$ is either 0 or $|x_{aj} - \overline{x_j}|$ based on the nature of the concept and attribute being considered and likewise with $\delta(x_{aj}, \overline{x'_j})$.

[8]See, for example, http://en.wikipedia.org/wiki/Generalized_mean.

Now the weighted psychological distance from the prototype is defined as follows:

$$d_u(a, \overline{b}) = \sqrt{\sum_{j=1}^{n} w_j(u)\ \delta(x_{aj}, \overline{x_j})^2}$$

$$d_u(a, \overline{b'}) = \sqrt{\sum_{j=1}^{n} w_j'(u)\ \delta(x_{aj}, \overline{x_j'})^2}$$

This in turn leads to similarity.

$$s_u(a, \overline{b}) = e^{-c(u)d_u(a,\overline{b})}$$

$$s_u(a, \overline{b'}) = e^{-c'(u)d_u(a,\overline{b'})}$$

And finally to the psychological probabilities for $\mathcal{A}$ as above.

$$P(bald \mid a; u) = \frac{s_u(a, \overline{b})}{s_u(a, \overline{b}) + s_u(a, \overline{b'})}$$

$$P(not\ \ bald \mid a; u) = \frac{s_u(a, \overline{b'})}{s_u(a, \overline{b}) + s_u(a, \overline{b'})}$$

As stated above, the key difference is that distances and similarities are measured with respect to a "summary representation," an *average* (or generalized mean) of all the exemplars.

Both models give us somewhat different ways to compute the same psychological probabilities $P(bald \mid a; u)$ and $P(not\ bald \mid a; u)$. I now put them to use.

Characterizing Vagueness

Intuitively, if a candidate is sufficiently similar to clear exemplars of both *bald* and *not bald*, it is reasonable to think it is a borderline case. This suggests the following definitions.

Definition 1 A candidate a is a borderline case of a concept C for an agent $\mathcal{A}$ in situation u if and only if $|P(C \mid a; u) - P(not\ C \mid a; u)| < \epsilon_u$ where $0 < \epsilon_u < 1$ is $\mathcal{A}$'s threshold in u.[9] If a is not a borderline case, then it is classified as clearly belonging to C if and only if $P(C \mid a; u) > P(not\ C \mid a; u)$ and as clearly belonging to *not* C if and only if $P(C \mid a; u) < P(not\ C \mid a; u)$.

[9] I have deliberately suppressed the agent in specifying the probabilities and threshold to avoid notational clutter.

Definition 2 A concept is vague for an agent $\mathcal{A}$ in situation u if and only if it has borderline cases. Otherwise, it is precise or "classical."

The decision to count an item as borderline or clear is *derived from* agent-relative psychological probabilities but is not itself probabilistic or belief-based (e.g. a is borderline with "subjective" probability q). So it is doubly immune to the charge against subjective probability views made by Schiffer (2003, Chap. 5): it is not based on beliefs and it is deterministic.

It also does not follow the Luce Choice Rule as described by Murphy (2004, p. 69) which enjoins the agent to make judgments that are themselves probabilistic. This rule has some empirical support for a task such as classifying an animal as a dog, cat, or burro, but has probably not been tested against my way of using these probabilities to get at complementary categories such as *bald* and *not bald.*

The threshold ϵ_u can be thought of in different ways. It can be conceived as a precise number or a fuzzy number.[10] It should also be seen as something the agent does not know for himself in different situations. That is, it operates sub-personally or non-consciously and the agent may lack a firm conviction about a particular decision to count an item as borderline or clear. Lastly, the situated threshold arises through communicative interactions and so agents in the same community tend to share it to a greater degree than intuition might suggest. Here, knowledge effects of the kind alluded to earlier when I described the knowledge approach to concepts may play an important role as such thresholds tend to partly arise also from the goals and interests of agents.

$\mathcal{A}$'s cognitive system determines when a case is borderline or clear. It is not entirely a conscious decision. This is confirmed by the familiar feeling of being stymied when we are asked to make a conscious judgment about a borderline case. There is simply no way to reason *decisively* about it based on the external facts. In a sense, the exemplars and the thresholds are the inaccessible "hidden variables" of the decision process. Indeed, when we scrutinize a judgment we have arrived at we may become uncertain about it as there is *no non-vague argument* available for the *feeling* that a is borderline. It is largely sub-personal. And all this is intensified when the threshold is a fuzzy number rather than a precise number.

An item a that is a borderline case for one agent need not be so for another agent. Likewise, the borderline cases of a concept shift with u for the same agent because all aspects of the definition depend on u, the probabilities as well as the threshold. The same agent $\mathcal{A}$ may choose to call Alex bald in one situation but not in another as the sentence ALEX IS BALD may be true for $\mathcal{A}$ in one situation, false

[10] A fuzzy number A is generally expressed as a function $A : \mathbb{R} \rightarrow [0, 1]$ such that:

$$A(x) = \begin{cases} f(x) & \text{for } x \in [a, b] \\ 1 & \text{for } x \in [b, c] \\ g(x) & \text{for } x \in [c, d] \\ 0 & \text{for } x < a \text{ and } x > d \end{cases}$$

where $a \leq b \leq c \leq d$, f is a continuous function that increases to 1 at point b, and g is a continuous function that decreases from 1 at point c. See Klir et al. (1997, p. 170).

for $\mathcal{A}$ in another, and indeterminate for $\mathcal{A}$ in a third. Consider the sentences SHE WON'T DATE ALEX—HE'S BALD and ALEX ISN'T BALD—HE NEEDS A HAIRCUT. In the first case, the situation u is such that a less stringent membership condition for *bald* is operative, either because relatively less bald exemplars are weighted more or because the magnitude of the threshold ϵ_u is relatively smaller making the penumbra, the region of borderline cases, correspondingly smaller as well, or possibly because both factors apply simultaneously. In the second case, the situation u is the opposite—a more stringent membership condition is used. This explains what Schiffer (2010) calls *penumbral shift* in a natural way.

In order to obtain "natural" concepts of the kind that would be useful in thinking and communication, we must assume that the positive exemplars (i.e. the set $\{b_i\}$ for *bald*) are so distributed that all members of their convex closure are instances of C. Otherwise, we would get strange and seemingly arbitrary outcomes for what belongs to C, what is borderline C, and what is not C. This requirement translates into a restriction on ϵ_u: it must be sufficiently small. In other words, if $b = (b_i)$, $b' = (b'_{i'})$, the latter in each case being the vectors of positive and negative exemplars of *bald*, then $0 < \epsilon_u < \kappa_u(b, b') < 1$ where $\kappa_u(b, b')$ is a function of the N positive exemplars and N' negative exemplars that derives from the convexity assumption. The same kind of condition is obviously not required for the negative exemplars (i.e. the set $\{b'_{i'}\}$ for *bald*) as they can, in general, lie anywhere outside the convex closure of the positive exemplars in the n-dimensional attribute space. A candidate a that is judged to belong to C *need not* lie within the convex closure. All that is required is that it be sufficiently close to it. Indeed, subsequently, a would become a positive exemplar itself and the convex closure could be correspondingly enlarged. This suggests a dynamic model of concept learning that results in possibly expanded convex closures as more exemplars are encountered. After a while, the category would converge to a convex polytope in the attribute space with somewhat different boundaries for different situations u.

This convexity assumption is very similar to the convexity assumption made by Gärdenfors (2000) and Warglien and Gärdenfors (2013). The approach I have adopted of using exemplars to derive concepts seems to allow a clearer development of these ideas from a more foundational starting point. Also, their decision to banish the *external* significance of language from their model seems unnecessary and raises too many problems (e.g. Barwise and Perry 1983, pp. 28–31; Putnam 1975). As I discuss briefly in the next section, conventional meanings are mental representations but referential meanings are external entities such as the individuals and properties that make up propositions. It is possible to have one's cake and eat it too.

What may be true or false or indeterminate for one agent may not be so for another. There is *no* agent-independent or objective truth value in other words. However, because the agents must belong to the same linguistic community and the exemplars each agent draws upon are often shared through communication, they may agree more often than expected. In fact, for a concept to be socially useful as most are, its exemplars must be *sufficiently* shared among the community. This points to a community model of interacting agents where concepts are constantly being revised to have sufficient overlap of the kind explored by Parikh (1994).

The definitions above suggest that if the threshold is a precise, non-fuzzy number there is no *higher-order vagueness*, the phenomenon that the borderline between clear and borderline cases is itself unclear, resulting in borderline borderline cases and so on ad infinitum. The foregoing implies an epistemic view of this as agents are typically unaware of their thresholds in particular situations.

If we wish to allow for higher-order vagueness, we can identify the threshold with a fuzzy number. In this case, there is no precise cut-off between clear and borderline cases and higher-order vagueness can be admitted. I believe my model is richer than Hampton (2007) because it allows for both graded membership in a category as well as fuzzy judgments of when a case is clear or borderline whereas Hampton (2007, p. 377) only allows for the former while pointing to the latter as important experimental evidence. Something akin to Definition 1 would have to be introduced into his model.

Much of the evidence seems to indicate that higher-order vagueness is real. This implies that the threshold ϵ_u must be a fuzzy number and not a precise number. However, for the purposes of this paper, I keep the matter open and continue to discuss both possibilities.

The mistake "epistemicists" such as Williamson (1994) seem to make is to assume the existence of sharp thresholds between clear cases of a concept and its complement. That is, they not only reject higher-order vagueness but also first-order vagueness, which is completely unrealistic. It is easy enough to introduce precise thresholds between clear and borderline cases and thus allow for first-order vagueness without higher-order vagueness; further, for greater realism, by taking the threshold to be a fuzzy number, we can accommodate higher-order vagueness as well.

Indeed, it is possible to *characterize* higher-order vagueness by treating Definition 1 as a base case for an inductive definition. The key idea is to identify exemplars at each level, and therefore, psychological distance, similarity, and psychological probabilities at each level. For example, Definition 1 provides a precise or fuzzy account of first-order borderline cases. Then we can identify positive and negative exemplars for what is clearly borderline and what is clearly not borderline which, in turn, gives rise to psychological distance and similarity and psychological probabilities at the next level. The latter can then be used in a manner analogous to Definition 1 to define borderline borderline cases or, in other words, second-order vagueness. And so on to higher-order vagueness for all n.

In the definition below, I assume that we have the exemplars for nth order vagueness and therefore the nth order psychological probabilities $P_n(C_n \mid a; u)$ and $P_n(\textit{not } C_n \mid a; u)$ where C_n is the nth order concept of being nth order borderline C and *not* C_n is the corresponding complementary concept. When $n = 0$, this is understood as just standing for the concepts C and *not* C.

Definition 3 A candidate a is an $(n+1)$st order borderline case of a concept C for an agent $\mathcal{A}$ in situation u if and only if $|P_n(C_n \mid a; u) - P_n(\textit{not } C_n \mid a; u)| < \epsilon_{n,u}$ where $0 < \epsilon_{n,u} < 1$ is $\mathcal{A}$'s threshold in u. If a is not an $(n+1)$st order borderline case, then it is classified as being clearly nth order borderline or clearly not nth order borderline according as $P_n(C_n \mid a; u)$ or $P_n(\textit{not } C_n \mid a; u)$ is greater.

Combining this definition with Definition 1 yields a characterization of higher-order vagueness for all n. Now, the threshold $\epsilon_{n,u}$ has to be understood as fuzzy and this gives rise to a fuzzy *fractal*[11] set with no crisp boundaries even in the limit. There is some indirect evidence for the fractal nature of higher-order vagueness in Hampton et al. (2012). In practice, of course, an agent will not actually possess the threshold $\epsilon_{n,u}$ for all n, only for the first and possibly second order borderline case as there is no practical utility in having such thresholds.

A slightly different way to approach vagueness is described below.[12]

Definition 4 $P(\textit{borderline } C \mid a;\, u) = 1 - |P(C \mid a;\, u) - P(\textit{not } C \mid a;\, u)|$.

Now we have $P(C \mid a;\, u)$, $P(\textit{not } C \mid a;\, u)$, and $P(\textit{borderline } C \mid a;\, u)$.[13] These psychological probabilities can be used to define borderline cases in different ways. One possible decision rule is that $\mathcal{A}$ judges a to be borderline $P(\textit{borderline } C \mid a;\, u) \times 100$ percent of the time. For example, if $P(\textit{borderline } C \mid a;\, u) = 0.6$ then $\mathcal{A}$ will judge a to be borderline C 60 % of the time.

Definition 5 A candidate a is a borderline case of a concept C for an agent $\mathcal{A}$ in situation u $P(\textit{borderline } C \mid a;\, u) \times 100$ percent of the time. If a is not judged to be a borderline case, then it is classified as clearly belonging to C if and only if $P(C \mid a;\, u) > P(\textit{not } C \mid a;\, u)$ and as clearly belonging to *not* C if and only if $P(C \mid a;\, u) < P(\textit{not } C \mid a;\, u)$.

This decision rule implies that there is no sharp cut-off between clear and borderline cases and so also leaves open the possibility of higher-order vagueness which can be defined in analogy with Definition 3. However, this definition faces certain problems as it may lead to odd results when a case that would ordinarily (e.g. by Definition 1) be judged to be borderline is not so judged owing to the decision rule being probabilistic. Incidentally, to the extent Definition 5 involves a probabilistic rule, it does follow something like the Luce Choice Rule referred to above.

So far, only vague concepts have been defined. Vague properties depend on the community's vague concepts.

Definition 6 A property is vague if and only if it is based on the community's corresponding vague concepts.

Definition 6 covers both types of property, the average kind and the individual kind. It is deliberately vague as there are somewhat messy issues relating to what happens if some members of the community have incorrect concepts and also if some members are vague and others are precise about the same concept. As should be obvious, the same sorts of observations, mutatis mutandis, follow for vague properties. However, aggregate vague properties are objectively true or false or indeterminate as they are derived from the concepts of all the individuals in a community. But there is no way

[11] A fractal is an object or quantity that displays "self-similarity" on all scales.

[12] The general idea was suggested to me by Gregory Murphy. I have fleshed out the details.

[13] There are overlaps among these probabilities so they do not sum to 1.

to know with certainty which of these truth values actually obtains as we have only approximate epistemic access to such a property.

The alert reader will have noticed that the apparatus of similarity and similarity-based probabilities is in fact unnecessary for these definitions. One could directly define borderline and clear cases once the notion of psychological distance is available. However, I do not do this in order to more closely mimic the psychological literature for greater realism and also to connect with other non-distance-based ways of capturing similarity as the latter seems like the fundamental idea. That is, instead of dispensing with similarity, we could dispense with distance, and define similarity on some different underlying basis.

We are now ready to apply these ideas to communication and to the sorites paradox.

2.3 Communication

As argued in Parikh (2010), communication and meaning involve four constraints: phonetic, syntactic, semantic, and flow. Here, I will confine myself to the semantic constraint that says that every word in an utterance is transformed by a conventional map into its conventional meaning(s) and then further transformed by an informational map into its referential meaning(s). This scheme is a generalization and refinement of Frege (1892/1980) system of sense and reference by which a word is converted first into its sense which is then converted into its reference. Each conventional meaning is the word's conventionally associated concept for the speaker and the addressee, and each referential meaning is the corresponding individuals and properties into which the former is mapped relative to the utterance situation u.

Pictorially:

$$\text{word} \longrightarrow \text{conventional meaning(s)} \stackrel{u}{\longrightarrow} \text{referential meaning(s)}$$

Symbolically:

$$\omega \longrightarrow C^{\omega} \stackrel{u}{\longrightarrow} P^{\omega}$$

assuming the concept C^{ω} is converted into the corresponding property P^{ω}.[14] The u on top of the second arrow implies that it is an argument of the informational map together with the conventional meaning. This schema combines the internal and external significance of language.

Consider now an utterance by $\mathcal{A}$ to $\mathcal{B}$ in u of the sentence $\varphi = $ ALEX IS BALD. Then the two maps above will apply to each of the three words in the utterance. As our interest is in BALD, we get the picture:

$$\text{BALD} \longrightarrow C^{\text{BALD}} \stackrel{u}{\longrightarrow} P^{\text{BALD}}$$

[14]The symbol P is being used for both the property and for probability but there should be no confusion.

Each concept and property can be understood as marked by the relevant agent $\mathcal{A}$ or $\mathcal{B}$ implying that there are actually *two* such arrow diagrams for the word. The vague concept C^{BALD} can be more or less any of the many situated concepts that $\mathcal{A}$ (or $\mathcal{B}$) has used in the past or it can be some average of these. Further, in the current situation u, it gets transformed via a new u-relative concept into its corresponding vague property. That is, the concept that is the conventional meaning shifts to a related concept relative to u and thence to the corresponding property.

The property in question can also be taken to be the *intersubjectively derived* property rather than the subjective property and it can be assumed that each agent has just a partial understanding of the content of the utterance. Also, in practice, there would be more than two diagrams as the word BALD is *ambiguous* besides being vague. For example, it can also mean *plain* or *blunt* as in A BALD STATEMENT. So there will be two or more conventional meanings, each of which will be mapped into their respective referential meanings. But we can ignore this complexity here.

For our purposes, we need to ask what made $\mathcal{A}$ utter φ in the first place. In order to do so, he would have had to determine that Alex *is* bald and to make that determination he would have resorted to the calculations of the previous section. That is, he would have ascertained that $|P(C^{\text{BALD}} \mid \text{Alex}; u) - P(\textit{not } C^{\text{BALD}} \mid \text{Alex}; u)| \geq \epsilon_u$ and that $P(C^{\text{BALD}} \mid \text{Alex}; u) > P(\textit{not } C^{\text{BALD}} \mid \text{Alex}; u)$ as Definition 1 requires. Because the threshold is more likely to be fuzzy, and fuzzy numbers involve membership functions, the calculation will involve some situated rule based on interval arithmetic for deciding what degree of membership is sufficient for counting someone bald. A similar approach would be followed if we used Definition 5. In other words, such calculations are an integral part of natural language generation. Since most words in language are vague, this shows that speakers have quite a bit to do and it is something of a psycholinguistic mystery how so much is accomplished so quickly. Perhaps the human brain's parallel processing just is very fast with such probabilistic comparisons.

On $\mathcal{B}$'s side, she simply has to access her concept C^{BALD} and then her property P^{BALD} but if she wants to decide whether to agree or disagree with $\mathcal{A}$, then she too has to go through the same arithmetic with her own threshold.

Penumbral shift, the change of truth value of the same utterance in different situations, does not seem to raise any special problems once the general context-sensitivity of language is accounted for as in my book cited above.

It turns out, therefore, that the conceptual difficulties vagueness poses do not overly complicate our semantic frameworks. There is, of course, a great deal more to communication and meaning than I have indicated here but many of those complexities occur with non-vague language as well. One somewhat new phenomenon that arises with vague concepts is how precise an agent needs to make his utterance to balance the conflicting demands of costs and benefits. This is similar but not identical to the issue of how much to explicitly disambiguate the lexical and structural ambiguities in a sentence. I address these game-theoretic issues in Parikh (2018) as they would take up too much space here. So once one has the right way to view vagueness, its apparent hurdles seem to melt away.

2.4 The Sorites Paradox

The sorites paradox can be formulated for any vague property. It consists of the following type of argument:

1. A hairless person is bald.
2. For all k, if a person with k hairs is bald then a person with $k + 1$ hairs is bald.
3. Therefore, all persons are bald.

Most proposed solutions to the paradox deny the second premise on either semantic or epistemic grounds. Schiffer (2003, Chap. 5) does an able job of dispelling such proposals. My resolution also denies the same premise but on *psychological* grounds. First, consider an agent-relative concept-based (or individual property-based) restatement of this premiss:

- For all k, if $\mathcal{A}$ judges a person with k hairs to be bald then he would judge a person with $k + 1$ hairs to be bald.

The key to the resolution is that such judgments are made on the basis of *multiple* exemplars or a category prototype which is also based on multiple exemplars. So it is quite possible for $\mathcal{A}$ to judge a person with k hairs to be bald and then judge a person with $k + 1$ hairs to be only borderline bald for some definite or fuzzy $k^\star$. This would be true even if we restricted our attention to all the exemplars in the sequence that have already been judged by $\mathcal{A}$ to be bald. That is, as the value of k increases, the distance from the early members of the series (or from the dynamically changing prototype) keeps growing, their similarity keeps dropping, and, at $k^\star$, $\mathcal{A}$ finds himself with a borderline case that is not clearly bald. This follows easily from Definition 1.

I am *not* saying that there is a definite cut-off between bald and not bald as many solutions do; I am saying there is either a precise or fuzzy cut-off between clearly bald and borderline bald (and also between borderline bald and not bald). Whether $k^\star$ is a precise or fuzzy number depends on whether $\mathcal{A}$'s threshold ϵ_u is a precise or fuzzy number. Moreover, $\mathcal{A}$'s own cut-off will change with the situation u in which he is asked to make the judgments because ϵ_u depends on u. Finally, different agents will have different cut-offs because their thresholds will generally be a little different.

The reason why the sorites argument seems plausible is that its formulation tricks us into consciously focusing on just a single exemplar: the previous case k in the second premise. Because our judgments about vague concepts are typically sub-personal and non-conscious, we are not aware of the multiple exemplars and thresholds that go into our judgments. Indeed, our judgments also lack the kind of firm conviction we have in judging that $2 + 2 = 4$ and we may waffle over the exact value of $k^\star$. So the sorites works by forcing us to make intermediate judgments in a conscious and unnatural way and then freeing us to judge the conclusion that all persons are bald in a non-conscious natural way.

Since properties are abstract social constructs built out of a community's individual concepts, it follows that the agent-independent property version of the sorites paradox will also have the same kind of resolution. That is, there will be some function of the individual $k^\star$ values for each agent that yields some social cut-off $K^\star$ although it will not be possible for anyone to know what its precise or fuzzy value is.

If Definition 5 is used instead, the argument is less smooth because $\mathcal{A}$ may judge a person with k hairs to be borderline and then the next person with $k + 1$ hairs to be bald. And $\mathcal{A}$ may never judge a candidate to be borderline because that decision is probabilistic and may go directly from baldness to non-baldness. But, barring these odd cases, the basic idea works because $P(\mathit{borderline}\ C \mid a; u)$ keeps growing as the number of hairs grows.

Since Definition 1 gives us a knockdown argument against the sorites and allows us to consider higher-order vagueness as well, it appears that it is to be preferred to Definition 5. In any case, the key point underlying both accounts is the same: new candidates are judged against multiple exemplars, not just the previous exemplar in the sequence.

2.5 Conclusion

I have construed the exemplar and prototype approaches of cognitive psychology in certain ways to characterize vagueness, describe certain aspects of communication, and approach the sorites paradox in a new way. The models appear to have some empirical support though I want to also emphasize the *kind* of reasoning that is involved in addressing these problems. More realistic models will doubtless become available but the underlying structure of explanation is not likely to change materially.

All three themes of Rohit Parikh's paper—vague communication, the lack of objectivity, and the centrality of agency—are echoed in the present paper.

References

Barwise, J., & Perry, J. (1983). *Situations and attitudes*. Cambridge, MA: The MIT Press.

Frege, G. (1892/1980). On sense and meaning. In: P. Geach & M. Black (Eds.), *Translations from the philosophical writings of Gottlob Frege* (pp. 56–78). Oxford: Basil Blackwell Ltd.

Gärdenfors, P. (2000). *Conceptual spaces: The geometry of thought*. Cambridge, MA: The MIT Press.

Hampton, J. A. (2007). Typicality, graded membership, and vagueness. *Cognitive Science*, *31*(3), 355–384.

Hampton, J. A., Estes, Z., & Simons, C. L. (2005). Comparison and contrast in perceptual categorization. *Journal of Experimental Psychology: Learning, Memory, and Cognition*, *31*, 1459–1476.

Hampton, J. A., Aina, B., Andersson, J. M., Mirza, H., & Parmar, S. (2012). The Rumsfeld effect: The unknown unknown. *Journal of Experimental Psychology: Learning, Memory, and Cognition*, *38*, 340–355.

Klir, G. J., Clair, U. S., & Yuan, B. (1997). *Fuzzy set theory: Foundations and applications*. New Jersey: Prentice Hall.

Medin, D. L., & Schaffer, M. M. (1978). Context theory of classification learning. *Psychological Review*, *85*, 207–238.
Mohanty, J. N. (2000). *Classical Indian Philosophy*. Maryland: Rowman and Littlefield Publishers.
Murphy, G. L. (2004). *The big book of concepts*. Cambridge, MA: The MIT Press.
Nosofsky, R. M. (1992). Exemplars, prototypes, and similarity rules. In A. Healy, S. Kosslyn, & R. Shiffrin, (Eds.), *From Learning Theory to Connectionist Theory: Essays in Honor of W. K. Estes* (vol. 1, pp. 149–168). Hillsdale, NJ: Erlbaum.
Nosofsky, R. M., & Palmeri, T. J. (1997). An exemplar-based random walk model of speeded categorization. *Psychological Review*, *104*, 266–300.
Parikh, P. (2010). *Language and equilibrium*. Cambridge, MA: The MIT Press.
Parikh, P. (2018). *Communication and content*. To be published.
Parikh, R. (1983). The problem of vague predicates. In R. S. Cohen & M. W. Wartofsky (Eds.), *Language, logic, and method* (pp. 241–261). Dordrecht: D. Reidel Publishing Company.
Parikh, R. (1994). Vagueness and utility: The semantics of common nouns. *Linguistics and Philosophy*, *17*, 521–535.
Putnam, H. (1975). The meaning of 'meaning'. In *Mind, language, and reality: Philosophical papers* (Vol. 2). Cambridge: Cambridge University Press.
Rosch, E. (1978). Principles of categorization. In E. Rosch & B. B. Lloyd (Eds.), *Cognition and categorization* (pp. 27–48). Hillsdale, NJ: Erlbaum.
Rosch, E., & Mervis, C. B. (1975). Family resemblance: Studies in the internal structure of categories. *Cognitive Psychology*, *7*, 573–605.
Schiffer, S. (2003). *The things we mean*. Oxford: Oxford University Press.
Schiffer, S. (2010). Vague properties. In R. Dietz & S. Moruzzi (Eds.), *Cuts and clouds: Vagueness, its nature, and its logic* (pp. 109–130). Oxford: Oxford University Press.
Shepard, R. N. (1987). Toward a universal law of generalization for psychological science. *Science*, *237*, 1317–1323.
Smith, E. E., & Medin, D. L. (1981). *Categories and concepts*. Cambridge, MA: Harvard University Press.
Warglien, M., & Gärdenfors, P. (2013). Semantics, conceptual spaces, and the meeting of minds. *Synthese*, *190*, 2165–2193.
Williamson, T. (1994). *Vagueness*. London: Routledge.
Wittgenstein, L. (1953/1968). *Philosophical investigations*, 3rd edn. New York: Macmillan Publishing Company, Inc.

Chapter 3
Nonmonotonicity and Knowability: As Knowable as Possible

Robert van Rooij

Abstract According to the anti-realistic theory of meaning, everything that is true is knowable. Fitch's (1963) paradox—based on very standard assumptions made in modal logic—is seen as a challenge to this theory. In this paper I argue that there is something wrong with Fitch's derivation of the absurdity. Assuming—for the sake of argument— the thesis of anti-realism, I argue with Beall (2000) that this doesn't lead to disaster in case we allow some contradictions to be true. By making use of a nonmonotonic consequence relation, I show that for all of Tennant's (1997) so-called 'Cartesian propositions' that are true, we can derive that it is true and not false that they are knowable.

Keywords Knowability paradox · Nonmonotonic consequence · Modal logic

3.1 Introduction: The Knowability Paradox

In recent years, Fitch's knowability (Fitch 1963) paradox received a lot of attention (cf. Salerno 2009). According to some, this paradox proves that verificationism, or more generally, Dummett's anti-realistic theory of meaning is impossible. According to others, there must be something wrong with the proof: how else can it be explained that Dummett's reasonable type of anti-realism reduces to absurdities like naive idealism and even lingualism?

According to anti-realism, if it is impossible to know something, this something cannot be true. In formal terms, $\neg\Diamond\mathbf{K}\phi \rightarrow \neg\phi$. Contrapositively, this means that if something is true, it is knowable, i.e., it can be known:

I would like to thank two anonymous reviewers for helpful comments and suggestions. This paper builds on joint work with Pablo Cobreros, Paul Egré and Dave Ripley, and I want to thank them as well. This work was done with main support from the Marie Curie Initial Training network ESSENCE-project, and the NWO-sponsored Language in Interaction-project.

R. van Rooij (✉)
Institute for Logic, Language and Computation, University of Amsterdam, Amsterdam, The Netherlands
e-mail: R.A.M.vanRooij@uva.nl

C. Başkent et al. (eds.), *Rohit Parikh on Logic, Language and Society*, Outstanding Contributions to Logic 11, DOI 10.1007/978-3-319-47843-2_3

- $\phi \rightarrow \Diamond \mathbf{K}\phi$. Anti-realism

The problem is that this seemingly innocent assumption gives rise to trouble when combined with some standard assumptions concerning the behavior of the involved modal operators '$\Diamond$' (or its dual '$\Box$') and '**K**'. The standard assumptions are the following:

$$\begin{array}{ll} \mathbf{N} & \models \phi \quad \Rightarrow \quad \models \Box\phi \\ \mathbf{K} & \models \mathbf{K}(\phi \rightarrow \psi) \rightarrow (\mathbf{K}\phi \rightarrow \mathbf{K}\psi). \\ \mathbf{T} & \models \mathbf{K}\phi \rightarrow \phi. \end{array}$$

From **N** and **K** one can straightforwardly prove that **C**, i.e. $\models \mathbf{K}(\phi \wedge \psi) \rightarrow \mathbf{K}\phi \wedge \mathbf{K}\psi$ holds. Let us now assume that we know that there are some unknown truths: $\exists\phi(\mathbf{K}(\phi \wedge \neg\mathbf{K}\phi))$. Let p be one of those. Thus $\mathbf{K}(p \wedge \neg\mathbf{K}p)$. The trouble is that from this assumption together with anti-realism and the standard modal assumptions **N**, **K**, and **T**, one can derive the seemingly absurd thesis that any truth is already known.

To show this, assume $\mathbf{K}(p \wedge \neg\mathbf{K}p)$. With **C** it follows that $\mathbf{K}p \wedge \mathbf{K}\neg\mathbf{K}p$, and thus $\mathbf{K}p$. But from the other conjunct $\mathbf{K}\neg\mathbf{K}p$ it follows with **T** that $\neg\mathbf{K}p$. Thus, from $\mathbf{K}(p \wedge \neg\mathbf{K}p)$ one can derive a contradiction, which means (according to classical logic) that $\mathbf{K}(p \wedge \neg\mathbf{K}p)$ cannot be true, i.e. $\models \neg\mathbf{K}(p \wedge \neg\mathbf{K}p)$. By the necessitation rule **N** it follows that $\Box\neg\mathbf{K}(p \wedge \neg\mathbf{K}p)$, which is equivalent to $\neg\Diamond\mathbf{K}(p \wedge \neg\mathbf{K}p)$.

According to anti-realism, all truths are knowable, formalized as $\phi \rightarrow \Diamond\mathbf{K}\phi$. Now take ϕ to be our assumption that p is an unknown truth: $\phi \equiv p \wedge \neg\mathbf{K}p$. With anti-realism it would follow that $\Diamond\mathbf{K}(p \wedge \neg\mathbf{K}p)$. But from our above reasoning we have concluded that $\neg\Diamond\mathbf{K}(p \wedge \neg\mathbf{K}p)$, which states the exact opposite. We must conclude that according to the anti-realist there can be no proposition p that is true but unknown $\neg\exists p(p \wedge \neg\mathbf{K}p)$. Stating it otherwise, this leads with **T** to the seemingly absurd proposition that truth and knowledge are indistinguishable, $\forall p(p \leftrightarrow \mathbf{K}p)$: **idealism**.

It is perhaps worth stating that the reasoning behind the paradox leads to **lingualism** as well, i.e. the thesis that all truths are truly expressed, $\forall p(p \leftrightarrow \mathbf{E}p)$, at least on the mild assumption that the factive modality of 'truly expressed', **E**, satisfies axiom **K** too. Suppose that we assume *the thesis of expressibility*: that if something is true, it can be expressed, and thus be truly expressed: $\phi \rightarrow \Diamond\mathbf{E}\phi$, for any ϕ. If we now make the natural assumption that the sentential operator **E** is a necessity operator and validates both (1) $\mathbf{E}\phi \rightarrow \phi$ and (2) $\mathbf{E}(\phi \wedge \psi) \rightarrow \mathbf{E}\phi \wedge \mathbf{E}\psi$, we can in exact analogy with the derivation of idealism immediately conclude that all truths are truly expressed: *lingualism*. Of course, the predicted requirement for any truth that it has to be expressed seems even more absurd than the prediction of (naive) idealism.

There are some well-known obvious reactions to the knowability paradox. First, one might simply bite the bullet, and accept the idealistic conclusion: there is no p that is both true and unknown. Second, one might argue that the paradox shows that anti-realism, just as the thesis of expressibility, is simply wrong. Both conclusions are radical, far-reaching, and not very popular. Indeed, isn't it more natural to assume that the absurd consequences show that there is something wrong with the proof?

Indeed, much more popular is Tennant's (1997) response: anti-realism should be limited to so-called *Cartesian* propositions, i.e., propositions that do not lead to contradictions if anti-realism is assumed. This response is completely correct: full-fletched anti-realism cannot escape the conclusion that some truths lead to inconsistency. [1] According to another response (Beall 2000), one should not worry too much about our knowledge being contradictory, because a closely related paradox (i.e., Montague's (1963) *knower paradox*) already shows that we do have contradictory knowledge, and such contradictory knowledge need not lead to as much disaster as classical logicians assumed. Also Beall seems to be correct: not allowing knowledge to be self-contradictory requires unnatural limitations on what can be expressed.

Still, even the latter two more moderate responses are not fully satisfying: Tennant's proposal seems more like an *escape* than a solution to the paradox, and although we agree with Beall's observation that the *knower*-paradox is relevant to the solution of the knowability paradox, Beall seems to *weaken* the anti-realistic thesis, i.e. $\phi \rightarrow \Diamond \mathbf{K}\phi$ for any ϕ, much more than desired. The reason is that if Beall uses the most popular paraconsistent logic that allows for sentences to be both true and false, i.e. the logic *LP*,[2] it follows that one can only conclude for any true ϕ that the formula $\Diamond \mathbf{K}\phi$ is also true, but not that $\Diamond \mathbf{K}\phi$ is *only true*, i.e., he allows a version of semantic anti-realism where $\Diamond \mathbf{K}\phi$ is false, for any arbitrary truth ϕ.

In this paper we will nevertheless build on the proposals of Tennant and Beall. We won't give up on anti-realism for the sake of argument, and we will assume that there are some unknown truths. In contrast to Tennant (1997), we will assume (again for the sake of argument) that the thesis of anti-realism, i.e. $\phi \rightarrow \Diamond \mathbf{K}\phi$, holds for any arbitrary ϕ, thus also for non-Cartesian ones. Like Beall (2000) we will make use of a logic that allows contradictions to be true, but our anti-realism will be stronger than the one stated by Beall: for most propositions ϕ we don't allow that if ϕ is true, $\Diamond \mathbf{K}\phi$ can still be false (as well as true). We will make use of Tennant's distinction between Cartesian and non-Cartesian propositions, and derive that for all Cartesian propositions ϕ that are true, $\Diamond \mathbf{K}\phi$ will be *only true*.

What is distinctive about our approach is that we will make use of pragmatic interpretation to account for this notion of 'only truth'. To make use of Tennant's distinction between Cartesian and non-Cartesian propositions, we will make use of a 3-valued semantics, and a non-monotonic consequence relation. I will discuss the prospects of using the non-monotonic logic LP^m that is based on LP to strengthen Beall's (2000) thesis of anti-realism. I will conclude that the logic won't do, because it would lead to the undesired consequence that there can be no unknown truths

[1] van Benthem (2004) argues that Tennant's limitation is too weak. The thesis of verificationalism should not be limited to propositions that can be known consistently, but rather to propositions that one can learn consistently. Van Benthem observes that there are sentences that can be known consistently, but that cannot be consistently learned: $\phi = (p \wedge \Diamond\neg p) \vee \mathbf{K}\neg p$.

[2] *LP* is so named by Priest (1979), who has advanced the logic widely in philosophy, though the logic was introduced, for the same reason of accommodating 'glutty theories' (true inconsistent but non-trivial theories), by Asenjo (1966) and Aseno and Tamburino (1975).

after all. In the next sections I will motivate the three-valued logic used by Cobreros et al. (2012, 2013), and introduce a new non-monotonic logic that accounts for conversational implicatures (Grice 1967) and for the pragmatic interpretation of vague language. Afterwards I will use these notions to do the required work about knowability.

3.2 3-Valued Semantics and Pragmatic Interpretation

3.2.1 3-Valued Semantics

Let $\mathcal{M} = \langle \mathcal{D}, \mathcal{I} \rangle$ be a three-valued model for predicate logic, where D is a domain, as usual, and where $\mathcal{I}$ is a total function from atomic sentences to $\{0, 1, \frac{1}{2}\}$. Now we can define the truth values of sentences as follows[3]:

- $\mathcal{V}_{\mathcal{M}}(\phi) = \mathcal{I}_{\mathcal{M}}(\phi)$, if ϕ is atomic
- $\mathcal{V}_{\mathcal{M}}(\neg\phi) = 1 - \mathcal{V}_{\mathcal{M}}(\phi)$
- $\mathcal{V}_{\mathcal{M}}(\phi \wedge \psi) = min\{\mathcal{V}_{\mathcal{M}}(\phi), \mathcal{V}_{\mathcal{M}}(\psi)\}$
- $\mathcal{V}_{\mathcal{M}}(\phi \vee \psi) = max\{\mathcal{V}_{\mathcal{M}}(\phi, \mathcal{V}_{\mathcal{M}}(\psi)\}$
- $\mathcal{V}_{\mathcal{M}}(\forall x\phi) = min\{\mathcal{V}_{\mathcal{M}}([\underline{d}/_x]\phi) : d \in D\}$

We say that ϕ is **strictly true** in $\mathcal{M}$ iff $\mathcal{V}_{\mathcal{M}}(\phi) = 1$, and that ϕ is **tolerantly true** iff $\mathcal{V}_{\mathcal{M}}(\phi) \geq \frac{1}{2}$. In terms of this semantics we can define some well-known logics: Kleene's $K3$ and *LP*. According to both logics, the consequence-relation is truth preserving. The only difference between the two is that while according to $K3$ only value 1 counts as true, according to *LP*, both 1 and $\frac{1}{2}$ do (while in $K3$ value $\frac{1}{2}$ stands for 'neither true nor false', in LP it denotes 'both true and false'). Thus, $\Gamma \models^{K3} \phi$ iff $\forall\mathcal{M}$: if $\forall\gamma \in \Gamma : \mathcal{V}_{\mathcal{M}}(\gamma) = 1$, then $\mathcal{V}_{\mathcal{M}}(\phi) = 1$, and $\Gamma \models^{LP} \phi$ iff $\forall\gamma \in \Gamma : \mathcal{V}_{\mathcal{M}}(\gamma) \geq \frac{1}{2}$, then $\mathcal{V}_{\mathcal{M}}(\phi) \geq \frac{1}{2}$. In some recent joint publications with Pablo Cobreros, Paul Egré, and David Ripley, we showed that a slight variant of $K3$ and LP can account for paradoxes of vagueness (Cobreros et al. 2012) and transparent truth (Ripley 2012; Cobreros et al. 2013) in an, arguably, more satisfying way than either $K3$ or LP can. The crucial idea of the analysis of vagueness and transparent truth is that, although we don't give up the idea that entailment is truth-preserving, we allow the 'strength' of truth of the conclusion to be weaker than the strength of truth of the premises. We say that a sentence ϕ is st**-entailed** by a set of premisses Γ, $\Gamma \models^{st} \phi$, iff $\forall\mathcal{M}$: if $\forall\gamma \in \Gamma : \mathcal{V}_{\mathcal{M}}(\gamma) = 1$, then $\mathcal{V}_{\mathcal{M}}(\phi) \geq \frac{1}{2}$. This analysis has two immediate consequences: (1) it interprets value $\frac{1}{2}$ as a notion of truth, just like *LP* does, and it thus allows for certain sentences to be both true *and* false, (2) the notion of consequence is *non-transitive*. One appealing feature of the logic is that in contrast to either $K3$ and LP, it is a conservative extension of classical logic: it only differs from classical logic if we extend the language with (1) a similarity relation '$\sim$'

[3]Notice that the semantics is just like that of Fuzzy Logic, but now limited to three truth values.

(Cobreros et al. 2012), so that the tolerance principle ($\forall x, y((Px \wedge x \sim_P y) \rightarrow Py))$ becomes valid, or with a truth-predicate '$\mathcal{T}$' (Cobreros et al. 2013) that behaves fully transparent ($\mathcal{V}_{\mathcal{M}}(\mathcal{T}(\phi)) = \mathcal{I}_{\mathcal{M}}(\phi)$ for any ϕ). And even in these cases the resulting logical differences are minimal, it is only in very special cases (i.e., when it gives rise to paradox) that transitivity fails. To illustrate this for vagueness, if we assume that $Px \wedge x \sim_P y \wedge y \sim_P z$ is strictly true, we can conclude using $\models^{st}_{\sim}$ that Py is at least tolerantly true. And if $Py \wedge y \sim_P z$ is, or were, strictly true, we could conclude with $\models^{st}_{\sim}$ that Pz would be at least tolerantly true. However, the two inferences cannot be joined together to give rise to the Sorites paradox: We cannot conclude from the strict truth of $Px \wedge x \sim_P y \wedge y \sim_P z$ to the tolerant truth of Pz.

3.2.2 Borderline Contradictions

In Cobreros et al. (2012) we argued that if Adam is a borderline case of a tall man, the sentence 'Adam is tall' is both true and false. We motivated this by a number of recent experiments (i.e., Alxatib and Pelletier 2011; Ripley 2011) that show that naive speakers find a logical contradiction like 'Adam is tall and Adam is not tall' acceptable exactly in case Adam is a borderline tall man. In Cobreros et al. (2012) we proposed that the explanation is that we always interpret a sentence pragmatically in the strongest possible way. This pragmatic interpretation accounts, on the one hand, for the intuition that if one *says* that Adam is tall, what is *meant* is that Adam is *only* tall, but, on the other, for the experimentally observed acceptability of contradictions at the border, because contradictions like '$Ta \wedge \neg Ta$' can only be interpreted as true when tolerant truth is at stake. In Cobreros et al. (2012) we show that such a pragmatic interpretation also accounts for the observed unacceptability (cf. Serchuk et al. 2011) of classical tautologies like '$Ta \vee \neg Ta$' if Adam is borderline tall. Unfortunately, the interpretation rule gives rise to trouble for more complex sentences. Alxatib et al. (2013) show that we wrongly predict that a sentence like 'Adam is tall and not tall, or John is rich' (of the form $(p \wedge \neg p) \vee q$) means that John is strictly rich, although it intuitively should mean that *either* Adam is borderline tall *or* John is strictly rich. Similarly, our analysis mispredict that a sentence like 'Adam is tall and Adam is not tall, and John is rich' (of the form $(p \wedge \neg p) \wedge q$) can be appropriately asserted if John is not strictly rich.

3.2.3 Pragmatic Interpretation

How should we account for pragmatic interpretation such that we can show what we cannot express, i.e. that a sentence is *only true*, and can solve the above problems with complex sentences involving borderline contradictions? Here is a proposal[4]: the

[4]For a more elaborate discussion of this proposal, see Cobreros et al. (2015).

pragmatic interpretation of ϕ makes (exactly) one minimal truth-maker of ϕ as true as possible. What are the minimal truth-makers of ϕ? and how to think of 'as true as possible'? As for the first question, a minimal truth-maker of ϕ will be thought of as a set of literals, and the set of minimal truth-makers of ϕ, $T(\phi)$, can easily be defined recursively as follows[5]:

- $T(\phi) = \{\{\phi\}\}$, if ϕ is a literal.
- $T(\phi \vee \psi) = T(\phi) \cup T(\phi)$.
- $T(\phi \wedge \phi) = T(\phi) \otimes T(\psi) = \{A \cup B | A \in T(\phi), B \in T(\psi)\}$.
- $T(\forall x \phi) = \bigotimes_{d \in D} T(\phi[^x/_d])$

Notice that according to these rules, $T(p) = \{\{p\}\}$, $T(\neg p) = \{\{\neg p\}\}$, $T(p \vee q)$, $= \{\{p\}, \{q\}\}$, $T(p \vee \neg p) = \{\{p\}, \{\neg p\}\}$, $T(p \wedge \neg p) = \{\{p, \neg p\}\}$, $T((p \wedge \neg p \vee q)q)\vee, = \{\{p, \neg p\}, \{q\}\}$, $T((p \wedge \neg p) \wedge q) = T(p \wedge (\neg p \wedge q)) = \{\{p, \neg p, q\}\}$, and $T((p \vee q) \wedge (r \vee s)) = \{\{p, r\}, \{p, s\}, \{q, r\}, \{q, s\}\}$.

To make sense of the notion 'as true as possible', we can define an ordering $<_S$ between models, with S a set of sentences (a minimal truth maker), defined as follows: $\mathcal{M} <_S \mathcal{N}$ iff$_{def}$ $\{\phi \in S : \mathcal{V}_{\mathcal{M}}(\phi) = 1\} \subset \{\phi \in S : \mathcal{V}_{\mathcal{N}}(\phi) = 1\}$. In terms of this we define the pragmatic interpretation of ϕ, *PRAG*(ϕ) (where $[\![S]\!]^t$ abbreviates $\bigcap_{\psi \in S}[\![\psi]\!]^t$, and where $[\![\psi]\!]^t = \{\mathcal{M} | \mathcal{V}_{\mathcal{M}}(\psi) \geq \frac{1}{2}\}$):

- $PRAG(\phi) = \{\mathcal{M} \in [\![S]\!]^t | \; S \in T(\phi) \;\&\; \neg\exists \mathcal{N} \in [\![S]\!]^t : \mathcal{M} <_S \mathcal{N}\}$.

Notice that for literals and conjunctive sentences, this pragmatic interpretation rule simply tries to make its minimal truth-maker as true as possible, i.e., strictly true. Thus, even if one cannot really *express* in language that a sentence like 'Adam is tall' is only true, it follows from the above pragmatic interpretation. But in general a sentence might have more than one minimal truth-maker, i.e., when the sentence is disjunctive: so the general pragmatic interpretation rule says that it is enough if one of its minimal truth-makers is as true as possible. As a result, *PRAG*$((p \wedge \neg p) \wedge q)$ singles out those models where p is tolerantly but not strictly true, *and* where q is strictly true, while *PRAG*$((p \wedge \neg p) \vee q)$ singles out those models where either p is tolerantly but not strictly true, *or* where q is strictly true.

Of course, the above interpretation rules cannot interpret all sentences translated in a propositional language. For one thing, propositional logic also contains (bi)conditionals, for another, it might be that a negation has scope over a complex sentence. However, both problems can be solved easily, either by also defining the minimal false-makers of a sentence,[6] or by stipulating that the pragmatic interpretation rule should be applied to a sentence only after it is put in its so-called 'Negative Normal Form'.[7]

---

[5]The definition of $T(\phi)$ is taken from the analysis of minimal truth-makers proposed by van Fraassen (1969).

[6]Cf. van Fraassen (1969). This path is taken in Cobreros et al. (2015).

[7]ϕ' is a sentence in Negative Normal Form of ϕ if (1) it is logically equivalent with ϕ, and (2) negations only stand in front of atomic formulas.

Let us see what this pragmatic analysis predicts for some examples involving vagueness: (1) 'p' is interpreted as being only true; (2) '$p \vee \neg p$' is pragmatically interpreted as saying that either p is strictly true, or $\neg p$ is; (3) '$p \wedge \neg p$' is predicted to be interpreted as saying that p is only tolerantly true; (4) '$(p \wedge \neg p) \vee q$' is meant as saying that either p is only tolerantly true, or q is strictly true, while (5) '$(p \wedge \neg p) \wedge q$' is predicted to be interpreted as saying that p is only tolerantly true, and that q is strictly true. All these predictions seems to be in accordance with the experimental results and intuition.

3.3 The Knowability Paradox in Light of the Knower

In Ripley (2012) really and Cobreros et al. (2013) it is shown that the logic $\models^{st}$ can account for the liar paradox, in case the language allows for self-reference and has a truth-predicate. But it is easy to show that the same logic can also account for Montague's *knower-paradox*, which is standardly taken to show the limitations of syntactic treatments of modality.[8] Suppose that we allow in our language both a knowledge-operator and for self-reference. Then we can express a sentence κ (known as *the knower*) that says of itself that it is unknown: κ = this sentence is not known. $\kappa := \neg\mathbf{K}(\kappa)$.

Assume that the law of excluded middle, LEM, is valid. Then $\mathbf{K}(\kappa) \vee \neg\mathbf{K}(\kappa)$. If $\mathbf{K}(\kappa)$, then κ (by factivity of **K**, i.e. **T**), and by the meaning of κ thus $\neg\mathbf{K}(\kappa)$. Obviously, if $\neg\mathbf{K}(\kappa)$, then $\neg\mathbf{K}(\kappa)$. In either case $\neg\mathbf{K}(\kappa)$, and thus κ. But this means that κ is a validity, $\models \kappa$. By necessitation (rule **N**) it follows that $\mathbf{K}(\kappa)$. Thus, there is a sentence, namely, κ, for which we can derive both $\mathbf{K}(\kappa)$ and $\neg\mathbf{K}(\kappa)$. By classical logic, it follows that we can derive everything: $\top \models \mathbf{K}(\kappa) \vee \neg\mathbf{K}(\kappa)$; $\mathbf{K}(\kappa) \vee \neg\mathbf{K}(\kappa) \models \mathbf{K}(\kappa) \wedge \neg\mathbf{K}(\kappa)$, and $\mathbf{K}(\kappa) \wedge \neg\mathbf{K}(\kappa) \models \bot$, the logic becomes trivial.

As always, there are various options open to solve this paradox. According to one option, we allow some sentences to be both true and false. If so, there are still various ways to go: either give up on explosion (as in *LP*), or give up on transitivity of consequence (e.g. by using our $\models^{st}_{T}$). In both cases, the only way to give a satisfactory semantics for κ is to say that it has value $\frac{1}{2}$. In fact, it has to have this value in all models/worlds. By the meaning of κ, this means that $\mathbf{K}(\kappa)$ has to have value $\frac{1}{2}$ in all models/worlds. What does this mean for the semantics of $\mathbf{K}(\phi)$? The most natural suggestion is to say the following (with $R_{\mathcal{M}}$ as the epistemic accessibility relation):

- $\mathcal{V}_{\mathcal{M},w}(\mathbf{K}(\phi)) = 1$ iff $\forall v \in R_{\mathcal{M}}(w) : \mathcal{V}_{\mathcal{M},v}(\phi) = 1$
  $\mathcal{V}_{\mathcal{M},w}(\mathbf{K}(\phi)) = 0$ iff $\exists v \in R_{\mathcal{M}}(w) : \mathcal{V}_{\mathcal{M},v}(\phi) = 0$
  $\mathcal{V}_{\mathcal{M},w}(\mathbf{K}(\phi)) = \frac{1}{2}$ otherwise.

[8] 'Thus, if necessity is to be treated syntactically, that is, as a predicate of sentences, as Carnap and Quine have urged, then virtually all of modal logic [...] must be scarified' (Montague, 1963, p. 9).

Notice that this means that $\mathcal{V}_{\mathcal{M},w}(\mathbf{K}(\phi)) = \frac{1}{2}$ iff (1) $\forall v \in R_{\mathcal{M}}(w) : \mathcal{V}_{\mathcal{M},v}(\phi) \geq \frac{1}{2}$ and (2) $\exists v \in R_{\mathcal{M}}(w) : \mathcal{V}_{\mathcal{M},v}(\phi) \neq 1$. Indeed, on this semantics it comes out that not only κ, but also both $\mathbf{K}(\kappa)$ and $\neg\mathbf{K}(\kappa)$ and thus $\mathbf{K}(\kappa) \vee \neg\mathbf{K}(\kappa)$ and $\mathbf{K}(\kappa) \wedge \neg\mathbf{K}(\kappa)$ have value $\frac{1}{2}$ in any world in any model. It is easy to see that one can solve Montague's knower paradox by giving up explosion. With respect to our consequence-relation $\models^{st}$, however, explosion is valid. But using $\models^{st}$ one can still solve the paradox, because the consequence relation is non-transitive. The tautology $\top$ st-entails $\mathbf{K}(\kappa) \vee \neg\mathbf{K}(\kappa)$. But the latter sentence only has value $\frac{1}{2}$. Each disjunct entails κ, and thus it follows that $\mathbf{K}(\kappa) \wedge \neg\mathbf{K}(\kappa)$. The latter, in turn, st-entails $\bot$. But nothing disastrous follows, because the two st-entailments cannot be put together.

Back to the knowability paradox. We have learned from the knower-paradox that there is at least one instance of ϕ, namely κ, for which we have inconsistent knowledge. Putting it otherwise, we know that the knower is valid, and by its meaning it holds that $\kappa \wedge \neg\mathbf{K}\kappa$ is true. In fact, this sentence *must* have value $\frac{1}{2}$ because both conjuncts have value $\frac{1}{2}$ in all worlds in all models. Let us assume that $\mathcal{V}_{\mathcal{M},w}(\Diamond(\phi)) = 1$ iff $\exists v : \mathcal{V}_{\mathcal{M},v}(\phi) = 1$ and $\mathcal{V}_{\mathcal{M},w}(\Diamond(\phi)) = 0$ iff $\forall v : \mathcal{V}_{\mathcal{M},v}(\phi) = 0$. We will also assume that the strength (strict or tolerant) of $\Box\phi$'s validity in rule **N**, for any necessary-modality '$\Box$' (and thus also '**K**'), depends on the strength of the validity of ϕ.

N $\quad \models^{s/t} \phi \quad \Rightarrow \quad \models^{s/t} \Box\phi$

But this means that not only $\kappa \wedge \neg\mathbf{K}\kappa$ is tolerantly valid, but that the same holds for $\mathbf{K}(\kappa \wedge \neg\mathbf{K}\kappa)$. In fact, also $\Diamond\mathbf{K}(\kappa \wedge \neg\mathbf{K}\kappa)$ can only have value $\frac{1}{2}$. But from this it follows that the thesis of anti-realism, -$\psi \rightarrow \Diamond\mathbf{K}\psi$, cannot be strictly valid, because $\kappa \rightarrow \Diamond\mathbf{K}(\kappa)$ can only have value $\frac{1}{2}$.[9]

The fact that the thesis of anti-realism cannot be strictly valid does not mean that we have to give up on anti-realism because of the knower paradox: Both *LP* and our logic based on $\models^{st}$ allow us to maintain the validity of $\phi \rightarrow \Diamond\mathbf{K}\phi$, for all ϕ. The reason is that both logics only demand that the conditional is tolerantly true. But notice with Beall (2000) that for exactly the same reason as for *the knower*, the LP- and ST-validity of $\phi \rightarrow \Diamond\mathbf{K}\phi$ is not a problem for any other sentence p that is (though perhaps only contingently) true and unknown. Our above semantics and interpretation of the thesis of anti-realism allows for $p \wedge \neg\mathbf{K}p$ to even have value 1. The reason is that although we might take the axioms **K** and **T** to hold both strictly and tolerantly,

K $\quad \models^{s,t} \mathbf{K}(\phi \rightarrow \psi) \rightarrow (\mathbf{K}\phi \rightarrow \mathbf{K}\psi)$,

T $\quad \models^{s,t} \mathbf{K}\phi \rightarrow \phi$,

from the tolerant truth of $\Diamond\mathbf{K}(p \wedge \neg\mathbf{K}p)$, and thus that of $\mathbf{K}(p \wedge \neg\mathbf{K}p)$ in a world, we cannot even derive the tolerant truth of $\mathbf{K}p \wedge \neg\mathbf{K}p$ in that world, because $\mathcal{V}_{\mathcal{M},w}(\phi \rightarrow \psi) \geq \frac{1}{2}$ if $\mathcal{V}_{\mathcal{M},w}(\phi) = \frac{1}{2}$

Although both *LP* and *ST* allow us, in contrast to Tennant (1997), to maintain the full thesis of anti-realism, we are still not there where we would like to be. The reason

[9]The thesis could still receive value 1, though, if we used Łucasiewics' conditional, which is like material implication, except that it assigns to $\phi \rightarrow \psi$ value 1 if both ϕ and ψ have value $\frac{1}{2}$.

is that we have generalized to the worst case: because there is at least one type of sentence (a sentence of the form $p \wedge \neg \mathbf{K}p$) for which $\phi \rightarrow \Diamond \mathbf{K}\phi$ cannot be strictly true, we cannot demand the thesis to hold strictly for *any* ϕ. What we would like to claim, instead, is that the thesis holds strictly, *except* for those types of sentences for which it gives rise to trouble, i.e., for Tennant's non-Cartesian propositions. It is only natural to try to account for this default behavior in terms of a non-monotonic logic.

3.4 Anti-realism and Non-monotonicity

3.4.1 Anti-realism and LP^m

We have argued above that in case Beall would use *LP* (and makes use of material implication), the thesis of anti-realism would become *too weak*: for any truth ϕ one can conclude at most that $\Diamond \mathbf{K}\phi$ is true, not that it is *only true*. One might try to strengthen this conclusion by making use of a stronger and non-monotonic logic that still allows for contradictions to be true.

There are many paraconsistent logics that are non-monotonic. One of them is Priest's (1991) LP^m, which is strictly stronger than LP. Its non-monotonicity is due to minimizing inconsistency: in LP^m one looks only at the set of *minimally inconsistent* models where the premisses are all tolerantly true. The set of minimally inconsistent worlds/models of ϕ as used in LP^m as $MI(\phi) = \{\mathcal{M} \in [\![\phi]\!]^t | \neg \exists \mathcal{N} \in [\![\phi]\!]^t : \mathcal{M} < \mathcal{N}\}$, where $\mathcal{M} < \mathcal{N}$ iff$_{df}$ $\{p \in ATOM | \mathcal{M} \in [\![p \wedge \neg p]\!]^t\} \subset \{p \in ATOM | \mathcal{M} \in [\![p \wedge \neg p]\!]^t\}$. If one would use LP to account for the knower and knowability paradox—as suggested by Beall (2000)—it seems that Priest's (1997) LP^m is a natural candidate to strengthen the force of the anti-realistic thesis as desired. Unfortunately, this candidate won't do.

An important (and, at least for the cases relevant to this paper, unfortunate) feature of LP^m is that p is an $\models^{LP^m}$-consequence of $p \vee \phi$—$p \vee \phi \models^{LP^m} p$, in case ϕ (but not p) is contradictory. Now consider the thesis of anti-realism: $\phi \rightarrow \Diamond \mathbf{K}\phi$. If '$\rightarrow$' is material implication, this is equivalent to $\neg \phi \vee \Diamond \mathbf{K}\phi$. Thus, using LP^m it follows from anti-realism that we can conclude that ϕ is strictly false, in case $\Diamond \mathbf{K}\phi$ is contradictory. Now take ϕ to be an unknown true proposition, e.g. $p \wedge \neg \mathbf{K}p$. In the introduction we have shown that by standard assumptions of modal logic one can prove that $\neg \mathbf{K}(p \wedge \neg \mathbf{K}p)$ is valid. Using a three-valued logic, one can live with this validity—in combination with the truth of $\Diamond \mathbf{K}(p \wedge \neg \mathbf{K}p)$—by assuming that the validity of $\neg \mathbf{K}(p \wedge \neg \mathbf{K}p)$ and thus (with **N**) of $\neg \Diamond \mathbf{K}(p \wedge \neg \mathbf{K}p)$ holds only tolerantly: both $\Diamond \mathbf{K}(p \wedge \neg \mathbf{K}p)$ and $\neg \Diamond \mathbf{K}(p \wedge \neg \mathbf{K}p)$ are tolerantly true, which means that $\Diamond \mathbf{K}(p \wedge \neg \mathbf{K}p)$ can at most be both true and false, i.e. is contradictory if true. Let us now try to *use* the validity of the thesis of anti-realism in our reasoning, by making the thesis as strong as possible by looking at the minimally inconsistent models of $(p \wedge \neg \mathbf{K}p) \rightarrow \Diamond \mathbf{K}(p \wedge \neg \mathbf{K}p)$. Unfortunately, because $\Diamond \mathbf{K}(p \wedge \neg \mathbf{K}p)$

is inconsistent if true, it follows that al minimally inconsistent models where $(p \wedge \neg\mathbf{K}p) \rightarrow \Diamond\mathbf{K}(p \wedge \neg\mathbf{K}p)$ is true are ones where $\neg(p \wedge \neg\mathbf{K}p)$ is strictly true, i.e., ones where the antecedent of the conditional is strictly false! Thus, by using LP^m as our non-monotonic logic in combination with the thesis of anti-realism, it would follow that there can be no unknown truths after all! Even though this idealistic conclusion can only be reached in an empty context, and it might be that $p \rightarrow \neg\mathbf{K}p$ does not hold anymore in case more is assumed (because the logic LP^m is non-monotonic), the fact that idealism is seen as the default case strikes us as unmotivated and undesired.

3.4.2 A Pragmatic Consequence Relation

The problem with using LP^m as the non-monotonic consequence relation was due to the use of the minimally inconsistent models. Observe that the use of minimally inconsistent models would also predict that a sentence of the form $(p \wedge \neg p) \vee q$ would be interpreted as q. But it was exactly to escape this prediction that we introduced our pragmatic interpretation rule in Sect. 3.2.3. Now we want to suggest that we should use this pragmatic interpretation to define a new non-monotonic consequence relation in terms of which we can strengthen the thesis of anti-realism.

There are various ways in which we can use our pragmatic interpretation rule to define a non-monotonic consequence relation. Two natural candidates are the following (where $PRAG(\Gamma)$ abbreviates $\bigcap_{\phi\in\Gamma} PRAG(\phi)$ and $PRAG(\Gamma, c)$ abbreviates $\bigcap_{\phi\in\Gamma} PRAG(\phi) \cap c$):

- $\Gamma \models^{prt} \psi$ iff $PRAG(\Gamma) \subseteq [\![\psi]\!]^t$
- $\Gamma \models^{prpr}_c \phi$ iff_{df} (for all c) $PRAG(\Gamma, c) \subseteq PRAG(\psi, PRAG(\Gamma))$

According to both consequence relations, we should look at the pragmatic interpretation of the premisses. As a result, using neither of the consequence relations we can conclude q from $(p \wedge \neg p) \vee q$. By looking at the pragmatic interpretation of the premisses, the new relations differ crucially from the consequence relation $\models^{st}$ that we introduced in Cobreros et al. (2012). With '$\models^{st}$' as our consequence relation, explosion (from a contradiction, everything can be derived) is predicted to be valid, even if the contradiction was meant to be expressed: $Ta \wedge \neg Ta$ can be interpreted as meaning that Adam is borderline tall. One unfortunate consequence of this picture was that the relation between assertion and inference was lost. Some of what was said (or, better, meant) was ignored in determining what can be inferred from it. The new consequence relations score better on this: If one claims that Adam is borderline tall by saying that he is tall *and* not tall, what one claims is taken seriously.

Another and related unfortunate consequence of using $\models^{st}$ as our consequence relation was that we were forced to make a distinction between the Sorites reasoning with and without the tolerance principle as explicit premise. Without the principle as explicit premise we predicted in Cobreros et al. (2012) that although each step in the argument is valid, the argument as a whole is *in*valid, because the arguments cannot be joined together. We felt, and still feel, that this is intuitively the correct diagnosis of the Sorites paradox. However, in Cobreros et al. (2012) we had to claim that with the tolerance principle as explicit premise, the argument is valid, but that one of the premisses (i.e., the tolerance principle) is not true enough to be used as a premise in an $\models^{st}$ inference. Making use of one of the new consequence-relations, we can also diagnose the Sorites reasoning with the tolerance principle as explicit premise as invalid, even though all the steps are valid. The fact that the tolerance principle $\forall x_i, x_j(Px_i \wedge x_i \sim_P x_j) \rightarrow Px_j)$ (with $1 \leq i, j \leq n$) cannot be strictly true if both Pa_1 and $\neg Pa_n$ are taken as premisses that are strictly true, does not rule out that it can be used appropriately in an inference where the premisses are interpreted pragmatically.

According to both notions of entailment conjunction-elimination and disjunction-introduction are valid. The fact that we now look at what was meant by a sentence means that, even though $\phi \wedge \neg\phi \models^{prt} \phi$, it does not hold that $\phi \wedge \neg\phi \models^{prt} \psi$. Thus, explosion is not valid. In this sense, both consequence-relations are a type of para-consistent entailment relations.

Interestingly, our new consequence relations are *non-monotonic*. For instance it will be the case on either consequence-relation that the Disjunctive Syllogism—$p, \neg p \vee q \models q$—is valid when p and q are atomic sentences, but when we strengthen the first premise from p to $(p \wedge \neg p)$ the conclusion fails to hold. Moreover, for the first consequence relation $\models^{prt}$, but not for the second ($\models^{prpr}$) it can be that although $\phi \models^{prt} \chi$, it holds that $\phi \wedge \psi \not\models^{prt} \chi$. This can be illustrated by the following example: $Tall(a) \wedge a \sim_T b \models^{prt} Tall(b)$ but $Tall(a) \wedge a \sim_T b \wedge \neg Tall(a) \not\models^{prt} Tall(b)$.

3.4.3 Pragmatic Entailment and Anti-realism

We naturally propose that instead of using LP^m to strengthen the thesis of anti-realism, we should make use of one of our pragmatic consequence-relations $\models^{prt}$ or $\models^{prpr}$.[10] Let us pretend that we can express the anti-realistic thesis in the object language by one sentence ϕ: $\phi \equiv \forall p(p \rightarrow \Diamond \mathbf{K}p)$. Notice that only by using the second consequence relation we can assume that the thesis of anti-realism is valid. However,—and what we feel is more important—on both accounts we can take the thesis of anti-realism as a substantial premise in our reasoning. Taking this thesis as a premise in reasoning results in the following set of models:

- $PRAG(\phi) = \{\mathcal{M} \in [\![S]\!]^t \mid S \in T(\phi) \;\&\; \neg\exists\mathcal{N} \in [\![S]\!]^t : \mathcal{M} <_S \mathcal{N}\}$.

[10] Recall that $\models^{prt}$ is neither transitive (if the language contains a similarity-relation or a transparent truth-predicate) nor does it validate explosion. It is by the latter feature, and not by non-transitivity, that it solves the *the knower*.

Notice that for arbitrary p for which $\Diamond\mathbf{K}p$ can be 'only true' (i.e., not also false) in some worlds, it is predicted that $\Diamond\mathbf{K}p$ *is* 'only true', in case p is true. This is the same as what would have been predicted by using LP^m, instead. But by using $\models^{prt}$ we now don't predict anymore that $p \rightarrow \neg\mathbf{K}p$ is the default case for any arbitrary p. By any logic that accepts the thesis of anti-realism and takes the conditional to be material implication, we predict $\neg(p \wedge \mathbf{K}p) \vee \Diamond\mathbf{K}(p \wedge \neg\mathbf{K}p)$ to be true. In contrast to LP^m, though, we don't predict that thus the first disjunct $p \rightarrow \neg\mathbf{K}p$ must be strictly true, because we allow the second disjunct $\Diamond\mathbf{K}(p \wedge \neg\mathbf{K}p)$ to be tolerantly true. Thus, idealism does not follow as the default case.

3.5 Conclusion

This paper follows a proposal of Beall (2000) to use a paraconsistent logic in order to save verificationalism from paradox. Even though Beall predicts $\phi \rightarrow \Diamond\mathbf{K}\phi$ to be valid, for any ϕ, this prediction turns out to be much weaker than desired. In this paper it is shown that the validity can be strengthened considerably by making use of a non-monotonic paraconsistent logic. But not any such non-monotonic paraconsistent logic will do. I argued for the use of a particular such logic and motivated its choice by looking at the pragmatic principles involved in the interpretation of vague expressions.

References

Alxatib, S., & Pelletier, F. R. (2011). The psychology of vagueness: Borderline cases and contradictions. *Mind and Language*, *26*, 287–326.

Alxatib, S., Pagin, P., Sauerland, U., et al. (2013). Acceptable contradictions: Pragmatics or semantics? A reply to Cobreros. *Journal of Philosophical Logic*, *42*, 619–634.

Asenjo, F. (1966). Calculus of antinomies. *Notre Dame Journal of Formal Logic*, *7*, 103–105.

Aseno, F., & Tamburino, J. (1975). Logic of antinomies. *Notre Dame Journal of Formal Logic*, *16*, 17–44.

Beall, Jc. (2000). Fitch's proof, verificationism, and the knower paradox'. *Australian Journal of Philosophy*, *78*(2), 241–247.

Cobreros, P., Egré, P., Ripley, D., & van Rooij, R. (2012). Tolerant, classical, strict. *Journal of Philosophical Logic*, *41*, 347–385.

Cobreros, P., Egré, P., Ripley, D., & van Rooij, R. (2013). Reaching transparent truth. *Mind*, *122*, 841–866.

Cobreros, P., Egré, P., Ripley, D., & van Rooij, R. (2015). Pragmatic interpretations of vague expressions: strong meaning and nonmonotonic consequence. *Journal of Philosophical Logic*.

Fitch, F. (1963). A logical analysis of some value concepts. *The Journal of Symbolic Logic*, *28*, 135–142.

Grice, H. P. (1967). Logic and conversation. *William James Lectures*. Harvard University, reprinted in *Studies in the Way of Words*, 1989. Cambridge, MA: Harvard University Press.

Montague, R. (1963). Syntactic treatment of modality. *Acta Philosophica Fennica*, *16*, 153–167.

Priest, G. (1979). The logic of paradox. *Journal of Philosophical Logic*, *8*, 219–241.

Priest, G. (1991). Minimally inconsistent LP. *Studia Logica, 50*, 321–331.
Ripley, D. (2011). Contradictions at the borders. In R. Nouwen, R. van Rooij, H.-C. Schmitz, & U. Sauerland (Eds.), *Vagueness in communication*. Berlin: LICS, Springer.
Ripley, D. (2012). Conservatively extending classical logic with transparent truth. *Review of Symbolic Logic, 5*, 354–378.
Salerno, J. (Ed.). (2009). *New essays on the knowability paradox*. Oxford: Oxford University Press.
Serchuk, P., Hargreaves, I., & Zach, R. (2011). Vagueness, logic and use: Four experimental studies on vagueness. *Mind and Language, 26*, 540–573.
Tennant, N. (1997). *The taming of the true*. Oxford: Oxford University Press.
van Benthem, J. (2004). What one may come to know. *Analysis, 64*, 95–105.
van Fraassen, B. (1969). Facts and tautological entailments. *Journal of Philosophy, 66*, 477–487.

Chapter 4
What is Money?

Jan van Eijck and Philip Elsas

Abstract Rohit Parikh advocates a collaboration between logicians, philosophers, computer scientists and game theorists, in an effort to use the techniques from their fields to shed light on social phenomena, in his well known plea for social software. So far, this enterprise has largely neglected a topic of key importance: money. What is money, how is it created, how does it disappear, and how does it shape our society? The paper takes on these questions in the form of a discourse where participants from a variety of backgrounds shine their lights on them.

Keywords Social software · Rules of the money game · Banking crisis · Construction of social reality · Artificial money · Functions of money

It is early Autumn 2014. A philosopher, an economist, an (ex-)banker and a computer scientist are having a meeting at CWI, the Dutch national centre for research in mathematics and computer science, in Amsterdam. Their plan is to arrive at a better understanding of the present international financial crisis by clarifying basic concepts and by learning from one another. They have decided to try to crack a difficult nut first: what is money?

Philosopher: Thanks for agreeing to take part in this discussion. In my invitation to you I mentioned Rohit Parikh's plea for an enterprise of *social software* (Parikh 2002), in which logicians, philosophers, and computer scientists collaborate with social scientists to gain a better understanding in the social mechanisms that make our societies tick, or sometimes cause them to break down. I challenged you to explain to me what money is. I have tried to prepare myself for this meeting by means of a bit of preliminary reading in economics textbooks, but if anything this left me more

---

J. van Eijck (✉)
CWI and ILLC, Amsterdam, The Netherlands
e-mail: jan.van.eijck@cwi.nl
URL: http://www.ComputationalAuditing.com

P. Elsas
Computational Auditing Com, Gatineau, Quebec, Canada

C. Başkent et al. (eds.), *Rohit Parikh on Logic, Language and Society*,
Outstanding Contributions to Logic 11, DOI 10.1007/978-3-319-47843-2_4

confused. It seems clear to me that money is an agreement in a community, just like a marriage agreement or an agreement about shared ownership. Money is part of the construction of social reality (Searle 1995). Yet, most textbooks I consulted treat money as some kind of 'thing.'

Computer Scientist: I suppose in the days when people still paid their debts in coins made of precious metal, it was quite easy to understand what money was. Just pieces of convenient stuff valued by everyone, and therefore suitable as a medium of exchange.

Economist: Money whose value comes from a commodity that it is made of, such as gold, silver or copper, is called commodity money. Gold coins have value in themselves as well as value in their use as medium of exchange.

Philosopher: But what makes gold precious or valuable in itself? Why do people want to have it? Let us see. It is easily divisible, it is scarce, it is reasonably easy to carry around, it does not deteriorate, it is hard to counterfeit, it can be used to make beautiful jewelry. Maybe these are all connected?

Economist: Well, does it matter? Let us say that the intrinsic value of a gold coin is what people are willing to trade it for. At some point the intrinsic value of coins will start to deviate from its nominal value. Usually there is a government that sets the nominal value of the coins.

Computer Scientist: Thus creating all kinds of difficulties. I remember a time in the Netherlands when the silver in a one guilder coin was worth more than the nominal value of the coin. These silver guilders quickly disappeared from circulation.

(Ex-)banker: That is a well-known phenomenon called Gresham's law. When a government undervalues one kind of money, that kind of money tends to disappear from circulation because it is hoarded.

Computer Scientist: Yes, these silver guilders were quickly replaced by nickel guilders. In normal circumstances, the metal value of coins is never more than its nominal value.

Philosopher: But if the government issues paper money, and rules that a paper guinea is worth the same as a guinea coin made of gold, then people are not hoarding the gold guineas. Why not?

Economist: Indeed, they will not, as long as they trust the statement of the government that the paper bills can be exchanged at any moment for the gold coins.

Philosopher: You seem to believe that what makes money valuable is this connection with precious metal. That's also the message I get from the textbooks I consulted. In primitive times, there was barter, but since it was inconvenient when you had to go to the baker with something the baker wanted in exchange for bread, cash was invented. Suddenly exchanges at armories, bakeries, breweries, became a lot easier.

Computer Scientist: That story is hard to believe, isn't it? Well, you have proposed to us to investigate the concept of *money* from the perspective of social software. I find it remarkable that a social software treatment of money has not already been done. The closest we got with methods from computer science was to simulate scrip systems to study how they can be optimized.

Philosopher: What is scrip?

Computer Scientist: Artificial money. Token money. Kash et al. (2012) give a lucid description, taking their lead from the story of the Capitol Hill Baby Sitting Co-op, which issued coupons that were good for baby sitting services. These coupons had to be earned by taking care of babies of other parents. It turns out that there is an optimum for the number of coupons that have to be in circulation for a given number of participants. The authors engage in a game-theoretical analysis, distinguishing various kinds of agents besides the standard ones: altruists, hoarders. Having altruists is the same as adding money. Having hoarders has the effect of removing money. Too many altruists or too much money make the system break down. So game theory can be used to analyze how money functions, to some extent.

Philosopher: But surely this is just one aspect of the role of money. It seems to me that real money, as opposed to token money, acts as a capability multiplier. The more money an agent has, the greater his or her powers in the game. Players without money have almost no power in the game. Players in debt to other players become virtually enslaved.

Computer Scientist: Ample scope for game theoretic modelling and logical analysis, but it seems to me that we still have to start.

Philosopher: And part of the problem with starting the analysis is confusion about the nature of money. In a paper Innes (1913) written in 1913 Alfred Mitchell Innes exposes the story of a barter economy that got replaced by a money economy as a myth invented by Smith (1982). It got repeated by econonomy textbook writers ever since, but Mitchell Innes believes that it is false. Instead of bartering, people just kept track of mutual obligations, and if they used money then only as a unit of accounting. "Money [...] is credit and nothing but credit. A's money is B's debt to him, and when B pays his debt, A's money disappears. This is the whole theory of money."

Computer Scientist: So there was hardly ever a need to exchange coins for goods?

Philosopher: This leads to a different or complementary theory of money. The way Georg Friedrich Knapp explains it, in a book written in the 1920s, money originated from the attempts of states to regulate economic activities (Knapp 1924). The state creates indebtedness of its subjects by levying tax on imports and on the produce of farming. Once these obligations exist, the state can rule that they can be paid off in the paper money of the state. This is what gives the paper money its value. As Knapp states in the first sentence of his book, "Money is a creature of law."

(Ex-)banker: Still, gold coins existed before paper money, and the coins were worth what the gold was worth, there is no way around it. And historically it was banks who first issued paper notes. As long as people trust the bankers, they will not hoard gold but use the much more convenient promissory notes issued by the bank.

Philosopher: It is not so clear what existed earlier, tally sticks for keeping track of debet and credit, or coins. But let's accept your story for now. Then this is how it goes on: the bankers soon discover that it is not necessary to have one million pounds of silver in their vaults to cover the worth of one million promissory notes "Good for one pound of silver."

(Ex-)banker: This was in fact discovered by goldsmiths who gave out notes for gold deposited with them for safekeeping. They observed that people do not all

reclaim their deposit at the same time, and they saw no objection to invest part of the gold that they had received for safekeeping.

Economist: Indeed, as Adam Smith observed, two hunderd thousand pounds of silver would be enough to cover the worth of promissory notes for one million pounds sterling. In fact, two hundred thousand government backed pound notes would also have been enough.

(Ex-)banker: The reserve requirement (or cash reserve ratio) is a regulation by the central bank setting the minimum fraction of deposits and notes that each commercial bank should hold as reserve. The 20 % that Adam Smith suggests is very high, by modern standards. Nowadays, Chinese banks have this. In the USA it is more like 10 %. In the Euro zone it is much less still.

Computer Scientist: And the converse of the cash reserve ratio, is that what is called the money multiplier?

Economist: That's right. But there is fierce debate among economists about whether the money multiplier exists (Carpenter and Demiralp 2010).

Computer Scientist: One way to think about the cash reserve ratio R is by saying that a bank is allowed to lend out $(1 - R)X$ of every deposit X that it receives. Next, this money can be put in deposit again. either with the same bank, or with another bank, it does not matter. Suppose for simplicity it gets deposited with the same bank. Then $1 - R$ of the new deposit can be loaned out again. This is $(1 - R)^2X$. And so on. This is a geometric series. Let's see. The original amount X gets multiplied to

$$X + (1 - R)X + (1 - R)^2X + \cdots$$

Applying the formula for a converging geometric series this gives

$$\frac{X}{1 - (1 - R)} = \frac{1}{R}X.$$

So the money multiplier is indeed the converse of the reserve ratio.

Economist: An easier way to see this is is as follows. The bank receives X in deposit. If the deposit consists of cash, it simply reasons, hey, we have to keep fraction R of our loans in reserve, so on the basis of this new cash deposit X in our vaults we can loan out $\frac{1-R}{R}X$. Together with the deposit X this gives $X + \frac{1-R}{R}X = \frac{1}{R}X$. So $\frac{1}{R}$ is the money multiplier. This is explained in many places, in books you can download from internet (Nichols 1961; Patman 1964).

Philosopher: The money multiplier explains why it is a severe threat to the financial system when a really large bank fails, I suppose. If the bank is worth one billion dollars, we may assume it has a cash reserve of about one billion dollars. Assuming a reserve ratio of 10 %, when it goes bankrupt ten billion dollars get destroyed. In fact much more gets destroyed, for the banks lend out much more than allowed by the cash reserve ratio.

(Ex-)banker: You are forgetting that the money was lent out in exchange for securities, an entitlement to your house, for instance. The net worth of a bank is not equal to its cash reserve ratio, but to the difference between its assets and its liabilities. When the bank goes bust the securities are still there. How much gets lost depends on the quality of the collateral. Often, the collateral cannot easily be turned

into money. So if there is a run on a bank, there is not enough money to pay the worried customers.

Computer Scientist: Here is a famous quote:

> Banking was conceived in iniquity and was born in sin. The Bankers own the earth. Take it away from them, but leave them the power to create deposits, and with the flick of the pen they will create enough deposits to buy it back again. However, take it away from them, and all the great fortunes like mine will disappear and they ought to disappear, for this would be a happier and better world to live in. But, if you wish to remain the slaves of Bankers and pay the cost of your own slavery, let them continue to create deposits.

Philosopher: Who said that?

Computer Scientist: Sir Josiah Stamp. He was president of the Bank of England in the 1920s, and the second richest man in Britain at that time.

Philosopher: And was he serious?

(Ex-)banker: There is no doubt he was. And right too. The only surprising thing about the quote is that it is unusually frank and lucid, for a banker. You should know that I decided to quit my profession some time ago.

Computer Scientist: I can see that with a scheme like this, backing up by some government assurance is important. But this government-backing has ceased, right? In 1971 Richard Nixon unilaterally cancelled the direct convertibility of dollar bills into the gold in Ford Knox.

Philosopher: And still, people continue to trust dollar bills.

Economist: Right now, the USA has an astronomical national debt of more that 17.6 trillion dollars, in the American sense of 'trillion'. That is 17.6 times 10^{12} dollars. Mind you, this is the debt of the government that is backing up the dollar bills.

Philosopher: So why do people still trust dollar bills? Can anyone explain this to me, please?

(Ex-)banker: If the figure of 17.6 trillion dollars scares you, let me remind you that the Gross Domestic Product of the US in 2014 is also roughly 17.6 trillion dollars.

Economist: Maybe we should talk a bit more about what money is, and then hope that gradually things will become clearer. For it is definitely not true that all money consists of either government-backed notes or bullion. In fact, most money is of a different kind.

(Ex-)banker: Money not backed by a guarantee of convertibility to anything else is called fiat money. What gives fiat money its value is the ruling that it is legal tender: it cannot be refused as payment for debts. Also, you can pay your taxes with it.

Philosopher: If it is the government that creates and circulates the money, why does it give it to the people first, and then takes some of it back later? Why not simply keep some of it in the first place? Why doesn't the central bank simply transfer the tax amount to the treasury each month?

(Ex-)banker: Something like a automated national tax payment from freshly created money? That would be cool. But when the central bank creates money, it does not simply *give* it to the taxpayers. Instead, it brings money into circulation by *buying* something with it, for example government bonds.

Philosopher: Still, if the state has the power to create cash, why not credit the new money as state revenue? Isn't this precisely what monetary reform activists propose (Huber and Robertson 2001)?

Economist: But cash is not the only kind of money. There is another kind that gets created by banks. Call this IOU money, or more precisely, spendable IOUs. These are created over and over. My colleague professor Charles Goodhart gives a nice explanation. See www.fractionalreserves.com.

(Ex-)banker: Goodhart is professor emeritus of banking and finance, London School of Economics. Not precisely your colleague.

Philosopher: Never mind. Can you summarize his explanation for us?

Economist: Goodhart explains it with a story. Mick wants to borrow 10 pounds from Jim. So he writes an IOU ("I owe you") for 10 pounds, signs it, hands it over to Jim, and receives his 10 pounds. When Mick repays the loan, Jim will hand back the IOU agreement and dispose of it. It has done its job. An IOU gets created, next there is a period of time when it has value, and finally, when the loan gets repaid, the IOU has no further use and expires.

Philosopher: You say that the IOU has value. But the IOU note is not money. If Jim takes it to his local pub to pay for his beer, the note will be refused.

Economist: That depends a bit on whether the publican knows Mick. If not, then it will be refused, yes.

(Ex-)banker: But that situation changes when a bank puts its stamp of approval. Then the bank says that the IOU note is as good as a 10 pound note.

Economist: These are spendable IOUs. When a spendable IOU gets created, it gets created out of nothing. It did not exist before, and it was not created from anything else, nor obtained from anybody else.

(Ex-)banker: See, this is what happens all the time. You go to the bank, sign an agreement that you owe them 200,000 euros, with the house you are going to buy as security against the loan, and you receive 200,000 euros on your account. The snag is that this is not money they have and lend to you.

Computer Scientist: By government regulation they only need to have 20,000 euros themselves against the 200,000 euros they give you. Still, they receive interest from you on 200,000 euros.

(Ex-)banker: What you are giving now is the textbook account. This is not exactly how it happens. If you go to the bank and ask for your mortgage, do you really think the desk clerk first looks at her computer to see if the bank still has 10 % of the money they are going to lend to you? Of course not. She just presses a button, and lo and behold, next time you log in to your bank account you see that you have 200,000 euros extra to your credit. And when you close your deal with the former owner of the house you buy, the 200,000 euros disappear from your bank account and make their appearance on his. Banks operate within an electronic clearing system that nets out multilateral payments at the end of each day. So they need to hold only a tiny proportion of central bank money to meet their payment requirements. Much less, in fact, than the cash reserve ratio.

Computer Scientist: The story brings to mind a quote from John Kenneth Galbraith:

> The study of money, above all other fields is one in which complexity is used to disguise truth or to evade truth, not to reveal it. The process by which banks create money is so simple that the mind is repelled. (Galbraith 1975)

Economist: Next, Goodhart explains the bathtub model of money circulation. If the bank's seal of approval can change a non-spendable IOU into a spendable IOU, then with every bank loan the bank is effectively creating money. If the bank grants you a loan, the bank gets you to sign an IOU. In return they give you a checkbook, or access to an elecronic account by means of a debit card. So the bank loan is effectively an IOU swapping exercise.

(Ex-)banker: Money is created when loans are made. Money disappears when loans get repaid. So money is continually being created and destroyed. It flows into the bathtub at the tap, as new loans being made. It flows out at the sink, as repayments of loans. So the supply of money in the system depends critically on the rate of flow in and out of the system. Banks make their money on the difference between what they earn making loans and what they spend in paying interest to depositors and to lenders like the central bank.

Philosopher: Well, banks are there to make money, I suppose. But it surprises me they do it in such a very literal sense. Your story makes lots of things clear. It explains why banks charge no money for taking care of our money, for instance. They lend out our money and earn interest on it as soon as we deposit it.

Economist: In the golden age of Dutch banking, in the Seventeenth Century, the Bank of Amsterdam charged a fee for deposits in silver or gold.

(Ex-)banker: Indeed, for a while the Bank of Amsterdam, maybe the first example of a central bank, functioned on a full reserve basis. But that was abandoned in 1657, when the bank started to allow depositors to overdraw their accounts. Also, it was providing large loans to the City of Amsterdam and to the Dutch East India Company. This was initially done in secret, and when it became public knowledge the bank got in trouble and had to be taken over by the City of Amsterdam, in 1791.

Philosopher: Well, it seems to me that this system where banks can lend out money they don't have is dangerous.

Computer Scientist: I don't mean the defend capitalist economy, but cash reserve ratio is one of the fundamental concepts that fueled the global economical development. I don't think that people invented this system to be evil. It simply makes sense. After all, even after all the economical meltdowns, still, global and national economies are in gain. They take some backwards steps from time to time, but it is always improving in a larger scale in capitalist societies.

Philosopher: You seem to agree with Mitchell Innes. Here is a quote from the paper I mentioned (Innes 1913) that I found so remarkable that I made a note of it: "To attempt the regulation of banking by limiting the note issue is to entirely misunderstand the whole banking problem, and to start at the wrong end. The danger lies not in the bank-note but in imprudent or dishonest banking. Once insure that banking shall be carried on by honest people under a proper understanding of the principles of credit and debt, and the note issue may be left to take care of itself."

Computer Scientist: Well, that sounds a bit naive, doesn't it? I think honest or dishonest is beside the point. The money trade is a game, so we should talk about the rules.

(Ex-)banker: Yes, there is a definite need to rethink the rules of the money game. One could argue that institutions like the IMF and the WTO are causing untold misery to people in third world countries, and are now helping to turn countries like Greece into places of misery too.

Philosopher: Let me ask you an important question. *(Takes a 10 Euro bill from his wallet.)* Does this 10 Euro bill represent wealth, or would it be more accurate to say that it represents debt? I am asking this because I recently read David Graeber's *Debt* (2012). Graeber is an anthropologist who argues that money turns personal obligation into impersonal debt: "…money [has the capacity] to turn morality into a matter of impersonal arithmetic — and by doing so, to justify things that would otherwise seem outrageous or obscene."

(Ex-)banker: Wow! Well, your bill is cash. You can keep it indefinitely, but it will lose its value by inflation. Or you can deposit it in a bank. And the instant you deposit your money, the bank starts playing musical chairs with all but a tiny fraction of it.

Philosopher: In itself, that is not a problem. For if the bank creates money, the created money always gets exchanged for an IOU of the lender, isn't that right?

(Ex-)banker: That is exactly right. And the big problem right now is that nobody knows what all these IOUs are really worth. For that you need to know who owes what to whom, and what else they possess. Right now, this is a great mess, and it needs to be disentangled. This is all clearly indicated in a blog by Bookstaber (2016).

Computer Scientist: Wait, wait. Let us go back to the 10 Euro question, for I think we can answer it now. The 10 Euro bill represents credit or debt, for that is the same thing, depending how you look at it. You can use it to fulfil an obligation of a very specific weight, to any stranger, for money quantifies obligation and makes it anonymous. And how can this magic occur? By the power of a state — or a community of states — ruled by law, that is behind all this.

Philosopher: Yes, and we can say more. The value of 10 Euros now is different from that of 10 Euros a year in the future. Economists call this time discounting, I believe. Also, the value of 100 Euros to a poor person is very different from what 1000 Euros offer an affluent one. Economists call this diminishing marginal utility of money. And the function of money in bankers' trading is very different from that in buying bread, or in saving for one's pension.

Economist: The function of money in buying bread is as a means of exchange. The function of money in saving for one's pension is as a store of value. The function of money in banker's trading, I am not so sure.

Philosopher: I propose to leave that for another occasion. A social software study should at least analyze three different functions of money: medium of exchange, store of value, unit of account. And it may well end with a plea for reform. Advocates of monetary reform usually start with an analysis of how money functions in society, and then move on to how it *should* function (Huber and Robertson 2001; Robertson 2012).

Acknowledgements The text owes much to the perceptive comments of two anonymous reviewers.

References

Bookstaber, R. (2016). *The risks of financial modeling: VaR and the economic meltdown—Testimony to the house*. http://rick.bookstaber.com/2009/09/risks-of-financial-modeling-var-and.html.
Carpenter, S. B., & Demiralp, S. (2010). *Money, reserves, and the transmission of monetary policy: Does the money multiplier exist? Technical report*. Federal Reserve Board, Washington, DC: Divisions of Research & Statistics and Monetary Affairs.
Galbraith, J. K. (1975). *Money: Whence it came, where it went*. Boston: Houghton Mifflin.
Graeber, D. (2012). *Debt*. Brooklyn: Melville House Publishing.
Huber, J., & Robertson, J. (2001). Creating new money: A monetary reform for the information age. *New Economics Foundation*.
Innes, A. M. (1913). What is money? *Banking Law Journal, 30*(5):377–408. http://www.community-exchange.org/docs/whatismoney.htm.
Kash, I. A., Friedman, E. J., & Halpern, J. Y. (2012). *Optimizing scrip systems: Crashes, altruists, hoarders, sybils and collusion*. Technical report, arXiv.
Knapp, G. F. (1924). *The state theory of money*. Macmillan. Translation of the German *Staatliche Theorie des Geldes*.
Nichols, D. M. (1961). *Modern money mechanics*. Technical report, Federal Reserve Bank of Chicago.
Parikh, R. (2002). Social software. *Synthese*, *132*, 187–211.
Patman, W. (1964). *A primer on money*. USA: Government Printing Office.
Robertson, J. (2012). *Future money: Breakdown or breakthrough?*. Green Books.
Searle, J. R. (1995). *The construction of social reality*. New York: Simon and Schuster.
Smith, A. (1982). *An inquiry into the nature and causes of the wealth of nations*. Indianapolis: Liberty Fund. Original edition: 1776.

Chapter 5
Focusing on Campaigns

Dominik Klein and Eric Pacuit

All errors in government and in society
are based on philosophic errors which in turn
are derived from errors in natural science.

Marquis de Condorcet

(This quote of Condorcet is taken from the recent book *Gaming the Vote* by William Poundstone [2008, p. 134.])

Abstract One of the important lessons to take away from Rohit Parikh's impressive body of work is that logicians and computer scientists have much to gain by focusing their attention on the intricacies of political campaigns. Drawing on recent work developing a theory of *expressive voting*, we study the dynamics of voters' opinions during an election. In this paper, we develop a model in which the relative importance of the different issues that concern a voter may change either in response to candidates' statements during a campaign or due to unforeseen events. We study how changes in a voter?s attention to the issues influence voting behavior under voting systems such as plurality rule and approval voting. We argue that it can be worthwhile for candidates to reshape public focus, but that doing so can be a complex and risky activity.

Keywords Decision theory · Voting systems · Expressive voting

---

D. Klein
Tilburg Institute for Logic and the Philosophy of Science, Tilburg, The Netherlands
e-mail: D.Klein@uvt.nl

E. Pacuit (✉)
Department of Philosophy, University of Maryland, College Park, MD, USA
e-mail: epacuit@umd.edu
URL: http://www.pacuit.org

C. Başkent et al. (eds.), *Rohit Parikh on Logic, Language and Society*,
Outstanding Contributions to Logic 11, DOI 10.1007/978-3-319-47843-2_5

5.1 Introduction

Rohit Parikh has noted that political elections are a "bonanza of data" for logicians and game theorists. Much of his recent work (see, especially, his contributions on *social software* Parikh 2002, 2001; Pacuit and Parikh 2006) uses the sophisticated mathematical tools developed by logicians and game theorists to provide penetrating analyses of political phenomena. Even Parikh's more technical work in logic is often motivated by an interest in reasoning about social phenomena (cf. his seminal contribution on game logic Parikh 1985).

In this paper, we follow up on some ideas presented in a recent paper by Dean and Parikh (2011) on the logic of campaigning. They begin their paper with a quote from the satirical newspaper *The Onion*, commenting on the primary race between Barack Obama and Hillary Clinton in 2008:

> After Sen. Barack Obama's comments last week about what he typically eats for dinner were criticized by Sen. Hillary Clinton as being offensive to both herself and the American voters, the number of acceptable phrases presidential candidate can now say is officially down to four. "At the beginning of 2007 there were 38 things candidates could mention in public that wouldn't be considered damming to their campaigns, but now they are mostly limited to 'Thank you all for coming' and 'God bless America"' ABC News chief Washington correspondent George Stephanapoulos said in Sunday's episode of *This Week.*
>
> (The Onion, 2008)

Parikh and Dean develop a formal framework to analyze the phenomenon satirized above.

Suppose that the main issues in an election are represented by a finite set of propositions. Formally, let $\mathcal{L}$ be a propositional language generated from a set At of atomic propositions. For example, the set $\mathsf{At} = \{p, q, r\}$ might represent policies on health care, nuclear energy, and the Israeli-Palestinian conflict. Each voter is assumed to have a preferred ideal world—a complete set of elements of At specifying which policies are enacted. In this setting, *worlds* are most naturally represented as propositional valuation functions $V : \mathsf{At} \rightarrow \{-1, 1\}$ where $V(p) = -1$ means that p is false (the policy is not enacted), and $V(p) = 1$ means that p is true (the policy is enacted). A voter i's preference is represented by a function $pref_i : \mathcal{L} \rightarrow \{-1, 0, 1\}$ where $pref_i(\varphi) = 1$ means that voter i prefers that φ is true; $pref_i(\varphi) = -1$ means that voter i prefers that φ is false; and $pref_i(\varphi) = 0$ means that voter i is neutral concerning φ. In addition, there is a weighing function for each voter i, $w_i : \mathsf{At} \rightarrow [0, 1]$, assigning weights to the different atomic propositions.[1] The weight of a formula $p \in \mathsf{At}$, $w_i(p)$, represents the relative importance of the proposition to voter i. Using these functions, each voter i can assign a value to each world V:

$$Val_i(V) = \sum_{q \in \mathsf{At}} V(q) \times pref_i(q) \times w_i(q).$$

[1] Dean and Parikh normalize the weight function so that $\sum_{p \in \mathsf{At}} w_i(p) = 1$. This simplification is not needed for this paper.

Then, $Val_i(V)$ is a measure of how far way the world V is from the voter's ideal world (which is assigned 1 provided the voters' weights sum to 1).[2]

During a campaign, voters develop theories about the candidates, based on their past utterances (and, perhaps, also on any preconceived ideas the voters may have about the candidates). For a candidate c, let $T_i(c) \subseteq \mathcal{L}$ denote voter i's theory of candidate c. The decision problem that a candidate faces is which statement(s) $\varphi \in \mathcal{L}$ maximizes support among the voters. Solving this decision problem involves two additional features of the voters. First, it depends on how a voter in question will update her theory in response to an utterance. Let $T_i(c) \circ \varphi$ denote i's theory about c that is *updated* with the statement φ. In general, $\circ$ may be any theory change operation, such as an AGM belief revision function (Alchourrón et al. 1985). The second important feature of a voter is how she evaluates theories of candidates. Dean and Parikh consider three different types of voters. Suppose that T is a theory and that $V \models T$ means that $V(\varphi) = 1$ for all $\varphi \in T$. Then, define the following utility functions for a voter i:

$$\begin{array}{lrl} \text{(pessimistic voter)} & U_i^{min}(T) &= \min\{Val_i(V) \mid V \models T\} \\ \text{(optimistic voter)} & U_i^{max}(T) &= \max\{Val_i(V) \mid V \models T\} \\ \text{(averaging voter)} & U_i^{ev}(T) &= \frac{\sum_{V \models T} Val_i(V)}{|\{V \mid V \models T\}|} \end{array}$$

When speaking to a block of voters (i.e., voters who share a theory of the candidate), a candidate is faced with a maximization problem: to choose a statement (which may or may not be consistent with the candidate's actual beliefs) that maximizes the overall utility of the updated theory for a group of voters. Dean and Parikh go on to discuss some intriguing connections with AGM belief revision theory (Alchourrón et al. 1985) (especially, Parikh's important work on splitting languages Parikh 1999).

In this paper, we are interested in studying voters' changing opinions during a campaign. In the model sketched above, the voters' theories of a candidate change in response to that candidate's statements. We take a different perspective in this paper. Instead of allowing voters to change their theory of a candidate during a campaign, we study voters who may *focus* on different issues throughout an election. That is, during a campaign, the relative importance of the different issues for a voter may change either in response to candidates' statements or due to unforeseen events.

The main contribution of this exploratory paper is to raise questions and point out interesting issues rather than to provide a fully worked-out theory. Such a theory will be left for future work. Section 5.2 introduces a framework for reasoning about how voters express themselves when voting. This framework is based on a recent article (Aragones et al. 2011) and has much in common with the Dean and Parikh model sketched above. In Sect. 5.3, we show how to model situations in which the voters' *focus* on issues shifts during an election. Finally, Sect. 5.4 offers some general conclusions and ideas for future work.

[2]This model is not only interesting for the theoretician. There are websites, such as www.isidewith.com, that use a variant of this model to rank candidates in upcoming elections according to how close they are to the voter's opinions about a number of relevant policy issues.

5.2 Expressive Voting

Dean and Parikh's model of a campaign is focused on the candidates' decision to make various statements during a campaign and how these statements change the voters' theories of the candidates. In this paper, we conceptualize the dynamics of a campaign differently, by adapting the approach in a paper by Aragones et al. (2011). They develop a model in which a voter's decision to vote is based solely on the need to express herself. To motivate their approach, they cite numerous psychological studies showing that people have an intrinsic need for their opinions to be heard. Of course, as already acknowledged in Aragones et al. (2011), this is an idealization. Voters have many ways to express themselves besides voting. However, for the purposes of this paper, we assume that a voter's only reason for voting is to express her political opinions.[3] Interestingly, this approach drastically changes the analysis of well-studied voting procedures.[4]

In this section, we introduce the model of an election used in Aragones et al. (2011), henceforth called the *AGW* model. Suppose that $T = \{1, \dots, m\}$ is a set of parties, or candidates. Each party $j \in T$ is characterized by its positions on the various issues of concern $I = \{1, \dots, n\}$. This is represented as follows: Each $j \in T$ is associated with a vector $\mathbf{p}^j \in [-1, 1]^n$ giving j's positions on each of the issues. The idea is that $p_i^j \in [-1, 1]$ is the degree to which candidate j supports issue i, where 1 denotes total support and -1 total opposition. To simplify the discussion, in this paper, we assume that candidates take extreme positions on each of the issues: for each $i \in I$ and candidate position vector $\mathbf{p} = (p_1, \dots, p_n)$, we have $p_i \in \{-1, 1\}$. Voters are represented is a similar way. Each voter's opinion about the different issues is represented by a vector $\mathbf{v}^i \in [-1, 1]^n$. For such a vector $\mathbf{v}$, the sign of v_i describes the voter's general attitude towards issue i (i.e., is the voter for or against issue i?), while the absolute value $|v_i|$ reflects the weight that the voter attaches to issue i. The greater the weight, the more important the issue is to the voter. To simplify the notation, we use $\mathbf{p}$ to denote an arbitrary candidate and $\mathbf{v}$ to denote an arbitrary voter.

The key idea behind *expressive voting* is that each ballot in an election is associated with a "statement" giving the amount of support for each issue. To that end, a **ballot** is a vector $\mathbf{x} = (x_1, \dots, x_m) \in \mathbb{R}_+^m$ (where $\mathbb{R}_+$ is the set of nonnegative real numbers), representing the amount of support for each candidate. For example, **abstention** corresponds to the zero vector (denoted $\mathbf{0}$ where $\mathbf{0}_j = 0$ for each $j = 1, \dots, m$). The statement made by the ballot $\mathbf{x} = (x_1, \dots, x_m)$, denoted $s(\mathbf{x})$, is the vector

$$\left(\sum_{j \in T} x_j \times \mathbf{p}_1^j, \dots, \sum_{j \in T} x_j \times \mathbf{p}_n^j\right) \in \mathbb{R}^n$$

Voter $\mathbf{v}$'s decision problem is to find a ballot that makes a statement as close as possible to her actual position—that is, to find the ballot $\mathbf{x}$ that minimizes the

[3]See, also, Brennan and Lomasky (1993, pp. 40–46) for a discussion of this point.

[4]For instance, the motivation to misrepresent one's position ceases altogether.

Euclidean distance from the statement made by $\mathbf{x}$ to the voter's own position $\mathbf{v}$. More precisely, if F is the set of *feasible* ballots (i.e., the ballots admitted by the voting rule), then voter $\mathbf{v}$ must solve the following minimization problem:

$$\arg\min_{\mathbf{x}\in F} \text{dist}(\mathbf{v}, s(\mathbf{x})),$$

where $\text{dist}(\mathbf{x}, \mathbf{y}) = \sqrt{\sum_i (x_i - y_i)^2}$ is the usual Euclidean distance. If the solution to this minimization problem is $\mathbf{x} = \mathbf{0} \in \mathbb{R}^m$, then the voter will abstain.[5] The main contribution from Aragones et al. (2011) is a rigorous analysis of the statements that can be made using plurality rule compared to approval voting.

A voting system consists of a set $F \subset \mathbb{R}^m_+$ of *feasible* ballots, together with an aggregation method for selecting a winner from a profile of ballots (one ballot for each voter).

Plurality rule: Each voter selects a single candidate, and the candidate with the most votes is declared the winner. Thus, the feasible ballots are $F^M = \{\mathbf{0}\} \cup \{\mathbf{e}^j\}_{j\leq m}$, where $\mathbf{e}^j$ is the vector with 1 in the jth position and 0 everywhere else.

Example 5.2.1 Suppose that there are three issues $I = \{1, 2, 3\}$ and two candidates $T = \{d, r\}$. Assume that the candidates take extreme opposing positions on the three issues: $\mathbf{p}^d = (1, 1, 1)$ and $\mathbf{p}^r = (-1, -1, -1)$. Consider a voter with the opinion vector $\mathbf{v} = (0, 0.7, -0.1)$. Under plurality rule, there are three possible ballots: $F^M = \{(0, 0), (1, 0), (0, 1)\}$. The statements associated with these ballots are the positions of the two candidates and the null vector:

$$\begin{aligned} s(1, 0) &= 1 * \mathbf{p}^d + 0 * \mathbf{p}^r = (1, 1, 1)\\ s(0, 1) &= 1 * \mathbf{p}^d + 0 * \mathbf{p}^r = (-1, -1, -1)\\ s(0, 0) &= 0 * \mathbf{p}^d + 0 * \mathbf{p}^r = (0, 0, 0) \end{aligned}$$

The voter must choose a ballot $\mathbf{x} \in F^M$ that minimizes $\text{dist}(\mathbf{v}, s(\mathbf{x}))$. The calculations are:

$$\begin{aligned} \text{dist}(\mathbf{v}, s(1, 0)) &= \sqrt{(0-1)^2 + (0.7-1)^2 + (-0.1-1)^2} = \sqrt{2.3}\\ \text{dist}(\mathbf{v}, s(0, 1)) &= \sqrt{(0+1)^2 + (0.7+1)^2 + (-0.1+1)^2} = \sqrt{4.7}\\ \text{dist}(\mathbf{v}, s(0, 0)) &= \sqrt{(0-0)^2 + (0.7-0)^2 + (-0.1-0)^2} = \sqrt{0.5} \end{aligned}$$

Thus, in this case, the voter's best statement is to abstain.

While the plurality rule asks voters to identify the best possible alternative (if one exists), **approval voting** asks voters to identify the candidates that are *approved* (Brams and Fishburn 1983). It turns out that there are two types of ballots that correspond to approval voting in this setting. Following Aragones et al. (2011), the first version defines a ballot by distributing a voter's support among all the candidates that are approved.

[5] Thus, a voter's choice to abstain is due to an inability to express herself in the voting system rather than any cost associated with voting.

Approval Voting, version 1 In approval voting, each voter selects a set of "approved" candidates. In the AGW model, the statement that such a ballot makes is an average over all of the approved candidates' positions. For each $S \subseteq T$, there is a ballot $\mathbf{x}^S$ defined as follows:

$$\mathbf{x}^S = \frac{1}{|S|} \sum_{j \in S} \mathbf{e}^j.$$

So, the total support is divided evenly among the approved candidates. Let F^A denote the set of feasible approval ballots.

Example 5.2.2 Suppose that there are three issues $I = \{1, 2, 3\}$ and three candidates $T = \{l, m, r\}$. Assume that the candidates take the following positions on the three issues: $\mathbf{p}^d = (1, -1, 1)$, $\mathbf{p}^m = (1, 1, 1)$, and $\mathbf{p}^r = (-1, 1, -1)$. Consider a voter with the opinion vector $\mathbf{v} = (0.55, -0.2, 0.8)$. Under plurality rule, there are four possible ballots:

$$F^M = \{(0, 0, 0), (1, 0, 0), (0, 0, 1), (0, 1, 0)\}.$$

As the reader is invited to check, the best statement for the voter in this case is to abstain:

$$\text{dist}(\mathbf{v}, s(0, 0, 0)) = \sqrt{(0.55 - 0)^2 + (-0.2 - 0)^2 + (0.8 - 0)^2} = \sqrt{0.9825}.$$

However, approving of $\{d, m\}$ is preferred to abstaining:

$$\text{dist}(\mathbf{v}, (1, 0, 1)) = \sqrt{(0.55 - 1)^2 + (0 - -0.2)^2 + (1 - 0.8)^2} = \sqrt{0.2825}.$$

According to the above definition, a voter evaluates a coalition of candidates by averaging the positions of its candidates. It is not hard to construct examples in which a voter with moderate positions on the issues may approve of a set of candidates with opposing positions on the same issues. More generally (dropping the assumption that candidates can take only extreme positions on the issues), a moderate voter may prefer a coalition of candidates with extreme opposing positions to a single candidate with relatively moderate positions on the issues. The soundness of the above definition of approval voting relies on voters believing that if the approved candidates are all elected, they will work together to implement more-moderate policies. Of course, examples abound in which the election of officials with opposing positions does not lead to moderate policies, but, rather, to deadlock. This discussion motivates the following variant of approval voting (Klein and Pacuit 2013).

We begin by considering each issue in the election individually. Since we assume that $\mathbf{p}_i \in \{-1, 1\}$ for each candidate $\mathbf{p}$ and issue i, the payoff for a voter $\mathbf{v} = (v_1, \ldots, v_n)$ choosing the candidate $\mathbf{p}$ on issue i is either $|v_i|$ or $-|v_i|$, depending on whether or not the signs of v_i and p_i agree. Formally, we have:

$$val_{\mathbf{v}}(i, \mathbf{p}) = \begin{cases} |v_i| & \text{iff } v_i \cdot p_i \geq 0 \\ -|v_i| & \text{iff } v_i \cdot p_i < 0 \end{cases}$$

Building on this idea, we can precisely define when a voter may approve of a candidate.

Definition 5.2.3 (*k-Approves*) Suppose that $k \in [-1, 1]$ (this is called the **approval coefficient**). A voter $\mathbf{v}$ k-approves of all parties $\mathbf{p}$ that satisfy:

$$\sum p_i v_i \geq k \cdot \sum |v_i|.$$

Since, $\sum p_i v_i$ is the standard scalar product $\mathbf{p} \cdot \mathbf{v}$, and $\sum |v_i|$ is the 1-norm, denoted $|\mathbf{v}|_1$, we have that a voter $\mathbf{v}$ k-approves of a candidate $\mathbf{p}$ provided that:

$$(1) \qquad \mathbf{p} \cdot \mathbf{v} \geq k|\mathbf{v}|_1.$$

Typically, we assume that $k \geq 0$. This means that if a voter k-approves of a candidate, then that voter agrees with the candidate on more issues than she disagrees on.

Definition 5.2.3 has an interesting geometric interpretation. For a vector $\mathbf{x} \in \mathbb{R}^n$ and some angle α, let $C(x, \alpha)$ be the cone of all vectors $\mathbf{y}$ in $\mathbb{R}^n - \{\mathbf{0}\}$ such that the angle between $\mathbf{x}$ and $\mathbf{y}$ is at most α. Then, we have the following proposition:

Proposition 5.2.4 *Let* $\mathbf{v}$ *be a voter and let* k *be as in the definition of approval voting. Then, there is some angle* α *depending upon* n, k *and* $\mathbf{v}$ *such that for each party* $\mathbf{p}$ *holds*

$$\mathbf{p} \in C(\mathbf{v}, \alpha) \Leftrightarrow \mathbf{p} \cdot \mathbf{v} \geq k|\mathbf{v}|_1.$$

Furthermore, α *satisfies* $\arccos(k) \leq \alpha \leq \arccos(\frac{k}{\sqrt{n}})$.

Proof For $\mathbf{x}, \mathbf{y} \in \mathbb{R}^n$, the angle α between $\mathbf{x}$ and $\mathbf{y}$ is described by the following well-known equation:

$$\frac{\mathbf{x} \cdot \mathbf{y}}{|\mathbf{x}|_2 |\mathbf{y}|_2} = \cos \alpha. \tag{5.1}$$

where $|\mathbf{x}|_2 = \sqrt{\sum \mathbf{x}_i^2}$ denotes the Euclidean length of $\mathbf{x}$. On the other hand, inequality (1) can be transformed to

$$\frac{\mathbf{v} \cdot \mathbf{p}}{|\mathbf{v}|_1} \geq k$$

Using a simple algebraic manipulation and multiplying both sides by $\frac{1}{|\mathbf{v}|_2 \sqrt{n}}$, we have

$$\frac{\mathbf{v} \cdot \mathbf{p}}{|\mathbf{v}|_2 \sqrt{n}} \geq \frac{k}{\sqrt{n}} \frac{|\mathbf{v}|_1}{|\mathbf{v}|_2}$$

Since $|\mathbf{p}|_2 = \sqrt{\sum_i 1} = \sqrt{n}$. This is exactly Eq. 5.1 for

$$\alpha = \arccos(\frac{k}{\sqrt{n}} \frac{|\mathbf{v}|_1}{|\mathbf{v}|_2}).$$

The last claim follows from the inequality $|\mathbf{x}|_2 \leq |\mathbf{x}|_1 \leq \sqrt{n}|\mathbf{x}|_2$, for all $\mathbf{x} \in \mathbb{R}^n$. □

Thus, if we interpret the vector $\mathbf{v}$ as providing a voter's general "direction of opinion", the above formula says that a voter k-approves of all parties that lie in roughly the same direction.

We conclude this section by showing that this alternative approach to approval voting is compatible with the definition of plurality rule given above. The following proposition shows that k-approving of a candidate boils down to plurality rule when there are only single-voter ballots. Of course, this depends on the value of the approval coefficient k. We stipulate that a voter $\mathbf{v}$ approves of the candidate $\mathbf{p}$ that minimizes[6] the angle between $\mathbf{v}$ and $\mathbf{p}$. That is, the voter approves of the candidate $\mathbf{p}$ that maximizes the quotient $\frac{\sum v_i \cdot p_i}{\sum |v_i|}$ $\mathbf{v}$.

Proposition 5.2.5 *Suppose that $\overline{T}$ is the set of position vectors for each candidate in T. For any voter $\mathbf{v}$, the voter k-approves of $\mathbf{p}$ with $k = \frac{\sum v_i \cdot p_i}{\sum |v_i|}$ if, and only if,* $\text{dist}(\mathbf{p}^*, \mathbf{v}) = \min_{\mathbf{p} \in \overline{T}} \text{dist}(\mathbf{p}, \mathbf{v})$.

Proof Suppose that $\mathbf{v}$ is a voter for a set of candidates T, and suppose that $\mathbf{p}^*$ is a vector of positions for a candidate. Then, the claim is a direct consequence of the following equivalence:

$$\text{dist}(\mathbf{p}^*, \mathbf{v}) = \min_{\mathbf{p} \in \overline{T}} \text{dist}(\mathbf{p}, \mathbf{v}) \iff \frac{\sum v_i p_i^*}{\sum |p_i^*|} = \max_{p \in P} \frac{\sum v_i p_i}{\sum |p_i|}.$$

Recall that $\mathbf{p}_i \in \{-1, 1\}$ for each topic $i \in I$. Fix a voter $\mathbf{v}$. For any party $\mathbf{p}$, let $U_{\mathbf{p}} \subseteq \{1 \ldots |I|\}$ be defined by:

$$i \in U_{\mathbf{p}} \Leftrightarrow v_i \cdot p_i < 0$$

Thus, $U_{\mathbf{p}}$ is the set of indices where the sign of $\mathbf{v}$ and $\mathbf{p}$ disagree. Now we have:

$$\begin{aligned} \text{dist}(\mathbf{v}, \mathbf{p}) &= \sqrt{\sum_i (v_i - p_i)^2} \\ &= \sqrt{n + \sum_i v_i^2 - 2\sum_i v_i p_i} \\ &= \sqrt{n + \sum_i v_i^2 - 2\sum_i |v_i| + 4\sum_{i \in U_{\mathbf{p}}} |v_i|} \end{aligned}$$

Observe that only the last term depends on $\mathbf{p}$. Therefore, we have for any $\mathbf{p}, \mathbf{p}' \in P$:

[6]The voter randomly selects a candidate if there is more than one such candidate.

$$\mathrm{dist}(\mathbf{p}, \mathbf{v}) \leq \mathrm{dist}(\mathbf{p}', \mathbf{v}) \Leftrightarrow \sum_{i \in U_{\mathbf{p}}} |v_i| \leq \sum_{i \in U_{\mathbf{p}'}} |v_i|$$

We also have:

$$\sum_i v_i p_i = \sum_i |v_i| - 2 \sum_{i \in U_{\mathbf{p}}} |v_i|.$$

Thus,

$$\frac{\sum v_i p_i}{\sum |p_i|} \geq \frac{\sum v_i p'_i}{\sum |p'_i|} \Leftrightarrow \sum_{i \in U_{\mathbf{p}}} |v_i| \leq \sum_{i \in U'_{\mathbf{p}}} |v_i|.$$

□

5.3 Focus

During the 2011 German state elections in Baden-Wuerttemberg, it appeared that the governing Christian Conservatives would easily remain in power. The party's position on nuclear energy did not quite match the majority opinion, but most voters were focused on different issues. Then, on March 11, a tsunami hit the Japanese province of Tohoku, causing a major nuclear incident at the Fukushima Daiichi Nuclear Power Plant. Suddenly, nuclear energy was on everyone's mind. This had a drastic effect on the elections: After nearly 40 years of governing, the Christian Conservatives were swept out of office by a Green left coalition (which strongly opposed against nuclear energy).

We can model the German 2011 election scenario using the framework discussed in the previous section. The set of issues is $I = \{i_1, i_2, i_3, i_4\}$ with[7]

i_1 : "We must support the car industry."
i_2 : "We should be conservative about public spending."
i_3 : "We ought to continue nuclear energy."
i_4 : "Do not increase funding for education."

In this framework, the conservatives are represented by a vector $\langle 1, 1, 1, 1 \rangle$, while their two main opponents have a -1 on i_3 and also on some of the other items. Say that the Social Democrats are represented by $\langle 1, -1, -1, -1 \rangle$ and the Greens by $\langle -1, 1, -1, -1 \rangle$.

A typical voter from the Southwest emphasized industry and/or education but displayed only a relatively small concern about nuclear energy. For instance, the following two voters represent a typical voter in this area: $\mathbf{v}_1 = (0.8, 0.9, -0.3, 0.4)$ or

[7]Some people claim that a significant number of voters originally based their decision on a fifth issue i_5: "This party has been in office for the last 40 years". We do not wish to comment on this claim here. We note, however, that our framework is rich enough to incorporate such considerations.

$\mathbf{v}_2 = (0.4, 0.8, -0.3, 0.9)$. Under normal circumstances, this would lead to a crushing victory for the Christian Conservatives, using any of the voting methods discussed in the previous section. However, as stated above, the Fukushima Power Plant incident changed the voters' focus.

Arguably, a change in focus does not necessarily change a voter's general attitude—i.e., the sign of a particular position. It does, however, change the magnitude $|v_i|$ of the entries. Thus, we think of a change in focus as a linear transformation of the space of positions for each voter.[8] This suggests the following definition:

Definition 5.3.1 A **focus matrix** is a diagonal matrix $A \in [0, 1]^{n \times n}$ (i.e., for all $1 \leq i, j \leq n$, if $i \neq j$, then $A_{ij} = 0$). Voter $\mathbf{v}$'s position after a focus change with A, denoted $\mathbf{v}A$, is calculated in the standard way using matrix multiplication.

The following is a possible focus change matrix triggered by the Fukushima incident:

$$A_{Fuku} = \begin{pmatrix} 0.05 & & & 0 \\ & 0.05 & & \\ & & 1 & \\ 0 & & & 0.05 \end{pmatrix}$$

Clearly, this will make nuclear energy the focus of attention for all voters. After applying this focus change to the two voters mentioned above, the resulting position vectors are $\mathbf{v}_1 A_{Fuku} = (0.04, 0.045, -0.3, 0.02)$ and $\mathbf{v}_2 A_{Fuku} = (0.02, 0.04, -0.3, 0.045)$. Such voters would end up supporting either the Social Democrats or the Green party.

The above example shows that redirecting the voters' focus is a powerful tool that can drastically change the outcome of an election. Indeed, as any political pundit will report, much of the rhetoric during an election is aimed at trying to focus the attention of voters on certain sets of issues. Recall Umberto Eco's famous quote from *Towards a Semiological Guerrilla Warfare* (1967):

> Not long ago, if you wanted to seize political power in a country, you had merely to control the army and the police. Today it is only in the most backward countries that fascist generals, in carrying out a coup d'etat, still use tanks. If a country has reached a high level of industrialization, the whole scene changes. The day after the fall of Khruschev, the editors of Pravda, Izvestiia, the heads of the radio and television were replaced; the army wasn't called out. Today, a country belongs to the person who controls communications.

We conclude this section with a number of examples that illustrate the subtleties involved in changing the focus of a group of voters.

Example 5.3.2 Suppose that there are two candidates $T = \{d, r\}$ competing in a two-topic election (i.e., $I = \{i_1, i_2\}$). The two parties have completely opposing views on

[8] Of course, this is not the only way to represent a change in focus. In general, any transformation (not necessarily linear) on the space of voters' positions can be used to describe a shift of focus during a campaign. A very interesting direction for future research is to explore these different modeling choices.

both topics, say $\mathbf{p}^d = (1, 1)$ and $\mathbf{p}^r = (-1, -1)$. Suppose that almost half of the voters are clearly in favor of the second candidate, $\mathbf{p}^r$. The rest of the voters are relatively undecided, not feeling that either of the parties is particularly close to their views. This example shows that there is a way to focus the voters so that the first candidate, d, is the winner.

To make things more concrete, suppose that there are three voters: $\mathbf{v}_1 = (-1, -0.8)$, $\mathbf{v}_2 = (-1, 0.7)$ and $\mathbf{v}_3 = (1, -0.7 + \epsilon)$. Clearly, d will lose the election given these voters. However, d can win a plurality election by changing the voters' focus using the following matrix:

$$\begin{pmatrix} 0.7 - \delta & 0 \\ 0 & 1 \end{pmatrix}$$

where $\delta \in (0, \epsilon)$. Note that candidate d cannot win the election by focusing on only one of the two issues.

Example 5.3.3 Suppose that there are three candidates $T = \{d, m, r\}$ and six issues $I = \{i_1, \ldots, i_6\}$. Assume that d is in favor of all the topics, $\mathbf{p}^d = (1, 1, 1, 1, 1, 1)$, and r opposes all the topics, $\mathbf{p}^r = (-1, -1, -1, -1, -1, -1)$. The candidates' campaign staffs have determined that d maximizes its share of votes if the voters focus on i_1, i_2 and i_3, while r receives the maximum support when the voters are focused on i_4, i_5 and i_6. In both cases, the maximum support among the voters is enough to win the election using plurality rule. Now, if both candidates d and r think about their public opinion campaigns, then they will try to direct the voters' focus to the issues that maximize their support. However, this may lead to a situation in which candidate m wins a plurality vote.

To fill in the remaining details, suppose that m supports only issues i_3 and i_6 ($\mathbf{p}^m = (-1, -1, 1, -1, -1, 1)$. There are three voters with $\mathbf{v}_1 = \mathbf{v}_2 = (-0.25, 0.3, 1, -0.1, -0.1, -0.1)$ and $\mathbf{v}_3 = (1, -1, 0.9, 1, 1, 1)$. Now, it is not hard to see that:

- In an election in which the voters are focused primarily on i_1, i_2 and i_3, d would win.
- In an election in which the voters are focused primarily on i_4, i_5 and i_6, r would win.
- In an election in which the voters are evenly focused on all the issues $i_1, \ldots, i_6$, m would win. However, if none of the voters focuses on i_1, then d would win the election.

5.4 Concluding Remarks

One of the important lessons to take away from Parikh's impressive body of work is that logicians and computer scientists have much to gain by focusing their attention on the intricacies of political campaigns. Drawing on some recent work developing a theory of *expressive voting*, we have provided some initial observations about the dynamics of voters' opinions during an election. The model of the voters' opinions and candidates' positions on the main issues of a campaign has much in common with Dean and Parikh's model introduced in Sect. 5.1. Our main contributions in this paper are to draw parallels with recent work on expressive voting (Aragones et al. 2011; Gilboa and Vielle 2004) and to stress the importance of the fact that the voters' attention to the main issues may shift during an election.

There are many avenues for future research. Except for some brief remarks when discussing the Dean and Parikh model of campaigns, we did not explicitly make use of any logical machinery. This raises a natural question about the type of logical framework that can naturally capture the phenomena discussed in this paper. We conclude by briefly discussing two additional directions for future work.

Subjective Focus Matrices In this paper, we have assumed that candidates are fully opinionated on the different issues. This modeling choice is an idealization (often, candidates cannot be described as being either fully in favor of or fully against a particular issue). However, the assumption can be justified provided that the issues are suitably fine-grained. There is a trade-off between the size of the set of issues and the richness of the candidates' positions on these issues. Indeed, there is no technical reason preventing us from allowing candidates to adopt positions strictly between -1 and 1 on the issues. Moving to such a model would allow us to represent an aspect of the voter that is present in the Dean and Parikh model but not in our framework.

In Sect. 5.2, we assumed that the candidates' position on each issue is commonly known among the voters. However, as Dean and Parikh note, voters often do not have access to the complete theory of each candidate. This may be due to a lack of information (e.g., the candidate has not stated her full position on the issues) or due to the fact that the voter may not fully trust the candidate (e.g., a candidate's positions represent how likely a voter thinks it is that the candidate will actually follow through on her promises). There are different ways to incorporate this observation into our framework. One approach is to follow Dean and Parikh and assume that each voter has her own theory of the candidates, represented by a set of the candidates' possible position vectors. Alternatively, we could describe the situation in terms of *subjective* focus matrices that depend on both the voters and the candidates.

Generalizing Approval Voting *Range voting* refers to a family of voting systems. The underlying idea behind all of these is that voters are asked to grade candidates using grades which are linearly ordered. The range voting systems differ in the aggregation method used to combine a profile of ballots.[9] Much of the analysis in this paper can be adapted to range voting systems. Suppose that there are n different grades. We can define the analogue of Definition 5.2.3. The n grades are associated with n graded approval coefficients $-1 = t_1 \leq \ldots \leq t_n \in [-1; 1]$. The grade that voter $\mathbf{v}$ assigns to candidate $\mathbf{p}$ is then given by

[9]For instance, majority judgement (Balinski and Laraki 2010) elects the candidate with the highest median grade. Score voting (Smith 2014) elects the candidate with the highest overall mean grade.

$$grade(\mathbf{v}, \mathbf{p}) := \max\{i \mid \mathbf{v} \cdot \mathbf{p} \geq t_i|\mathbf{v}|_1\}.$$

Translated into the language of cones, this means that a voter is associated with a sequence $\mathcal{C}(\mathbf{v}, \alpha_1) \supseteq \ldots \supseteq \mathcal{C}(\mathbf{v}, \alpha_n)$ of narrower and narrower cones, where the first is the entire space. The grade of a party is then given by the highest index i such that $\mathbf{p} \in \mathcal{C}(\mathbf{v}, \alpha_i)$, or equivalently by the number of cones containing $\mathbf{p}$. Approval voting can be seen as a special case of range voting with the grade requirements $t_{disapproval} = -1 \leq t_{approval} = k$.

References

Alchourrón, C. E., Gärdenfors, P., & Makinson, D. (1985). On the logic of theory change: Partial meet contraction and revision functions. *Journal of Symbolic Logic*, *50*(2), 510–530.

Aragones, E., Gilboa, I., & Weiss, A. (2011). Making statements and approval voting. *Voting Theory and Decission*, *71*, 461–472.

Balinski, M., & Laraki, R. (2010). *Majority judgment: Measuring, ranking, and electing*. Cambridge: MIT Press.

Brams, S. J., & Fishburn, P. C. (1983). *Approval voting*. Boston: Birkhauser.

Brennan, G., & Lomasky, L. (1993). *Democracy and decision—The pure theory of electoral choice*. Cambridge: Cambridge University Press.

Dean, W., & Parikh, R. (2011). The logic of campaigning. In *Logic and its applications* (pp. 38–49). Berlin: Springer.

Gilboa, I., & Vielle, N. (2004). Majority vote following a debate. *Social Choice and Welfare*, *23*, 115–125.

Klein, D., & Pacuit, E. (2013). Expressive voting: Modeling a voter's decision to vote. Manuscript, 2013.

Pacuit, E., & Parikh, R. (2006). Social interaction, knowledge, and social software. In D. Goldin, S. Smolka, & P. Wegner (Eds.), *Interactive computation* (pp. 441–461). Berlin: Springer.

Parikh, R. (1985). The logic of games and its applications. In M. Karplnski & J. van Leeuwen (Eds.), *Topics in the theory of computation selected papers of the international conference on 'foundations of computation theory'* (Vol. 102, pp. 111–139). Amsterdam: North-Holland.

Parikh, R. (1999). Beliefs, belief revision, and splitting languages. In L. Moss, J. Ginzburg, & M. de Rijke (Eds.), *Logic, language and computation* (Vol. 2, pp. 266–278). Stanford: Center for the Study of Language and Information.

Parikh, R. (2001). Language as social software. In *Future pasts: The analytic tradition in twentieth century philosophy*, pp. 339–350.

Parikh, R. (2002). Social software. *Synthese*, *132*(3), 187–211.

Poundstone, W. (2008). *Gaming the vote*. New York: Hill and Wang.

Smith, W. D. (2014). Rangevoting.org.

Chapter 6
A Non-classical Logical Approach to Social Software

Can Başkent

Abstract The term *social software* was coined by Rohit Parikh in 2002. Social software can be viewed as a research program which studies the construction and verification of social procedures by using tools in logic and computer science. However, to the best of my knowledge, social software has not been considered from a non-classical logical perspective. In this paper, I argue how non-classical logical approaches can enrich, broaden and support the agenda of social software.

Keywords Social software · Non-classical logics · Paraconsistent logic

6.1 Introduction and Motivation

The term *social software* was coined by Rohit Parikh in his 2002 paper (Parikh 2002). Social software can be viewed as a research program which studies the construction and verification of social procedures by using tools in logic and computer science. By definition, it relates closely to a variety of neighboring fields including game theory, social choice theory and behavioral economics. However, to the best of my knowledge, social software has not been considered from a non-classical logical perspective. In this paper, I argue how non-classical logical approaches can enrich, broaden and support the agenda of social software. Additionally, I will claim that incorporating non-classical elements to the program of social software aligns very well with its initial motivation.

Parikh himself does not seem to commit himself to the classical logic in the original paper, yet the de facto logic he utilizes in his work is classical. On the other hand, classical logic does not seem to be an essential element of the program of social software:

> I want to argue that (...) no doubt we shall never have social procedures which work ideally, we can nonetheless have a theory of social procedures which is analogous to the formal

C. Başkent (✉)
University of Bath, Bath, England, UK
e-mail: can@canbaskent.net
URL: http://www.canbaskent.net/logic

C. Başkent et al. (eds.), *Rohit Parikh on Logic, Language and Society*, Outstanding Contributions to Logic 11, DOI 10.1007/978-3-319-47843-2_6

> theories for computer algorithms which exist in computer science. I am referring here to a whole group of theories, some of which have come into existence during the early seventies and some are newer.
>
> (Parikh 2002)

I argue that the above quoted claim and, in general, the research program of social software, suggest the inclusion of formal tools beyond classical logic to study social software. Plurality of social procedures and their various anomalies (such as lies, jokes and speech acts) necessitate a pluralistic approach. Moreover, truth *gaps* and truth *gluts* are ordinary parts of logical formalisms which can be used to give a formal account of a variety of social phenomena. In fact, this is one of the main motivations behind logical pluralism: classical Boolean logic suffers from various restrictions which render it rather problematic in explaining human behavior and reasoning. In some cases, we can have different notions of logical consequence; in some cases, we may need more truth values; in some cases, we may have to reevaluate and redefine the logical connectives. Furthermore, it is not entirely clear how exactly people reason in social situations, and to which logical framework they are usually committed (Kahneman 2011).

In this work, in terms of logical pluralism and non-classical logics, I mainly focus on paraconsistent logics. I use the term *paraconsistency* for the logical systems in which the rule of explosion fails. In such systems, for some φ, ψ, we have $\varphi, \neg\varphi \nvdash \psi$. In short, paraconsistent systems are the logical frameworks that allow non-trivial inconsistent theories where a trivial theory is a set of sentence from which everything logically follows. Paraconsistent logic, therefore, allows us build inconsistency-tolerant models. I believe this is a key notion in understanding social software.

There are various reasons for that. First of all, contradictions occur in social phenomena. People lie, cheat, make mistakes, and misunderstand each other, they happen to be wrong in their thoughts and actions, and all of these situations (and possibly many more) require an inconsistency-friendly framework for expressive power and normative predictions. Moreover, various data from behavioral economics indicate that people usually do not reason in the way that the classical logic predicts (Kahneman 2011; Ariely 2008, 2010). This observation, by no means, entails that people always reason in a non-classical logical way. However, it casts doubt on the soundness of classical logical tools and encourages us to consider non-classical logical apparatus. Also, there can be found a variety of situations in social software that seem to fit well with non-classical logical reasoning. For instance, when people make an error in reasoning that can cause an inconsistency, the very existence of the inconsistency does not render the formalism trivial. People keep reasoning in their inconsistent model in a sound way. Sometimes they revise their beliefs, sometimes they reason non-monotonically, sometimes they ignore the inconsistency.

Yet, there also exist some other sort of inconsistencies in human reasoning and social procedures. Perhaps, a canonical example for such cases comes from normativity. The problem is how people *should* act under the presence of contradictory obligations. This is a social software issue as well as a problem in legal philosophy.

There are various approaches to obligations and normativity, yet very few of them mention the above critical point (Başkent et al. 2012; Priest 2006).

Priest, for example, describes inconsistent obligations as the "situations where someone is obliged both to do x and not to do x" (Priest 2006, p. 182). He further elaborates on inconsistent obligations as follows.

> The source of contradictory obligations need not be different contracts, but may be one and the same contract. Of course, in practice it is rare for a contract per se to be blatantly inconsistent, but it is not unusual for a contract plus contingent circumstances to give someone inconsistent obligations. Suppose, for example, that I contract to do z under condition X, but refrain from doing z under condition Y. We may suppose that X and Y are events not under the control of the parties of the contract, and that there is no reasonable likelihood of X and Y both occurring. Suppose that, despite this, both do occur. Can I then be held in breach for whichever of the actions I do not perform?
>
> (Priest 2006, p. 183)

The issue, raised by Priest here, has some ontological commitments to the contradictory co-existence of X and Y, and the philosophical implications of this situation go beyond the scope of the current article. Nevertheless, it shows that paraconsistent approach to social situations presents itself as an important perspective, and there seems to be no reason why it should not be included within the agenda of social software.

Another example of a contradictory situation comes from one of Parikh's recent papers. In my opinion, the "Kitty Genovese" case that Parikh and his coauthors discussed illustrates similar concerns.

> [I]n the Kew Gardens section of Queens, New York City, Catherine Genovese began the last walk of her life in the early morning hours of March 13, 1964. As she locked her car door, she took notice of a figure in the darkness walking towards her. She became immediately concerned as soon as the stranger began to follow her.
>
> As she got of the car she saw me and ran, the man told the court later, I ran after her and I had a knife in my hand... I could run much faster than she could, and I jumped on her back and stabbed her several times, the man later told the cops.
>
> Many neighbours saw what was happening, but no one called the police. Mr. Koshkin wanted to call the police but Mrs. Koshkin thought otherwise. I didnt let him, she later said to the press, I told him there must have been 30 calls already.
>
> When the cops finished polling the immediate neighbourhood, they discovered at least 38 people who had heard or observed some part of the fatal assault on Kitty Genovese.
>
> Some 35 min passed between Kitty Genovese being attacked and someone calling the police. Why?
>
> (Pacuit et al. 2006)

In this case, the classical logic oriented analysis that the authors suggested is of deontic and epistemic logical in nature. That is, the witnesses did not call the police, thus did not fulfill their moral duty as they did not possess the full information of the event and their agency in relation to each other. Simply put, witnesses thought that some other people might have called the police already. This analysis is plausible. Yet, some other analyses can also be given for the Genovese case underlining that people may behave inconsistently in a non-trivial way.

One of the descriptive analysis of the situation calls for a paraconsistent framework. It is assumed that the witnesses are morally obliged to call the police, and these moral obligations are generally assumed to be factual and truthful, and they must be fulfilled. Yet, they sometimes are not. Based on these deontic presuppositions, this is what we have:

$$\mathsf{WitnessAMurder} \rightarrow \mathbf{Obliged}(\mathsf{CallPolice}), \mathsf{WitnessAMurder} \nvdash \mathsf{CallPolice}$$

If we endorse modus ponens, and assume that moral obligations need to be fulfilled (that is if $\mathbf{Obliged}(\mathsf{CallPolice}) \rightarrow \mathsf{CallPolice}$), then we derive both $\mathsf{CallPolice}$ and $\neg\mathsf{CallPolice}$ which is contradictory and incoherent under the classical negation and consequence relation. As widely known, there are various ways to modalize the above formulation by using deontic operators that stand for obligations, yet we will not dwell on the matter by entering into deontic logical debates and their paradoxes here.

It should be noted that the paradoxical situation in this case is avoidable. As the standard analysis for the Genovese example explicates, if the witnesses knew individually that none of the other witnesses called the police, they could have fulfilled their moral obligation.

Additionally, a simple game theoretical analysis of the situation can be considered. If a witness calls the police to report the incident, the cost to the witness for the call is less than a dollar and couple minutes which is by far negligible compared to the possible benefit that the call might bring about: saving the life of Kitty Genovese. Simply put, even if a moral agent i assumes that 1000 people saw the incident, and the chances that i will be the one who will report the incident first to the police is 0.1 %, it is still the rational move to make, since a person's life (nearly universally) is more valuable than the troubles that i needs to go through to report the incident—yielding a much higher expected utility for the call. Therefore, regardless of attaining the full knowledge of the case, I maintain that the witnesses have the obligation *to the best of their knowledge* to report the incident. Perhaps, they would be the 999th person to report it, which is perfectly acceptable, but maybe they would be the first. Additional irregularities can also be the case in the Genovese example. For the legal authorities, receiving multiple calls for the same incident may backfire or perhaps can be ignored, and consequently trigger a higher cost for the individuals when they consider making the call. Moreover, as *The New York Times* article reporting the event quoted, witnesses were afraid to get involved and thought that it was a "lover's quarrel".[1] Such reasons are perfectly justifiable but may not rationally be the best move for not making the call. In either of these cases, game theoretical reasoning dictates that rational agents should make the call. Yet, they did not. Then, the classical analysis suggests that the agents in this case are not rationa—perhaps excluding the ones with imperfect information. Paraconsistent analysis, on the other hand, prevents this over-reaching revision. It can very well be the case that agents could be perfectly rational (and most likely they were), had perfect information about the incident (they were the witnesses), yet they still did not make the call.

[1] *37 Who Saw Murder Didn't Call the Police*, Martin Gansberg, The New York Times, 27.03.1965.

Therefore, adopting a paraconsistent view point helps us construct a broader formal framework where we do not need to revise the initial assumptions of the theory just because an inconsistency occurred. In paraconsistent social software, we can very well have perfectly rational agents that can make *mistakes*. This simple example, in my opinion, argues that social software can easily allocate non-classical logics, and can be enriched by it. This seems as a powerful research direction within the domain of social software by extending the discussions on rationality, action-based models and utility theory.

The Kitty Genovese example and many others from law[2] illustrate the possibility of applying non-classical logical methods to social software. The central claim of this paper is to argue that social software lies within the interesting intersection of logical, moral and economical pluralisms, and it can further benefit from incorporating non-classical logical methods into the theory. Additionally, apart from the descriptive perspective it provides, non-classical logical theories in social software can depict normative theories. In this work, we will not go into the details of this distinction.

In short, in order to analyze a variety of interesting social procedures and phenomena, we may need to use a variety of different logics. And social software, in all its richness, seems to provide an ideal domain to test the strengths (and weaknesses) of different formalisms. Similar to logical pluralism, I will argue for pluralism in social software, and this clearly falls within the agenda of social software as a research program. Rich formalisms in non-classical logics, the extensive research in behavioral economics and the way it discusses the pluralities in rational and social behavior, and finally alternative economic theories open up new avenues for social software and relate it to a broader audience. I will argue for these points by suggesting a pluralistic framework—logically, morally and economically.

6.2 A Broader Social Base for Social Software

The recent rise of behavioral economics in both popular literature and academic research points out a well-known missing link between formal logic, and social and individual human behavior: people do not reason or behave as normatively as manifested by the classical logic. People often times make various deductions that diverge from the classical logic, hinting at the possibility of adopting logical pluralism to address the *logic of society*.

There is a rich literature that discusses real-life examples combining various logical issues in game and decision theory (Ariely 2008, 2010; Brafman and Brafman 2008; Gigerenzer 2008; Harford 2009; Kahneman 2011; Smith 2010). What is relevant for our purposes here is the immediate observation that classical logic falls short when analyzing individual and social human reasoning and interaction. If a logic claims to be *the* system of correct reasoning, there seems to be a problem here.

[2] A canonical example from law is *civil disobedience* where agents deliberately break the law and create an inconsistent situation where moral obligations and legal duties clash. Yet, still we obtain a *non-trivial* and coherent inconsistent situation.

More precisely, at a descriptive level, classical logic is not sufficient to explicate and analyze many interesting social phenomena—but perhaps is sufficient in some others. This, by no means, entails that non-classical logics are normatively the only logics that the epistemic and rational agents need to employ—this is exactly the opposite of logical pluralistic view point. Granted, it is not only non-classical logic that may help us understand human behavior within the domain of social software. Decision and game theory and formal epistemology are among the formal sciences that attempt at analyzing similar issues. Nevertheless, from a logical perspective, it can be argued that social software not only can enrich itself by focusing on such pluralistic cases, but also, perhaps more importantly, can help us analyze the examples in behavioral economics. As an illustration, let us consider a very simple example, the two horsemen, that Parikh also discussed.

Example 6.2.1 (Parikh 2002) Two horsemen are on a forest path chatting about something. A passerby, the mischief maker, comes along and having plenty of time and a desire for amusement, suggests that they race against each other to a tree a short distance away and he will give a prize of $100. However, there is an interesting twist. He will give the $100 to the owner of the slower horse.

I maintain that the way negation (or game duality) treated in this puzzle is not strong enough to generalize, and more importantly can be limiting for the overall agenda of social software. The idea of switching to the dual role (which is obtained by using the classical negation) is not a universal strategy that can apply to other similar games. In general, players do not necessarily deal with negated statements in this fashion. The dual game in this example possesses some simple properties: it is easier to determine, and the negation of *slow* is clear to decide. Yet, such properties do not exist in all games. Can we play checkers in this way? Can we play football as such?

For example, for the games with three players, computing the dual game and permuting the roles for the players are not trivial (Olde Loohuis and Venema 2010). If we modify the Example 6.2.1 by allowing a third player, then we can have 2 "dual" games—the permutations of horsemen and horses where nobody rides their own horse. The number of "dual" games increases if we consider even more players and additional intermediate states besides slow / fast.

First and foremost, Example 6.2.1 shows that the formal analysis of social phenomena has traditionally restricted itself to some well-defined and well-behaved subset of the society. On the other hand, it is quite evident that Parikh's recent work incorporates a broad variety of examples and phenomena to social software, somehow supporting my point by expanding the domain of social software (Parikh 2014).

It is possible to take one more step and use this perspective that social software provides to question the basic tenets of traditional game theory and rational choice theory. These two theories construct a utilitarian understanding of society with a quite restricted notion of rationality. Yet, utilitarian rationality is a very controversial assumption—which largely remains unearthed. Recently, some authors in economics and finance communities criticized this approach heavily. For instance, Yves Smith, the author of the popular book *ECONned*, remarks the following.

> The dominant economic paradigm, neoclassical economics, became ascendant in part because it offered a theory of behavior that could be teased out in elegant formulation. Yet it rests on assumptions that are patently ridiculous: that individuals are rational and utility-maximizing (which has become a slippery notion as to be meaningless), that buyers and sellers have perfect information, that there are no transaction costs, that capital flows freely.
>
> (Smith 2010)

Hartford argues along similar lines.

> Fundamental to von Neumanns approach was the assumption that both players were as clever as von Neumann himself. (...) The second problem is that game theory becomes less useful if your opponent is fallible. If player two is not an expert, player one should play to exploit his mistakes rather than defend against brilliant strategies that will never be found. The worse the opponent, the less useful the theory is.
>
> (Harford 2009)

Such limitations of the classical theory of games and rationality do not directly carry over to social software. As Parikh himself has been underlining more and more, social software encompasses a broader outlook to society, and it seems to me that it emerges as a more grounded theory to analyze social behavior and procedures, compared to the classical and traditional theory of games. I will now argue that social software can also be broadened if an alternative understanding of utilities are considered, as an alternative to the traditional utilities of von Neumann–Morgenstern.

Even though such a direct game theoretical and ideological influence is difficult to trace in Parikh's works, some utilitarianism based understanding of rationality and semantics seems to be the one that Parikh endorses (Parikh 1994). Parikh argues that "Roughly speaking, if an agent has a choice among several actions, we would expect that the agent will carry out that (pure) action which the agent thinks will bring the maximal benefit (utility) to the agent" (Parikh 2002). As widely known, ordinal utilities can be translated to preference orderings at the cost of losing some information, yet, the main problems of the traditional von Neumann–Morgenstern framework still remain.

It can be argued that a *deontological* approach can live side-by-side with the utilitarianism within the domain of social software. The deontological approach suggests that utility based moral analysis does not fully consider the deontological commitments of the moral agents. My suggestion here can be considered as an instance of moral pluralism for social software. Such a pluralism argues that utilitarianism based understanding of rationality can be a perfect fit for some social phenomena, but also deontological analysis can be beneficial in understanding some other social situations.

Let me illustrate this point with an example. In his Parikh (2002), Parikh mentions the well-known theorem of Gibbard and Satterthwaite that suggests that any social choice function which takes preference orderings of the voters as inputs, and returns a social preference ordering for the society, will be vulnerable to manipulation in the form of strategic voting. Here, Parikh discusses the United States presidential election of 2004 as an example of Gibbard and Satterthwaite theorem, and concludes that "this is murky territory and I shall not venture further into it." (Parikh 2002).

I believe that strategic voting and manipulations in elections constitute a very interesting focal point for social software, and they can be helpful illustrating the need to expand the agenda of social software. If we consider voting as a form of utilitarian calculus, and take strategic voting as a legal and permissible strategy in it (which it is), then we will be puzzled with the results like Gibbard and Satterthwaite or Arrow's Impossibility Theorem or even Sen's result on the Impossibility of Pareto Liberal. One of the main reasons for negative results in the social choice theory is that the theory does not generally take the moral and ethical compass of the society into account.[3] Moreover, such considerations are not even representable in most social choice theories. The reason why people did not vote strategically in the 2004 US elections is not only epistemic, and perhaps epistemic reasons do not even count among the main reasons.[4] One of the real reasons, in my opinion, is that many people (if not most) people consider strategic voting as a *betrayal* to their political conviction for understandable reasons.[5] For many people, voting represents their commitment and loyalty, and honoring their own opinions, and even if they feel that the party/candidate they support will clearly not win, they do not switch to another one for the aforementioned reasons.

Some disagree with my perspective (Brennan 2011; Chisholm 1963). Chisholm discusses those imperatives which are "telling us what we ought to do if we neglect certain of our duties", and argues that the deontic logic (with its deontic modality O) is not sufficient to formalize them (Chisholm 1963). He argues as follows.

> Ordinarily the rules of a game do not tell us how to proceed with the game after the rules have been violated. In such a case, we may: (1) go back to the point at which the rule was broken, correct the mistake, and resume the game; (2) call off the game; or (3) conclude that since one rule has been broken, others may now be broken, too. But these possibilities are not open to us when we have broken a rule of morality. Instead we are required to consider the familiar duties associated with blame, confession, restoration, reparation, punishment, repentance, and remedial justice, in order to be able to answer the question: 'I have done something I should not have done-so what should I do now?' (Or even: 'I am going to do something I shouldn't do-so what should I do after that?') For most of us need a way of deciding, not only what we ought to do, but also what we ought to do after we fail to do some of the things we ought to do.
>
> (Chisholm 1963)

This argument is interesting in-itself. For my purposes it is a valid example to justify strategic voting. For this reason, this subject becomes even more interesting for social software, especially once it is supplemented by a logical framework (classical or non-classical) that can formalize contrary-to-duty actions (Carmo and Jones 2002; da Costa and Carnielli 1986; Hansen 2006).

---

[3]Other reasons being, no cost of information, no reference to the actual society, etc.

[4]It is generally argued that, in the 2004 US elections, if Greens—which is a very small political party in the US—had voted strategically against Bush, he might not have been reelected.

[5]Brennan mentions Habermas who argued that "strategic voting is disrespectful to other citizens" (Brennan 2011).

Chisholm's argument can possibly be suggested as a counter-argument to my point. Contrary-to-duty obligations can be viewed as "second degree obligations" that may seem to by-pass the social and individual moral preferences. Nevertheless, notice that the second degree obligations (the ones you are supposed to do, after violating your initial duties) still depend on social, political and economical morality and the ethics of the individual and his priors. Moreover, the first violation of moral duties (which come from the very definition of "contrary-to-duty actions") calls for a inconsistency-friendly framework, at least for expressive strength. Because, after a violation of a moral obligation, not any proposition but only contrary-to-duty actions become obligatory. Thus, Chisholm formulated how the agents reason under some moral inconsistencies in a sound and non-trivial way. This is nothing but a paraconsistent reasoning.

Similarly, if an individual decides to vote strategically, then, the candidate he is going to vote for strategically reflects his "second degree" duties which are also shaped by his own individual and social moral compass. An individual voting for his second best choice to block the worst candidate based on his preferences, is still reflecting his own preferences. In the 2004 US elections, Greens were expected to vote strategically, because, in a broader perspective, it can be argued that they prefer the Democratic candidate to the Republican one, and this choice between the two major parties reflects their own choices.

Brennan discusses various forms of strategic voting and concludes that, for him, "there is no objection in principle to strategic voting, so long as strategic voting does not impose too much risk and tends to produce better outcomes than one justifiedly believes otherwise would occur" (Brennan 2011). Yet, the logical and mathematical complications of strategic voting make it a philosophically interesting subject for social software—broadly construed. On the other hand, discussions on "ethical voters" have been initiated by Harsanyi in late 70s (Harsanyi 1977) and it can be seen as a relatively new field. In short, incorporating some of the ideas suggested here only enriches the field and relates it to some current debates.

Also, a deontological extension of social software, which we account for as moral pluralism, can be suggested based on similar motivations. For this purpose of mine, let us reconsider the Kitty Genovese example which was discussed in Sect. 6.1. Besides its epistemic analysis, there is also a strong deontic component in this case. Namely, it is fair to assume that people are obliged to help others when they do not risk anything comparable, and we can take this principle as our deontological commitment. Regardless of the cost of the phone bill or even of the risk of overwhelming the authorities or whether it was a lover's quarrel, it can very well be argued from a deontological perspective that the witnesses are morally obliged to call the police. Moreover, the questions that whether others made a similar call or whether the police would make it on time to the crime scene do not exclude anyone from following their deontological moral obligations. In short, possibly undesired consequences of making the call do not constitute enough of a reason to nullify the *a priori* deontological commitments. It may minimize or economize an individual's personal duty if it is shared by the others, yet the individual is still obliged to help according to the deontological commitments we have formulated above.

The dichotomy between deontological morality and utilitarian calculus to evaluate the moral actions is an exciting research field. The discussion in this field extends to many major social debates ranging from capital punishment to abortion. A formal and analytical approach to such problems falls within the scope of social software. More precisely, developing a deontic logic for Kantian priors and incorporating it to social choice and game theories appear as a rather big research agendas within social software.

Similarly, language games, a special favorite of Parikh's, exhibit similar issues. Parikh himself argued that language has a utilitarian element (Parikh 1994). We choose the nouns (or in general any other language elements) that work. Parikh develops this thought and connects it to vagueness in a very interesting way in the aforementioned work. This explains a significant part of semantics, except perhaps literature and poetry, where ambiguity and vagueness in meaning are intentional and even desirable. Therefore, on some occasions, language games can take another form in art where the utility based analysis of semantics becomes complicated if not impossible.

As another example of a non-utilitarian social phenomena, the case of having children can be considered. As is widely known, numerous reasons can be given to bear and have children. Let us consider them in two main categories following Overall: deontological and consequentialist (Overall 2012). Deontological reasons include carrying on the family line and name, duty towards the society and the family, whereas the consequentialist ones include the traditional economical benefit (of the children) to the family, and psychological benefits to the parents. Overall goes ahead and argues from a moral perspective that none (and more) of these reasons cannot be ethically justified as a reason to have children. Nevertheless, the same issue can be approached from a social software point of view. As we all know, having children has a lot of difficulties as well. They increase the stress level of the parents, and in many societies, it is very expensive to raise them. In short, the quantitative and measurable cost of having children (for instance, increased stress hormones and diminishing bank accounts) needs to be compared with the qualitative and unmeasurable benefit of having children (happiness and all that), according to the traditional game theoretical approach. Yet, the traditional approach appears to be not very fruitful in this direction. That is, how can we compare the utility value of having a baby with the university tuition that the parents will need to pay for the child.[6] For such examples, where self-sacrifice and deontological commitments play a central role, we need a broader understanding of social software that goes beyond the traditional consequentialist method of game theory and formal epistemology.

Another interesting argument towards some game theoretical concepts can be found in Graeber (Graeber 2011). He questions Hobbes's use of "self-interest" to describe human motivation. Graeber comments on "self-interest" as follows.

---

[6]Clearly, putting all of the financial and emotional burden of raising a child (with tuition and care cost etc.) to couples is a socio-economical decision taken by governements, and does not apply equivalently to all countries and societies. We leave such issues aside in this paper, even if they constitute an interesting direction for social software as well.

> Part of the terms appeal was that it derived from bookkeeping. It was mathematical. This made it seem objective, even scientific. Saying we are all really pursuing our own self-interest provides a way to cut past the welter of passions and emotions that seem to govern our daily existence, and to motivate most of what we actually observe people to do (not only out of love and amity, but also envy, spite, devotion, pity, lust, embarrassment, torpor, indignation, and pride) and discover that, despite all this, most really important decisions are based on the rational calculation of material advantage which means that they are fairly predictable as well.

Graeber does not only skeptically argue that the utilitarian social choice based analysis of games in society are missing an important component, but also implies that such calculations, if possible at all, are more complicated than they look.

I see no reason as to why social software cannot be approached from a perspective that can allocate the cases I mentioned so far. Parikh himself seems to agree with his point when he considered even a broader set of examples in his recent work (Parikh 2014).

In short, I believe that society exhibits many interesting cases which reflect a wide variety of cultural and historical pluralism, and recent anthropological work argues that people exhibit a broad variety of behavior when it comes to rational decision making (Graeber 2011). Incorporating such observations to the agenda of social software will result in a mutually beneficial and stronger cooperation between these research areas and the formal sciences of social software.

6.3 A Broader Logical Formalism for Social Software

Logical pluralism is a "*pluralism* about logical consequence" asserting that there can be more than one logical consequence relation (their emphasis Beall and Restall 2006). Logical pluralists endorse the view that from a given set of sentences, it is possible (whenever a formalism can be given) to deduce various conclusions. I maintain that logical pluralism is essential to social software, and it can be viewed as the counter-part of moral and social pluralism.

As widely known, in an intuitionistic universe, law of excluded middle does not hold as it can be ontologically possible that there are propositions which are *neither* true nor false. Similarly, in a paraconsistent (or dialetheic) universe, the law of non-contradiction is not valid. Because, it is thought that there are propositions which are *both* true and false. The important point here is the fact that non-classical logics are motivated not only by logical and mathematical observations, but also by various social, epistemological and ontological phenomena.

For instance, quantum physics provide us with various ontological and epistemological examples with undetermined truth values such as the Pauly Indeterminacy Principle. Similarly, law raises various issues where dialetheism and paraconsistent consequence relations can be put in use, as we have discussed earlier (Priest 2006). Moreover, there are various other situations where paradoxes appear in social and game theoretical contexts.

Take *Parrondo's Paradox*. Consider the following two games: Game 1 and Game 2. In Game 1, you lose \$1 every time you play. In Game 2, if you have left an even number of dollars, you win \$3, if you have an odd number of dollars left, you lose \$5. Say, you start playing this game with \$50. If you play Game 1, you will lose all your money in 50 rounds. If you play Game 2, you will still lose all your money in 50 rounds following the sequence:

$$50 - 53 - 48 - 51 - 46 - 49 - 44 - \dots$$

However, the catch point is, if you play the games in the order of "Game 2–Game 1–Game 2–Game 1–....", then you will always win following the sequence:

$$50 - 53 - 52 - 55 - 54 - 57 - \dots$$

The *paradoxical* result here is the fact that by combining two losing strategies, it is possible to obtain a winning strategy which is somehow surprising and unintuitive. Non-classical logical elements in this analysis are quite striking.

Another major example is *dialectic*. Consider an agent, let us call him Karl the CEO, struggling to make a decision. Assume that he has been suggested two opposing points of view: φ and $\neg\varphi$, the thesis and the antithesis respectively. Then, any rational agent would not give up his logical system or decision procedure as there are contradictory and plausible statements expressed formally in the system and therefore rendering it inconsistent. In this case, we would expect Karl to reach a conclusion, a *synthesis*, say ψ, after a dialectical procedure. Thus, we will have for Karl, $\varphi, \neg\varphi \vdash \psi$ yet $\varphi, \neg\varphi \nvdash \neg\psi$. For Karl, the decision ψ follows from the given contradictory evidence $\{\varphi, \neg\varphi\}$ whereas the decision $\neg\psi$ simply does not. In short, the system of dialectic reasoning is not *explosive*: there is a statement (namely, $\neg\psi$) which did not follow from a contradiction. As this example illustrates, there is an interesting relation between dialectic, dialetheism and paraconsistency, and to do justice to the subject, we refer the reader to the following work in the subject for a broader treatment (Ficara 2013; Priest 1989, 2006). Nevertheless, taken as a decision procedure, it is clear to see how dialectic can fall within the domain of social software and how it carries along non-classical logic with it to social software.

Finally, as I underlined earlier, case studies from behavioral economics provide rich examples which demonstrate that people do not usually reason in the way that the classical logic predicts (Ariely 2008, 2010; Gigerenzer 2008; Harford 2009; Smith 2010; Stenning and van Lambalgen 2008). Such examples direct us towards logical pluralism where the logical consequence of what is given can be, to say the least, unexpected and surprising from a classical logical perspective. People do not end up with trivial theories when they encounter paradoxical situations. They simply work their way through it—usually in a sound and rational way. Non-classical logics can be viewed as formalisms attempting to give a formal account for such situations. Let us see it in more examples.

6.3.1 More Examples

I will now examine some more cases starting with some examples from Parikh's original paper, and observe how they can be inspiring and motivating to introduce non-classical logical elements into social software.

Example 6.3.1 (*Carousel Example,* Parikh 2002) In order to prevent the overcrowded carousels at the airports, Parikh mentions a simple solution. The airport authorities should paint a line a certain distance from the carousel and post signs that say "Do not cross the line until you see your suitcase".

Let us assume that this solution is implemented. Without doubt, there will still be people who approach the carousel before seeing their luggage.

An intuitionistic approach to this anomaly suggests that there are people who neither know nor do not know that they should approach the carousel. In other words, these are the people *who did not care about the sign*. It does not mean that they disagree with the solution procedure. It also does not mean that they agree with it. They are simply indifferent to this solution concept. Thus, the solution, taken as a proposition, has no truth value in those people's mental models. Anyone who has observed people breaking some simple rules carelessly might agree that this is a very common phenomenon.

Paraconsistent logicians might argue that some people, even if they approve of the solution, would still not wait until they see their luggage. This clearly creates an incoherent if not inconsistent situation within the mental model of the agents. The agent agrees with the solution, and thinks that she should wait. Nevertheless, she simply does not wait. Those are perhaps the agents who intentionally break the rules. For such situations, which arguably happens quite often, paraconsistency suggests an inconsistency-tolerant framework for those agents with inconsistent mental models.

Clearly, one can also unify the above approaches, at least formally, in the framework of First-Degree-Entailment (Dunn 1976; Routley and Routley 1972). We refrained ourselves from using the First-Degree Entailment for this formalization as the incompleteness- and inconsistency-tolerant logics may separately provide a clearer understanding of the phenomenon.

Example 6.3.2 (*Russellian Barbers*) The example of two horsemen (Example 6.2.1, also in Parikh (2002) suggests that sometimes it is wiser to switch to the dual game with dualized strategies. A non-classical variation of the puzzle presents an interesting approach. Let us consider the following situation which we call *Two Russellian Barbers*. Take two Russellian barbers who can only cut the hair of the people who cannot cut their own hair themselves.

Assume that in the case of Russellian barbers, they were asked to compete in a game where the one who gets his hair cut fastest wins. Let us apply the solution concept which we mentioned for Two Horsemen example. If the barbers switch to the dual game and cut each other's hair, they will be slow, and not even cut the hair. Then, it seems, then each barber should cut his own hair. If they commit themselves cutting their own hair, then they can compete to be the fastest, it seems. Yet, recall that these barbers are Russellian who only cut the hair of the people who cannot cut their own hair themselves. Thus, the strategy of switching to the dual game does not directly work for Russellian barbers. Perhaps, it can be argued that the solution concept of Two Horsemen applied to the Russellian Barbers call for an auxiliary agent who would be willing to cut the barbers' hair. The strategy that is entailed by this solution concept require auxiliary players, which in-itself an interesting concept.

The logical implications of this problems aside, this example illustrates how non-classical ideas can introduce interesting cases to social software.

Example 6.3.3 (*King Solomon Example,* Parikh 2002; van Eijck and Verbrugge 2009) This example is a very old Biblical story about King Solomon. In the story, two women claim a baby and ask the King's help to resolve the issue. The King threatens to cut the baby into two and share it between the women, thinking that the real mother would not allow it, and let the other woman take the baby. Therefore, the woman who denies the motherhood under the presence of this procedure is indeed the real mother, according to the solution concept of this procedure.

As it is pointed out in van Eijck and Verbrugge (2009), the surprise element in Solomon's procedure is essential—this is what prevents the players from playing strategically. Yet, logically, surprises seem to be difficult to formalize. This point begs the question whether surprises are the focal points that require a non-classical analysis.

The surprise element here involves a component that renders the problem and the solution void. In other words, in this puzzle, a hidden assumption requires the puzzle solver to keep the baby alive—otherwise there would be no need to determine the motherhood. Suggesting that the baby will be killed is not actually a surprise at the moment of it being suggested. It is deemed as a surprise *later* when the solution is fully introduced by the King. Therefore, *when* it is suggested, it creates an inconsistency and requires a logical framework that can tolerate it.

Some further discussion on the King Solomon example can be found in (van Eijck and Verbrugge 2009). This problem can also be analyzed from the view point of counterfactual conditionals, yet we shall not delve into that aspect here—even though it also supports our claim that non-classical analysis can enrich our understanding of social phenomena.

Example 6.3.4 (*Game Semantics as a Language Game*) Conceived as a meaning-construing procedure by Hintikka, game semantics provides a very interesting perspective on formal semantics. *Semantic verification game* is played by two players, *falsifier* and *verifier* which we call Abelard and Heloise respectively. The goal of Heloise in the game is to verify the truth of the formula whereas for Abelard it is to falsify it. The rules of the semantic verification game are specified syntactically based on the form of the formula. During the game, the given formula is broken into subformulas step by step by the players. The game terminates when it reaches the propositional literals and when there is no more moves to make. If we end up with a propositional literal which is true in the model in question, then Heloise wins the game. Otherwise, Abelard wins. When the main connective is a conjunction, it is

Abelard's turn to choose, and similarly, disjunction yields a choice for Heloise. The negation operator switches the roles of the players. The major result of this approach states that Heloise has a winning strategy if and only if the given formula is true in the given model. For an overview of the field and its relation to various epistemic and scientific topics, we refer the reader to (Pietarinen 2003). Moreover, (Pietarinen and Sandu 2000; Hintikka and Sandu 1997) provide expositions of game theoretical semantics and its relevance to philosophy.

Not much argument is needed to show the relevance of game semantics to language games which was Parikh's starting point in his article (Parikh 2002). However, a non-classical analysis of game semantics reveals that semantic verification games can formalize non-classical behavior. By using non-classical logics as the underlying formal framework, it is possible to have verification games with additional players, concurrent play, and variable sum games where more than one player can win, or one's loss does not entail the opponent's win (Pietarinen 2000).

Example 6.3.5 (*Law*) Real-life paradoxes in social situations are not easy to pin point. However, law provides a unique playground both for paraconsistency and social software. Almost without exceptions, every legal system contains inconsistencies, and one way or the other, they still function. Priest gives various examples of legal dialetheias and inconsistent obligations, and considers the following simple example.

> Suppose that there is a certain country which has a constitutional parliamentary system of government. And suppose that its constitution contains the following clauses:
>
> In a parliamentary election:
>
> (1) no person of the female sex shall have the right to vote;
>
> (2) all property holders shall have the right to vote.
>
> We may also suppose that it is part of common law that women may not legally possess property. As enlightenment creeps over the country, this part of common law is revised to allow women to hold property. We may suppose that a de facto right is eventually recognized as a *de jure* one. Inevitably, sooner or later, a woman, whom we will call Jan, turns up at a polling booth for a parliamentary election claiming the right to vote on the ground that she is a property holder. A test case ensues. Patently, the law is inconsistent. Jan, it would seem, both does and does not have the right to vote in this election.
>
> (Priest, 2006, pp. 207–8)

This shows that legal systems can be very well viewed as non-trivial inconsistent theories, exemplifying paraconsistent reasoning. Moreover, in many real-life cases, the point is not genuinely to create a legal system with no inconsistencies or incompletenesses whatsoever. Yet, the real focus is to make this system work—whatever the phrase *working system* entails. By itself, such situations call for a social software based analysis.

Similar examples can be multiplied especially when *dialogues* are considered (Carlson 1983; Rahman and Carnielli 2000). Taken as a formal model about a social situation, dialogues contain inconsistencies, and are genuine examples for paraconsistency. Since language and communication are essential parts of social interaction, thus of social software, formal models of dialogues constitute interesting case studies which fall within the intersection of social software and paraconsistent reasoning.

The examples which we have discussed so far show that various social procedures call for various different logical and computational paradigms. By itself, this constitutes a quite central meta-problem: which logical framework would be an ideal fit for which social phenomenon? Such problems and meta-problems suggest further extensions of social software and illustrate how logical and social issues interact.

6.4 A Broader Economics for Social Software

The real-world economics movement, which was born in Paris in 2000, heavily criticizes the foundations of neoclassical economics (Fullbrook 2008; Reardon 2009). Even if we accept the assumption of *homo economicus*, we will stumble upon many problems when we consider the *markets* as described by the neoclassical economics and game theory (Benicort and Guerrien 2008). The proponents of real-world economics argue that focusing on equilibrium points, which are not even predictive of the future outcomes, misses the point. They argue that "in an uncertain world, making sophisticated calculations before making each decision is nonsense", and conclude provocatively with the observation that "to understand the real world, one has to forget microeconomics" (Benicort and Guerrien 2008).

Similarly, there is an increasing number of works that criticize the heavy mathematical machinery used in economics. As Milton Friedman put it "... economics has become increasingly an arcane branch of mathematics rather than dealing with real economic problems" (ibid). Solow argues that "economics as taught in America's graduate schools ... bears testimony to a triumph of ideology over science" (ibid). Moreover, Guerrien and Jallais argue that:

> Game theory does not resolve concrete problems or make predictions about player choices. It focuses on the complexity of the decision interactions of persons conscious of being in interaction. As the renowned game theorist Ariel Rubinstein explains,
>
> > game theory is a fascinating and abstract discussion that is closer to philosophy than to the economics pages of the newspaper. It has no direct applications, and if it has any practical utility (which I doubt), then it is in the winding and inscrutable way that our minds absorb ideas and use them when the time comes for real action. And this too must be proved.[7]
>
> (Reardon 2009 pp. 37–8)

Without much effort, we can find similar approaches, albeit not as radical and explicit, in various other works as well (Ariely 2008, 2010; Gigerenzer 2008; Har-

[7] November 17, 2000 in Israeli daily *Haaretz*. In a similar way, his 2009 (2012, in English) book is called *Economic Fables* to underline the fact that game theorists are tellers of fables.

ford 2009; Kahneman 2011). I believe there is some truth in this criticism towards micro-economics, and this body of criticism easily carries over to game theory and social choice theory. More importantly, for our purposes here, social software and more generally mathematical and logical analysis of social algorithms and social phenomena become a target of the aforementioned criticism as both game theory and social software make similar (perhaps, implicit) assumptions.

The central claim of above approaches of the real-world economics is the fact that neoclassical economics fails to address a broad spectrum of social and economical phenomena. Behavioral economics and real-world economics suggest some sound conceptual alternatives within the field, and I believe that logical approaches to such problems should address those concerns—either affirmatively or negatively. I argue, *economic pluralism*, that is crystallized in the real-world economics, may help illustrate the use of logical pluralism in social software. Furthermore, I claim that logical pluralism can be the key point to address different economical paradigms within logic and game theory, and hence in social software. The limitations of the traditional game theory can therefore be an opportunity for social software to expand its domain.

One of the central demands of the real-economics movement is to have a pluralistic understanding of economics (Fullbrook 2008; Reardon 2009). This provides the field of economics with different theories for different economical phenomena, as opposed to a monist and monolithic methodology that strives to explain and predict it all. Simply put, different social situations presuppose different logical reasoning. Therefore, plurality in social norms and rules and interactions presuppose a pluralistic view of logic. Real-world economics, in this manner, provides examples and ideas from the same domain that social software attempts to analyze. A theory of economics which rely on pluralistic foundations will therefore be pluralistic.

An interesting argument in favor of real-world economics perspective is that the economics should not be approached from an individual-centric perspective. In terms of social software theory, this translates into the thesis that agents-based analysis of epistemic, doxastic and deontic situations lack an important component of social interaction: the society itself. Clearly, it can be argued that "markets" can play the role of the society, and be expressed as an agent. Nevertheless, this gives rise to the notion of "invisible hand" which can be considered as one of the problematic points of the theory. Such different approaches can easily be discussed within the domain of social software.

Reconsider the "Kitty Genovese" case. A real-economics oriented "real-social software" would approach this example by considering the social dynamics of the community in which the incident took place. For instance, if it was a small town in a closely knit society where the incident took place instead of a New York City neighborhood where social interaction among the neighbors are much looser or perhaps non-existent, the analysis of the case would be much different—both epistemically and deontologically. The histories of the agents would have much more in common, they would share a larger common language, supposedly they would share a common moral background and moral priors, and the cost of not calling the police would be much higher. In cosmopolitan New York City, the social dynamics are obviously much different, and "the real-world social software" should reflect this parameter in its analysis of the case.

Interesting enough, the reason that such considerations have not received their fare share of analytical treatment within social software is not due the initial assumptions or restrictions of the program. The broad outlook of social software seems to be eager to analyze, both computationally and logically, such variants of different micro-economical theories.

6.5 Conclusion

In this paper, I managed to avoid game theoretical and social choice theoretical formalisms. My pragmatic goal was to clarify some foundational ideas in social software, and to investigate the connection between non-classical logical, specifically paraconsistent perspectives and social software. If, I argued, social software has a strong emphasis on logic and computation, then different logical traditions might provide it with a broader outlook.

On the other hand, perhaps at a more personal level, based on my long discussions with him, I am more or less convinced that my thoughts put together in this paper simply complement what Parikh had in mind when he originally suggested the idea of social software. The theory should be comprehensive and powerful, and based on this conviction, the ideas put forward in this work do not suggest otherwise, but provides a broader outlook and paradigm for social software.

References

Ariely, D. (2008). *Predictably irrational: The hidden forces that shape our decisions*. New York, NY: HarperCollins.

Ariely, D. (2010). *The upside of irrationality*. New York: Harper.

Başkent, C., Olde Loohuis, L., & Parikh, R. (2012). On knowledge and obligation. *Episteme*, *9*(2), 171–188.

Beall, J. C., & Restall, G. (2006). *Logical pluralism*. Oxford: Clarendon Press.

Benicort, E., & Guerrien, B. (2008). Is anything worth keeping in microeconomics? *Review of Radical Political Economics*, *40*(3), 317–323.

Brafman, O., & Brafman, R. (2008). *Sway*. New York: Doubleday.

Brennan, J. (2011). *The ethics of voting*. Princeton: Princeton University Press.

Carlson, L. (1983). *Dialogue games*. Boston: D. Reidel Publishing.

Carmo, J., & Jones, A. J. I. (2002). Deontic logic and contrary-to-duties. In D. Gabbay & F. Guenthner (Eds.), *Handbook of philosophical logic* (Vol. 8, pp. 265–343). Berlin: Kluwer.

Chisholm, R. M. (1963). Contrary-to-duty imperatives and deontic logic. *Analysis*, *24*(2), 33–36.

da Costa, N. C. A., & Carnielli, W. A. (1986). On paraconsistent deontic logic. *Philosophia*, *16*(3–4), 293–305.

Dunn, J. M. (1976). Intuitive semantics for first-degree entailments and 'coupled trees'. *Philosophical Studies*, *29*(3), 149–168.

Ficara, E. (2013). Dialectic and dialetheism. *History and Philosophy of Logic*, *34*(1), 35–52.

Fullbrook, E. (Ed.). (2008). *Pluralist economics*. London: Zed Books.
Gigerenzer, G. (2008). *Gut feelings*. London: Penguin.
Graeber, D. (2011). *Debt: The first 5000 years*. Brooklyn: Melville House.
Hansen, J. (2006). The paradoxes of deontic logic: Alive and kicking. *Theoria*, *72*(3), 221–232.
Harford, T. (2009). *Logic of life*. New York: Random House.
Harsanyi, J. (1977). Morality and the theory of rational behavior. *Social Research*, *44*(4), 623–656.
Hintikka, J., & Sandu, G. (1997). Game-theoretical semantics. In J. van Benthem & A. ter Meulen (Eds.), *Handbook of logic and language* (pp. 361–410). Amsterdam: Elsevier.
Kahneman, D. (2011). *Thinking, fast and slow*. New York: Farrar, Straus and Giroux.
Olde Loohuis, L., & Venema, Y. (2010). Logics and algebras for multiple players. *The Review of Symbolic Logic*, *3*(3), 485–519.
Overall, C. (2012). *Why have children?*. Cambridge: MIT Press.
Pacuit, E., Parikh, R., & Cogan, E. (2006). The logic of knowledge based obligation. *Synthese*, *149*(2), 311–341.
Parikh, R. (1994). Vagueness and utility: The semantics of common nouns. *Linguistics and Philosophy*, *17*, 521–535.
Parikh, R. (2002). Social software. *Synthese*, *132*(3), 187–211.
Parikh, R. (2014). Epistemic reasoning in life and literature. In S. O. Hansson (Ed.), *David Makinson on classical methods for non-classical problems. Outstanding Contributions to Logic* (pp. 143–153). Berlin: Springer.
Pietarinen, A. (2000). Logic and coherence in the light of competitive games. *Logique et Analyse*, *43*, 371–391.
Pietarinen, A., & Sandu, G. (2000). Games in philosophical logic. *Nordic Journal of Philosophical Logic*, *4*(2), 143–173.
Pietarinen, A.-V. (2003). Games as formal tools versus games as explanations in logic and science. *Foundations of Science*, *8*(4), 317–364.
Priest, G. (1989). Dialectic and dialetheic. *Science & Society*, *53*(4), 388–415.
Priest, G. (2006). *In contradiction* (2nd ed.). Oxford: Oxford University Press.
Rahman, S., & Carnielli, W. A. (2000). The dialogical approach to paraconsistency. *Synthese*, *125*, 201–231.
Reardon, J. (Ed.). (2009). *The handbook of pluralist economics education*. London: Routledge.
Routley, R., & Routley, V. (1972). The semantics of first degree entailment. *Noûs*, *6*(4), 335–359.
Smith, Y. (2010). *Econned*. Basingstoke: Palgrave Macmillan.
Stenning, K., & van Lambalgen, M. (2008). *Human reasoning and cognitive science*. Cambridge: MIT Press.
van Eijck, J., & Verbrugge, R. (Eds.). (2009). *Discourses on social software*. Amsterdam: Amsterdam University Press.

Chapter 7
Why Bother with Syntax?

Joseph Y. Halpern

Abstract When economists represent and reason about knowledge, they typically do so at a *semantic* (or set-theoretic) level. In this paper, it is argued that there are also benefits in using a syntactic representation.

Keywords Syntax, Semantics, Epistemic reasoning

Rohit and I go back a long way. We started talking about Dynamic Logic back when I was a graduate student, when we would meet at seminars at MIT (my advisor Albert Meyer was at MIT, although I was at Harvard, and Rohit was then at Boston University). Right from the beginning I appreciated Rohit's breadth, his quick insights, his wit, and his welcoming and gracious style. Rohit has been interested in the interplay between logic, philosophy, and language ever since I've known him. Over the years, both of us have gotten interested in game theory. I would like to dedicate this short note, which discusses issues at the intersection of all these areas, to him.[1]

7.1 Introduction

When economists represent and reason about knowledge, they typically do so at a *semantic* (or set-theoretic) level. Events correspond to sets and the knowledge operator maps sets to sets. On the other hand, in the literature on reasoning about knowledge in philosophy or logic, there is an extra layer of what may be viewed as unnecessary overhead: syntax. There have been papers in the economics literature that have argued for the importance of syntax. For example, Feinberg (2000, p. 128)

[1]Some material in this note appears in Halpern (2003, Sect. 7.2.4).

Supported in part by NSF Grants IIS-0812045, IIS-0911036, and CCF-1214844, AFOSR Grants FA9550-08-1-0266 and FA9550-12-1-0040, and ARO Grant W911NF-09-1-0281.

J. Halpern (✉)
Computer Science Department, Cornell University, Ithaca, NY 14853, USA
e-mail: halpern@cs.cornell.edu
URL: http://www.cs.cornell.edu/home/halpern

C. Başkent et al. (eds.), *Rohit Parikh on Logic, Language and Society*,
Outstanding Contributions to Logic 11, DOI 10.1007/978-3-319-47843-2_7

says "The syntactic formalism is the more fundamental and—intuitively—the more descriptive way to model economic situations that involve knowledge and belief ... It is fine to use the semantic formalism, as long as what we say semantically has a fairly clear intuitive meaning—that it can be said *in words*. This amounts to saying that it can be stated syntactically."

In this brief note, I point out some technical advantages of using syntax. Roughly speaking, they are the following:

1. Syntax allows us to make finer distinctions than semantics; a set may be represented by the two different expressions, and an agent may not react to these expressions in the same way. Moreover, different agents may react differently to the same expression, that is, the expression may represent different sets according to different agents.
2. Syntax allows us to describe in a model-independent way notions such as "rationality". This enables us to identify corresponding events (such as "agent 1 is rational") in two different systems.
3. The structure of the syntax provides ways to reason and carry out proofs. For example, many technical results proceed by induction on the structure of formulas. Similarly, formal axiomatic reasoning typically takes advantage of the syntactic structure of formulas.

In the rest of this note, I briefly review the semantic and syntactic approaches, explain the advantages listed above in more detail, and point out ways in which the economics literature has not exploited the full power of the syntactic approach thus far. In another paper (Halpern 1999), I have made essentially the opposite argument, namely, that computer scientists and logicians have not exploited the full power of the semantic approach, pointing out that we can often dispense with the overhead of syntax and the need to define a $\models$ relation by working directly with the semantics. I believe both arguments! This just goes to show that each community can learn from the approaches used by the other community. I return to this point in the concluding section of this paper.

7.2 Semantics Versus Syntax: A Review

The standard approach in the economic literature on knowledge starts with what is called in Fagin et al. (1995) an *Aumann structure* $A = (\Omega, \mathcal{P}_1, \ldots, \mathcal{P}_n)$, where Ω is a set of states of the world and $\mathcal{P}_i$, $i = 1, \ldots, n$ are partitions, one corresponding to each agent. Intuitively, worlds in the same cell of partition i are indistinguishable to agent i. Given a world $\omega \in \Omega$, let $\mathcal{P}_i(\omega)$ be the cell of $\mathcal{P}_i$ that contains ω.

Knowledge operators K_i, $i = 1, \ldots, n$ mapping events in (subsets of) Ω to events are defined as follows:

$$\mathsf{K}_i(A) = \{\omega \colon \mathcal{P}_i(\omega) \subseteq A\}.$$

We read $\mathsf{K}_i(A)$ as "agent i knows A". Intuitively, it includes precisely those worlds ω such that A holds at all the worlds that agent i cannot distinguish from ω.

The logical/philosophical approach adds an extra level of indirection to the set-theoretic approach. The first step is to define a *language* for reasoning about knowledge, that is, a set of well-formed formulas. We then associate an event with each formula. We proceed as follows.

We start with a set Φ of primitive propositions $p_1, p_2, \ldots$ representing propositions of interest. For example, p_1 might represent "agent 1 is rational" and p_2 might represent "agent 2 is following the strategy of always defecting (in a game of repeated Prisoner's Dilemma)". We then close off this set under conjunction, negation, and the *modal operators* $K_1, \ldots, K_n$. Thus, if ϕ and ψ are formulas, then so are $\phi \wedge \psi$, $\neg\phi$, and $K_i\phi$, for $i = 1, \ldots, n$. We typically write $\phi \vee \psi$ as an abbreviation for $\neg(\neg\phi \wedge \neg\psi)$ and $\phi \Rightarrow \psi$ as an abbreviation for $\neg\phi \vee \psi$. Thus, we can write formulas such as $K_1K_2p_1 \wedge \neg K_2K_1p_2$, which could be read as "agent 1 knows that agent 2 knows that agent 1 is rational and agent 2 does not know that agent 1 knows that agent 2 is following the strategy of always defecting". It should be stressed that a formula like $K_1K_2p_1 \wedge \neg K_2K_1p_2$ is just a string of symbols, not a set.

A *Kripke structure* is a tuple $M = (\Omega, \mathcal{P}_1, \ldots, \mathcal{P}_n, \pi)$, where $(\Omega, \mathcal{P}_1, \ldots, \mathcal{P}_n)$ is an Aumann structure, and π is an *interpretation*, that associates with each primitive proposition an event in Ω.[2] We can then extend π inductively to associate with each formula an event as follows:

- $\pi(\phi \wedge \psi) = \pi(\phi) \cap \pi(\psi)$
- $\pi(\neg\phi) = \overline{\pi(\phi)}$ (where $\overline{E}$ denotes the complement of E)
- $\pi(K_i\phi) = \mathsf{K}_i(\pi(\phi))$.[3]

Notice that not every subset of Ω is necessarily definable by a formula (even if Ω is finite). That is, for a given subset $E \subseteq \Omega$, there may be no formula ϕ such that $\pi(\phi) = E$. The set of events definable by formulas form an algebra.

If all that is done with a formula is to translate it to an event, why bother with the overhead of formulas? Would it not just be simpler to dispense with formulas and interpretations, and work directly with events? It is true that often there is no particular advantage in working with syntax. However, sometimes it does come in handy. Here I discuss the three advantages mentioned above in some more detail:

1. The two events E and $(E \cap F) \cup (E \cap \overline{F})$ denote the same set. Hence, so do $\mathsf{K}_i(E)$ and $\mathsf{K}_i((E \cap F) \cup (E \cap \overline{F}))$. Using the set-theoretic approach, there is no way to distinguish these events. Even if we modify the definition of K, and move to non-partitional definitions of knowledge (Bacharach 1985; Fagin et al.

[2]In the literature, π is often taken to associate a *truth value* with each primitive proposition at each world; that is, $\pi \colon \Phi \times \Omega \to \{\textbf{true}, \textbf{false}\}$. Using this approach, we can then associate with each primitive proposition p the event $\{\omega \colon \pi(p, \omega) = \textbf{true}\}$. Conversely, given a mapping from primitive propositions to events, we can construct a mapping from $\Phi \times \Omega$ to truth values. Thus, the two approaches are equivalent.

[3]In the literature, one often sees the notation $(M, \omega) \models \phi$, which is read as "the formula ϕ is true at world ω in Kripke structure M". The definition of $\models$ recapitulates that just given for π, so that $\omega \in \pi(\phi)$ iff $(M, w) \models \phi$.

1995; Geanakoplos 1989; Samet 1990), we still cannot distinguish these formulas. In propositional logic, the formula p is equivalent to $(p \wedge q) \vee (p \wedge \neg q)$. (Of course, these formulas were obtained from the events by substituting p for E, q for F, and replacing $\cap$, $\cup$, and - by $\wedge$, $\vee$, and $\neg$, respectively.) In the standard semantics defined above, the formulas $K_i p$ and $K_i((p \wedge q) \vee (p \wedge \neg q))$ are also equivalent. However, an agent may not recognize that the formulas p and $(p \wedge q) \vee (p \wedge \neg q)$ are equivalent, and may react differently to them. There are approaches to giving semantics to knowledge formulas that allow us to distinguish these formulas (Fagin et al. 1995, Chap. 9). This issue becomes important if we are trying to model resource-bounded notions of knowledge. We may, for example, want to restrict the agent's knowledge to formulas that are particularly simple, according to some notion of simplicity; thus, it may be the case that $K_i p$ holds, while $K_i((p \wedge q) \vee (p \wedge \neg q))$ does not. Syntactic approaches to dealing with *awareness* (Fagin and Halpern 1988; Halpern and Rêgo 2013) can capture this intuition by allowing agent i to be aware of p but not of $(p \wedge q) \vee (p \wedge \neg q)$. This issue also figures prominently in a recent approach to decision theory (Blume et al. 2006) that takes an agent's object of choice to be a syntactic program, which involves tests (which are formulas). Again, an agent's decision can, in principle, depend on the form of the test. A yet more general approach, where an agent's utility function is explicitly defined on formulas in the agent's language, is considered in Bjorndahl et al. (2013).

A slightly more general notion of Kripke structure allows us to deal with (at least one form of) *ambiguity*. The standard definition of Kripke structure has a single interpretation π. But, in practice, agents often interpret the same statement differently. What one agent calls "red" might not be "red" to another agent. It is easy to deal with this; we simply have a different interpretation π_i for each agent i (see Halpern and Kets 2012 for the implications of allowing such ambiguity for game theory).

2. Language allows us to describe notions in a model-independent way. For example, the typical approach to defining rationality is to define the event that agent i is rational in a particular Aumann structure. But suppose we are interested in reasoning about two related Aumann structures, A_1 and A_2, at the same time. Perhaps each of them has the same set Ω of possible worlds, but different partitions. We then want to discuss "corresponding" events in each structure, for example, events such as "agent 1 is rational" or "agent 2 is following the strategy of always defecting". This is not so easy to do if we simply use the events. The event "agent 1 is rational" corresponds to different sets in A_1 and A_2. In the set-based approach, there is no way of relating them. Using syntax, an event such as "agent 1 is rational" would be described by a formula (the exact formula depends, of course, on the definition of rational, a matter of some controversy; see, for example, Binmore 2009; Blume and Easley 2008; Gilboa 2010). The same formula may well correspond to different events in the two structures. For example, if the formula involves K_1, then if different partitions characterize agent 1's knowledge in A_1 and A_2, the formula would correspond to different subsets of Ω. Although the formula describing a notion like "agent 1 is rational" would correspond to different

events in different structures, we may well be able to prove general properties of the formula that are true of all events corresponding to the formula. For example, if ϕ is the formula that says that agent 1 is rational, we may be able to show that $\phi \Rightarrow K_i\phi$ is *valid* (true in every state of every Aumann structure); this says that if agent 1 is rational, then agent 1 knows that he is rational.

3. There are standard examples of when the use of formulas can be useful in proofs. For one thing, it allows us to prove results by induction on the structure of formulas. Examples of such proofs can be found in Fagin et al. (1992, 1995). Secondly, the syntactic structure of a formula we would like to prove can suggest a proof technique. For example, a disjunction can often be proved by cases. Thirdly, when proving that certain axioms completely characterize knowledge, the standard proof involves building a *canonical model*, whose worlds can be identified with sets of formulas (Aumann 1999; Fagin et al. 1995). This approach is also implicitly taken by Heifetz and Samet (1998). They construct a *universal type space*, a space that contains all possible types of agents (where a *type*, roughly speaking, is a complete description of an agent's beliefs about the world and about other agents' types). Their construction involves certain events that they call *expressions*; these expressions are perhaps best thought of as syntactic expressions in a language of belief.

At some level, the economics community is already aware of the advantages of syntax, and syntactic expressions are used with varying levels of formality in a number of papers. To cite one example, in Balkenborg and Winter (1997) the notion of an *epistemic expression* is defined, that is, a function from Aumann spaces to events in that Aumann space. An epistemic expression such as R_i, which can be thought of as representing the proposition "agent i is rational" then becomes a mapping from an Aumann structure to the event consisting of all worlds where i is rational.[4] More complicated epistemic expressions are then allowed, such as $\mathsf{K}_1\mathsf{K}_2(R_3)$. Perhaps not surprisingly, some proofs then proceed by induction on the structure of event expressions. It should be clear that event expressions are in fact syntactic expressions, with some overloading of notation. Epistemic expressions are just strings of symbols. The K_i in a syntactic expression is not acting as a function from events to events. It is precisely because they are strings of symbols that we can do induction on their length. Indeed, in Fagin et al. (1992), epistemic expressions are introduced in the middle of a proof (under the perhaps more appropriate name *(event) descriptions*, since in fact they are used even to describe non-epistemic events), precisely to allow a proof by induction.

7.3 Discussion

Expressive power In the semantic approach, we can apply the knowledge operator to an arbitrary subset of possible worlds. In the syntactic approach, we apply the

[4]Not all papers are so careful. For example, Aumann (1995) calls R_i an event. Of course, his intention is quite clear.

knowledge operator to formulas. While formulas are associated with events, it is not necessarily the case that every event is definable by a formula. Indeed, one of the major issues of concern to logicians when considering a particular syntactic formalism is its *expressive power*, that is, what events are describable by formulas in the language.

Using a more expressive formalism has both advantages and disadvantages. The advantages are well illustrated by considering Feinberg's (2000) work on characterizing the common prior assumption (CPA) syntactically. Feinberg considers a language that has knowledge operators and, in addition, belief operators of the form $p_i^\alpha(f)$, which is interpreted as "according to agent i, the probability of f is at least α". Feinberg does not have common knowledge in his language, nor does his syntax allow expressions of the form $1/2p_i(f) + 2/3p_i(f') \geq 1$, which can be interpreted as "according to agent i, 1/2 of the probability of f plus 2/3 of the probability of f' is at least 1" or, more generally, expressions of the form $\alpha_1 p_i(f_1) + \cdots \alpha_k p_i(f_k) \geq \beta$. Expressions of the latter form are allowed, for example, in Fagin et al. (1990) and Fagin and Halpern (1994). The fact that he does not allow common knowledge causes some technical problems for Feinberg (which he circumvents in an elegant way). Linear combinations of probability terms allow us to make certain statements about expectations of random variables (at least, in the case of a finite state space). Feinberg has an elegant semantic characterization of CPA in the finite case: Roughly speaking, he shows that CPA holds if and only if it is not the case that there is a random variable X whose expectation agent 1 judges to be positive and agent 2 judges to be negative. Since Feinberg cannot express expectations in his language, he has to work hard to find a syntactic characterization of CPA. With a richer language, it would be straightforward.

This is not meant to be a criticism of Feinberg's results. Rather, it points out one of the features of the syntactic approach: the issue of the exact choice of syntax plays an important role. There is nothing intrinsic in the syntactic approach that prevents us from expressing notions like common knowledge and expectation. Rather, just as in the semantic approach, where the modeler must decide exactly what the state space should be, in the syntactic approach, the modeler must decide on the choice of language. By choosing a weaker language, certain notions become inexpressible.

Why would we want to choose a weaker language? There are three obvious reasons. One is aesthetic: Just as researchers judge theories on their elegance, we have a notion of elegance for languages. We expect the syntax to be natural (although admittedly "naturalness", like "elegance", is in the eye of the beholder) and to avoid awkward expressions. A second is more practical: to the extent that we are interested in doing inference, simpler languages typically admit simpler inferences procedures. This intuition can be made formal. There are numerous results characterizing the difficulty of determining whether a formula in a given language is *valid* (that is, true in every state in every structure) (Fagin et al. 1995, Chap. 3). These results demonstrate that more complicated languages do in fact often lead to more complex decision procedures. However, there are also results showing that, in this sense, there is no cost to adding certain constructs to the language. For example, results of Fagin et al. (1990) and Fagin and Halpern (1994) show that allowing linear combinations of probability terms does not increase the complexity over just allowing operators such as p_i^{α}. The third reason is that a weaker language might give us a more appropriate level at which to study what we are most interested in. For example, if we are interested only in qualitative beliefs (A is more likely than B), we may not want to "clutter up" the language with quantitative beliefs; that is, we may not want to be able to talk about the exact probability of A and B. A simpler language lets us focus on the key issues being examined. Indeed, roughly speaking, we can think of different languages as providing us with different notions of "sameness"; in other words, "isomorphism" is language-dependent, so any result that holds "up to isomorphism" is really a language-dependent result.[5]

One other issue: the focus in the semantic approach has been on finding operators that act on events. Not all syntactic expressions correspond in a natural way to operators. For example, recall that in Fagin et al. (1990), expressions such as $\alpha_1 p_i(f_1) + \cdots \alpha_k p_i(f_k) \geq \beta$ are considered. This could be viewed as corresponding to a k-ary operator $B^{\alpha_1,\ldots,\alpha_k,\beta}$ such that $\omega \in B^{\alpha_1,\ldots,\alpha_k,\beta}(E_1,\ldots,E_k)$ if $\alpha_1 p_i(E_1) + \cdots + p_k(E_k) \geq \beta$, however, this is not such a natural operator. It seems more natural to associate an event with this syntactic expression directly in terms of a probability distribution on worlds, without bothering to introduce such operators.

Models One of the advantages of the syntactic approach is that it allows us to talk about properties without specifying a model. This is particularly important both if we do not have a fixed model in mind (for example, when discussing rationality, we may be interested in general properties of rationality, independent of a particular model) or when we do have a model in mind, but we do not wish to (or cannot) specify it completely. In general, a formula may be true in many models, not just one.

To the extent that models for formulas have been considered in the economics literature, the focus has been on one special model, the so-called *canonical model* (Aumann 1999). This canonical model has the property that if a (possibly infinite) collection of formulas is true at some world in some model, then it is true at some world of the canonical model. Thus, in a sense, the canonical model can be thought of as including all possible models.

While the canonical model is useful for various constructions, it also has certain disadvantages. For one thing, it is uncountable. A related problem is that it is far too complicated to be useful as a tool for specifying simple situations. We can think of a formula ϕ as specifying a collection of structure-world pairs, namely, all the pairs (M, w) such that $(M, w) \models \phi$. It is well known that, at least in the case of epistemic logic, if a formula ϕ is satisfiable at all (that is, if $(M, w) \models \phi$ for some (M, w)), then there is a finite structure M and a world w in M such that $(M, w) \models \phi$. Instead of focusing on an uncountable canonical structure, it is often much easier to focus on a finite structure.

Thinking in terms of the set of models that satisfy a formula (rather than just one canonical model) leads us to consider a number of different issues. In many cases we do not want to construct a model at all. Instead, we are interested in the logical consequences of some properties of a model. A formula ϕ is a *logical consequence* of

[5]Thanks to Adam Bjorndahl for stressing this point.

a collection Ψ of formulas if, whenever $(M, w) \models \psi$ for every formula $\psi \in \Psi$, then $(M, w) \models \phi$. For example, we may want to know if some properties of rationality are logical consequences of other properties of rationality; again, this is something that is best expressed syntactically.

7.4 Conclusion

The point of this short note should be obvious: there are times when syntax is useful, despite its overhead. Moreover, like Molière's M. Jourdain, who discovers he has been speaking prose all his life (Molière 2013), game theorists have often used syntax without realizing it (or, at least, without acknowledging its use explicitly). However, they have not always taken full advantage of it.

That said, as I have pointed in Halpern (1999), there are times when semantics is useful, and the overhead of syntax is not worth it. Semantic proofs of validity are typically far easier to carry out than their syntactic counterparts (as anyone who has tried to prove the validity of even a simple formula like $K(\phi \wedge \psi) \equiv K\phi \wedge K\psi$ from the axioms knows). I believe that the finer expressive power of syntax is particularly useful when dealing with notions of awareness, and trying to capture how agents react differently to two different representations of the same event. This is something that arguably cannot be done by a semantic approach without essentially incorporating the syntax in the semantics. But when considering an approach where these issues do not arise, a semantic approach may be the way to go. The bottom line here is that it is useful to have both syntactic and semantic approaches in one's toolkit!

Acknowledgements Thanks to Adam Bjorndahl for and the two reviewers of this paper for very useful comments.

References

Aumann, R. J. (1995). Backwards induction and common knowledge of rationality. *Games and Economic Behavior*, *8*, 6–19.

Aumann, R. J. (1999). Interactive epistemology I: Knowledge. *International Journal of Game Theory*, *28*(3), 263–300.

Bacharach, M. (1985). Some extensions of a claim of Aumann in an axiomatic model of knowledge. *Journal of Economic Theory*, *37*, 167–190.

Balkenborg, D., & Winter, E. (1997). A necessary and sufficient epistemic condition for playing backward induction. *Journal of Mathematical Economics*, *27*, 325–345.

Binmore, K. (2009). *Rational decisions*. Princeton, NJ: Princeton University Press.

Bjorndahl, A., Halpern, J. Y., & Pass, R. (2013). Language-based games. In *Theoretical aspects of rationality and knowledge: Proceedings of the fourteenth conference (TARK 2013)* (pp. 39–48).

Blume, L., & Easley, D. (2008). Rationality. In L. Blume & S. Durlauf (Eds.), *The new Palgrave: A dictionary of economics*. New York: Palgrave Macmillan.

Blume, L., Easley, D., & Halpern, J. Y. (2006). Redoing the foundations of decision theory. In *Principles of knowledge representation and reasoning: Proceedings of the tenth international conference (KR '06)* (pp. 14–24). A longer version, entitled "Constructive decision theory", can be found at http://www.cs.cornell.edu/home/halpern/papers/behfinal.pdf.

Fagin, R., J. Geanakoplos, Halpern, J. Y., & Vardi, M. Y. (1992). The expressive power of the hierarchical approach to modeling knowledge and common knowledge. In *Theoretical aspects of reasoning about knowledge: Proceedings of the fourth conference* (pp. 229–244).

Fagin, R., & Halpern, J. Y. (1988). Belief, awareness, and limited reasoning. *Artificial Intelligence, 34*, 39–76.

Fagin, R., & Halpern, J. Y. (1994). Reasoning about knowledge and probability. *Journal of the ACM, 41*(2), 340–367.

Fagin, R., Halpern, J. Y., & Megiddo, N. (1990). A logic for reasoning about probabilities. *Information and Computation, 87*(1/2), 78–128.

Fagin, R., Halpern, J. Y., Moses, Y., & Vardi, M. Y. (1995). *Reasoning about knowledge*. Cambridge, MA: MIT Press. A slightly revised paperback version was published in 2003.

Feinberg, Y. (2000). Characterizing common priors in the form of posteriors. *Journal of Economic Theory, 91*, 127–179.

Geanakoplos, J. (1989). *Game theory without partitions, and applications to speculation and consensus*. Cowles Foundation Discussion Paper #914, Yale University.

Gilboa, I. (2010). Questions in decision theory. *Annual Reviews in Economics, 2*, 1–19.

Halpern, J. Y. (1999). Set-theoretic completeness for epistemic and conditional logic. *Annals of Mathematics and Artificial Intelligence, 26*, 1–27.

Halpern, J. Y. (2003). *Reasoning about uncertainty*. Cambridge, MA: MIT Press.

Halpern, J. Y. & Kets, W. (2012). Ambiguous language and differences in beliefs. In *Principles of knowledge representation and reasoning: Proceedings of the thirteenth international conference (KR '12)* (pp. 329–338).

Halpern, J. Y., & Rêgo, L. C. (2013). Reasoning about knowledge of unawareness revisited. *Mathematical Social Sciences, 66*(2), 73–84.

Heifetz, A., & Samet, D. (1998). Typology-free typology of beliefs. *Journal of Economic Theory, 82*, 324–341.

Molière, J. B. P. (2013). *Le Bourgeois Gentilhomme*. Paris: Gallimard.

Samet, D. (1990). Ignoring ignorance and agreeing to disagree. *International Journal of Game Theory, 52*, 190–207.

Chapter 8
Talking About Knowledge

Johan van Benthem

Abstract In current studies of knowledge at the interface of logic and epistemology, philosophical positions and logical systems lore meet in new ways. In this little piece, a programmatic sequel to van Benthem (2011) and a prequel to Baltag et al. (2015), I add some further perspectives and issues to this mix from dynamic-epistemic logics of information and inquiry. My aim is to show that we can have a yet richer agenda of epistemic themes, and a richer view of the interplay of logic and epistemology, when we make epistemic action a major focus on a par with knowledge or belief per se.

Keywords Epistemology · Dynamic-epistemic logic · Relevant alternatives theory

---

Rohit Parikh is an inspirational leader at the interface of logic, epistemology, mathematics, computer science and game theory whose influence (some might say, grace) has touched so many people. This piece does no justice to the depth of his thinking, but I hope it is in his spirit of free inquiry.
I thank Shi Chenwei and, especially, Wesley Holliday for valuable comments on earlier versions of this paper. Two anonymous referees for this volume provided helpful feedback at a later stage.

J. van Benthem (✉)
University of Amsterdam, Amsterdam, The Netherlands
e-mail: J.vanBenthem@uva.nl

J. van Benthem
Stanford University, Stanford, CA, USA

J. van Benthem
Changjiang Scholar Program, Tsinghua University, Beijing, China

C. Başkent et al. (eds.), *Rohit Parikh on Logic, Language and Society*,
Outstanding Contributions to Logic 11, DOI 10.1007/978-3-319-47843-2_8

8.1 Background: Epistemic Logic and Notions of Knowledge

Modern encounters of logic and epistemology started with epistemic logic (Hintikka 1962).[1] This system reads its key modality $K\varphi$ as truth of the proposition φ in the current range of epistemically accessible worlds for the relevant agent. The connection of this notion to knowledge in the philosopher's sense has long been a focus of discussion (cf. Stalnaker 1985; Williamson 2000). It may be viewed as an account of implicit ideal knowledge of ordinary agents, of explicit knowledge of ideal agents, or as a mere statement of the content of an agent's knowledge, without any claim to define knowledge itself in that way (Holliday 2012b). Be this as it may, among many epistemic logicians, $K\varphi$ has come to be viewed as standing for yet a different notion, equally important in philosophy and the sciences, namely the 'semantic information' that the agent has available. Especially in large open-ended spaces of epistemic options, this often seems too demanding to allow for significant knowledge, and current philosophical accounts of knowledge have taken other roads.

Indeed, there has been a veritable wave of creativity in the literature since the 1960s. Many philosophers analyze knowledge as belief, a notion that scans less than the full range of options for semantic information, and then upgrade belief with extra requirements of various kinds, such as tracking the truth (Nozick 1981), robustness under update with new information (Stalnaker 2006), temporal convergence to true belief (Kelly 1996), or defensibility in challenge games (Lehrer 1990).[2] The resulting styles of reasoning about knowledge diverge considerably from classical epistemic logic, in line with philosophical intuitions that have been much discussed in the literature. In particular, there is no automatic closure of knowledge under simple logical inferences such as weakening the known proposition, or combining two known propositions to knowing their conjunction.[3]

8.2 Knowledge and Modeling Relevant Alternatives

These richer philosophical accounts allow for significant senses in which one can know that φ at the present stage of inquiry even when not all possible $\neg\varphi$-worlds (alternatives, states, scenarios) have been ruled out: knowledge ventures beyond the semantic information. In what follows I will be looking at the influential proposals of Dretske (1970, 1981), and Lewis (1996) that read knowledge of φ as a state of having ruled out all relevant potential counter-examples to the proposition φ. This

---

[1] Hintikka's views on the interface of logic and epistemology are still highly relevant, and they have kept evolving, witness Hintikka (1973) and the 'Socratic epistemology' of Hintikka (2007).

[2] Cf. Parikh et al. (2013) for another take on the striking entanglements of knowledge and games.

[3] These are not the usual omniscience failures of bounded rationality that are often bandied against epistemic logic. As Dretske points out, they would even occur for "ideally astute logicians".

is the view that I will be discussing as the running theme of this paper, since it seems the best suited for making my general points.[4]

Recently, logicians have looked in new ways at what makes such relevant alternatives ('RA', for short) theories tick, going beyond the basics of epistemic logic. A trail-blazing study is Holliday (2012b) that presents a joint analysis of relevant-alternative and tracking theories of knowledge, and determines the complete valid closure principles governing reasoning about knowledge in these styles. Further analyses, using ideas from dynamic-epistemic logic (van Benthem 2011), are found in Cornelisse (2011), Xu (2010), Shi (2013) and Holiday (2012a). I will not go into details of these systems here. The technical discussion of RA theory to follow is in the spirit of these papers, but the general issues I will raise may be all my own.[5]

Relevant Alternatives Models For the purpose of this paper a minimal modeling suffices. We use worlds for alternative situations, with the current accessible worlds those that have not been ruled out yet as options for what the actual world is like, and thus potentially inviting further inspection.[6] Equally crucially to the analysis, worlds come with a primitive order of relevance R that may have many plausible intuitive origins [cf. Hawke (2014, 2015) for a critical survey of current philosophical approaches]. Putting together all these basic ingredients, we will work with simple pointed epistemic models

$$\boldsymbol{M} = (W, \sim, R, V, s), \quad \text{with a designated actual world } s.$$

In what follows, we will assume that $\sim$ is an equivalence relation, while the relevance order is transitive and well-founded, allowing no loops or infinite upward sequences.[7]

Authors sometimes rush to the inferences validated by such models, but we will go slowly, more in model-theoretic mode. Prior to any specific definition of knowledge, and perhaps even more importantly than that, models such as these are not just semantic structures, they also embody an implicit proposal for a *conceptual framework* of notions, a core family that constitutes an epistemic practice of stating claims and rejoinders, and based on that, reasoning. It is a substantive view in itself that the epistemic core notions are the ones chosen here, and of course, it is not an entirely uncontroversial choice.

The present conceptual framework suggests a logical base language with two modalities:

---

[4]Personally, I am also deeply attracted to accounts of knowledge that involve belief and truth-tracking, but this second love should not be a serious bias in the following presentation.

[5]Analogies with the cited work may still be worth pursuing for their independent interest.

[6]Another view of alternatives would use 'possibilities models' (Humberstone 1980; Holliday 2013).

[7]This is highly simplified. A key choice point (cf. Holliday 2012b) is whether the relevance relation R is uniform, as in Lewis (1996), or depends on the proposition φ for which knowledge is claimed, as in Dretske (1970). Moreover, in a more sophisticated analysis, it becomes mandatory to make the epistemic and relevance ordering world-dependent instead of uniform across the models. An even more general framework of modal neighborhood models and 'selection functions' for relevant worlds to be inspected and alternatives already ruled out is developed in Holliday (2012b). The points to be made in this paper carry over to such richer settings, and even acquire more force there.

(a) $[\sim]\varphi$ for the semantic information of the agent, that we will write $I\varphi$,
(b) $[\uparrow]\varphi$ for truth in all worlds that are more relevant than the current one.[8]

Defining Knowledge This is arguably the simplest setting in which a philosophically defensible notion of knowledge $K\varphi$ becomes an, if you wish, more subtle defined logical operator stating that all relevant counterexamples to φ have been ruled out. We define it as follows, in close analogy with Holliday (2012a):

$$K\varphi := I\,(\varphi \vee <\uparrow>\neg\varphi)$$

This definition requires some explanation. It says that all epistemic alternatives are either φ, or if not, then there exists some more relevant $\neg\varphi$-alternative. Why does this capture the above account? What I have in mind here are finite, or at least upward well-founded models in the ordering R, where the following is going to happen.

Fact On models with well-founded transitive relevance orders, $I\,(\varphi \vee <\uparrow>\neg\varphi)$ holds iff all the most relevant $\neg\varphi$-worlds are epistemically inaccessible.

Proof This follows from analyzing $I\,(\varphi \vee <\uparrow>\neg\varphi)$. If I see a live $\neg\varphi$-alternative x, then there is a more relevant $\neg\varphi$-alternative y. Now either y is not epistemically accessible, and we are done (y has been ruled out), or it is. In the latter case, we repeat the argument. By well-foundedness, this process must stop short of creating an infinite upward chain, and so, by transitivity of the order R, some more relevant $\neg\varphi$-alternative for x was ruled out.[9] ■

Remark Veridicality.
This definition does not require the presence of φ-worlds in the epistemic range of the current world. Hence, in order to make sure that $K\varphi$ implies φ, authors in the area make assumptions such as the actual world being the most relevant one in the model. This raises some interesting issues about the justification for this. On externalist views of knowledge, where relevance order reflects real features of the actual situation, this may be just right. On internalist views of knowledge, the preceding assumption may be somewhat more ad-hoc. I will sidestep these issues here, and just enforce veridicality by adding a conjunct that φ is true, while the main body $I\,(\varphi \vee <\uparrow>\neg\varphi)$ says that the claim that φ is 'defensible'. In what follows, I use veridicality, assuming that it has been introduced in some plausible manner.[10] There is more to the RA view of knowledge than meets the eye. In particular, it tolerates a lot of $\neg\varphi$-worlds contradicting the claim. Depending on the relevance order, one may know that φ even when the overwhelming majority of the current alternatives fails to satisfy φ.

---

[8]We will also use the dual existential modalities occasionally in the principles to follow.

[9]One can also find definitions that are more suitable to infinite or non-wellfounded models.

[10]More should probably be said here. The fact that Veridicality is a tag-on in the setting of this paper worries me a bit, and it makes me wonder where the relevant-alternatives paradigm is really intimately tied to knowledge, rather than to a wider range of further epistemic attitudes.

But there are other, less obvious points as well. For instance, knowledge now depends essentially on making relevance comparisons with worlds that have already been ruled out. If I recalibrate my epistemic horizon to just the accessible worlds as my current model, new most urgent counterexamples to φ would arise. In this model, it is easy to show that, with the same relevance ordering as above, the above reading of $K\varphi$ collapses into semantic information $I\varphi$. This contrast deserves some emphasis: the context where we make the relevance comparison is essential, and past effort still counts today.[11] In particular, then, even for something as simple as a single S5-agent with an epistemic equivalence relation, we cannot assume that models consist of one connected epistemic component: there may be different components, reflecting earlier stages of the process of ruling out.

Other Epistemic Notions Our framework does not exclude other legitimate intuitions about knowledge, and in fact, it facilitates their exploration. In particular, it may not be completely obvious in RA theory why we only 'compensate' counterexamples to the proposition φ by more relevant counterexamples to φ that have been ruled out. Given that our knowledge claim is in favor of the proposition φ, why not compensate a counter-example by some more relevant φ-world (epistemically accessible or not)? Taking the latter line would change the definition of knowledge from $I(\varphi \vee <\uparrow>\neg\varphi)$ to another notion studied in Holliday (2012a):

$$I(\varphi \vee <\uparrow>T)$$

stating that all most relevant epistemically accessible worlds in the model are φ.[12] The formula $I(\varphi \vee <\uparrow> T)$ seems one more serious notion of knowledge with its own applications, though weaker than our first definition, from which it follows. It might fit another intuition about knowledge, viz. that one's epistemic state is 'in favor' of the known proposition qua worlds supporting it, whence situations that are even more in favor do not disrupt this. Reasoning with this second notion differs from that with the first: for instance, it supports both weakening (upward monotonicity) and conjunction of known propositions. I think that a logical analysis should bring out such options and study them together.[13]

Summary We have identified two different components so far in RA theory. The first was a conceptual framework, consisting of basic notions of semantic information and relevance, that come with their own language. Next, within that framework, there was a proposal for a particular definition of knowledge. This distinction matters.

---

[11] Similar contrasts occur in logics of games when defining 'non-dominated' strategies (van Benthem 2007), or in the logic of knowledge and preference when defining 'regret' (van Benthem and Liu 2007) where objects that have been ruled out still play an essential role in justifying current moves. Compare how reference to a lost past is essential to calling someone an emeritus professor.

[12] This clause looks like a truth condition for absolute belief on plausibility orderings (see below).

[13] While this may be a logician's tendency towards tolerance, philosophers might object that this leads to easy relativism concerning the notion of knowledge. However, one can also have a benign view on the plurality. Instead of forcing a choice between different views of knowledge, a logical framework allows for communication and cooperation between different philosophical positions.

One may endorse a conceptual framework for basic epistemics, even if one rejects the specific view of knowledge—but one might also reject the framework altogether. Next, prior to any account of valid and invalid inferences, definability in the resulting language turned out to be of interest in its own right. For instance, we saw how plausible alternative views of knowledge could be formulated.[14]

8.3 The Logic of Relevant Alternatives-Based Knowledge

Once we have our models and formal language representing what we took as the essential features of the RA-paradigm for knowledge, we can investigate the logic that is validated: if you wish, the induced style of epistemic reasoning. For our system, we start with a standard bimodal logic for the two static modalities I for semantic information and $[\uparrow]$ for relevance.

Static Base Logic Depending on what constraints we put on the accessibility and relevance orders in our models, this will be a combination of known separate modal logics for these two modalities, plus perhaps bridge axioms reflecting connections between $\sim$ and $<$.[15] Inside this axiomatizable bimodal logic, there will then be an axiomatizable logic for any of its defined notions of knowledge, in particular, one for the RA notion $K\varphi = I(\varphi \vee <\uparrow>\neg\varphi)$.

Closure Principles Still, this does not imply that there is a perspicuous base logic of 'closure principles' for formulas with K and Boolean connectives only. That is why the completeness results in Holliday (2012b) are such a combinatorial tour de force. But even without an explicit axiomatization in a closure principle format, we can understand some basic properties of inference with K merely in virtue of its syntactic form. The RA-formula for $K\varphi$ is neither upward not downward monotonic with respect to φ, as it contains both a positive and a negative occurrence of φ. This fits the intuitions about closure principles stated in the RA literature.[16] But there are further interesting inferential features to RA knowledge.[17] Of these, we merely mention one curious observation that seems to have gone unnoticed in the literature. A knowledge claim may sometimes be strengthened in surprising ways.

---

[14]There are also interesting more technical issues about the order of notions. For instance, we have defined RA knowledge in terms of semantic information. Conversely, can we also define semantic information in terms of RA knowledge? I think not, but a proof may be somewhat complicated.

[15]For instance, we may want to insist that the relevance order is known to the agent in the sense expressed by the axiom $<\uparrow>\varphi \rightarrow I <\uparrow>\varphi$ (cf. van Benthem and Liu 2007).

[16]By contrast, the alternative notion of knowledge $I(\varphi\vee <\uparrow> T)$ given above has a syntax whose occurrence of φ predicts the validity of monotonicity and conjunction inferences automatically.

[17]One sign of complexity is that standard *conditionalizations* such as $I(\varphi \rightarrow \psi)$ for $I^{\varphi}\ \psi$ do not work. Conditional $K^{\varphi}\ \psi = I(\varphi \rightarrow (\psi \vee <\uparrow>(\varphi \wedge \neg\psi))$ has no reduction by substituting in the absolute notion.

Example Knowledge claims may 'self-strengthen'.

Consider the following model with actual world 1, whose only epistemic alternatives are 1, 2 and 3, while world 4 is more relevant than both 2 and 3:

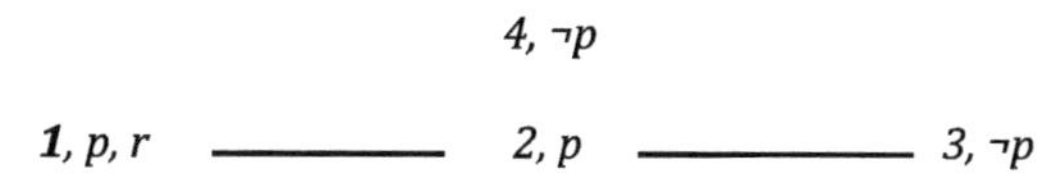

This supports the knowledge claim *Kp* in 1. But it also supports the stronger $K(p \wedge r)$ true only at 1: the $\neg(p \wedge r)$-world 2 has a more relevant $\neg(p \wedge r)$-world 4 that was ruled out. ■

Stated more generally, we have the following observation:

Fact The formula $K\varphi \rightarrow K(\varphi \wedge \neg{<}\uparrow{>}\neg\varphi)$ is valid.

Proof This follows since, even just in the minimal modal logic, $I(\varphi \vee {<}\uparrow{>}\neg\varphi)$ implies $I((\varphi \wedge \neg{<}\uparrow{>}\neg\varphi) \vee {<}\uparrow{>}\neg\varphi)$, which in its turn implies $I((\varphi \wedge \neg{<}\uparrow{>}\neg\varphi) \vee {<}\uparrow{>}\neg(\varphi \wedge \neg {<}\uparrow{>}\neg\varphi))$.[18] ■

The principle may look like something contorted that one would never say. But in a given model, $\varphi \wedge \neg{<}\uparrow{>}\neg\varphi$ may well denote a set of worlds that also has a more concrete definition. We leave it to the reader to judge whether this closure principle, or others in the same vein, is a desirable or an undesirable effect of the relevant alternatives view of knowledge.

Remark Closure principles versus richer logic.
Closure principles are not necessarily the most revealing about the behavior of a knowledge operator. Consider an analogy with a more standard operator, Boolean conjunction. Taken by itself, this supports very few valid inference patterns, all of them rather trivial. But it would be a mistake to conclude from this fact that conjunction has no interesting logic. The point is just that the latter will show only in interactions with other operators, such as disjunction and negation, witness the distribution and De Morgan laws. Likewise, the full inferential power of knowledge may only show in its inferential interactions with our other modalities. As an illustration, notice that, though Conjunction fails, we do have this

Fact The formula $(K\varphi \wedge I\psi) \rightarrow K(\varphi \wedge \psi)$ is valid.

It is easy to prove this, and related principles that govern the behavior of K-claims. ■

Much more can be said about the bimodal logic of RA models and its relation to philosophical intuitions about epistemic reasoning[19]—but in this paper, our main concern is with an extension of the logical horizon to which we will turn now.

[18] This principle may fail in more general RA models than our simple setting with uniform relations.

[19] In particular, some critics see the above language, not as a natural medium for knowledge claims, but as epistemologists' 'meta-speak' about knowledge. I disagree: talking about and reasoning with

8.4 Dynamic Logic of Semantic Information and Knowledge

Epistemic Actions The system of the preceding section may seem the logical core of RA theory. However, we are not done yet. Propositional attitudes are not the only driver in the functioning of knowledge as discussed here. There is a third epistemically important component, too, that still needs to be made explicit for a complete logical picture. Our analysis so far has left out what is perhaps the most striking aspect of RA theory. Intuitively, ruling out is an *activity* that we engage in, and a task that must be performed in order to deserve the appellation of knowledge for claims that we are making. The very flavor of RA theory is therefore dynamic: current knowledge claims are based on things that we have done, or perhaps, that we still need to do. Thus the framework comes with one more crucial ingredient: this time, an explicit proposal for also modeling what are the basic epistemic actions that create or modify knowledge.[20]

In what follows, I will explore this theme of epistemic action and show where it takes us. There will be two basic activities of ruling out and relevance raising, but the point made above is much more general. There is an amazingly rich repertoire of actions that make up epistemic practices, and ruling out and relevance raising are just a small slice.

Dynamic Superstructure: Ruling Out as Model Update What dynamic superstructure fits the standard epistemic logic introduced in the preceding section? We need an explicit account of what it means to rule out a world or a set of worlds, where we assume these are defined by propositions φ. To do this we employ the framework of dynamic-epistemic logic (van Ditmarsch et al. 2007; van Benthem 2011), where actions are treated by defining the corresponding model transformation, viewed as a change in epistemic state.

A first candidate for ruling out proposition φ would be the well-known transformation of 'public announcement' $(\neg\varphi)!$ that turns any current pointed model $(\boldsymbol{M}, s)$ into its submodel $(\boldsymbol{M}|\neg\varphi, s)$ consisting of all worlds satisfying $\neg\varphi$.[21] Public announcement is natural, and it may well be the common sense model of information growth as shrinking a current range.

However, this update mechanism cannot quite work for us here, because of a point that we have already noted earlier. The spirit of RA theory makes essential reference

(Footnote 19 continued)
relevance seems a natural part of our natural language. Incidentally, if theoretical origins are viewed as a problem, then many natural language expressions are in trouble (think "objective", "subjective", "rational") as ghosts of departed philosophical theories, even though they have somehow managed to transcend this, becoming useful and well-understood members of our common vocabulary.

[20]This dynamics is not confined to RA theory, it is the main theme of my own work at the interface of logic and epistemology (van Benthem 1996, 2011; Baltag et al. 2015).

[21]This is best understood more abstractly as a form of update with hard information that could be triggered by many events: public announcement, totally reliable observation, and so on.

to worlds that have been ruled out. These should be kept around, like the vanquished in a Roman triumph. Therefore, we will work with a slight variant:

Definition Test operation
The *test* $!\alpha$ cuts all links between α-worlds and $\neg\alpha$-worlds in the current model. ■

Though not overly faithful to our operational intuitions about testing complex assertions in terms of their syntax, this simple semantic operation occurs in many places in the literature, for instance, as a way of describing what happens after performing an experiment, or when seeing a question answered.[22] However, if the reader would prefer another account of the informal notion of ruling out, the spirit of this paper is in its approach, rather than its details. The techniques to follow work for a broad variety of model-changing operations.

Dynamic Logic of Testing The test operation changes models, and hence assertions of knowledge that held before may not hold afterwards: knowledge may be acquired, or lost. It is straightforward to axiomatize this in a modal language that enriches our earlier setting with action expressions $!\alpha$, where α is any formula of the language. We add new action modalities $[!\alpha]\,\varphi$ stating that φ holds at the distinguished world of the updated model:

$$(\boldsymbol{M}, s)| = [!\alpha]\varphi \text{ iff } (!\alpha(\boldsymbol{M}), s)| = \varphi$$

In this way, we get a logic that can talk about semantic information, relevance, and RA-style knowledge, but also about actions of ruling out and what these do to the preceding notions.

To axiomatize the reasoning supported by this dynamic setting, we need to find the basic recursion laws telling us how the basic static operators change under testing. These are the driving equations of the information dynamics at work. It is an application of standard dynamic-epistemic techniques to find these laws for a simple operation like testing:

Fact The following recursion laws are valid for semantic information and relevance:

(a) $[!\alpha]I\varphi \leftrightarrow ((\alpha \wedge I(\alpha \rightarrow [!\alpha]\varphi)) \vee (\neg\alpha \wedge I(\neg\alpha \rightarrow [!\alpha]\varphi)))$
(b) $[!\alpha][\uparrow]\varphi \leftrightarrow [\uparrow][!\alpha]\varphi$

Principle (a) reflects precisely how link cutting works, while the commutation in (b) shows that link cutting does not change the relevance relation—a topic that will return later.

Theorem *The dynamic-epistemic logic of semantic information, relevance, and test is completely axiomatizable.*

---

[22]There is an issue of whether the test operation respects the constraints that we have imposed on the orderings in our models. Such technical issues make sense, but we will ignore them here.

Proof By standard techniques from the literature, we can use the recursion laws to push dynamic modalities systematically inside until they hit atoms.[23] In the latter position, we use the evident validity of the equivalence $[!\alpha]p \leftrightarrow p$ to remove them altogether. The result is a provably equivalent dynamic-free formula in the base language for which we already had a complete logic. This reduction giving the dynamics an almost free ride on the statics works for many dynamic-epistemic logics. But we will discuss a generalization to 'protocol models' later on where no such reduction is possible—though completeness remains. ■

This dynamic logic still needs unpacking to see what it does with defined modalities. In particular, using the above recursion laws, we can calculate step by step how RA-style knowledge changes under testing, or equivalently, under ruling out:

$$[!\alpha]K\varphi \leftrightarrow [!\alpha]I(\varphi \vee \langle\uparrow\rangle\neg\varphi) \leftrightarrow$$
$$((\alpha \wedge I(\alpha \rightarrow [!\alpha](\varphi \vee \langle\uparrow\rangle\neg\varphi))) \vee (\neg\alpha \wedge I(\neg\alpha \rightarrow [!\alpha](\varphi \vee \langle\uparrow\rangle\neg\varphi)))) \leftrightarrow$$
$$((\alpha \wedge I(\alpha \rightarrow ([!\alpha]\varphi \vee [!\alpha]\langle\uparrow\rangle\neg\varphi))) \vee (\neg\alpha \wedge I(\neg\alpha \rightarrow ([!\alpha]\varphi \vee [!\alpha]\langle\uparrow\rangle\neg\varphi)))) \leftrightarrow$$
$$((\alpha \wedge I(\alpha \rightarrow ([!\alpha]\varphi \vee \langle\uparrow\rangle[!\alpha]\neg\varphi))) \vee (\neg\alpha \wedge I(\neg\alpha \rightarrow ([!\alpha]\varphi \vee \langle\uparrow\rangle[!\alpha]\neg\varphi)))) \leftrightarrow$$
$$((\alpha \wedge I(\alpha \rightarrow ([!\alpha]\varphi \vee \langle\uparrow\rangle\neg[!\alpha]\varphi))) \vee (\neg\alpha \wedge I(\neg\alpha \rightarrow ([!\alpha]\varphi \vee \langle\uparrow\rangle\neg[!\alpha]\varphi))))$$

This is not a simple recursion law, and so we see that RA knowledge is not a trivial notion in a dynamic setting: something that also showed, of course, in its static inferential behavior.

However, if we focus on base-level knowledge only, given by purely propositional 'factual' formulas φ, a simplification occurs. It is easy to show that in that case:

Fact $[!\alpha]\varphi$ is equivalent to φ itself for all factual formulas φ.

Proof Ruling out does not change atomic facts in worlds, or their Boolean combinations. ■

The earlier complex recursion law for RA-knowledge then simplifies to something much closer to standard ways of conditionalizing knowledge:

$$[!\alpha]K\varphi \leftrightarrow ((\alpha \wedge I(\alpha \rightarrow (\varphi \vee \langle\uparrow\rangle\neg\varphi))) \vee (\neg\alpha \wedge I(\neg\alpha \rightarrow (\varphi \vee \langle\uparrow\rangle\neg\varphi))))$$

8.5 Epistemic Actions in General

Dynamic Perspectives on Knowledge My main interest here is not the technical dynamic machinery per se. I want to show what new themes are facilitated by this new analysis tool in a discussion about knowledge. I will consider a few topics that extend the usual focus on closure principles or other static properties that knowledge is supposed to satisfy. In one form or another, all of these themes can be found in the

[23] What we also need to proceed inside the formula are two simple valid recursion laws for the Boolean operations: $[!\alpha]\neg\varphi \leftrightarrow \neg[!\alpha]\varphi$, $[!\alpha](\varphi \wedge \psi) \leftrightarrow ([!\alpha]\varphi \wedge [!\alpha]\psi)$.

philosophical literature, but I will show that they can now be brought in to the open, and treated in an exact logical style.

What Actions Generate Knowledge? Knowledge is good to have, but how does it arise? Let us first consider this for semantic information. An immediate intuition might be this. If a proposition φ is true, then a test $!\varphi$ gives us the semantic information that φ is true:

$$\varphi \to [!\varphi]I\varphi$$

One might view this as a basic intuition, on a par with the usual closure principles. However, the situation is delicate, and in fact, the principle is not valid in general. To see this, consider an evergreen: so-called 'Moore sentences' of the form $p \wedge \neg Ip$. It may be true right now that p while I do not have this information. But after the test, I will have the information that p, and hence the conjunction has become false: a Moore sentence 'refutes itself'. Of course, there is no inconsistency here, just a surprise. The recursion laws of our dynamic logic tell us precisely what Moore sentences achieve—and $\varphi \to [!\varphi]I\ \varphi$ is definitely not one of its valid principles for all formulas φ. Admittedly, there is a price to be paid for this sophistication: for, it is the reason why proofs usually have to be sprinkled with $[!\varphi]$ modalities penetrating inside formulas. As it happens, the principle $\varphi \to [!\varphi]I\ \varphi$ does hold for special classes of statement in our language, such as factual formulas containing no I-operators at all.[24,25]

The Role of Intuitions What should we make of this? We must be careful with our intuitions: their range of validity may have hidden restrictions. For instance, common intuitions about knowledge may not hold when we move beyond factual propositions to propositions that themselves refer to knowledge and ignorance. In that case, we could view our logic as an instrument for 'bootstrapping': starting from initial intuitions, we respect them as far as they go, and then refine them for larger realms. But we may also view the extension to epistemic statements in a language that is closed under epistemic and even all dynamic modalities as logical 'system pressure', that imports irrelevant concerns. Logicians might opt for the exciting dangerous road, but does it lead anywhere important?[26]

I conclude this theme with a simpler question. Since our notion of knowledge $K\varphi$ is weaker than semantic information $I\varphi$, what act would generate it, restricting ourselves to the safe realm of empirical propositions φ? The answer is simple: we want to get $K = I(\varphi \vee <\uparrow>\neg\varphi)$ and by the preceding, a test for a weaker formula than φ will do it, viz. $!(\varphi \vee <\uparrow>\neg\varphi)$.[27]

---

[24]For a precise determination of all valid instances of our principle, see Holliday and Icard (2010).

[25]It is an interesting question if $!\varphi$ is the *minimal* action inducing $I\varphi$: I will not pursue this here.

[26]Here are two interesting technical problems in this setting. First, for every valid principle for factual formulas in this paper which possibly larger classes of modal formulas preserve its validity? But also, can we axiomatize the richer set of schematic valid principles that hold for factual formulas?

[27]Caveat: Testing for a weaker formula is not necessarily easier than testing for a stronger one.

Knowledge and Persistence Under Update Another basic aspect of knowledge, once in our possession, is what happens to it as we learn new things. RA theory gives a past-looking account: we have knowledge if we have made the relevant inquiries. But there is also a common intuition that knowledge gives us a guarantee for the future: it will persist. This, too, can be made sense of in our framework. We merely make one observation among many:

Fact The formula $K\varphi \rightarrow [!\alpha]K\varphi$ is valid for all factual φ and all α.

Proof Consider $I(\varphi \vee <\uparrow>\neg\varphi)$ and apply the test $!\alpha$. This can only diminish the number of epistemic links from the current world: and the disjunction will still be true there, since factual formulas do not change their truth values under testing. For arbitrary formulas φ, the situation is more complex, though again the recursion laws are up to their analysis. ■

The preceding observation is reminiscent of the view of knowledge as stable belief that survives arbitrary new information (Stalnaker 2006; Rott 2005; Baltag and Smets 2008). Of course, as we saw with inference, many other questions make sense, of a more mixed type. For instance, we could ask what 'upgrades' knowledge to semantic information: and the above principle in fact gives a stronger mixed principle, using veridicality of $K\varphi$:

Fact The formula $K\varphi \rightarrow [!\varphi]I\varphi$ is valid for all factual φ.

My point is not that this is a deep difficult principle. It is not. But again it reflects a basic intuition, and one could investigate (as with the earlier Moore-type problems) how far it can be pushed for more complex propositions than merely factual ones.

From Closure to Epistemic Work I conclude by returning to the closure principles that often serve to frame epistemological discussions. RA theory leads to a variety of closure failures, which tell us something important about the proposed account of knowledge. But in a dynamic setting, this is not the end of the story. A natural further question emerges. Closure principles say that, if we know certain things, then we also know other things *automatically*. While this sort of gain may suffice for the idle rich, I would submit that a natural broader accompanying question would be what an ordinary working class agent would have to *do* in order to get new knowledge from available known propositions.

Van Benthem (2011) analyzes the resulting *dynamic closure principles* in terms of a 'gap' for actions: the correct epistemic principles should not be, say, $K\varphi \rightarrow K(\varphi \vee \psi)$, but

$$K\varphi \rightarrow [act]K(\varphi \vee \psi), \text{ where } act \text{ is some appropriate epistemic action.}$$

This approach has merits for analyzing omniscience, when *act* might stand for an act of drawing an inference, or engaging in introspection. However, such scenarios of bounded rationality are ruled out here, since RA theory assumes Dretske's 'ideally astute logicians'. But even enviable beings like that may still profit from other, non-inferential actions, such as receiving new information. Here is one relevant observation:

Fact The formula $K\varphi \rightarrow [!\psi]K(\varphi \wedge \psi)$ is valid.[28]

Given that many such facts hold in a dynamic setting, a lack of standard closure principles may lose its sting, in RA theory or in even less closure-tolerant accounts of knowledge.

Digression I even feel the pull of an opposite intuition, an 'Anti-Closure Principle' that makes knowledge a supremely delicate epistemic state, whose high status is as fragile as that of Andersen's heroine in The Princess and the Pea. More concretely, on this intuition, an assertion $K\varphi$ should only imply knowledge of propositions that are equivalent to φ in the current model. In particular, this would make knowledge so delicate that it admits no weakening or strengthening at all.[29] On such a view, *all* genuine extensions of available knowledge would have to come by epistemic actions.

Summary I have shown how adding a dynamic component to epistemic logics is easy to do, and gets us a much richer view of intuitions that might hold for knowledge. But I have only illustrated this by means of a few epistemic actions, many more could be considered. So, let us see whether this proposal also works in extended settings beyond the bare minimum.

8.6 Dynamic Logic of Knowledge: Relevance

The dynamic logic defined so far is not in balance. We had two static base modalities: I for semantic information and $[R]$ for relevance, but only one of them was 'dynamified' in the sense of adding explicit test actions $!\alpha$ that affect semantic information to the logic. But relevance, too, is a dynamic entity that can change as people engage in conversation about knowledge claims. In particular, the relation R can be affected by raising an issue—the way a critic might draw attention to some overlooked class of worlds in a conversation about a knowledge claim. This seems crucial to many epistemic scenarios, such as encounters with skeptics who try to change our view of what should count as relevant possibilities. In fact, it is this dynamics of making, defending, or modifying knowledge claims that seems the most interesting aspect of many epistemic scenarios in the literature (cf. van Benthem 2006).

Again this new action can be made precise as a model transformation in several ways, and there are different ways of raising an issue, with different force. We will explore only one.

Changing Relevance Let us define an operation $+\alpha$ that transforms a current model so as to make α relevant. This operation will change the given relevance order, and it might do so in many ways. We borrow a proposal made for preference change in van Benthem and Liu (2007):

---

[28] I have not been able to find an interesting action *act* that validates $K\varphi \rightarrow [act]\ K(\varphi \vee \psi)$. But see the later reference to dynamic logics of inference, van Benthem and Velazquez-Quesada (2010).

[29] Intuitions like this place a well-defined technical constraint on the proof power of epistemic logics, be it of a somewhat unusual sort. It would be interesting to see what modal logics satisfy this.

Definition Suggestion

A *relevance upgrade* $+\alpha$ changes the relevance ordering R in the current epistemic model by deleting all upward links running from α-worlds to $\neg\alpha$-worlds.[30] ■

Dynamic Logic of Relevance This way of making α an issue is again a simple relation transformer whose complete dynamic logic can be determined using our earlier methods:

Theorem *The dynamic logic of semantic information, relevance, test, and issue raising is completely axiomatizable using recursion laws.*

Here are the basic recursion laws, showing how relevance is affected:

(c) $[+\alpha][\uparrow]\varphi \leftrightarrow ((\alpha \wedge [\uparrow](\alpha \rightarrow [+\alpha]\varphi)) \vee (\neg\alpha \wedge [\uparrow][+\alpha]\varphi)))$
(d) $[+\alpha]I\varphi \leftrightarrow I[+\alpha]\varphi$

This full system seems much more congenial to analyzing epistemic scenarios. Information can change, but so can relevance, and the laws of the logic describe how the basic static modalities are modified. Again there are also many interesting technical issues here, some of them new. For instance, there can now be an interplay of the two basic actions, and the reader may want to ponder the potentially different effects of orders $!\varphi; +\psi$ versus $+\psi; !\varphi$.

Epistemic Dynamics Once More Recall the epistemic dynamics that we discussed before. It is easy to see that all our earlier issues still make sense here. We merely consider two examples. One is persistence of knowledge under epistemic events. We first note that knowledge does not change if we raise the issue of whether its proposition is true.

Fact The formula $K\varphi \rightarrow [+\varphi]\mathrm{K}\varphi$ is valid for all factual φ.

Proof The reason is that the 'compensation' clause in our definition of RA knowledge runs inside the set of $\neg\varphi$-worlds where relevance order is unaffected by the model transformation of issue raising as defined above. Moreover, factual formulas have their truth value unaffected by issue raising, and the same is even true in the presence of I modalities. ■

But unlike the case of tests, knowledge is not preserved under arbitrary relevance changes—something well realized by all those skeptics prowling our streets looking for victims.

Example Knowledge may be lost after a suggestion…

Consider the following model with actual world 1, whose only epistemic alternatives are 1, 2, while world 3 is more relevant than 2:

[30] Another option, out of many, might be adding the set of worlds satisfying α to the current range of the relation R, creating the new relation $R \cup (\{s\} \times \alpha^{M})$.

The knowledge claim Kp is true in world 1. However, after a relevance change $+q$, there will be no upward relevance link any more from 2 to 3, and hence Kp is no longer true at 1. ■

Relevance changes make reasoning much more delicate. One way in which this shows are relations between knowledge and semantic information. For instance, in a given model $(\boldsymbol{M}, s)$, let I_s be the proposition that defines the set of all worlds that are epistemically accessible from s. This encodes the semantic information that the agent possesses. Merely making this relevant leads to the following connection between our two notions. Under the right issue change, achieving RA-style knowledge amounts to having semantic information:

Fact $I\varphi$ is equivalent with $[+I_s]K\varphi$ for factual propositions φ.

Proof By earlier observations, $I\varphi$ implies $[+I_s]I\varphi$ and hence $[+I_s]K\varphi$. Conversely, once I_s has been made relevant in our sense, no compensation links can run from it into its complement, and hence $K\varphi$ can only be true because all epistemic alternatives satisfy φ. ■

This is only one of several dynamic connections between epistemic notions, and it shows the entanglements all around. Thus, even Dretske's ideally astute logician can still learn a lot under epistemic actions that arise naturally in the RA framework.[31] And more generally, we see how the earlier picture of epistemic statics and dynamics gets even more appealing when we consider the full range of actions supported by our conceptual framework.

Summary We have defined a dynamic version of RA theory, and we have shown how this enriches the sort of logical-epistemological discussions one can have about epistemic practices.

8.7 Dynamic Logic of Knowledge: Issues

We said at the start of this paper that our style of analysis is not restricted to RA theory. This section is a digression adding one more pilot for the above style of logical analysis for epistemic attitudes and actions, now in a somewhat deviant setting going beyond the basics of what was defined above, viz. dynamic logics of inquiry and 'issue management'.[32]

Inquiry and Issues Just postulating a primitive relevance relation R seems ad-hoc: can we say more? Major RA theorists have in fact said a lot more (cf. Dretske 1981;

---

[31] What I have not investigated so far is the question whether and how relevance upgrades, in addition to tests, could help validate, or invalidate closure principles. What is easy to see, however, is that further relevance transformers in our framework have this effect. For instance, an operation that adds, rather than removes, relevance links can link a current relevant alternative to a more relevant counterexample that has already been ruled out, thereby introducing knowledge.

[32] What follows here is a brief précis of van Benthem (2012), building on van Benthem and Minica (2012).

Lewis 1996–Hawke 2014 is an investigation of a wide range of possible approaches). In this section we explore just one very simple technical idea, viz. that relevance for knowledge claims is created by a *current issue*, that can be created and modified through typical acts of raising issues, in particular, by asking *questions*. While the idea of relating knowledge to questions is well known in itself (cf. Hintikka 1973; Stalnaker 2006), the idea to be stated now seems more in the spirit of Schaffer's issue-based account of knowledge (cf. Schaffer and Szabó 2014), though our implementation also differs in its form and substance.[33]

Issue Models It is common to model an issue as a partition of the set of possible worlds (Groenendijk and Stokhof 1984; Kelly 1996). The purpose of inquiry is to find out in which cell the actual world is located. But this suggests the following sense of relevance. The issue tells us which alternatives are relevant to the claim that we know a proposition φ. We do not need to inspect all $\neg\varphi$-worlds to rule them out, we only need to rule out those alternative hypotheses (entire partition cells in the current issue relation) that contradict φ.[34]

Questions and Issue Change Epistemic models $\boldsymbol{M} = (W, \sim, \approx, V)$ with an added issue relation $\approx$ support test dynamics as before, but they also have crucial actions of issue management. The current issue may change in the course of inquiry, or a conversation. The most basic action of this sort may well be asking a Yes/No question. Its obvious dynamic transformation resembles the earlier $!\alpha$-style link cutting for the epistemic relation:

Definition Issue change by a question

The model transformation $?\,\alpha$ takes the current issue relation $\approx$ to a new relation $\approx|\alpha$ where all links between α-worlds and $\neg\alpha$-worlds have been cut.[35] ■

Again one can introduce a modal base language here,[36] and add dynamic superstructure.

Theorem *With explicit modalities added for the actions* $[!\alpha]$ *and* $[?\alpha]$, *the dynamic logic of semantic information and issues can be axiomatized completely.*

Issue-Based Knowledge A notion of knowledge close to RA theory is as what can be claimed on the basis of our semantic information *without being contradicted by a live alternative hypothesis for the current issue*. While not the only option from

---

[33] Admittedly, this is an agent- and inquiry-related approach that may not sit well with, for instance, Dretske's more realist construal of epistemic relevance relations.

[34] Issues can be much more structured than partitions, often with *priorities*. 'Priority graphs' in the style of Liu (2011) can model such issues and their more fine-grained dynamics. One can also structure the *evidence* underlying what is ruled out in similar ways, keeping a record of earlier tests, as is done in van Benthem and Pacuit (2011) in terms of modal neighborhood models.

[35] Note that this operation is just like the earlier $!\alpha$-style link cutting for the epistemic relation.

[36] A useful set of base modalities for issue models has $I\varphi$ for semantic information, $Q\varphi$ for what is true in the current cell of the issue partition, plus a modality $R\varphi$ for the intersection of $\sim$ and $\approx$, stating what holds when the issue is *resolved* in combination with what we know.

a philosophical point of view,[37] this pilot case is easy to state in logical terms. To ground it in reality, we add a proviso: if we were to resolve the current issue with our current information, the claim holds.

This intuitive description can be expressed compactly in our language by the formula

$$K\varphi := R\varphi \wedge I{<}R{>}\varphi$$

Its inferential properties include upward monotonicity, but not conjunction of known propositions. Moreover, the above dynamic logic has recursion laws that track its dynamics.

Dynamic Features Of the new dynamic phenomena that arise in this setting, we list a few. Recall our earlier idea of deserving closure through epistemic action. Consider conjunction of knowledge that failed here: $K\varphi$ and $K\psi$ together did not imply $K(\varphi \wedge \psi)$. In issue cells where φ and ψ are both possible, obviating a need for ruling them out for $K\varphi$ and $K\psi$ separately, their conjunction need not be true for any world. Thus, to know the conjunction, we have to rule out such trouble zones, and here is one action achieving this:

Fact The formula $(K\varphi \wedge K\psi) \rightarrow [!((<R>\varphi \wedge <R>\psi) \rightarrow <R>(\varphi \wedge \psi))]\, K(\varphi \wedge \psi)$ is valid.

Further interesting phenomena arise with issue change. Merely asking a question can tho-roughly reshuffle the knowledge picture as the partition cells change. This need not always achieve closure: even raising what looks like a crucial question may not help. For instance, it is easy to see that the implication $K\varphi \rightarrow [?\psi]\ K(\varphi \vee \psi)$ is not valid. Even more startlingly, our sense of knowledge is not preserved under raising the issue whether it holds:

Fact The implication $K\varphi \rightarrow [?\varphi]K\varphi$ is not valid.[38]

But a slight variation on this observation does something that is perhaps more interesting: it captures a basic notion in dynamic terms. Our next technical observation expresses the following equivalence. If knowledge in the above issue-based sense persists under all ways of raising issues, then it can only be the earlier notion of semantic information I.

Fact The principle $[?\varphi]K\varphi \leftrightarrow I\varphi$ is valid in the dynamic logic of issues.

Summary These are just sketchy illustrations,[39] but we hope that the reader has seen how the dynamics of questions and answers, too, systematically affects epistemic notions.

---

[37] The notion to follow may be closer to what an agent can reasonably claim than to what she knows.

[38] It is easy to see how this can fail. Consider any issue zone inside which the epistemically accessible worlds include both φ-worlds and $\neg\varphi$-worlds. After raising the issue whether φ, this zone splits into two, one of which now needs to be ruled out in order to have knowledge of φ by our definition.

[39] The earlier-cited van Benthem and Minica (2012) discusses many further topics, such as non-question-like acts of issue management, and temporal protocols with constraints on available acts of inquiry.

8.8 Adding Further Structure

Despite the new level of epistemic actions, our analysis is still quite austere in terms of basic epistemic notions and models. It is entirely possible to add further structure that matches the intuitive presentation of relevant alternatives theory, or other epistemic frameworks. In this section, we list a few directions that can be taken with the machinery we have set up.

Richer Core Families One can wonder whether knowledge should be studied in isolation, or whether we should start at once by identifying a richer core family of epistemic attitudes that govern our behavior.[40] One obvious candidate for this is *belief*, itself an ingredient of many proposed definitions of knowledge, but also clearly an important epistemic notion in its own right. The notions and techniques of the dynamic logics in this paper extend easily to belief, for instance, in the dynamics of plausibility models found in Baltag and Smets (2008) and van Benthem (2011). Basic actions then include softer forms of information change where no worlds are eliminated, but the plausibility order gets modified.[41,42] Plausibility dynamics might be a good way of construing natural weaker ways of ruling out relevant alternatives. Similar techniques work for other notions tied to knowledge, such as the counterfactuals used in tracking accounts of knowledge (Baltag et al. 2015).

Richer Epistemic Actions But still more important to the special focus of this paper is the choice of basic dynamics. There are many more epistemically relevant actions than testing, modifying relevance, or asking questions. Just to go one step further than we have done: possibilities can be ruled out, but later also 'ruled in'—and issues may be raised in inquiry, but it is highly beneficial that they can also be 'dropped'. If one steps back and looks at what really drives epistemic scenarios in the philosophical literature, one finds actions of detecting mistakes, repairing beliefs, raising doubts and objections, and many others. I do not have anything like a taxonomy of our basic repertoire, but it is certainly quite rich.

One particular type of action that falls outside the purely semantic scope of this paper is *inference* and reasoning in general. One might think that the valid principles of the earlier logical systems take care of this, but they are the meta-laws that govern test or issue raising, not actual acts of inferring. To deal with the subtler inferential aspects of epistemic activity, we need to move to another notion of information, more fine-grained than the possible worlds modeling we have provided. Dynamic logics of inference and other actions involving syntax and code are studied in van Benthem and Velazquez-Quesada (2010), but there are other paradigms for fine-grained epistemic

---

[40] Interestingly, Dretske (1970) itself contains many more notions than those that have made it into RA theory. In particular, he emphasizes that propositions come with 'normal criteria' determining what are their relevant alternatives. This suggests further model structure representing such criteria, but also an explicit use of normality as a sort of default assumption summarizing past experience with knowledge claims.

[41] Plausibility is even one way of reading RA relevance orderings, but of course, not the only one.

[42] For a recent example of soft plausibility changing operators on models, applied to describe reasoning with epistemic modals, see Hawke and Steinert-Threlkeld (2015).

dynamics, some closer to argumentation theory. In terms of closure principles, this is the realm where one can study what actions close knowledge under inference, or what forms of reflection validate introspection principles.[43]

Knowledge, Decision, Action One can even wonder to which extent the purely informational focus of this paper, and much of the epistemological literature, is appropriate. Many authors point out that knowledge functions in a context of action, but then drop this theme. Relevant alternatives theory itself provides an obvious angle here. Relevance as used in daily discourse seems a relational notion, since our way of construing relevant options usually depends on something else. In particular, that something else is often reflection on possible outcomes before taking a *decision* or performing an action.[44] It would be easy to add this in the above models by giving the agent a set of possible actions a that can be performed in worlds s to produce new worlds sa that can then be compared as to the agent's preferences. One way in which an alternative might be relevant then is if it taking it into account would change the optimal action that the agent would perform. In this way, justifiable knowledge claims would be tied intimately to possible action: they 'matter'.

Social Aspects A striking feature of many epistemic scenarios in the philosophical literature is their multi-agent character. Encounters with skeptics are about people getting into a discussion about knowledge claims and relevant alternatives. And in Socratic dialogues, knowledge is what one can defend against others having perhaps different information. This multi-agent character usually gets lost in logical treatments, but it should be there. Now dynamic-epistemic logic was originally designed to deal with multi-agent scenarios, so there is no obstacle in principle. We can make all orderings agent-dependent and study levels of interaction, and forms of group knowledge. I think it would be interesting to give such types of analysis for skeptical scenarios, where there are two relevance orders, and negotiation has to take pace to change either. Or perhaps differently, there is social choice involved: the two discussants may merge their individual relevance orders into an ordering for the group consisting of the persons involved (effectively, a third epistemic agent), which can then serve as a basis for their further discussion. The multi-agent aspect may also help make something visible that tends to be ignored when focusing on success features of knowledge for agents: the fact that disagreement, not assent, is often the engine of epistemic progress.[45]

Time, Protocols, and Games Another important epistemic aspect ignored in this paper is the nature of the process where knowledge functions. Epistemic actions are single steps in something bigger. There may be a procedure or *protocol* allowing us to perform certain actions, and not others—as is true in any process of inquiry,

---

[43]Closure or closure failure hardly does justice to the subtleties of legal reasoning. One of the judges in the novel Summerscale (2012), when told that he cannot let the man in an affair go free, while convicting the woman involved of having had that affair, answers that he can convict the women of having had an affair *with someone* while absolving the man of having had an affair with *her*.

[44]The view that knowledge is intimately tied to potential for action is widespread in the literature.

[45]It should be emphasized though that this social agent-dependent perspective is not that of the founding fathers of relevant alternatives theory, who see relevance as something objective in nature.

experimentation, or conversation (Hoshi 2009). A major aspect of such protocols is the regimentation of available tests: we can only rule out alternatives by some legitimate means of inquiry that is at our disposal. But a process limitation can apply equally well to questions that can be asked, or to admissible acts of changing relevance. If we encode this into our models, the laws of the dynamic logics that we had before will change in interesting ways (in particular, a static base reduction will no longer be possible), but the general style of analysis remains the same.

Another major feature of processes is their global *temporal horizon*: what is now also important epistemically is what happens in the long run. Perhaps the best known example of this is formal learning theory (Kelly 1996, 2014) and its notions of inductive knowledge in the limit. Another relevant procedural framework, combining temporal horizon with social structure, is that of game theory where strategic interactive aspects of knowledge come to the fore. The knowledge games of Lehrer (1990) and Hintikka (2007) show these at work in epistemology (Fiutek 2013; Hamami 2014). Learning theory and game theory fit well with the dynamic logics in this paper (van Benthem 2014) and add further concerns of their own.

Other Levels of Representing Information Finally, the pervasive slant of this paper has been semantic. But there are clearly many other levels for representing information that are of interest to epistemology. One richer level is that of *syntax*, proof, and combinatorial practice, which fits with major reason-based strands in epistemology, not discussed here. We have ignored this line to keep some focus in an already overloaded agenda, but cf. van Benthem (2009), van Benthem and Velazquez-Quesada (2010, 2014) for relevant studies of acts of inference with their epistemic impact for agents who can be aware of what they are doing in some sense that goes beyond classical semantic modeling. Likewise, *probability* affords richer views of agents' epistemic attitudes and their dynamics, and the fact that we have omitted this research tradition here is not to deny its importance. For a systematic view of how these various levels of representing informational attitudes and informational action are connected, we refer the reader to van Benthem (2015).

8.9 Conclusion

This paper is a sample of a particular logical style of looking at epistemology. I started with a simple model of relevant alternatives theory and then gradually added forms of epistemic dynamics. In doing so, I viewed epistemic theories as having three different components that can be evaluated separately: a conceptual framework of the proposed basic epistemic notions, a definition of knowledge, and often implicit, but crucial here: a repertoire of basic epistemic actions that constitute an epistemic practice. By providing a family of definitions, systems, and formal results, I hope to have shown how dynamic-epistemic logics can help think about all these aspects of knowledge in a coherent and suggestive manner. Finally, I have pointed at broader social and temporal dimensions where all this fits.

I see this as proposing an extended agenda for investigations at the interface of logic and epistemology, and for eliciting a wider range of philosophical intuitions about knowledge. To me, this involves cooperation between partners bringing different assets. Logic and epistemology are not the same mindset, and this can be an engine of progress. For instance, on my own agenda of intelligent interactive agency (van Benthem 2011, 2014), the core target are our rich information-driven epistemic practices, involving an array of different epistemic core notions, and different perspectives on knowledge,[46] and it is a major task of logical analysis to bring out their fruitful co-existence. But so far, most of the really innovative guiding ideas about knowledge that are around these days have come from the philosophers, and their intuitions may well come to affect how we develop epistemic logic. This paper just follows a few threads in this encounter.

The results of teaming up epistemology and logic are not entirely predictable, and hence their collaboration becomes even more alluring.

References

Baltag, A., & Smets, S. (2008). A qualitative theory of dynamic interactive belief revision. In G. Bonanno, W. van der Hoek, & M. Wooldridge (Eds.), *Texts in logic and games* (Vol. 3, pp. 9–58). Amsterdam: Amsterdam University Press.

Baltag, A., van Benthem, J., & Smets, S. (2015). *The music of knowledge: Dynamic epistemic logic meets epistemology*. Institute for Logic, Language and Computation, University of Amsterdam.

Cornelisse, I. (2011). *Context dependence of epistemic operators in dynamic evidence logic*. Master of Logic Thesis ML-2011-09. Institute for Logic, Language and Computation, University of Amsterdam.

Dretske, F. (1970). Epistemic operators. *The Journal of Philosophy*, *67*(24), 1007–1023.

Dretske, F. (1981). The pragmatic dimension of knowledge. *Philosophical Studies*, *40*, 363–378.

Fine, K. (2013). Truthmaker semantics for intuitionistic logic. *Journal of Philosophical Logic*. doi:10.1007/s10992-013-9281-7.

Fiutek, V. (2013). *Playing with knowledge and belief. Dissertation*. Institute for Logic, Language and Computation, University of Amsterdam.

Groenendijk, J., & Stokhof, M. (1984). *Studies on the semantics of questions and the pragmatics of answers*. Philosophical Institute, University of Amsterdam.

Hamami, Y. (2014). *The interrogative model of inquiry meets dynamic epistemic logic*. Brussels: Centre for Logic and Science-Philosophy. To appear in Synthese.

Hawke, P. (2015). *The problem of epistemic relevance*, Dissertation. Department of Philosophy, Stanford University.

Hawke, P., & Steinert-Threlkeld, S. (2015). *Epistemic modality and the dynamics of discourse*. Berkeley-Stanford Graduate Conference, Department of Philosophy, University of California at Berkeley.

Hawke, P. (2014). Relevant alternatives in epistemology and logic. In S. Rahman & J. Redmond (Eds.), *The dynamics of knowledge: From reasoning to epistemology and back*. Dordrecht: Springer.

Hintikka, J. (1962). *Knowledge and belief*. Ithaca, NY: Cornell University Press.

Hintikka, J. (1973). *Logic, language games and information*. Oxford: Clarendon Press.

[46]For instance, my own recent work approaches knowledge in terms of invariants and equilibria in longer-term epistemic activity (cf. van Benthem 2011; Baltag et al. 2015).

Hintikka, J. (2007). *Socratic epistemology: Explorations of knowledge-seeking by questioning*. Cambridge: Cambridge University Press.
Holliday, W. (2012a). Epistemic logic, relevant alternatives, and the dynamics of context. In *New directions in logic, language, and computation*, Springer Lecture Notes in Computer Science (Vol. 7415, pp. 109–129).
Holliday, W. (2012b). *Knowing what follows. Epistemic closure and epistemic logic*, Dissertation. Department of Philosophy, Stanford University & Dissertation DS-2012-09, ILLC, University of Amsterdam. A substantial part relevant to this paper is to appear in the Journal of Philosophical Logic under the title 'Epistemic Logic and Epistemic Closure I: Relevant Alternatives and Subjunctivism'.
Holliday, W. (2013). *From worlds to possibilities for knowledge and belief.* Berkeley: Department of Philosophy, University of California. Follow-up versions through 2015.
Holliday, W., & Icard, Th. (2010). Moorean phenomena in epistemic logic. In L. Beklemishev, V. Goranko, & V. Shehtman (Eds.), *Advances in modal logic* (Vol. 8, pp. 178–199). London: College Publications.
Hoshi, T. (2009). *Epistemic dynamics and protocol information*, Ph.D. thesis, Department of Philosophy, Stanford University (Dissertation ILLC-DS-2009-08).
Humberstone, L. (1980). From worlds to possibilities. *Journal of Philosophical Logic*, *10*, 313–339.
Kelly, K. (1996). *The logic of reliable inquiry*. Oxford: The Clarendon Press.
Kelly, K. (2014). A computational learning semantics for inductive empirical knowledge. In A. Baltag & S. Smets (Eds.), *Van Benthem outstanding contributions volume*. Dordrecht: Springer.
Lehrer, K. (1990). *Theory of knowledge*. Boulder: Westview Press.
Lewis, D. (1996). Elusive knowledge. *Australasian Journal of Philosophy*, *74*, 549–567.
Liu, F. (2011). *Dynamic logics of preference change*. Dordrecht: Springer.
Nozick, R. (1981). *Philosophical explanations*. Princeton, NJ: Princeton University Press.
Parikh, R., Taşdemir, Ç., & Witzel, A. (2013). The power of knowledge in games. *International Game Theory Review*, *15*, 1340030. (28 pages).
Rott, H. (2005). Stability, strength and sensitivity: Converting belief into knowledge. *Erkenntnis*, *61*, 327–351.
Schaffer, J., & Szabó, Z. (2014). Epistemic comparativism: A contextualist semantics for knowledge. *Philosophical Studies*, *168*, 491–543.
Shi, C. (2013). *Logic of evidence-based knowledge*. Lecture Notes in Computer Science (Vol. 8196, pp. 347–351). Berlin: Springer.
Stalnaker, R. (1985). *Inquiry*. Cambridge, MA: The MIT Press.
Stalnaker, R. (2006). On logics of knowledge and belief. *Philosophical Studies*, *128*(1), 169–199.
Summerscale, K. (2012). *Mrs Robinson's disgrace*. London: Bloomsbury.
van Benthem, J. (1996). *Exploring logical dynamics*. Stanford: CSLI Publications.
van Benthem, J. (2006). Epistemic logic and epistemology: The state of their affairs. *Philosophical Studies*, *128*, 49–76.
van Benthem, J. (2007). Rational dynamics. *International Game Theory Review*, *9*(1), 13–45.
van Benthem, J. (2009). The information in intuitionistic logic. *Synthese*, *167*(2), 251–270.
van Benthem, J. (2011). *Logical dynamics of information and interaction*. Cambridge: Cambridge University Press.
van Benthem, J. (2014). *Logic in games*. Cambridge MA: The MIT Press.
van Benthem, J. (2015). Tracking information at different levels. In K. Bimbo (Ed.), *Outstanding contributions volume for Mike Dunn*. Dordrecht: Springer.
van Benthem, J., Girard, P., & Roy, O. (2009). Everything else being equal: A modal logic for ceteris paribus preferences. *Journal of Philosophical Logic*, *38*, 83–125.
van Benthem, J., & Liu, F. (2007). Dynamic logics of preference upgrade. *Journal of Applied Non-Classical Logics*, *17*, 157–182.
van Benthem, J., & Minica, S. (2012). Toward a dynamic epistemic logic of questions. *Journal of Philosophical Logic*, *41*, 633–669.

van Benthem, J., & Pacuit, E. (2011). Dynamic logics of evidence-based beliefs. *Studia Logica, 99*, 61–92.
van Benthem, J., & Velazquez-Quesada, F. (2010). The dynamics of awareness. *Synthese (Knowledge, Rationality & Action), 177*(1), 5–27.
van Ditmarsch, H., van der Hoek, W., & Kooi, B. (2007). *Dynamic epistemic logic*. Dordrecht: Springer.
van Ormondt, P. (2012). *Finite narrative modeling, contextual dynamic semantics and elusive knowledge*. Master of Logic Thesis, Institute for Logic, Language and Computation, University of Amsterdam.
Velázquez-Quesada, F. (2014). Dynamic epistemic logic for implicit and explicit beliefs. *Journal of Logic, Language and Information, 23*(2), 107–140.
Williamson, T. (2000). *Knowledge and its limits*. Oxford: Oxford University Press.
Xu, Z. (2010). *Knowledge, evidence and relevant alternatives: A logical dynamical approach*, Dissertation. Department of Philosophy, Peking University.
Yablo, S. (2014). *Aboutness*. Princeton, NJ: Princeton University Press.

Chapter 9
The Pleasure of Gossip

Maduka Attamah, Hans van Ditmarsch, Davide Grossi and Wiebe van der Hoek

Abstract Rohit Parikh has written on levels of knowledge (Parikh and Krasucki in Sadhana 17(1):167–191, 1992). Levels of knowledge are relevant for the analysis of gossip protocols. Gossip protocols describe the dissemination of information over a network. We present some examples of epistemic gossip protocols, wherein the agents or processes communicate with each other by peer-to-peer contact (telephone calls), as in the usal gossip protocols, but wherein the decision to contact another agent is based on the calling agent's information only. This is, as far as we know, unusual in gossip protocols. In this we wish to honour Rohikh Parikh's long career and many contributions to logic and computer science.

Keywords Dynamic epistemic logic · Modal logic · Networks · Gossip protocols

9.1 Gossip Protocols

> Six friends each know a secret. They can call each other. In each call they exchange all the secrets they know. How many calls are needed for everyone to know all secrets?

We thank a reviewer for insightful comments. Hans is also affiliated to IMSc, Chennai, as research associate, and he acknowledges support from ERC project EPS 313360.

M. Attamah · D. Grossi · W. van der Hoek
Computer Science, University of Liverpool, Liverpool, UK

H. van Ditmarsch (✉)
LORIA/CNRS, University of Lorraine, Nancy, France
e-mail: hans.van-ditmarsch@loria.fr

C. Başkent et al. (eds.), *Rohit Parikh on Logic, Language and Society*,
Outstanding Contributions to Logic 11, DOI 10.1007/978-3-319-47843-2_9

We solve this for n friends, and let us work our way upwards bottom-up. For one friend, no calls need to be made, and for two friends, a single call between them is sufficient. For three friends, any two need to call each other first, then the friend who did not make a call yet needs to call either of the two who just called, and finally either of those who made the second call, call the friend not involved in that call. That makes three calls.

Let there now be four friends Amal, Bharat, Chandra, and Devi (a, b, c, d) who hold, respectively, secrets A, B, C, D. First, some terminological and modelling assumptions. We see a secret as a propositional variable such that 'knowing the secret' means knowing the truth value of that variable. 'Amal knows secret A' means that Amal knows whether A, i.e., Amal knows that A is false or Amal knows that A is true. We represent by ab a call from a to b. The informative consequences of a call (i.e., which secrets are exchanged) are independent from who initiates a call, so in that sense a call ab is the same as a call ba. We will define protocols wherein ab is allowed but not ba, but in our examples we tend to respect lexicographic order for convenience of presentation (for example, when both ab and ba are possible). Given the setting of our contribution we prefer to talk about agents instead of friends (just consider 'multi-friend epistemic logic'). However, we will continue to talk about secrets and not about propositions, propositional variables, or facts. Knowing a proposition or a fact tends to mean that you know that it is true, whereas knowing a secret tends to mean that you know whether it is true. That is what we want.

Back now to the four agents Amal, Bharat, Chandra, and Devi (a, b, c, d). The four calls ab; cd; ac; bd distribute all secrets. The underlying protocol, of which this call sequence is an execution, is as follows.

Protocol 1 (Four agents) *Any two agents make the first call; the second call is then between the remaining two agents; the third call is then between an agent who made the first call and an agent who made the second call; and the fourth call is between the two who were not chosen in the third call.*

The distribution of secrets given the four calls is as follows. The rows list the distribution of secrets after a particular call took place.

	a	b	c	d
	A	B	C	D
ab	AB	AB	C	D
cd	AB	AB	CD	CD
ac	$ABCD$	AB	$ABCD$	CD
bd	$ABCD$	$ABCD$	$ABCD$	$ABCD$

No other protocol of sequential calls solves this in four calls, and that less than four calls is insufficient to distribute all secrets. We can easily show this:

In an execution of any other protocol one of the first callers will also make the second call. So, it has to start like this.

	a	b	c	d
	A	B	C	D
ab	AB	AB	C	D
ac	ABC	AB	ABC	D
	...			

How will this continue? For the third call, let us distinguish between the case that Devi (d) is not involved and the case that she is involved. If Devi is not involved, then another call ac does not result in more information, and the cases ab and bc are symmetric. Take the first, then we get

	a	b	c	d
	A	B	C	D
ab	AB	AB	C	D
ac	ABC	AB	ABC	D
ab	ABC	ABC	ABC	D
	...			

We now have to make three more calls in order for all agents to know all secrets: for example, one by one, Devi now calls Amal, Bharat, and Chandra. That makes *six* calls altogether.

But if the third call involves Devi, there also will always remain two agents who do not know d yet. Again, two or three further calls are needed, so that we need at least *five* calls altogether.

This also demonstrates that less than four calls is insufficient to distribute all secrets, because any execution starts with either ab; ac (at least five calls to termination) or ab; cd (at least four calls to termination), modulo a permutation of agents.

For $n = 4$, $2n - 4$ calls are sufficient to distribute all the secrets. Let there now be $n > 4$ agents. Then $2n - 4$ is also sufficient. Suppose the agents are $a, b, c, d, e, f, \ldots$.

Protocol 2 (Fixed schedule) *First, a makes a call to all agents* $e, f, \ldots$ *except* b, c, d. *Then, the calls* ab, cd, ac, bd *are made (this is an execution of Protocol 1). Finally a makes, again, a call to all agents* $e, f, \ldots$ *except* b, c, d.

This adds up to $(n - 4) + 4 + (n - 4) = 2n - 4$ calls. It will be clear that all secrets are then distributed over all agents.

Let us do this for $n = 6$, such that we get $2n - 4 = 8$ calls. Given are six agents Amal, Bharat, Chandra, Devi, Ekram, and Falguni (a, b, c, d, e, f) who hold secrets A, B, C, D, E, F. Amal starts by calling Ekram and then Falguni, etc.

	a	b	c	d	e	f
	A	B	C	D	E	F
ae	AE	B	C	D	AE	F
af	AEF	B	C	D	AE	AEF
ab	$ABEF$	$ABEF$	C	D	AE	AEF
cd	$ABEF$	$ABEF$	CD	CD	AE	AEF
ac	$ABCDEF$	$ABEF$	$ABCDEF$	CD	AE	AEF
bd	$ABCDEF$	$ABCDEF$	$ABCDEF$	$ABCDEF$	AE	AEF
ae	$ABCDEF$	$ABCDEF$	$ABCDEF$	$ABCDEF$	$ABCDEF$	AEF
af	$ABCDEF$	$ABCDEF$	$ABCDEF$	$ABCDEF$	$ABCDEF$	$ABCDEF$

Less than $2n - 4$ calls are insufficient to distribute all secrets. This has been shown in Tijdeman (1971), as also related in Hedetniemi et al. (1988), Hurkens (2000).

This is not the only protocol to distribute the secrets in $2n - 4$ calls. For example, in Protocol 2 some calls are made more than once. For the depicted $n = 6$ execution, these are ae and af. The following also achieves distribution of all secrets over all agents but in *all different calls*.

	a	b	c	d	e	f
	A	B	C	D	E	F
ab	AB	AB	C	D	E	F
cd	AB	AB	CD	CD	E	F
ef	AB	AB	CD	CD	EF	EF
ac	$ABCD$	AB	$ABCD$	CD	EF	EF
de	$ABCD$	AB	$ABCD$	$CDEF$	$CDEF$	EF
af	$ABCDEF$	AB	$ABCD$	$CDEF$	$CDEF$	$ABCDEF$
bd	$ABCDEF$	$ABCDEF$	$ABCD$	$ABCDEF$	$CDEF$	$ABCDEF$
ce	$ABCDEF$	$ABCDEF$	$ABCDEF$	$ABCDEF$	$ABCDEF$	$ABCDEF$

Not all sequences of eight different calls distribute the secrets over all agents. For example, when we change the sixth call from af into bf, Amal will only know the secrets A, B, C, D after those eight calls.

Maximum number of calls If gossip is the goal, prolonging gossip is better! As long as two agents who call each other still exchange all the secrets that they know and at least one of them learns something new from the call, what is the *maximum* number of calls to distribute all secrets?

The maximum number of calls to distribute all secrets is $\binom{n}{2} = \frac{n \cdot (n-1)}{2}$. This is also the maximum number of different calls between n agents/agents. For six agents a, b, c, d, e, f the following calls can be made such that in every call at least one agent learns one secret—for convenience we generate the execution sequence in lexicographic order again.

$$ab;\ ac;\ ad;\ ae;\ af;\ bc;\ bd;\ be;\ bf;\ cd;\ ce;\ cf;\ de;\ df;\ ef$$

For four agents we get

$$ab;\ ac;\ ad;\ bc;\ bd;\ cd$$

Let us be explicit and give the detailed distribution of secrets for four agents:

	a	b	c	d
	A	B	C	D
ab	AB	AB	C	D
ac	ABC	AB	ABC	CD
ad	$ABCD$	AB	ABC	$ABCD$
bc	$ABCD$	ABC	ABC	$ABCD$
bd	$ABCD$	$ABCD$	ABC	$ABCD$
cd	$ABCD$	$ABCD$	$ABCD$	$ABCD$

Rounds of calls Instead of consecutive telephone calls wherein all secrets are exchanged between both parties, several calls between two agents might as well take place at the same time, to speed up the exchange of information. A number of simultaneous telephone calls is called a *round*. What is the minimum number of rounds to communicate all secrets between n agents? This is analyzed in an elegant one-page journal contribution Knödel (1975). The answer is: $\lceil log_2 n \rceil$ for n even, and $\lceil log_2 n \rceil + 1$ for n odd. We will not show the proof, except for the obvious case, but instead give some examples.

First consider the case that $n = 2^m$. (This is the obvious case.) Then we need exactly m rounds. In this case, let the n agents be named $1, \dots, n$. We count modulo 2^m. The first round consists of 2^{m-1} parallel calls between two agents: for $i = 1$ to $i = 2^{m-1}$, all agents $2i$ (simultaneously) call their neighbour $2i + 1$ (i.e., for future convenience, $2i + 2^1 - 1$). The second round also consists of 2^{m-1} parallel calls but now between agents that were not paired in the first round. A way to implement this, is for all agents $2i$ (simultaneously) to call agents $2i + 3$, i.e., $2i + 2^2 - 1$. (And nobody will find the line engaged!) We continue to do so m times altogether, namely until in the mth round all $2i$ (simultaneously) call $2i + 2^m - 1$. For example, for eight agents a, b, c, d, e, f, g, h (i.e., $1, 2, \dots, 8$) the three rounds are $\{ab, cd, ef, gh\}$; $\{ac, bd, eg, fh\}$; $\{ae, bf, cg, dh\}$. Let us be explicit again.

	a	b	c	d	e	f	g	h
	A	B	C	D	E	F	G	H
i	AB	AB	CD	CD	EF	EF	GH	GH
ii	$ABCD$	$ABCD$	$ABCD$	$ABCD$	$EFGH$	$EFGH$	$EFGH$	$EFGH$
iii	$ABCDEFGH$	...	...	...	...	...	...	...

When the number of agents is not a power of 2 this requires some more work, e.g., for $n = 5$ the minimal number of rounds is $\lceil log_2 5 \rceil + 1 = 4$. Such a four round parallel call sequence is: $\{ab, cd\}$; $\{ac, be\}$; $\{ae, bc\}$; $\{ad\}$. Isn't it strange that the case $n = 5$ needs more calls than the case $n = 8$? One can easily verify from the table below that less than four is indeed impossible.

	a	b	c	d	e
	A	B	C	D	E
$\{ab, cd\}$	AB	AB	CD	CD	E
$\{ac, be\}$	$ABCD$	ABE	$ABCD$	CD	ABE
$\{ae, bc\}$	$ABCDE$	$ABCDE$	$ABCDE$	CD	$ABCDE$
$\{ad\}$	$ABCDE$	$ABCDE$	$ABCDE$	$ABCDE$	$ABCDE$

Another configuration for the first two rounds starts with $\{ab, cd\}$; $\{ac, bd\}$; .... But then we even need three more rounds, and therefore five in total. A minimal completion of that is $\{ab, cd\}$; $\{ac, bd\}$; $\{ae\}$; $\{ab, ce\}$; $\{de\}$. Note that in the third round, there is nothing else to do but to make the single call between e and another agent.

Knowledge-Based Gossip Protocols So far, we assumed that the agents can coordinate their actions before making any calls. In the Fixed Schedule protocol the individual called Amal has been assigned to make the first call, and to make this call to the individual named Bharat. Then, the individual named Chandra calls the individual named Devi. And so on. A better way to define a protocol seems that, e.g., any two agents can make the first call, and then any two other agents can make the second call. So it doesn't have to be between these exact four individuals. The question then becomes if the protocol can be rephrased such that these scheduling decisions can be made by the agents themselves, at execution time, and based on what they know. There are many realistic settings wherein we cannot assume the existence of a global scheduler to assign optimal behaviour to all agents, but wherein agents (or nodes) have to figure this out by themselves: by way of what they know. Consequently, such protocols may not be those with the shortest possible executions, but merely those with the best under such more restricted circumstances.

All agents only know their own secret initially. They cannot distinguish between the other agents by their knowledge. So for the first call we can choose any two. One can easily justify that the first two callers are selected non-deterministically. One of them is simply the agent getting through before the others, in making a call, and the recipient of that call can be any other agent. But for the second call we have a problem. Now there are two agents who know two secrets, and the remaining agents only know one secret. In other words, they have different knowledge. We may pick any agent who only knows one secret to initiate a call, this choice is knowledge-based (and anyone fulfilling the condition can be chosen), and this rules out those who made the call in the first round. In our attempt to generalize the Fixed Schedule protocol, this agent now has to call another agent who only knows one secret. But the agent initiating that second call cannot choose such a one-secret-only agent *based on its knowledge*. If Chandra initiates the second call, she has no reason to prefer Devi over any other agent, if she were ignorant about who made the first call. It seems not unreasonable to assume that she only knows that she was not involved herself in that first call. That means that, from Chandra's point of view, the first call could have been between Amal and Bharat, or between Amal and Devi, or between Bharat and Devi. (And from each pair, either agent initiating the call.) She does not know which one really happened! We can also say that Chandra cannot distinguish different *histories* of calls, as in Parikh and Ramanujam (1985, 2003).

Learn New Secrets protocol Let us now consider an epistemic protocol wherein an agent makes a decision about whom to call based on its knowledge only, and such that any agent fulfilling the knowledge condition is chosen non-deterministically. This knowledge may be about the secrets that it knows, it may be about the protocol that it is engaged in, and about its call history. In the protocols we present, we only use knowledge of secrets. The agent names remain Amal, Bharat, …($a, b, \dots$), and the corresponding secrets $A, B, \dots$; but agent variables will be $x, y, \dots$ and propositional variables will be $p_x, p_y, \dots$; where p_x is the secret of agent x, and so on.

Protocol 3 (Learn new secrets) *Until all agents know all secrets: choose an agent x who does not know all secrets, let x choose an agent y whose secret it does not know, and let x call y.*

It is easy to see that this protocol will achieve the epistemic goal that everybody knows every secret. No call sequence obtained from the Fixed Schedule protocol can be obtained by the Learn New Secrets protocol, because in the final two calls (ae; af) from the Fixed Schedule protocol agent a contacts agents of which she already knows the secret. But the same information transitions can be achieved by an execution sequence of Learn New Secrets: instead of final calls ae; ef, make final calls from Ekram and Falguni to Bharat (or to Chandra, or to Devi): eb; fb. This is legal, as Ekram and Falguni do not know Bharat's secret at the time of that call. The Learn New Secrets protocol also allows for longer execution sequences than the Fixed Schedule protocol. The longest possible execution of length $n \cdot (n-1)/2$ already mentioned above is a possible execution sequence of Learn New Secrets. For example, for $n = 4$: ab; ac; ad; bc; bd; cd. One can easily show that any length of call sequence between the minimum of $2n - 4$ and the maximum of $n \cdot (n-1)/2$ can be realized by the Learn New Secrets protocol. (On the assumption that any non-deterministic choice between available callers is uniformly random, what would then be the expected length of an execution sequence? Some examples are given in Attamah et al. (2014).)

Expected and Known Information Growth protocols There are many such knowledge-based protocols. In the (not knowledge-based) Fixed Schedule protocol, Amal calls the same agents at the end as at the beginning, because she has learnt something new in the intervening calls that she wishes to tell them. This feature can also be used in a knowledge-based protocol.

The reason for Amal to call Ekram again is that Ekram will learn something new in that call, because Amal has learnt new secrets in the intervening calls. When Amal first called Ekram, Amal learnt E and Ekram learnt A. When Amal calls Ekram again, Ekram learns B, C, D, and F from Amal. Amal learns nothing from Ekram in the second call. Instead of Amal calling Ekram, the fifth call could also have resulted from Ekram calling Amal. Let us define 'p_z is learnt in the call xy' as 'before the call, x (exclusive) or y did not know whether p_z, but after the call x and y know whether p_z'. Now consider these protocols. (Below, there is no probabilistic meaning associated to 'consider x possible'. It merely means that x is in the collection of objects that cannot be ruled out.)

Protocol 4 (Expected Information Growth) *Until all agents know all secrets: choose an agent x, let x choose an agent y such that x considers it possible that there is a secret p_z that would be learnt in the call xy, and let x call y.*

Protocol 5 (Known Information Growth) *Until all agents know all secrets: choose an agent x, let x choose an agent y such that x knows that there is a secret p_z that would be learnt in the call xy, and let x call y.*

The Expected Information Growth protocol may loop and therefore termination is not guaranteed! For example, for four agents, the following is an infinite execution sequence: ab; cd; ab; cd; ab; ... (i.e., $(ab; cd)^*$ is an initial call sequence for all finite iterations). In the third call, a considers it possible that b has learnt something new, namely if the second call had been one of bc, bd, cb, db. Therefore, after ab; cd, call ab can be chosen according to the protocol. We could also say that at the moment of the third call Amal is unable to distinguish the call sequences / histories *ab*; *cd*, *ab*; *dc*, *ab*; *bc*, *ab*; *cb*, *ab*, *bd*, *ab*, *db*.

At first sight, the Expected Information Growth protocol seems to have an advantage over the Known Information Growth protocol. Maybe there are situations wherein after a certain number of calls, due to the uncertainty about who called who, no agent knows for certain that calling any other agent will result in information growth. That would cause a deadlock in the Known Information Growth protocol, from which an agent can still escape when using the Expected Information Growth protocol.

But on second sight, such a situation cannot occur. Consider any situation wherein it is not yet the case that all agents know all secrets. Then there is an agent x who does not know some secret p_y. That agent *knows* that when it calls agent y, it will learn p_y. So the knowledge condition that x knows that there is a p_z that is learnt in the call xy, is fulfilled, namely for $p_z = p_y$.

A crucial moment is when exactly the secret is chosen that the agent who initiates the call will learn. Consider the following 'de re' variation of the Known Information Growth protocol—that for the purpose of making it different should be considered de 'de dicto' Known Information Growth. A similar variation exists for the Expected Information Growth protocol. 'De re' and 'De dicto' knowledge are considered in Jamroga and Hoek (2004), van Benthem (2001), Ågotnes and Ditmarsch (2011).

Protocol 6 (Known Information Growth (de re)) *Until all agents know all secrets: choose an agent* x*, let* x *choose an agent* y *and choose a secret* p_z *such that* x *knows that* p_z *is learnt in the call* xy*, and let* x *call* y.

Consider the sequence

$$ab; cd; ac; ab$$

and let us assume that the system is synchronous: calls are made at regular intervals. Clearly, this is not a possible execution of the Learn New Secrets protocol, because two agents will never call each other twice in that protocol. But it is a possible execution of the Known Information Growth (de re) protocol. Consider the point of view of agent b, Bharat. After the initial call ab, Bharat is not involved in the two subsequent calls (as calls are at regular intervals, he *knows* that he missed two calls), and is then called again. The two intervening calls must therefore have been between a, c and d. The following histories are then possible—here, we only take into account the informational consequences and identify calls xy with yx:

$$ab; ac; ad; ab \ (i)$$
$$ab; ac; cd; ab \ (ii)$$
$$ab; ad; ac; ab \ (iii)$$
$$ab; ad; cd; ab \ (iv)$$
$$ab; cd; ac; ab \ (v)$$
$$ab; cd; ad; ab \ (vi)$$

In all of those cases b will learn new secrets from agent a in the fourth call: namely C, D in (i), C in (ii), C, D in (iii), D in (iv), C, D in (v), and C, D in (vi). Therefore, Bharat knows that there is a secret (namely C or D) that he will learn by calling Amal again. Therefore, the Known Information Growth (de dicto) protocol applies. However, Bharat does not know that he will learn secret C from calling Amal and Bharat also does not know that he will learn secret D from calling Amal. So, it is not the case that there is a secret such that Bharat knows that he will learn that secret from calling Amal. Therefore, the Known Information Growth (de re) protocol does not apply ...to Bharat. But, it applies to Amal initiating a call at that point and making the call to Bharat. Because, for Amal only the actual call sequence is considered possible, given the sequence ab; cd; ac; ab (in the third call ac, Amal learns that the second call was cd). Therefore, there is a secret that Amal knows that Bharat will learn from the fourth call, namely C (or D).

Many more variations are possible, e.g., (i) knowledge-based gossip protocols with uncertainty over the number of calls that have taken place (asynchronous systems), such that in sequence ab; cd; ac; ab Bharat learns that two intervening calls must have taken place when he is called again; (ii) knowledge-based gossip protocols for rounds of parallel calls, wherein, for five agents a, b, c, d, e, Ekram learns that two calls must be taking place in a given round when he finds every other agent engaged (so he is unable to distinguish rounds $\{ab, cd\}$, $\{ac, bd\}$, $\{ad, bc\}$), and (iii) knowledge-and-history-based gossip protocols, e.g., when Amal, after she has called someone, is not allowed to initiate the next call (this is not a knowledge condition).

Variations such as between rounds of calls and sequential calls are standard in the gossip protocol community, where we based ourselves on the combinatorial survey (Hedetniemi et al. 1988). The absence of a global scheduler is common in this community (related to network theory and signal processing), where the method to make up for this absence is to assume random scheduling instead (Kempe et al. 2003; Boyd et al. 2006). We do not know of approaches where one makes up for the absence of a scheduler by using the knowledge of the processes and let them schedule instead, i.e., knowledge-based protocols, that are based on information available in individual network nodes. It would be interesting to find out how the speed of dissemination in knowledge-based scheduling compares to that in random scheduling. It may well be that knowledge-based protocols are only computationally feasible in small networks. The cost of determining whom to call, and only then to make a call, may exceed that of just calling anyone such that eventually you learn something—doesn't this sound like real life!

9.2 Knowledge and Gossip

What sort of knowledge do the agents obtain in these protocols? This becomes interesting if we do not only consider what agents know about the secrets but also consider what they know about each other. We systematically overview knowledge in the initial state of information (wherein every agent only knows its own secret), the change of knowledge due to a call between two agents, and the knowledge conditions after termination of a protocol consisting of such calls.

We will informally use ‘agent x knows something’ ($K_x\varphi$), ‘the agents in B have general knowledge of φ’ ($E_{\mathsf{B}}\varphi$), and ‘the agents in B have common knowledge of φ’ ($C_{\mathsf{B}}\varphi$). But not imprecisely! We will merely use the English paraphrases. The formal language, the Kripke models, and the semantics of calls will only be defined in the next section.

Initial state of information We can represent the uncertainty of the agents about their secrets in a multi-agent Kripke model. In that case we consider Amal’s secret A as a proposition A of which the value is initially only known by Amal (a). For four agents a nice depiction would be already a four-dimensional model, so let us depict the one for three agents. Below, a node like 011 stands for ‘A is false and B is true and C is true’ (in the order a, b, c the digits 0 and 1 stand for the value of the propositions A, B, C, respectively).

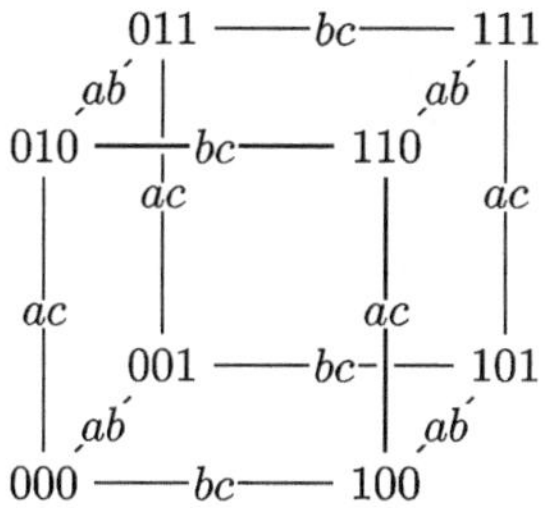

This is a most standard kind of situation. The secret of an agent is its local state, and every agent only knows its local state (and this is common knowledge). (So, it is an interpreted system (Fagin et al. 1995.) For example, in state 011 we have that Amal knows that A is false (because A is false in 011 and 111, the two states considered possible by a), and that Bharat knows that B is true and that Chandra knows that C is true.

The distributions of secrets over agents that we already considered in the previous section correspond in a precise way to such a Kripke model. We represent the distribution of secrets over agents as a list (or, if you want, as a function from agents to subsets of the set of all secrets), as we already have done in the previous section. The one above is succinctly represented by $A.B.C$. The situation $AB.AB.C$ is represented by

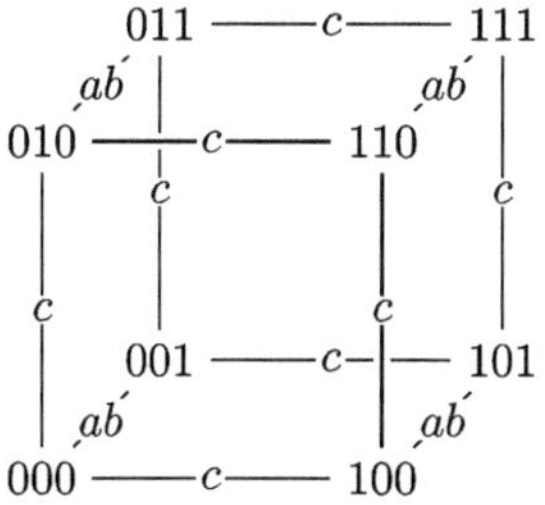

As we are only interested in whether agents know secrets, and as any agent knows the same number of secrets in any state of such a model, it is sufficient to stick to this model perspective in the logic. For example 'Amal knows whether A and whether B' is a model validity of the above model. A list like $A.B.C$ and $AB.AB.C$ we call a gossip state. In a gossip state, the agents have common knowledge of the distribution of secrets, i.e., each agent knows for all agents how many secrets those agents know, and it knows its own secrets.

Executing a phone call Let us now execute a telephone call in this setting. We get from $A.B.C$ to $AB.AB.C$ by executing the call ab. What sort of dynamics is a telephone call? A telephone call is a very different form of communication than an announcement in the presence of other agents. An announcement is *public*. This means that, after Amal says 'The old name of Chennai is Madras' in the presence of Bharat and Chandra, then Bharat knows that the old name of Chennai is Madras, but Chandra also knows that Bharat knows that, and Amal knows that Chandra knows that Bharat knows that, and so on. The information that the old name of Chennai is Madras, is common knowledge between the three agents. But if first Amal calls Bharat to tell him that, and then Bharat calls Chandra, all three know that the old name of Chennai is Madras, but it is not common knowledge. It is even impossible that this becomes common knowledge if nothing is known about the timing of the phone calls.

From the left-most information transition below it is clear that a telephone call is not public announcement. No worlds are eliminated and no links are completely cut. (We could still conceive a public announcement of 'the truth about A' as a non-deterministic announcement $!A \cup !\neg A$, resulting in cutting all bc labelled links of the initial cube.) Instead of the mere transition for the call ab we list those for the sequence ab; ac; bc.

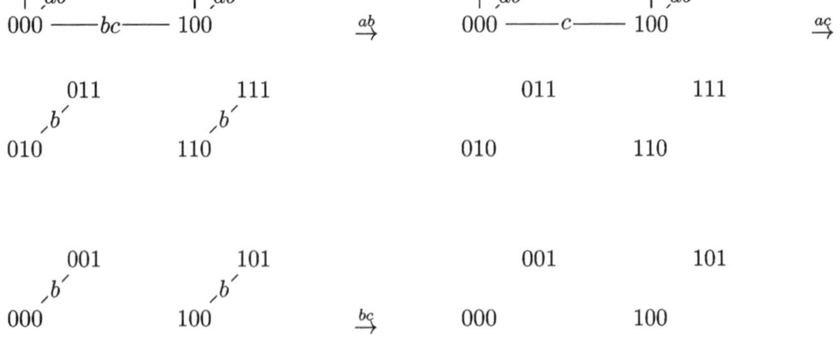

The corresponding transitions between the gossip states (the list of who knows what secrets) are as follows.[1]

$$A.B.C \stackrel{ab}{\rightarrow} AB.AB.C \stackrel{ac}{\rightarrow} ABC.AB.ABC \stackrel{bc}{\rightarrow} ABC.ABC.ABC$$

Now here is an obvious but surprising observation. Having first explained that calls do not create common knowledge of the secrets, after all, at the end, there is common knowledge that all three agents know all the secrets. We can understand this as follows: the agents have common knowledge what protocol is being carried out. In this case, this could be a Fixed Schedule protocol, but also an execution sequence of the Learn New Secrets protocol. On the assumption of synchronization, if there are three agents and the second call is ac, then Bharat knows that the call ac—or ca—is taking place, because that is the only call that he is not involved in. The agents know that after three steps all agents know all secrets. In each step there is some change in common knowledge, that finally results in common knowledge of all secrets.

For any Fixed Schedule Protocol this remains the case for more than four agents, under conditions of synchronicity. We then assume that there is common knowledge which call sequence is executed, and with what time interval between calls. We could imagine the agents sitting around a table and making the calls from there, in view of each other, but whispering, so that any other person only notices that a call is made, but not what is said.

Executing an epistemic protocol Now consider four agents, and a call between Amal and Bharat such that Chandra and Devi consider any other call possible that does not involve them. Although the real transition is $A.B.C.D \stackrel{ab}{\rightarrow} AB.AB.C.D$, Chandra considers it possible that the transition was $A.B.C.D \stackrel{ab}{\rightarrow} AD.B.C.AD$, or $A.B.C.D \stackrel{ab}{\rightarrow} A.BD.C.BD$. As we model the information change resulting from a call—which secrets are learnt—we abstract from who initiates the call and who receives it, so ab and ba are treated on a par. We now get the transition resulting in the following model.

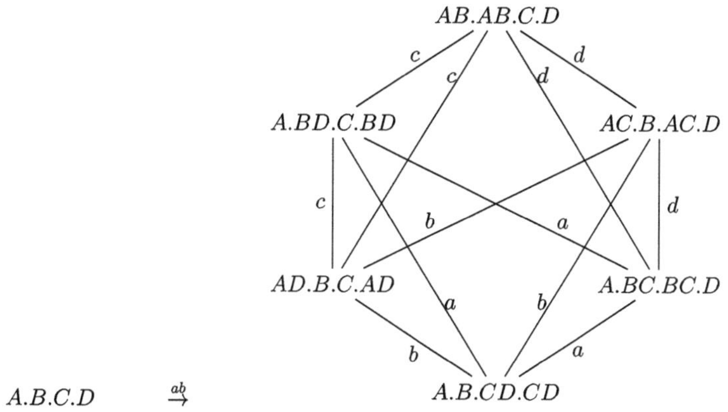

[1]We will later define a restricted logical language which justifies to use only that succinct representation, there is no need for the more complex representation.

We call this a gossip model. In fact, the unit of interpretation is the combination of the gossip model and a designated gossip state: what really happened, namely the result $AB.AB.C.D$. Such a gossip model also represents a multi-agent Kripke model, namely one wherein we replace each gossip state by the corresponding Kripke model, take their disjoint union, and add accessibility links appropriately (two worlds in gossip states are indistinguishable for an agent, even across gossip states, if it knows the same secrets in both).

Postconditions of protocol execution Clearly the agents have *general knowledge of all secrets* after the execution of such a protocol, i.e., every agent x knows the value of all secrets $A, B, \ldots$—wherein we consider such secrets as propositional variables with value true or false. Do they know more than that? This depends somewhat on further assumptions. If they do not know the protocol, not much more can be said. If we suppose that a Fixed Schedule protocol is known to the agents, the last agents to call, now know that everybody knows all secrets. This is already a bit stronger. Under such conditions we can also obtain higher order knowledge: let subsequently to the last call, an agent involved in that call, call all other agents. For example, following $ae; af; ab; cd; ac; bd; ae; af$ let a now call everybody else once more, in five calls ab, ac, ad, ae, af. We have now obtained general knowledge of general knowledge of all secrets. (And a and f have general knowledge of that.) And so on.

So, instead of $2n - 4$ calls to achieve general knowledge of the secrets, we need $2n - 4 + (n - 1) = 3n - 5$ calls to achieve general knowledge of general knowledge of the secrets. And so on... Still, it remains out of reach to make this common knowledge.

Let us additionally assume that calls take place at regular intervals, like one every 10 min, and that this is also known to the agents. In other words, we assume synchronization. Then the secrets are *common knowledge* after termination of a Fixed Schedule protocol! For example, for six agents, this will take only 1 h and 40 min.

But for epistemic protocols it is harder to achieve common knowledge. Take the Learn New Secrets protocol, and four agents. We have seen that the executions consists of between four and six calls (between $2n - 4$ and $n \cdot (n - 1)/2$). What if general knowledge is already obtained after four calls? The two agents not involved in that call do not necessarily know that. In this case, 'stuffing' the protocol with a number of a skip actions will still achieve common knowledge, after six time steps: 1 h.

9.3 Outline of Gossip Logic

Let a finite set of n *agents* (agents) $\mathsf{A} = \{a, b, \ldots\}$ and a corresponding set of *secrets* (propositional variables) $\mathsf{P} = \{A, B, \ldots\}$ be given (i.e.: A is the secret of a, B of b, etc.). We let $x, y, z, \ldots$ be variables over A, and $p_x, p_y, p_z \ldots$ the corresponding variables for their secrets (where p_x is the secret of x, etc.).

Definition 1 (*Language*) The language $\mathcal{L}$ is defined as

$$\mathcal{L} \ni \varphi ::= Kw_x p_y \mid \neg\varphi \mid (\varphi \wedge \varphi) \mid Kw_x \varphi \mid [\pi]\varphi$$
$$\pi ::= ?\varphi \mid xy \mid (\pi \,;\, \pi) \mid (\pi \cup \pi) \mid \pi^*$$

where $x, y \in \mathsf{A}$, $x \neq y$, and $p_y \in \mathsf{P}$.

Disjunction and implication are defined as usual. We read $Kw_x\varphi$ as 'agent x knows whether φ'. 'Agent x knows φ' is defined by abbreviation as $K_x\varphi := \varphi \wedge Kw_x\varphi$, and the dual of $K_x\varphi$ is written $\hat{K}_x\varphi$. (As we also have that $Kw_x\varphi$ is definable as $K_x\varphi \vee K_x\neg\varphi$, this is merely basic modal logic but with atomic formulas $Kw_x p_y$ instead of p_y.) For any $\mathsf{Q} \subseteq \mathsf{P}$, $Kw_x\mathsf{Q} := \bigwedge_{p_x \in \mathsf{Q}} Kw_x p_x$, which means that agent x knows the value of all secrets in Q. Formula $[xy]\varphi$ stands for 'after a call from agent x to y, φ (is true)'. For $(?\neg\varphi \,;\, \pi)^* \,;\, ?\varphi$ we may write 'until φ do π'. For $(?\varphi \,;\, \pi) \cup (?\neg\varphi \,;\, \pi')$ we may write 'if φ then π else π''. Epistemic protocols will be defined as such programs π but with additional constraints. Informally, a protocol is a program that intends to get all agents to know all secrets.

The *extension* of a program π is the set of its execution sequences of calls.

Definition 2 (*Extension of a program*) Defined by induction of the structure of programs. $\Sigma(xy) = \{xy\}$, $\Sigma(?\varphi) = \emptyset$, $\Sigma(\pi \,;\, \pi') = \{\sigma \,;\, \sigma' \mid \sigma \in \Sigma(\pi) \text{ and } \sigma' \in \Sigma(\pi')\}$, $\Sigma(\pi \cup \pi') = \Sigma(\pi) \cup \Sigma(\pi')$, $\Sigma(\pi^*) = \{\sigma^* \mid \sigma \in \Sigma(\pi)\}$.

A gossip state assigns a subset of all propositional variables to each agent. But not in any way, it has to be done properly. We can also see a gossip state as a partition on the set $\{0, 1\}^{\mathsf{P}}$ of valuations of propositional variables, which defines a Kripke model.

Definition 3 (*Gossip state and gossip model*) We define *gossip states* by induction. The function $g : \mathsf{A} \to \mathsf{P}$ with $g(x) = p_x$ is a gossip state. Let g be a gossip state and $x, y \in \mathsf{A}$. Consider $h : \mathsf{A} \to \mathsf{P}$ defined as: $h(x) = h(y) = g(x) \cup g(y)$, and for all $z \neq x, y$: $h(z) = g(z)$. Then h is a gossip state. We also write $g^{\{x,y\}}$ for h. For each agent, knowing the same number of secrets induces an equivalence relation on the set of gossip states: for $g(x) = g'(x)$ we write $g \approx_x g'$. A *gossip model* is a pair $G = (\mathcal{S}, \approx)$, where $\mathcal{S}$ is a set of gossip states and for each agent x, $\approx_x$ is such an equivalence relation. A *pointed gossip model* is a pair (G, g), where $g \in \mathcal{S}$.

The *initial gossip state* is the gossip state g with $g(x) = p_x$ for all agents. The *initial gossip model* is the singleton gossip model $((\{g\}, \approx), g)$ for the initial gossip state g, where $\approx_x = \{(g, g)\}$ for all agents.

Given a set of agents and propositional variables, a multi-agent Kripke model is a triple (S, R, V) (with R a function assigning to each agent a binary relation on S, and V a function assigning to each state in S a valuation of propositional variables; for $R(x)$ we write R_x and for $V(p)$, for such a propositional variable, we write V_p). If all accessibility relations R_x are equivalence relations $\sim_x$ we call this an *epistemic model*. A gossip state and a gossip model correspond to an epistemic model for the set of agents A and the set of propositional variables P. This is how to construct these epistemic models.

Every gossip state $g : \mathsf{A} \to \mathcal{P}(\mathsf{P})$ induces an epistemic model $M(g) = (S, \sim, V)$ such that $S = \{0, 1\}^{\mathsf{P}}$, where for every $x \in \mathsf{A}$, $s \sim_x t$ iff for all $p_y \in g(x)$, $s(p_y) = t(p_y)$, and where valuation $V : S \to \mathsf{P} \to \{0, 1\}$ is such that $V_s = s$. (As the domain consists of the set of all valuations it is superfluous to list the valuation function separately, as we do.)

Every gossip model $G = (\mathcal{S}, \approx)$ induces a multi-pointed epistemic model $M(G) = (S', \sim', V')$ such that: $S' = \{s_g | g \in \mathcal{S}$ and $s \in M(g)\}$; $s_g \sim'_x t_h$ iff $g \approx_x h$ and, in $g, s \sim_x t$; and $V'_{s_g} = s$ (observe that s and t are valuations and that the domain of any gossip state consists of all valuations). To a pointed gossip model (G, g) corresponds a multi-pointed epistemic model $(M(G), M(g))$.

Definition 4 (*Semantics on gossip models*) Let $G = (\mathcal{S}, \approx)$ be a gossip model. We inductively define the interpretation of a formula $\varphi \in \mathcal{L}$ on a gossip state $g \in \mathcal{S}$.

$$
\begin{array}{ll}
G, g \models Kw_x p_y & \textit{iff } p_y \in g(x) \\
G, g \models \neg\varphi & \textit{iff } G, g \not\models \varphi \\
G, g \models \varphi \wedge \psi & \textit{iff } G, g \models \varphi \text{ and } G, g \models \psi \\
G, g \models Kw_x\varphi & \textit{iff } \text{for every } h \approx_x g : G, h \models \varphi \text{ iff } G, g \models \varphi \\
G, g \models [\pi]\psi & \textit{iff } \text{for all } (G', g') \text{ such that } (G, g)[\![\pi]\!](G', g') : G', g' \models \psi
\end{array}
$$

where

$$
\begin{array}{ll}
[\![xy]\!] & = \{((G, g), (G^{\mathsf{call}}, g^{\{x,y\}}))\} \\
[\![?\varphi]\!] & = \{((G, g), (G, g))\} \text{ whenever } G, g \models \varphi \\
[\![\pi; \pi']\!] & = [\![\pi]\!] \cdot [\![\pi']\!] \\
[\![\pi \cup \pi']\!] & = [\![\pi]\!] \cup [\![\pi']\!] \\
[\![\pi^*]\!] & = [\![\pi]\!]^*
\end{array}
$$

and where $G^{\mathsf{call}} = (\mathcal{S}', \approx')$ such that $\mathcal{S}' = \{g^{\{z,w\}} \mid g \in \mathcal{S}$ and $z \neq w \in \mathsf{A}\}$, and for any $g', h' \in \mathcal{S}$: $g' \approx_z h'$ iff $g'(z) = h'(z)$.

The advantage of a logical language wherein $Kw_x p_y$ is an atomic proposition instead of the usual propositional variable p_y, now appears from the semantics: this less expressive logic allows us to have gossip models as primitive semantic objects instead of (more complex) epistemic models. Anything we do in gossip models can still be done at the level of epistemic models (but not vice versa, for example, we cannot evaluate p_y in a gossip model).

Instead of the semantics of $Kw_x p_y$ as given, we could alternatively have defined it more closely resembling the usual semantics of 'knowing whether', on epistemic models. Let $M(g) = (S, \sim, V)$, then

$$M(G), M(g) \models Kw_x p_y \text{ iff for all } s \in S, \text{ for all } t \sim_x s : V_s(p_y) = V_t(p_y).$$

Instead of the semantics of $Kw_x\varphi$ as given, we could alternatively have defined knowing that φ, or have stipulated $K_x\varphi$ as a primitive in the language. They are interdefinable: $Kw_x\varphi \leftrightarrow (K_x\varphi \vee K_x\neg\varphi)$, and $K_x\varphi \leftrightarrow (\varphi \wedge Kw_x\varphi)$. With $Kw_x p_y$ as an atomic proposition it seemed more elegant to have $Kw_x\varphi$ as a primitive construct than $K_x\varphi$. For $K_x\varphi$ we get:

$$G, g \models K_x\varphi \text{ iff for every } h \approx_x g : G, h \models \varphi.$$

When two agents call each other in a given gossip state, then instead of taking the union of the secrets they know, we can alternatively take the intersection of the equivalence relations for those agents in the induced epistemic models: given a gossip state g with induced epistemic model $M(g) = (S, \sim, V)$, we have that $M(g^{\{x,y\}}) = (S, \sim', V)$ with $\sim'_x = \sim'_y = \sim_x \cap \sim_y$ and for all $z \neq x, y$, $\sim'_z = \sim_z$.

The informative effect of x calling y is the same as the informative effect of y calling x. In other words, $[\![xy]\!] = [\![yx]\!]$. For that reason we write $g^{\{x,y\}}$ and not g^{xy} for the gossip state resulting from a call between x and y in g.

Although a gossip state induces an epistemic model consisting of different states (worlds), in our logic we never need to evaluate formulas in a given state. This is because the basic formulas are expressions $Kw_x p_y$ that are either valid on such an epistemic model, or their negation is valid.

We now define epistemic gossip protocols as programs $\pi \in \mathcal{L}$ satisfying some additional constraints. If a formula ψ contains a subformula $Kw_x\varphi$ or $Kw_y p_x$ we say that x *occurs in* ψ. We write $\psi(x)$ to denote all (possibly 0) occurrences of x in ψ.

Definition 5 (*Epistemic gossip protocol*) An *epistemic gossip protocol* is program π of form

$$\texttt{until} \bigwedge_{x \in \mathsf{A}} Kw_x \mathsf{P} \ \texttt{do} \bigcup_{x,y \in \mathsf{A}} (?K_x\psi(y) \ ; \ xy),$$

where $\psi(y) \in \mathcal{L}$.

An assumption that the agents know the protocol needs to be hardwired in the semantics of a call, by restricting the construction of the gossip model G^{call} resulting from a call xy. Instead of the $G^{\mathsf{call}} = (\mathcal{S}', \approx')$ with $\mathcal{S}' = \{g^{\{z,w\}} \mid g \in \mathcal{S} \text{ and } z \neq w \in \mathsf{A}\}$ we then need a $\mathcal{S}' = \{g^{\{z,w\}} \mid g \in \mathcal{S} \text{ and } zw \in \pi(g)\}$, where $\pi(g)$ is the set of calls possible in g according to π (defined in the obvious way). Details are omitted.

These are some epistemic gossip protocols, as they have already featured informally in prior sections. In their description we implicitly use the validities $\neg Kw_x\varphi \leftrightarrow K_x\neg Kw_x\varphi$ and $\hat{K}_x\varphi \leftrightarrow K_x\hat{K}_x\varphi$: these are really knowledge conditions. In all protocols below there is a single occurrence of the agent variable y in the knowledge condition $?K_x\psi(y)$.

Definition 6 (*Knowledge-based Protocols in* $\mathcal{L}$) Below, ∇ denotes exclusive disjunction.

- $\texttt{until} \bigwedge_{x \in \mathsf{A}} Kw_x\mathsf{P} \ \texttt{do} \bigcup_{x,y \in \mathsf{A}}(?\neg Kw_x p_y \ ; \ xy)$
  Learn New Secrets (Protocol 3)
- $\texttt{until} \bigwedge_{x \in \mathsf{A}} Kw_x\mathsf{P} \ \texttt{do} \bigcup_{x,y \in \mathsf{A}}(?\hat{K}_x \bigvee_{p_z \in \mathsf{P}}(Kw_x p_z \nabla Kw_y p_z) \ ; \ xy)$
  Expected Information Growth (Protocol 4)

- $\texttt{until} \bigwedge_{x \in \mathsf{A}} Kw_x \mathsf{P}\ \texttt{do} \bigcup_{x,y \in \mathsf{A}} (?K_x \bigvee_{p_z \in \mathsf{P}} (Kw_x p_z \nabla Kw_y p_z)\ ;\ xy)$
  Known Information Growth (de dicto) (Protocol 5)
- $\texttt{until} \bigwedge_{x \in \mathsf{A}} Kw_x \mathsf{P}\ \texttt{do} \bigcup_{x,y,z \in \mathsf{A}} (?K_x (Kw_x p_z \nabla Kw_y p_z)\ ;\ xy)$
  Known Information Growth (de re) (Protocol 6)

Proposition 7 (Protocol 3 $\subset$ Protocol 5 $\subset$ Protocol 4) *Every execution (call sequence) of the Learn New Secrets protocol (Protocol* 3*) is an execution of the Known Information Growth protocol (Protocol* 5*), and every execution of the Known Information Growth protocol (Protocol* 5*) is an execution of the Expected Information Growth protocol (Protocol* 4*). The inclusion is strict.*

We recall that the Expected Information Growth protocol (Protocol 4) has executions of infinite length.

Proposition 8 (Protocol 6 $\subset$ Protocol 5) *Every execution (call sequence) of the* (de dicto*) Known Information Growth protocol (Protocol* 5*) is an execution of the* de re *Known Information Growth protocol (Protocol* 6*).*

Not surprisingly, in this logic of knowledge and change of knowledge, the dynamics of calls in gossip logic can be described in other, more general, dynamic epistemic logics. Gossip calls have been modelled in van Ditmarsch (2000, Section 6.6) in a dialect of dynamic epistemic logic, they have been modelled in a logic of messages proposed in Sietsma (2012), and a modelling in action model logic is found in Attamah et al. (2014).

We recall the difference between the *extension* $\Sigma(\pi)$ of a protocol according to Definition 1, and the *meaning* $[\![\pi]\!]$ of a protocol, according to Definition 4. The extension is the number of its executions, the number of branches of the execution tree. The meaning is the number of different information transitions, i.e., the number of branches of the tree induced by the initial gossip model and π.[2] For example, call sequences *ab*; *ac*; *bc* and *ab*; *ac*; *ab* are different in the extension, but have the same meaning: in both cases they induce the transitions $A.B.C \rightarrow AB.AB.C \rightarrow ABC.AB.ABC \rightarrow ABC.ABC.ABC$.

For example, for the Known Information Growth protocol, for three agents, the extension consists of 96 different call sequences, and the meaning consists of 6 different transformation chains of gossip models. For the Learn New Secrets protocol, these figures for three agents are 24, and 6. The extension of a protocol allows to determine its average execution length (as a function of the number of agents), and that would allow us to compare different epistemic gossip protocols. Surely, we are interested in a minimum expected execution length.

It is interesting for all agents to know all secrets. But for a given agent, it may even be more interesting to be *the first* to know all secrets. Epistemic gossip protocols may also play a role in 'gossip games', say, in a setting similar to some of Rohit Parikh's work, such as Parikh et al. (2013), Chopra et al. (2004).

---

[2]In other words, the meaning is the set of pointed gossip model sequences, where the points are the actual gossip states. Note that this is different from a set of gossip state sequences. Different pointed gossip model sequences may have the same induced gossip state sequence; i.e., if you just look at their points, they are the same.

References

Ågotnes, T., & van Ditmarsch, H. (2011). What will they say?—Public announcement games. *Synthese*, *179*(S.1), 57–85.

Attamah, M., van Ditmarsch, H., Grossi, D., & van der Hoek, W. (2014). Knowledge and gossip. In *Proceedings of the 21st ECAI* (pp. 21–26). IOS Press.

Boyd, S., Ghosh, A., Prabhakar, B., & Shah, D. (2006). Randomized gossip algorithms. *IEEE/ACM Transactions on Networks*, *14*(SI), 2508–2530.

Chopra, S., Pacuit, E., & Parikh, R. (2004). Knowledge-theoretic properties of strategic voting. In *Proceedings of the 9th JELIA, LNCS 3229* (pp. 18–30).

Fagin, R., Halpern, J. Y., Moses, Y., & Vardi, M. Y. (1995). *Reasoning about knowledge*. Cambridge, MA: MIT Press.

Hedetniemi, S. M., Hedetniemi, S. T., & Liestman, A. L. (1988). A survey of gossiping and broadcasting in communication networks. *Networks*, *18*, 319–349.

Hurkens, C. A. J. (2000). Spreading gossip efficiently. *Nieuw Archief voor Wiskunde*, *5/1*(2), 208–210.

Jamroga, W., & van der Hoek, W. (2004). Agents that know how to play. *Fundamenta Informaticae*, *63*, 185–219.

Kempe, D., Dobra, A., & Gehrke, J. (2003). Gossip-based computation of aggregate information. In *Proceedings of the 44th FOCS* (pp. 482–491). IEEE Computer Society.

Knödel, W. (1975). New gossips and telephones. *Discrete Mathematics*, *13*, 95.

Parikh, R., & Krasucki, P. (1992). Levels of knowledge in distributed systems. *Sadhana*, *17*(1), 167–191.

Parikh, R., & Ramanujam, R. (1985). Distributed processing and the logic of knowledge. In *Logic of programs, LNCS 193* (pp. 256–268). Berlin: Springer. Similar to *JoLLI 12*, 453–467, 2003.

Parikh, R., & Ramanujam, R. (2003). A knowledge based semantics of messages. *Journal of Logic, Language and Information*, *12*, 453–467.

Parikh, R., Tasdemir, C., & Witzel, A. (2013). The power of knowledge in games. *IGTR*, 15(4).

Sietsma, F. (2012). *Logics of communication and knowledge*. PhD thesis, University of Amsterdam. ILLC Dissertation Series DS-2012-11.

Tijdeman, R. (1971). On a telephone problem. *Nieuw Archief voor Wiskunde*, *3*(19), 188–192.

van Benthem, J. (2001). Games in dynamic epistemic logic. *Bulletin of Economic Research*, *53*(4), 219–248.

van Ditmarsch, H. (2000). *Knowledge games*. PhD thesis, University of Groningen. ILLC Dissertation Series DS-2000-06.

Chapter 10
Modeling Monty Hall in If Logic

Gabriel Sandu and Silviu Velica

Abstract We provide a game-theoretical solution to the Monty Hall problem, and its representation in IF-logic, a system of logic which is an extension of first-order logic with Independent Quantifiers. This solution has been originally proposed by Allen Mann, but here we focus more on the game-theoretical aspects and the philosophical significance.

Keywords Monty Hall · IF-logic · Game-theoretical semantics

10.1 Monty Hall: Formulation of the Problem

Let us start by considering two formulations of the Monty Hall problem (Grinstead and Snell, *Introduction to Probabilities*).

The first formulation is:

> Suppose you are on a Monty Hall's *Let's Make a Deal*! You are given the choice of three doors, behind one door is a car, the others goats. You pick up a door, say 1, Monty Hall opens another door, say 3, which has a goat. Monty says to you "Do you want to pick door 2?" Is it to your advantage to switch your choice of doors? (nd Snell, *Introduction to Probabilities*, Example 4.6, p. 136)

The second formulation is more general:

> We say that C is using the "stay" strategy if she picks a door, and, if offered a chance to switch to another door, declines to do so (i.e., he stays with his original choice). Similarly, we say that C is using the "switch" strategy if he picks a door, and, if offered a chance to switch to another door, takes the offer. Now suppose that C decides in advance to play the "stay" strategy. Her only action in this case is to pick a door (and decline an invitation to switch, if one is offered). What is the probability that she wins a car? The same question can be asked about the "switch" strategy. (Idem, p. 137)

G. Sandu (✉) · S. Velica
Helsinki, Finland
e-mail: sandu@mappi.helsinki.fi

C. Başkent et al. (eds.), *Rohit Parikh on Logic, Language and Society*,
Outstanding Contributions to Logic 11, DOI 10.1007/978-3-319-47843-2_10

Grinstead and Snell remark that the first formulation of the problem "asks for the conditional probability that C wins if she switches doors, *given that* she has chosen door 1 and that Monty Hall has chosen door 3" whereas the second formulation is about the comparative probabilities of two kinds of strategies for C, the "switch" strategy and the "stay" strategy. They point out that using the "stay" strategy, the contestant C will win the car with probability 1/3, since 1/3 of the time the door he picks will have the car behind it. But if *C* plays the "switch" strategy, then he will win whenever the door he originally picked does not have the car behind it, which happens 2/3 of the time. (Idem, p. 13).

Grinstead and Snell give a solution to the first problem using conditional probabilities. van Benthem (2003) gives a solution to the same problem in terms of product updates and probabilites.

Sandu (2015b) a modelling of the Monty Hall problem is given using a two player win-lose extensive game of imperfect information. The idea comes from Mann (2010). The emphasis in that paper is on the comparison between different solutions to the Monty Hall problem: on one side the solutions to the first problem above in terms of conditional probabilities and product updates in dynamic logic, and the game-theoretical solution to the second problem, on the other.

The emphasis of the present paper is on the game-theoretical solution to the second problem. We consider several variants of the second problem and model them in IF logic endowed with a probabilistic semantics. The modelization is essentially the same one as in Sandu (2015b) but the overall purpose of the paper is to offer a justification for probabilistic IF logic.

10.2 Monty Hall: A Game-Theoretical Modelization

10.2.1 Monty Hall as an Extensive Game of Imperfect Information

The second formulation of the Monty Hall problem mentioned above refers to strategies. This makes it natural to bring in game-theoretical conceptualizations. We think of Monty Hall and the Contestant as two players, the latter, call her C, trying to identify the door with the prize, whereas her opponent, call him MH, tries to deceive her. The tree which constitutes the extensive form of the game has maximal histories (plays of the game) of the form (x, y, z, t) where:

- x represent the door where MH hides the prize
- y is the door chosen by C
- z is the door opened by MH
- t is the second choice of C

The rules of the game dictate that

- z must be distinct from x and y, and
- t must be distinct from z.

Thus the sequence $(1, 1, 2, 1)$ represents the possible play:

> MH hides the prize behind door 1; C makes a guess: door 1; MH opens the empty door 2; C guesses again: door 1.

There are 24 plays (maximal histories). The specification of the player function P which associates with each history other than maximal ones the player whose turn is to move should be quite obvious.

Some of the histories are indistinguishable (informationally equivalent). This holds only for those histories where player C is to move:

C1 Any histories (x) and (x') are equivalent for player C.
C2 Any histories (x, y, z) and (x', y', z') where $y = y'$ and $z = z'$ are equivalent for player C.

(C1) expresses the fact that C does not know the door where the prize is hidden when making her first guess. And (C2) epresses the fact that she does not know the door where the prize is hidden, when she makes her second guess.

Each play is winning either for MH or for C. C wins (and MH looses) every play (x, y, z, t) in which she identifies the door which hides the prize, i.e. $t = x$. MH wins (and C looses) the remaining ones.

This ends up the description of the game. The result is a 2 player, win-lose finite extensive game of imperfect information.

10.2.2 Strategies

A strategy for player C is any function F which gives her a choice for any history where she is to move. Thus F will give her, for every choice of x by MH, a value for y, and for every sequence (x, y, z), a value for t. Imperfect information will impose the following restriction (uniformity) on any strategy F:

- If the histories h and h' are equivalent, then $F(h) = F(h')$.

We prefer to decompose any F into two "local" strategies, that is, two functions f and f' such that f yields a value for y and f' yields a value for t. Given (C1), f will be have to be a constant function, that is, a door y. Given (C2), f' will take only y and z as arguments.

Thus player C's set S_C of strategies will consist of pairs (i, h_i), where i stands for a door and h_i for a function of two arguments (y, z). A strategy (i, h_i) is winning if C wins every play where she follows it. We focus on two kinds of strategies for player C.

- The first kind, denoted byS_C^{Stay}, groups together all the "stay" strategies, where C choose a door, and then stick to her initial guess no matter what MH does:

$$S_C^{Stay} = \{(i, h_i) : i = 1, 2, 3\},$$

where

$$h_i(y, z) = i$$

Each strategy (i, h_i) is followed in the play

$$(x, i, z, h_i(i, z))$$

for any x and z. There are three "stay" strategies:

$$\begin{array}{ll} h_1(1,2) = 1 & h_1(1,3) = 1 \\ h_2(2,3) = 2 & h_2(2,1) = 2 \\ h_3(3,2) = 3 & h_3(3,1) = 3 \end{array}$$

It should be obvious that each "stay" strategy is winning all the plays where C's initial guess is correct (i.e., $i = x$) and losing the remaining ones.

- The second kind, denoted by S_C^{Switch} groups together all the "switch" strategies, where C choose a door, and then, after MH opens a door, switch doors:

$$S_C^{Switch} = \{(1, f_1), (2, f_2), (3, f_3)\}$$

where

$$\begin{array}{ll} f_1(1,2) = 3 & f_1(1,3) = 2 \\ f_2(2,3) = 1 & f_2(2,1) = 3 \\ f_3(3,2) = 1 & f_3(3,1) = 2 \end{array}$$

Given that there are only three doors, each of the three strategies wins all the plays in which the initial choice is incorrect, $i \neq x$, and loses the remaining ones. For illustration, suppose C follows the strategy $(2, f_2)$. Suppose MH chooses (hides the prize behind) door 1. C's first choice (guess) is door 2. By the rules of the game, MH's next choice can be only door 3. C's final choice is $f_2(2, 3) = 1$. C wins.

MH's strategies consist also of pairs (j, g): the first corresponds to a value for x; the function g associates to each argument (x, y) a value for z.

The only strategy available to MH (given the rules of the game) is: "hide the prize behind a door, and after C chooses a door, open any other door". Here are all the strategy pairs she can follow:

$(1, g_1)$:	$g_1(1, 1) = 2$	$g_1(1, 2) = 3$	$g_1(1, 3) = 2$
$(1, g_1')$:	$g_1'(1, 1) = 3$	$g_1'(1, 2) = 3$	$g_1'(1, 3) = 2$
$(2, g_2)$:	$g_2(2, 1) = 3$	$g_2(2, 2) = 1$	$g_2(2, 3) = 1$
$(2, g_2')$:	$g_2'(2, 1) = 3$	$g_2'(2, 2) = 3$	$g_2'(2, 3) = 1$
$(3, g_3)$:	$g_3(3, 1) = 2$	$g_3(3, 2) = 1$	$g_3(3, 3) = 1$
$(3, g_3')$:	$g_3'(3, 1) = 2$	$g_3'(3, 2) = 1$	$g_3'(3, 3) = 2$

Each of the strategy pair (j, g_j) is followed in every play of the form

$$(j, y, g_j(j, y), t)$$

for any y and t. It is winning whenever $j \neq t$ and losing otherwise. None of these strategies is winning all the plays in which it is followed.

Let us get more dynamics into the game. Let S_{MH} denote the set of strategies of MH and S_C the set of strategies of player C. Whenever MH follows one of his strategies in S_{MH}, and$C$ follows one of her strategies in S_C, a play of the extensive game is generated which is a win for either one of the players. For instance, when MH follows $(3, g_3)$ and C follows $(1, h_1)$, the result is the play $(3, 1, 2, 1)$ which is a win for MH.

The following table registers the payoffs of the players for all the strategy pairs the players might play:

	$(1, g_1)$	$(1, g_1')$	$(2, g_2)$	$(2, g_2')$	$(3, g_3)$	$(3, g_3')$
$(1, h_1)$	$(1, 0)$	$(1, 0)$	$(0, 1)$	$(0, 1)$	$(0, 1)$	$(0, 1)$
$(2, h_2)$	$(0, 1)$	$(0, 1)$	$(1, 0)$	$(1, 0)$	$(0, 1)$	$(0, 1)$
$(3, h_3)$	$(0, 1)$	$(0, 1)$	$(0, 1)$	$(0, 1)$	$(1, 0)$	$(1, 0)$
$(1, f_1)$	$(0, 1)$	$(0, 1)$	$(1, 0)$	$(1, 0)$	$(1, 0)$	$(1, 0)$
$(2, f_2)$	$(1, 0)$	$(1, 0)$	$(0, 1)$	$(0, 1)$	$(1, 0)$	$(1, 0)$
$(3, f_3)$	$(1, 0)$	$(1, 0)$	$(1, 0)$	$(1, 0)$	$(0, 1)$	$(0, 1)$

We shall call the game described by the matrix the Monty Hall game. It is a finite, two players, win-lose *strategic* game

$$\Gamma_{MH} = (S_C, S_{MH}, u_C, u_{MH})$$

where u_{MH} and u_C are the payoffs of the two players as shown in the matrix.

We can now try to "solve" the game. A natural solution concept to be applied is that of equilibrium. But there is none in the Monty Hall game and we must conclude that the game is indeterminate.

10.3 Variants of the Monty Hall Game

We could consider a variant of the Monty Hall game in which the use of the "switch" strategy is built into the winning conditions of the game. The extensive game will be exactly as above, except that the plays (x, y, z, t) won by C are exactly those for which $x = t$ and $t \neq y$. That is, compared to the previous game, C wins less plays and MH wins more in the present variant of the Monty Hall game. The sets S_C, S_{MH} of strategies of the two players are exactly the same as in the previous game. However, it is not difficult to see that the "switch" strategies win the same plays as before, whereas the "stay" strategies never win:

	$(1, g_1)$	$(1, g_1')$	$(2, g_2)$	$(2, g_2')$	$(3, g_3)$	$(3, g_3')$
$(1, h_1)$	(0, 1)	(0, 1)	(0, 1)	(0, 1)	(0, 1)	(0, 1)
$(2, h_2)$	(0, 1)	(0, 1)	(0, 1)	(0, 1)	(0, 1)	(0, 1)
$(3, h_3)$	(0, 1)	(0, 1)	(0, 1)	(0, 1)	(0, 1)	(0, 1)
$(1, f_1)$	(0, 1)	(0, 1)	(1, 0)	(1, 0)	(1, 0)	(1, 0)
$(2, f_2)$	(1, 0)	(1, 0)	(0, 1)	(0, 1)	(1, 0)	(1, 0)
$(3, f_3)$	(1, 0)	(1, 0)	(1, 0)	(1, 0)	(0, 1)	(0, 1)

Like the previous game, this game, let us call it the first variant of the MH game, is also indeterminate.

Finally, we could consider a second variant of the Monty Hall game in which the use of the "stay" strategy is built into the winning conditions of the game. The extensive game will be exactly as in the two earlier games, except that the plays (x, y, z, t) won by C are such that $x = t$ and $t = y$. The strategies of the two players are exactly the same as in the previous games. However, in this case the "switch" strategies never win whereas the "stay" strategies win the same plays as in the Monty Hall game:

	$(1, g_1)$	$(1, g_1')$	$(2, g_2)$	$(2, g_2')$	$(3, g_3)$	$(3, g_3')$
$(1, h_1)$	(1, 0)	(1, 0)	(0, 1)	(0, 1)	(0, 1)	(0, 1)
$(2, h_2)$	(0, 1)	(0, 1)	(1, 0)	(1, 0)	(0, 1)	(0, 1)
$(3, h_3)$	(0, 1)	(0, 1)	(0, 1)	(0, 1)	(1, 0)	(1, 0)
$(1, f_1)$	(0, 1)	(0, 1)	(0, 1)	(0, 1)	(0, 1)	(0, 1)
$(2, f_2)$	(0, 1)	(0, 1)	(0, 1)	(0, 1)	(0, 1)	(0, 1)
$(3, f_3)$	(0, 1)	(0, 1)	(0, 1)	(0, 1)	(0, 1)	(0, 1)

This game is indeterminate too.

10.4 Logics with Imperfect Information: Independence-Friendly Logic (If Logic)

Hintikka and Sandu (1989) introduce Independence-Friendly logic (IF logic) in order to express more patterns of dependencies (and independencies) of quantifiers than those allowed by ordinary first-order logic. The syntax of IF logic contains quantifers of the form $(\exists x/W)$, and $(\forall x/W)$ where W is a finite set of variables. The intended interpretation of e.g. $(\exists x/W)$ is that the choice of a value for x is independent of the choices of the values for the variables in W. When $W = \varnothing$ we recover the standard quantifiers. Alternatively, in terms of the idea of dependence: the choice of a value for x depends on all the values of the variables in whose scope $(\exists x/W)$ is, except for those in W.

IF logic is useful for modeling certain puzzles or phenomena of imperfect information. Here is a well known example (Matching Pennies).

Two players, Abelard (the universal player) and Eloise (the existential player) play the following game: Abelard hides one euro in his left or right hand without Eloise seeing it. Eloise has to guess it. If she guesses correctly, she wins; otherwise Abelard wins.

We can model this game in terms of the IF sentence φ_{MP}: $\forall x(\exists y/\{x\})x = y$ and the model set $\mathbb{M} = \{L, R\}$. They determine a semantical game $G(\mathbb{M}, \varphi_{MP})$ which is played like this: Abelard chooses $a \in \{L, R\}$ to be the value of the variable x. Eloise chooses $b \in \{L, R\}$ to be the value of y without "seeing" the choice of Abelard. Eloise wins the play (a, b) if the assignment $x = a$, $y = b$ satisfies the quantifier free formula $x = y$, that is, if a is the same as b. Otherwise Abelard wins the play.

What is important is not to win a particular play but to have a winning strategy for the whole game. A strategy for a player is a sequence of functions, one for each of his or her move defined on all the possible "known" or "seen" earlier moves. In our semantical game, a strategy for Abelard reduces to an individual of the universe and so does a strategy for Eloise. The set $S_\forall$ of strategies for Abelard is thus the same as the set $S_\exists$ of strategies for Eloise which is $\mathbb{M}$. Truth and falsity of φ_{MP} in $\mathbb{M}$ are defined as expected:

- $\mathbb{M} \models^+ \varphi_{MP}$ iff there is a winning strategy for Eloise in $G(\mathbb{M}, \varphi_{MP})$ iff there is $b \in \{L, R\}$ such that for all $a \in \{L, R\} : a = b$
- $\mathbb{M} \models^- \varphi_{MP}$ iff there is a winning strategy for Abelard in $G(\mathbb{M}, s, \varphi_{MP})$ iff there is $a \in \{L, R\}$ such that for all $b \in \{L, R\} : a \neq b$

As we see, we have indeterminacy: $\mathbb{M} \not\models^+ \varphi_{MP}$ and $\mathbb{M} \not\models^- \varphi$.

Notice that were the informal game to have been one of perfect information, that is, were Eloise to see in which hand Abelard hides the coin, we would have modelled it by the semantical game $G(\mathbb{M}, \varphi)$ where φ is the ordinary FOL sentence $\forall x \exists y x = y$. In that case $S_\forall$ would be, as before, identical to $\mathbb{M}$ but $S_\exists$ would consist of four functions (i.e. (a, a), (b, b), (a, b), (b, a)). Given that Eloise has more strategies than in the earlier game, $G(\mathbb{M}, \varphi)$ is determined with (a, b) a winning strategy for Eloise. In other words, $\mathbb{M} \models^+ \varphi$.

If negation occurs in positions other than in front of atomic formulas, then the rules of the semantical game would be changed. In such cases, each occurrence of the negation sign $\neg$ would prompt a swap of the roles of the two players: The moves made "normally" by Eloise would be now made by Abelard, and *vice versa*. Also the rules for winning and losing would be reversed. One can easily show that the following properties hold for every IF sentence φ and model $\mathbb{M}$:

- $\mathbb{M} \models^+ \neg\varphi$ if and only if $\mathbb{M} \models^- \varphi$
- $\mathbb{M} \models^- \neg\varphi$ if and only if $\mathbb{M} \models^+ \varphi$.

There is a counterpart of the game-theoretical semantics sketched above which underlies better strategic, game-theoretical phenomena. It is the Skolem semantics. Let us sketch it briefly by way of an example.

With every IF formula φ in negation normal form in a given vocabulary L we associate its Skolem form $Sk(\varphi)$, which is a first-order formula in an extended vocabulary. Essentially each existentially quantified subformula $(\exists x/W)\psi$ introduces a new Skolem term $f(y_1, \ldots, y_n)$ where $y_1, \ldots, y_n$ are all the free variables of $(\exists x/W)\psi$ minus the variables in W. The new function symbol f is intended to codify the "local" strategy of Eloise and the value of the term $f(y_1, \ldots, y_n)$ is the individual which is the value of the variable x when $y_1, \ldots, y_n$ have received appropriate values earlier in the game. Again we prefer to illustrate with an example. Let φ be the IF sentence

$$\exists x\forall y(\exists z/\{y\})\forall w R(x, y, z, w).$$

Its Skolem form $Sk(\varphi)$ is

$$\forall y\forall w R(c, y, f(x), w)$$

where c is a new 0-place function symbol (constant) and f is a new unary function symbol. The truth of the sentence φ on a given model $\mathbb{M}$ in the Skolem semantics is defined as

- $\mathbb{M} \models^+_{Sk} \varphi$ if and only if there exist an individual c and a function g in the universe of $\mathbb{M}$ to be the interpretations of the new symbols in $Sk(\varphi)$ such that

$$\mathbb{M}, c, g \models Sk(\varphi).$$

  The functions a and g are called Skolem functions. We should realize at this point that the Skolem functions are the counterpart of the strategies of Eloise in the semantical game $G(\mathbb{M}, \varphi)$.

Note also that, for an arbitrary IF sentence φ in negation normal form, its Skolem form $Sk(\varphi)$ and any arbitrary model $\mathbb{M}$ determine the set $S_\exists$ of strategies of Eloise in the semantical game $G(\mathbb{M}, \varphi)$: it is the set of all sequences of Skolem functions which may be the possible interpretations in $\mathbb{M}$ of the new function symbols of $Sk(\varphi)$.

The strategies of Abelard in $G(\mathbb{M}, \varphi)$ have also a natural counterpart as Kreisel counterexamples. Again, the details have been given elsewhere Mann et al. (2011).

With every IF formula φ in negation normal form in a given vocabulary L we associate its Kreisel form $Kr(\varphi)$, which is also a first-order formula in an extended vocabulary. This procedure is the dual of the procedure of Skolemization. Essentially each universally quantified subformula $(\forall x/W)\psi$ introduces a new Kreisel term $h(y_1, \ldots, y_m)$ where $y_1, \ldots, y_m$ are all the free variables of $(\forall x/W)\psi$ minus the variables in W. In our example the Kreisel form of φ is

$$\forall x \forall z \neg R(x, h_1(x), z, h_2(x, f(x), z))$$

noindentThe falsity of φ on a given model $\mathbb{M}$ in the Skolem semantics is defined as:

- $\mathbb{M} \models^{-}_{Sk} \varphi$ if and only if there exist function g_1 and g_2 in $\mathbb{M}$ to be the interpretations of the new function symbols in $Kr(\varphi)$ such that

$$\mathbb{M}, g_1, g_2 \models Kr(\varphi)$$

We call g_1 and g_2 Kreisel counterexamples.

Mann et al. (2011) gives a detailed introduction to game-theoretical semantics, Skolem semantics and the compositional semantics and shows their equivalence.

10.5 Some Basic Results on If Prefixes

In this section we describe some of the results in Sevenster (2014) which help us to evaluate the expressive power of IF quantifier prefixes.

An IF prefix is a series $(Q_1x_1/U_1)\ldots(Q_nx_n/U_n)$ of quantifiers, where Q_i is either $\exists$ or $\forall$, u_i a variable and U_i is a finite set of variables disjoint from u_i. We require that all quantified variables x_i in an IF prefix be distinct.

A prefix $(Q_1x_1/U_1)\ldots(Q_nx_n/U_n)$ is sentential if $U_i \subseteq \{u_1, \ldots, u_{i-1}\}$ for every $1 \leq i \leq n$. An IF formula $\Pi\varphi$ is a sentence if the prefix Π is sentential and all variables appearing in φ are contained in the set of variables quantified in Π.

Let φ and ψ be two IF sentences. We say that φ and ψ are (truth) equivalent, if for every suitable structure $\mathbb{M}$ we have,

$$\mathbb{M} \models^{+} \varphi \quad \texttt{iff} \quad \mathbb{M} \models^{+} \psi.$$

Let Π and Φ be IF prefixes. Then, Π and Φ are equivalent if $\Pi\varphi$ and $\Phi\varphi$ are equivalent for every quantifier-free formula φ.

The IF formalism introduces a difference between syntactical scope and semantical scope (dependence). Let Qx/X and Qy/Y be two quantifiers in an IF formula such that Qx/X is in the (syntactical) scope of Qy/Y. We say that Qx/X depends on Qy/Y if $y \notin X$. Notice that if $X = \varnothing$, then Qx depends on Qy/Y. Two patterns of dependence turn out to be important for expressive power:

- An IF prefix Π is a *Henkin prefix* if it contains four quantifiers

$$(\forall x/X), (\forall y/Y), (\exists u/U), (\exists v/V)$$

  such that $(\exists u/U)$ depends on x and not on y; and $(\exists v/V)$ depends on y and not on x and u. Notice that the first condition requires $(\exists u/U)$ to be in the syntactical scope of $(\forall x/X)$ and the second condition requires $(\exists v/V)$ to be in the syntactical scope of $(\forall y/Y)$. For example

$$\forall x \forall y(\exists u/\{y\})(\exists v/\{x, u\})$$

  is a Henkin prefix and so is

$$\forall x \exists u \forall y(\exists v/\{x, u\}).$$

- An IF prefix Π is a *signaling prefix* if it contains three quantifiers

$$(\forall u/U), (\exists v/V), (\exists w/W)$$

  syntactically ordered as above such that $(\exists v/V)$ depends on u and $(\exists w/W)$ depends on v but not on u. For example

$$\forall u \exists v(\exists w/\{u\})$$

  is a signaling prefix.

Sevenster (2014) considers three operations on prefixes that preserve truth equivalence.

The first operation is *quantifier swapping*. Every two prefixes

$$\Pi(Qu/U)(Qv/V \cup u)\Pi'$$

and

$$\Pi(Qv/V - u)(Qu/U \cup v)\Pi'$$

of IF quantifiers are equivalent.

The second operation is *emptying the slash set from an universal quantifier*. Every two prefixes

$$(\forall v/V)\Pi \textit{ and } \forall v\Pi$$

of IF quantifiers are equivalent.

The third operation consists of *emptying an existentially quantified slash set.* Every two sentential prefixes

$$\Pi(Qv/V)\Pi'$$

and

$$\Pi Qv\Pi'$$

of IF quantifiers are equivalent whenever V contains variables that are quantified only by existential quantifiers in Π.

A prefix Π of IF quantifiers is called *primary* if it is neither Henkin nor signaling.

Lemma (Sevenster 2014) *Let Π be a IF prefix and let Π_0 be the IF prefix that is the result of swapping two quantifiers in (as in the first operation above); or emptying the slash set of a universal quantifier in Π (as in the second operation above); or emptying the slash set of a quantifier in Π that contains only existentially quantified variables (as in the third operation above). Then, Π_0 is primary, whenever Π is.*

Lemma (Sevenster 2014) *Every primary, sentential IF prefix $\Pi(\exists v/V)\Pi'$ in which Π is an FOL prefix, is equivalent to a primary, sentential IF prefix $\Sigma\Pi'$ in which Σ is an FOL prefix.*

Using these two Lemmas, Sevenster (2014) shows:

Theorem *Every prefix which is neither signaling nor Henkin is (truth) equivalent to an ordinary FOL prefix.*

10.6 Defining the Monty Hall Games in If Logic

We shall consider three IF sentences,

$$\begin{aligned}
\varphi_{MH} &=: \forall x(\exists y/\{x\})\forall z(\exists t/\{x\})[x \neq z \wedge y \neq z \rightarrow t = x] \\
\varphi_{MH1} &=: \forall x(\exists y/\{x\})\forall z(\exists t/\{x\})[x \neq z \wedge y \neq z \rightarrow (t \neq y \wedge x = t)] \\
\varphi_{MH2} &=: \forall x(\exists y/\{x\})\forall z(\exists t/\{x\})[x \neq z \wedge y \neq z \rightarrow (t = y \wedge x = t)].
\end{aligned}$$

We can think of the Contestant, C, as the existential quantifier and of MH as the universal quantifier. Let $\mathbb{M}$ be a model set which consists of three doors. Any of the three sentences $\varphi \in \{\varphi_{MH}, \varphi_{MH1}, \varphi_{MH2}\}$ and the model $\mathbb{M}$ determine a semantical game $G(\mathbb{M}, \varphi)$, as we have seen. In such a game, the two players choose individuals from the set $\mathbb{M}$ to be the values of the variables x, y, z and t. A play of any of these games is thus a sequence of four individuals in $\mathbb{M}$. For instance, the sequence $(1, 2, 3, 1)$ is intended to represent the play:

MH hides the prize behind door 1; C makes a guess: door 2; MH opens door 3; C guesses again: door 1.

The games $G(\mathbb{M}, \varphi_{MH})$, $G(\mathbb{M}, \varphi_{MH1})$ and $G(\mathbb{M}, \varphi_{MH2})$ have the same set of histories. They differ only in the winning conditions (payoffs) of the maximal histories (plays) that are determined in each case by the relevant quantifier free subformula. For instance, the play $(1, 2, 3, 1)$ determines the assignment $x = 1, y = 2, z = 3$ and $t = 1$ which satisfies the quantifier free subformula of φ_{MH}. We conclude that the play is a win for C. As the assignment determined by any play of any of the games either satisfies the relevant quantifier-free subformula or not, it is obvious that $G(\mathbb{M}, \varphi_{MH})$, $G(\mathbb{M}, \varphi_{MH1})$ and $G(\mathbb{M}, \varphi_{MH3})$ are 2 player, win-lose extensive game of imperfect information. We shall take them to "define" in IF logic the three games discussed in an earlier section.

Strictly speaking, the semantical games do not correspond exactly to the earlier Monty Hall game and its two variants. But they are "equivalent" to them. To see this, let us consider the Monty Hall game and the semantical game $G(\mathbb{M}, \varphi_{MH})$ that we take to define it. The definitory rules of the Monty Hall game are built into the winning conditions of the semantical game and the rationality assumptions behind it. Take the rule that z must be distinct from x and y. In the semantical game, if, for instance, MH chooses to open either the door x where he hid the price or the door y chosen by C, he will lose right away. Therefore if MH is a rational player (and we assume he is), he will choose to open a door z such that $z \neq x$ and $z \neq y$. Similarly for the other definitory rule which prescribes that t must be distinct than z. In the semantical game nothing in the rules of the game prevents C from choosing t to be the same as z. But we already pointed out that it is in MH's interest to choose z distinct from x. And as C is rational herself, and she knows that MH is rational, it is not in her interest to choose t to be the same as z. All these considerations have a precise game-theoretical counterpart: such "irrational" strategies turn out to be weakly dominated by other strategies of the same player (cf. below.)

The information sets of the two players will be determined by the same principles (C1) and (C2) that we discussed earlier.

The strategies of the two players in a semantical game are determined by the Skolem and Kreisel form of the relevant IF sentence and the underlying model set, as we pointed out earlier. For instance the Skolem form of φ_{MH} is

$$\forall x \forall z[x \neq z \wedge c \neq z \rightarrow f(y, z) = x]$$

and its Kreisel form is

$$\forall y \forall t[d \neq g(x, y) \wedge y \neq g(x, y) \wedge t \neq d]$$

Therefore we shall take C's strategies to consist of all pairs (i, f_i), such that $i \in M$ and $f_i : M^2 \rightarrow M$. And the strategies of Monty Hall will be also pairs (j, g_j) such that $j \in M$ and $g_j : M^2 \rightarrow M$. These are also the strategies of C and respectively MH in the two other semantical games.

10.6.1 The Semantical Indeterminacy of the Monty Hall Sentences

We will show that φ_{MH}, φ_{MH1} and φ_{MH2} are semantically indeterminate on the model set $\mathbb{M}$ with three elements. There are at least three ways to do this.

One way would be to reason as follows. We have established that $G(\mathbb{M}, \varphi_{MH})$, $G(\mathbb{M}, \varphi_{MH1})$ and $G(\mathbb{M}, \varphi_{MH2})$ define the Monty Hall game and its variants. Given that the latter are indeterminate in the game theoretical sense, that is, there is no equilibrium in the games, then it also follows that that the former are indeterminate in the semantical sense. This argument would require that we further establish an equivalence between the lack of an equilibrium in the latter games and the lack of a winning strategy for either player in the semantical game. Such an argument for the general case may be found in Barbero and Sandu (2014).

A second way would be to use the definitions of truth and falsity in Skolem semantics to show that φ_{MH}, φ_{MH1} and φ_{MH2} are semantically indeterminate in $\mathbb{M}$. The definitions tells us that:

- $\mathbb{M} \models^{+}_{Sk} \varphi_{MH}$ iff there exist $i \in M$ and a function f_i in $\mathbb{M}$ such that

$$\mathbb{M}, i, f_i \models \forall x \forall z[x \neq z \wedge c \neq z \rightarrow f(y, z) = x]$$

and

- $\mathbb{M} \models^{-}_{Sk} \varphi_{MH}$ iff there exist $j \in M$ and a function g_j in $\mathbb{M}$ such that

$$\mathbb{M}, j, g_j \models \forall y \forall t[d \neq g(x, y) \wedge y \neq g(x, y) \wedge t \neq d]$$

We could then show that none of the conditions on the right side of the equivalences holds.

We prefer to apply a simpler, third alternative procedure based on Sevenster's result described earlier. We first notice that each of the three prefixes is neither branching nor signaling. By Sevenster's theorem, each of the three sentences is truth equivalent with an ordinary FOL sentence. In the case of φ_{MH1}

$$\forall x(\exists y/\{x\})\forall z(\exists t/\{x\})[x \neq z \wedge y \neq z \rightarrow (t \neq y \wedge x = t)]$$

we obtain its first-order equivalent by the following steps.

- First we swap the first two quantifiers:

$$\exists y(\forall x/\ \{x)\})\forall z(\exists t/\{x\})[x \neq z \wedge y \neq z \rightarrow (t \neq y \wedge x = t)].$$

- We then empty the slash set of the universal quantifier in the resulting sentence:

$$\exists y \forall x \forall z(\exists t/\{x\})[x \neq z \wedge y \neq z \rightarrow (t \neq y \wedge x = t)].$$

- Next we swap the two universal quantifiers:

$$\exists y \forall z \forall x (\exists t/\{x\})[x \neq z \wedge y \neq z \rightarrow (t \neq y \wedge x = t)]$$

- We swap the last two quantifiers:

$$\exists y \forall z \exists t (\forall x / \{t\})[x \neq z \wedge y \neq z \rightarrow (t \neq y \wedge x = t)].$$

- Finally we empty the slashed set of the universal quantifier:

$$\exists y \forall z \exists t \forall x [x \neq z \wedge y \neq z \rightarrow (t \neq y \wedge x = t)].$$

Let us denote this sentence by φ^+_{MH1}. By Sevenster's result, φ_{MH1} and φ^+_{MH1} are truth equivalent on all models. In particular, they are truth equivalent on our model set $\mathbb{M}$ with three elements. However it is easy to show that the first-order sentence φ^+_{MH1} is not true in $\mathbb{M}$ (i.e. it is false in $\mathbb{M}$). We conclude that

$$\mathbb{M} \not\models^+ \varphi_{MH1}.$$

On the other side, from the game-theoretical semantics of IF logic, we know that

$$\mathbb{M} \models^- \varphi_{MH1} \quad \texttt{iff} \quad \mathbb{M} \models^+ \neg\varphi_{MH1}.$$

But $\neg\varphi_{MH1}$ is truth equivalent with the IF sentence which is obtained by pushing negation infront of the atomic formulas in such a way that every quantifier and connective is changed into its dual. In our case this sentence, let us denote it by φ^*_{MH1}, is

$$\exists x (\forall y/\{x\}) \exists z (\forall t/\{x\})[x \neq z \wedge y \neq z \wedge (x \neq t \vee t = y)].$$

Applying Sevenster's transformations, φ^*_{MH1} can be shown to be equivalent to the first-order sentence

$$\exists x \forall y \exists z \forall t [x \neq z \wedge y \neq z \wedge (x \neq t \vee t = y)].$$

This sentence is false in $\mathbb{M}$. Thus

$$\mathbb{M} \not\models^+ \varphi_{MH1} \quad \texttt{and} \quad \mathbb{M} \not\models^- \varphi_{MH1}.$$

A similar argument also shows that

$$\mathbb{M} \not\models^+ \varphi_{MH2} \quad \texttt{and} \quad \mathbb{M} \not\models^- \varphi_{MH2}$$

and

$$\mathbb{M} \not\models^{+} \varphi_{MH} \quad \text{and} \quad \mathbb{M} \not\models^{-} \varphi_{MH}.$$

(We are indebted to Barbero for the above argument). We have reached the conclusion that φ_{MH}, φ_{MH1} and φ_{MH2} are neither true nor false on $\mathbb{M}$. Of course this semantic indeterminacy mirrors the lack of pure strategy equilibria in the three Monty Hall games discussed earlier.

The question which interests us now is: Is their a way to distinguish between φ_{MH}, φ_{MH1} and φ_{MH2}? Of course the distinction cannot be spelled out in terms of truth-functional content, that is, by evaluating what the three sentences "assert" or "exclude": for, as the above argument showed, what these sentences assert is false, and what they exclude is also false. Still, there are differences between them. Consider, for instance, φ_{MH1} and φ_{MH2}: the condition $(t \neq y \wedge x = t)$ makes the "switch" strategy weakly dominant, whereas the condition $(t = y \wedge x = t)$ makes the "stay" strategy to be so. Is there a way to incorporate this difference into the semantical values of the two sentences?

The framework which yields the desired result has been worked out for the first time in Sevenster (2006), developed in Sevenster and Sandu (2010) and Mann et al. (2011). The application of this framework to Monty Hall has been given for the first time by Mann (2010), and developed in Sandu (2015a).

10.7 Mixed Strategy Equilibria: von Neumann's Minimax Theorem

We shall overcome the indetermincy of the three games above by an appeal to a technique which is well known in classical game theory: von Neumann's Minimax Theorem.

Let

$$\Gamma = (S_{\exists}, S_{\forall}, u_{\exists}, u_{\forall})$$

be a two player, win-lose, finite strategic game, where S_p is the set of pure strategies of player p. A mixed strategy ν for player p is a probability distribution over S_p, that is, a function $\nu : S_p \to [0, 1]$ such that $\sum_{\tau \in S_i} \nu(\tau) = 1$. ν is uniform over $S_i' \subseteq S_i$ if it assigns equal probability to all strategies in S_i' and zero probability to all the strategies in $S_i - S_i'$. The support of ν is the set of strategies to which ν assigns nonzero probability. Obviously we can simulate a pure strategy σ with a mixed strategy ν such that ν assigns σ probability 1. Given a mixed strategy μ for player $\exists$ and a mixed strategy ν for player $\forall$, the expected utility for player p is given by:

$$U_p(\mu, \nu) = \sum_{\sigma \in S_{\exists}} \sum_{\tau \in S_{\forall}} \mu(\sigma)\nu(\tau)u_p(\sigma, \tau).$$

Let μ be a mixed strategy for player $\exists$ and ν a mixed strategy for player $\forall$. The pair (μ, ν) is an equilibrium if the following two conditions hold:

1. for every mixed strategy μ', of Eloise, $U_\exists(\mu, \nu) \geq U_\exists(\mu', \nu)$
2. for every mixed strategy ν' of Abelard, $U_\forall(\mu, \nu) \geq U_\forall(\mu, \nu')$.

The following two results are well known.

Theorem (von Neuman's Minimax Theorem; von Neumann 1928) *Every finite, two-person, constant-sum game has an equilibrium in mixed strategies.*

Corollary *Let* (μ, ν) *and* (μ', ν') *be two mixed strategy equlibria in a constant sum game. Then* $U_p(\mu, \nu) = U_p(\mu', \nu')$

These two results guarantee that we can talk about *the value* $V(\Gamma)$ *of a strategic* game Γ: it is the expected utility returned to player $\exists$ by any equilibrium in the relevant strategic game.

The next results will help us to identify equilibria. Their proof may be found in Mann et al. (2011, Chap. 7).

We start with a couple of definitions.

Let $\Gamma = (S_\exists, S_\forall, u_\exists, u_\forall)$ be a strategic game where S_p is the set of pure strategies of player p. For $\sigma, \sigma' \in S_\exists$, we say that σ' *weakly dominates* σ if the following two conditions hold:

(i) For every $\tau \in S_\forall : u_\exists(\sigma', \tau) \geq u_\exists(\sigma, \tau)$
(ii) For some $\tau \in S_\forall : u_\exists(\sigma', \tau) > u_\exists(\sigma, \tau)$.

A similar notion is defined for Abelard.

We say that σ' is *payoff equivalent to* σ if for every $\tau \in S_\forall : u_\exists(\sigma', \tau) = u_\exists(\sigma, \tau)$.

A similar notion is defined for Abelard.

We list three results from the game-theoretical literature. The first one enables one to identify equilibria in mixed strategies. It is given here in order to illustrate how indeterminate IF sentences can be distinguished by their probabilistic semantics. The other two enable us to eliminate weakly dominated strategies and payoff equivalent ones. They will be applied to the Monty Hall sentence.

For the first result we recall that we can simulate a pure strategy with a mixed strategy which assigns 1 to that pure strategy and 0 to the rest.

Then for $\sigma \in S_\exists$ and ν a mixed strategy for Abelard we have:

$$U_p(\sigma, \nu) = \sum_{\tau \in S_\forall} \nu(\tau) u_p(\sigma, \tau).$$

Similarly, for $\tau \in S_\forall$ and μ a mixed strategy for Eloise, we have

$$U_p(\mu, \tau) = \sum_{\sigma \in S_\exists} \mu(\sigma) u_p(\sigma, \tau).$$

Proposition *Let μ^* be a mixed strategy for Eloise and ν^* a mixed strategy for Abelard in a finite, two player strategic game Γ. The pair (μ^*, ν^*) is an equilibrium in Γ if and only if the following conditions hold:*

1. $U_\exists(\mu^*, \nu^*) = U_\exists(\sigma, \nu^*)$ for every $\sigma \in S_\exists$ in the support of μ^*
2. $U_\exists(\mu^*, \nu^*) = U_\forall(\mu^*, \tau)$ for every $\tau \in S_\forall$ in the support of ν^*
3. $U_\exists(\mu^*, \nu^*) \geq U_\forall(\sigma, \nu^*)$ for every $\sigma \in S_\exists$ outside the support of μ^*
4. $U_\forall(\mu^*, \nu^*) \geq U_\forall(\mu^*, \tau)$ for every $\tau \in S_\forall$ outside the support of ν^*.

Proof See Osborne (2003, p. 116).

Here is a well known example in the IF literature (see e.g. Sevenster 2006), where the above proposition is applied. We compare two strategic IF games, $\Gamma(\mathbb{M}, \varphi_{MP})$ and $\Gamma(\mathbb{M}, \varphi_{IMP})$ where $\mathbb{M} = \{1, 2, 3, 4\}$, φ_{MP} is the Matching Pennies sentence $\forall x(\exists y/\{x\})x = y$ and φ_{IMP} is the Inverted Matching Pennies sentence $\forall x(\exists y/\{x\})x \neq y$. The first game is represented on the left and the second on the right:

	τ_1	τ_2	τ_3	τ_4
σ_1	(1, 0)	(0, 1)	(0, 1)	(0, 1)
σ_2	(0, 1)	(1, 0)	(0, 1)	(0, 1)
σ_3	(0, 1)	(1, 0)	(1, 0)	(0, 1)
σ_4	(0, 1)	(0, 1)	(0, 1)	(1, 0)

	τ_1	τ_2	τ_3	τ_4
σ_1	(0, 1)	(1, 0)	(1, 0)	(1, 0)
σ_2	(1, 0)	(0, 1)	(1, 0)	(1, 0)
σ_3	(1, 0)	(1, 0)	(0, 1)	(1, 0)
σ_4	(1, 0)	(1, 0)	(1, 0)	(0, 1)

We show that the strategy pair (μ^*, ν^*) where both μ^* and ν^* are uniform strategies with suport $M = \{1, 2, 3, 4\}$. First notice that in the first game $U_\exists(\mu^*, \nu^*) = 1/4$ and thus $U_\forall(\mu^*, \nu^*) = 3/4$. Notice that clauses (3) and (4) of the above Proposition are trivially satisfied. For clause (1), consider σ_1 and recall the equation

$$U_\exists(\sigma_1, \nu^*) = \sum_{\tau \in S_{II}} \nu^*(\tau) u_I(\sigma_1, \tau)$$

to get $U_\exists(\sigma_1, \nu^*) = 1/4$. A similar computation establishes clause (2). The argument for the second game is similar.

Proposition *Let $\Gamma = (S_\exists, S_\forall, u_\exists, u_\forall)$ be a 2 player, finite, win-lose strategic game. Then Γ has a mixed strategy equilibrium (μ, ν) such that for each player p none of the pure strategies in the support of his mixed strateggy is weakly dominated in Γ. (The proof requires the finiteness of the game).*

Proposition *Let $\Gamma = (S_\exists, S_\forall, u_\exists, u_\forall)$ be a 2 player, finite, win-lose strategic IF game. Then Γ has a mixed strategy equilibrium (μ, ν) such that for each player there are no pure strategies in the support of his mixed strategy which are payoff equivalent.*

We apply these results to the Monty Hall game. The second proposition allows us to reduce the game to the smaller one:

	$(1, g_1)$	$(2, g_2)$	$(3, g_3)$
$(1, h_1)$	$(1, 0)$	$(0, 1)$	$(0, 1)$
$(2, h_2)$	$(0, 1)$	$(1, 0)$	$(0, 1)$
$(3, h_3)$	$(0, 1)$	$(0, 1)$	$(1, 0)$
$(1, f_1)$	$(0, 1)$	$(1, 0)$	$(1, 0)$
$(2, f_2)$	$(1, 0)$	$(0, 1)$	$(1, 0)$
$(3, f_3)$	$(1, 0)$	$(1, 0)$	$(0, 1)$

We notice that each "stay" strategy is weakly dominated by some "switch" strategy. We apply the first proposition and reduce the game to:

	$(1, g_1)$	$(2, g_2)$	$(3, g_3)$
$(1, f_1)$	$(0, 1)$	$(1, 0)$	$(1, 0)$
$(2, f_2)$	$(1, 0)$	$(0, 1)$	$(1, 0)$
$(3, f_3)$	$(1, 0)$	$(1, 0)$	$(0, 1)$

Let μ be the uniform probability distribution $\mu(1, f_i) = \frac{1}{3}$ and ν the uniform probability distribution $\nu(j, g_j) = \frac{1}{3}$. It may be shown that this is an equilibrium in the game and that the expected utility of player C for this equilibrium is $2/3$.

We consider the first variant of the MH game. By applying the two propositions above we obtain the game described in the last paragraph. Hence it's value is the same, $2/3$.

As for the last game, after applying the two propositions we reduce it to

	$(1, g_1)$	$(2, g_2)$	$(3, g_3)$
$(1, h_1)$	$(1, 0)$	$(0, 1)$	$(0, 1)$
$(2, h_2)$	$(0, 1)$	$(1, 0)$	$(0, 1)$
$(3, h_3)$	$(0, 1)$	$(0, 1)$	$(1, 0)$

Let μ be the uniform probability distribution $\mu(1, h_i) = \frac{1}{3}$ and ν the uniform probability distribution $\nu(j, g_j) = \frac{1}{3}$. It may be shown that this is an equilibrium in the game and that the expected utility of player C for this equilibrium is $1/3$.

10.8 Some General Remarks

We shall take the probabilistic value of an IF sentence φ on a finite model $\mathbb{M}$ to be the value of the 2 player, win-lose, finite strategic game which can be obtained from $G(\mathbb{M}, \varphi)$ along the procedure illustrated in the preceding section. It is the expected utility returned to Eloise by any of the mixed strategy equilibria in this game. The existence of such an equilibrium is guaranteed by von Neumann's Minimax theorem. We denote the probabilistic value by $P(\varphi, \mathbb{M})$. Mann et al. (2011) collect the basic results on probabilistic IF logic. They prove, among other things, the following:

P1 $P(\varphi \vee \psi, \mathbb{M}) = max(P(\varphi, \mathbb{M}), P(\psi, \mathbb{M}))$.
P2 $P(\varphi \wedge \psi, \mathbb{M}) = min(P(\varphi, \mathbb{M}), P(\psi, \mathbb{M}))$.
P3 $P(\neg\varphi, \mathbb{M}) = 1 - P(\varphi, \mathbb{M})$.
P4 $P(\exists x\varphi, \mathbb{M}) = max\,\{P(\varphi, b, \mathbb{M}) : b \in M\}$.
P5 $P(\forall x\varphi, \mathbb{M}) = min\,\{P(\varphi, b, \mathbb{M}) : b \in M\}$.

In P4 and P5 $\exists x\varphi$ and $\forall x\varphi$ are sentences and $P(\varphi, b, \mathbb{M})$ is the value of strategic game $\Gamma(\mathbb{M}, s, \varphi)$ where s is the assignment which assigns b to x.

Let us give an example which illustrates how we compute the probabilistic value of an IF sentence form the probabilistic values of its subformulas. The example if from Mann et al. (2011, p. 168). Consider the IF sentence ψ

$$\exists u \forall w (u \neq w \vee \varphi_{MP})$$

on a finite model-set $\mathbb{M}$ with n elements. Here φ_{MP} is the Matching Pennies sentence.

P4 and P5 tell us that

$$NE(\psi, \mathbb{M}) = max_a min_b \{NE(u \neq w \vee \varphi_{MP}, (u, a), (w, b), \mathbb{M}) : a, b \in M\}.$$

By P1, we know that

$$NE(u \neq w \vee \varphi_{MP}, (u, a), (w, b), \mathbb{M}) = max(NE(u \neq w, (u, a), (w, b), \mathbb{M}), NE(\varphi_{MP}, (u, a), (w, b), \mathbb{M})).$$

We know already that $NE(\varphi_{MP}, (u, a), (w, b), \mathbb{M})$ is $1/n$. By P1 we get

$$NE(u \neq w \vee \varphi_{MP}, (u, a), (w, b), \mathbb{M}) = max(NE(u \neq w, (u, a), (w, b), \mathbb{M}), 1/n).$$

>From (E1) and (E2) we know that $NE(u \neq w, (u, a), (w, b), \mathbb{M})$ is 1 if $a \neq b$ and 0 otherwise. Thus for a fixed a, $min_b \{NE(u \neq w \vee \varphi_{MP}, (u, a), (w, b), M) : b \in M\}$ is reached when w is a and this minimum is $1/n$. We conclude that $max_a min_b$ is $1/2$.

On the basis of P1-P3 it is straightforward to show that the following hold for any IF sentence φ and finite model $\mathbb{M}$:

Ax1 $P(\varphi, \mathbb{M}) \geq 0$
Ax2 $P(\varphi, \mathbb{M}) + P(\neg\varphi, \mathbb{M}) = 1$
Ax3 $P(\varphi, \mathbb{M}) + P(\psi, \mathbb{M}) \geq P(\varphi \vee \psi, \mathbb{M})$
Ax4 $P(\varphi \wedge \psi, \mathbb{M}) = 0 \rightarrow P(\varphi, \mathbb{M}) + P(\psi, \mathbb{M}) = P(\varphi \vee \psi, \mathbb{M})$.

The axioms are known as *Kolmogorov axioms of probabilities*.

It has been pointed out that there is another principle which may be considered natural in the context of logic:

S. (Substitutivity) If φ and ψ are truth equivalent, then $P(\varphi, \mathbb{M}) = P(\psi, \mathbb{M})$ for every finite model $\mathbb{M}$.

The IF sentence φ_{MH} and the first-order sentence $\exists y \forall z \exists t \forall x[x \neq z \wedge y \neq z \rightarrow x = t]$ which is equivalent with it provide a counter-example to (S). On any model $\mathbb{M}$ with at least three elements, $P(\varphi_{MH}, \mathbb{M}) \neq 0$, $P(\varphi_{MH}, \mathbb{M}) \neq 1$ but

$$P(\exists y \forall z \exists t \forall x[x \neq z \wedge y \neq z \rightarrow x = t], \mathbb{M}) = 0.$$

For a more detailed comparison between IF probabilities and other kinds of probabilistic semantics we refer to Sandu (forthcoming, 2015b).

References

Barbero, F. (2013). On existential declarations of independence in IF logic. *The Review of Symbolic Logic*, *6*, 254–280.

Barbero, F., & Sandu, G. (2014). Signaling in independence-friendly logic. *Logic Journal of IGPL*, *22*, 638–664. doi:10.1093/jigpal/jzu004.

Grinstead, Ch. M., & Snell, L. (1998). *Introduction to probabilities* (2nd edn.). Providence: American Mathematical Society.

Hintikka, J., & Sandu, G. (1989). Informational Independence as a semantic phenomenon. In J. E. Fenstad, et al. (Eds.), *Logic, methodology and philosophy of science* (Vol. 8, pp. 571–589). Amserdam: Elsevier.

Mann, A. (2010). *A logical analysis of the Monty Hall problem*. Varna: Logic Colloquium. http://www.pps.univ-paris-diderot.fr/logic2010/contributed_talks.

Mann, A. I., Sandu, G., & Sevenster, M. (2011). *Independence-friendly logic: A game-theoretic approach*. Cambridge: Cambridge University Press.

Sandu, G. (2015a). Languages with imperfect information. In S. Gosh, R. Verbrugge, & J. van Benthem (Eds.), *Modeling strategic reasoning, texts in logic and games*. Berlin: Springer.

Sandu, G. (2015b). Probabilistic IF logic. In I. Parvu, G. Sandu, & I. Toader (Eds.), *Romanian studies in philosophy of science, Boston studies* (forthcoming).

Sevenster, M. (2006). *Branches of imperfect information: Logic, games, and computation*. PhD Thesis, Amsterdam: University of Amsterdam.

Sevenster, M. (2014). Dichotomy results for independence-friendly prefixes of generalized quantifiers. *Journal of Symbolic Logic*, forthcoming.

Sevenster, M., & Sandu, G. (2010). Equilibrium semantics of languages of imperfect information. *Annals of Pure and Applied Logic*, *161*, 618–631.

van Benthem, J. K. (2003). Conditional probability meets update logic. *Journal of Logic, Language and Information*, *12*, 409–421.

Von Neumann, J. (1928). Zur Theorie der Gesellschaftsspiele. *Mathematische Annalen*, *100*, 295–320.

Chapter 11
Solving for Best Responses and Equilibria in Extensive-Form Games with Reinforcement Learning Methods

Amy Greenwald, Jiacui Li and Eric Sodomka

Abstract We present a framework to solve for best responses and equilibria in extensive-form games (EFGs) of imperfect information by transforming a game together with an other-agent policy into a set of Markov decision processes (MDPs), one per player, and then applying simulation-based reinforcement learning (RL) to the ensuing MDPs. More specifically, we first transform a turn-taking partially observable Markov game (TT-POMG) into a set of partially observable Markov decision processes (POMDPs), and we then transform that set of POMDPs into a corresponding set of Markov decision processes (MDPs). Next, we observe that EFGs are a special case of TT-POMGs, and hence can be transformed as described. Furthermore, because each transformation preserves the strategically-relevant information of the model to which it is applied, an optimal policy in one of the ensuing MDPs corresponds to a best response to the given other-agent policy in the original EFG. We then go on to prove that our RL algorithm finds a near-optimal policy (and therefore a near-best response in the original EFG) in finite time, although the sample complexity is lower bounded by a function with an exponential dependence on the horizon. Nonetheless, we apply this algorithm iteratively to search for equilibria in an EFG. When the iterative procedure converges, the resulting MDP policies comprise an approximate weak perfect Bayesian equilibrium. Although this procedure is not guaranteed to converge, it frequently did in numerical experiments with sequential auctio.

Keywords Reinforcement learning · Partially observable Markov games · Bayesian equillibrium

A. Greenwald (✉) · E. Sodomka
Department of Computer Science, Brown University, Providence, RI 02912, USA
e-mail: amy_greenwald@brown.edu

E. Sodomka
e-mail: sodomka@gmail.com

J. Li
Graduate School of Business, Stanford University, Stanford, CA 94305-7298, USA
e-mail: jiacui@stanford.edu

C. Başkent et al. (eds.), *Rohit Parikh on Logic, Language and Society*,
Outstanding Contributions to Logic 11, DOI 10.1007/978-3-319-47843-2_11

11.1 Introduction

Real-world interactions between individuals and organizations are commonly modeled as extensive-form games (EFGs), in which players act in sequence, and payoffs to one player depend on the actions of all players. Under standard game-theoretic assumptions, each player in an EFG is predicted to play her *best response*, which is a strategy that maximizes her payoff, given fixed other-agent strategies. This notion of best response is central to game theory, and forms the basis of Nash Equilibrium (NE), the canonical game-theoretic solution concept: at NE, all players simultaneously play best responses to one another (Nash 1951). Best responses also underpin several classic game-theoretic learning schemes, such as *best-reply dynamics* and *fictitious play* (Fudenberg and Levine 1998). But best responses are difficult to calculate in general; and NE (and many other game-theoretic solutions) are more difficult to find, still (Daskalakis 2009).

In this paper, we develop a simulation-based reinforcement learning (RL) algorithm that approximates best responses in EFGs. Our approach relies on transforming EFGs into more computationally-friendly representations. We first show that EFGs are a special case of turn-taking partially-observable Markov games (TT-POMGs), so that any EFG can be viewed as a TT-POMG. Next, from the perspective of a single player searching for a best response, a TT-POMG can be seen as a partially-observable Markov decision process (POMDP). We endeavor to solve such a POMDP by further transforming it into a belief-state MDP. Chaining this whole line of reasoning together, we find that to solve for a best response in an EFG, we need only to compute an optimal policy in the corresponding belief-state MDP.

Our RL algorithm is then specifically designed to solve for optimal policies in the belief-state MDPs that represent transformed EFGs. We derive upper and lower sample complexity bounds for this algorithm, showing that it is guaranteed to find a near-optimal policy in finite time, although there is an unavoidable exponential dependence on the horizon. Relative to model-free reinforcement learning methods, such as POMCP (Silver and Veness 2010), our model-based approach is more efficient in games where players' types (i.e., their payoffs, essentially) are private and independent, two properties which are typical of many sequential auctions of interest (e.g., Katzman 1999; Menezes and Monteiro 2003; Weber 1983).

Our algorithm can be also be applied iteratively to search for equilibria in EFGs, and if it converges, the resulting strategy profile is an approximate weak perfect Bayesian equilibrium. To evaluate the performance of this iterative search procedure, we ran simulation experiments. Specifically, we looked at three sequential auction models that had been analyzed in the literature to see whether our algorithm could faithfully recover known equilibria. Our algorithm not only discovered close approximations of the known equilibria, but it also discovered heretofore unknown ones. A complete description of these experiments can be found in Greenwald et al. (2012).

The rest of this paper proceeds as follows. Section 11.2 spells out generic transformations from TT-POMGs to POMDPs, and then on to MDPs. Section 11.3 then presents EFGs as a special case of TT-POMGs, and proceeds to apply these transformations to EFGs. Section 11.4 presents our simulation-based reinforcement-learning algorithm and its performance guarantees. This algorithm, when applied to MDPs that are actually transformed EFGs, is our best-response finding algorithm. Section 11.5 describes an iterative procedure for searching for equilibria, which employs our best-response finding algorithm as a subroutine. Section 11.7 surveys related literature, and Sect. 11.8 concludes. Technical details are relegated to appendices.

11.2 From TT-POMGs to POMDPs to MDPs

To begin our story, we review a few definitions of standard models of sequential decision making, in both multi-agent and single agent environments. We begin with turn-taking partially-observable Markov games (TT-POMGs). These games are very much like the more widely-studied simultaneous-move partially-observable Markov games (see, for example, Hansen et al. 2004), except that the players take turns.

For ease of exposition, and so that we can more easily apply our transformations to EFGs, we assume all models include fully-observable terminal states. Furthermore, rewards are associated with states only, not with state-action pairs. Neither of these assumptions is essential.

Definition 1 (*Turn-taking partially-observable Markov game*) A turn-taking partially-observable Markov game Θ is defined as a tuple $\langle N, S, A, T, O, Q, r, \zeta \rangle$ where:

- $N = \{1, ..., n\}$ is the set of players.
- S is a (finite) set of states. Further, let $\langle S_1, \ldots, S_n, Z \rangle$ be a partition over S, where $S_i \subseteq S$ is the set of states at which it is player i's turn to move, and $Z \subseteq S$ is a designated set of terminal states.
  (We write $\texttt{player}(s)$ to denote the player whose turn it is at state s.)
- $A = \bigcup_{ij} A_{ij}$, where A_{ij} denotes the (finite) set of actions available to player i at state $j \in S_i$. At all states $z \in Z$, and for all players i, $A_{iz} = \emptyset$ (i.e., there are no actions available at terminal states).
- $O = \langle O_1, \ldots, O_n \rangle$ is a tuple of (finite) observation sets, where O_i is the set of observations for player i. Each player makes an observation when he enters a state in which it is his turn to play. Upon entering a terminal state (in which it is no player's turn), all players observe that the game is over.
- $P : \bigcup_i S_i \times A \times S \times \bigcup_i O_i \rightarrow [0, 1]$ is a function that expresses the joint probability over the next state and accompanying observation. In particular, $P(j, k, j', h)$ is the joint probability of reaching state j' and observing observation $h \in O_{\texttt{player}(j')}$, conditional on the player whose turn it is taking action k in state j. There are no transitions out of terminal states. (**Note:** P is Markov, by definition.)
  (We write $\texttt{obs}(s)$ to denote the observation observed at state s.)
  The following abbreviations will come in handy:

$$T[j' \mid j,k] \equiv \frac{\Pr[j,k,j']}{\Pr[j,k]} \tag{11.1}$$

$$= \frac{\sum_h P(j,k,j',h)}{\sum_{j''}\left(\sum_h P(j,k,j',h)\right)} \tag{11.2}$$

$$Q_{\texttt{player}(j')}[h \mid j'] \equiv \frac{\Pr[j',h]}{\Pr[j']} \tag{11.3}$$

$$= \frac{\sum_{j,k} P(j,k,j',h)}{\sum_{j,k}\left(\sum_h P(j,k,j',h)\right)} \tag{11.4}$$

- $r = \langle r_1, \ldots, r_n \rangle$ is a tuple of reward functions where $r_i : S \to \mathbb{R}$ is player i's reward function. That is, $r_i(j)$ denotes player i's reward upon reaching state $j \in S$.
- $\zeta : S \to [0, 1]$ is a distribution over initial states. Specifically, $\zeta(j)$ is the probability that the initial state is $j \in S$.

For all players i and for all observations $h \in O_i$, if there exist j and j' s.t. $Q_i[h \mid j] > 0$ and $Q_i[h \mid j'] > 0$, then there must be a 1-to-1 correspondence between A_{ij} and $A_{ij'}$, and it must be the case that $r_i(j) = r_i(j')$. Otherwise, players would be able to distinguish states j and j' from one another. With this in mind, we let $A_i(h)$ denote the set of actions available to player i given observation h (i.e., $A_i(h) = A_{ij} = A_{ij'}$, for all such j and j'). Likewise, we write $r_i(h)$ to denote player i's rewards given observation h (i.e., $r_i(h) = r_i(j) = r_i(j')$, for all such j and j').

Here is how the play of a game unfolds: An initial game state is drawn from the distribution ζ. The player designated to move at that state does not observe the state; rather, he observes h_1 and then takes action a_1. Next, the game transitions according to T to a new state, where it is another player's turn to move. Again, that player does not observe the state, but rather, he observes h_2 and then takes action a_2. The players accrue rewards along the way, and receive further rewards upon transitioning to a terminal state.

A *history* μ^τ in a turn-taking partially-observable Markov game is a sequence $\langle s^1, h^1, a^1, \ldots, s^\tau, h^\tau, a^\tau \rangle$ of τ states, observations, and actions, where each observation is observed by exactly one player, and that is the player whose turn it is at the corresponding state. Hence, each game history μ^τ can be restricted to a subsequence μ_i^τ of player i's states, observations, and actions.

Generally speaking, a *policy* for a player in a TT-POMG is a mapping from the set of that player's histories to the set of available actions (or a probability distribution over actions). We restrict our attention to a subclass of these general policies. Specifically, we define a policy θ_i for player i as a mapping from a single observation $h \in O_i$ to an action $a \in A_i(h)$ (or a probability distribution over actions $a \in \Delta(A_i(h))$). Such policies are called *memoryless*, because they depend on the most recent observation only.[1]

[1] It turns out that restricting to memoryless policies is not a limitation in our setting. Ultimately, we will represent extensive-form games, which are trees, as TT-POMGs. In this representation,

Let $\langle r_i^1, \ldots, r_i^\tau \rangle$ denote player i's sequence of rewards through time τ. Then player i's expected reward through time τ is given by $\mathbb{E}\left[\sum_{t=1}^{\tau} r_i^t\right]$, where the expectation is taken over all possible state sequences of length τ. The probabilities of the various sequences are dictated by the players' (possibly randomized) policies $\boldsymbol{\theta}$ and the game's transition and observation probabilities. We abbreviate this expectation as $R_i^\tau(\boldsymbol{\theta})$.

Definition 2 (*Best response*) Given a profile of other-agent policies $\boldsymbol{\theta}_{-i}$, a τ-*best response* for player i in a TT-POMG is a policy θ_i such that $R_i^\tau(\theta_i, \boldsymbol{\theta}_{-i}) \geq R_i^\tau(\theta_i', \boldsymbol{\theta}_{-i})$, for all policies θ_i'.

Definition 3 (*Partially-observable Markov decision process*) A partially-observable Markov decision process Ψ is a TT-POMG with only one player, who is called the "agent".

Definition 4 (*Optimal policy*) An τ-*optimal policy* in a POMDP is a policy ψ such that $R^\tau(\psi) \geq R^\tau(\psi')$, for all policies ψ'.

Definition 5 (*Markov decision process*) A Markov decision process Φ is defined as a tuple $\langle S, A, T, r, \zeta \rangle$, where:

- S is a (finite) set of states, with $Z \subseteq S$ a designated set of terminal states.
- $A = \bigcup_j A_j$, where A_j denotes the set of possible actions at state $j \in S$. At all states $z \in Z$, $A_z = \emptyset$ (i.e., there are no actions available at terminal states).
- $T : S \times A \times S \rightarrow [0, 1]$ is a possibly stochastic transition function. More specifically, $T[s' \mid s, a]$ denotes the probability of reaching state $s' \in S \backslash Z$, conditioned on taking action $a \in A_s$ in state $s \in S$. (**Note:** T is Markov, by definition.)
- $r : S \rightarrow \mathbb{R}$ is a reward function. That is, $r(s)$ denotes the agent's reward upon reaching state $s \in S$.
- $\zeta : S \rightarrow [0, 1]$ is a distribution over initial states. More specifically, $\zeta(s)$ denotes the initial probability of state $s \in S$.

Here is how the dynamics of an MDP unfold: Initially, a state is drawn from the distribution ζ. The agent observes this state s_1, and then chooses an action a_1. Next, the game transitions according to T to a new state. Again, that agent observes the state s_2, and then chooses an action a_2. The agent accrues rewards along the way, and receive further rewards upon transitioning to a terminal state.

A *history* μ^t in an MDP is a sequence of t states and actions: e.g., $\langle s^1, a^1, \ldots, s^t, a^t \rangle$. Generally speaking, a *policy* in an MDP is a mapping from the set of histories to the set of actions (or a probability distribution over the set of actions). Still WLOG (see, for example, Puterman 1994), we define a policy ϕ as a mapping from a state $j \in S$ to an action $a \in A_j$ (or a probability distribution over actions $a \in \Delta(A_j)$).

(Footnote 1 continued)
we assume the states in the TT-POMG are the paths in the tree. Thus, a memoryless policy in the TT-POMG is in fact a mapping from a history to an action in the EFG. Moreover, since EFGs are trees, defining states as paths is no more expensive than defining them as nodes. See Footnote 4.

Let $\langle r^1, \ldots, r^\tau \rangle$ denote the agent's sequence of rewards through time τ. Then the expected reward through time t is given by $\mathbb{E}\left[\sum_{t=1}^{t} r^t\right]$, where the expectation is taken over all possible state sequences of length τ. The probabilities of the various sequences are dictated by the agent's (possibly randomized) policy and the MDP's transition probabilities. We abbreviate this expectation as $R^\tau(\phi)$.

Definition 6 (*Optimal policy*) An τ-optimal policy in an MDP is a policy ϕ such that $R^\tau(\phi) \geq R^\tau(\phi')$, for all policies ϕ'.

With these definitions of various Markov models in hand, we are ready to present our transformations. First, we transform TT-POMGs into POMDPs, from the point of view of a single player. Second, we transform POMDPs into MDPs. Taken together, the two transformations yield a procedure for converting TT-POMGs to MDPs.

11.2.1 TT-POMG to POMDP Transformation

The main idea of our transformation from TT-POMGs to POMDPs is to collapse a game into a single-agent decision process by folding given other-agent strategies into the transition probabilities and the initial probability distribution of the TT-POMG. Our goal is to carry out this transformation so that a best-response in the TT-POMG corresponds exactly to an optimal policy in the POMDP.

Before we present the transformation, we make the following important observation, which follows directly from our assumptions (state and observation probabilities are Markov, and policies are memoryless):

$$\begin{aligned}
&\mathtt{Pr}_{\boldsymbol{\theta}}^{\mathrm{POMG}}[h^y, a^y, s^{y+1} \mid \mu^{y-1}, s^y] && (11.5)\\
&= \mathtt{Pr}_{\boldsymbol{\theta}}^{\mathrm{POMG}}[s^{y+1} \mid \mu^{y-1}, s^y, h^y, a^y]\, \mathtt{Pr}_{\boldsymbol{\theta}}^{\mathrm{POMG}}[a^y \mid \mu^{y-1}, s^y, h^y]\, \mathtt{Pr}_{\boldsymbol{\theta}}^{\mathrm{POMG}}[h^y \mid \mu^{y-1}, s^y] && (11.6)\\
&= T^{\mathrm{POMG}}[s^{y+1} \mid s^y, a^y]\, (\theta_{\mathtt{player}(s^y)}(h^y))(a^y)\, Q^{\mathrm{POMG}}_{\mathtt{player}(s^y)}[h^y \mid s^y] && (11.7)
\end{aligned}$$

Now based on this observation, we derive the probability $\mathtt{Pr}_{\boldsymbol{\theta}}^{\mathrm{POMG}}$ of transitioning from one state, say s^1, to another, say $s^{\tau+1}$, assuming players are abiding by policy profile $\boldsymbol{\theta}$:

$$\mathtt{Pr}_{\boldsymbol{\theta}}^{\mathrm{POMG}}[s^1] = \zeta^{\mathrm{POMG}}(s^1) \qquad (11.8)$$

$$\begin{aligned}
\mathtt{Pr}_{\boldsymbol{\theta}}^{\mathrm{POMG}}[s^{\tau+1} \mid s^1] &= \sum_{\mu^\tau_{-s_1}} \mathtt{Pr}_{\boldsymbol{\theta}}^{\mathrm{POMG}}[h^1, a^1, \ldots, s^\tau, h^\tau, a^\tau, s^{\tau+1} \mid s^1] && (11.9)\\
&= \sum_{\mu^\tau_{-s_1}} \prod_{y=1}^{\tau} \mathtt{Pr}_{\boldsymbol{\theta}}^{\mathrm{POMG}}[h^y, a^y, s^{y+1} \mid \mu^{y-1}, s^y] && (11.10)\\
&= \sum_{\mu^\tau_{-s_1}} \prod_{y=1}^{\tau} T^{\mathrm{POMG}}[s^{y+1} \mid s^y, a^y]\; \underbrace{(\theta_{\mathtt{player}(s^y)}(h^y))(a^y)}_{\text{probability } \mathtt{player}(s^y) \text{ plays } a^y}\; Q^{\mathrm{POMG}}_{\mathtt{player}(s^y)}[h^y \mid s^y] && (11.11)
\end{aligned}$$

In words, this probability is the sum over all possible histories $\mu^{\tau}_{-s_1} = \langle h^1, a^1, \ldots, s^{\tau}, h^{\tau}, a^{\tau}\rangle$ of the product of the probability of observing each h^y, the probability (dictated by the players' policies) of playing each a^y, and the probability of transitioning to each s^{y+1}.

In our transformation from a multi-player TT-POMG to a single-agent POMDP, of particular concern are histories in which it is never a given player, say i's, turn to play. In such cases, we denote the corresponding probability as follows: $\mathtt{Pr}^{\text{POMG}}_{\boldsymbol{\theta}_{-i}}[s^{\tau+1} \mid s^1]$, indicating dependence on only $\boldsymbol{\theta}_{-i}$ not the full policy profile $\boldsymbol{\theta}$. This probability is also calculated using Eq. (11.11), but the sum over histories $\langle h^1, a^1, \ldots, s^{\tau}, h^{\tau}, a^{\tau}\rangle$ only includes those histories in which $s^2, \ldots, s^{\tau} \in S^{\text{POMG}}_{-i} \equiv \cup_{i' \neq i} S^{\text{POMG}}_{i}$. (N.B. $s_1 \in S^{\text{POMG}}_{-i}$.)

Our first transformation is encapsulated by the following definition of iPOMDPs, which are TT-POMGs from the point of view of a select player i.

Definition 7 (*iPOMDP*) Given a TT-POMG Θ, and an other-agent profile of memoryless policies $\boldsymbol{\theta}_{-i}$, we define an iPOMDP $\Psi_i(\Theta, \boldsymbol{\theta}_{-i})$, from the point of view of player i, as follows:

- $S^{i\text{POMDP}} = S_i \cup Z$, where $S_i = S^{\text{POMG}}_i$ and $Z = Z^{\text{POMG}}$. (Observe that $S^{i\text{POMDP}} = S^{\text{POMG}} \backslash S^{\text{POMG}}_{-i}$.)
- $A^{i\text{POMDP}} = \bigcup_j A_j$, where $A_j = A^{\text{POMG}}_{ij}$ is the set of i's actions at state $j \in S_i$.
- $T^{i\text{POMDP}}$ defines the probability of transitioning from state $j \in S_i$ to state $j' \in S^{i\text{POMDP}}$ when action $k \in A_j$ is chosen. This probability is calculated as you would expect—it is the sum over all other-player next states s in Θ of the probability of transitioning to s times the probability of transitioning from s to j', as determined by the other-agent policy profile $\boldsymbol{\theta}_{-i}$:

$$T^{i\text{POMDP}}[j' \mid j, k] = \sum_{s \in S^{\text{POMG}}_{-i}} T^{\text{POMG}}[s \mid j, k]\ \mathtt{Pr}^{\text{POMG}}_{\boldsymbol{\theta}_{-i}}[j' \mid s] \tag{11.12}$$

- $O^{i\text{POMDP}} = O^{\text{POMG}}_i$.
- $Q^{i\text{POMDP}}$ is an observation function defined such that $Q^{i\text{POMDP}}[h \mid j] = Q^{\text{POMG}}_i[h \mid j]$, for all $j \in S^{i\text{POMDP}}$ and $h \in O^{i\text{POMDP}}$.
- $r^{i\text{POMDP}}$ is a reward function defined such that $r^{i\text{POMDP}}(j) = r^{\text{POMG}}_i(j)$, for all $j \in S^{i\text{POMDP}}$.
- Recall that $\mathtt{Pr}^{\text{POMG}}_{\boldsymbol{\theta}_{-i}}[j \mid s]$ is the total probability of all histories through the TT-POMG that start at state s, and lead to state $j \in S^{i\text{POMDP}}$, without encountering a state in which it is i's turn to move. So, the total probability of all histories through the TT-POMG that lead to state $j \in S^{i\text{POMDP}}$ without encountering a state in which it is i's turn to move is given by:

$$\zeta^{i\text{POMDP}}(j) = \sum_{s \in S^{\text{POMG}}_{-i}} \zeta^{\text{POMG}}(s)\ \mathtt{Pr}^{\text{POMG}}_{\boldsymbol{\theta}_{-i}}[j \mid s] \tag{11.13}$$

We have defined iPOMDPs so that, from the point of view of player i, a TT-POMG Θ and an iPOMDP $\Psi_i(\Theta, \boldsymbol{\theta}_{-i})$ are identical. First, a policy for the agent in $\Psi_i(\Theta, \boldsymbol{\theta}_{-i})$ is a mathematically equivalent object to a policy for player i in Θ. This is because a policy in a POMDP depends on a history—a sequence of observations and actions. Likewise, in a TT-POMG, a policy for a player depends only on that player's observations and actions. In other words, player i observes exactly the same history in the TT-POMG and the corresponding iPOMDP, and hence makes decisions on exactly the same basis in both models.

Second, by construction, the probability of an i-history (i.e., $\langle s_i^1, h_i^1, a_i^1, \ldots s_i^T, h_i^T, a_i^T \rangle$) is equal in both models, so that $R^{i\text{POMDP},T}(\psi) = R^{\text{POMG},T}(\psi)$, for all policies ψ. From these two observations, it follows that a best-response to a memoryless policy profile $\boldsymbol{\theta}_{-i}$ in Θ is an optimal policy in $\Psi_i(\Theta, \boldsymbol{\theta}_{-i})$, and vice versa.

Theorem 1 *An optimal (not necessarily memoryless) policy ψ in an iPOMDP $\Psi_i(\Theta, \boldsymbol{\theta}_{-i})$ is a best-response to an other-agent profile of memoryless policies $\boldsymbol{\theta}_{-i}$ in the original TT-POMG Θ, and vice versa.*

A proof of this theorem appears in "Appendix 1".

11.2.2 POMDP to MDP Transformation

The next transformation we present is a variation of one that is commonly used in the literature (e.g., Kaelbling 1998): from POMDPs to *belief-state* MDPs. Like our first transformation, this transformation preserves optimal policies: i.e., an optimal policy in the ensuing belief-state MDP is an optimal policy in the original POMDP, and vice versa.

Although this transformation applies more generally, we restrict our attention to POMDPs with *perceptual aliasing* (Whitehead and Ballard 1991)—namely, those in which multiple states give rise to a single observation—because they are sufficient to represent EFGs. Specifically, we assume the set of observations O is a partition of the underlying state space. Furthermore, we assume that each state $j \in h$ gives rise precisely to observation $h \in O$. With this specific structure in mind, we describe a belief state not as a probability distribution over the entire state space, but as a conditional probability distribution over a restricted set of states (i.e., an observation). We call these MDPs *observation MDPs*, which we abbreviate OMDPs.

Definition 8 (*OMDP*) Given a POMDP Ψ, we construct an OMDP $\Phi(\Psi)$ as follows:

- S^{OMDP} is the set of all possible pairs $\langle h, \beta \rangle$ consisting of an observation $h \in O^{\text{POMDP}}$ together with beliefs $\beta \in \Delta(h)$ (i.e., a probability distribution over the set of states contained in the observation h).
- $A^{\text{OMDP}} = \bigcup_{h \in O^{\text{POMDP}}} A(h)$, where $A(h)$ denotes the set of actions available to the agent at observation $h \in O^{\text{POMDP}}$.
- The probability of transitioning from state $\langle h, \beta \rangle$ to state j' in Ψ upon taking action $k \in A(h)$ is computed in expectation, where the expectation is taken with respect to beliefs β:

$$T^{\mathrm{POMDP}}[j' \mid \langle h, \beta \rangle, k] = \sum_{j \in h} \beta[j \mid h] \, T^{\mathrm{POMDP}}[j' \mid j, k] \tag{11.14}$$

Likewise, the probability of transitioning from state $\langle h, \beta \rangle$ to observation h' in Φ is:

$$T^{\mathrm{OMDP}}[h' \mid \langle h, \beta \rangle, k] = \sum_{j' \in S^{\mathrm{POMDP}}} T^{\mathrm{POMDP}}[j' \mid \langle h, \beta \rangle, k] \, Q^{\mathrm{POMDP}}[h' \mid j'] \tag{11.15}$$

$$= \sum_{j' \in h'} T^{\mathrm{POMDP}}[j' \mid \langle h, \beta \rangle, k] \tag{11.16}$$

Equation (11.16) follows from the assumption of perceptul aliasing, which implies that $Q^{\mathrm{POMDP}}[h' \mid j']$ is an indicator function: i.e., its evaluates to 1 when $j' \in h'$, and 0 otherwise.

Conditioned on h', beliefs transition deterministically according to Bayes' rule: for $j' \in S^{\mathrm{POMDP}}$,

$$\beta'[j' \mid h'] = \frac{\Pr[h' \mid j'] \Pr[j']}{\sum_{j''} \Pr[h' \mid j''] \Pr[j'']} \tag{11.17}$$

$$= \frac{Q^{\mathrm{POMDP}}[h' \mid j'] \Pr[j']}{\sum_{j'' \in S^{\mathrm{POMDP}}} Q^{\mathrm{POMDP}}[h' \mid j''] \Pr[j'']} \tag{11.18}$$

Assuming perceptual aliasing, and then simplifying accordingly yields: for $j' \notin h'$, $\beta'[j' \mid h'] = 0$, while for $j' \in h'$,

$$\beta'[j' \mid h'] = \frac{\Pr[j']}{\sum_{j'' \in h'} \Pr[j'']} \tag{11.19}$$

$$= \frac{T^{\mathrm{POMDP}}[j' \mid \langle h, \beta \rangle, k]}{T^{\mathrm{POMDP}}[h' \mid \langle h, \beta \rangle, k]} \tag{11.20}$$

Thus, $T^{\mathrm{OMDP}}[\langle h', \beta' \rangle \mid \langle h, \beta \rangle, k] = T^{\mathrm{POMDP}}[h' \mid \langle h, \beta \rangle, k]$.

- Like transition probabilities, the reward at observation h is an expectation, where the expectation is computed with respect to beliefs β:

$$r_{\beta}^{\mathrm{POMDP}}(h) = \sum_{j \in h} \beta[j \mid h] \, r^{\mathrm{POMDP}}(j) \tag{11.21}$$

The reward at state $\langle h, \beta \rangle$ is defined in exactly this way: $r^{\mathrm{OMDP}}(\langle h, \beta \rangle) = r_{\beta}^{\mathrm{POMDP}}(h)$.

- The initial probability of observation h is given by (assuming perceptual aliasing):

$$\zeta^{\mathrm{POMDP}}(h) = \sum_{j \in S^{\mathrm{POMDP}}} \zeta^{\mathrm{POMDP}}(j) \, Q^{\mathrm{POMDP}}[h \mid j] = \sum_{j \in h} \zeta^{\mathrm{POMDP}}(j) \tag{11.22}$$

Conditioned on h, the initial beliefs β_0 are deterministic: for $j \in h$,

$$\beta_0[j \mid h] = \frac{\zeta^{\texttt{POMDP}}(j)}{\zeta^{\texttt{POMDP}}(h)} \tag{11.23}$$

Thus, $\zeta^{\texttt{OMDP}}(\langle h, \beta_0\rangle) = \zeta^{\texttt{POMDP}}(h)$, and $\zeta^{\texttt{OMDP}}(\langle h, \beta\rangle) = 0$ for all $\beta \neq \beta_0$.

We have defined OMDPs so that, from the point of view of the agent, a POMDP Ψ and an OMDP $\Phi(\Psi)$ are identical. First, a policy for the agent in Ψ is a mathematically equivalent object to a policy for the agent in $\Phi(\Psi)$. This is because a policy in an OMDP is a function from observations together with beliefs to actions. A (general, not memoryless) policy in a POMDP is a function from histories to actions. But beliefs in an OMDP encode histories in a POMDP! (Beliefs are updated as actions are taken and observations are made.) Hence, the agent makes decisions on exactly the same basis in both models.

Second, by construction, the probability of a history $\langle s^1, h^1, a^1, \ldots, a^\tau, s^{\tau+1}, h^{\tau+1}\rangle$ in the POMDP is precisely the belief $\beta^{\tau+1}$ that arises in the history $\langle\langle h^1, \beta^1\rangle, a^1, \ldots, a^\tau, \langle h^{\tau+1}, \beta^{\tau+1}\rangle\rangle$ in the OMDP, so that $R^{\texttt{OMDP},\tau}(\phi) = R^{\texttt{POMDP},\tau}(\phi)$, for all policies ϕ. From these two observations, it follows that an optimal policy in Ψ is an optimal policy in $\Phi(\Psi)$, and vice versa.

Theorem 2 *An optimal policy ϕ in an OMDP $\Phi(\Psi)$ is an optimal policy in the original POMDP Ψ, and vice versa.*

Observe that beliefs in an OMDP are deterministic. Consequently, it is not actually necessary to represent beliefs explicitly, nor define policies in an OMDP as dependent on them. Instead, it is sufficient to define an OMDP policy as s function from observations to actions. Likewise, it is sufficient to model transitions between observations alone, provided $T^{\texttt{POMDP}}[h' \mid \langle h, \beta\rangle, k]$ can be computed accurately. Indeed, the key to our reinforcement learning approach is to estimate transition probabilities $T^{\texttt{OMDP}}[h' \mid h, k]$ by approximating $T^{\texttt{POMDP}}[h' \mid \langle h, \beta\rangle, k]$ without explicitly computing β.

11.3 From EFGs to OMDPs

Thus far, we have described how to interpret a TT-POMG Θ from the point of view of player i, together with a strategy profile $\boldsymbol{\theta}_{-i}$ for all players other than i, as a partially-observable Markov decision process (POMDP), $\Psi(\Theta, \boldsymbol{\theta}_{-i})$, such that an optimal policy in $\Psi(\Theta, \boldsymbol{\theta}_{-i})$ is a best-response to $\boldsymbol{\theta}_{-i}$ in Θ. We achieved this result by folding the other-agent strategies into the transition probabilities and the initial probability distribution of the TT-POMG. We also argue that, by taking observations and beliefs together as states, and by defining transition probabilities, rewards, and initial probabilities appropriately, a POMDP can be viewed as an equivalent OMDP—an observation MDP—such that an optimal policy in the OMDP is an optimal policy in the POMDP. Combining these observations, we can solve for the best-response in a TT-POMG by solving for an optimal policy in the equivalent MDP.

In this section, we apply these insights to extensive-form games, which we view as a special kind of TT-POMG. After doing so, we obtain the following theorem:

Theorem 3 *If we view a given EFG* Γ *as a TT-POMG, and transform* Γ *and a given strategy profile* $\boldsymbol{\theta}_{-i}$ *into an OMDP* $\Phi(\Psi(\Gamma, \boldsymbol{\theta}_{-i}))$ *via the aforementioned transformations, then an optimal policy in* $\Phi(\Psi(\Gamma, \boldsymbol{\theta}_{-i}))$ *is a best-response to* $\boldsymbol{\theta}_{-i}$ *in* Γ.

11.3.1 Extensive-Form Games of Imperfect Information

An extensive-form game (EFG) is a model of dynamic multi-agent interactions. It is most frequently used to represent a non-cooperative sequential decision-making setting.

Technically, an EFG is a game *tree* comprised of *nodes* and *directed edges*, as illustrated in Fig. 11.1. Each non-terminal node is a decision point, controlled by exactly one player, who chooses from the available actions, represented by outgoing edges. A *trajectory*, which arises as the game is played, is a sequence of nodes from the root to a leaf (i.e., a terminal node), with *payoffs* specified at the leaves. Consequently, a game trajectory and ensuing payoffs are jointly determined by all the players' actions.

In an EFG of *imperfect information*, each player's decision nodes are partitioned into *information sets*. When choosing actions, players know only their information

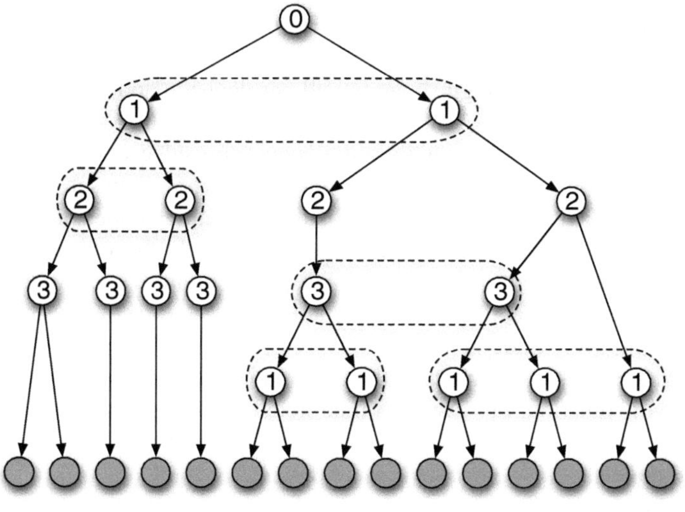

Fig. 11.1 Sample EFG of imperfect information: *circles* are nodes, *arrows* are edges, and *dashed ovals* are non-singleton information sets. Rewards are not specified. Each node is labeled with the identity of its controlling player. Nature is player 0

set, not the precise decision node within that set, and therefore make decisions under uncertainty. In Fig. 11.1, nodes in the same information set are encircled with dashed lines; any node that is not so encircled is its own singleton information set.

To model non-strategic uncertainty (i.e., uncertainty due to forces other than the strategic play of other agents), some games include a special player called nature, who chooses actions at *chance nodes* according to a distribution commonly known to all (other) players. In Fig. 11.1, chance nodes are labeled 0. Player 1 does not observe nature's action before her first decision, while player 2 does observe player 1's action if nature chooses the right branch, but not if she chooses the left branch.

With this picture in mind, we now proceed to formally define EFGs of imperfect information. Our definition is based on that of Hart (1992).

Definition 9 (*Extensive-form game of imperfect information*) An extensive-form game of imperfect information Γ is defined as a tuple $\langle N, T, P, H, u\rangle$ where:

- $N = \{0, 1, ..., n\}$ is the set of players, with a select (non-strategic) player 0 called nature.
- T is a tree, called the game tree, with nodes V (including a select root node), directed edges E, and terminal nodes $Z \subset V$.
- P is a partition of non-terminal nodes $V \backslash Z$ into subsets $P_0, P_1, \ldots, P_n$, where P_i is the set of player i's decision nodes: i.e., those in which it is player i's turn to choose an action. (The nodes in P_0 are the chance nodes.)
  We denote by A_{ij} the set of actions available to player i at node j. Note that this set is isomorphic to the set of outgoing edges leaving node $j \in P_i$. (We write $\texttt{action}(e)$ to denote the action associated with edge e.)
- $H = \langle H_0, H_1, \ldots, H_n\rangle$ is a tuple of partitions, one for each player i. Each H_i is a partition of the corresponding P_i. Each subset of nodes $h \in H_i$ is called an *information set*. Intuitively, a player cannot distinguish among the decision nodes in one of his information sets. However, H_0 is assumed to consist of only singletons, so nature can always distinguish among all her decision nodes.
  Note: For any two nodes j and j' in the same information set, there must be a 1-to-1 correspondence between j's and j''s outgoing edges: i.e., exactly the same set of actions must be available at all nodes in an information set. If this were not the case, then players would be able to distinguish among nodes in an information set, but information sets are meant to model precisely those nodes that players *cannot* distinguish from one another. With this in mind, we let $A_i(h)$ denote the set of possible actions at player i's information set $h \in H_i$ (i.e., the set of outgoing edges leaving every node in h).
  (We write $\texttt{infoset}(j)$ to denote the information set associated with state j.)
- u is a payoff function. For each terminal node $z \in Z$, $u(z) = \langle u_1(z), \ldots, u_n(z)\rangle$ is an n-dimensional vector of payoffs, where $u_i(z)$ denotes player i's payoff at terminal node z.
- $\mathcal{P}$ is a set of probability distributions, one for each node $j \in P_0$ (or equivalently, each information set $h \in H_0$). Each $p \in \mathcal{P}$ is a probability distribution over the outgoing edges (i.e., nature's possible actions): i.e., $p(\texttt{action}(e))$ is the probability of the action in $A_0(h)$ associated with edge e.

A *pure strategy* for player i is a function s_i on H_i such that $s_i(h) \in A_i(h)$; that is, s_i specifies an action for player i at each information set. Let S_i be the set of all possible pure strategies. A *mixed strategy* is a randomization over pure strategies; that is, a mixed strategy is an element of $\Delta(S_i)$.[2]

A *behavioral strategy* for player i is a function σ_i on H_i such that $\sigma_i(h) \in \Delta(A_i(h))$; that is, σ_i specifies a distribution over actions for player i at each information set. Let Σ_i be the set of player i's possible behavioral strategies.

We assume a game of *perfect recall*, in which agents do not forget information they have previously acquired (e.g., players do not forget their own past actions). By Kuhn's theorem (Kuhn 1953), in a game of perfect recall, every mixed strategy has a corresponding payoff-equivalent behavioral strategy. Thus, as is common in the game-theoretic literature, we restrict our attention to behavioral strategies.

A *strategy profile* $\sigma = (\sigma_1, \ldots, \sigma_n)$ dictates a behavioral strategy for each player. Let $U_i(\sigma)$ be player i's *expected* utility when strategy profile σ is played, where the expectation is taken over game trajectories induced by σ (the players' behavioral strategies) and over nature's random actions.

Let $\sigma_{-i} = (\sigma_1, \ldots, \sigma_{i-1}, \sigma_{i+1}, \ldots, \sigma_n)$ dictate a strategy for every player except player i.

Definition 10 (*Best response*) Given an other-agent strategy profile σ_{-i}, a *best response* for player $i \neq 0$ in an EFG Γ of imperfect information is a strategy σ_i such that $U_i(\sigma) \geq U_i(\sigma_i', \sigma_{-i}), \forall \sigma_i' \in \Sigma_i$.

Definition 11 (*Bayes-Nash equilibrium*) A strategy profile σ is a *Bayes-Nash Equilibrium* (NE) of an EFG Γ of imperfect information if σ consists of only best responses: i.e., $U_i(\sigma) \geq U_i(\sigma_i', \sigma_{-i})$, for all players $i \neq 0$ and for all $\sigma_i' \in \Sigma_i$. Note that nature, whose actions are specified by $\mathcal{P}$, need not best-respond.

Having defined EFGs, we will now go on to show that EFGs are a special kind of turn-taking partially-observable Markov game.

11.3.2 EFGs as TT-POMGs

In this section, we instantiate our definitionn of TT-POMGs with EFGs, viewing the latter as an instance of the former. Indeed, TT-POMGs generalize EFGs in the following ways:

1. TT-POMGs may distribute rewards at intermediate states, while EFGs distribute payoffs at terminal nodes only. Furthermore, TT-POMG rewards can depend on state-action pairs, while EFG payoffs are typically associated with states alone.
2. EFGs are typically trees of finite horizon, with designated terminal nodes, while TT-POMGs can be more general graphs.

---

[2]Given a discrete set X, the notation $\Delta(X)$ is commonly used to represent the set of all probability distributions over X: i.e., $\Delta(X) = \{\langle p_1, \ldots, p_{|X|}\rangle \mid \sum_{i=1}^{|X|} p_i = 1 \text{ and } p_i \geq 0, \forall i \in \{1, \ldots, |X|\}\}$.

3. TT-POMGs may have more general informational structures. In EFGs (with perfect recall), there is a unique path to each information set; the same cannot be said for TT-POMGs.

To simplify the exposition, we restrict our attention to EFGs in which nature plays just once, and first. In this case, we can easily fold nature's play into the initial probability distribution of the TT-POMG, and eliminate nature as a player in the game. Modulo notational complications, our results extend immediately to EFGs, such as backgammon, in which nature plays intermittently.

Notation: Given a set X, we let $X^* = \{\{x\} \mid x \in X\}$.

Observation 4 *Every EFG Γ can be viewed as a TT-POMG Θ by taking the set of states to be the set of paths in the EFG game tree,*[3] *and taking the set of observations to be the set of (paths to) information sets.*

Proof

- $N^{\text{POMG}} = N^{\text{EFG}} \backslash \{0\}$: i.e., nature is not a player in this game.
- $S^{\text{POMG}} = \langle S_1, \ldots, S_n, Z \rangle$, where for all $i \in N^{\text{POMG}}$, $S_i = \{\text{the path to } p \mid \forall p \in P_i^{\text{EFG}}\}$, and $Z = \{\text{the path to } z \mid \forall z \in Z^{\text{EFG}}\}$.[4]
- A^{POMG} includes all edges in E^{EFG} except those emanating from the root: i.e., $A^{\text{POMG}} = \bigcup_{i \in N^{\text{POMG}}} A_i^{\text{POMG}}$, where $A_i^{\text{POMG}} = \bigcup_{j \in S_i} A_{ij}^{\text{EFG}}$. The edges emanating from the root represent nature's moves, which are folded into the initial probabilities.
- $T^{\text{POMG}}[j' \mid j, k]$ is a deterministic function that specifies whether or not the game transitions to state j' when the player whose turn it is in state j takes action k.
- $O^{\text{POMG}} = \langle O_1, \ldots, O_n \rangle$, where $O_i = H_i^{\text{EFG}} \cup Z^*$. In other words, (paths to) information sets in the EFG are encoded as observations in the TT-POMG, as are the terminal states.
- $Q^{\text{POMG}} = \langle Q_1, \ldots, Q_n \rangle$, where Q_i is an observation function such that

$$Q_i[h \mid j] = \begin{cases} 1 & \text{if } j \in h \\ 0 & \text{otherwise} \end{cases} \tag{11.24}$$

  for all $j \in S_i$ and $h \in O_i$. The players observe termination.
- $r^{\text{POMG}} = \langle r_1, \ldots, r_n \rangle$, where r_i is a reward function such that

$$r_i(j) = \begin{cases} u_i(j) & \text{if } j \in Z \\ 0 & \text{otherwise} \end{cases} \tag{11.25}$$

  for all $j \in S^{\text{POMG}}$.

---

[3]Consider, for example, a representation of the simultaneous-move Prisoners' Dilemma in the extensive form. This game, in the TT-POMG framework, would not be Markov unless the states in the TT-POMG are the paths to the nodes in the EFG. Otherwise, transitions would have to depend not only on the current state, but on past actions as well.

[4]Note that defining states as paths rather than nodes does not lead to any increase in the size of the representation, because there are the same number of nodes and paths in a tree, and paths can be stored efficiently.

- The initial probability $\zeta^{\text{POMG}}(j) = p^{\text{EFG}}(\texttt{action}(\text{root}, j))$, for all $j \in S^{\text{POMG}}$.

In the next two sections, we apply the transformations described in Sects. 11.2.1 and 11.2.2 to the TT-POMG defined in Observation 4, namely to an EFG. The result is an MDP that is strategically equivalent from the point of view of a single player to an EFG (together with an other-agent strategy profile).

11.3.3 EFGs as iPOMDPs

After transforming an EFG, viewed as a TT-POMG, together with other-agent policies $\boldsymbol{\theta}_{-i}$, this is the iPOMDP that ensues (see Fig. 11.2):

- $S^{i\text{POMDP}} = S_i \cup Z$, where $S_i = P_i^{\text{EFG}}$ and $Z = Z^{\text{EFG}}$.
- $A^{i\text{POMDP}}$ includes all edges in E^{EFG} emanating from states at which it is i's turn to play. Let A_j denote the actions available at state $j \in S_i$.
- $T^{i\text{POMDP}}$ defines the probability of transitioning from state $j \in S_i$ to state $j' \in S^{i\text{POMDP}}$ when action $k \in A_j$ is chosen. More specifically, $T^{i\text{POMDP}}[j' \mid j, k]$ is the total probability of all histories (in which i has no moves) to state j', originating at state j with action k:

$$
\begin{aligned}
T^{i\text{POMDP}}[j' \mid j, k] &= \sum_{s \in S_{-i}^{\text{POMG}}} T^{\text{POMG}}[s \mid j, k]\, \texttt{Pr}_{\boldsymbol{\theta}_{-i}}^{\text{POMG}}[j' \mid s] \qquad (11.26) \\
&= \sum_{\mu_{-s_1}^{\mathcal{T}} \in H_{-i}^{\mathcal{T}}} \prod_{y=1}^{\mathcal{T}} T^{\text{POMG}}[s^{y+1} \mid s^y, a^y] \underbrace{[(\theta_{\texttt{player}(s^y)}(h^y))(a^y)]}_{\text{probability } \texttt{player}(s^y) \text{ plays } a^y} Q_{\texttt{player}(s^y)}^{\text{POMG}}[h^y \mid s^y]
\end{aligned}
$$

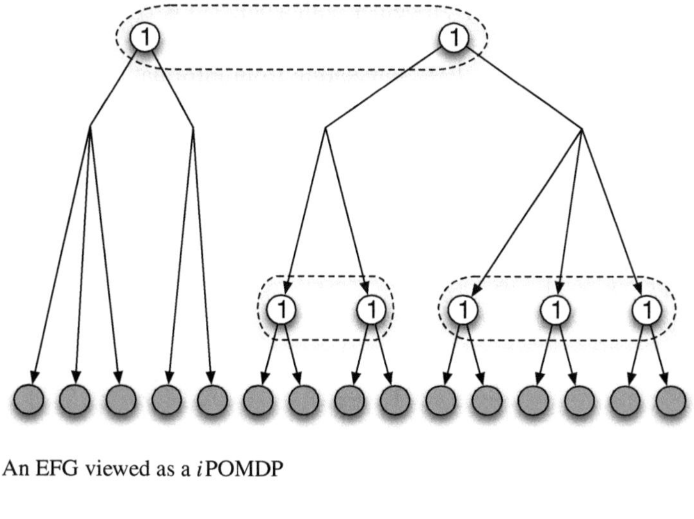

Fig. 11.2 An EFG viewed as a iPOMDP

- $O^{i\text{POMDP}} = H_i^{\text{EFG}} \cup Z^*$. In other words, information sets in the EFG are encoded as observations in the iPOMDP, as are the terminal states.
- $Q^{i\text{POMDP}}$ is an observation function such that

$$Q^{i\text{POMDP}}[h \mid j] = \begin{cases} 1 & \text{if } j \in h \\ 0 & \text{otherwise} \end{cases} \tag{11.27}$$

for all $j \in S^{i\text{POMDP}}$ and $h \in O^{i\text{POMDP}}$. The agent observes termination.
- $r^{i\text{POMDP}}$ is a reward function such that

$$r^{i\text{POMDP}}(j) = \begin{cases} u_i(j) & \text{if } j \in Z \\ 0 & \text{otherwise} \end{cases} \tag{11.28}$$

for all $j \in S^{i\text{POMDP}}$.
- The initial probability of a state $j \in S^{i\text{POMDP}}$ is the total probability of all paths in the TT-POMG that lead to j without encountering a state in which it is i's turn to move:

$$\zeta^{i\text{POMDP}}(j) = \sum_{s \in S_{-i}^{\text{POMG}}} p^{\text{EFG}}(\texttt{action}(\text{root}, s)) \, \texttt{Pr}_{\boldsymbol{\theta}_{-i}}^{\text{POMG}}[j \mid s] \tag{11.29}$$

11.3.4 EFGs as OMDPs

After transforming an EFG, viewed as a iPOMDP, this is the OMDP that ensues (see Fig. 11.3):

- S^{OMDP} is the set of all possible pairs $\langle h, \beta \rangle$ consisting of an information set or terminal state $h \in H_i^{\text{EFG}} \cup (Z^{\text{EFG}})^*$, together with a belief $\beta \in \Delta(h)$ (i.e., a probability distribution over the set of states contained in h).
- A^{OMDP} includes all edges in E^{EFG} emanating from information sets at which it is i's turn to play. Let $A(h)$ denote the actions available at information set $h \in H_i^{\text{EFG}}$.
- The probability of transitioning from state $\langle h, \beta \rangle$ to state j' in the POMDP upon taking action $k \in A(h)$ is computed in expectation, where the expectation is taken with respect to beliefs β:

$$T^{i\text{POMDP}}[j' \mid \langle h, \beta \rangle, k] = \sum_{j \in h} \beta[j \mid h] \, T^{i\text{POMDP}}[j' \mid j, k] \tag{11.30}$$

$$= \sum_{j \in h} \beta[j \mid h] \left(\sum_{s \in S_{-i}^{\text{POMG}}} T^{\text{POMG}}[s \mid j, k] \, \texttt{Pr}_{\boldsymbol{\theta}_{-i}}^{\text{POMG}}[j' \mid s] \right) \tag{11.31}$$

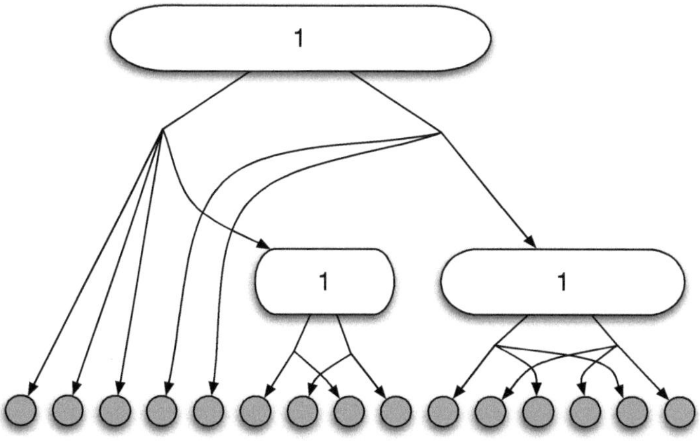

Fig. 11.3 An EFG viewed as an OMDP

Likewise, the probability of transitioning from state $\langle h, \beta \rangle$ to observation h' in Φ is:

$$T^{\text{OMDP}}[h' \mid \langle h, \beta \rangle, k] = \sum_{j' \in h'} T^{i\text{POMDP}}[j' \mid \langle h, \beta \rangle, k] \tag{11.32}$$

$$= \sum_{j' \in h'} \sum_{j \in h} \beta[j \mid h] \left(\sum_{s \in S_{-i}^{\text{POMG}}} T^{\text{POMG}}[s \mid j, k] \, \text{Pr}_{\boldsymbol{\theta}_{-i}}^{\text{POMG}}[j' \mid s] \right) \tag{11.33}$$

Conditioned on h', beliefs transition deterministically: for $j' \notin h'$, $\beta'[j' \mid h'] = 0$, while for $j' \in h'$,

$$\beta'[j' \mid h'] = \frac{T^{i\text{POMDP}}[j' \mid \langle h, \beta \rangle, k]}{T^{i\text{POMDP}}[h'' \mid \langle h, \beta \rangle, k]} \tag{11.34}$$

Thus, $T^{\text{OMDP}}[\langle h', \beta' \rangle \mid \langle h, \beta \rangle, k] = T^{i\text{POMDP}}[h' \mid \langle h, \beta \rangle, k]$.

- Like transition probabilities, the reward at observation h is an expectation, where the expectation is computed with respect to beliefs β:

$$r_{\beta}^{i\text{POMDP}}(h) = \sum_{j \in h} \beta[j \mid h] \, r^{i\text{POMDP}}(j) \tag{11.35}$$

The reward at state $\langle h, \beta \rangle$ is defined in exactly this way: $r^{\text{OMDP}}(\langle h, \beta \rangle) = r_{\beta}^{i\text{POMDP}}(h)$.

- The initial probability of observation h is given by

$$\zeta^{i\mathtt{POMDP}}(h) = \sum_{j \in h} \zeta^{i\mathtt{POMDP}}(j) \tag{11.36}$$

Conditioned on h, the initial beliefs β_0 are deterministic: for $j \in h$,

$$\beta_0[j \mid h] = \frac{\zeta^{i\mathtt{POMDP}}(j)}{\zeta^{i\mathtt{POMDP}}(h)} \tag{11.37}$$

Thus, $\zeta^{\mathtt{OMDP}}(\langle h, \beta_0\rangle) = \zeta^{i\mathtt{POMDP}}(h)$, and $\zeta^{\mathtt{OMDP}}(\langle h, \beta\rangle) = 0$ for all $\beta \neq \beta_0$.

11.3.5 Beliefs

Beliefs calculated via Bayesian updating, as in Eq. (11.34), are called *consistent*. However, it can happen that the denominator (and numerator) in Eq. (11.34) evaluates to 0. In this case, it is not possible to compute β', so transition probabilities out of state $\langle h, \beta\rangle$ in the OMDP are left unspecified. We will encounter this issue in Sect. 11.5, when we search for equilibria in EFGs. For now, we simply note that our OMDPs have not yet been fully specified.

Setting aside the problem of zero-probability events, there is another problem with computing beliefs, and that is their sheer complexity. Even representing beliefs is expensive. In the next section, we leverage reinforcement learning methods to solve OMDPs, without representing beliefs explicitly, nor computing them via Eq. 11.34. Specifically, we learn only the transition probabilities between information sets. Because beliefs are deterministic, these probabilities are sufficient for finding an optimal policy in an OMDP, and, in turn, an equilibrium in an EFG.

11.4 RL Approach to Solving OMDPs

Recall that our goal is to develop a best-response finding algorithm for EFGs. Having reduced EFGs to OMDPs, our present subgoal, then, is to develop an algorithm that solves for an optimal policy in an OMDP. We take a reinforcement learning (RL) approach, and assume a black box capable of simulating the (O)MDP (or, equivalently, the EFG, given an other-agent strategy profile).

At a high level, our approach is straightforward: first, we learn an approximation of the MDP via simulation; then, we solve for an optimal policy in the approximate (i.e., learned) MDP. We do not specify a means of solving the learned MDP; this can be accomplished via dynamic programming, for example Bellman (1957). Our RL method primarily concerns the learning of the MDP via simulation.

Our main theorems quantify the number of samples necessary to learn an MDP, and thereby learn a near-optimal policy, to varying degrees of accuracy (with high probability). As noted above, this learning is accomplished without specific reference to the beliefs in the state space of the OMDP, but only to the transition probabilities. Consequently, our algorithm and its analysis are applicable to any tree-shaped MDP.

Our approach can be compared with model-free reinforcement learning methods, such as POMCP (Silver and Veness 2010), that learn optimal policies directly, without first learning an explicit model of the MDP. We opt for a model-based approach so that we can more easily exploit an independence structure that is typical of EFGs. In particular, when nature moves first and chooses a private type t_i for each player i independently, her move factors the game into N independent strategic components (which we transform into MDPs), one per player.[5] Specifically, from player i's perspective, nature's move factors $\boldsymbol{t}$ into (t_i, t_{-i}), where t_i influences only player i's (by assumption, terminal) rewards, and likewise for t_{-i}. Consequently, in player i's MDP, the transition probabilities do not depend on t_i (only the terminal rewards do), and the terminal rewards do not depend on t_{-i} (only the transition probabilities—and any intermediate rewards[6]—do). Our approach efficiently exploits this independence structure, while a typical model-free method would be forced to learn a different policy for all other-agent type profiles $\boldsymbol{t}_{-i}$.

11.4.1 Sampling Algorithm

We describe our sampling algorithm twice, first as simulating repeated play of an EFG, and second as simulating repeated trajectories through an iPOMDP.

Given an EFG, to learn player i's MDP while the other agents are playing $\boldsymbol{\sigma}_{-i}$:

- Repeatedly play the EFG, with:
  - nature playing its prescribed strategy
  - other-agents playing their prescribed strategies, $\boldsymbol{\sigma}_{-i}$
  - player i exploring all actions (e.g., with uniform probability)
- After collecting histories of many game trajectories, approximate i's OMDP as follows:

[5]For example, consider a one-shot first-price auction with *independent private values* Krishna (2009). At the start of the game, nature draws a vector of independent types, one per player: $t_i \sim F_i$. Each player i then bids according to strategy $\sigma_i : t_i \rightarrow$ BidSet. The highest bidder wins, paying her bid, and getting a payoff of t_i less her bid, while the others get 0.

[6]Our approach also applies when any intermediate rewards earned by player i depend on t_{-i}, but not t_i, which is the case in sequential first- and second-price auctions when payments are modeled as intermediate (negative) rewards.

- the initial probabilities $\zeta(h)$, by the empirical frequency of initializing the OMDP at information set h, meaning nature's and the other agents' moves before player i moves lead to information set h
- the transition probabilities $T[h' \mid h, k]$, by the empirical frequency of transitioning to information set h' when player i chooses action k at information set h, and the other agents play according to $\boldsymbol{\sigma}_{-i}$
- the expected rewards $r(h)$, by the average of all rewards received at information set h

This algorithm is restated in terms of Markov models in Algorithm 1.

Algorithm 1 Sampling Algorithm

Input: Γ together with $\boldsymbol{\sigma}_{-i}$ represented as an iPOMDP, σ_i, L
Output: an estimate of T^{OMDP} and r^{OMDP}

$T[h' \mid h, k] \leftarrow 0$, for all h, k, h'
$\zeta(h) \leftarrow 0$, for all h
$r(h) \leftarrow 0$, for all h
for $l \leftarrow 1$ to L **do**
  $j \leftarrow$ sample from $\zeta^{i\text{POMDP}}(\cdot)$
  $h \leftarrow \texttt{infoset}(j)$
  Increment $\zeta(h)$
  Increment $r(h)$
  while $j \notin Z$ **do**
    $k \leftarrow$ sample from $\sigma_i(h)$ ▷ If $\sigma_i(h)$ is uniform, all actions are sampled with equal probability.
    $j' \leftarrow$ sample from $T^{i\text{POMDP}}(j, k, \cdot)$
    $h' \leftarrow \texttt{infoset}(j')$
    Increment $T[h' \mid h, k]$
    Increment $r(h')$
    $j \leftarrow j'$
    $h \leftarrow h'$
  end while
end for
Normalize the Ts
Average the rs
return T
return r

11.4.2 Sample Complexity

Two particular properties inherited from EFGs pose particular challenges to learning OMDPs effectively. First of all, because EFGs are trees, the number of states in the corresponding OMDPs is exponential in the horizon. Specifically, since the player remembers her own past moves (perfect recall), even without observing her opponents' or nature actions, the number of information sets after H actions is at least A^H, where A is the number of available actions.

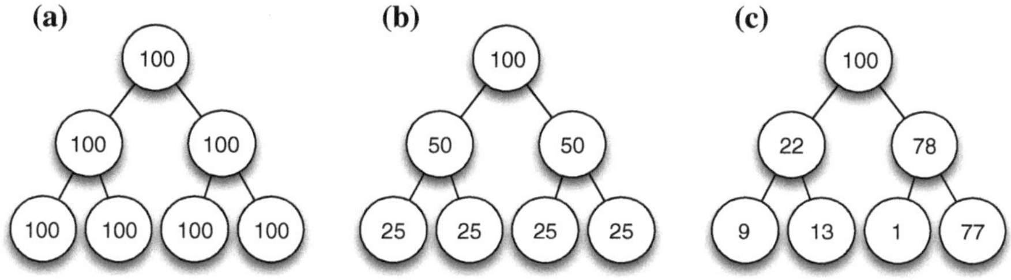

Fig. 11.4 The number of times states were sampled in a hypothetical OMDP. **a** The case assumed achievable in the sparse sampling algorithm of Kearns et al. (2002). **b** The best case that can possibly occur in learning an OMDP. **c** A typical black box simulation of an OMDP. The states deep in the tree can be poorly learned even if the total number of simulations is large

Second, each simulation consists of a single uninterrupted trajectory from the root to some leaf, so the nodes deep in the tree can only be sampled very infrequently and their dynamics poorly learned, even after numerous simulations. Even among nodes at the same level in the tree, the distribution can be very uneven. Kearns et al. (2002) designed a sparse sampling algorithm to efficiently plan in large MDPs with exponential state spaces, but their algorithm only works when it is possible to resample actions at any point along a trajectory. That way, the simulator can make the distribution of sample points look like that in Fig. 11.4a, instead of what it could easily look like in our case, namely Fig. 11.4c.

These two challenges—that the state space is exponential, and that all sample trajectories must originate at the root—more or less ensure that there will always be some states whose transitions are poorly learned, rendering it difficult to derive optimality guarantees for learned policies. Nonetheless, Theorem 4 states that our simple RL algorithm is guaranteed to learn a near-optimal policy with high probability in finite time.

Theorem 4 *Consider a tree-shaped MDP with finite horizon H, branching factor S, at most A actions per state, and total rewards in the range $[0, 1]$. Further, suppose transitions are unknown, but that we are given a black box that simulates the MDP. Assuming undiscounted rewards, there exists an exploration strategy for learning the MDP's transitions such that, for any $\epsilon, \delta > 0$, the following sample size guarantees that the learned policy (i.e., the optimal policy in the learned MDP) is ϵ-optimal with probability at least $1 - \delta$:*

$$N = \mathcal{O}\left(\left(\frac{2ASH}{\epsilon}\right)^{2H}\left(\log\frac{1}{\delta} + H\log S\right)\right) \tag{11.38}$$

Although this bound is exponential in the horizon, it is linear in both the branching factor S and the number of actions A. In many applications, such as K sequential auctions, the number of actions is quite large (it includes all combinations of bids, one per bidder, over some discretized bid space; or worse, over a continuous bid space), while the horizon is only K.

Nonetheless, this bound can perhaps still be improved by considering selective sampling schemes, such as UCT (Kocsis et al. 2006), or by better proof techniques. However this bound is not "too unsharp" either. We also prove a lower sample complexity bound which says it is impossible to get rid of an exponential dependence on the horizon H:

Theorem 5 (Lower bound) *For any sampling algorithm, even adaptive, there exists an MDP such that the algorithm must generate at least* $\Omega\left(\frac{A^H}{\epsilon}(1 - H\epsilon^{1/H})(1-\delta)\right)$ *samples to guarantee that the learned policy be* ϵ*-optimal with probability at least* $1 - \delta$.

The proofs of both theorems are constructive. For the upper bound, we examine convergence of the learned MDP policy under a *balanced wandering* exploration strategy: at every state, choose the least sampled action, and break ties uniformly randomly. For the lower bound, we explicitly construct an MDP that is difficult to learn. The full proofs are relegated to "Appendices 2" and "3".

Most sample complexity bounds in the literature apply to infinite-horizon discounted MDPs. In contrast, we are dealing with MDPs of finite horizon. But a casual look at such bounds confirms the difficulty in dealing with exponentially large state spaces. For instance, directly applying the result in Azar et al. (2012) yields a *lower bound* that is also linear in AS^H and $\log\frac{1}{\delta} + H\log(S)$.

11.5 An Iterative Search for Equilibrium

While our RL approach only computes one player's best response, we can apply it iteratively, to each player in turn, to search for an equilibrium in an EFG. This idea is motivated by the method of best-reply dynamics, commonly studied in game theory (Fudenberg and Levine 1998). Specifically, we propose the following iterative procedure:

1. Initialize with an arbitrary strategy profile $\boldsymbol{\sigma}^{(0)} = (\sigma_1^{(0)}, \sigma_2^{(0)}, \ldots, \sigma_n^{(0)})$.
2. For each iteration $l = 1, \ldots, L$ and for each player $i = 1, \ldots, N$:
   (a) Run Algorithm 1 given Γ and $\sigma_{-i}^{(l-1)}$ represented as an iPOMDP. Return player i's OMDP.
   (b) Update strategy $\sigma_i^{(l)}$ to be an optimal policy in the learned OMDP: i.e., a best response to $\boldsymbol{\sigma}_{-i}^{(l-1)}$.
3. Terminate with success if $\boldsymbol{\sigma}^{(l)}$ is the same as, or very similar to, $\boldsymbol{\sigma}^{(l-1)}$.

This algorithm is restated in terms of Markov models in Algorithm 2.

Algorithm 2 Equilibrium Search Algorithm

Input: Γ represented as a TT-POMG, L
Output: near-equilibrium policy profile $\boldsymbol{\theta}$
Initialize a (memoryless) policy profile $\boldsymbol{\theta}^{(0)}$
$l \leftarrow 0$
repeat
  $l \leftarrow l + 1$
  for $i \leftarrow 1$ to n **do**
    Build player i's iPOMDP, $\Psi^{(l)}$, given $\boldsymbol{\theta}_{-i}^{(l-1)}$
    Run Algorithm 1 on $\Psi^{(l)}$ to learn an OMDP, $\Phi^{(l)}$
    $\boldsymbol{\theta}_i^{(l)} \leftarrow$ an optimal policy in $\Phi^{(l)}$
  end for
until $\boldsymbol{\theta}_{-i}^{(l)}$ is similar to $\boldsymbol{\theta}_{-i}^{(l-1)}$ or $l \geq L$
return policy profile $\boldsymbol{\theta}^{(l)}$

It is desirable to initialize the (behavioral) strategy profile to be totally mixed, so that players choose all actions with positive probability, because doing so makes it more likely that all parts of the EFG will be explored. But regardless of initialization, after the first iteration, we can no longer be guaranteed that all information sets will be reached with positive probability.

Definition 12 Given a strategy profile $\boldsymbol{\sigma}$, information sets that cannot be reached with positive probability are called *off-path*, because they are guaranteed to be off the path of play. Information sets that can be reached with positive probability are called *on-path*.

Consider the example in Fig. 11.5, which involves two players, P1 and P2. Suppose beliefs are initialized so that the rightmost state occurs with probability $\alpha > 1/(2 - \epsilon)$, while the leftmost states in P2's information set occurs with probability $1 - \alpha$. Suppose further that P2 updates first. Given the aforementioned beliefs, his best response is to play r (because $\alpha(1 - \epsilon) > 1 - \alpha$). Now it is P1's turn to update; his best response to P2's move is L (because $1 > 0$). The baton is passed

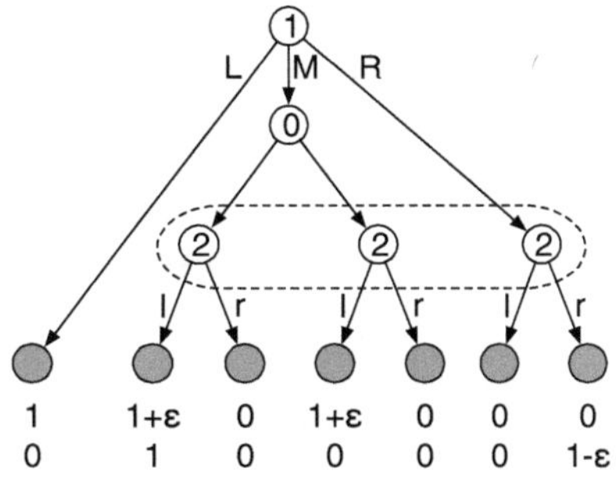

Fig. 11.5 An example EFG. The *top* and *bottom rows* of numbers below the terminal nodes are the payoffs for P1 and P2, respectively. Nature chooses among her branches with equal probability

back to P2 to update, but when P1 plays L, there is zero probability on P2's sole information set.

There is nothing for P2 to do when his sole information set is off-path, so let's allow P2 to invent some arbitrary beliefs. If he chooses uniform, then his best response is l. P1's best response when P2 plays l is M (because $1 + \epsilon > 1$). And now, P2's information set is no longer off-path. The two leftmost states each occur with probability 0.5. P2's best response to these beliefs remains l, and the dynamics have converged. Furthermore, they have converged to a Bayes-Nash equilibrium: P1 is best-responding to P2's choice of l; likewise, P2 is best-responding to P1's choice of M.

At this point, we would be remiss not to fully specify a means of constructing transition probabilities in an OMDP. One possibility is to proceed as described above: to simply allow players to invent arbitrary beliefs about the states in off-path information sets. This possibility has the drawback that it is unlikely to lead to convergence, particularly if a player's choice of beliefs differs from iteration to iteration.

Another possibility is to initialize all actions uniformly (so that there are no off-path information sets), and then to insist that if ever in our search for an equilibrium, the dynamics lead to a situation in which an information set is off-path, then the beliefs about the states in that information set should be inherited from the previous iteration. Returning to our example, if, when his sole information set is off-path, P2 instead operates under the assumption that his beliefs are unchanged, he continues to play r, so P1 chooses to play L again. At this point, the dynamics have converged, once again to a Bayes-Nash equilibrium (P1 is best-responding to P2's choice of r; likewise, P2 is best-responding to P1's choice of L).

At this point, it should be clear why we deferred fully specifying OMDPs until now. We simply wanted to integrate their full specification into our equilibrium search algorithm, which we do presently:

1. Initialize with an arbitrary *mixed* strategy profile $\boldsymbol{\sigma}^{(0)} = (\sigma_1^{(0)}, \sigma_2^{(0)}, \ldots, \sigma_n^{(0)})$ *with full support.*
2. *Initialize each player with beliefs at each of his information sets (i.e., observations).*
3. For each iteration $l = 1, \ldots, L$ and for each player $i = 1, \ldots, N$:
   (a) Run Algorithm 1 given Γ and $\sigma_{-i}^{(l-1)}$ represented as an iPOMDP. Return player i's OMDP.
   (b) *At any information set (i.e., observation) off the path of play, define beliefs to be whatever they were during the previous iteration.*[7]
   (c) Update strategy $\sigma_i^{(l)}$ to be an optimal policy in the learned OMDP: i.e., a best response to $\boldsymbol{\sigma}_{-i}^{(l-1)}$.
4. Terminate with success if $\boldsymbol{\sigma}^{(l)}$ is the same as, or very similar to, $\boldsymbol{\sigma}^{(l-1)}$.

[7]There are other equally reasonable (and perhaps less arbitrary) means of updating off-path information sets, but we did not try any others, as this method worked sufficiently well in our experiments (see Greenwald et al. 2012).

At first glance, this modified search procedure might appear confusing, because beliefs are not represented or learned explicitly in our reinforcement learning approach to solving games. But beliefs are still implicit in the transition probabilities and rewards learned in Algorithm 1. Consequently, we can implement these changes with only a few minor modifications to Algorithms 1 and 2. First, Algorithm 1 should be generalized to take one additional input, transition probabilities. These default transition probabilites can then be returned as the transition probabilities of off-path information sets, while the transition probabilities to on-path information sets can be learned as usual. Likewise, Line 7 in Algorithm 2 should be extended so that the current transition probabilities are passed as a parameter in the call to Algorithm 1. Finally, Line 1 of Algorithm 2 should be initialized to a (memoryless) *mixed* policy profile *with full support.*

Theorem 6 *If our equilibrium search algorithm terminates with success, it finds an approximate Bayes-Nash equilibrium.*

If our iterative search algorithm terminates with success, and $\boldsymbol{\sigma}^{(l)} = \boldsymbol{\sigma}^{(l-1)}$, then for all players i, $\sigma_i^{(l)}$ is an approximate best response (because of the nature of our best-response finding algorithm) to $\boldsymbol{\sigma}_{-i}^{(l-1)} = \boldsymbol{\sigma}_{-i}^{(l)}$. Since all players are near-best responding to one another's strategies, their play is an approximate Bayes-Nash equilibrium.

Another potential source of error in the approximation could arise if play does not converge precisely, but if instead $\boldsymbol{\sigma}^{(l)}$ is very similar to $\boldsymbol{\sigma}^{(l-1)}$. Indeed, in practice, our iterative search procedure requires a user-specified measure of dissimilarity between $\boldsymbol{\sigma}^{(l)}$ and $\boldsymbol{\sigma}^{(l-1)}$. One such dissimilarity measure is the proportion of information sets at which the two strategy profiles prescribe different actions. This measure is simple to calculate, but it does not take into account the fact that some information sets are more likely to be encountered in equilibrium than others, so should perhaps be weighted more heavily. (Indeed some information sets may be off-path entirely!)

A more sophisticated measure estimates player i's *expected* loss in payoffs, $\hat{U}$, for playing $\sigma_i^{(l-1)}$ rather than $\sigma_i^{(l)}$ when other agents play $\boldsymbol{\sigma}_{-i}^{(l-1)}$: i.e., $\left|\hat{U}(\sigma_i^{(l)}, \boldsymbol{\sigma}_{-i}^{(l-1)}) - \hat{U}(\boldsymbol{\sigma}^{(l-1)})\right|$. Taking the maximum of this measure across all players yields the *ϵ-factor* that makes $\boldsymbol{\sigma}^{(l-1)}$ an ϵ-BNE. For more details about how to effectively implement our iterative search procedure, including information about approximating and interpreting ϵ-factors, see Greenwald et al. (2012).

11.6 Beyond Bayes-Nash Equilibria

One common dissatisfaction with the concept of Bayes-Nash equilibrium is that BNE may contain unrealistic strategy profiles, in the sense that they are non-credible threats (see, for example, Myerson 1991). What this means is that a BNE may consist of strategies that are not actually everywhere best responses to other strategies in the profile. In particular, they may not be best responses *at off-path information sets*. But our overall approach is robust to this concern. In this section, we argue that, when our iterative procedure converges, it finds a stronger equilibrium concept than BNE, namely *weak perfect Bayesian equilibrium (wPBE)*, a refinement of BNE that rules out non-credible threats off the equilibrium path of play.

An informal definition of wPBE follows. See Mas-Colell et al. Mas-Colell et al. (1995) for a more formal treatment.

Definition 13 (*weak Perfect Bayesian Equilibrium*) A strategy profile $\boldsymbol{\sigma}$ together with a profile of beliefs $\boldsymbol{\beta}$ comprise a weak Perfect Bayesian Equilibrium (wPBE) of an EFG iff

1. $\boldsymbol{\sigma}$ is sequentially rational given $\boldsymbol{\beta}$: i.e., for all players i, for all information sets h, $\sigma_i(h)$ maximizes expected utility, given $\boldsymbol{\sigma}_{-i}$, where the expectation is taken with respect to β_i, and
2. $\boldsymbol{\beta}$ is consistent given $\boldsymbol{\sigma}$, whenever possible: i.e., on the equilibrium path, beliefs are determined using Bayes' rule and $\boldsymbol{\sigma}$.

Recall that we do not represent or learn beliefs explicitly in our reinforcement learning approach to solving EFGs. But beliefs are calculated explicitly (via Eq. 11.34) in the definition of OMDPs. Given such beliefs, an optimal policy in an OMDP is a sequentially-rational strategy. Moreover, these beliefs are consistent whenever possible—it is possible to apply Eq. 11.34 on the equilibrium path, where there are no zeros in the denominator, but it is impossible to apply it off the equilibrium path.

It is straightforward to envision a variant of our sampling algorithm that estimates transition probabilities from observations to game tree nodes rather than between observations: estimate $T^{\mathrm{OMDP}}[j' \mid h, k]$ by approximating $T^{\mathrm{POMDP}}[j' \mid \langle h, \beta \rangle, k]$ (without explicitly computing β). With such estimates, "learned" beliefs $\hat{\beta}$ can again be calculated in a consistent manner whenever possible via Eq. 11.34. Therefore:

Lemma 7 *An optimal policy in a learned OMDP is a sequentially rational strategy given implicit learned beliefs $\hat{\beta}$, which are calculated according to Eq. 11.34, and consistent whenever possible.*

If Algorithm 2 (modified as described) converges, this lemma applies to all players, so that the resulting strategy profile is sequentially rational given learned beliefs $\hat{\boldsymbol{\beta}}$, which are consistent whenever possible. So:

Theorem 8 *If our iterative procedure terminates with success, it finds an approximate weak Perfect Bayesian Equilibrium.*

11.7 Related Work

The problem of finding a Nash equilibrium or even an approximate Nash equilibrium in a general-sum normal-form game is PPAD-complete (Daskalakis 2009). Regardless, various approaches have been taken to search for such equilibria. Black-box game solvers such as GAMBIT (McKelvey et al. 2010) provide easy off-the-shelf solutions for finding equilibria in games, but they are limited by the size of games in which they can tractably solve.

Generalizing to one-shot games of incomplete information, Rabinovich and Naroditskiy (2013) extend fictitious play to incomplete-information games with finite actions, and thereby compute approximate equilibria in environments with utilities expressible as linear functions over a one-dimensional type space. Reeves and Wellman (2004) describe an algorithm for computing a best response in two-player games of incomplete information with infinite action spaces, using a two-player sealed-bid auction as a motivating example.

Generalizing to zero-sum extensive-form games, Koller et al. (1996) proposed the first efficient equilibrium-finding algorithms. Nowadays, the most widespread algorithm for large-scale zero-sum equilibrium computation is based on the idea of counterfactual regret minimization (Lanctot et al. 2009). This algorithm has been applied successfully in practice in the poker competition by Brown et al. (2015). Oliehoek et al. (2005) also studied a version of poker, and developed a best-response finding algorithm much like ours. Their model is a special case of ours—a two-player zero-sum game in which the other player's actions are fully observable.

One line of research advocates solving large games only after restricting the strategy space, for reasons of tractability. Armantier et al. (2008) study the problem of approximating equilibrium in Bayesian games by searching for constrained strategic equilibrium (CSE). A strategy profile is a CSE if no agent is better off by deviating to a different strategy within the constrained strategy set, typically represented by some parametrization of the full strategy space. Mostafa and Lesser (2008) describe an anytime algorithm for approximating equilibria in incomplete-information games that reduces the size of the game tree by collapsing nodes with "obvious" decisions.

Specifically in the auction domain, the iterative method *self-confirming price prediction* employed in Osepayshvili et al. (2005) is very similar to the best-response dynamics we employ here, as is the approach taken in Naroditskiy and Greenwald (2007).

In terms of applying reinforcement learning methods to games, many variants of Q-learning (Littman 2001; Hu and Wellman 2003; Greenwald and Hall 2003) have been used to approximate game-theoretic equilibrium. While we consider EFGs, others have applied their methods to stochastic games. Sample complexity bounds for reinforcement learning have been studied by many; a summary appears in Strehl et al. (2009).

11.8 Conclusion

In this paper, we purport to make three main contributions. First, we clarify the theoretical relationships between four game-theoretic and decision-theoretic representations (EFGs, TT-POMGs, POMDPs with perceptual aliasing, and MDPs), and provide general recipes from transforming one to another. Second, we design an algorithm that can solve for a near-optimal policy in finite time in tree-shaped MDPs that do not admit resampling part-way through a trajectory, and hence near-best responses in an EFG. Finally, we use our best-response finding algorithm to iteratively search for equilibria, and argue that our iterative search procedure, when it converges, finds approximate weak perfect Bayesian equilibria.

Although our iterative search procedure is not guaranteed to converge, we found that it often converges to near-equilibrium in experiments with sequential auctions (Greenwald et al. 2012). One limitation, however, is that we can only search for pure-strategy equilibria, not mixed-strategy equilibria. This is an inherent difficulty in applying reinforcement learning methods to games, because there is no principled way to break ties between equally attractive actions, and exploring the space of all possible mixed strategies is computationally prohibitive. The good news is that pure-strategy equilibria exist in many games with general properties (Proposition 8.D.3 in Mas-Colell et al. 1995), while mixed equilibria have frequently been regarded as unnatural (Aumann 1985).

Acknowledgements This work would not have been possible without the guidance and ongoing mentorship of Rohit Parikh, who has been a constant source of inspiration, and a friend, to the first author since she first met him in 1995. We are also grateful to two anonymous reviewers who provided a plethora of constructive criticism that led to improvements which undoubtedly strengthened the quality of this work. Finally, this work was funded by NSF Grant #1217761 and Microsoft Research.

Appendix 1: Proof of Theorem 1

In this appendix, we show that, from the point of view of player i, a TT-POMG Θ and an iPOMDP $\Psi_i(\Theta, \boldsymbol{\theta}_{-i})$ are identical. That is, the probability of all histories are equal in both models. From this, it follows that a best-response to a memoryless policy profile $\boldsymbol{\theta}_{-i}$ in Θ is an optimal policy in $\Psi_i(\Theta, \boldsymbol{\theta}_{-i})$, and vice versa.

Lemma 9 *Given a TT-POMG Θ together with an other-agent memoryless policy profile $\boldsymbol{\theta}_{-i}$ and a corresponding iPOMDP $\Psi(\Theta, \boldsymbol{\theta}_{-i})$ with policy ψ, the probability of a history $\langle \mu^{\tau-1}, s^{\tau} \rangle$ through Ψ is equal to the probability of that same history (denoted $\langle \mu_i^{\tau-1}, s_i^{\tau} \rangle$) through Θ: i.e., $\mathtt{Pr}_{\psi}^{iPOMDP}[\mu^{\tau-1}, s^{\tau}] = \mathtt{Pr}_{\psi,\boldsymbol{\theta}_{-i}}^{POMG}[\mu_i^{\tau-1}, s_i^{\tau}]$, for all $\tau = 1, \ldots$.*

Proof The proof is by induction on t.

Basis: In the base case, when $t = 1$, the history $\mu^{\tau-1}$ through Ψ is of length 0. Correspondingly, the only histories of interest in Θ are those leading to state s^1 in which it is never i's turn to move. Consequently, $\mathtt{Pr}_{\psi}^{i\text{POMDP}}[s^1] = \zeta^{i\text{POMDP}}(s^1) = \sum_{s \in S_{-i}^{\text{POMG}}} \zeta^{\text{POMG}}(s) \mathtt{Pr}_{\boldsymbol{\theta}_{-i}}^{\text{POMG}}[s^1 \mid s] = \mathtt{Pr}_{\psi,\boldsymbol{\theta}_{-i}}^{\text{POMG}}[s^1]$.

Step: Assume the induction hypothesis, namely, $\mathtt{Pr}_{\psi}^{i\mathrm{POMDP}}[\mu^{\tau-1}, s^{\tau}] = \mathtt{Pr}_{\psi,\boldsymbol{\theta}_{-i}}^{\mathrm{POMG}}[\mu_i^{\tau-1}, s_i^{\tau}]$, for a history of length $\tau - 1$ through Ψ. We must show the same for a history of length τ.

First, note how the dynamics of Ψ and Θ play out:

$$\begin{aligned}\mathtt{Pr}_{\psi}^{i\mathrm{POMDP}}[h^{\tau}, a^{\tau}, s^{\tau+1} \mid \mu^{\tau-1}, s^{\tau}] &= \mathtt{Pr}_{\psi}^{i\mathrm{POMDP}}[s^{\tau+1} \mid \mu^{\tau-1}, s^{\tau}, h^{\tau}, a^{\tau}]\, \mathtt{Pr}_{\psi}^{i\mathrm{POMDP}} \\ &\quad [a^{\tau} \mid \mu^{\tau-1}, s^{\tau}, h^{\tau}]\, \mathtt{Pr}_{\psi}^{i\mathrm{POMDP}}[h^{\tau} \mid \mu^{\tau-1}, s^{\tau}] && (11.39)\\ &= T^{i\mathrm{POMDP}}(s^{\tau}, a^{\tau}, s^{\tau+1})\, (\psi(\mu^{\tau-1}, s^{\tau}, h^{\tau}))(a^{\tau})\, Q^{i\mathrm{POMDP}}(s^{\tau}, h^{\tau}) && (11.40)\\ &= T^{\mathrm{POMG}}(s_i^{\tau}, a_i^{\tau}, s_i^{\tau+1})\, (\psi(\mu^{\tau-1}, s^{\tau}, h_i^{\tau}))(a_i^{\tau})\, Q_i^{\mathrm{POMG}}(s_i^{\tau}, h_i^{\tau}) && (11.41)\\ &= \mathtt{Pr}_{\psi,\boldsymbol{\theta}_{-i}}^{\mathrm{POMG}}[s_i^{\tau+1} \mid \mu_i^{\tau-1}, s_i^{\tau}, h_i^{\tau}, a_i^{\tau}]\, \mathtt{Pr}_{\psi,\boldsymbol{\theta}_{-i}}^{\mathrm{POMG}} \\ &\quad [a_i^{\tau} \mid \mu_i^{\tau-1}, s_i^{\tau}, h_i^{\tau}]\, \mathtt{Pr}_{\psi,\boldsymbol{\theta}_{-i}}^{\mathrm{POMG}}[h_i^{\tau} \mid \mu_i^{\tau-1}, s_i^{\tau}] && (11.42)\\ &= \mathtt{Pr}_{\psi,\boldsymbol{\theta}_{-i}}^{\mathrm{POMG}}[h_i^{\tau}, a_i^{\tau}, s_i^{\tau+1} \mid \mu_i^{\tau-1}, s_i^{\tau}] && (11.43)\end{aligned}$$

Equation (11.40) follows immediately via the *i*POMDP construction. Equation 11.41 follows from the fact that Θ is Markov in both transition and observation probabilities. The claim now follows via the induction hypothesis (Eq. (11.46)):

$$\begin{aligned}\mathtt{Pr}_{\psi}^{i\mathrm{POMDP}}[\mu^{\tau}, s^{\tau+1}] &= \mathtt{Pr}_{\psi}^{i\mathrm{POMDP}}[\mu^{\tau-1}, s^{\tau}, h^{\tau}, a^{\tau}, s^{\tau+1}] && (11.44)\\ &= \mathtt{Pr}_{\psi}^{i\mathrm{POMDP}}[h^{\tau}, a^{\tau}, s^{\tau+1} \mid \mu^{\tau-1}, s^{\tau}]\, \mathtt{Pr}_{\psi}^{i\mathrm{POMDP}}[\mu^{\tau-1}, s^{\tau}] && (11.45)\\ &= \mathtt{Pr}_{\psi,\boldsymbol{\theta}_{-i}}^{\mathrm{POMG}}[h_i^{\tau}, a_i^{\tau}, s_i^{\tau+1} \mid \mu_i^{\tau-1}, s_i^{\tau}]\, \mathtt{Pr}_{\psi,\boldsymbol{\theta}_{-i}}^{\mathrm{POMG}}[\mu_i^{\tau-1}, s_i^{\tau}] && (11.46)\\ &= \mathtt{Pr}_{\psi,\boldsymbol{\theta}_{-i}}^{\mathrm{POMG}}[\mu_i^{\tau-1}, s_i^{\tau}, h_i^{\tau}, a_i^{\tau}, s_i^{\tau+1}] && (11.47)\\ &= \mathtt{Pr}_{\psi,\boldsymbol{\theta}_{-i}}^{\mathrm{POMG}}[\mu_i^{\tau}, s_i^{\tau+1}] && (11.48)\end{aligned}$$

□

Theorem 10 *An optimal (not necessarily memoryless) policy ψ in an iPOMDP $\Psi_i(\Theta, \boldsymbol{\theta}_{-i})$ is a best-response to an other-agent profile of memoryless policies $\boldsymbol{\theta}_{-i}$ in the original TT-POMG Θ, and vice versa.*

Proof We claim that for all times τ and for all policies ψ, $R^{i\mathrm{POMDP},\tau}(\psi) = R^{\mathrm{POMG},\tau}(\psi, \boldsymbol{\theta}_{-i})$. Therefore, since ψ is an optimal policy in $\Psi_i(\Theta, \boldsymbol{\theta}_{-i})$, it follows that ψ is a best-response to $\boldsymbol{\theta}_{-i}$ in Θ.

This claim follows from Lemma 9, which states that, under any policy ψ, the probability of a history through Ψ is equal to the probability of that same history through Θ, together with the fact that player i's rewards in Ψ are by definition the rewards in Θ.

$$R^{i\mathrm{POMDP},\tau}(\psi) = \mathbb{E}_{\langle\mu^{\tau-1}, s^{\tau}\rangle \sim \mathtt{Pr}_{\psi}^{i\mathrm{POMDP}}}\left[\sum_{t=1}^{\tau} r^{i\mathrm{POMDP}}(s^{t})\right] \quad (11.49)$$

$$= \mathbb{E}_{\langle \mu_i^{\tau-1}, s_i^{\tau} \rangle \sim \Pr^{\text{POMG}}_{\psi, \boldsymbol{\theta}_{-i}}} \left[\sum_{t=1}^{\tau} r_i^{\text{POMG}}(s_i^t) \right] \tag{11.50}$$

$$= R_i^{\text{POMG},\tau}(\psi, \boldsymbol{\theta}_{-i}) \tag{11.51}$$

□

Appendix 2: Proof of Theorem 4

In our proof, we assume the simulation process follows a *balanced wandering* policy π_0: during each iteration, one of the actions sampled least often so far is chosen, with ties broken uniformly. A desirable feature of balanced wandering [8] is that if a state is reached mA times, then all actions at that state are guaranteed to have been chosen m times.

Note: To simplify exposition, we prove the theorem for undiscounted MDPs with deterministic rewards, assuming the maximum reward along each path is bounded by [0, 1], but the result can be extended to handle stochastic and arbitrarily bounded non-negative rewards.

Proof Plan

Recall that our proof pertains only to tree-shaped MDPs of finite horizon H. We use s^h to denote a generic MDP state at depth h, and $p^h = (s^1, s^2, ..., .s^h)$ to denote a path of length h. Then $r(s)$ from (resp., $r(p)$) denotes the reward upon reaching state s (resp., following path p). We let $P_h^{\pi}(s^h)$ (resp., $P_h^{\pi}(p^h)$) denote the probability of reaching state s^h (resp., following path p^h) under policy π, omitting the subscript π when the choice of policy is clear. Estimators appear with hats over their true counterparts: e.g., $\hat{T}(s^{h-1}, a, s^h)$, $\hat{P}_h(p^h)$.

Because of the MDP's tree structure, each state uniquely defines a path, so we can equivalently associate rewards with paths, and then define the expected rewards $R(\pi)$ earned by policy π as path-probability weighted rewards. Furthermore, inspired by Kearns and Singh (2002), we divide up the set of paths through the MDP into a set of *rare* paths, and its complement set of *regular* paths:

$$R(\pi) = \sum_{p^H} P_H^{\pi}(p^H) r(p^H) \tag{11.52}$$

[8] Desirable in terms of facilitating the proof. In our simulation experiments, we observe that adaptive sampling methods frequently produce better MDP approximations, and hence policies with higher rewards, under the same computational budget.

$$= \sum_{p^H \text{ rare}} P_H^{\pi}(p^H) r(p^H) + \sum_{p^H \text{ regular}} P_H^{\pi}(p^H) r(p^H) \quad (11.53)$$

$$\stackrel{\text{def}}{=} R^{\text{rare}}(\pi) + R^{\text{regular}}(\pi) \quad (11.54)$$

Denote the optimal policy by π^* and our learned policy by $\hat{\pi}^*$. Then the loss in rewards experienced by our simulation algorithm is:

$$R(\pi^*) - R(\hat{\pi}^*) = \underbrace{(R^{\text{rare}}(\pi^*) - R^{\text{rare}}(\hat{\pi}^*))}_{\texttt{Err}_{\text{rare}}} + \underbrace{(R^{\text{regular}}(\pi^*) - R^{\text{regular}}(\hat{\pi}^*))}_{\texttt{Err}_{\text{regular}}} \quad (11.55)$$

Our proof strategy targets rare and regular paths separately:

- Rare paths will be visited insufficiently often in the simulation process, so transition probabilities along these paths are not guaranteed to be learned well. But $\texttt{Err}_{\text{rare}}$ is still small precisely because these paths will be visited rarely, so choosing suboptimal actions along them has limited impact on total rewards.
- $\texttt{Err}_{\text{regular}}$ will also be small (with high probability), because the transitions along regular paths will be sampled sufficiently often and thus well learned, so our learned policy will fare well in the regular MDP.
  - Lemmas 13, 14, and 15 show that with sufficient sampling, all regular path probabilities will be estimated with low error.
  - Lemma 16 shows that when path probabilities are well estimated, $\texttt{Err}_{\text{regular}}$ is small.

Rare MDP

Theorem 12 shows that $\texttt{Err}_{\text{rare}}$ is small if we choose the parameter β to be small.

Definition 14 (*β-rare paths*) We call a state s^h *β-rare* if the probability of transitioning to s^h from its parent state s^{h-1} is bounded above by a small constant $\beta > 0$ under any policy: $\frac{1}{A}\sum_a T(s^{h-1}, a, s^h) < \beta$. Likewise, we call a path *β-rare* if any of its states are β-rare.

Given an MDP and a policy π, let $P^{\text{rare}}(\pi) = \sum_{p^H \text{ rare}} P_H^{\pi}(p^H)$. That is, $P^{\text{rare}}(\pi)$ is the probability of traversing a rare path in the MDP under policy π. Then the probability of traversing a rare path (regardless of policy) is bounded above by $\max_\pi P^{\text{rare}}(\pi)$. The following lemma upper bounds this latter probability.

Lemma 11 *The probability of traversing a path that is β-rare will be small if β is small:*

$$\max_{\pi} P^{\,rare}(\pi) \leq \beta ASH \tag{11.56}$$

Proof Define $\mathcal{E}_h$ as the event of transitioning to a β-rare state on the hth step under an arbitrary policy π. Then:

$$P^{\pi}(\mathcal{E}_h) = \sum_{s^h \text{ rare}} T(s^{h-1}, \pi(s^{h-1}), s^h) \tag{11.57}$$

$$\leq \sum_{s^h \text{ rare}} \max_a T(s^{h-1}, a, s^h) \tag{11.58}$$

$$\leq \sum_{s^h \text{ rare}} \sum_a T(s^{h-1}, a, s^h) \tag{11.59}$$

$$\overset{\text{Definition of } \beta\text{-rare}}{\leq} \sum_{s^h \text{ rare}} \beta A \tag{11.60}$$

$$\overset{\text{Branching factor = } S}{\leq} \beta AS \tag{11.61}$$

Now, to traverse a β-rare path under π requires visiting at least one β-rare state:

$$P^{\,\text{rare}}(\pi) = P^{\pi}(\cup_{h=1}^{H} \mathcal{E}_h) \tag{11.62}$$

$$\overset{\text{Union bound}}{\leq} \sum_{h=1}^{H} P^{\pi}(\mathcal{E}_h) \tag{11.63}$$

$$\overset{\text{Equation (62)}}{\leq} \sum_{h=1}^{H} \beta AS \tag{11.64}$$

$$= \beta ASH \tag{11.65}$$

Finally, since π was arbitrary, the result follows.

Theorem 12 *The error along rare paths is small if β is small. In particular, $\texttt{Err}_{rare} \leq \beta ASH$.*

Proof

$$\texttt{Err}_{\text{rare}} = R^{\,\text{rare}}(\pi^*) - R^{\,\text{rare}}(\hat{\pi}^*) \tag{11.66}$$

$$\overset{\text{Rewards}\geq 0}{\leq} R^{\,\text{rare}}(\pi^*) \tag{11.67}$$

$$= \max_{\pi} R^{\,\text{rare}}(\pi) \tag{11.68}$$

$$= \max_{\pi} \sum_{p^H \text{ rare}} P_H^{\pi}(p^H) r(p^H) \tag{11.69}$$

$$\overset{\text{Rewards}\leq 1}{\leq} \max_{\pi} \sum_{p^H \text{ rare}} P_H^{\pi}(p^H) \tag{11.70}$$

$$\overset{\text{Lemma 3}}{=} \beta ASH \tag{11.71}$$

Regular MDP

A state that is *not* β-rare is called *regular*. Likewise, a *regular* path is one that visits no β-rare states. The following lemma gives a lower bound on how frequently a regular path will be sampled.

Lemma 13 *Assuming N simulations and balanced wandering, all regular paths will be sampled at least* $\frac{\beta^H N}{2}$ *times with probability at least* $1 - S^H \exp\left(-\frac{\beta^{2H} N}{2}\right)$.

Proof Under balanced wandering, the probability of transitioning to a regular state s^h from its parent state s^{h-1} given action a is $\frac{1}{A} T(s^{h-1}, a, s^h)$. Therefore, the total probability of transitioning to a regular state s^h from its parent s^{h-1} (summing up over all actions) is at least β, because s^h is regular. It follows that $P(p^H) \geq \beta^H$ for regular path p^H, because regular paths visit no β-rare states.

Denote by $N(p^H)$ the number of times p^H is sampled during N simulations. Then a direct application of the Chernoff bound gives:

$$P\left(N(p^H) < \frac{\beta^H N}{2}\right) < \exp\left(-\frac{\beta^{2H} N}{2}\right) \tag{11.72}$$

Now since there are at most S^H paths:

$$P\left(\exists \text{ a regular path } \bar{p}^H \text{ s.t. } N(\bar{p}^H) < \frac{\beta^H N}{2}\right) = P\left(\bigcup_{p^H} \left\{N(p^H) < \frac{\beta^H N}{2}\right\}\right) \tag{11.73}$$

$$\leq \sum_{p^H} P\left(N(p^H) < \frac{\beta^H N}{2}\right) \tag{11.74}$$

$$\leq \sum_{p^H} \exp\left(-\frac{\beta^{2H} N}{2}\right) \tag{11.75}$$

$$\leq S^H \exp\left(-\frac{\beta^{2H} N}{2}\right) \tag{11.76}$$

Therefore, $P\left(\forall \text{ regular paths } \bar{p}^H, N(\bar{p}^H) \geq \frac{\beta^H N}{2}\right) \geq 1 - S^H \exp\left(-\frac{\beta^{2H} N}{2}\right)$. □

Once again, Lemma 13 guarantees that regular paths will be sampled sufficiently often. But then, under balanced wandering, all state-action pairs along regular paths will also be sampled sufficiently often. The next lemma, Lemma 14, shows that, as a consequence, the error in learned transition probabilities will be small. We calculate errors in terms of L1-norms:

$$||\hat{T}(s,a,\cdot) - T(s,a,\cdot)||_1 \stackrel{\text{def}}{=} \sum_{s'} |\hat{T}(s,a,s') - T(s,a,s')| \tag{11.77}$$

$$||\hat{P}_h(\cdot) - P_h(\cdot)||_1 \stackrel{\text{def}}{=} \sum_{p^h} |\hat{P}_h(p^h) - P_h(p^h)| \tag{11.78}$$

The following lemma, which lower bounds the number of times $N(s,a)$ that state action pair (s,a) should be sampled, is due to Weissman et al. (2003):

Lemma 14 *For each state-action pair* (s,a), *to ensure that* $P(||\hat{T}(s,a,\cdot) - T(s,a,\cdot)||_1 \leq \epsilon_0) \geq 1 - \delta_0$, (s,a) *must be sampled at least* N_0 *times: i.e.,*

$$N(s,a) \geq N_0 \stackrel{def}{=} \frac{2}{\epsilon_0^2} \ln\left(\frac{2^S}{\delta_0}\right) \tag{11.79}$$

Next, we bound the sampling errors in estimating regular path probabilities by summing up the sampling errors in estimating one-step transition probabilities.

Lemma 15 *If the state-action pairs along all regular paths obey* $||\hat{T}(s,a,\cdot) - T(s,a,\cdot)||_1 \leq \epsilon_0$ *as in Lemma* 14, *then under any policy* π, *it follows that* $||\hat{P}_H^\pi(\cdot) - P_H^\pi(\cdot)||_1 \leq H\epsilon_0$.

Proof Recall that $P_h(\cdot)$ is the probability distribution over length-h paths (under π). The proof is by induction on the path length h.

Basis: $||\hat{P}_1(\cdot) - P_1(\cdot)||_1 = ||\hat{T}(s^0,a,\cdot) - T(s^0,a,\cdot)||_1$, which, by assumption, is bounded above by ϵ_0.

Step: Assume the induction hypothesis: $||\hat{P}_{H-1}(\cdot) - P_{H-1}(\cdot)||_1 \leq (H-1)\epsilon_0$. Then:

$$||\hat{P}_H(\cdot) - P_H(\cdot)||_1 = \sum_{p^{H-1}} \sum_{s^H} \left|\hat{P}_H(p^{h-1} \cup s^H) - P_H(p^{H-1} \cup s^H)\right| \tag{11.80}$$

$$= \sum_{p^{H-1}} \sum_{s^H} \left|\hat{P}_{H-1}(p^{H-1})\hat{T}(s^{H-1},a,s^H) - P_{H-1}(p^{H-1})T(s^{H-1},a,s^H)\right| \tag{11.81}$$

$$\stackrel{\text{Simplify notation: } p \stackrel{\text{def}}{=} p^{H-1}, s \stackrel{\text{def}}{=} s^H, P(\cdot) \stackrel{\text{def}}{=} P_H(\cdot), T(\cdot) \stackrel{\text{def}}{=} T(s,a,\cdot)}{=} \sum_p \sum_s \left|\hat{P}(p)\hat{T}(s) - P(p)T(s)\right| \tag{11.82}$$

$$= \sum_p \sum_s \left|\hat{P}(p)\hat{T}(s) - \hat{P}(p)T(s) + \hat{P}(p)T(s) - P(p)T(s)\right| \tag{11.83}$$

$$\leq \sum_p \sum_s \left(\left|\hat{P}(p)\hat{T}(s) - \hat{P}(p)T(s)\right| + \left|\hat{P}(p)T(s) - P(p)T(s)\right| \right) \tag{11.84}$$

$$= \sum_p \sum_s \left(\hat{P}(p) \left|\hat{T}(s) - T(s)\right| + T(s) \left|\hat{P}(p) - P(p)\right| \right) \tag{11.85}$$

$$= \sum_p \hat{P}(p) \sum_s \left|\hat{T}(s) - T(s)\right| + \sum_s T(s) \sum_p \left|\hat{P}(p) - P(p)\right| \tag{11.86}$$

$$\leq (1)\epsilon_0 + (1)(H-1)\epsilon_0 \tag{11.87}$$

$$= H\epsilon_0 \tag{11.88}$$

If all regular path probabilities have low estimation errors, this leads to low estimation error in rewards. More specifically, the following *simulation lemma*[9] states that an estimate of a policy's rewards cannot differ by more than $H\epsilon_0$ from the true value of the policy.

Lemma 16 *If the estimation errors of length-H path probabilities are at most $H\epsilon_0$, as in Lemma* 15, *then the error in the estimated expected reward of any policy π is bounded above by $H\epsilon_0$.*

Proof

$$|\hat{R}(\pi) - R(\pi)| = \left| \sum_{p^H} \hat{P}_H(p^H) r(p^H) - \sum_{p^H} P_H(p^H) r(p^H) \right| \tag{11.89}$$

$$\leq \sum_{p^H} \left| \hat{P}_H(p^H) - P_H(p^H) \right| r(p^H) \tag{11.90}$$

$$\overset{\text{Rewards bounded above by 1}}{\leq} \sum_{p^H} \left| \hat{P}_H(p^H) - P_H(p^H) \right| \tag{11.91}$$

$$= ||\hat{P}_H(\cdot) - P_H(\cdot)||_1 \tag{11.92}$$

$$\leq H\epsilon_0 \tag{11.93}$$

Equation (11.93) follows from Lemma 15. □

Theorem 17 *The learned policy will be within $2H\epsilon_0$ of the true optimal.*

Proof Denote the true optimal policy as π^* and the learned policy as $\hat{\pi}^*$. If they are not the same policy, it must be that $\hat{\pi}^*$ is estimated to have higher rewards than π^* in the learned MDP. Therefore, we can infer the following lower bound on $\hat{\pi}^*$'s reward in the true MDP:

$$R(\hat{\pi}^*) \geq \hat{R}(\hat{\pi}^*) - H\epsilon_0 \tag{11.94}$$

$$\geq \hat{R}(\pi^*) - H\epsilon_0 \tag{11.95}$$

$$= R(\pi^*) - 2H\epsilon_0 \tag{11.96}$$

[9]There are several versions of the simulation lemma, the earliest of which appears in Kearns and Singh (2002).

Equations (11.94) and (11.96) follow from our simulation lemma, Lemma 16. Equation (11.95) follows from the assumption that $\hat{\pi}^*$ is an optimal policy in the learned MDP.

Choosing Parameters

Let's sum up the story so far. The total reward contribution of rare-paths is bounded (Lemma 11), so Err_{rare} is small. With high probability, all states-action pairs in the regular MDP will be sampled sufficiently often (Lemma 13), so all regular paths will be learned well (Lemmas 14, 15), hence the learned policy will be near-optimal in the regular MDP (Theorem 17).

All that is left to do is express the free parameters β, ϵ_0, N_0, δ_0 as functions of input variables A, S, H, δ, ϵ, and then to solve the sample size N.

1. **Reward loss in rare MDP**: We choose $\beta = \frac{\epsilon}{2ASH}$, so that the maximum loss in rewards from choosing suboptimal actions on rare paths is bounded by $\epsilon/2$ according to Lemma 11. That is, in the worst case, and in the rare MDP, our learned policy achieves 0 reward while the optimal policy achieves $\epsilon/2$. This result is deterministic, meaning it is independent of N.
   All of the following steps are for the regular MDP.
2. **Reward loss in regular MDP**: We choose $\epsilon_0 = \frac{\epsilon}{4H}$, so that, according to Theorem 17, the maximum reward loss from regular paths is also bounded by $\epsilon/2$ with high probability. Therefore, total reward loss (in both the rare and the regular MDP) is bounded by ϵ with high probability.
3. In the regular MDP, bound P **(sampled sufficiently)**: Lemma 13 states that all regular paths, and therefore all state-action pairs along regular paths, will be sampled $N_0 = \frac{\beta^H N}{2}$ times (i.e., "sufficiently") with probability at least $1 - S^H \exp\left(-\frac{\beta^{2H} N}{2}\right)$.
4. In the regular MDP, bound $P(\exists$ **a transition that is not well learned | it is sampled sufficiently)**: A state-action pair (s, a) is sampled sufficiently if $N(s, a) \geq N_0$. The transition probabilities of a particular state-action pair (s, a) are "not well learned" if $||\hat{T}(s, a, \cdot) - T(s, a, \cdot)||_1 > \epsilon_0$. By Lemma 14, P $((s, a)$ well learned| it is sampled sufficiently) $\geq 1 - \delta_0$. So:

$$P\left((s,a) \text{ not well learned} \mid \text{it is sampled sufficiently}\right) \leq \delta_0 \tag{11.97}$$

$$\overset{\text{Inverting Equation (80)}}{=} 2^S \exp\left(-\frac{\epsilon_0^2}{2} N_0\right) \tag{11.98}$$

Applying a union bound, and noting that there are at most $AS + AS^2 + \ldots + AS^H \leq AHS^H$ state-action pairs yields:

$$P\left(\exists\, (s,a) \text{ not well learned} \mid \text{all } (s,a) \text{ are sampled sufficiently}\right) \tag{11.99}$$

$$= P\left(\bigcup_{(s,a)} \{(s,a) \text{ not well learned}\} \mid \text{all } (s,a) \text{ are sampled sufficiently}\right) \tag{11.100}$$

$$\leq \sum_{(s,a)} P\left((s,a) \text{ not well learned} \mid (s,a) \text{ is sampled sufficiently}\right) \tag{11.101}$$

$$\leq AHS^H \delta_0 \tag{11.102}$$

5. **Putting it all together**: Recall that the loss in rewards obtained in the rare MDP is no greater than $\epsilon/2$ *deterministically*, so we need only bound the probability of failing to achieve $\epsilon/2$-optimality in the regular MDP. Putting everything together yields:

$$\delta \stackrel{\text{def}}{=} P(\text{learned policy not } \epsilon\text{-optimal}) \tag{11.103}$$

$$= P(\text{learned policy not } \epsilon/2\text{-optimal in the regular MDP}) \tag{11.104}$$

$$= P(\text{failed to learn the regular MDP well}) \tag{11.105}$$

$$\begin{aligned} &= P(\text{didn't sample sufficiently}) \\ &\quad + P(\text{sampled sufficiently}) P(\exists \text{ a transition that is not well learned} \mid \\ &\text{it is sampled sufficiently}) \end{aligned} \tag{11.106}$$

$$\begin{aligned} &\leq P(\text{didn't sample sufficiently}) + P(\exists \text{ a transition that is not well learned} \mid \\ &\text{it is sampled sufficiently}) \end{aligned} \tag{11.107}$$

$$\leq S^H \exp\left(-\frac{\beta^{2H}}{2} N\right) + AHS^H \delta_0 \tag{11.108}$$

$$= S^H \left[\exp\left(-\frac{\beta^{2H}}{2} N\right) + AH\delta_0\right] \tag{11.109}$$

$$\stackrel{\text{Plugging in } \delta_0}{=} S^H \left[\exp\left(-\frac{\beta^{2H}}{2} N\right) + AH2^S \exp\left(-\frac{\epsilon_0^2}{2} N_0\right)\right] \tag{11.110}$$

$$\stackrel{\text{Plugging in } \epsilon_0}{=} S^H \left[\exp\left(-\frac{\beta^{2H}}{2} N\right) + AH2^S \exp\left(-\frac{\epsilon^2}{32H^2} N_0\right)\right] \tag{11.111}$$

$$\stackrel{\text{Plugging in } N_0}{=} S^H \left[\exp\left(-\frac{\beta^{2H}}{2} N\right) + AH2^S \exp\left(-\frac{\epsilon^2 \beta^H}{64AH^2} N\right)\right] \tag{11.112}$$

$$\stackrel{\text{Plugging in } \beta}{=} S^H \left[\exp\left(-\underbrace{\frac{\epsilon^{2H}}{2(2ASH)^{2H}}}_{\gamma_1} N\right) + AH2^S \exp\left(-\underbrace{\frac{\epsilon^{H+2}}{64AH^2(2ASH)^H}}_{\gamma_2} N\right)\right] \tag{11.113}$$

The two terms in Eq. (11.113) are both exponentially decaying in N, but for small enough values of ϵ, and when $H > 2$, $\gamma_1 < \gamma_2$. This means that the second term will vanish much faster than the first, asymptotically, so the first will dominate. Therefore, when N is large, Eq. (11.113) simplifies to:

$$\delta \leq 2S^H e^{-\gamma_1 N} \tag{11.114}$$

$$= 2S^H \exp\left[-\frac{1}{2}\left(\frac{\epsilon}{2ASH}\right)^{2H} N\right] \tag{11.115}$$

Rearranging yields:

$$N \leq 2\left(\frac{2ASH}{\epsilon}\right)^{2H} \ln\left(\frac{2S^H}{\delta}\right) \tag{11.116}$$

$$= 2\left(\frac{2ASH}{\epsilon}\right)^{2H}\left(\ln 2 + H\ln S + \ln\frac{1}{\delta}\right) \tag{11.117}$$

Ignoring constants and simplifying expressions proves the theorem.

Appendix 3: Proof of Theorem 5

To prove this theorem, we construct a specific MDP which achieves this bound.

Suppose there is only one terminal state with reward 1 while all other states have reward 0. Further suppose that from each state there is only one optimal action that successfully transitions to the desired next-round state with probably $\epsilon^{1/H}$; all other actions will only lead to "wrong" states: i.e., states that can only lead to zero reward.

In the remainder of this proof, we rely on the following terminology:

- Choose the right action: choosing the only optimal action, call it a^*, at a state.
- Transition successfully: choosing a^* *and* transitioning to the desired next round state.
- Getting a hit: reaching a rewarding terminal state.

The optimal policy, which chooses the right action at each state, is the only policy that can possibly get a hit, and its expected reward is 1 times the probability of a hit, which is ϵ:

$$P(\text{hit}) = P(\text{transition 1 successful, transition 2 successful, ..., transition } H \text{ successful}) \tag{11.118}$$

$$\overset{\text{Independence of transitions}}{=} P(\text{transition 1 successful}) \cdot P(\text{transition 2 successful}) \cdot \ldots \cdot$$

$$P(\text{transition } H \text{ successful}) \tag{11.119}$$

$$= (\epsilon^{1/H})^H = \epsilon \tag{11.120}$$

All other policies generate an expected reward of 0, so to learn an ϵ-optimal policy in this MDP is equivalent to discovering the optimal policy, which in turn requires continued simulation until a hit occurs. Therefore our task becomes bounding the number of samples N needed until a hit occurs. We will now prove the theorem by showing that even under the optimal exploration policy, the probability of getting a hit during each simulation will be small, and hence N needs to be large.

Lemma 18 *In this specially constructed MDP, balanced wandering (choosing the least-sampled action and breaking ties uniformly at random) is the optimal exploration policy.*

Proof Balanced wandering is optimal because at any point in the simulation, it maximizes the probability of choosing the right action.

- When all actions have been sampled the same number of times, all actions are equally likely to be the correct one, so breaking ties uniformly at random is optimal.
- When an action a has been sampled less often than any of its alternatives, as long as there has not been a hit yet, a is more likely than its alternatives to be right.

Next, we bound the probability of getting a hit during a simulation. □

Lemma 19 *Under balanced wandering, the probability of a successful transition is bounded by* $\frac{\epsilon^{1/H}}{A(1-\epsilon^{1/H})}$*, and the probability of getting a hit is bounded by* $\bar{p} = \frac{\epsilon}{A^H(1-H\epsilon^{1/H})}$.

Proof When all actions at a state have been sampled the same number of times, the probability of a successful transition under balanced wandering is the probability of choosing the right action, $1/A$, times the probability of a successful transition upon the correct choice, $\epsilon^{1/H}$. This yields a success probability of $\frac{\epsilon^{1/H}}{A}$, at each step, and since there are H steps, the probability of a hit is $(\frac{\epsilon^{1/H}}{A})^H = \frac{\epsilon}{A^H}$.

But we can obtain a tighter bound by considering the situation in the midst of repeated simulations, when some actions have been sampled less than others, balanced wandering will choose the right action with probability greater than $1/A$. This is because actions that have been chosen more often and have failed more often are less likely than the others to be the right action.

Denote by $\mathcal{A}_i(m)$ the event that action a_i has been tried m times, but that none of those tries led to a successful transition. The most favorable case for an action, say a_1 WLOG, is the event $\mathcal{A} \stackrel{\text{def}}{=} \mathcal{A}_1(n-1) \cup \bigcup_{i \neq 1} \mathcal{A}_i(n)$. In other words, this event generates the highest probability that a_1 is a^*.[10] Balanced wandering will choose a_1

[10]That this is the most favorable case (i.e., that this situation yields the highest probability a_1 is the right action) is easy to see by example. Suppose there are three actions, that have all been tried (and failed) the same number of times. Then the probability distribution over the right action is $(\frac{1}{3}, \frac{1}{3}, \frac{1}{3})$. Now suppose action 2 is again tried without success; then the situation becomes $(\frac{1}{3} + \frac{\eta}{2}, \frac{1}{3} - \eta, \frac{1}{3} + \frac{\eta}{2})$, where η is a small probability calculated using Bayesian updating. Still more probability will be shifted to action 1 if action 3 is then tried without success. This is the highest probability that action 1 is the right action.

in this case, and $P(a^* = a_1)$ is indeed greater than $1/A$, but it cannot be too much greater:

$$P(a^* = a_1 \mid \mathcal{A}) = \frac{P(\mathcal{A} \mid a^* = a_1)P(a^* = a_1)}{P(\mathcal{A})} \tag{11.121}$$

$$= \frac{P(\mathcal{A} \mid a^* = a_1)P(a^* = a_1)}{P(\mathcal{A} \mid a^* = a_1)P(a^* = a_1) + \sum_{i \neq 1} P(\mathcal{A} \mid a^* = a_i)P(a^* = a_i)} \tag{11.122}$$

$$= \frac{P(\mathcal{A}_1(n-1) \mid a^* = a_1)P(a^* = a_1)}{P(\mathcal{A}_1(n-1) \mid a^* = a_1)P(a^* = a_1) + \sum_{i \neq 1} P(\mathcal{A}_i(n) \mid a^* = a_i)P(a^* = a_i)} \tag{11.123}$$

$$= \frac{(1-\epsilon^{1/H})^{n-1} \cdot \frac{1}{A}}{(1-\epsilon^{1/H})^{n-1} \cdot \frac{1}{A} + (1-\epsilon^{1/H})^{n} \cdot \frac{A-1}{A}} \tag{11.124}$$

$$= \frac{1}{1 + (1-\epsilon^{1/H})(A-1)} \tag{11.125}$$

$$= \frac{1}{A(1-\epsilon^{1/H}) + \epsilon^{1/H}} \tag{11.126}$$

$$< \frac{1}{A(1-\epsilon^{1/H})} \tag{11.127}$$

Equation (11.123) follows because for all $j \neq 1$, $P(\mathcal{A}_j(n) \mid a^* = a_1) = 1$, for all n.

Therefore, the probability of a successful transition is actually bounded by $\frac{1}{A(1-\epsilon^{1/H})} \cdot \epsilon^{1/H} = \frac{\epsilon^{1/H}}{A(1-\epsilon^{1/H})}$. It follows that the probability of getting a hit is bounded by:

$$P(\text{hit}) < \left(\frac{\epsilon^{1/H}}{A(1-\epsilon^{1/H})}\right)^H \tag{11.128}$$

$$= \frac{\epsilon}{A^H(1-\epsilon^{1/H})^H} \tag{11.129}$$

$$\leq \frac{\epsilon}{A^H(1-H\epsilon^{1/H})} \tag{11.130}$$

$$\stackrel{\text{def}}{=} \bar{p} \tag{11.131}$$

□

Putting Lemmas 18 and 19 together, we see that to guarantee P(learned policy is ϵ-optimal) $\geq 1-\delta$ requires that $\delta \geq 1 - P$(learned policy is ϵ-optimal) $= P$ (learned policy is *not* ϵ-optimal). In other words, we require that

$$\delta \geq P(\text{learned policy is } \textit{not } \epsilon\text{-optimal}) \tag{11.132}$$

$$= P(\text{no hit in } N \text{ simulations}) \tag{11.133}$$

$$> (1-\bar{p})^N \tag{11.134}$$

$$\geq 1 - N\bar{p} \tag{11.135}$$

But then, it is necessary that

$$N > \frac{1-\delta}{\bar{p}} \tag{11.136}$$

$$\overset{\text{plug in } \bar{p} \text{ as in Equation 132}}{=} \frac{A^H}{\epsilon}\left(1 - H\epsilon^{1/H}\right)(1-\delta) \tag{11.137}$$

References

Armantier, O., Florens, J.-P., & Richard, J.-F. (2008). Approximation of Nash equilibria in Bayesian games. *Journal of Applied Econometrics, 23*(7), 965–981.

Aumann, R. J. (1985). What is game theory trying to accomplish? In K. Arrow & S. Honkapohja (Eds.), *Frontiers of economics*. Citeseer.

Azar, M. G., Munos, R., & Kappen, H. J. (2012). On the sample complexity of reinforcement learning with a generative model.

Bellman, R. E. (1957). *Dynamic programming*. Princeton, NJ: Princeton University Press.

Brown, N., Ganzfried, S., & Sandholm, T. (2015). Hierarchical abstraction, distributed equilibrium computation, and post-processing, with application to a champion no-limit texas hold'em agent. In *Proceedings of the 2015 international conference on autonomous agents and multiagent systems, AAMAS 2015*, Istanbul, Turkey, May 4–8, 2015, pp. 7–15. http://dl.acm.org/citation.cfm?id=2772885.

Daskalakis, C., Goldberg, P. W., & Papadimitriou, C. H. (2009). The complexity of computing a Nash equilibrium. *SIAM Journal on Computing, 39*(1), 195–259.

Fudenberg, D., & Levine, D. K. (1998). *The theory of learning in games* (Vol. 2). Cambridge: MIT Press.

Greenwald, A., & Hall, K. (2003). Correlated Q-learning. In *Twentieth international conference on machine learning* (pp. 242–249). Washington, DC.

Greenwald, A., Li, J., & Sodomka, E. (2012). Approximating equilibria in sequential auctions with incomplete information and multi-unit demand. In *Advances in neural information processing systems* (Vol. 25, pp. 2330–2338).

Hansen, E. A., Bernstein, D. S., & Zilberstein, S. (2004). Dynamic programming for partially observable stochastic games. In *AAAI* (pp. 709–715).

Hart, S. (1992). Games in extensive and strategic forms. *Handbook of Game Theory with Economic Applications, 1*, 19–40.

Hu, J., & Wellman, M. P. (2003). Nash Q-learning for general-sum stochastic games. *Journal of Machine Learning Research, 4*, 1039–1069.

Kaelbling, L. P., Littman, M. L., & Cassandra, A. R. (1998). Planning and acting in partially observable stochastic domains. *Artificial Intelligence, 101*(1), 99–134.

Katzman, B. (1999). A two stage sequential auction with multi-unit demands. *Journal of Economic Theory, 86*(1), 77–99.

Kearns, M., Mansour, Y., & Ng, A. Y. (2002). A sparse sampling algorithm for near-optimal planning in large Markov decision processes. *Machine Learning, 49*, 193–208.

Kearns, M. J., & Singh, S. P. (2002). Near-optimal reinforcement learning in polynomial time. *Machine Learning, 49*(2), 209–232.

Kocsis, L., & Szepesvári, C. (2006). Bandit Based Monte-Carlo planning. In *Seventeenth European conference on machine learning* (pp. 282–293). Berlin.

Koller, D., Megiddo, N., & von Stengel, B. (1996). Efficient computation of equilibria for extensive two-person games. *Games and Economic Behavior, 14*(2), 247–259.

Krishna, V. (2009). *Auction theory*. Cambridge: Academic Press.

Kuhn, H. W. (1953). Extensive games and the problem of information. *Contributions to the Theory of Games*, *2*(28), 193–216.
Lanctot, M., Waugh, K., Zinkevich, M., & Bowling, M. (2009). Monte Carlo sampling for regret minimization in extensive games. In Y. Bengio, D. Schuurmans, J. D. Lafferty, C. K. I. Williams & A. Culotta (Eds.), *Advances in neural information processing systems* (Vol. 22, pp. 1078–1086). Curran Associates, Inc., 2009. http://papers.nips.cc/paper/3713-monte-carlo-sampling-for-regret-minimization-in-extensive-games.pdf.
Littman, M. L. (2001). Friend-or-Foe Q-learning in general-sum games. In *Eighteenth international conference on machine learning* (pp. 322–328). Williamstown.
Mas-Colell, A., Whinston, M. D., & Green, J. R. (1995). *Microeconomic theory*. New York: Oxford University Press.
McKelvey, R. D., McLennan, A. M., & Turocy, T. L. (2010). *Gambit: Software tools for game theory*. Technical report, Version 0.2010.09.01. http://www.gambit-project.org.
Menezes, F. M., & Monteiro, P. K. (2003). Synergies and price trends in sequential auctions. *Review of Economic Design*, *8*(1), 85–98.
Mostafa., H., & Lesser, V. (2008). Approximately solving sequential games with incomplete information. In *AAMAS-08 workshop on multi-agent sequential decision making in uncertain multi-agent domains* (pp. 92–106). Estoril.
Myerson, R. B. (1991). *Game theory: Analysis of conflict*. Cambridge, MA: Harvard University.
Naroditskiy, V., & Greenwald, A. (2007). Using iterated best-response to find symmetric Bayes–Nash equilibria in auctions. In *Twenty-second national conference on artificial intelligence* (pp. 1894–1895). Vancouver.
Nash, J. (1951). Non-cooperative games. *Annals of Mathematics*, *54*(2), 286–295. ISSN 0003486X. doi:10.2307/1969529. http://www.jstor.org/stable/1969529.
Oliehoek, F., Spaan, M. T. J., & Vlassis, N. (2005). Best-response play in partially observable card games. In *Benelearn 2005: Proceedings of the 14th annual machine learning conference of Belgium and the Netherlands* (pp. 45–50).
Osepayshvili, A., Wellman, M. P., Reeves, D. M., & MacKie-Mason, J. K. (2005). Self-confirming price prediction for bidding in simultaneous ascending auctions. In *Twenty-first conference on uncertainty in artificial intelligence* (pp. 441–449). Edinburgh.
Puterman, M. L. (1994). *Markov decision processes: Discrete stochastic dynamic programming*. Hoboken: Wiley-Interscience.
Rabinovich, Z., Naroditskiy, V., Gerding, E. H., & Jennings, N. R. (2013). Computing pure Bayesian Nash Equilibria in games with finite actions and continuous types. *Artificial Intelligence*, *195*, 106–139.
Reeves, D. M., & Wellman, M. P. (2004). Computing best-response strategies in infinite games of incomplete information. In *Twentieth conference on uncertainty in artificial intelligence* (pp. 470–478). Banff.
Silver, D., & Veness, J. (2010). Monte-Carlo planning in large pomdps. In *Advances in neural information processing systems* (pp. 2164–2172).
Strehl, A. L., Li, L., & Littman, M. L. (2009). Reinforcement learning in finite MDPs: PAC analysis. *Journal of Machine Learning Research*, *10*, 2413–2444.
Weber, R. J. (1983). Multiple-object auctions. In R. Engelbrecht-Wiggans, R. M. Stark, & M. Shubik (Eds.), *Competitive bidding, auctions, and procurement* (pp. 165–191). New York: New York University Press.
Weissman, T., Ordentlich, E., Seroussi, G., Verdu, S., & Weinberger, M. J. (2003). *Inequalities for the l1 deviation of the empirical distribution*. Hewlett-Packard Labs, Tech. Rep.
Whitehead, S. D., & Ballard, D. H. (1991). Learning to perceive and act by trial and error. *Machine Learning*, *7*, 45–83.

Chapter 12
The Logic of Approximate Dependence

Jouko Väänänen

Abstract In my joint paper (Parikh and Väänänen in Ann Pure Appl Log 134(1):83–93, 2005) with Rohit Parikh we investigate a logic arising from finite information. Here we consider another kind of limited information, namely information with a small number of errors, and prove a related completeness theorem. We point out that this approach naturally leads to considering multi-teams in the team semantics that lies behind (Parikh and Väänänen 2005).

Keywords Team semantics · Dependence logic · Approximation

The idea of *finite information logic* (Parikh and Väänänen 2005). is that when quantifiers, especialy the existential quantifiers, express choices in social context, the choices are based on *finite* information about the parameters present. In this paper we consider a different kind of restriction. We do not restrict the information available but we allow a small number of errors. In *social sofware* a few errors can perhaps be allowed, especially if there is an agreement about it.

Consider the sentences

"On these flights I have an exit-row seat, apart from a few exceptions.",

"Apart from a few, the participants are logicians.",

One way to handle such expressions is the introduction of generalized quantifiers, such as "few x", "most x", "all x but a few", etc. The approach of this paper is

---

This paper was written while the author was visiting the Computer Science Department of the University of California, Santa Cruz. The author is grateful to his host Phokion Kolaitis for the invitation and for the hospitality. The author is grateful for helpful discussions on this topic with P. Galliani, L. Hella, and P. Kolaitis. The author also thanks J. Kivinen, Jixue Liu, H. Toivonen and M. Warmuth for helpful suggestions. Research partially supported by Grant 40734 of the Academy of Finland.

J. Väänänen (✉)
Department of Mathematics and Statistics, University of Helsinki, Helsinki, Finland
e-mail: jouko.vaananen@helsinki.fi

J. Väänänen
Institute for Logic, Language and Computation, University of Amsterdam, Amsterdam, The Netherlands

C. Başkent et al. (eds.), *Rohit Parikh on Logic, Language and Society*,
Outstanding Contributions to Logic 11, DOI 10.1007/978-3-319-47843-2_12

different, or at least on the surface it looks different. We use the *team semantics* of Väänänen (2007).

In team semantics the main tool for defining the meaning of formulas is not that of an assignment but that of a *set* of assignments. Such sets are called in Väänänen (2007) teams. Intuitively a team can represent (or manifest) different kinds of things, such as

uncertainty
belief
plays of a game
data about a scientific experiment
possible voting profiles
database
dependence
independence
inclusion
exclusion
etc.

The obvious advantage of considering meaning in terms of teams rather than single assignments is that teams can indeed manifest change and variation unlike static single assignments. For example, Table 12.1 can tells us e.g. that $y = x^2$ apart from one exception, and z is a constant zero apart from one exception. Likewise, Table 12.2 tells us that an employee's salary depends only on the department except for one person. In medical data about causes and effects of treatment there can often be exceptions although there may be compelling evidence of a causal relationship otherwise.

While team semantics is suitable for many purposes, we focus here on the concept of dependence, the main concept of the paper (Parikh and Väänänen 2005), too. Dependence is used throughout science and humanities. In particular it appears in database theory in the form of functional dependence.

Table 12.1 A team with an error

x	y	z
2	4	0
5	25	0
3	9	1
2	3	0

Table 12.2 A dependence with an exception

Employee	Department	Salary
John	I	120 000
Mary	II	130 000
Ann	I	120 000
Paul	I	120 000
Matt	II	130 000
Julia	I	130 000

In Väänänen (2007) the following concept was introduced:

Definition 1 A team is any set of assignments for a fixed set of variables. A team X is said to satisfy the dependence atom

$$=(x, y), \tag{12.1}$$

where x and y are finite sequences of variables, if any two assignments s and s' in X satisfy

$$s(x) = s'(x) \to s(y) = s'(y). \tag{12.2}$$

Dependence logic (Väänänen (2007)) arises from first order logic by the addition of the dependence atoms (12.1). The logical operations $\neg$, $\wedge$, $\vee$, $\forall$, $\exists$ are defined in such a way that dependence logic is a conservative extension of classical first order logic. The exact expressive power of dependence logic is existential second order logic.

With the purpose in mind to capture a concept of dependence logic which is more realistic in the sense that a couple of errors are allowed, we now define[1]:

Definition 2 Suppose p is a real number, $0 \leq p \leq 1$. A finite team X is said to satisfy the approximate dependence atom

$$=_p(x, y)$$

if there is $Y \subseteq X$, $|Y| \leq p \cdot |X|$, such that the team $X \setminus Y$ satisfies $=(x, y)$. We then write $X \models =_p(x, y)$. For arbitrary teams (finite or infinite) X we say that X satisfies the atom $=(x, y)$ *mod finite*, if there is a finite Y such that $X \setminus Y$ satisfies $=(x, y)$. In symbols $X \models =^*(x, y)$.

In other words, a finite team of size n satisfies $=_p(x, y)$ if it satisfies $=(x, y)$ after we delete a portion measured by the number p, of the assignments of X. More exactly, we delete up to $p \cdot n$ assignments from the team. Hence the word "approximate".

The emphasis in approximate dependence $=_p(x, y)$ is on small p but the general concept is defined for all p. It is difficult to imagine any practical importance for, say, $=_{.95}(x, y)$. This is the proposition that the team has a 5 % portion where y is functionally determined by x. However, if we suppose that the relationship between x and y is totally random, then it may be significant in a big dataset of millions of rows, to observe that the $=_{.95}(x, y)$ holds and violates total randomness.

For a trivial example, every finite team satisfies $=_1(x, y)$, because the empty team always satisfies $=(x, y)$. On the other hand, $=_0(x, y)$ is just the old $=(x, y)$. Since singleton teams always satisfy $=(x, y)$, a team of size n always satisfies $=_{1-\frac{1}{n}}(x, y)$. A finite team trivially satisfies $=^*(x, y)$, whatever x and y, so the "mod finite" dependence is only interesting in infinite teams.

[1] An essentially same, as well as related approximate functional dependences, were introduced already in Kivinen and Mannila (1995).

The team of Table 12.1 satisfies $=_{\frac{1}{4}}(x, y)$ and the team of Table 12.2 satisfies

$$=_{\frac{1}{6}}(\texttt{Department}, \texttt{Salary}).$$

We claim that approximate dependence $=_p(x, y)$ is a much more common phenomenon in science and humanities than full dependence $=(x, y)$. Any database of a significant size contains errors for merely human reasons or for reasons of errors in transmission. Any statistical data of medical, biological, social, etc. information has exceptions partly because of the nature of the data. One rarely if ever encounters absolute dependence of the kind $=(x, y)$ in practical examples. The dependencies we encounter in practical applications have exceptions, the bigger the data the more there are exceptions. For the dependence $=_{.1}(x, y)$ we allow an error value in 10% of the cases. This may be unacceptable for some applications but overwhelming evidence of functional dependence in others.

A different kind of approximate functional dependence arises if we think of the individual *values* of variables as being slightly off. For example, we can consider a functional dependence in which values of y are *almost* the same whenever the values of x are almost the same. This direction is pursued in Bělohlávek and Vychodil (2006).

We have emphasized the relevance of $=_p(x, y)$ over and above $=(x, y)$. So how does dependence logic change if we allow $=_p(x, y)$ in addition to $=(x, y)$, that is, if we allow dependance with errors in addition to dependence without errors?

One of the first results about database dependencies is the so called Armstrong Completeness Theorem (Armstrong 1974). It has as its starting point a set of axioms for the dependance $=(x, y)$. We now adapt the axioms from (Armstrong 1974) to the more general case of approximate dependence atoms. Concatenation of two finite sequences of variables, x and y, is denoted xy. Such finite sequences can be also empty.

Definition 3 The *axioms* of approximate dependence are:

A1 $=_0(xy, x)$ (Reflexivity)
A2 $=_1(x, y)$ (Totality)

The *rules* of approximate dependence are:

A3 If $=_p(x, yv)$, then $=_p(xu, y)$ (Weakening)
A4 If $=_p(x, y)$, then $=_p(xu, yu)$ (Augmentation)
A5 If $=_p(xu, yv)$, then $=_p(ux, yv)$ and $=_p(xu, vy)$ (Permutation)
A6 If $=_p(x, y)$ and $=_q(y, v)$, where $p + q \leq 1$, then $=_{p+q}(x, v)$ (Transitivity)
A7 If $=_p(x, y)$ and $p \leq q \leq 1$, then $=_q(x, y)$ (Monotonicity)

These axioms are always satisfied in finite teams. As to (A1), we observe that the empty team $\emptyset$ satisfies $=(x, y)$ and hence we can take $Y = X$ in Definition 3. As to (A2) we observe that the empty team satisfies $=(x, y)$ and so we can take $Y = X$ in Definition 2. The axiom (A3) can be verified as follows. Suppose $X \setminus Y \models =(x, yz)$, where $Y \leq p \cdot X$, and the domain of X (and of Y) includes $xuyz$ so that both

$=(x, yz)$ and $=(xu, y)$ can be meaningfully checked for satisfiability in X. Suppose $s, s' \in X\backslash Y$ such that $s(xu) = s'(xu)$. Then $s(x) = s'(x)$. Hence $s(yz) = s'(yz)$, whence finally $s(y) = s'(y)$. Let us then verify the validity of (A6). Suppose $X\backslash Y \models =(x, y)$, $X\backslash Z \models =(y, z)$, where $|Y| \leq p \cdot |X|$ and $|Z| \leq q \cdot |X|$. Then $|Y \cup Z| \leq |Y| + |Z| \leq (p+q) \cdot |X|$ and $X\backslash(Y \cup Z) \models =(x, z)$. Finally, (A7) is trivial.

The above axioms and rules are designed with finite derivations in mind. With infinitely many numbers p we can have infinitary logical consequences (in finite teams), such as

$$\{=_{\frac{1}{n}}(x, y) : n = 1, 2, \ldots\} \models =_0(x, y),$$

which do not follow by the axioms and rules (A1)–(A6).[2] We now focus on finite derivations and finite sets of approximate dependences. We prove the following Completeness Theorem[3]:

Theorem 4 *Suppose Σ is a finite set of approximate dependence atoms. Then $=_p(x, y)$ follows from Σ by the above axioms and rules if and only if every finite team satisfying Σ also satisfies $=_p(x, y)$.*

We first develop some auxiliary concepts and observations for the proof.

Let τ be a pair $(\Sigma, =_p(x, y))$, where Σ is a finite set of approximate dependencies. For such τ let Z_τ be the finite set of all variables in $\Sigma \cup \{=_p(x, y)\}$. Let C_τ be the smallest set containing Σ and closed under the rules $(A1)$–$(A6)$ (but not necessarily under (A7)) for variables in Z_τ. Note that C_τ is finite.

Lemma 5 $\Sigma \vdash =_t(u, v)$ *iff* $\exists r \leq t(=_r(u, v) \in C_\tau)$.

Proof The implication from right to left is trivial. For the converse it suffices to show that the set

$$\Sigma' = \{=_t(u, v) : \exists r \leq t(=_r(u, v) \in C_\tau)\}$$

is closed under (A1)–(A7). □

Definition 6 Suppose $\tau = (\Sigma, =_p(x, y))$. For any variable y let

$$d_\tau(y) = \min\{r \in [0, 1] : =_r(x, y) \in C_\tau\}.$$

This definition makes sense because there are only finitely many $=_r(u, v)$ in C_τ. Note that $d_\tau(x) = 0$ by axiom (A1). By Lemma 5,

$$d_\tau(y) = \min\{r \in [0, 1] : \Sigma \vdash =_r(x, y)\}.$$

Lemma 7 *If $\Sigma \vdash =_p(u, v)$, then $d_\tau(v) - d_\tau(u) \leq p$.*

---

[2]We can use this example to encode the Halting Problem to the question whether a recursive set of approximate dependence atoms logically implies a given approximate dependence atom.

[3]Proposition A.3 of Kivinen and Mannila (1995) is a kind of completeness theorem in the spirit of the below theorem for one-step derivations involving approximate dependence atoms.

Fig. 12.1 The team X_τ

	x	$\dots$	u	$\dots$
s_0	0	$\dots$	0	$\dots$
s_1	0	$\dots$	1	$\dots$
s_2	0	$\dots$	2	$\dots$
$\vdots$				
s_m	0	$\dots$	m	$\dots$
$\vdots$				
s_{n-1}	0	$\dots$	m	$\dots$

Proof Suppose $d_\tau(u) = r$, $d_\tau(v) = t$, $\Sigma \vdash =_r(x, u)$ (r minimal) and $\Sigma \vdash =_t(x, v)$ (t minimal). Now $\Sigma \vdash =_r(x, u)$ and $\Sigma \vdash =_p(u, v)$. Hence $\Sigma \vdash =_{r+p}(x, v)$. By the minimality of t, $t \leq r + p$. Hence $t - r \leq p$. □

For a given Σ there are only finitely many numbers $d_\tau(u)$, $u \in Z_\tau$, because C_τ is finite. Let A_τ consist of p and the set of $d_\tau(u)$ such that $u \in Z_\tau$. Let $n = 1 + \max\{\lceil 2/(a - b)\rceil : a, b \in A - \tau, a \neq b\}$. We define a team X_τ of size n as follows (Fig. 12.1):

$$X_\tau = \{s_0, \dots, s_n\},$$

where for $\frac{m}{n} \leq d_\tau(u) < \frac{m+1}{n}$ we let

$$s_i(u) = \begin{cases} i, & \text{if } i \leq m \\ m, & \text{if } i > m \end{cases}$$

Lemma 8 *Suppose $X_\tau \models =_p(x, y)$. Then $\Sigma \vdash =_p(x, y)$.*

Proof Suppose $X_\tau \models =_p(x, y)$ but $\Sigma \nvdash =_p(x, y)$. Now $d_\tau(y) > p$. Let $\frac{m}{n} \leq d_\tau(y) < \frac{m+1}{n}$. One has to take all the assignments s_i, $i \leq m - 1$, away from X_τ in order for the remainder to satisfy $=(x, y)$. Hence $p \cdot n \geq m$ i.e. $p \geq \frac{m}{n}$. But we have chosen n so that $1/n < d_\tau(y) - p$. Hence

$$p < d_\tau(y) - \frac{1}{n} \leq \frac{m+1}{n} - \frac{1}{n} = \frac{m}{n},$$

a contradiction. □

Lemma 9 *Suppose $\Sigma \vdash =_q(u, v)$. Then $X_\tau \models =_q(u, v)$.*

Proof We know already $d_\tau(v) - d_\tau(u) \leq q$. If $d_\tau(v) \leq d_\tau(u)$, then $X_\tau \models =(u, v)$, and hence all the more $X_\tau \models =_q(u, v)$. Let us therefore assume $d_\tau(v) > d_\tau(u)$. Since $2/n < d_\tau(v) - d_\tau(u)$, there are m and k such that

$$\frac{m}{n} \leq d_\tau(u) < \frac{m+1}{n} < \frac{k}{n} \leq d_\tau(v) < \frac{k+1}{n}.$$

In order to satisfy $=(x, y)$ one has to delete $k - m$ assignments from X_τ. But this is fine, as $qn \geq (d_\tau(v) - d_\tau(u))n > k - d_\tau(u)n \geq k - m$. □

Fig. 12.2 Non-locality of approximation

x	y	z
0	0	0
0	0	1
0	1	1

Y

x	y
0	0
0	1

$Y \upharpoonright xy$

Lemmas 8 and 9 finish the proof of Theorem 4.

A problematic feature of the approximate dependence atom is that it is **not local**, that is, the truth of $X \models =_p(x, y)$ may depend on the values of the assignments in X on variables u not occurring in x or y. To see this, consider the team Y of Fig. 12.2. Now Y satisfies $=_{\frac{1}{3}}(x, y)$. Let Z be the team $Y \upharpoonright xy$. Now Z does not satisfy $=_{\frac{1}{3}}(x, y)$, as Fig. 12.2 shows.

This problem can be overcome by the introduction of **multi-teams**:

Definition 10 A *multi-team* is a pair (X, τ), where X is a set and τ is a function such that

1. $\text{Dom}(\tau) = X$,
2. If $i \in X$, then $\tau(i)$ is an assignment for one and the same set of variables. This set of variables is denoted by $\text{Dom}(X)$.

An ordinary team X can be thought of as the multi-team (X, τ), where $\tau(i) = i$ for all $i \in X$. When approximate dependence is developed for multi-teams the non-locality phenomenon disappears (see Fig. 12.3). Moreover, the above Theorem 4 still holds.

The canonical example of a team in dependence logic is the set of plays where a player is using a fixed strategy. Such a team satisfies certain dependence atoms reflecting commitments the player has made concerning information he or she is using. If such dependence atoms hold only approximatively, the player is allowed to make a small number of deviations from his or her commitments. Let us suppose the player is committed to y being a function of x during the game. Typically y is a move of this player and x is the information set available for this move. When we look at a table of plays where the player is following his or her strategy, we may observe that indeed y is functionally determined by x except in a small number of plays. To evaluate the amount of such exceptional plays we can look at the table of all possible plays where the said strategy is used and count the numerical proportion of plays that have to be omitted in order that the promised functional dependence holds.

Fig. 12.3 Multi-teams

x	y	z
0	0	0
0	0	1
0	1	1

Y

x	y
0	0
0	0
0	1

$Y \upharpoonright xy$

We have here merely scratched the surface of approximate dependence. When approximate dependence atoms are added to first order logic we can express propositions such as "the predicate P consists of half of all elements, give or take 5 %" or "the predicates P and Q have the same number of elements, with a 1 % margin of error". To preserve locality we have to introduce multi-teams. On the other hand that opens the door to probabilistic teams, teams where every assignment is associated with a probability with which a randomly chosen element of the team is that very assignment. We will not pursue this idea further here.

References

Armstrong, W. W. (1974). Dependency structures of database relationships. *IFIP Congress 1974*, 580–583.

Bělohlávek, R., & Vychodil, V. (2006). Data tables with similarity relations: Functional dependencies, complete rules and non-redundant bases. In M. Lee, K.-L. Tan, & V. Wuwongse (Eds.), *Database systems for advanced applications* (Vol. 3882, pp. 644–658)., of Lecture Notes in Computer Science Berlin: Springer.

Kivinen, J., & Mannila, H. (1995). Approximate inference of functional dependencies from relations. *Theoretical Computer Science, 149*(1), 129–149. Fourth International Conference on Database Theory (ICDT '92).

Parikh, R., & Väänänen, J. (2005). Finite information logic. *Annals of Pure and Applied Logic, 134*(1), 83–93.

Väänänen, J. (2007). *Dependence logic, volume of London Mathematical Society Student Texts*. Cambridge: Cambridge University Press.

Chapter 13
On Height and Happiness

Melvin Fitting

Abstract Predicate abstraction provides very simple formal machinery that represents some of the ideas connecting intension and extension. It has been investigated and applied for many years now. Here we do two things with it. First, we present some natural examples, and show how predicate abstraction can be applied to them. Second, we give a formal proof procedure for predicate abstraction, based on *nested sequents*. Soundness and completeness proofs are sketched.

Keywords Intensional logic · Modal logic · Nested sequents · Non-rigidity · Predicate abstraction

13.1 Introduction

One simple and natural way of enhancing propositional modal logic is via *predicate abstraction*, which goes back to Stalnaker and Thomason (1968) and Thomason and Stalnaker (1968). It disambiguates many of the perceived syntactic and semantic problems with modal logic, and does so with a certain amount of intuitive appeal. The basic idea behind predicate abstraction in modal logics is elementary, but like first-order logic itself, there are many concrete versions of it. With first-order logic(s) one can have constant symbols, or not, function symbols, or not, multiple sorts, or not. A similar range of choices is available for predicate abstraction. Here we introduce the minimal machinery to handle the kinds of problems we discuss.

We begin with some examples of commonly perceived modal problems, and discuss how predicate abstraction deals with them. Though there are domains, predicate symbols, and function symbols, there is no quantification. Domains are constant. Type complexity is minimal. Think of this presentation as something that can be expanded on, as needed. All the really important ideas are already present. After the examples, there is a formal presentation of the syntax and semantics. Then we go

M. Fitting (✉)
Department of Philosophy, CUNY Graduate Center,
365 Fifth Avenue, New York, NY 10016, USA
e-mail: melvin.fitting@gmail.com

C. Başkent et al. (eds.), *Rohit Parikh on Logic, Language and Society*,
Outstanding Contributions to Logic 11, DOI 10.1007/978-3-319-47843-2_13

on to formulate a nice nested sequent calculus for predicate abstraction. Up to this point the work presented is not new, though we hope the discussion is simple and enlightening. The nested sequent calculus is both new and not new—a remark that will be elaborated then and not now.

In 2010, Kai Wehmeier gave a talk at the CUNY Graduate Center. That talk (and presumably many others) crystalized into *Subjunctivity and Cross-World Predication* (Wehmeier 2012). His presentation, and the discussion it initiated, were the origins of the present paper. His way of handling the problems discussed here is very different than ours. In a sense, ours is more traditional. His approach involves the introduction of symbols that are designed to represent relations between objects across worlds. It is very natural, especially from a linguistic point of view, and his paper is highly recommended. The two approaches are not in competition, but are rather complementary. It would not be surprising if there were natural embeddings between the two versions, but an investigation must wait until another day.

13.2 Examples

Suppose we are talking about the King (or Queen) of Sweden. We can think of the monarch as represented by a non-rigid constant symbol, say m, designating different persons at different times, or nobody at some times too. This could be elaborated into a definite description, but such machinery isn't really needed for what is discussed here, and would require the introduction of quantifiers anyway. Suppose also that we wish to express the idea that someday the monarch might be taller than now. We can give a modal operator a temporal reading, so that $\Diamond X$ asserts that X will be true at some future time. Also suppose we have a two-place predicate symbol T whose intended reading is: both arguments are people, and the first is taller than the second. With standard modal machinery, the best formalization we can manage is the absurd $\Diamond T(m, m)$.

It isn't m, the monarchical role or office, that we want to measure for height, it's the particular monarch at a particular time. We want to compare what m designates now with what m designates at some future time. In $\Diamond T(m, m)$, the first m must be understood as designating a person now; the second as designating at a possible future. Predicate abstraction gives us precisely that ability. A formal definition will come later, but the idea is that $\langle \lambda x \cdot X \rangle(c)$ should be true at a state if X is true at that state when x is given the value that c designates at that state (we'll worry about possible non-designation later on). Then we can express that someday the monarch will be taller than now as follows.

$$\langle \lambda y \cdot \Diamond \langle \lambda x \cdot T(x, y) \rangle(m) \rangle(m) \tag{13.1}$$

This says it is true of the person who is presently King or Queen (the value of y) that at some possible future state the taller-than predicate will hold between the person who then holds the position (the value of x) and the present person.

This seems straightforward. But now suppose we want to say about some person, Alice, that she might have been taller than she is (say given different nourishment as a baby than she actually had). We can shift our intended reading of $\Diamond$ so that $\Diamond X$ asserts that X is true in some other possible world, whatever we may mean by possible. And we can represent Alice by a constant symbol, say a, similarly to our use of m above. But the following sentence, corresponding to (13.1), won't work.

$$\langle \lambda y \cdot \Diamond \langle \lambda x \cdot T(x, y) \rangle (a) \rangle (a) \tag{13.2}$$

The problem is that 'Alice' is a name and so a is commonly understood to be a *rigid* constant symbol. We are talking about the same individual a in the present world and in alternative circumstances, and in effect we are saying a person might be taller than themselves. More technically, if a is rigid we have the truth of the following.

$$\langle \lambda y \cdot \Box \langle \lambda x \cdot x = y \rangle (a) \rangle (a) \tag{13.3}$$

And from (13.2) and (13.3) it follows that

$$\langle \lambda x \cdot \Diamond T(x, x) \rangle (a) \tag{13.4}$$

and this is clearly wrong (we will give a derivation of (13.4) from (13.2) and (13.3) in Sect. 13.8, after we have introduced nested sequents for predicate abstraction).

There is a way around the problem however, one that applies to the two cases just discussed, though it is overkill for the King of Sweden. When we say that Alice is represented by a rigid designator, a, what does that mean? In some sense a designates the same person in different possible worlds, of course, but in different possible worlds that person might have different inessential properties. In particular, the height of a might be different in different possible worlds, even though a is the same person.

Suppose we introduce a non-rigid function symbol, h, intended to map a person to that person's height, thought of as a number. What h assigns to a depends on the possible world in which $h(a)$ is evaluated (of course we need to assume some constancy in our standard of measuring, but let us do so for purposes of the present discussion). The point is that even though a is rigid, $h(a)$ can vary from possible world to possible world. This happens because an inessential attribute of a can vary from world to world, and this is reflected in the non-rigidity of h, rather than of a. If we assume h is a non-rigid function symbol, and G is the two-place greater-than relation on numbers, our assertion that Alice might have been taller than she is, formalizes as follows.

$$\langle \lambda y \cdot \Diamond \langle \lambda x \cdot G(x, y) \rangle (h(a)) \rangle (h(a)) \tag{13.5}$$

For the height example above, we don't need the whole real number system as part of our model. In practice we can only distinguish a finite set of height measurements, and these can certainly be built into formal *finite* models. But now, suppose we move from height to happiness. We can say that someday the King of Sweden will be happier than the King is now, rather similarly to (13.1). But how can we express that Alice might have been happier than she is? If we introduce a function symbol as we did above, we would have to think of it as designating a mapping of individuals to some kind of degrees of happiness, and assume that these degrees can (often) be compared. The ordering might not be linear—a partial order seems more plausible. Perhaps psychologists really do something like this, but for most of us it seems to expand our horizons a bit. On the other hand, since we generally can make sense out of saying that Alice is happier than Bob, we do seem to be implicitly making use of such machinery anyway. At any rate, there is certainly no bar to a formalization along these lines.

13.3 Predicate Abstraction Syntax

We begin the precise formulation of predicate abstraction. It is, in a sense, intermediate between propositional and first-order machinery. It makes use of relation symbols, variables, constant and function symbols, but quantifiers need not be involved. A formulation can be given in a variety of ways that differ in some technical details, but these details don't matter all that much. Below we will say something about what could be modified without changing anything essential.

The basic distinction is between *objects* and *intensions*. Very loosely, objects (or extensions) are the entities we bump into: a person, the number 4 (which we bump into in a soft sort of way), and so on. Intensions, on the other hand, are non-rigid designators, often specified by definite descriptions: the number of nations in the UN, my best friend, and so on. Semantically, intensions pick out objects in possible worlds, though perhaps not in all of them. This is actually part of Carnap's method of intension and extension (Carnap 1947). As formulated here, equality relates objects, while synonymy relates intensions (though we will not discuss this now). This is enough for us to get started.

There is a family of *object variables*, typically $x, y, \ldots$, and *intension constants*, $a, b, \ldots$ [in Fitting (2006) we allowed intension variables, but did not allow any constants—it makes no deep difference]. We also have *intension function symbols*, $a, b, \ldots$, of various arities, which take object variables as arguments. Finally, we have *relation symbols*, $P, Q, \ldots$ of various arities, also taking object variables as arguments in the usual way [in Fitting (2004) we allowed relation symbols to have intension variables as arguments as well, and positions in relations were typed—see also Fitting (2006, 2007). A similar generalization could be extended to the arguments of function symbols as well]. We will make use of a special two-place relation symbol, $=$, intended to denote equality. For ease in reading, we will write $x = y$ instead of $=(x, y)$.

An *intension function term* is $a(x_1, \ldots, x_n)$ where $x_1, \ldots, x_n$ are object variables and a is an n-ary intension function symbol. It is convenient for us to assume that intension constants are 0-place intension function symbols, so they are also function terms. Note that intension functions are not allowed to be nested—arguments are object variables—but there is a way around this, discussed below.

An *atomic formula* is $P(x_1, \ldots, x_n)$ where $x_1, \ldots, x_n$ are object variables and P is an n-ary relation symbol. Note that intension function terms don't come into things here. There is a way around this too, also discussed below.

Formulas (and *predicate abstracts*) are defined by recursion. Similarly for *free variable occurrences*.

- An atomic formula is a formula. All variable occurrences are free occurrences.
- If X and Y are formulas, so are $(X \wedge Y)$, $(X \vee Y)$, $(X \supset Y)$. Free variable occurrences in these formulas are the free variable occurrences of X and of Y.
- If X is a formula, so are $\neg X$, $\Box X$, and $\Diamond X$. Free variable occurrences in these formulas are the free variable occurrences of X.
- If X is a formula, x is an object variable, and t is an intension function term, $\langle \lambda x \cdot X \rangle(t)$ is a formula, called a predicate abstract. Free variable occurrences are those of X, except for occurrences of x, together with any free variable occurrences in t.

We finish with some comments and examples. Both $\langle \lambda x \cdot \Diamond P(x) \rangle(a)$ and $\Diamond \langle \lambda x \cdot P(x) \rangle(a)$ are formulas, where a is an intension constant and $P(x)$ is atomic. Neither formula has any free variable occurrences. Predicate abstracts impose a scope distinction, which is something that goes all the way back to Russell (1905) and Whitehead and Russell (1927). For him a crucial issue was non-designating terms, in particular definite descriptions. If $X(x)$ is a formula with object variable x free, and a is an intension constant, how should $\neg X(a)$ be read? Does it say that the object designated by a has the $\neg X$ property, or that it fails to have the X property? If a doesn't designate, these come out differently, as the present King of France would know. In predicate abstract notation the difference is syntactically represented by $\langle \lambda x \cdot \neg X(x) \rangle(a)$ and $\neg \langle \lambda x \cdot X(x) \rangle(a)$. Loosely the first says the a object, if any, has the $\neg X$ property, and the second says it fails to have the X property. We will assume that nothing can be correctly predicated of the object designated by a non-designating term, and so $\langle \lambda x \cdot \neg X(x) \rangle(a)$ is false if a does not designate. But then $\langle \lambda x \cdot X(x) \rangle(a)$ is also false, and hence $\neg \langle \lambda x \cdot X(x) \rangle(a)$ is true.

We did not allow intension function symbols to appear in atomic formulas. This simplifies some things, without restricting expressiveness. Suppose a is an intension constant. Instead of $P(a)$, which is not legal, we can write $\langle \lambda x \cdot P(x) \rangle(a)$, which is. We will consider the first as a convenient abbreviation for the second. But we must be careful, since this does not extend to formulas generally, or we would run into an ambiguity with, say, $\Diamond P(a)$. The familiar broad scope/narrow scope distinction is at work here, but by itself it is not sufficient—consider $\Diamond\Diamond P(a)$ for instance, where broad scope and narrow scope are not all the readings there are. Generally, we will not abbreviate unless an explicit proper version is also specified.

We also did not allow nesting of function symbols. This is not a problem because we can understand something line $P(f(g(a)))$ as an abbreviation for

$$\langle \lambda z \cdot \langle \lambda y \cdot \langle \lambda x \cdot P(x) \rangle (f(y)) \rangle (g(z)) \rangle (a)$$

where substitution is made explicit using λ. Note that the two formulas

$$\Diamond\langle\lambda y \cdot \langle\lambda x \cdot P(x)\rangle(f(y))\rangle(a) \tag{13.6}$$
$$\langle\lambda y \cdot \Diamond\langle\lambda x \cdot P(x)\rangle(f(y))\rangle(a) \tag{13.7}$$

are not syntactically the same, and will be seen to differ semantically as well. We might abbreviate (13.6) as $\Diamond P(f(a))$, using a narrow scope reading, but (13.7) has no corresponding abbreviation.

Finally, in the interest of readability we will sometimes abbreviate $\langle\lambda x \cdot \langle\lambda y \cdot X\rangle(b)\rangle(a)$ as $\langle\lambda x, y \cdot X\rangle(a, b)$, and similarly for larger numbers of nested abstracts.

13.4 Predicate Abstract Semantics

Domains, as in first-order models, must be part of the semantic machinery since we have relation and function symbols to model. We are not considering quantifiers here, so a distinction between constant and varying domain models really cannot be seen. We confine things to constant domain models for simplicity (and other reasons as well). Think of members of domains as the objects—extensions—that intensions pick out at possible worlds.

A *model* is a structure $\mathcal{M} = \langle\mathcal{G}, \mathcal{R}, \mathcal{D}, \mathcal{I}\rangle$ meeting the following conditions.

- $\langle\mathcal{G}, \mathcal{R}\rangle$ is a *frame*, with $\mathcal{G}$ the set of possible worlds and $\mathcal{R}$ a binary accessibility relation on $\mathcal{G}$.
- $\mathcal{D}$ is a non-empty set called the *object domain*.
- $\mathcal{I}$ is an *interpretation* function. It must cover relation and intension function symbols.
  If P is an n-ary relation symbol then $\mathcal{I}(P)$ is a mapping from $\mathcal{G}$ to subsets of $\mathcal{D}^n$. It is required that $\mathcal{I}(=)$ be the constant function mapping each world to the identity relation on $\mathcal{D}$ (thus equality is interpreted rigidly, and is total).
  If a is an n-ary intension function symbol then $\mathcal{I}(a)\colon S \to (\mathcal{D}^n \to \mathcal{D})$, for some $S \subseteq \mathcal{G}$ (the idea is that $\mathcal{I}(a)$ provides an n-ary function from $\mathcal{D}$ to itself for some, but perhaps not all, of the possible worlds in $\mathcal{G}$. Such a function is provided for those worlds in S, but not for worlds outside S). As a special case, a 0-ary intension function symbol, or intension constant, is simply a partial function from $\mathcal{G}$ to $\mathcal{D}$. If $\mathcal{I}(a) : S \to (\mathcal{D}^n \to \mathcal{D})$, we say *a designates* at the worlds in S.

A *valuation* in model $\mathcal{M}$ is a map assigning to each object variable a member of $\mathcal{D}$. We write $\mathcal{M}, w \Vdash_v X$ to symbolize that formula X is true at possible world w of model $\mathcal{M}$ with respect to valuation v. The conditions for this are as follows.

Atomic $\mathcal{M}, w \Vdash_v P(x_1, \ldots, x_n) \Leftrightarrow \langle v(x_1), \ldots, v(x_n)\rangle \in \mathcal{I}(P)(w)$.
Propositional $\mathcal{M}, w \Vdash_v X \supset Y \Leftrightarrow \mathcal{M}, w \not\Vdash_v X$ or $\mathcal{M}, \Gamma \Vdash_v Y$
And similarly for other propositional connectives.
Necessity $\mathcal{M}, w \Vdash_v \Box X \Leftrightarrow \mathcal{M}, w' \Vdash_v X$ for all $w' \in \mathcal{G}$ such that $w\mathcal{R}w'$.

Possibility $\mathcal{M}, w \Vdash_v \Diamond X \Leftrightarrow \mathcal{M}, w' \Vdash_v X$ for some $w' \in \mathcal{G}$ such that $w\mathcal{R}w'$.

Predicate Abstraction, n-ary Function Symbol

If intension function symbol a designates at w, $\mathcal{M}, w \Vdash_v \langle\lambda y \cdot X\rangle(a(x_1, \ldots, x_n))$ if
$\mathcal{M}, w \Vdash_{v'} X$ where v' is like v except that $v'(y) = \mathcal{I}(a)(w)(v(x_1), \ldots, v(x_n))$.
If a does not designate at w, $\mathcal{M}, w \nVdash_v \langle\lambda x \cdot X\rangle(a(x_1, \ldots, x_n))$.

The special case of predicate abstraction where a is 0-place, that is, an intension constant, is of particular interest, so we state it explicitly.

Predicate Abstraction, intension constant

If a designates at w, $\mathcal{M}, w \Vdash_v \langle\lambda y \cdot X\rangle(a)$ if
$\mathcal{M}, w \Vdash_{v'} X$ where v' is like v except that $v'(y) = \mathcal{I}(a)(w)$.
If a does not designate at w, $\mathcal{M}, w \nVdash_v \langle\lambda x \cdot X\rangle(a)$.

A formula X is valid in a model $\mathcal{M} = \langle\mathcal{G}, \mathcal{R}, \mathcal{D}, \mathcal{I}\rangle$ if $\mathcal{M}, w \Vdash_v X$ for every $w \in \mathcal{G}$ and every valuation v, and is valid if it is valid in every model. We have placed no restrictions on accessibility, so this is a version of K with predicate abstracts. Reflexivity, transitivity, and so on, can be imposed in the familiar way.

The first predicate abstraction condition says that $\mathcal{M}, w \Vdash_v \langle\lambda x \cdot X\rangle(t)$ if the object designated by t at w has the property specified by X at w. The second condition says that no property can correctly be ascribed to the object designated by a term that does not designate. Then intension symbols enter into things via what they designate, if anything. Other things are possible as well, and were considered in Fitting (2004).

13.5 Examples Continued

In (13.1) we gave a formula intended to express the idea that someday the King/Queen of Sweden might be taller than now. We repeat the formula here for convenience.

$$\langle\lambda y \cdot \Diamond\langle\lambda x \cdot T(x, y)\rangle(m)\rangle(m)$$

Now we use this formula to illustrate the way the semantics works. Here are a few historical facts we can make use of.

The list of Swedish royalty commonly begins with Eric the Victorious, who died around 995. But to be on the safe side, the first vague reference to a Swedish King is in Tacitus, around the year 100. So we can plausibly assume that in the year 50 there was no such King.

It is hard to get data about heights before modern times, but an examination of skeletons shows that, in the thirteenth century the average adult male Swede was 174.3 cm (68.6 in.) (Ward 2013). King Magnus III Barnlock (who we will represent as mb) died in 1290, and we will assume his height was exactly average.

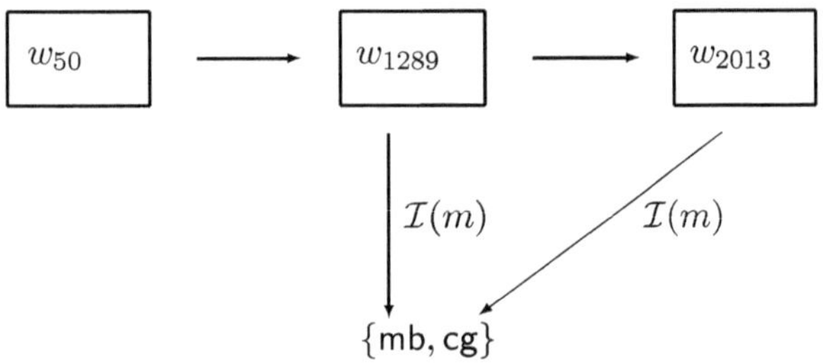

Fig. 13.1 Model $\mathcal{M}$ with Swedish Kings

Finally, King Carl XVI Gustaf, king at the time of writing this, is 179 cm (70.5 in.) tall. We will use cg to represent him. It is a small difference, but he is taller than Magnus III.

Figure 13.1 graphically shows a model, $\mathcal{M}$. For this model the set of possible worlds is $\mathcal{G} = \{w_{50}, w_{1289}, w_{2013}\}$, intended to represent states of the world in the years 50, 1289, and 2013. Accessibility is shown using right pointing arrows, and is intended to represent passage to a future state (we may assume transitivity, but it is not shown in the figure and is not needed for the points we wish to make). The object domain is $\mathcal{D} = \{\mathsf{mb}, \mathsf{cg}\}$, intended to represent the two Swedish kings discussed above. The interpretation function $\mathcal{I}$ is such that $\mathcal{I}(T)(w_{50}) = \mathcal{I}(T)(w_{1289}) = \mathcal{I}(T)(w_{2013}) = \{(\mathsf{cg}, \mathsf{mb})\}$ (thus T is interpreted rigidly, and informally says that cg is taller than mb). We assume m designates at $\{w_{1289}, w_{2013}\}$, and $\mathcal{I}(m)(w_{1289}) = \mathsf{mb}$ and $\mathcal{I}(m)(w_{2013}) = \mathsf{cg}$. That is, $\mathcal{I}(m)$ picks out the King of Sweden at each time state we consider, at which a king exists.

We check that $\mathcal{M}, w_{1289} \Vdash_v \langle\lambda y \cdot \Diamond\langle\lambda x \cdot T(x, y)\rangle(m)\rangle(m)$, thus showing that (13.1) is formally true under the expected circumstances (v is an arbitrary valuation. In fact, the formula has no free variables, so a choice of v won't matter). Now, $\mathcal{M}, w_{1289} \Vdash_v \langle\lambda y \cdot \Diamond\langle\lambda x \cdot T(x, y)\rangle(m)\rangle(m)$ just in case $\mathcal{M}, w_{1289} \Vdash_{v'} \Diamond\langle\lambda x \cdot T(x, y)\rangle(m)$, where $v'(y) = \mathcal{I}(m)(w_{1289}) = \mathsf{mb}$. This holds just when $\mathcal{M}, w_{2013} \Vdash_{v'} \langle\lambda x \cdot T(x, y)\rangle(m)$, and in turn we have this just in case $\mathcal{M}, w_{2013} \Vdash_{v''} T(x, y)$ where $v''(y) = v'(y) = \mathsf{mb}$ and $v''(x) = \mathcal{I}(m)(w_{2013}) = \mathsf{cg}$. And this is the case, because $(\mathsf{cg}, \mathsf{mb}) \in \mathcal{I}(T)(w_{2013})$.

Next $\mathcal{M}, w_{50} \not\Vdash_v \langle\lambda y \cdot \Diamond\langle\lambda x \cdot T(x, y)\rangle(m)\rangle(m)$, because m does not designate at w_{50}. Also $\mathcal{M}, w_{2013} \not\Vdash_v \langle\lambda y \cdot \Diamond\langle\lambda x \cdot T(x, y)\rangle(m)\rangle(m)$ because w_{2013} is a possible world with no future. The first is what we would expect. The second really reflects an inadequacy in our model—we have not represented time correctly (we dearly hope).

We move on to the example in which it is said, of Alice, that she might have been taller than she is, initially formalized in (13.5) which we repeat here.

$$\langle\lambda y \cdot \Diamond\langle\lambda x \cdot G(x, y)\rangle(h(a))\rangle(h(a))$$

The immediate difficulty is that this is not a legal formula as we have set things up—see the discussion at the end of Sect. 13.3. What takes its place is the appalling

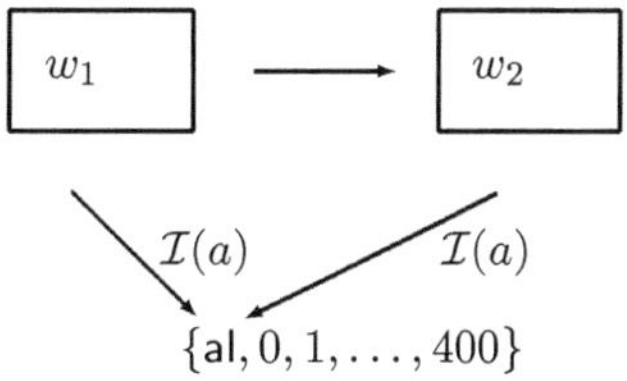

Fig. 13.2 Model $\mathcal{N}$ with Alice

looking formula (13.8), which is actually simpler in a sense, because there is no function/constant symbol nesting.

$$\langle \lambda w \cdot \langle \lambda y \cdot \Diamond \langle \lambda z \cdot \langle \lambda x \cdot G(x, y) \rangle (h(z)) \rangle (a) \rangle (h(w)) \rangle (a) \tag{13.8}$$

We want the interpretation of a to be rigid, but the function interpreting h to be non-rigid. For simplicity, let us suppose we measure heights to the nearest centimeter, and we only need heights from, say, 0–400. This keeps things finite, though nothing essential depends on it.

Figure 13.2 shows a model, $\mathcal{N}$. The set of possible worlds $\mathcal{G}$ consists of w_1 and w_2 (how things are, and how they might have been), with $w_1 \mathcal{R} w_2$. The object domain $\mathcal{D}$ is $\{\mathsf{al}, 0, 1, \ldots, 400\}$, where al is intended to be the 'Alice object.'

Next we have the interpretation function $\mathcal{I}$. We set $\mathcal{I}(G)(w_1) = \mathcal{I}(G)(w_2) = \{(x, y) \mid x, y \text{ are integers from 0 to 400 and } x > y\}$. Intension constant a designates at both w_1 and w_2, and $\mathcal{I}(a)(w_1) = \mathcal{I}(a)(w_2) = \mathsf{al}$. Thus a is interpreted rigidly. h is interpreted non-rigidly, though it designates at both possible worlds; $\mathcal{I}(h) : \mathcal{G} \rightarrow (\mathcal{D} \rightarrow \mathcal{D})$. We set $\mathcal{I}(h)(w_1)(\mathsf{al}) = 165$ and $\mathcal{I}(h)(w_2)(\mathsf{al}) = 180$ (values on other members of $\mathcal{D}$ are quite irrelevant, so we won't bother to specify them).

It can now be verified that we have

$$\mathcal{N}, w_1 \Vdash_v \langle \lambda w \cdot \langle \lambda y \cdot \Diamond \langle \lambda z \cdot \langle \lambda x \cdot G(x, y) \rangle (h(z)) \rangle (a) \rangle (h(w)) \rangle (a)$$

and so (13.8) is satisfiable, pretty much as expected.

Equality is a relation on objects, and not on intensions, and is independent of the particular possible world. We have the validity of $(x = y) \supset \Box(x = y)$ and, $\neg(x = y) \supset \Box\neg(x = y)$. On the other hand, we do *not* have validity of $\langle \lambda x, y \cdot (x = y) \rangle (a, b) \supset \Box \langle \lambda x, y \cdot (x = y) \rangle (a, b)$, which says that if a and b designate the same object in the present world, they must designate the same object in all accessible worlds. We do, however, have the validity of $\langle \lambda x, y \cdot (x = y) \rangle (a, b) \supset \langle \lambda x, y \cdot \Box(x = y) \rangle (a, b)$. Think about it.

Suppose a is a one-place intension function symbol and b is zero-place, that is, an intension constant. The expression $a(b) = a(b)$ is not a legal formula, and there is more than one way of unabbreviating it, but all versions have similar behavior. Let us examine the simplest.

$$\langle \lambda x \cdot \langle \lambda y \cdot y = y \rangle (a(x)) \rangle (b) \tag{13.9}$$

Suppose $\mathcal{M} = \langle \mathcal{G}, \mathcal{R}, \mathcal{D}, \mathcal{I} \rangle$ is a model, $w \in \mathcal{G}$, and v is a valuation. Under what circumstances do we have $\mathcal{M}, w \Vdash_v$ (13.9)?

1. Suppose b does not designate at w. Then directly, $\mathcal{M}, w \not\Vdash_v$ (13.9).
2. Now suppose b designates at w. Let v' be like v except that $v'(x) = \mathcal{I}(b)(w)$. Then $\mathcal{M}, w \Vdash_v$ (13.9) just in case $\mathcal{M}, w \Vdash_{v'} \langle \lambda y \cdot y = y \rangle (a(x))$. If a does not designate at w, again directly $\mathcal{M}, w \not\Vdash_{v'} \langle \lambda y \cdot y = y \rangle (a(x))$, and hence $\mathcal{M}, w \not\Vdash_v$ (13.9).
3. Now suppose b designates, and also a designates. Let v'' be like v' except that $v''(y) = \mathcal{I}(a)(w)(v'(x))$. Then $\mathcal{M}, w \Vdash_{v'} \langle \lambda y \cdot y = y \rangle (a(x))$ just in case $\mathcal{M}, w \Vdash_{v''} y = y$, and this is so.

Conclusion: (13.9) is true at a possible world exactly when both a and b designate at that world—essentially what we would expect.

13.6 Propositional Nested Sequents

Nested sequent systems are proof systems that allow sequents to be nested inside of other sequents. They were invented several times with some variation (Brünnler 2009, 2010; Kashima 1994; Poggiolesi 2009). Usually they are for modal logics and generalize one-sided sequents, though there are exceptions. There is, for instance, a two-sided version for intuitionistic logic in Fitting (2014), including constant domain logic. It turns out that nested sequents bear the same relationship to prefixed tableaus that Gentzen sequents do to ordinary tableaus, namely one is the other 'upside down,' (Fitting 2012). Nested sequent systems are forward reasoning, while tableaus are backward reasoning systems.

We sketch a nested sequent system for the modal logic K with predicate abstraction and equality. Ultimately it derives from a prefixed tableau system for the same logic, a relationship discussed in Fitting (2012), though there are significant modifications. We think of a sequent as a set, so there are no structural rules. We do not allow the empty sequent, though it would not change things much if we did. The definition of nested sequent that we use is a recursive one.

Definition 13.6.1 A *nested sequent* is a non-empty finite set of formulas and nested sequents.

A one-sided sequent can be thought of as a disjunction, while nesting corresponds to necessitation. Formally, let $\Gamma = \{X_1, \ldots, X_n, \Delta_1, \ldots, \Delta_k\}$ be a nested sequent, where each X_i is a formula and each Δ_j is a nested sequent. This translates to an ordinary modal formula $\Gamma^\dagger$, which can be thought of as the 'meaning' of Γ.

$$\Gamma^\dagger = X_1 \vee \cdots \vee X_n \vee \Box\Delta_1^\dagger \vee \cdots \vee \Box\Delta_k^\dagger$$

Axioms $\Gamma(A, \neg A)$,

A an atomic formula

Double Negation Rule

$$\frac{\Gamma(X)}{\Gamma(\neg\neg X)}$$

α Rule

$$\frac{\Gamma(X) \quad \Gamma(Y)}{\Gamma(X \wedge Y)} \quad \frac{\Gamma(\neg X) \quad \Gamma(\neg Y)}{\Gamma(\neg(X \vee Y))} \quad \frac{\Gamma(X) \quad \Gamma(\neg Y)}{\Gamma(\neg(X \supset Y))}$$

β Rule

$$\frac{\Gamma(X, Y)}{\Gamma(X \vee Y)} \quad \frac{\Gamma(\neg X, \neg Y)}{\Gamma(\neg(X \wedge Y))} \quad \frac{\Gamma(\neg X, Y)}{\Gamma(X \supset Y)}$$

ν Rule

$$\frac{\Gamma([X])}{\Gamma(\Box X)} \quad \frac{\Gamma([\neg X])}{\Gamma(\neg\Diamond X)}$$

π Rule

$$\frac{\Gamma(\Diamond X, [X, \ldots])}{\Gamma(\Diamond X, [\ldots])} \quad \frac{\Gamma(\neg\Box X, [\neg X, \ldots])}{\Gamma(\neg\Box X, [\ldots])}$$

Fig. 13.3 Nested sequent rules for K

We follow common notational conventions for nested sequents. Enclosing outer set brackets are generally omitted. A nested sequent that is a member of another nested sequent has its members listed in square brackets, and is called a *boxed sequent*. For example, $A, B, [C, [D, E], [F, G]]$, is the conventional way of writing $\{A, B, \{C, \{D, E\}, \{F, G\}\}\}$. We use Γ, Δ, ... for nested sequents, boxed or top level.

Suppose that $\Gamma(P)$ is a nested sequent in which propositional letter P (technically, a zero-place relation symbol) occurs exactly once, as a direct member of some subsequent of $\Gamma(P)$ and not as part of a more complex formula. Then we may write $\Gamma(X)$ to denote the result of replacing P in $\Gamma(P)$ with X. Similarly for $\Gamma(X, Y)$, $\Gamma(\Delta)$, and so on. Think of $\Gamma(P)$ as supplying a 'context.' In other presentations the role of P is played by a 'hole,' but the idea is the same.

The rules for propositional connectives and modal operators are given in Fig. 13.3. These are standard, and are for the logic K. In stating them we assume that $\Gamma(P)$ is some nested sequent with one occurrence of propositional letter P. Also we use $[\ldots]$ to stand for a *non-empty* nested sequent, and $[Z, \ldots]$ is $[\ldots]$ but with Z added.

Sequent proofs start with axioms and end with the nested sequent being proved. More precisely, they are trees with axioms at the leaves, the item being proved at the root, with each non-leaf node labeled with a nested sequent that follows from the nested sequents labeling its child nodes using one of the nested sequent rules. Proof of a formula is a derivative notion: a proof of the nested sequent consisting of just the formula X is taken to be a proof of X itself. For illustration, Fig. 13.4 displays a proof of $\Box(A \supset B) \supset (\Box A \supset \Box B)$. Formulas A and B are atomic. Later, atomic formulas will have internal structure, but the details don't matter now. The two subsequents shown at the top are axioms, and otherwise reasons are displayed for each inference.

$$\dfrac{\dfrac{\dfrac{\dfrac{\dfrac{\dfrac{\neg\Box(A\supset B),\neg\Box A,[\neg A,A,B] \quad \neg\Box(A\supset B),\neg\Box A,[\neg A,\neg B,B]}{\neg\Box(A\supset B),\neg\Box A,[\neg A,\neg(A\supset B),B]}\ \alpha\ \textbf{Rule}}{\neg\Box(A\supset B),\neg\Box A,[\neg(A\supset B),B]}\ \pi\ \textbf{Rule}}{\neg\Box(A\supset B),\neg\Box A,[B]}\ \pi\ \textbf{Rule}}{\neg\Box(A\supset B),\neg\Box A,\Box B}\ \nu\ \textbf{Rule}}{\neg\Box(A\supset B),\Box A\supset\Box B}\ \beta\ \textbf{Rule}}{\Box(A\supset B)\supset(\Box A\supset\Box B)}\ \beta\ \textbf{Rule}$$

Fig. 13.4 Proof of $\Box(A \supset B) \supset (\Box A \supset \Box B)$

13.7 Nested Sequents and Predicate Abstraction

From this point on relation and function symbols play a significant role. They were suppressed in the previous propositional section. It is common in proof systems for first-order logic to introduce special symbols, just for use in proofs—new free variables, special constant symbols, parameters. They play the role of existential witnesses. We do a similar thing here. Just for use in proofs, we introduce into the formal language *extension function symbols*. We might have allowed such things in formulas from the start, but in the present treatment their use is confined to proofs—they will not appear in formulas being proved. For convenience, we refer to them as parameters. We also introduce a new kind of atomic formula, connecting these new extension function symbols with the intension function symbols we have been using.

Definition 13.7.1 (*Parameters and Object Terms*) *Parameters* are a new family of symbols, typically written as $p,\ q,\ r,\ \ldots$, each having an *arity*, 0-place, 1-place, and so on.

A new kind of atomic formula is introduced, $a \nrightarrow p$, where a is an intension function symbol and p is a parameter, both having the same *arity*.

An *object term* is an expression built up entirely from parameters (but not from free variables). Object terms are allowed to appear in proofs directly as arguments of intension function symbols and of atomic formulas.

Informally, a parameter represents the extensional function picked out at a particular possible world by an intension function symbol. The new atomic formula $a \nrightarrow p$ is intended to assert that, at a particular possible world, *a does not* designate extensional function p (it is more convenient to introduce $\nrightarrow$ directly, than to introduce $\rightarrow$ and combine it with negation. The point is a minor one). As to allowing object terms to appear directly as arguments, parameters are to represent functions that are not world-dependent, so the sort of ambiguity that can arise with nested intension function symbols cannot arise with parameters.

Only closed formulas may appear in proofs where a closed formula has no free variable occurrences, though parameters may be present. We only prove closed formulas not containing parameters.

Positive Abstract Rule
$$\frac{\Gamma(X(p(t_1,\ldots,t_n)), a \not\twoheadrightarrow p)}{\Gamma(\langle\lambda x.X(x)\rangle(a(t_1,\ldots,t_n)), a \not\twoheadrightarrow p)}$$

Negative Abstract Rule 1
$$\frac{\Gamma(\neg X(p(t_1,\ldots,t_n)), a \not\twoheadrightarrow p)}{\Gamma(\neg\langle\lambda x.X(x)\rangle(a(t_1,\ldots,t_n)), a \not\twoheadrightarrow p)}$$

Negative Abstract Rule 2
$$\frac{\Gamma(\neg X(p(t_1,\ldots,t_n)), a \not\twoheadrightarrow p)}{\Gamma(\neg\langle\lambda x.X(x)\rangle(a(t_1,\ldots,t_n)))}$$
p not in conclusion
$a \not\twoheadrightarrow q$ not in conclusion for any q

In the rules above $t_1, \ldots, t_n$ are object terms, and n could be 0.

Fig. 13.5 Predicate abstract rules

Fig. 13.6 Proof of $\langle\lambda x \cdot \neg P(x)\rangle(a) \supset \neg\langle\lambda x \cdot P(x)\rangle(a)$

$$\cfrac{\cfrac{\cfrac{\cfrac{P(p), a \not\twoheadrightarrow p, \neg P(p)}{\neg\neg P(p), a \not\twoheadrightarrow p, \neg P(p)}\ \neg\neg\ \textbf{Rule}}{\neg\neg P(p), a \not\twoheadrightarrow p, \neg\langle\lambda x.P(x)\rangle(a)}\ \textbf{Neg Abs 1}}{\neg\langle\lambda x.\neg P(x)\rangle(a), \neg\langle\lambda x.P(x)\rangle(a)}\ \textbf{Neg Abs 2}}{\langle\lambda x.\neg P(x)\rangle(a) \supset \neg\langle\lambda x.P(x)\rangle(a)}\ \beta\ \textbf{Rule}$$

In Fig. 13.5 we give rules for predicate abstracts. In them $X(x)$ is a formula that may contain free occurrences of variable x, and $X(p(t_1, \ldots, t_n))$ is the result of replacing all free occurrences of x with occurrences of object term $p(t_1, \ldots, t_n)$. For Negative Abstract Rule 2 two restrictive conditions are imposed. The first is needed to ensure rule soundness. The second somewhat simplifies our completeness proof. Of course we would still have completeness without it since any proof meeting a restriction is still a proof if the restriction is relaxed.

Figure 13.6 displays a proof of $\langle\lambda x \cdot \neg P(x)\rangle(a) \supset \neg\langle\lambda x \cdot P(x)\rangle(a)$ where P is a one-place relation symbol and a is an intension constant, that is, a 0-place function symbol. The converse, $\neg\langle\lambda x \cdot P(x)\rangle(a) \supset \langle\lambda x \cdot \neg P(x)\rangle(a)$ is not valid. One can easily construct a model in which a does not designate at a world, and so $\langle\lambda x \cdot P(x)\rangle(a)$ fails there, and hence the antecedent is true, but of course the consequent is false. Once soundness is established, this tells us the converse is not provable. The reader may find it instructive to see what goes wrong with a direct attempt at constructing a proof. On the other hand, $[\neg\langle\lambda x \cdot P(x)\rangle(a) \wedge \langle\lambda x \cdot Q(x)\rangle(a)] \supset \langle\lambda x \cdot \neg P(x)\rangle(a)$ is provable, and constructing a proof is a good exercise.

13.8 Nested Sequents and Equality

In the semantics from Sect. 13.4, equality is a relation on the object domain, $\mathcal{D}$, of a model. This is reflected in nested sequent proofs, in which the symbol $=$ may appear

Equality Axioms $$\Gamma(t = t)$$

Equality Up Rule $$\frac{\Gamma([\neg t = u, \ldots], \neg t = u)}{\Gamma([\ldots], \neg t = u)}$$

Equality Down Rule $$\frac{\Gamma([\neg t = u, \ldots], \neg t = u)}{\Gamma([\neg t = u, \ldots])}$$

Equality Left-Right Substitution Rule $$\frac{\Gamma(\neg t = u, X(u), X(t))}{\Gamma(\neg t = u, X(u))}$$

Equality Right-Left Substitution Rule $$\frac{\Gamma(\neg t = u, X(u), X(t))}{\Gamma(\neg t = u, X(t))}$$

In the rules above, t and u are object terms,
X is a simple formula
and $\ldots$ is allowed to be empty in the Down Rule,
but not in the Up Rule.

Fig. 13.7 Equality rules

between object terms as defined in the previous section. Now we give rules for $=$, and this can be done in two ways—reflexivity is the key point. We could introduce a rule as follows.

$$\frac{\Gamma(\neg t = t, \ldots)}{\Gamma(\ldots)}$$

This gives a complete system when combined with the rules from Fig. 13.7. It is easier to prove completeness than it is for the system actually adopted here. But it has the annoying property of possibly letting a relation symbol (equality) and function terms (those in t) disappear during a rule application. To avoid this, we adopt an axiom instead, loosely basing our approach on Sect. 9.8 of Fitting (1990).

Definition 13.8.1 A formula is *simple* if it is the negation of an atomic formula, or it is an atomic formula whose relation symbol is $=$.

Figure 13.7 displays the nested sequent rules for equality. In stating the Substitution Rule $X(t)$ is a (closed, simple) formula and $X(u)$ is the result of replacing *some* (as opposed to *every*) occurrence of object term t in it with an occurrence of object term u.

In Sect. 13.2 we noted that formula (13.4) followed from (13.2) and (13.3). We now show this using a nested sequent proof, in Fig. 13.8.

Formula (13.10) is a related example one might practice on. Part of the antecedent is $\langle \lambda x \cdot \Box \langle \lambda y \cdot x = y \rangle(a) \rangle(a)$. In Fitting and Mendelsohn (1999) this was called a *local rigidity* condition—it expresses that the intension constant a designates in the current world, and must designate the same value in all related worlds. The second

$$
\begin{array}{cl}
[\neg T(q,p), \neg T(p,p), a \not\twoheadrightarrow q, \neg q = p, T(p,p)], a \not\twoheadrightarrow p, \neg\Box\langle\lambda x.x = p\rangle(a), \Diamond T(p,p) & \\ \hline
[\neg T(q,p), a \not\twoheadrightarrow q, \neg q = p, T(p,p)], a \not\twoheadrightarrow p, \neg\Box\langle\lambda x.x = p\rangle(a), \Diamond T(p,p) & \textbf{Eq R-L Subs} \\ \hline
[\neg T(q,p), a \not\twoheadrightarrow q, \neg\langle\lambda x.x = p\rangle(a), T(p,p)], a \not\twoheadrightarrow p, \neg\Box\langle\lambda x.x = p\rangle(a), \Diamond T(p,p) & \textbf{Neg Abs 1} \\ \hline
[\neg\langle\lambda x.T(x,p)\rangle(a), \neg\langle\lambda x.x = p\rangle(a), T(p,p)], a \not\twoheadrightarrow p, \neg\Box\langle\lambda x.x = p\rangle(a), \Diamond T(p,p) & \textbf{Neg Abs 2} \\ \hline
[\neg\langle\lambda x.T(x,p)\rangle(a), \neg\langle\lambda x.x = p\rangle(a)], a \not\twoheadrightarrow p, \neg\Box\langle\lambda x.x = p\rangle(a), \Diamond T(p,p) & \pi\ \textbf{Rule} \\ \hline
[\neg\langle\lambda x.T(x,p)\rangle(a)], a \not\twoheadrightarrow p, \neg\Box\langle\lambda x.x = p\rangle(a), \Diamond T(p,p) & \pi\ \textbf{Rule} \\ \hline
\neg\Diamond\langle\lambda x.T(x,p)\rangle(a), a \not\twoheadrightarrow p, \neg\Box\langle\lambda x.x = p\rangle(a), \Diamond T(p,p) & \nu\ \textbf{Rule} \\ \hline
\neg\Diamond\langle\lambda x.T(x,p)\rangle(a), a \not\twoheadrightarrow p, \neg\Box\langle\lambda x.x = p\rangle(a), \langle\lambda x.\Diamond T(x,x)\rangle(a) & \textbf{Pos Abs} \\ \hline
\neg\Diamond\langle\lambda x.T(x,p)\rangle(a), a \not\twoheadrightarrow p, \neg\langle\lambda y.\Box\langle\lambda x.x = y\rangle(a)\rangle(a), \langle\lambda x.\Diamond T(x,x)\rangle(a) & \textbf{Neg Abs 1} \\ \hline
\neg\langle\lambda y.\Diamond\langle\lambda x.T(x,y)\rangle(a)\rangle(a), \neg\langle\lambda y.\Box\langle\lambda x.x = y\rangle(a)\rangle(a), \langle\lambda x.\Diamond T(x,x)\rangle(a) & \textbf{Neg Abs 2} \\ \hline
\neg 2, \neg 3, 4 & \textit{Abbreviating} \\ \hline
\neg(2 \wedge 3), 4 & \alpha\ \textbf{Rule} \\ \hline
(2 \wedge 3) \supset 4 & \alpha\ \textbf{Rule}
\end{array}
$$

Fig. 13.8 Equality proof example

condition, $\Box\langle\lambda x \cdot \langle\lambda y \cdot y = y\rangle(h(x))\rangle(b)$ can only be true if h designates a function in all related worlds. Of course it also requires that b designates in all related worlds, but the values designated play no significant role.

$$
\begin{aligned}
&\left[\langle\lambda x \cdot \Box\langle\lambda y \cdot x = y\rangle(a)\rangle(a) \wedge \Box\langle\lambda x \cdot \langle\lambda y \cdot y = y\rangle(h(x))\rangle(b)\right] \\
&\supset \\
&\langle\lambda x \cdot \Box\langle\lambda y \cdot \langle\lambda z \cdot \langle\lambda w \cdot z = w\rangle(h(y))\rangle(h(x))\rangle(a)\rangle(a)
\end{aligned}
\tag{13.10}
$$

13.9 Soundness

We begin with an obvious problem that affects our completeness argument as well. Parameters can occur in proofs, but were not mentioned in the discussion of semantics in Sect. 13.4. They must be given a semantic meaning. Similarly for the atomic formula $a \not\twoheadrightarrow p$.

Let $\mathcal{M} = \langle \mathcal{G}, \mathcal{R}, \mathcal{D}, \mathcal{I} \rangle$ be a model. We extend the definition of interpretation so that values are supplied for parameters too. If p is an n-argument parameter, $\mathcal{I}(p) : \mathcal{D}^n \to \mathcal{D}$. Note that this is not world dependent, unlike with intension function symbols. This in turn gives a natural semantic meaning to each *object term*, Definition 13.7.1. We write $\mathcal{I}(t)$ for the member of $\mathcal{D}$ assigned to object term t. Definition is by recursion: $\mathcal{I}(p(t_1, \ldots, t_n)) = \mathcal{I}(p)(\mathcal{I}(t_1), \ldots, \mathcal{I}(t_n))$, with $\mathcal{I}(p)$ defined directly as part of the condition for being an interpretation now.

We have a new atomic formula allowed in proofs. $\not\twoheadrightarrow$ is interpreted as a two-place relation symbol, across types, so that $\mathcal{M}, w \Vdash_v a \not\twoheadrightarrow p$ if either a does not designate at w, or else a does designate, but $\mathcal{I}(a)(w) \neq \mathcal{I}(p)$. More formally, but less intelligibly, $\mathcal{I}(\not\twoheadrightarrow)(a, p) = \{w \in \mathcal{G} \mid a \text{ does not designate at } w\} \cup \{w \in \mathcal{G} \mid a \text{ designates at } w \text{ and } \mathcal{I}(a)(w) \neq \mathcal{I}(p)\}$.

Finally, the interpretation of predicate abstraction must be modified a bit, from Sect. 13.4. We now want the following, which agrees with the earlier definition in which no parameters are present.

Predicate Abstraction, n-ary Function Symbol Suppose u_i is either a variable or an object term. Let $\overline{u_i}$ be $v(u_i)$ if u_i is a variable, and $\mathcal{I}(u_i)$ if u_i is an object term.

> If intension function symbol a designates at w, $\mathcal{M}, w \Vdash_v \langle\lambda y \cdot X\rangle(a(u_1, \ldots, u_n))$ if
> $\mathcal{M}, w \Vdash_{v'} X$ where v' is like v except that $v'(y) = \mathcal{I}(a)(w)(\overline{u_1}, \ldots, \overline{u_n})$.
> If a does not designate at w, $\mathcal{M}, w \not\Vdash_v \langle\lambda x \cdot X\rangle(a(x_1, \ldots, x_n))$.

Detailed arguments are lengthy, but the basic ideas are rather straightforward. We sketch them, and leave it to the reader to work out the specifics.

In Sect. 13.6 is given the usual translation of nested sequents into formulas. If one shows the translate of each axiom is valid, and the rules preserve validity of translations, then soundness is an immediate consequence. There are just a few general cases. First, we have the following.

$$\text{Let } A \text{ be a closed formula, allowing object terms and } \not\twoheadrightarrow. \text{If } A \text{ is valid, so is } \Gamma(A)^{\dagger}. \tag{13.11}$$

Item (13.11) is shown by induction on the nesting level of A in $\Gamma(A)$. We leave this to the reader. Once it has been established, validity of the nested sequent axioms follows.

Next is another general result whose proof also can be shown by induction on nesting depth.

$$\begin{array}{l}\text{Let } A \text{ and } B \text{ be closed formulas, allowing object terms and } \not\twoheadrightarrow. \\ \text{If} A \supset B \text{ is valid, so is } \Gamma(A)^{\dagger} \supset \Gamma(B)^{\dagger}.\end{array} \tag{13.12}$$

This allows easy treatment of most of the single-premise rules. Consider, for example, the Positive Abstract Rule. First one shows validity of the following formula. We omit the verification.

$$\left[X\left(p(t_1, \ldots, t_n)\right) \vee a \not\twoheadrightarrow p\right] \supset \left[\langle\lambda x \cdot X(x)\rangle(a(t_1, \ldots, t_n)) \vee a \not\twoheadrightarrow p\right]$$

Then soundness of the Positive Abstract Rule follows using (13.12). In a similar way one handles Double Negation, β, ν, π, Positive Abstract, Negative Abstract 1, and all the Equality Rules.

Double premise rules can be handled once the following is shown.

$$\begin{array}{l}\text{Let } A, B, \text{ and } C \text{ be closed formulas allowing object terms and } \not\twoheadrightarrow. \\ \text{If } (A \wedge B) \supset C \text{ is valid, so is } [\Gamma(A)^{\dagger} \wedge \Gamma(B)^{\dagger}] \supset \Gamma(C)^{\dagger}.\end{array} \tag{13.13}$$

Using (13.13) one easily gets soundness of the β rules.

We still have not verified soundness of Negative Abstract Rule 2, which is more complicated. For this we need something that is related to the usual treatment of quantification. We say a model $\mathcal{M}' = \langle \mathcal{G}, \mathcal{R}, \mathcal{D}, \mathcal{I}' \rangle$ is a *p-variant* of model $\mathcal{M} = \langle \mathcal{G}, \mathcal{R}, \mathcal{D}, \mathcal{I} \rangle$ just in case $\mathcal{I}'$ agrees with $\mathcal{I}$ except possibly on parameter p.

Assume $\Gamma(\neg\langle\lambda x \cdot (x)\rangle(a(t_1, \ldots, t_n)))$ does not contain parameter p. Let $\mathcal{M}$ be a model, and w be an arbitrary member of $\mathcal{G}$. If $\mathcal{M}', w \Vdash_v \Gamma(\neg X(p(y_1, \ldots, y_n)) \vee a \nrightarrow p)$ for every p variant $\mathcal{M}'$ of $\mathcal{M}$, then $\mathcal{M}, w \Vdash_v \Gamma(\neg\langle\lambda x \cdot X(x)\rangle(a(t_1, \ldots, t_n)))$. (13.14)

As with earlier items in this section, the proof of (13.14) is by induction on nesting depth. For the ground case where there is no nesting of sequents, what needs to be shown is: if $\mathcal{M}', w \Vdash_v A_1 \vee \ldots \vee A_k \vee \neg X(p(t_1, \ldots, t_n)) \vee a \nrightarrow p$ for every p variant $\mathcal{M}'$ of $\mathcal{M}$, then $\mathcal{M}, w \Vdash_v A_1 \vee \ldots \vee A_k \vee \neg\langle\lambda x \cdot X(x)\rangle(a(t_1, \ldots, t_n))$, where $A_1, \ldots, A_k$, $X(x)$ do not contain p, and k might be 0.

This is most easily argued in the contrapositive direction. Assume $\mathcal{M}, w \Vdash_v \neg A_i$ for $i = 1, \ldots, k$, and $\mathcal{M}, w \Vdash_v \langle\lambda x \cdot X(x)\rangle(a(t_1, \ldots, t_n))$. Then a must designate at w, and $\mathcal{M}, w \Vdash_{v'} X(x)$ where $v'(x) = \mathcal{I}(a)(w)(\mathcal{I}(t_1), \ldots, \mathcal{I}(t_n))$, and otherwise v' and v agree. Now introduce a new parameter, p, and define $\mathcal{I}'$ to be like $\mathcal{I}$ except that $\mathcal{I}'(p) = \mathcal{I}(a)(w)$, and let $\mathcal{M}' = \langle \mathcal{G}, \mathcal{R}, \mathcal{D}, \mathcal{I}' \rangle$. Then $\mathcal{M}', w \Vdash_v X(p(t_1, \ldots, t_n))$, and $\mathcal{M}', w \Vdash_v \neg(a \nrightarrow p)$. Of course $\mathcal{M}', w \Vdash_v \neg A_i$ since A_i does not contain p so $\mathcal{M}$ and $\mathcal{M}'$ behave alike on it.

We have now finished the argument for the ground case. We leave the induction step to the reader. With (13.14) available, soundness of Negative Abstract Rule 2 follows, completing the proof of the soundness of the nested sequent system.

13.10 Completeness

Let Z be a closed parameter-free formula, fixed for this section. Assume Z is unprovable; we show a counter model for Z exists.

It is likely that this could be done using maximal consistency as described in Fitting (2011), but here we follow a more traditional route. With sequent systems one often reasons backwards from the formula to be proved, in order to discover a proof. A backward search for a proof should be done systematically, but there is much variation possible. A search must be *fair*, in that any applicable rule must eventually be brought into the proof. Beyond this we omit specifics, and simply assume some fair search procedure can be formulated.

Apply a fair proof search procedure, attempting to prove formula Z. This constructs a tree, with the formula being proved at the bottom, the tree branching upward, with branching associated with α Rule applications. Maximum nesting depth of nested sequents cannot exceed the modal depth of the formula being proved, as a look at the ν and π rules in Fig. 13.3 makes clear. If a proof is found during the course of a search, each branch terminates with an Axiom, either $\Gamma(A, \neg A)$ or $\Gamma(t = t)$. If a proof is not found there must be a branch—call it *open*—without an axiom occurrence. Let us say $\mathcal{B}(Z)$ is a specific such branch, arising during a fair search for a proof of Z. Like Z, this too is fixed for the rest of this section.

Suppose, as an example, that on branch $\mathcal{B}(Z)$ the following step appears:

$$\begin{array}{c} \vdots \\ A, [B, C], [D, X], [E, [F]] \\ A, [B, C], [D, \neg\neg X], [E, [F]] \\ \vdots \end{array}$$

Most of the structure remains the same in this step, but $[D, X]$ 'turns into' $[D, \neg\neg X]$. We can think of these as the *same* boxed sequent, but we are seeing different stages of its history. Then a boxed sequent, dynamically, is really a sequence of boxed sequents appearing in $\mathcal{B}(Z)$, where the terms of the sequence are the stages through which the sequent passes. We will call such a sequence a *dynamic sequent*. This can be made mathematically precise, but an informal approach should suffice for our purposes: for the example above, we say that D, X, and $\neg\neg X$ all appear in the same dynamic sequent, though X and $\neg\neg X$ appear at different stages. If w is a dynamic sequent, we use the special notation $A \dot{\in} w$ to indicate that A is a member at some stage. Members of a nested sequent can be other nested sequents, which have their own histories, but the general idea should be clear enough for present purposes.

We are about to use $\mathcal{B}(Z)$ to construct a model. We need a few results first, and it is easiest to state them if we say at this point what the possible worlds of our model will be. Making use of the discussion about stage-of-construction above, *let $\mathcal{G}$ be the set of dynamic boxed sequents that appear on $\mathcal{B}(Z)$, together with the top level dynamic sequent.*

Lemma 13.10.1 *Suppose A and B are closed atomic or negated atomic formulas, including those of the form $a \twoheadrightarrow p$. Also suppose $w \in \mathcal{G}$. If $A \dot{\in} w$ and $B \dot{\in} w$, then both A and B are present simultaneously at some stage of the construction of w.*

Proof An inspection of the nested sequent rules shows that (applied forward) none introduces either an atomic formula or its negation. Then when rules are applied backwards during a proof search, atomic formulas and their negations are not eliminated. It follows that during a proof search, if A and B are both present at different stages of the construction of w, both must be present at whichever stage is uppermost. ■

Lemma 13.10.2 *If P is atomic we cannot have both $P \dot{\in} w$ and $\neg P \dot{\in} w$ for any $w \in \mathcal{G}$.*

Proof Immediate from Lemma 13.10.1 and the fact that $\mathcal{B}(Z)$ is open. ■

Lemma 13.10.3 *Suppose $w \in \mathcal{G}$. Then for each intension function symbol a, $a \twoheadrightarrow p \dot{\in} w$ for at most one parameter p.*

Proof Suppose both $a \twoheadrightarrow p \dot{\in} w$ and $a \twoheadrightarrow q \dot{\in} w$, where $p \neq q$. Then both must be present at the same stage of the construction of w, by Lemma 13.10.1. But during the fair proof search that constructs $\mathcal{B}(Z)$, working from Z upward, the only rule that can introduce an expression $a \twoheadrightarrow p$ into a nested sequent is Negative Abstract Rule 2, from Fig. 13.5. This rule can only be applied if $a \twoheadrightarrow q$ is not already present, for any q, so it is impossible to have both present at the same stage. ■

The remaining items all have to do with the behavior of equality.

Lemma 13.10.4 *Suppose t_1 and t_2 are object terms, and $\neg t_1 = t_2 \dot{\in} w$ for some $w \in \mathcal{G}$. Then $\neg t_1 = t_2 \dot{\in} w$ for every $w \in \mathcal{G}$.*

Proof This is an easy consequence of the Equality Up Rule and the Equality Down Rule. ■

Lemma 13.10.5 *Suppose $w \in \mathcal{G}$ and t_1, t_2, and t_3 are object terms.*

1. *If $\neg t_1 = t_2 \dot{\in} w$ then $\neg t_2 = t_1 \dot{\in} w$.*
2. *If $\neg t_1 = t_2 \dot{\in} w$ and $\neg t_2 = t_3 \dot{\in} w$, then $\neg t_1 = t_3 \dot{\in} w$.*

Proof We show 1; item 2 is similar. Suppose $\neg t_1 = t_2 \dot{\in} w$. Recall, we are conducting a backwards search for a proof. Now, $\neg t_1 = t_2$ can be inferred (in any context), from $\neg t_1 = t_2, \neg t_2 = t_2$ using the Equality Right-Left Substitution Rule (and deleting duplicates, since we are dealing with sets). Since our proof search procedure is a fair one, at some point this rule must be tried. In turn, $\neg t_1 = t_2, \neg t_2 = t_2$ can be inferred from $\neg t_1 = t_2, \neg t_2 = t_2, \neg t_2 = t_1$ using the Equality Left-Right Substitution Rule, and fairness says this will be tried. But then $\neg t_2 = t_1 \dot{\in} w$. ■

Definition 13.10.6 We say object term t_1 $\mathcal{B}(Z)$*-rewrites* to object term t_2 if t_2 results from the replacement of some subterm u_1 of t_1 with u_2, where $\neg u_1 = u_2 \dot{\in} w$ for some (any) $w \in \mathcal{G}$. We also say t_1 and t_2 are $\mathcal{B}(Z)$*-equivalent* if t_1 can be turned into t_2 via a sequence of $\mathcal{B}(Z)$-rewrites (possibly 0).

Lemma 13.10.7 *Suppose X_1 is a closed simple formula containing an occurrence of object term t_1, and X_2 is like X_1 but with some occurrence of t_1 replaced with an occurrence of object term t_2. If $X_1 \dot{\in} w$ for some $w \in \mathcal{G}$, and t_1 and t_2 are $\mathcal{B}(Z)$-equivalent, then $X_2 \dot{\in} w$.*

Proof A direct consequence of the Equality Substitution Rules and our fair search procedure. ■

Lemma 13.10.8 *Suppose t_1 is $\mathcal{B}(Z)$-equivalent to t_2. Then it is not the case that $t_1 = t_2 \dot{\in} w$ for any $w \in \mathcal{G}$.*

Proof Suppose $t_1 = t_2 \dot{\in} w$ and t_1 is $\mathcal{B}(Z)$-equivalent to t_2. Then using Lemma 13.10.7, $t_2 = t_2 \dot{\in} w$, contradicting the fact that $\mathcal{B}(Z)$ is not closed and so contains no axiom occurrence. ■

Now we are ready to define our model. Actually we construct two models because, as in most approaches to equality, it is first modeled by an equivalence relation, then equivalence classes are introduced.

Construction of Model One

We have already said what our possible worlds are, but we repeat the definition for convenience. Let $\mathcal{G}$ be the set of dynamic boxed sequents that appear on $\mathcal{B}(Z)$, together with the top level dynamic sequent. Continuing, if $w_1, w_2 \in \mathcal{G}$ set $w_1 \mathcal{R} w_2$ just in case $w_2 \dot{\in} w_1$. And let $\mathcal{D}$ be the Herbrand universe consisting of object terms (terms built up from parameters, Definition 13.7.1). We thus have $\mathcal{G}$, $\mathcal{R}$, and $\mathcal{D}$. The interpretation $\mathcal{I}$ is a bit more complicated.

Since parameters are involved, $\mathcal{I}$ must cover them. We do the thing usual with Herbrand universes. If p is 0-place, $\mathcal{I}(p) = p$, and if p is n-ary and $h_1, \ldots, h_n \in \mathcal{D}$, $\mathcal{I}(p)(h_1, \ldots, h_n)$ is the object term $p(h_1, \ldots, h_n)$. Then $\mathcal{I}(t) = t$ for any object term t.

Let a be an n-ary intension function symbol. We take a to designate at $w \in \mathcal{G}$ if $a \not\twoheadrightarrow p \dot{\in} w$ for some p (p is unique by Lemma 13.10.3). Then we set $\mathcal{I}(a) : S \to (\mathcal{D}^n \to \mathcal{D})$ where S is the subset of $\mathcal{G}$ at which a designates, and if $w \in S$, then $\mathcal{I}(a)(w)(h_1, \ldots, h_n) = p(h_1, \ldots, h_n)$, where p is the unique parameter such that $a \not\twoheadrightarrow p \dot{\in} w$, and $p(h_1, \ldots, h_n)$ is a Herbrand term.

Suppose P is an n-ary relation symbol other than $=$, and $w \in \mathcal{G}$. Set $\mathcal{I}(P)(w) = \{\langle h_1, \ldots, h_n \rangle \mid \neg P(h_1, \ldots, h_n) \dot{\in} w\}$. For equality a little more is needed. We set $\mathcal{I}(=)(w)$ to be $\{(h_1, h_2) \mid \neg(h_1 = h_2) \dot{\in} w\}$ together with $\{(h_1, h_2) \mid h_1 \text{ is } \mathcal{B}(Z)$ equivalent to $h_2\}$. Finally we model $a \not\twoheadrightarrow p$ as we did in the Soundness section, so that $\mathcal{M}, w \Vdash_v a \not\twoheadrightarrow p$ if a does not designate at w, or else a designates at w, but doesn't designate p.

This completes the definition of a model, $\mathcal{M} = \langle \mathcal{G}, \mathcal{R}, \mathcal{D}, \mathcal{I} \rangle$. The only thing that still needs checking is that equality is interpreted rigidly—the same at all members of $\mathcal{D}$. But this is an easy consequence of the Equality Up and Equality Down Rules.

Now we prove the following key result. It is the opposite of what one might have expected.

Lemma 13.10.9 (Truth Lemma) *Let $w \in \mathcal{G}$ and let A be any closed formula, allowing parameters. If $A \dot{\in} w$ then $\mathcal{M}, w \not\Vdash_v A$ (choice of v is not important because A is closed. In the proof v will be arbitrary unless otherwise specified).*

Proof By induction on the complexity of A. We show the result simultaneously for A and $\neg A$. Assume the result is known for formulas simpler than A. There are several cases to consider.

Atomic A is atomic, so there are no simpler formulas.

We begin with the conventional non-equality case. Suppose $\neg P(h_1, \ldots, h_n)\dot{\in} w$. Then $\langle h_1, \ldots, h_n \rangle \in \mathcal{I}(P)(w)$, so $\mathcal{M}, w \Vdash_v P(h_1, \ldots, h_n)$, and of course $\mathcal{M}, \not\Vdash_w \neg P(h_1, \ldots, h_n)v$.

Suppose $P(h_1, \ldots, h_n)\dot{\in} w$. Then we do not have $\neg P(h_1, \ldots, h_n)\dot{\in} w$ by Lemma 13.10.2. But then $\langle h_1, \ldots, h_n \rangle \notin \mathcal{I}(P)(w)$, so $\mathcal{M}, w \not\Vdash_v P(h_1, \ldots, h_n)$.

Next we consider equality. If $\neg h_1 = h_2 \dot{\in} w$ the argument is as in the previous case. Now suppose $h_1 = h_2 \dot{\in} w$. Since $\mathcal{B}(Z)$ is not closed, we do not have $\neg(h_1 = h_2)\dot{\in} w$. By Lemma 13.10.8, h_1 is not $\mathcal{B}(Z)$-equivalent to h_2. But then $(h_1, h_2) \notin \mathcal{I}(=)(w)$ and so $\mathcal{M}, w \not\Vdash_v h_1 = h_2$.

Finally we have the rather special case of $a \twoheadrightarrow p$. Now $\mathcal{M}, w \not\Vdash_v a \twoheadrightarrow p$ exactly when a designates at w, and it is p that it designates. But a look at the definitions earlier shows that these conditions obtain exactly when $a \twoheadrightarrow p \dot{\in} w$.

Propositional We consider only $A = X \wedge Y$, the other propositional cases are similar.

Suppose $X \wedge Y \dot{\in} w$. Since a fair proof search was made, at some point an α rule application was introduced, and so either $X \dot{\in} w$ or $Y \dot{\in} w$. Then by the induction hypothesis, either $\mathcal{M}, w \not\Vdash_v X$ or $\mathcal{M}, w \not\Vdash_v Y$, and hence $\mathcal{M}, w \not\Vdash_v X \wedge Y$.

The $\neg A = \neg(X \wedge Y)$ argument is along the same lines, but using the β rule.

Modal Suppose $A = \Box X$, the case of $\Diamond X$ is similar.

Suppose $\Box X \dot{\in} w$. Since a fair proof search was made, at some point a ν Rule application was introduced, and a new dynamic nested box, call it w', was created using X. Then $w' \in \mathcal{G}$, $w\mathcal{R}w'$, and $X \dot{\in} w'$. By the induction hypothesis, $\mathcal{M}, w' \not\Vdash_v X$, and so $\mathcal{M}, w \not\Vdash_v \Box X$.

The $\neg A = \neg\Box X$ case is similar, but using the π Rule.

Predicate Abstract Suppose $A = \langle \lambda x \cdot X(x) \rangle (a(t_1, \ldots, t_n))$.

This time we examine the negated case. Suppose $\neg\langle \lambda x \cdot X(x) \rangle (a(t_1, \ldots, t_n))\dot{\in} w$. At some point in our fair proof search an application of one of the Negative Abstract Rules 1 or 2 is introduced concluding this. Either way, we have a unique p so that both $a \twoheadrightarrow p \dot{\in} w$ and $\neg X(p(t_1, \ldots, t_n))\dot{\in} w$. By the induction hypothesis, $\mathcal{M}, w \not\Vdash_v \neg X(p(t_1, \ldots, t_n))$, and so $\mathcal{M}, w \Vdash_v X(p(t_1, \ldots, t_n))$. Since $a \twoheadrightarrow p \dot{\in} w$, we have that a designates at w, and $\mathcal{I}(a)(w)(\mathcal{I}(t_1), \ldots, \mathcal{I}(t_n)) = \mathcal{I}(a)(w)(t_1, \ldots, t_n) = p(t_1, \ldots, t_n)$. Let v' be like v except that $v'(x) = p(t_1, \ldots, t_n)$; then $\mathcal{M}, w \Vdash_{v'} X(x)$, and so $\mathcal{M}, w \Vdash_v \langle \lambda x \cdot X(x) \rangle (a(t_1, \ldots, t_n))$, and hence $\mathcal{M}, w \not\Vdash_v \neg\langle \lambda x \cdot X(x) \rangle (a(t_1, \ldots, t_n))$. The case where $A = \langle \lambda x \cdot X(x) \rangle (a(t_1, \ldots, t_n))$ is similar.

■

Let t be the top level sequent of $\mathcal{B}(Z)$. Then $Z \dot{\in} t$, so by the Truth Lemma, $\mathcal{M}, t \not\Vdash_Z v$. Thus $\mathcal{M}$ is a counter-model.

Construction of Model Two

The interpretation of the equality symbol in the model $\mathcal{M}$, constructed above, is not by the equality relation. However, it turns out that $\mathcal{I}(=)(w)$ is the same for all w in $\mathcal{G}$, is an equivalence relation on $\mathcal{D}$, and is a congruence with respect to all relation and function symbols. Once this is established, a 'factor' model can be constructed with an object domain consisting of equivalence classes from $\mathcal{D}$. This construction is quite standard, and we omit all details. It is necessary, however, to establish the facts about $\mathcal{I}(=)$ just mentioned.

First we show that $\mathcal{I}(=)(w)$ is the same for all $w \in \mathcal{G}$. Suppose $(h_1, h_2) \in \mathcal{I}(=)(w)$. The definition of interpretation for equality has two parts. It could be that $\neg h_1 = h_2 \dot{\in} w$. Then Lemma 13.10.4 gives us what we need. Or else it could be that h_1 is $\mathcal{B}(Z)$-equivalent to h_2. But Definition 13.10.6 does not depend on a particular choice of possible world, relying on Lemma 13.10.4.

Next we show that the interpretation of equality, which we now know is world independent, is an equivalence relation. It is reflexive because h is $\mathcal{B}(Z)$-equivalent to h since it $\mathcal{B}(Z)$-rewrites to itself with 0 rewritings. Now consider symmetry. Suppose $(h_1, h_2) \in \mathcal{I}(=)(w)$. There are two possibilities. First, $\neg h_1 = h_2 \dot{\in} w$, and we appeal to Lemma 13.10.5. Second, h_1 is $\mathcal{B}(Z)$-equivalent to h_2, and clearly $\mathcal{B}(Z)$-rewriting is bidirectional, making use of Lemma 13.10.5 again. We leave transitivity to the reader.

We need to establish we have a congruence with respect to all relation symbols. Let us assume P is not the equality symbol, and leave the equality case to the reader. Suppose $(h_1, \ldots, h_n) \in \mathcal{I}(P)(w)$ and $(h_1, h_1') \in \mathcal{I}(=)(w)$. By the first condition, and the definition of $\mathcal{I}$, $\neg P(h_1, \ldots, h_n) \dot{\in} w$. As usual, there are two possibilities for the second condition. We could have $\neg h_1 = h_1' \dot{\in} w$. In this case, by Lemma 13.10.1, both $\neg P(h_1, \ldots, h_n)$ and $\neg h_1 = h_1'$ must be present at some same stage in the construction of w, and a fair construction procedure will eventually bring in $\neg P(h_1', \ldots, h_n)$ using the Equality Left-Right Substitution Rule. The other possibility is that h_1 is $\mathcal{B}(Z)$ equivalent to h_1'. In this case we apply Lemma 13.10.7.

Finally we need to show congruence properties with respect to function symbols. To keep notation simple we work with a unary function symbol. Suppose $(h_1, h_2) \in \mathcal{I}(=)(w)$; we show $(f(h_1), f(h_2)) \in \mathcal{I}(=)(w)$. There are the usual two cases. We might have that $\neg h_1 = h_2 \dot{\in} w$, or we might have that h_1 is $\mathcal{B}(Z)$-equivalent to h_2. But either way, h_1 is $\mathcal{B}(Z)$-equivalent to h_2, and it follows from the definition that $f(h_1)$ is $\mathcal{B}(Z)$-equivalent to $f(h_2)$. giving us what we need.

This establishes what was needed, and ends our discussion of Model Two.

13.11 Conclusion

All formal work here has been for systems based on the simplest normal modal logic K. Propositional nested sequent rules for other modal logics can be found in Brünnler (2009); Fitting (2012) and elsewhere. Using rules for other modal logics, together with the Predicate Abstract and Equality Rules, gives appropriate formal systems for a family of modal logics containing the machinery discussed here. Our completeness and soundness methods carry over.

Predicate Abstract machinery is intermediate between propositional and first-order, in an obvious sense. Which way it leans is not always obvious. For instance, it is shown in Fitting (2002) that logics between K4 and S5, with predicate abstraction and equality are undecidable, and for S5, equality is not needed for undecidability. For K, <, and other simpler modal logics with predicate abstraction and equality, we have decidability if no function symbols are present—a fair proof search procedure must terminate. What the effect of function symbols might be is not known.

In Sects. 13.2 and 13.5 we discussed a few natural examples, and saw how predicate abstraction machinery dealt with them. This is not the only formalization possible. Quantifiers can be made use of. Cross-world predication can be used, as in Wehmeier (2012). A multiplicity of methods is not a drawback. Each approach has its own virtues and provides its own insights. Relationships between approaches needs investigation, and we encourage readers to think about this.

References

Brünnler, K. (2009). Deep sequent systems for modal logic. *Archive for Mathematical Logic, 48*(6), 551–577.

Brünnler, K. (2010). *Nested sequents*. Habilitation thesis, Institut für Informatik und angewandte Mathematik, Universität Bern.

Carnap, R. (1947). *Meaning and necessity*. Chicago: University of Chicago Press, enlarged edition 1956 edition.

Fitting, M. C. (1996). *First-order logic and automated theorem proving*. Berlin: Springer. First edition, 1990. Errata at http://comet.lehman.cuny.edu/fitting/errata/errata.html.

Fitting, M. C. (2002). Modal logics between propositional and first-order. *Journal of Logic and Computation, 12*, 1017–1026.

Fitting, M. C. (2004). First-order intensional logic. *Annals of Pure and Applied Logic, 127*, 171–193.

Fitting, M. C. (2006). FOIL axiomatized. *Studia Logica, 84*(1), 1–22. doi:10.1007/s11225-006-9000-2. See Correction Fitting (2007).

Fitting, M. C. (2007). Correction to FOIL axiomatized. *Studia Logica, 85*(2), 275. doi:10.1007/s11225-007-9032-2.

Fitting, M. C. (2011). Proving completeness for nested sequent calculi. In J.-Y. Béziau & M. E. Coniglio (Eds.), *Logic without frontiers, Festschrift for Walter Alexandre Carnielli on the occasion of his 60th Birthday*, Tributes (pp. 145–154). London: College Publications.

Fitting, M. C. (2012). Prefixed tableaus and nested sequents. *Annals of Pure and Applied Logic, 163*, 291–313. doi:10.1016/j.apal.2011.09.004.

Fitting, M. C. (2014). Nested sequents for intuitionistic logics. *Notre Dame Journal of Formal Logic, 55*(1), 41–61.

Fitting, M. C., & Mendelsohn, R. (1999). *First-order modal logic*. Kluwer. Paperback. Errata at http://comet.lehman.cuny.edu/fitting/errata/errata.html.

Kashima, R. (1994). Cut-free sequent calculi for some tense logics. *Studia Logica, 53*, 119–135.

Poggiolesi, F. (2009). The tree-hypersequent method for modal propositional logic. In D. Makinson, J. Malinowski, & H. Wansing (Eds.), *Trends in logic: Towards mathematical philosophy* (pp. 31–51). Berlin: Springer.

Russell, B. (1905). On denoting. *Mind, 14* (new series), 479–493.

Stalnaker, R., & Thomason, R. (1968). Abstraction in first-order modal logic. *Theoria, 34*, 203–207.

Thomason, R., & Stalnaker, R. (1968). Modality and reference. *Nous, 2*, 359–372.

Ward, C. (Gunnvôr). (2013). *The viking answer lady*. http://www.vikinganswerlady.com. Online; Accessed August 2, 2013.

Wehmeier, K. F. (2012). Subjunctivity and cross-world predication. *Philosophical Studies*, *159*(1), 107–122.

Whitehead, A. N., & Russell, B. (1927). *Principia mathematica* (2nd ed., p. 1910). Cambridge: Cambridge University Press, First edition.

Chapter 14
Epistemic Conditionals and the Logic of Subsets

Konstantinos Georgatos

Abstract This paper proposes a formalization of conditional reasoning using Moss and Parikh's logic of subsets so that a reasoner can express both conditional assertions about beliefs, as well as beliefs about conditional assertions. We present a complete axiomatization of the logic and show that it is decidable. A version of the Ramsey test is found to be compatible with this logic and provides a correspondence between conditionals and belief contraction.

Keywords Subset logic · Conditional logic · Belief revision · Ramsey test

14.1 Introduction

Conditionals have long been known to encode the reasoner's epistemic state. Knowing what is currently true (the facts) is not enough to describe what the agent knows about the world. The agent may be deemed intelligent if she or he has a plan for a case in which the current state of affairs does not apply. The only way to encode and predict our behavior when things are not as they currently are is by using conditionals. For example, consider the following:

Had it rained, I would have brought an umbrella.

A material conditional, whose antecedent is false, is always true. However, the subjunctive conditional above can be false even if its antecedent is false. Such conditionals are important because they can be used to model change and guide the agent to the new (updated) state of affairs. We will call these conditionals update conditionals.

More often than not, what the agent believes is false. We recognize that the current epistemic state is not only made out of true beliefs. Information we receive might

This research was supported by a PSC-CUNY grant. I am grateful to Bernhard Heinemann for his comments on an earlier draft of this paper.

K. Georgatos (✉)
Department of Mathematics and Computer Science, John Jay College CUNY,
514 West 59th Street, New York, NY 10019, USA
e-mail: kgeorgatos@jjay.cuny.edu

C. Başkent et al. (eds.), *Rohit Parikh on Logic, Language and Society*,
Outstanding Contributions to Logic 11, DOI 10.1007/978-3-319-47843-2_14

invalidate other information we received in the past and which formed the basis of our beliefs. Again, conditionals can encode which beliefs will be discarded and which will remain in view of the new information. For example, consider the following (Adams 1975):

If Oswald didn't kill Kennedy, then someone else did.

In this case, the conditional does not describe an alternate state of affairs but rather an alternative epistemic state that would result from a belief revision, when we believe that Oswald didn't kill Kennedy. Ramsey (Braithwaite 1931) made that explicit with the following test: a conditional $a > c$ in our belief set B, add the antecedent a to the set of beliefs and see whether c follows. In other words:

$$a > c \in B \quad \text{iff} \quad c \in B * a,$$

where $*$ is a revision operator. We will call such a conditional *epistemic*.

Conditionals have been studied systematically since the nineteen-sixties (Stalnaker 1968; Lewis 1973). Several formal systems called conditional logics have been proposed, and associated semantics, such as the system-of-spheres model, have been developed (see Nute 1984 for a survey). Unfortunately, this is not the case for epistemic conditionals. Epistemic conditionals have not been studied as extensively, although they recently became the center of attention because of research in belief change. In particular, there is no logic that satisfactorily formalizes the notion of epistemic conditional. At the center of this discrepancy rests a theorem that shows that this task is impossible. Gärdenfors has shown (Gärdenfors 1985, 1986) that the Ramsey test mentioned above is only possible if the revision operation is trivial. In other words, there cannot be interesting revision theories based on conditionals (when the latter are fined by the Ramsey test).

Several theories have been developed to resolve the dead end and break the correspondence that Gärdenfors has established that leads to triviality. Some have weakened or reinterpreted the theorem (Rott 1986; Lindström and Rabinowicz 1995), others replaced AGM revision with a weaker theory (Giordano et al. 2001) or even update (Grahne 1991; Ryan and Schobbens 1997), or used fragments of conditional logics (Friedman and Halpern 1994).

In a previous paper (Georgatos 2010), where we studied conditional operators on subsets, we characterized classes of conditional operators that satisfied two forms of the Ramsey test. The first was

$$B \subseteq A > C \quad \text{iff} \quad B * A \subseteq C.$$

One may notice that this is the set theoretic analogue of the Ramsey test mentioned above. We showed that it corresponds to update conditional operators. In a sense, this is nothing but the Gärdenfors' triviality result as we showed that the conditionals defined using the Ramsey test cannot be based on revision. This idea is hardly new as it is long known, at least as early as Grahne (1991), that the logic of conditionals defined by the Ramsey test is compatible with update operators.

At the same paper, we offered an alternative form of the Ramsey test:

$$B \in A > C \quad \text{iff} \quad B * A \subseteq C.$$

We showed that the conditional operators defined by this test are compatible with a revision operator. Notice the left side of the test: a conditional is not an element of the belief state, as in the first form of the Ramsey test. Rather the opposite: a belief set is an element of the conditional. Therefore, the extension of the conditional is comprised of belief states (that is, subsets of possible worlds). In other words, the meaning of the conditional is not determined on the present state of the agent but rather on the subset that contains the states that the agent considers possible (the belief state). The meaning of a conditional based on revision is contextual. This points to a treatment of conditionals on a higher level, within a language that can express beliefs.

This paper puts forth a different approach to epistemic conditionals. We propose to augment Moss and Parikh's epistemic logic of subsets logic with conditionals, so we are able to reason about conditional beliefs and beliefs about conditionals. We also give a standard straightforward semantics based on the system-of-spheres. Although $a > b$ will be a formula belonging to a belief set, this formula will itself be a belief whose truth will be contextual. The truth of a conditional will not depend on the state the agent is but rather on the beliefs that the agents has at that state. It is known (Lindström 1996) that a such a view of a conditional does not validate monotonicity, which is one of the requirements that leads to the triviality result. We will not use the standard conditional logic of Lewis (VC) or other variants which is to be expected. Instead we will express conditionals through a process of minimization very similar to the operator UNTIL of temporal logic and therefore we will use a linear temporal logic (LTL). In a sense, our solution rests on embedding conditionals to a more expressive framework, a framework that was enhanced with epistemic logic so that conditionals may acquire a contextual meaning. Therefore, there are two main novel features in our logic:

- conditionals are interpreted on subsets of possible states (belief states) rather than single states and,
- the process of minimization is explicit as the logic encodes the steps to reach the belief state that determines the truth of a conditional.

In the next section, we present subset logic (Sect. 14.2.1), its semantics and our modifications to accommodate epistemic conditionals (Sect. 14.2.2), the semantics based on system-of-spheres and the axiomatization. We show completeness and decidability in Sect. 14.3. We conclude with a discussion on the form of the Ramsey test the logic satisfies, as well as, on the ways this research work can be extended.

14.2 Syntax and Semantics

14.2.1 The Logic of Subsets

Subset logic is a bimodal logic that combines two modal operators, one corresponding to knowledge, and one corresponding to effort, and models the increase of knowledge after a larger amount of resources has been spent in acquiring it. Subset logic has been introduced by Moss and Parikh who also established the basic results (Moss and Parikh 1992; Dabrowski et al. 1996). A great deal of further research has been devoted to characterizing the underlying structure of subsets using axioms of this logic. For example, the system "topologic" has been found complete with respect to topological spaces (Georgatos 1993). Variants of this logic have also been developed to address knowledge after program termination and time passing (Heinemann 1999b, 2007).

The main novelty of subset logic is its semantics, where, after fixing a space of subsets of a set of worlds, sentences are interpreted over a pair (x, U), where x is the actual world the agent resides in, and U is the view the agent has. The agent's view consists of those worlds the agent considers possible. We can represent effort by restricting the agent's view. Restricting the view means that the agent cancels out some of the alternatives, and, as a result, increase of knowledge occurs.

Then the *language* $\mathcal{L}$ is the least set such that $\mathsf{Atom} \subseteq \mathcal{L}$ and closed under the following rules:

$$\frac{\phi, \psi \in \mathcal{L}}{\phi \wedge \psi \in \mathcal{L}} \quad \frac{\phi \in \mathcal{L}}{\neg\phi, \Box\phi, \mathsf{K}\phi \in \mathcal{L}}$$

The interpretation of the language using subsets follows:

Definition 1 Let X be a set, and $\mathcal{O}$ a subset of the powerset of X, i.e. $\mathcal{O} \subseteq \mathcal{P}(X)$ such that $X \in \mathcal{O}$. We denote the set $\{(x, U) : x \in X, U \in \mathcal{O}, \text{ and } x \in U\} \subseteq X \times \mathcal{O}$ by $X \dot{\times} \mathcal{O}$. Then the pair $\langle X, \mathcal{O} \rangle$ will be called a *subset frame*. A *model* is a triple $\langle X, \mathcal{O}, i \rangle$, where $\langle X, \mathcal{O} \rangle$ is a subset frame and i a map from Atom to $\mathcal{P}(X)$ with $i(\top) = X$ and $i(\bot) = \emptyset$ called *initial interpretation*.

Definition 2 The *satisfaction relation* $\models_{\mathcal{M}}$, where $\mathcal{M}$ is the model $\langle X, \mathcal{O}, i \rangle$, is a subset of $(X \dot{\times} \mathcal{O}) \times \mathcal{L}$ defined recursively by (we write $x, U \models_{\mathcal{M}} \phi$ instead of $((x, U), \phi) \in \models_{\mathcal{M}}$):

$$\begin{array}{ll}
x, U \models_{\mathcal{M}} A & \text{iff } x \in i(A), \text{ where } A \in \mathsf{Atom} \\
x, U \models_{\mathcal{M}} \phi \wedge \psi & \text{iff } x, U \models_{\mathcal{M}} \phi \text{ and } x, U \models_{\mathcal{M}} \psi \\
x, U \models_{\mathcal{M}} \neg\phi & \text{iff } x, U \not\models_{\mathcal{M}} \phi \\
x, U \models_{\mathcal{M}} \mathsf{K}\phi & \text{iff for all } y \in U,\ y, U \models_{\mathcal{M}} \phi \\
x, U \models_{\mathcal{M}} \Box\phi & \text{iff for all } V \in \mathcal{O} \text{ such that } V \subseteq U \text{ and } x \in V,\ x, V \models_{\mathcal{M}} \phi.
\end{array}$$

14.2.2 Epistemic Conditional Logic

The language of Epistemic Conditional Logic will be the language of Subset Logic augmented with the conditional operator $>$ and the next operator $\bigcirc$, i.e.:

$$\frac{\phi, \psi \in \mathcal{L}}{\phi > \psi, \bigcirc\phi \in \mathcal{L}}$$

We modify the original subset semantics in two ways: first we make use of expanding rather than decreasing subsets and, second, the expanding sets form a sequence which we will call a discrete system of spheres. The next operator is needed to describe the sequential expansion, that is, next is not temporal as it is in Heinemann (1998), Battigalli and Bonanno (1997), or Bonanno (2007).

Definition 3 Let X be a set, then a *(discrete) system of spheres* S is a sequence of non-decreasing subsets $\{U_i\}_{i=1}^{\infty}$ of X such that $\bigcup_{i=1}^{\infty} U_i = X$.

Let S be a system of spheres, as above, a *model* is a triple $\langle X, S, v\rangle$, where v is a map from Atom to $\mathcal{P}(X)$ with $v(\top) = X$ and $v(\bot) = \emptyset$ called *initial interpretation*.

The *satisfaction relation* $\models_{\mathcal{M}}$, where $\mathcal{M}$ is the model $\langle X, S, v\rangle$, is a subset of $(X \dot{\times} S) \times \mathcal{L}$ defined recursively by:

$$
\begin{array}{ll}
x, U_i \models_{\mathcal{M}} A & \text{iff } x \in i(A), \text{ where } A \in \mathsf{Atom} \\
x, U_i \models_{\mathcal{M}} \phi \wedge \psi & \text{iff } x, U_i \models_{\mathcal{M}} \phi \text{ and } x, U_i \models_{\mathcal{M}} \psi \\
x, U_i \models_{\mathcal{M}} \neg\phi & \text{iff } x, U_i \not\models_{\mathcal{M}} \phi \\
x, U_i \models_{\mathcal{M}} \bigcirc\phi & \text{iff } x, U_{i+1} \models_{\mathcal{M}} \phi. \\
x, U_i \models_{\mathcal{M}} \Box\phi & \text{iff for all } j \geq i\ x, U_j \models_{\mathcal{M}} \phi. \\
x, U_i \models_{\mathcal{M}} \mathsf{K}\phi & \text{iff for all } y \in U_i,\ y, U_i \models_{\mathcal{M}} \phi \\
x, U_i \models_{\mathcal{M}} \phi > \psi & \text{iff if } x, U_i \not\models_{\mathcal{M}} \Box\mathsf{K}\neg\phi \text{ then } x, U_j \models_{\mathcal{M}} \mathsf{K}(\phi \to \psi), \text{ where} \\
& j = \min\{k\colon k \geq i \text{ and there exists } z \in U_i \text{ with } z, U_k \models_{\mathcal{M}} \phi\}.
\end{array}
$$

If $x, U_i \models_{\mathcal{M}} \phi$ for all (x, U_i) belonging to $X \dot{\times} S$ then ϕ is *valid* in $\mathcal{M}$, denoted by $\mathcal{M} \models \phi$.

We describe the process of reasoning when the epistemic state grows continuously as it is encompassing other possibilities. This is also the central idea behind the system-of-spheres semantics for conditional logic. Knowing a proposition on a given epistemic state is modeled using necessity. This correspondence allow us to express the minimization of the distance of an epistemic state, where the antecedent is consistent, for the interpretation of conditionals.

The axiom system **ECL** consists of axiom schemes 1 through 16 and rules of Table 14.1 (see page xxx). We will write $\vdash_{\mathbf{ECL}} \phi$ iff ϕ is a theorem of **ECL**.

Axiom 2 stipulates that the non-epistemic facts true in the world of an agent will remain true as the only change we allow is epistemic. The actual state of an agent remains always the same although the agent's view may change.

Example 4 Consider the set $X = \{x_1, x_2, x_3, x_4\}$ with the system of spheres $\{U_i\}_{i=1}^{\infty}$, where $U_i = \{x_j\colon j \leq i\}$ for $i \leq 4$ and $U_j = U_4 = X$ for $i > 4$. Interpret atomic formulas A_1, A_2, A_3, A_4 and B_1, B_2, B_3 with $v(A_i) = \{x_j\colon i \leq j\}$ and $v(B_i) = U_i$.

We have $x_1 \in U_1$ and $x_1, U_1 \models A_1 \wedge B_1$ but $x_1, U_1 \models \neg A_2 \wedge B_2$. Also, $x_2, U_2 \models \mathsf{K}A_1$ but $x_2, U_2 \not\models \mathsf{K}A_2$ because $x_1, U_2 \not\models A_2$. We have $x, U_1 \models A_2 > B_2$ because U_2 is the smallest superset of U_1 that contains an element of A_2. This is explicitly encoded in the language as we have $x_1, U_1 \models \mathsf{K}\neg A_2 \wedge \bigcirc\mathsf{L}A_2$. In contrast $x, U_1 \models A_3 > \mathsf{L}\neg B_2$ because, similarly U_3 is the smallest superset of U_1 that contains an element of A_3. Nested conditionals can be easily verified: we have $x, U_1 \models (B_3 >$

Table 14.1 Axioms for ECL

Axioms

1. All propositional tautologies
2. $(A \rightarrow \Box A) \wedge (\neg A \rightarrow \Box\neg A)$, for all $A \in$ Atom
3. $\bigcirc(\phi \rightarrow \psi) \rightarrow (\bigcirc\phi \rightarrow \bigcirc\psi)$
4. $\bigcirc\neg\phi \leftrightarrow \neg\bigcirc\phi$
5. $\Box(\phi \rightarrow \psi) \rightarrow (\Box\phi \rightarrow \Box\psi)$
6. $\Box\phi \rightarrow \phi$
7. $\Box\phi \rightarrow \Box\Box\phi$
8. $\Box\phi \rightarrow \Box\bigcirc\phi$
9. $\Box(\phi \rightarrow \bigcirc\phi) \rightarrow (\phi \rightarrow \Box\phi)$
10. $\mathsf{K}(\phi \rightarrow \psi) \rightarrow (\mathsf{K}\phi \rightarrow \mathsf{K}\psi)$
11. $\mathsf{K}\phi \rightarrow \phi$
12. $\mathsf{K}\phi \rightarrow \mathsf{KK}\phi$
13. $\phi \rightarrow \mathsf{KL}\phi$
14. $\bigcirc\mathsf{K}\phi \rightarrow \mathsf{K}\bigcirc\phi$
15. $\Box\mathsf{K}\neg\phi \rightarrow \phi > \psi$
16. $\phi > \psi \leftrightarrow \mathsf{K}(\phi \rightarrow \psi) \wedge (\mathsf{L}\phi \vee \bigcirc(\phi > \psi))$

Rules

$$\frac{\phi \rightarrow \psi, \phi}{\psi}\ \text{MP}$$

$$\frac{\phi}{\mathsf{K}\phi}\ \text{K-Necessitation} \quad \frac{\phi}{\Box\phi}\ \Box\text{-Necessitation} \quad \frac{\phi}{\bigcirc\phi}\ \bigcirc\text{-Necessitation}$$

$\mathsf{L}A_3) > \bigcirc\mathsf{L}A_4$ because $\mathsf{K}(B_3 \wedge \neg A_3)$ is valid at U_1 and U_2, and so U_3 becomes the smallest superset of U_1 where $\mathsf{L}A_3$ is satisfied.

Axioms 3–9 is just the axiomatization of LTL (linear temporal logic) with next and always. Axioms 10–13 are the axioms of the modal logic **S5**. Axiom 14 is the reverse axiom of the Cross Axiom of the original logic of subsets, the "no learning" of Schmidt and Tishkovsky (2008) for a single modality. The effect of this axiom is that the sequence of subsets is non-decreasing. Axioms 15 and 16 are the axioms for epistemic conditionals. Axiom 15 says that, if the antecedent of a conditional can never be true, then the conditional holds vacuously. Axiom 16 is the inductive definition of the conditional, akin to the inductive definition of until in linear temporal logic. Axioms 15 and 16 show that the epistemic conditional we defined is a *knowledge* formula, that is, it is equivalent to a formula of the form $\mathsf{K}a$ or $\neg\mathsf{K}a$ because the following is a theorem

$$\phi > \psi \quad \leftrightarrow \quad \mathsf{K}(\phi > \psi).$$

This means that if $\phi > \psi$ is true the agent believes it. The extension $ex(\phi)$ of all knowledge formulas ϕ can be identified with a set of subsets

$$ex(\phi) = \{U_i \colon y, U_i \models \phi\}.$$

Therefore, the meaning of the epistemic conditional $ex(\phi > \psi)$ is contextual as it depends on the epistemic view of the agent, an important desideratum we expressed in the introduction. Also, spheres will represent the belief state of the agent as they determine the satisfaction of knowledge formulas. It will be useful to relate knowledge formulas with spheres with:

$$th(U_i) = \{\phi \colon \phi \text{ is a knowledge formula and } x, U_i \models \phi \text{ for some } x \in U_i\}.$$

Heinemann has combined the logic of subsets and linear temporal logic in several papers. He studied the temporal logic of decreasing subsets in Heinemann (1998) and added *until* in Heinemann (1999b). He then studied the temporal logic of increasing subsets (Heinemann 1999a) without next and a combination of alternating increasing and decreasing subsets in Heinemann (2000). Although the combination of the temporal *next* and the epistemic K has been studied extensively in his work, the semantics and axiomatization of the conditional, does not relate to his constructs and therefore our techniques differ. This should be evident from the proofs.

The following holds:

Theorem 5 *The axioms and rules of* **ECL** *are sound with respect to discrete systems of spheres.*

Proof The proof is straightforward and we show only soundness for Axiom 16. Suppose $x, U_i \models a > b$. If $x, U_i \models \Box \mathsf{K}\neg a$ then $x, U_i \models \bigcirc\Box \mathsf{K}\neg a$ and therefore $x, U_i \models \bigcirc(a > b)$ by Axiom 15. Now suppose that $x, U_i \models \Diamond \mathsf{L}a$. There is a least $j \geq i$ such that $x, U_j \models \mathsf{L}a$ and $x, U_j \models \mathsf{K}(a > b)$. If $i = j$ then we have $x, U_i \models \mathsf{K}(a > b)$. If $i \neq j$ then $i + 1 \leq j$ and therefore $x, U_{i+1} \models a > b$ so $x, U_i \models \bigcirc(a > b)$. The other direction is similar. ■

14.3 Completeness

The proof of completeness runs along the lines of the linear temporal logic. There are differences though which we will spell out in the proofs. We will define a canonical structure (Sect. 14.3.1), which is not a model of the logic but will provide the basis for one after we define a filtration (Sect. 14.3.2). The filtration will be unwound in to a model that can be converted to a discrete system of spheres (Sect. 14.3.3). We will state but not prove the results that we can transfer from the completeness theorem of LTL.

14.3.1 Canonical Structure

We will perform a filtration on the canonical structure of **ECL**. By canonical structure we mean the structure $(S, \overset{\circ}{\rightarrow}, \overset{\Diamond}{\rightarrow}, \overset{\mathsf{L}}{\rightarrow})$ where S is the set of maximal consistent sets of $\mathcal{L}$ and $\overset{\circ}{\rightarrow}, \overset{\Diamond}{\rightarrow}, \overset{\mathsf{L}}{\rightarrow}$ the accessibility relations corresponding to the modalities $\bigcirc, \Box$ and K respectively. That is, $s \overset{\circ}{\rightarrow} t$ iff $\bigcirc\phi \in s$ implies $\phi \in t$. Similarly for the other relations.

We will now prove some properties of the canonical structure

Lemma 6 *The following hold*

1. $\overset{\mathsf{L}}{\rightarrow}$ *is an equivalence*
2. $\overset{\circ}{\rightarrow}$ *is functional*
3. $\overset{\Diamond}{\rightarrow}$ *is reflexive, transitive and weakly connected.*

Proof $\overset{\mathsf{L}}{\rightarrow}$ is an equivalence because K is **S5**. $\overset{\circ}{\rightarrow}$ is functional because of the *Fun* axiom. $\overset{\Diamond}{\rightarrow}$ is reflexive, transitive and weakly connected because $\Box$ is **S4** and axiom 8. It becomes connected when the model is generated. ■

Lemma 7 *If* $s \overset{\Diamond}{\rightarrow} t$ *and* $s \overset{\circ}{\rightarrow} s'$ *then there exists* $t' \in S$ *such that* $t \overset{\circ}{\rightarrow} t'$ *and* $s' \overset{\Diamond}{\rightarrow} t'$ (*see Fig.* 14.1).

Proof Suppose $s \overset{\Diamond}{\rightarrow} t$ and $s \overset{\circ}{\rightarrow} s'$. Let $\{a\colon \bigcirc a \in t\} \cup \{b\colon \Box b \in s'\}$. Suppose this set is inconsistent. Without loss of generality we can assume that there are single a, b with $b \vdash \neg a$. We have $\bigcirc\Box b \vdash \bigcirc\Box\neg a$. We have $\Box b \in s'$ and therefore $\bigcirc\Box b \in s$ which implies that $\bigcirc\Box\neg a \in s$. By Axiom 8, we have $\Box \bigcirc \neg a \in s$ so $\bigcirc\neg a \in t$. Therefore, $\neg \bigcirc a \in t$, a contradiction. ■

Lemma 8 *The following hold*

1. *If* $s \overset{\mathsf{L}}{\rightarrow} t$ *and* $t \overset{\circ}{\rightarrow} t'$ *then there exists* $s' \in S$ *such that* $s \overset{\circ}{\rightarrow} s'$ *and* $s' \overset{\mathsf{L}}{\rightarrow} t'$ (*see Fig.* 14.2).
2. *If* $s \overset{\mathsf{L}}{\rightarrow} t$ *and* $t \overset{\circ}{\rightarrow} t'$ *and* $s \overset{\circ}{\rightarrow} s'$ *then* $s' \overset{\mathsf{L}}{\rightarrow} t'$ (*see Fig.* 14.3).

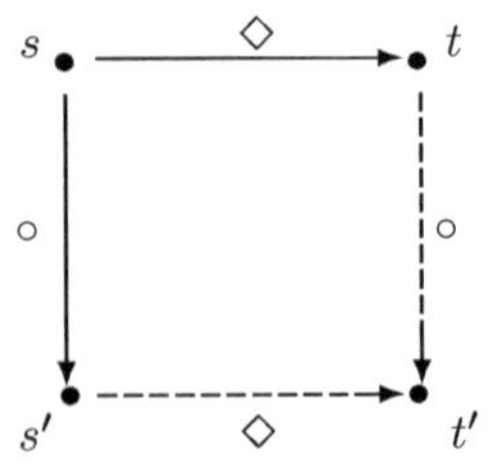

Fig. 14.1 Commutative diagram for Lemma 7

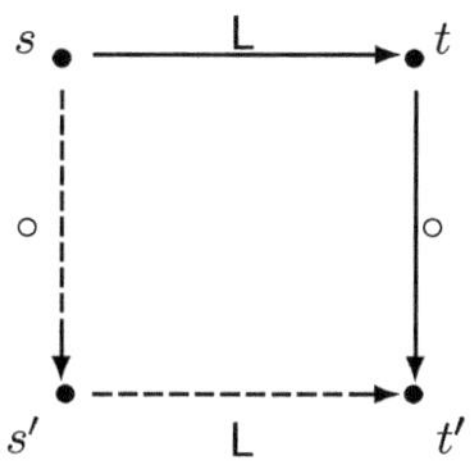

Fig. 14.2 Commutative diagram for Lemma 8.1

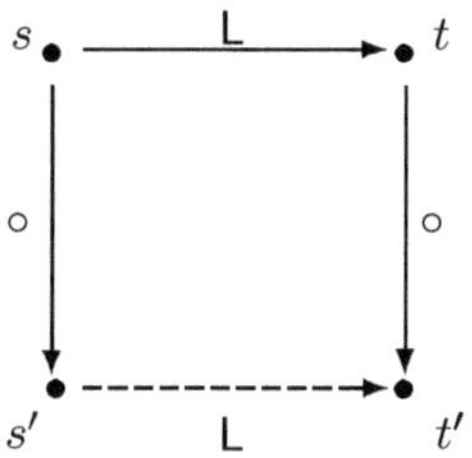

Fig. 14.3 Commutative diagram for Lemma 8.2

Proof Suppose $s \xrightarrow{\mathsf{L}} t$ and $t \xrightarrow{\circ} t'$. Let $\{a\colon \bigcirc a \in s\} \cup \{b\colon \mathsf{K}b \in t'\}$. Suppose this set is inconsistent. Without loss of generality we can assume that there are single a, b with $b \vdash \neg a$. We have $\bigcirc \mathsf{K}b \vdash \bigcirc \mathsf{K}\neg a$. We have $\mathsf{K}b \in t'$ and therefore $\bigcirc \mathsf{K}b \in t$ which implies that $\bigcirc \mathsf{K}\neg a \in t$. By Axiom 14, we have $\mathsf{K} \bigcirc \neg a \in t$ so $\bigcirc \neg a \in s$. Therefore, $\neg \bigcirc a \in s$, a contradiction.

For the second, let $a \in t'$ then $\bigcirc a \in t$ so $\mathsf{L} \bigcirc a \in s$ and by Axiom 14 $\bigcirc \mathsf{L}a \in s$. Therefore $\mathsf{L}a \in S'$. ■

Lemma 9 *The following hold*

1. *If* $s \xrightarrow{\mathsf{L}} t$ *and* $t \xrightarrow{\Diamond} t'$ *then there exists* $s' \in S$ *such that* $s \xrightarrow{\Diamond} s'$ *and* $s' \xrightarrow{\mathsf{L}} t'$ *(see Fig.* 14.4).
2. *If* $s \xrightarrow{\mathsf{L}} t$ *and* $s \xrightarrow{\Diamond} t$ *than* $s = t$.

Proof Suppose $s \xrightarrow{\mathsf{L}} t$ and $t \xrightarrow{\Diamond} t'$. Let $\{a\colon \Box a \in s\} \cup \{b : \mathsf{K}b \in t'\}$. Suppose this set is inconsistent. Without loss of generality we can assume that there are single a, b with $b \vdash \neg a$. We have $\Diamond b \vdash \Diamond \neg a$. We have $\mathsf{K}b \in t'$ and therefore $\Diamond \mathsf{K}b \in t$ which implies that $\mathsf{K}\Diamond b \in t$, by Axiom 14. Hence $\Diamond b \in s$ so $\Diamond \neg a \in s$, a contradiction.

For the second, we will show that for all pairs s, t such that $s \xrightarrow{\mathsf{L}} t$ and $s \xrightarrow{\Diamond} t$, we have $a \in s$ if and only if $a \in t$. We will show it by induction on a. The cases of atomic, negation, conjunction, K and $>$ formulas are straightforward. Suppose $\Box a \in s$. Then $\Box\Box a \in s$ and therefore $\Box a \in t$. Let $\Box a \in t$ and suppose $\Diamond \neg a \in S$. There exists s' with $s \xrightarrow{\Diamond} s'$ such that $\neg a \in s'$. By the first part, there exists t' with

$t \overset{\Diamond}{\to} t'$ and $s' \overset{\mathsf{L}}{\to} t'$. Observe that $s \overset{\Diamond}{\to} t'$ as well so from connectedness we have either $s' \overset{\Diamond}{\to} t'$ or $t' \overset{\Diamond}{\to} s'$. In both cases the induction hypothesis applies and since $a \in t'$ we must have $a \in s'$ which is a contradiction.

For the $\bigcirc a$ case, let $s \overset{\circ}{\to} s'$ and $t \overset{\circ}{\to} t'$. By the previous lemma (second case) we have $s' \overset{\mathsf{L}}{\to} t'$. Also, $s \overset{\Diamond}{\to} t'$ so $s' \overset{\Diamond}{\to} t'$ (by connectedness and without loss of generality). Now we have $\bigcirc a \in s$ iff $a \in s'$ iff (by IH) $a \in t'$ iff $\bigcirc a \in t$. ■

14.3.2 Filtration

To define the desired filtration of the canonical structure, we will combine the filtration of temporal logic and the filtration of subset logic. Fix a formula a and let

$$\Gamma' = Sf(a) \cup \{\bigcirc\Box b : \Box b \in Sf(a)\} \cup \{\bigcirc(b > c), \Box\neg c, \bigcirc\Box\neg c : b > c \in Sf(a)\}.$$

$$\Gamma^{\neg} = \Gamma' \cup \{\neg b : b \in \Gamma'\},$$

$$\Gamma^{\wedge} = \Gamma^{\neg} \text{ together with all finite conjunctions of distinct elements of } gG^{\neg},$$

$$\Gamma^{\mathsf{L}} = \{\mathsf{L}b : b \in \Gamma^{\wedge}\},$$

$$\Gamma = \Gamma^{\wedge} \cup \Gamma^{\mathsf{L}}.$$

We have that

$$a \in \Gamma$$

$$\Gamma \text{ is closed under subformulas},$$

$$\Box b \in \Gamma \text{ implies } \bigcirc\Box b \in \Gamma$$

$$b > c \in \Gamma \text{ implies } \bigcirc(b > c), \Box\neg c \in \Gamma$$

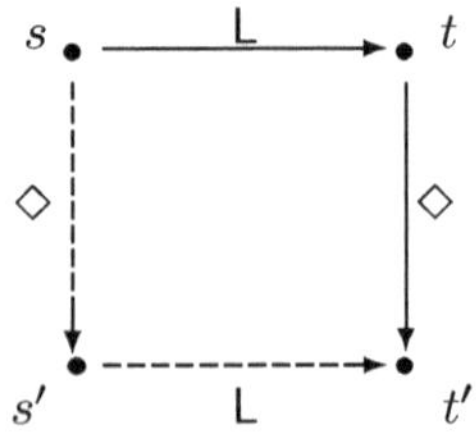

Fig. 14.4 Commutative diagram for Lemma 9.1

Now, let

$$\Gamma_s = s \cap \Gamma,$$

$$s \sim_\Gamma t \text{ iff } \Gamma_s = \Gamma_t,$$

$$\bar{s} = \{t : s \sim_\Gamma t\},$$

$$\bar{S} = \{\bar{s} : s \in S\}.$$

Observe that $b \in \Gamma_s$ than up to provable equivalence $\mathsf{L}b \in \Gamma_s$.

To define a model, let $x, y \in \bar{S}$ then define by $x \overset{\circ}{\rightarrow} y$ if there is an $s \in x$ and $t \in y$ such that $s \overset{\circ}{\rightarrow} t$. (We abuse notation by using the same symbols for the accessibility relation on both the canonical structure and its filtration.) Similarly, for $\overset{\mathsf{L}}{\rightarrow}$. One can show that $\overset{\circ}{\rightarrow}$ and $\overset{\mathsf{L}}{\rightarrow}$, as defined, they are minimal. In other words, they are the smallest relations that satisfy the following two properties (F1) $s \overset{\circ}{\rightarrow} t$ implies $\bar{s} \overset{\circ}{\rightarrow} \bar{t}$ and (F2) $\bar{s} \overset{\circ}{\rightarrow} \bar{t}$ implies that if $\Box b \in \Gamma^s$ then $b \in \Gamma^t$, and similarly for L. We can show the following

Lemma 10 *The following are equivalent:*

1. $\bar{s} \overset{\mathsf{L}}{\rightarrow} \bar{t}$ *iff* $\Gamma_s^{\mathsf{L}} = \Gamma_t^{\mathsf{L}}$
2. $\bar{s} \overset{\circ}{\rightarrow} \bar{t}$ *iff* $\Gamma_s \cup \bigcirc\Gamma_t$ *is consistent*

One of the significant consequences of employing a filtration is that one can show the following

Lemma 11 *The ancestral relation* $\overset{\circ *}{\rightarrow}$ *of* $\overset{\circ}{\rightarrow}$ *(on* $\bar{S}$*) is a filtration of* $\overset{\Diamond}{\rightarrow}$.

Also the following lemma [akin to the *Fun* Lemma of Goldblatt (1992) at page 94] is true.

Lemma 12 *The following are equivalent:*

1. $\bigcirc b \in s$
2. $\bar{s} \overset{\circ}{\rightarrow} \bar{t}$ *implies* $b \in \Gamma_t$
3. *there exists t such that* $\bar{s} \overset{\circ}{\rightarrow} \bar{t}$ *and* $b \in \Gamma_t$.

We will now prove a series of properties of the filtration that will be used for the completeness.

First the cross axioms hold true for the filtration:

Lemma 13 *We have*

1. *If* $\bar{s} \overset{\mathsf{L}}{\rightarrow} \bar{t}$ *and* $\bar{t} \overset{\circ}{\rightarrow} \bar{t}'$ *then there exists* $\bar{s}' \in S$ *such that* $\bar{s} \overset{\circ}{\rightarrow} \bar{s}'$ *and* $\bar{s}' \overset{\mathsf{L}}{\rightarrow} \bar{t}'$.
2. *If* $\bar{s} \overset{\mathsf{L}}{\rightarrow} \bar{t}$ *and* $\bar{t} \overset{\circ *}{\rightarrow} \bar{t}'$ *then there exists* $\bar{s}' \in S$ *such that* $\bar{s} \overset{\circ *}{\rightarrow} \bar{s}'$ *and* $\bar{s}' \overset{\mathsf{L}}{\rightarrow} \bar{t}'$.

Lemma 14 *If* $\bar{s} \overset{\circ *}{\rightarrow} \bar{t}_1$, $\bar{s} \overset{\circ *}{\rightarrow} \bar{t}_2$ *and* $\bar{t}_1 \overset{\mathsf{L}}{\rightarrow} \bar{t}_2$ *then* $\bar{t}_1 = \bar{t}_2$.

Proof Suppose $\bar{s} \stackrel{\circ^*}{\to} \bar{t}_1, \bar{s} \stackrel{\circ^*}{\to} \bar{t}_2$ and $\bar{t}_1 \stackrel{\mathsf{L}}{\to} \bar{t}_2$. We will show that $\phi \in \bar{t}_1$ if and only if $\phi \in \bar{t}_2$ by induction on ϕ.

- Suppose ϕ is atomic and $\phi \in \mathcal{L}_{\bar{t}_1}$. Since $\bar{s} \stackrel{\circ^*}{\to} \bar{t}_1$ there exist $x \in \bar{s}$ and $y \in \bar{t}_1$ with $x \stackrel{\Diamond}{\to} y$. We have that $\phi \in y$ and therefore $\Diamond\phi \in x$ so $\phi \in x$ and therefore $\phi \in \mathcal{L}_{\bar{s}}$. Similarly, there exist $x' \in \bar{s}$ and $y' \in \bar{t}_2$ with $x' \stackrel{\Diamond}{\to} y'$. We have that $\phi \in x'$ so $\Box\phi \in x'$ and therefore $\phi \in y$ which proves that $\phi \in \mathcal{L}_{\bar{t}_2}$. The other direction is symmetric.
- The cases of boolean connectives is straightforward.
- Suppose ϕ is of the form $\mathsf{K}\psi$ or $\psi > \chi$. In both cases we have that ϕ is equivalent to $\mathsf{L}\phi \in \mathcal{L}^{\mathsf{L}}$ and therefore $\phi \in \bar{t}_1$ if and only if $\phi \in \bar{t}_2$.
- Suppose now that ϕ is of the form $\Box\psi$ and $\Box\psi \in \bar{t}_1$. Towards contradiction suppose that $\Diamond\neg\psi \in \bar{t}_2$. For some $x \in \bar{t}_2$ we have $\Diamond\neg\psi \in x$. So there is $y \in C$ such that $x \stackrel{\Diamond}{\to} y$ and $\neg\psi \in y$. Therefore $\bar{t}_2 \stackrel{\circ^*}{\to} \bar{y}$ and $\neg\psi \in \bar{y}$. By Lemma 13, there exists $\bar{z}$ with $\bar{t}_1 \stackrel{\circ^*}{\to} \bar{z}$ and $\bar{y} \stackrel{\mathsf{L}}{\to} \bar{z}$. The former implies that $\psi \in \bar{z}$. Since $\stackrel{\circ^*}{\to}$ is transitive we have $\bar{s} \stackrel{\circ^*}{\to} \bar{z}, \bar{s} \stackrel{\circ^*}{\to} \bar{y}$ and by applying the induction hypothesis we have $\neg\psi \in \bar{z}$, a contradiction. The other direction is symmetric.
- Suppose now that ϕ is of the form $\bigcirc\psi$ and $\bigcirc\psi \in \bar{t}_1$. Towards contradiction suppose that $\bigcirc\neg\psi \in \bar{t}_2$. There exists y such that $\neg\psi \in y$ and $\bar{t}_2 \stackrel{\circ}{\to} \bar{y}$, by the *Fun* lemma. By Lemma 13, there exists $\bar{z}$ with $\bar{t}_1 \stackrel{\circ}{\to} \bar{z}$ and $\bar{y} \stackrel{\mathsf{L}}{\to} \bar{z}$. The former implies that $\psi \in \bar{z}$. As in the previous case we have $\bar{s} \stackrel{\circ^*}{\to} \bar{z}, \bar{s} \stackrel{\circ^*}{\to} \bar{y}$ and by applying the induction hypothesis we have $\neg\psi \in \bar{z}$, a contradiction. ■

Lemma 15 *If $\bar{s} \stackrel{\circ^*}{\to} \bar{t}$ and $\bar{s} \stackrel{\mathsf{L}}{\to} \bar{t}$ then $\bar{s} = \bar{t}$.*

Proof Suppose $\bar{s} \stackrel{\circ^*}{\to} \bar{t}$ and $\bar{s} \stackrel{\mathsf{L}}{\to} \bar{t}$. We will show that $\phi \in \bar{t}_1$ if and only if $\phi \in \bar{t}_2$ by induction on ϕ.

- The cases of atomic, boolean connectives, K and $>$ connectives are as above.
- Suppose now that ϕ is of the form $\Box\psi$ and $\Box\psi \in \bar{t}$. Towards contradiction suppose that $\Diamond\neg\psi \in \bar{s}$. For some $x \in \bar{s}$ we have $\Diamond\neg\psi \in x$. So there is $y \in C$ such that $x \stackrel{\Diamond}{\to} y$ and $\neg\psi \in y$. Theretofore $\bar{s} \stackrel{\circ^*}{\to} \bar{y}$ and $\neg\psi \in \bar{y}$. By the cross axiom Lemma 13, there exists $\bar{z}$ with $\bar{t} \stackrel{\circ^*}{\to} \bar{z}$ and $\bar{y} \stackrel{\mathsf{L}}{\to} \bar{z}$. The former implies that $\psi \in \bar{z}$. Since $\stackrel{\circ^*}{\to}$ is transitive we have $\bar{s} \stackrel{\circ^*}{\to} \bar{z}, \bar{s} \stackrel{\circ^*}{\to} \bar{y}$ and by the previous Lemma we have $\bar{y} = \bar{z}$, a contradiction. The other direction is straightforward. The case $\bigcirc\phi$ is similar. ■

Definition 16 Let $x, y \in \bar{S}$. We define a relation $x \stackrel{C}{\to} y$ iff for all $s \in x$ there is $t \in y$ such that $s \stackrel{\Diamond}{\to} t$.

The following is proved as in Goldblatt (1992)

Lemma 17 *If $\Box\phi \in \mathcal{L}$ and $\Box\phi \notin s \in X$ then there exists $t \in S$ with $\phi \notin t$ and $\bar{s} \stackrel{C}{\to} \bar{t}$.*

The cross axiom also holds for $\xrightarrow{c}$:

Lemma 18 *If $\bar{s} \xrightarrow{L} \bar{t}$ and $\bar{t} \xrightarrow{c} \bar{t'}$ then there exists $s' \in S$ such that $\bar{s} \xrightarrow{c} \bar{s'}$ and $\bar{s'} \xrightarrow{L} \bar{t'}$.*

Proof Pick $x \in \bar{s}$. Since $\bar{s} \xrightarrow{L} \bar{t}$ there exists $y \in \bar{t}$ with $x \xrightarrow{L} y$. Also, $\bar{t} \xrightarrow{c} \bar{t'}$ implies there is $y' \in \bar{t'}$ with $y \xrightarrow{\Diamond} y'$. By the cross axiom there exists x' with $x \xrightarrow{\Diamond} x'$ and $x' \xrightarrow{L} y'$. Therefore, $\bar{s} \xrightarrow{\circ^*} \bar{x'}$ and $\bar{x'} \xrightarrow{L} \bar{t'}$. To show that $\bar{s} \xrightarrow{c} \bar{c'}$, let $z \in \bar{s}$. As before we have z' with $z \xrightarrow{\Diamond} z'$ and $\bar{s} \xrightarrow{\circ^*} \bar{z'}$ and $\bar{z'} \xrightarrow{L} \bar{t'}$. By Lemma 15 $\bar{x'} = \bar{z'}$ which shows that $\bar{x} \xrightarrow{c} \bar{x'}$. ■

Lemma 19 *If $\bar{s} \xrightarrow{c} \bar{s'}$, $\bar{t} \xrightarrow{c} \bar{t'}$ and $\bar{s} \xrightarrow{L} \bar{t}$ then $\bar{s'} \xrightarrow{c} \bar{s}$ implies $\bar{t'} \xrightarrow{c} \bar{t}$.*

Proof Suppose $\bar{s} \xrightarrow{c} \bar{s'}$, $\bar{t} \xrightarrow{c} \bar{t'}$, $\bar{s} \xrightarrow{L} \bar{t}$ and $\bar{s'} \xrightarrow{c} \bar{s}$. By previous lemma, there exists $s'' \in C$ such that $\bar{s'} \xrightarrow{c} \bar{s''}$ and $\bar{s''} \xrightarrow{L} \bar{t'}$. We have $\bar{s} \xrightarrow{c} \bar{s''}$ and $\bar{s} \xrightarrow{c} \bar{s''}$, so by above lemma $\bar{s} = \bar{s''}$. ■

A subset U of S will be called $\bigcirc^*$-maximal if for any pair of elements $\bar{s}$ and $\bar{t}$ of U we have either $\bar{s} \xrightarrow{\circ^*} \bar{t}$ or $\bar{t} \xrightarrow{\circ^*} \bar{s}$ and U is maximal with respect to this property. One can show as in Goldblatt (1992) that this set can be unwound to a sequence of $\xrightarrow{c}$ clusters where all clusters but the last cluster are simple (a cluster is simple when it contains a single element).

Call a subset of $\bar{S}$ generated if there exists an element x such that for all y there exists z such that $x \xrightarrow{\circ^*} z$ and $z \xrightarrow{L} x$. A generated set can be partitioned into $\bigcirc^*$-maximal sets $U_1, U_2, \ldots, U_n$. Let U be the $\bigcirc^*$-maximal set containing x and let $x_1, x_2, \ldots, x_{n-1}, C_n$ be the unwind to $\xrightarrow{c}$ clusters where $C_n = \{x_{n_1}, \ldots, x_{n_k}\}$ is the last not necessarily simple cluster. This unwinding gives rise to an unwinding of all other $\bigcirc^*$-maximal sets as follows: Let U' be any other $\bigcirc^*$-maximal set and x_k be the earliest element of U such that there exists an element $y \in U'$ such that $x_k \xrightarrow{L} y$. Then by Lemma 18 there exists a sequence of $\xrightarrow{c}$ clusters $y_k = y, y_{k+1}, \ldots, y_{n-1}, \{y_{n_1}, \ldots, y_{n_k}\}$ which is an unwinding of U_i using Lemma 19. The following holds:

Lemma 20 *The last cluster of a $\bigcirc^*$-maximal set is simple.*

Proof Let $C_n = \{x_{n_1}, \ldots, x_{n_k}\}$ be the last cluster of U, as above. Let $|x_{n_1}|, \ldots, |x_{n_k}|$ be the corresponding $\xrightarrow{L}$ equivalence classes. All $\bigcirc^*$-maximal subsets have a last cluster $C = \{y_1, \ldots, y_k\}$ comprised exactly by an element from each of the $|x_{n_1}|, \ldots, |x_{n_k}|$. We will show that $y_1 = \ldots = y_k$ by induction on the structure of the formulas:

- The cases of atomic formulas (due to Axiom 2) and boolean connectives are straightforward.

- Suppose ϕ is of the form $\mathsf{K}\psi$ or $\psi > \chi$. In both cases we have that ϕ is equivalent to $\mathsf{L}\phi \in \mathcal{L}^{\mathsf{L}}$. Let $\mathsf{L}\phi \in y_j$ for some j, then there exists z_j such that $\phi \in z_j$. By induction hypothesis $\phi \in z_i$, for all $i = 1, \ldots, k$, and therefore $\mathsf{L}\phi \in y_i$, for all $i = 1, \ldots, k$.
- Suppose now that ϕ is of the form $\Box\psi$ and $\Box\psi \in y_j$, for some $j \in \{1, \ldots, k\}$. Since $y_j \overset{\bigcirc^*}{\rightarrow} y_i$ for all $i \in \{1, \ldots, k\}$ and $\overset{\bigcirc^*}{\rightarrow}$ is a filtration of $\overset{\Diamond}{\rightarrow}$ we have that $\psi \in y_i$ for all $i \in \{1, \ldots, k\}$. Suppose towards a contradiction that there exists $j \in \{1, \ldots, k\}$ such that $\Box\psi \notin y_j$. By Lemma above, there exists y with $\psi \notin y$ and $y_j \overset{c}{\rightarrow} y$. We necessarily have that $y = y_l$ for some $l = 1, \ldots, k$ as we have that $y_j \overset{c}{\not\rightarrow} y$, if y is outside the cluster. Therefore, $\psi \notin y_l$, a contradiction.
- Suppose now that ϕ is of the form $\bigcirc\psi$ and $\bigcirc\psi \in y_j$, for some $j \in \{1, \ldots, k\}$. Since $y_j \overset{\bigcirc^*}{\rightarrow} y_i$ for all $i \in \{1, \ldots, k\}$ there exists $l \in \{1, \ldots, k\}$ with $y_j \overset{\bigcirc}{\rightarrow} y_l$. By the *Fun* lemma we have that $\psi \in y_l$ and by induction hypothesis $\psi \in y_i$ for all $i \in \{1, \ldots, k\}$. Pick $m \in \{1, \ldots, k\}$ and let $m' \in \{1, \ldots, k\}$ be such that $y_m \overset{\bigcirc}{\rightarrow} y_{m'}$. We have that $\psi \in y_{m'}$ and by the *Fun* lemma $\bigcirc\psi \in y_m$. ■

Using the above lemma we can now proceed to the construction of a subset model based on the filtration of the generated structure.

14.3.3 Subset Model

Let X be the set of $\bigcirc^*$-maximal sets. Let U_0 be a $\bigcirc^*$-maximal set with maximal length with the following unwinding

$$x_1 \overset{\bigcirc}{\rightarrow} x_2 \overset{\bigcirc}{\rightarrow} \cdots \overset{\bigcirc}{\rightarrow} x_n.$$

The union of the $\overset{\mathsf{L}}{\rightarrow}$ equivalence classes for each element of U_0, i.e.

$$|x_1|, |x_2|, \ldots, |x_n|$$

equals the union of all $\bigcirc^*$-maximal sets. It is straightforward to show that if U is any other $\bigcirc^*$-maximal set then $U \cap |x_i| \neq \emptyset$ implies $U \cap |x_j| \neq \emptyset$ for all $j = i, i+1, \ldots, n$. Therefore we can define a system of spheres $\{S_i\}_{i=1}^{\infty}$ in X with

$$S_i = \begin{cases} \{U : U \cap |x_i| \neq \emptyset\} & \text{if } i \leq n \\ S_n & \text{if } n < i \end{cases}$$

Let v be the initial interpretation defined by

$$v(A) = \{U:\ \text{for all } \bar{s} \in U,\ A \in s\}.$$

Notice that $U \in v(A)$ follows from the existence of $\bar{s} \in U$ with $a \in S$ as the U is $\overset{\bigcirc^*}{\rightarrow}$ connected.

We will show the following

Lemma 21 *For all* $U \in X$, $i = 1, \dots n$, *and* $\phi \in \mathcal{L}$ *we have*

$$U, S_i \models \phi \text{ iff } \phi \in x,$$

where $U \cap |x_i| = \{\bar{x}\}$.

Proof By induction on ϕ.

- Suppose ϕ is atomic then the statement follows from the definition of the initial interpretation.
- The cases of boolean connectives is straightforward.
- Suppose ϕ is of the form $\mathsf{K}\psi$. Let $U, S_i \models \mathsf{K}\psi$ then there exists U' such that $U', S_i \models \psi$. Let $U \cap |x_i| = \{\bar{x}\}$. By induction hypothesis, $\psi \in y$ where $U' \cap |x_i| = \{\bar{y}\}$. We have that $\bar{x}$ and $\bar{y}$ belong to $|x_i|$ so $\bar{x} \xrightarrow{\mathsf{L}} \bar{y}$, This implies that there exists $y' \in \bar{y}$ such that $x \xrightarrow{\mathsf{L}} y'$. We have that $\psi \in y'$ and therefore $\mathsf{L}\psi \in x$. The other direction is similar.
- Suppose now that ϕ is of the form $\Box\psi$ and $\Box\psi \notin x$, where $U \cap |x_i| = \{\bar{x}\}$. By Lemma 17, then there exists $t \in S$ with $\phi \notin t$ and $\bar{x} \xrightarrow{c} \bar{t}$. Unwinding ensures that $|x_{i-1}| \xrightarrow{c} |x_i|$ but $|x_i| \not\xrightarrow{c} |x_{i-1}|$, for $i = 2, \dots, n$. Therefore $\{t\} = U \cap |x_j|$ where $i \leq j$. By induction hypothesis we have $U, S_j \models \neg\phi$ which implies $U, S_i \not\models \Box\phi$. If $\Box\psi \in x$, where $U \cap |x_i| = \{\bar{x}\}$ then there exists $y_j \in |x_j|$ for all $i \leq j$ such that $x \xrightarrow{\Diamond} y_j$ because $|x_i| \xrightarrow{c} |x_j|$. We have that $\psi \in y_j$ and by induction hypothesis $U, S_j \models \psi$ therefore $U, S_i \models \Box\psi$.
- Suppose now that ϕ is of the form $\bigcirc\psi$ and $U, S_i \models \bigcirc\psi$. We have $U, S_{i+1} \models \psi$. If $U \cap |x_i| = \{\bar{x}\}$ and $U \cap |x_{i+1}| = \{\bar{y}\}$ then $\bar{x} \xrightarrow{\circ} \bar{y}$. By induction hypothesis, $\psi \in y$ and therefore $\bigcirc\psi \in x$, using the *Fun* Lemma.
  Now suppose $\bigcirc\psi \in x$. We have that for all z such that $\bar{x} \xrightarrow{\circ} \bar{z}$, $\psi \in z$. In particular this holds for y and by induction hypothesis $U, S_{i+1} \models \psi$. Therefore, $U, S_i \models \bigcirc\psi$.
- Suppose now that ϕ is of the form $\psi > \chi$ and $\psi > \chi \in x$ where $U \cap |x_i| = \{\bar{x}\}$. If $\Box\mathsf{K}\neg\psi \in x$ then by induction hypothesis and a few applications of the above cases we have $U, S_i \models \Box\mathsf{K}\neg\psi$ and therefore $U, S_i \models \psi > \chi$. So assume that $\Diamond\mathsf{L}\psi \in x$ and again by induction hypothesis $U, S_i \models \Diamond\mathsf{L}\psi$. We have that there exists $j \geq i$ such that $U, S_j \models \mathsf{L}\psi$ and assume that it is the least such. Observe that $\psi > \chi \in x$ implies that $\mathsf{K}(\phi \to \chi) \in x$ so by induction hypothesis $U, S_i \models \mathsf{K}(\phi \to \chi)$.
  If $i = j$ than $U, S_i \models \mathsf{K}(\phi \to \chi) \wedge \mathsf{L}\psi$ and therefore $U, S_i \models \psi > \chi$.
  If $i < j$ then we will show that $\mathsf{K}(\psi \to \chi) \in x^j$ by induction. Suppose that $\mathsf{K}(\psi \to \chi) \in x^k$ for $i \leq k < j$ and we will show that it holds for $k + 1$. We have that $U, S_k \models \mathsf{K}\neg\psi$ because j is the least index that it does not hold, therefore, by induction hypothesis, $\mathsf{K}\neg\psi \in x^k$ so $\bigcirc(\psi > \chi) \in x^k$ using Axiom 16.

We have that $x^k \overset{\circ}{\to} x^{k+1}$ therefore $\psi > \chi \in x^{k+1}$. Using Axiom 16 we have that $\mathsf{K}(\psi \to \chi) \in x^{k+1}$. Since $\mathsf{K}(\psi \to \chi) \in x^j$ we have $U, S_j \models \mathsf{K}(\psi \to \chi)$ using the induction hypothesis and few application of other cases. This implies that $U, S_i \models \psi > \chi$.
For the other direction suppose $U, S_i \models \psi > \chi$. If $U, S_j \models \mathsf{K}\neg\psi$ for $i \leq j$ then $U, S_i \models \Box\mathsf{K}\neg\psi$ and by induction hypothesis $\Box\mathsf{K}\neg\psi \in x$ where $U \cap |x_i| = \{\bar{x}\}$. We have $\phi > \psi \in x$. Now suppose that $U, S_j \models \mathsf{L}\psi$ for some $j \geq i$ and j is the minimal such index. We will show that $\phi > \psi \in x$ by backwards induction. We have that $U, S_j \models \mathsf{L}\psi \wedge \mathsf{K}(\psi \to \chi)$. By induction hypothesis and a few applications of the other cases $\mathsf{L}\psi \wedge \mathsf{K}(\psi \to \chi) \in x^j$ where $U \cap |x_i| = \{\bar{x^j}\}$ and therefore $\psi > \chi \in x^j$ using Axiom 16. Suppose now that there exist l such that $i \leq l < j$ and it holds for all k such that $l < k \leq j$. We will show that it is true for l as well. Observe that $U, S_l \models \mathsf{K}\neg\phi$ and therefore $U, S_l \models \mathsf{K}(\phi \to \chi)$. By induction hypothesis, we have that $\mathsf{K}(\phi \to \chi) \in x^l$. Since $\psi > \chi \in x^{j+1}$ we have $\bigcirc(\psi > \chi) \in x^j$. Applying Axiom 16, we have $\psi > \chi \in x^j$. ■

The following holds:

Corollary 22 *The axioms and rules of ESC are complete with respect to discrete systems of spheres.*

Corollary 23 **ECL** *has the finite model property and therefore it is decidable.*

14.4 The Ramsey Test Revisited

We will now show what version of Ramsey test our logic validates. The formulation of the Ramsey test requires a set of theories $\mathcal{B}$ and a revision operator $*$ of a theory with a formula such that the result is again a theory of in $\mathcal{B}$, that is, $\mathcal{B}$ is $*$-closed, such that

$$\text{(RT)} \qquad a > c \in B \;\text{ iff }\; c \in B * a.$$

As we mentioned above, because of the Gärdenfors' impossibility result, there is no non-trivial revision operator $*$ such that the above holds in **ECL**. Instead we will need to reformulate a more appropriate test.

Notice that the first problem we face is that a system of spheres is not closed under revisions. If $a > b \in th(U_i)$ then we should look for a sphere U_j such that $th(U_j) = th(U_i) * a$. The AGM Reflexivity postulate requires that $a \in Th(U_j)$ which implies that $x, U_j \models \mathsf{K}a$. However, this might never happen as for the satisfaction of $a > b$ we only require the existence of $x \in U_j$ such that $x, U_j \models \mathsf{L}a$. Taking those considerations into account we can formulate the following similar test

$$\text{(RTK)} \qquad a > c \in B \;\text{ iff }\; \mathsf{K}(a \to c) \in B * \mathsf{L}a,$$

where the revision operator $*$ returns the theory of the smallest sphere where $\mathsf{L}a$ is true, if such a sphere exists, or $\mathcal{L}$ otherwise. The above form of Ramsey test cannot be compared directly with the AGM style revision as it specifies only revision with the modal formula $\mathsf{L}a$. To remedy that, we will appeal to the meaning of the revision with $\mathsf{L}a$. The meaning of the formula $\mathsf{L}a$ is the belief on the possibility of a and revision with this possibility translates to drop the belief on $\neg a$. So revision with $\mathsf{L}a$ points directly to contraction $\ominus$ with $\neg a$. At first sight, we only need to translate the AGM contraction postulates to our language. For example, the postulate $K \ominus a \subseteq K$ translates to $B * \mathsf{L}a \subseteq B$. However, the latter does not hold in our framework. Belief sets are complete with respect to knowledge formulas, i.e. $\mathsf{L}a \in B$ or $\neg\mathsf{L}a \in B$. So, if, say, the latter holds, then $\mathsf{L}a \in B * \mathsf{L}a \subseteq B$, which is a contradiction, if B is consistent. The reason for this discrepancy is that the AGM contraction postulates were not formulated for a language that can express beliefs (or conditionals). Belief is only determined indirectly by acceptance in a belief set. Our way to get around this discrepancy is to interpret the contraction operator semantically, in our case with a set-theoretic operator. This way AGM postulates acquire a meaning independently of a particular language that can cause paradoxes as Gardenförs impossibility theorem. A translation appears in Table 14.2. The set theoretic contraction operator $\ominus$ is from $S \times \mathcal{P}(S)$ to S, where S is the system of spheres, and it is defined with

$$U_i \ominus A = \begin{cases} U_j \text{ if } \uparrow U_i \not\subseteq A \\ U_i \text{ if otherwise} \end{cases}$$

where $j = \min\{k : k \geq i \text{ and } U_k \notin A\}$ and $\uparrow U = \{U_j : U \subseteq U_j\}$.

It is easy to see that the subset contraction operator we defined satisfies the properties of Table 14.2. Those properties correspond to the standard AGM contraction postulates (see Table 14.3). Closure and Extensionality are assured by our framework. Inclusion corresponds to Rule 1. Vacuity corresponds to Rule 2. Success corresponds to Rule 3. Intersection and Conjunction to Rules 5 and 6, respectively. Unfortunately, Recovery does not hold in our framework. Recovery has been the most disputed AGM postulate for contraction and is not relevant to the Ramsey test.

Table 14.2 Set-theoretic contraction rules

1. $U \subseteq U \ominus A$
2. If $U \notin A$, then $U \ominus A = U$
3. If $\uparrow U \not\subseteq A$, then $U \ominus A \notin A$
4. $U \ominus (A \cap B) \subseteq U \ominus A \cap U \ominus B$
5. If $U \ominus (A \cup B) \not\subseteq A$ then $U \ominus A \subseteq U \ominus (A \cup B)$

Table 14.3 AGM Contraction postulates

$K \div p = \mathrm{Cn}(K \div p)$	Closure
$K \div p \subseteq K$	Inclusion
If $p \notin K$, then $K \div p = K$	Vacuity
If $p \notin \mathrm{Cn}(\emptyset)$, then $p \notin \mathrm{Cn}(K \div p)$	Success
If $p \leftrightarrow q \in \mathrm{Cn}(\emptyset)$, then $K \div p = K \div q$	Extensionality
$K \subseteq \mathrm{Cn}((K \div p) \cup \{p\})$	Recovery
$K \div p \cap K \div q \subseteq K \div (p \wedge q)$	Intersection
If $p \notin K \div (p \wedge q)$, then $K \div (p \wedge q) \subseteq K \div p$	Conjunction

Recall that for a given a knowledge formula a, we have $ex(a) \in \mathcal{P}(S)$. Now, the following Lemma follows from the above.

Lemma 24 *Let* (X, S, i) *a model of* **ECL**, *then*

1. *The system of spheres* S *is closed under contraction* $\ominus$,
2. *the contraction operator* $\ominus$ *satisfies the rules of Table* 14.2
3. *the following holds*

$$a > b \in th(U_i) \quad \textit{iff} \quad \mathsf{K}(a \to c) \in th(U_i \ominus ex(\neg a)).$$

We showed that it is possible to incorporate epistemic conditionals within the language and axiomatize them in a manner consistent with the Ramsey test. This shows that the Gärdenfors impossibility theorem can be avoided once conditionals are axiomatized within an epistemic framework and acquire a contextual meaning. The fact that the impossibility of a Ramsey correspondence does not arise in a concrete logical system, that was developed without any special efforts to avoid it, is particularly encouraging to an investigation of similar such conditional logics. A salient direction is the one that allows the incorporation of a revision operator within the language in ways that a Ramsey test can be a theorem of the system. We find that ECL is the right step toward this direction.

References

Adams, E. W. (1975). *The logic of conditionals: An application of probability to deductive logic*. Synthese Library. Berlin: Springer.

Battigalli, P., & Bonanno, G. (1997). The logic of belief persistence. *Economics and Philosophy*, *13*(01), 39–59.

Bonanno, G. (2007). Axiomatic characterization of the AGM theory of belief revision in a temporal logic. *Artificial Intelligence*, *171*(2–3), 144–160.

Braithwaite, R. B. (Ed.), (1931). *The foundations of mathematics and other logical essays*. International Library of Psychology, Philosophy, and Scientific Method. London: K. Paul, Trench, Trubner & Co., Ltd.

Dabrowski, A., Moss, L. S., & Parikh, R. (1996). Topological reasoning and the logic of knowledge. *Annals of Pure and Applied Logic*, *78*(1–3), 73–110.

Friedman, N., & Halpern, J. Y. (1994). Conditional logics of belief change. In B. Hayes-Roth, & R. E. Korf (Eds.), *AAAI* (pp. 915–921). Menlo Park, CA: AAAI Press/The MIT Press.

Gärdenfors, P. (1985). *Knowledge in flux*. Cambridge, MA: MIT Press.

Gärdenfors, P. (1986). Belief revision and the Ramsey test for conditionals. *Philosophical Review*, *95*, 81–93.

Georgatos, K. (1993). *Modal logics for topological spaces*. Ph.D. thesis. City University of New York.

Georgatos, K. (2010). Conditioning by minimizing accessibility. In G. Bonanno, B. Löwe, & W. Hoek (Eds.), *Logic and the foundations of game and decision theory LOFT 8, lecture notes in computer science* (Vol. 6006, pp. 20–33). Berlin: Springer.

Giordano, L., Gliozzi, V., & Olivetti, N. (2001). Belief revision and the ramsey test: A solution. In F. Esposito (Ed.), *AI*IA, Lecture Notes in Computer Science* (Vol. 2175, pp. 165–175). Berlin: Springer.

Goldblatt, R. I. (1992). *Logics of time and computation*. CSLI lecture notes. Center for the Study of Language and Information.

Grahne, G. (1991). Updates and counterfactuals. In *Proceedings of the second international conference on principles of knowledge representation and reasoning* (pp. 269–276). Morgan Kaufmann.

Heinemann, B. (1998). Topological nexttime logic. In H. Wansing M. Kracht, M. de Rijke, M. Zakharyaschev (Eds.), Advances in modal logic '96 (pp. 99–113). Stanford: Center for the Study of Language and Information.

Heinemann, B. (1999a). On sets growing continuously. In C. Pandu Rangan, V. Raman, R. Ramanujam (Eds.), *Foundations of software technology and theoretical computer science, lecture notes in computer science* (Vol. 1738, pp. 420–431). Berlin: Springer.

Heinemann, B. (1999b). Temporal aspects of the modal logic of subset spaces. *Theoretical Computer Science*, *224*(1–2), 135–155.

Heinemann, B. (2000). Extending topological nexttime logic. In *Proceedings of the seventh international workshop on temporal representation and reasoning (TIME'00)* (pp. 87–94). Washington, DC: IEEE Computer Society.

Heinemann, B. (2007). A PDL-like logic of knowledge acquisition. In V. Diekert, M. V. Volkov, & A. Voronkov (Eds.), *CSR, lecture notes in computer science* (Vol. 4649, pp. 146–157). Berlin: Springer.

Lewis, D. (1973). *Counterfactuals*. Cambridge, MA: Harvard University Press.

Lindström, S. (1996). The Ramsey test and the indexicality of conditionals: A proposed resolution of Gärdenfors' paradox. In A. Fuhrmann & H. Rott (Eds.), *Logic, action, and information: Essays on logic in philosophy and artificial intelligence* (pp. 208–228). Berlin: Walter de Gruyter.

Lindström, S., & Rabinowicz, W. (1995). The ramsey test revisited. In G. Crocco, L. Fariñas Del Cerro, & A. Herzig (Eds.), *Conditionals: From philosophy to computer science* (pp. 147–192). Oxford: Oxford University Press.

Moss, L. S., & Parikh, R. (1992). Topological reasoning and the logic of knowledge. In Y. Moses (Ed.), *Proceedings of the fourth conference (TARK 1992)* (pp. 95–105).

Nute, D. (1984). Conditional logic. In D. Gabbay & F. Guenthner (Eds.), *Handbook of philosophical logic: Extensions of classical logic* (Vol. II, pp. 387–439). Dordrecht: Reidel.

Rott, H. (1986). Ifs, though, and because. *Erkenntnis*, *25*(3), 345–370.

Ryan, M., & Schobbens, P.-Y. (1997). Counterfactuals and updates as inverse modalities. *Journal of Logic, Language and Information*, *6*(2), 123–146.

Schmidt, R. A., & Tishkovsky, D. (2008). On combinations of propositional dynamic logic and doxastic modal logics. *Journal of Logic, Language and Information*, *17*(1), 109–129.

Stalnaker, R. (1968). A theory of conditionals. In N. Rescher (Ed.), *Studies in logical theory*. Oxford: Oxford University Press.

Chapter 15
On the Coalgebraic Theory of Kleene Algebra with Tests

Dexter Kozen

In honor of Rohit Parikh

Abstract We develop a coalgebraic theory of Kleene algebra with tests (KAT) along the lines of Rutten (1998) for Kleene algebra (KA) and Chen and Pucella (Electron Notes Theor Comput Sci 82(1), 2003) for a limited version of KAT, resolving some technical issues raised by Chen and Pucella. Our treatment includes a simple definition of the Brzozowski derivative for KAT expressions and an automata-theoretic interpretation involving automata on guarded strings. We also give a complexity analysis, showing that an efficient implementation of coinductive equivalence proofs in this setting is tantamount to a standard automata-theoretic construction. It follows that coinductive equivalence proofs can be generated automatically in *PSPACE*. This matches the bound of Worthington (2008) for the automatic generation of equational proofs in KAT.

Keywords Kleene algebra · Kleene algebra with tests · Coalgebra · Verification

15.1 Introduction

Kleene algebra (KA) is the algebra of regular expressions. The operations $+$, $\cdot$, and * of Kleene algebra can be used to model nondeterministic choice, sequential composition, and iteration, respectively, on a set of actions.

Kleene algebra with tests (KAT) is an extension of KA obtained by identifying a subset of *tests*, which must satisfy the axioms of Boolean algebra as well. The

---

D. Kozen (✉)
Computer Science Department, Cornell University, Ithaca, NY 14853-7501, USA
e-mail: kozen@cs.cornell.edu

C. Başkent et al. (eds.), *Rohit Parikh on Logic, Language and Society*,
Outstanding Contributions to Logic 11, DOI 10.1007/978-3-319-47843-2_15

two sorts, actions and tests, interact seamlessly: on tests, nondeterministic choice becomes join and sequential composition becomes meet.

The presence of tests allows KAT to model basic programming and verification constructs such as conditional tests, while loops, and Hoare triples. Thus KAT gives a simple equational approach to partial correctness. By now KAT has a well-developed theory, including an extensive model theory and results on expressiveness, deductive completeness, and complexity. It has been applied successfully in a number of areas, including the verification of compiler optimizations and communication protocols and various other program analysis tasks.

Traditionally, KAT is axiomatized equationally (Kozen 1997). In Chen and Pucella (2003), Chen and Pucella develop a coalgebraic theory of KAT inspired by Rutten's coalgebraic theory of KA based on deterministic automata (Rutten 1998). Remarking that "the known automata-theoretic presentation of KAT (Kozen 2003) does not lend itself to a coalgebraic theory," and that "the notion of derivative, essential to the coinduction proof principle in this context, is not readily definable for KAT expressions as defined in Kozen (1997)," Chen and Pucella develop a new interpretation of KAT expressions and a corresponding automata theory differing from Kozen (2003) in several respects. They give a coinductive proof principle and show how it can be used to establish equivalence of expressions. This gives an alternative to equational proofs using the standard axiomatization (Kozen 1997) or by minimization of deterministic automata (Kozen 2003).

The ability to generate equivalence proofs automatically has important implications for proof-carrying code. Chen and Pucella argue that the coalgebraic approach makes this possible, since proofs can be generated purely mechanically via repeated application of the Brzozowski derivative, whereas classical equational logic "requires creativity" (Chen and Pucella 2003). This is not strictly true, as Worthington (Worthington 2008) has shown that equational proofs can also be generated automatically. However, it is fair to say that the coinductive approach does provide a more natural mechanism.

Still unresolved is the issue of proof complexity in the coinductive setting. Chen and Pucella claim that coinduction can give shorter proofs, but they give no supporting evidence. Worthington's technique is known to require *PSPACE* and to produce exponential-size proofs in the worst case. This worst-case bound is unlikely to be significantly improved, as the equational theory of KAT is *PSPACE*-complete (Cohen et al. 1996).

Chen and Pucella's treatment has a few technical shortcomings, as they themselves point out. In their words:

> The "path independence" of a mixed automaton gives any mixed automaton a certain form of redundancy. This redundancy persists in the definition of bisimulation... An open question is to cleanly eliminate this redundancy; a particular motivation for doing this would be to make proofs of expression equivalence as simple as possible. Along these lines, it would be of interest to develop other weaker notions of bisimulation that give rise to bisimulations; pseudo-bisimulations require a sort of "fixed variable ordering" that does not seem absolutely necessary...

> Another issue for future work would be to give a class of expressions wider than our mixed expressions for which there are readily understandable and applicable rules for computing derivatives. In particular, a methodology for computing derivatives of the KAT expressions defined by Kozen (1997) would be nice to see. Intuitively, there seems to be a tradeoff between the expressiveness of the regular expression language and the simplicity of computing derivatives (in the context of KAT). Formal tools for understanding this tradeoff could potentially be quite useful (Chen and Pucella 2003).

This paper addresses these issues. We develop a coalgebraic theory of KAT, which we call KCT, along the lines of Chen and Pucella (2003); Rutten (1998). Our treatment includes a new definition of the Brzozowski derivative, but in the context of the original automata-theoretic formulation of KAT involving automata on guarded strings (Kozen 2003). The syntactic form of the Brzozowski derivative applies to all KAT expressions as defined in Kozen (1997). The somewhat artificial concepts of path independence, fixed variable ordering, and pseudo-bisimulation do not arise in this setting. This treatment places KCT within the general coalgebraic framework described by Bonsangue et al. (2007, 2009).

We also give a complexity analysis of the coinductive proof principle. We show that an efficient implementation is tantamount to the construction of nondeterministic automata from the given expressions by a Kleene construction, determinizing the two automata by a standard subset construction, and constructing a bisimulation on states of the resulting deterministic automata. It follows that coinductive equivalence proofs can be generated automatically in *PSPACE*. This matches Worthington's bound (Worthington 2008) for equational proofs.

15.2 KA and KAT

15.2.1 Kleene Algebra

Kleene algebra (KA) is the algebra of regular expressions (Conway 1971; Kleene 1956). The axiomatization used here is from Kozen (1994). A *Kleene algebra* is a structure $(K, +, \cdot, ^*, 0, 1)$ such that K is an idempotent semiring under $+$, $\cdot$, 0, and 1 and satisfies the axioms

$$1 + pp^* \le p^* \qquad q + pr \le r \Rightarrow p^*q \le r$$
$$1 + p^*p \le p^* \qquad q + rp \le r \Rightarrow qp^* \le r$$

for *. There is a natural partial order $p \le q \overset{\text{def}}{\Longleftrightarrow} p + q = q$.

Standard models include the family of regular sets over a finite alphabet, the family of binary relations on a set, and the family of $n \times n$ matrices over another Kleene algebra. Other more unusual interpretations include the min,+ algebra, also known as the *tropical semiring*, used in shortest path algorithms, and models consisting of convex polyhedra used in computational geometry.

The completeness result of Kozen (1994) says that the algebra of regular sets of strings over a finite alphabet Σ is the free Kleene algebra on generators Σ. The axioms are also complete for the equational theory of relational models.

15.2.2 Kleene Algebra with Tests

A *Kleene algebra with tests* (KAT) (Kozen 1997) consists of a Kleene algebra K with an embedded Boolean algebra B such that the semiring structure on B is a subalgebra of the semiring structure on K. Elements of B are called *tests*. The Boolean negation operator is defined only on tests.

Like KA, KAT has language and relational models and is deductively complete over these interpretations (Kozen and Smith 1996). The chief language-theoretic models are the regular sets of *guarded strings* over alphabets Σ and T of primitive action and test symbols, respectively (see Sect. 15.3.1). This is the free KAT on generators Σ, T. The set of guarded strings represented by a KAT expression e is denoted $\mathsf{GS}(e)$.

KAT can code elementary programming constructs and Hoare partial correctness assertions and subsumes propositional Hoare logic (PHL). It is deductively complete over relational models, whereas PHL is not. Moreover, KAT is no more difficult to decide, as PHL, KA, and KAT are all *PSPACE*-complete.

For KAT expressions e_1, e_2, we write $e_1 \leq e_2$ if this inequality holds in the free KAT on generators Σ, T; that is, if it is a consequence of the axioms of KAT.

See Kozen (1994, 1997, 2000) for a more detailed introduction.

15.3 Automata on Guarded Strings

Automata on guarded strings (AGS), also known as automata with tests, were introduced in Kozen (2003). They are a generalization of ordinary finite-state automata to include tests. An ordinary automaton with null transitions is an AGS over the two-element Boolean algebra.

15.3.1 Guarded Strings

Guarded strings were first introduced in Kaplan (1969). A *guarded string* over Σ, T is an alternating sequence $\alpha_0 p_1 \alpha_1 p_2 \alpha_2 \cdots p_n \alpha_n$, $n \geq 0$, where $p_i \in \Sigma$ and the α_i are atoms (minimal nonzero elements) of the free Boolean algebra B generated by T. The set of atoms is denoted At. The elements of At can be regarded either as conjunctions of literals of T (elements of T or their negations) or as truth assignments to T. A *guarded string* is thus an element of $\mathsf{At} \cdot (\Sigma \cdot \mathsf{At})^*$. The set of all guarded strings is denoted GS. Guarded strings represent the join-irreducible elements of the free KAT on generators Σ, T.

15.3.2 Nondeterministic Automata

A nondeterministic AGS consists of a labeled directed graph with two types of transitions, *action transitions* labeled with elements of Σ and *test transitions* labeled with elements of B. There is a set start of *start states* and a set accept of *accept states*.

An input to an AGS is a guarded string $\alpha_0 p_1 \alpha_1 p_2 \alpha_2 \cdots p_n \alpha_n$. Intuitively, it operates as follows. It starts with a pebble on a nondeterministically chosen start state and its input head scanning α_0. In the course of the computation, say the pebble is occupying state s and the input head is scanning α_i. If $i < n$, it may read the next action symbol p_{i+1} from the input string and move the pebble to any nondeterministically chosen state t such that there is an action transition from s to t with label p_{i+1}. The input head is advanced past p_{i+1} in the input string and is now scanning α_{i+1}. Also while scanning α_i, it may slide the pebble along an enabled test transition at any time without advancing the input head. A test transition is *enabled* if $\alpha_i \leq b$, where b is the label of the transition. The automaton accepts if it is ever scanning α_n while the pebble is on an accept state. Thus the automaton accepts a guarded string x if there is a directed path π from start to accept such that $x \leq e$, where e is the product of the labels of the edges along π.

Formally, a (nondeterministic) *automaton on guarded strings* (AGS) over Σ and T is a tuple

$$M = (Q, \Delta, \mathsf{start}, \mathsf{accept}),$$

where Q is a set of *states*, $\mathsf{start} \subseteq Q$ are the *start states*, $\mathsf{accept} \subseteq Q$ are the *accept states*, and Δ is the *transition function*

$$\Delta : (\Sigma + \mathsf{At}) \to Q \to 2^Q,$$

where $+$ denotes disjoint (marked) union.

The definition of acceptance involves the Kleisli composition $\bullet$ and asterate † operations on maps $Q \to 2^Q$ defined by:

$$R \bullet S \stackrel{\text{def}}{=} s \mapsto \bigcup_{t \in S(s)} R(t) \qquad R^0 \stackrel{\text{def}}{=} s \mapsto \{s\}$$

$$R^\dagger \stackrel{\text{def}}{=} \bigcup_{n \geq 0} R^n \qquad R^{n+1} \stackrel{\text{def}}{=} R \bullet R^n.$$

The map Δ generates a map

$$\hat{\Delta} : (\Sigma + \mathsf{At}) \to Q \to 2^Q \qquad \hat{\Delta}_\alpha \stackrel{\text{def}}{=} \Delta_\alpha^\dagger \qquad \hat{\Delta}_p \stackrel{\text{def}}{=} \Delta_p.$$

Intuitively, $\hat{\Delta}_\alpha(s)$ accumulates all states accessible from s by a sequence of test transitions enabled under α. The map $\hat{\Delta}$ extends further to a monoid homomorphism

$$\hat{\Delta} : (\Sigma + \mathsf{At})^* \to Q \to 2^Q \qquad \hat{\Delta}_\varepsilon \stackrel{\text{def}}{=} s \mapsto \{s\} \qquad \hat{\Delta}_{xy} \stackrel{\text{def}}{=} \hat{\Delta}_y \bullet \hat{\Delta}_x$$

from the free monoid $(\Sigma + \mathsf{At})^*$ to the monoid $Q \to 2^Q$ under Kleisli composition. The guarded strings $\mathsf{GS} = \mathsf{At} \cdot (\Sigma \cdot \mathsf{At})^*$ form a submonoid of $(\Sigma + \mathsf{At})^*$. The automaton M *accepts* $x \in \mathsf{GS}$ if there exist $s \in \mathsf{start}$ and $t \in \mathsf{accept}$ such that $t \in \hat{\Delta}_x(s)$. The set of guarded strings accepted by M is denoted $\mathsf{GS}(M)$.

15.3.3 Deterministic Automata

The definition of deterministic AGS here differs from that of Kozen (2003) so as to conform to the coalgebraic structure to be introduced in Sect. 15.4, but the difference is inessential. In Kozen (2003) the set of states of a deterministic AGS is separated into disjoint sets of *action states* and *test states*, whereas here we have elided the action states.

A *deterministic automaton on guarded strings* (AGS) over Σ and T is a structure

$$M = (Q, \delta, \varepsilon, \mathsf{start}),$$

where Q is a set of *states*, $\mathsf{start} \in Q$ is the *start state*, and

$$\delta : \mathsf{At} \cdot \Sigma \to Q \to Q \qquad \varepsilon : \mathsf{At} \to Q \to 2$$

with components

$$\delta_{\alpha p} : Q \to Q \qquad \varepsilon_\alpha : Q \to 2$$

for $\alpha \in \mathsf{At}$ and $p \in \Sigma$. The components ε_α play the same role as the accept states in a nondeterministic automaton.

Define the function $L : Q \to \mathsf{GS} \to 2$ inductively as follows:

$$L(u)(\alpha) \stackrel{\text{def}}{=} \varepsilon_\alpha(u) \qquad L(u)(\alpha p y) \stackrel{\text{def}}{=} L(\delta_{\alpha p}(u))(y), \tag{15.1}$$

where $y \in \mathsf{GS}$, $\alpha \in \mathsf{At}$, and $p \in \Sigma$. The machine is said to *accept* $x \in \mathsf{GS}$ if $L(\mathsf{start})(x) = 1$. The set of guarded strings accepted by M is denoted $\mathsf{GS}(M)$. Identifying a subset of GS with its characteristic function $\mathsf{GS} \to 2$, we can write $\mathsf{GS}(M) = L(\mathsf{start})$.

The map δ extends to a monoid homomorphism $\hat{\delta} : (\mathsf{At} \cdot \Sigma)^* \to Q \to Q$ from the free monoid $(\mathsf{At} \cdot \Sigma)^*$ to the monoid $Q \to Q$ under functional composition.

15.3.4 *Determinization*

Determinization is effected by a subset construction similar to that for ordinary automata. Given a nondeterministic AGS

$$N = (Q, \Delta, \mathsf{start}, \mathsf{accept}),$$

there is an equivalent deterministic AGS

$$M = (2^Q, \delta, \varepsilon, \mathsf{start}),$$

where for $A \subseteq Q$,

$$\varepsilon_\alpha(A) \stackrel{\text{def}}{=} \begin{cases} 1, & \text{if } \exists s \in A\ \exists t \in \mathsf{accept}\ t \in \hat{\Delta}_\alpha(s), \\ 0, & \text{otherwise} \end{cases} \qquad \delta_{\alpha p}(A) \stackrel{\text{def}}{=} \bigcup_{s \in A} \hat{\Delta}_{\alpha p}(s).$$

One can show by a straightforward induction on the length of $x \in \mathsf{GS}$ that for all $A \subseteq Q$,

$$L(A)(x) = \begin{cases} 1, & \text{if } \exists s \in A\ \exists t \in \mathsf{accept}\ t \in \hat{\Delta}_x(s), \\ 0, & \text{otherwise;} \end{cases}$$

in particular,

$$L(\mathsf{start})(x) = 1 \Leftrightarrow \exists s \in \mathsf{start}\ \exists t \in \mathsf{accept}\ t \in \hat{\Delta}_x(s).$$

As these are exactly the acceptance criteria for M and N respectively, the two machines accept the same set of guarded strings.

15.4 Kleene Coalgebra with Tests (KCT)

A Kleene coalgebra with tests (KCT) is very much like Kleene coalgebra (KC) (Rutten 1998), but with the addition of Boolean tests. Formally, a *Kleene coalgebra with tests* (KCT) over Σ and T is a structure

$$M = (Q, \delta, \varepsilon),$$

where Q is a set of *states* and

$$\delta : \mathsf{At} \cdot \Sigma \to Q \to Q \qquad\qquad \varepsilon : \mathsf{At} \to Q \to 2$$

for $\alpha \in \mathsf{At}$ and $p \in \Sigma$, exactly as in deterministic automata on guarded strings. Thus we can view a KCT as simply a deterministic AGS without a designated start state.

A KCT *morphism* $h : (Q, \delta, \varepsilon) \to (Q', \delta', \varepsilon')$ is a set map $h : Q \to Q'$ that commutes with δ, δ' and $\varepsilon, \varepsilon'$; that is,

$$\delta'_{\alpha p}(h(u)) = h(\delta_{\alpha p}(u)) \qquad\qquad \varepsilon'_{\alpha}(h(u)) = \varepsilon_{\alpha}(u).$$

We denote the category of KCTs and KCT morphisms over Σ and T also by KCT.

15.4.1 The Brzozowski Derivative, Semantic Form

There is a natural KCT over Σ and T defined in terms of the Brzozowski derivative on sets of guarded strings. The traditional Brzozowski derivative (Brzozowski 1964) is a kind of residuation operator on sets of ordinary strings. The current form is quite similar, except that we extend the definition to accommodate tests.

We define two maps

$$D : \mathsf{At} \cdot \Sigma \to 2^{\mathsf{GS}} \to 2^{\mathsf{GS}} \qquad\qquad E : \mathsf{At} \to 2^{\mathsf{GS}} \to 2,$$

where for $R \subseteq \mathsf{GS}$,

$$D_{\alpha p}(R) \stackrel{\text{def}}{=} \{y \in \mathsf{GS} \mid \alpha p y \in R\} \qquad\qquad E_{\alpha}(R) \stackrel{\text{def}}{=} \begin{cases} 1, & \text{if } \alpha \in R, \\ 0, & \text{if } \alpha \notin R. \end{cases}$$

It is clear that the structure

$$\mathsf{Brz} \stackrel{\text{def}}{=} (2^{\mathsf{GS}}, D, E)$$

forms a KCT. Indeed, it is the final object in the category KCT: for any KCT $M = (Q, \delta, \varepsilon)$, the function $L : Q \to 2^{\mathsf{GS}}$ defined in (15.1) is the unique KCT morphism $L : M \to \mathsf{Brz}$.

15.4.2 The Brzozowski Derivative, Syntactic Form

As with Brzozowski's original formulation (Brzozowski 1964), there is also a syntactic form of the Brzozowski derivative defined on KAT expressions. Let Exp denote the set of KAT expressions over Σ and T. We define a family of derivative operators

$$D : \mathsf{At} \cdot \Sigma \to \mathsf{Exp} \to \mathsf{Exp} \qquad\qquad E : \mathsf{At} \to \mathsf{Exp} \to \mathsf{2}$$

consisting of components

$$D_{\alpha p} : \mathsf{Exp} \to \mathsf{Exp} \qquad\qquad E_\alpha : \mathsf{Exp} \to \mathsf{2}$$

defined inductively as follows. For $\alpha \in \mathsf{At}$, $p, q \in \Sigma$, and $b \in B$,

$$\begin{aligned} D_{\alpha p}(e_1 + e_2) &\stackrel{\text{def}}{=} D_{\alpha p}(e_1) + D_{\alpha p}(e_2) \\ D_{\alpha p}(e_1 e_2) &\stackrel{\text{def}}{=} D_{\alpha p}(e_1) \cdot e_2 + E_\alpha(e_1) \cdot D_{\alpha p}(e_2) \\ D_{\alpha p}(e^*) &\stackrel{\text{def}}{=} D_{\alpha p}(e) \cdot e^* \end{aligned} \qquad \begin{aligned} D_{\alpha p}(q) &\stackrel{\text{def}}{=} \begin{cases} 1, & \text{if } p = q, \\ 0, & \text{otherwise,} \end{cases} \\ D_{\alpha p}(b) &\stackrel{\text{def}}{=} 0. \end{aligned}$$

$$\begin{aligned} E_\alpha(e_1 + e_2) &\stackrel{\text{def}}{=} E_\alpha(e_1) + E_\alpha(e_2) \\ E_\alpha(e_1 e_2) &\stackrel{\text{def}}{=} E_\alpha(e_1) \cdot E_\alpha(e_2) \\ E_\alpha(e^*) &\stackrel{\text{def}}{=} 1 \end{aligned} \qquad \begin{aligned} E_\alpha(b) &\stackrel{\text{def}}{=} \begin{cases} 1, & \text{if } \alpha \le b, \\ 0, & \text{otherwise,} \end{cases} \\ E_\alpha(q) &\stackrel{\text{def}}{=} 0. \end{aligned}$$

These operators on KAT expressions are collectively called the *syntactic Brzozowski derivative*.

The map E_α is just the evaluation morphism that for any KAT expression substitutes 0 for any $p \in \Sigma$, 1 for any $b \in T$ such that $\alpha \le b$, and 0 for any $b \in T$ such that $\alpha \le \bar{b}$, then simplifies the resulting expression over the two-element Kleene algebra 2. It is easily shown that for any KAT expression e,

$$E_\alpha(e) = \begin{cases} 1, & \text{if } \alpha \le e, \\ 0, & \text{if } \alpha \not\le e \end{cases} = \begin{cases} 1, & \text{if } \alpha \in \mathsf{GS}(e), \\ 0, & \text{if } \alpha \notin \mathsf{GS}(e). \end{cases}$$

The structure (Exp, D, E) is a KCT in the sense of Sect. 15.4, thus there is a unique KCT morphism $L : \mathsf{Exp} \to \mathsf{Brz}$ to the final coalgebra Brz defined in (15.1). We will show that $L(e) = \mathsf{GS}(e)$, where GS is the traditional interpretation of KAT expressions mentioned in Sect. 15.2.2.

Lemma 15.4.1 *For all* $\alpha \in \mathsf{At}$, $p \in \Sigma$, *and* $e, e' \in \mathsf{Exp}$,

$$\alpha p e' \le e \Leftrightarrow e' \le D_{\alpha p}(e).$$

Proof For the forward implication,

$$D_{\alpha p}(\alpha p e') = D_{\alpha p}(\alpha) p e' + E_\alpha(\alpha) D_{\alpha p}(p) e' + E_\alpha(\alpha) E_\alpha(p) D_{\alpha p}(e') = e'.$$

By monotonicity of $D_{\alpha p}$,

$$\alpha pe' \le e \Rightarrow e' = D_{\alpha p}(\alpha pe') \le D_{\alpha p}(e).$$

For the reverse implication, it suffices to show $\alpha p D_{\alpha p}(e) \le e$. We proceed by induction on the structure of e. For $p \in \Sigma$,

$$\alpha p D_{\alpha p}(p) = \alpha p \le p.$$

For the case $e_1 e_2$,

$$\begin{aligned}\alpha p D_{\alpha p}(e_1 e_2) &= \alpha p D_{\alpha p}(e_1) e_2 + \alpha p E_\alpha(e_1) D_{\alpha p}(e_2)\\ &= \alpha p D_{\alpha p}(e_1) e_2 + \alpha E_\alpha(e_1) \alpha p D_{\alpha p}(e_2)\\ &\le e_1 e_2.\end{aligned}$$

For the case e^*,

$$\alpha p D_{\alpha p}(e^*) = \alpha p D_{\alpha p}(e) e^* \le e e^* \le e^*.$$

All other cases are equally straightforward. □

Theorem 15.4.2 *For all* KAT *expressions* e, $\mathsf{GS}(e) = L(e)$. *Thus the set accepted by the automaton* (Exp, D, E, e) *is* $\mathsf{GS}(e)$.

Proof We wish to show that for all $x \in \mathsf{GS}$, $x \in \mathsf{GS}(e)$ iff $L(e)(x) = 1$. By the completeness theorem for KAT (Kozen and Smith 1996), we have $x \in \mathsf{GS}(e)$ iff $x \le e$, so it suffices to show that $x \le e$ iff $L(e)(x) = 1$. We proceed by induction on the length of x. The basis for x an atom α is immediate from the definition of E_α. For $x = \alpha p y$, by Lemma 15.4.1,

$$\alpha p y \le e \Leftrightarrow y \le D_{\alpha p}(e) \Leftrightarrow L(D_{\alpha p}(e))(y) = 1 \Leftrightarrow L(e)(apy) = 1.$$

□

15.5 Completeness

15.5.1 *Bisimulation on* KCTs

A *bisimulation* between two KCTs $M = (Q, \delta, \varepsilon)$ and $M' = (Q', \delta', \varepsilon')$ is a binary relation $\equiv$ between Q and Q' such that if $s \in Q$, $t \in Q'$, and $s \equiv t$, then for all $\alpha \in \mathsf{At}$ and $p \in \Sigma$,

(i) $\varepsilon_\alpha(s) = \varepsilon'_\alpha(t)$; and
(ii) $\delta_{\alpha p}(s) \equiv \delta'_{\alpha p}(t)$.

Lemma 15.5.1 *The relation*

$$s \mathrel{\hat{\equiv}} t \stackrel{\text{def}}{\Longleftrightarrow} L(s) = L(t)$$

is the unique maximal bisimulation between M and M'.

Proof It is easily shown that $\hat{\equiv}$ satisfies (i) and (ii). Moreover, if $\equiv$ is any relation satisfying (i) and (ii), one can show by a straightforward inductive argument that $\equiv$ refines $\hat{\equiv}$, thus $\hat{\equiv}$ is the unique maximal relation satisfying (i) and (ii). □

An *autobisimulation* is a bisimulation between M and itself. Bisimulations are closed under relational composition and arbitrary union, and the identity relation is an autobisimulation. Thus the reflexive, symmetric, and transitive closure of an autobisimulation is again an autobisimulation. An autobisimulation that is so closed is called a KCT-*congruence*. KCT-congruences are exactly the kernels of KCT-morphisms.

A KCT is bisimilar to its quotient by any KCT-congruence under the map $\{(s, [s]) \mid s \in Q\}$, where $[s]$ is the KCT-congruence class of s. The quotient by the unique maximal autobisimulation is a sub-coalgebra of Brz, the final coalgebra.

15.5.2 Bisimulation on Deterministic Automata

For deterministic automata, we add an extra condition. A *bisimulation* between two deterministic AGS $M = (Q, \delta, \varepsilon, \mathsf{start})$ and $M' = (Q', \delta', \varepsilon', \mathsf{start}')$ is a bisimulation $\equiv$ between the underlying KCTs (Q, δ, ε) and $(Q', \delta', \varepsilon')$ such that $\mathsf{start} \equiv \mathsf{start}'$. Two automata are *bisimilar* if there exists a bisimulation between them.

Lemma 15.5.2 *M and M' are bisimilar iff $\mathsf{GS}(M) = \mathsf{GS}(M')$.*

Proof Let $\hat{\equiv}$ be the relation defined in the proof of Lemma 15.5.1. If $\mathsf{GS}(M) = \mathsf{GS}(M')$, then $L(\mathsf{start}) = L(\mathsf{start}')$ by the definition of acceptance, therefore $\mathsf{start} \mathrel{\hat{\equiv}} \mathsf{start}'$. Then M and M' are bisimilar under $\hat{\equiv}$.

Conversely, if there exists a bisimulation $\equiv$ between M and M', then $\mathsf{start} \equiv \mathsf{start}'$, and by Lemma 15.5.1, $\equiv$ refines $\hat{\equiv}$, therefore $\mathsf{start} \mathrel{\hat{\equiv}} \mathsf{start}'$. Thus $\hat{\equiv}$ is a bisimulation of automata. □

The quotient of an automaton by its unique maximal autobisimulation gives the unique minimal equivalent automaton (ignoring inaccessible states).

Theorem 15.5.3 (Completeness) *The following are equivalent:*

(i) the automata (Exp, D, E, e) *and* (Exp, D, E, e') *are bisimilar;*
(ii) $L(e) = L(e')$;
(iii) $\mathsf{GS}(e) = \mathsf{GS}(e')$;

(iv) e and e′ are equivalent modulo the axioms of KAT.

Proof The equivalence of (i)–(iii) follows from Theorem 15.4.2 and Lemma 15.5.2. The equivalence of (iii) and (iv) are just the soundness and completeness of KAT for the guarded string model (Kozen and Smith 1996). □

15.6 Complexity

Let Exp_e denote the subautomaton of (Exp, D, E, e) consisting of those expressions that are accessible from e; that is, those expressions of the form $\hat{D}_x(e)$ for some $x \in (\mathsf{At} \cdot \Sigma)^*$. Theorem 15.5.3 by itself is not very useful as a deductive system or decision procedure for equivalence, because Exp_e is not a finite system in general. However, equivalent finite systems exist. In particular, by Theorem 15.5.3, KAT equivalence is the maximal congruence on Exp. The quotient with respect to this relation, ignoring inaccessible states, gives the minimal deterministic AGS accepting $\mathsf{GS}(e)$, which is finite since $\mathsf{GS}(e)$ is regular.

Unfortunately, to construct this automaton directly, we would need an independent algorithm to decide KAT equivalence. However, we can obtain finite automata with finer congruences that are easier to decide than full KAT equivalence. Chen and Pucella (2003) use equivalence modulo additive associativity, commutativity, and idempotence (ACI-equivalence). Here we consider equivalence modulo the axioms of idempotent commutative monoids for $+$, 0 and the axioms

$$1 \cdot x = x \qquad 0 \cdot x = 0 \qquad (x + y) \cdot z = xz + yz. \tag{15.2}$$

Multiplicative associativity is not assumed, nor is left distributivity. We might call structures satisfying these axioms *right presemirings*. We denote by $\approx$ the KAT-congruence on terms generated by these axioms. We will show that $\mathsf{Exp}_e/{\approx}$ has finitely many accessible classes. It is a coarser relation than ACI-equivalence, therefore has fewer classes, but is still easy to decide, as there are normal forms up to additive commutativity. Of course, it makes the most sense to use the coarsest relation possible that is easily decidable, because coarser relations give smaller automata.

Because there are only finitely many $\approx$-classes accessible from e, the quotient automaton $\mathsf{Exp}_e/{\approx}$ is finite, and we can use it to obtain finite coinductive equivalence proofs. More interestingly, we will also show that $\mathsf{Exp}_e/{\approx}$ is a homomorphic image of a deterministic automaton M_e obtained by creating a nondeterministic AGS N_e from the expression e by a Kleene construction, then determinizing N_e by a subset construction as described in Sect. 15.3.4. This characterization gives a bound on the size of $\mathsf{Exp}_e/{\approx}$, which we can then use to argue that coinductive equivalence proofs can be generated automatically in *PSPACE*.

Lemma 15.6.1 *The relation* $\approx$ *is a* KCT*-congruence on* Exp.

Proof We must show that if $e \approx e'$, then $E_\alpha(e) = E_\alpha(e')$ and $D_{\alpha p}(e) \approx D_{\alpha p}(e')$. The first conclusion follows from Theorem 15.5.3 and the fact that $\approx$ refines KAT-equivalence.

For the additive axioms of idempotent commutative monoids, the second conclusion follows from the additivity of $D_{\alpha p}$.

For the axioms (15.2),

$$\begin{aligned}
D_{\alpha p}(1x) &= D_{\alpha p}(1)x + E_\alpha(1)D_{\alpha p}(x) = 0x + 1D_{\alpha p}(x) \approx D_{\alpha p}(x) \\
D_{\alpha p}(0x) &= D_{\alpha p}(0)x + E_\alpha(0)D_{\alpha p}(x) = 0x + 0D_{\alpha p}(x) \approx D_{\alpha p}(0) \\
D_{\alpha p}((x+y)z) &= (D_{\alpha p}(x) + D_{\alpha p}(y))z + (E_\alpha(x) + E_\alpha(y))D_{\alpha p}(z) \\
&\approx D_{\alpha p}(x)z + E_\alpha(x)D_{\alpha p}(z) + D_{\alpha p}(y)z + E_\alpha(y)D_{\alpha p}(z) \\
&= D_{\alpha p}(xz + yz).
\end{aligned}$$

Finally, we must show that if $e_1 \approx e_2$, then $D_{\alpha p}(e_1 + e_3) \approx D_{\alpha p}(e_2 + e_3)$, $D_{\alpha p}(e_1 e_3) \approx D_{\alpha p}(e_2 e_3)$, $D_{\alpha p}(e_3 e_1) \approx D_{\alpha p}(e_3 e_2)$, and $D_{\alpha p}(e_1^*) \approx D_{\alpha p}(e_2^*)$. These arguments are all quite easy. For example,

$$\begin{aligned}
D_{\alpha p}(e_1 e_3) &= D_{\alpha p}(e_1)e_3 + E_\alpha(e_1)D_{\alpha p}(e_3) \\
&\approx D_{\alpha p}(e_2)e_3 + E_\alpha(e_2)D_{\alpha p}(e_3) = D_{\alpha p}(e_2 e_3)
\end{aligned}$$

and

$$D_{\alpha p}(e_1^*) = D_{\alpha p}(e_1)e_1^* \approx D_{\alpha p}(e_2)e_2^* = D_{\alpha p}(e_2^*).$$

□

15.6.1 Closure

To establish the finiteness of the quotient automaton $\mathsf{Exp}_e/{\approx}$ and explain its relationship to the Kleene construction, we derive a formal relationship between the set of accessible $\approx$-classes of derivatives $\{\hat{D}_x(e)/{\approx} \mid x \in (\mathsf{At} \cdot \Sigma)^*\}$ and certain sets of terms derived from e.

For KAT term e, we define the *closure* of e, denoted $\mathrm{cl}(e)$, to be the smallest set of terms containing e and 1 and closed under the following rules:

$$\frac{e \in \mathrm{cl}(e_1)}{e \in \mathrm{cl}(e_1 + e_2)} \qquad \frac{e \in \mathrm{cl}(e_1)}{ee_2 \in \mathrm{cl}(e_1 e_2)} \qquad \frac{e \in \mathrm{cl}(e_1)}{ee_1^* \in \mathrm{cl}(e_1^*)}$$
$$\frac{e \in \mathrm{cl}(e_2)}{e \in \mathrm{cl}(e_1 + e_2)} \qquad \frac{e \in \mathrm{cl}(e_2)}{e \in \mathrm{cl}(e_1 e_2)} \qquad \frac{e \in \mathrm{cl}(b)}{e \in \mathrm{cl}(\bar{b})} \tag{15.3}$$

Lemma 15.6.2 *The set* $\mathrm{cl}(e)$ *contains at most* $|e|+1$ *elements, where* $|e|$ *is the number of subterms of* e.

Proof We show by induction on e that $\mathrm{cl}'(e)$ contains at most $|e|$ elements, where $\mathrm{cl}'(e) = \mathrm{cl}(e) - \{1\}$. For $e \in \Sigma \cup T$, $\mathrm{cl}'(e) = \{e\}$. For the other operators, from the rules (15.3) we have

$$\begin{aligned}
\mathrm{cl}'(\bar{b}) &= \{\bar{b}\} \cup \mathrm{cl}'(b), \\
\mathrm{cl}'(e_1 + e_2) &= \{e_1 + e_2\} \cup \mathrm{cl}'(e_1) \cup \mathrm{cl}'(e_2), \\
\mathrm{cl}'(e_1 e_2) &= \{e_1 e_2\} \cup \{e e_2 \mid e \in \mathrm{cl}'(e_1)\} \cup \mathrm{cl}'(e_2), \\
\mathrm{cl}'(e_1^*) &= \{e_1^*\} \cup \{e e_1^* \mid e \in \mathrm{cl}'(e_1)\}.
\end{aligned}$$

The result follows. □

15.6.2 Set Representation of Derivatives

We now construct a nondeterministic transition function Δ on the set of states $\mathsf{Exp} + (\mathsf{At} \times \mathsf{Exp})$ as follows. The elements of Exp are called *test states* and the elements of $\mathsf{At} \times \mathsf{Exp}$ are called *action states*. The test transitions go only from test states to action states, and the action transitions go only from action states to test states. Thus for $\alpha \in \mathsf{At}$ and $p \in \Sigma$,

$$\Delta_\alpha : \mathsf{Exp} \to 2^{\mathsf{At} \times \mathsf{Exp}} \qquad\qquad \Delta_p : \mathsf{At} \times \mathsf{Exp} \to 2^{\mathsf{Exp}}.$$

The test transitions are deterministic: $\Delta_\alpha(e) \stackrel{\text{def}}{=} \{(\alpha, e)\}$. The action transitions are defined inductively:

$$\begin{aligned}
\Delta_p(\alpha, q) &\stackrel{\text{def}}{=} \begin{cases} \{1\}, & \text{if } q \in \Sigma \text{ and } q = p, \\ \emptyset, & \text{if } q \in \Sigma \text{ and } q \neq p \text{ or } q \in B, \end{cases} \\
\Delta_p(\alpha, e_1 + e_2) &\stackrel{\text{def}}{=} \Delta_p(\alpha, e_1) \cup \Delta_p(\alpha, e_2), \\
\Delta_p(\alpha, e_1 e_2) &\stackrel{\text{def}}{=} \begin{cases} \{e e_2 \mid e \in \Delta_p(\alpha, e_1)\} \cup \Delta_p(\alpha, e_2), & \text{if } E_\alpha(e_1) = 1, \\ \{e e_2 \mid e \in \Delta_p(\alpha, e_1)\}, & \text{if } E_\alpha(e_1) = 0, \end{cases} \\
\Delta_p(\alpha, e_1^*) &\stackrel{\text{def}}{=} \{e e_1^* \mid e \in \Delta_p(\alpha, e_1)\}.
\end{aligned}$$

Due to the bipartite structure of the states, we have $\hat{\Delta}_{\alpha p} = \Delta_p \bullet \Delta_\alpha$, where $\hat{\Delta}$ is the extension of Δ defined in Sect. 15.3.2. Then

$$\hat{\Delta}_{\alpha p}(e) = (\Delta_p \bullet \Delta_\alpha)(e) = \bigcup \{\Delta_p(\alpha, e)\} = \Delta_p(\alpha, e). \tag{15.4}$$

We thus have

$$\hat{\Delta}_{\alpha p}(q) \stackrel{\text{def}}{=} \begin{cases} \{1\}, & \text{if } q \in \Sigma \text{ and } q = p, \\ \emptyset, & \text{if } q \in \Sigma \text{ and } q \neq p \text{ or } q \in B, \end{cases}$$

$$\hat{\Delta}_{\alpha p}(e_1 + e_2) \stackrel{\text{def}}{=} \hat{\Delta}_{\alpha p}(e_1) \cup \hat{\Delta}_{\alpha p}(e_2),$$

$$\hat{\Delta}_{\alpha p}(e_1 e_2) \stackrel{\text{def}}{=} \begin{cases} \{ee_2 \mid e \in \hat{\Delta}_{\alpha p}(e_1)\} \cup \hat{\Delta}_{\alpha p}(e_2), & \text{if } E_\alpha(e_1) = 1, \\ \{ee_2 \mid e \in \hat{\Delta}_{\alpha p}(e_1)\}, & \text{if } E_\alpha(e_1) = 0, \end{cases}$$

$$\hat{\Delta}_{\alpha p}(e_1^*) \stackrel{\text{def}}{=} \{ee_1^* \mid e \in \hat{\Delta}_{\alpha p}(e_1)\}.$$

Lemma 15.6.3 *For all* KAT *terms* e *and* $x \in (\mathsf{At} \cdot \Sigma)^*$, $\hat{\Delta}_x(e) \subseteq \mathrm{cl}(e)$.

Proof We first show that for $\alpha \in \mathsf{At}$ and $p \in \Sigma$, $\hat{\Delta}_{\alpha p}(e) \subseteq \mathrm{cl}(e)$ by induction on the structure of e. The cases $e \in \Sigma$ or $e \in B$ are easy. For the other operators,

$$\hat{\Delta}_{\alpha p}(e_1 + e_2) = \hat{\Delta}_{\alpha p}(e_1) \cup \hat{\Delta}_{\alpha p}(e_2) \subseteq \mathrm{cl}(e_1) \cup \mathrm{cl}(e_2) \subseteq \mathrm{cl}(e_1 + e_2)$$

$$\hat{\Delta}_{\alpha p}(e_1 e_2) = \begin{cases} \{ee_2 \mid e \in \hat{\Delta}_{\alpha p}(e_1)\} \cup \hat{\Delta}_{\alpha p}(e_2), & \text{if } E_\alpha(e_1) = 1 \\ \{ee_2 \mid e \in \hat{\Delta}_{\alpha p}(e_1)\}, & \text{if } E_\alpha(e_1) = 0 \end{cases}$$

$$\subseteq \{ee_2 \mid e \in \mathrm{cl}(e_1)\} \cup \mathrm{cl}(e_2) \subseteq \mathrm{cl}(e_1 e_2)$$

$$\hat{\Delta}_{\alpha p}(e_1^*) = \{ee_1^* \mid e \in \hat{\Delta}_{\alpha p}(e_1)\} \subseteq \{ee_1^* \mid e \in \mathrm{cl}(e_1)\} \subseteq \mathrm{cl}(e_1^*).$$

For arbitrary $x \in (\mathsf{At} \cdot \Sigma)^*$, we proceed by induction on the length of x. The base case $x = \varepsilon$ is easy and the case $x = \alpha p$ is given by the previous argument. For $x \neq \varepsilon$ and $y \neq \varepsilon$,

$$\hat{\Delta}_{xy}(e) = (\hat{\Delta}_y \bullet \hat{\Delta}_x)(e) = \bigcup \{\hat{\Delta}_y(d) \mid d \in \hat{\Delta}_x(e)\}$$

$$\subseteq \bigcup \{\mathrm{cl}(d) \mid d \in \mathrm{cl}(e)\} = \mathrm{cl}(e).$$

□

Lemma 15.6.4 *For all* KAT *terms* e *and* $x \in (\mathsf{At} \cdot \Sigma)^*$, $\hat{D}_x(e) \approx \sum \hat{\Delta}_x(e)$.

Proof We first show that for $\alpha \in \mathsf{At}$ and $p \in \Sigma$, $D_{\alpha p}(e) \approx \sum \hat{\Delta}_{\alpha p}(e)$ by induction on the structure of e. For $q \in \Sigma$, we have

$$D_{\alpha p}(q) = \begin{cases} 1, & \text{if } p = q \\ 0, & \text{if } p \neq q \end{cases} = \begin{cases} \sum\{1\}, & \text{if } p = q \\ \sum \emptyset, & \text{if } p \neq q \end{cases} = \sum \hat{\Delta}_{\alpha p}(q).$$

For $b \in B$,

$$D_{\alpha p}(b) = 0 = \sum \emptyset = \sum \hat{\Delta}_{\alpha p}(b).$$

For the other operators,

$$\begin{aligned}D_{\alpha p}(e_1+e_2) &= D_{\alpha p}(e_1)+D_{\alpha p}(e_2) \approx \sum \hat{\Delta}_{\alpha p}(e_1) + \sum \hat{\Delta}_{\alpha p}(e_2)\\ &\approx \sum (\hat{\Delta}_{\alpha p}(e_1)\cup \hat{\Delta}_{\alpha p}(e_2)) = \sum \hat{\Delta}_{\alpha p}(e_1+e_2),\end{aligned}$$

$$\begin{aligned}D_{\alpha p}(e_1e_2) &= D_{\alpha p}(e_1)e_2 + E_\alpha(e_1)D_{\alpha p}(e_2)\\ &\approx \left(\sum \hat{\Delta}_{\alpha p}(e_1)\right) e_2 + E_\alpha(e_1)\sum \hat{\Delta}_{\alpha p}(e_2)\\ &\approx \begin{cases}\sum\{ee_2 \mid e\in \hat{\Delta}_{\alpha p}(e_1)\} + \sum \hat{\Delta}_{\alpha p}(e_2), & \text{if } E_\alpha(e_1)=1\\ \sum\{ee_2 \mid e\in \hat{\Delta}_{\alpha p}(e_1)\}, & \text{if } E_\alpha(e_1)=0\end{cases}\\ &\approx \begin{cases}\sum(\{ee_2 \mid e\in \hat{\Delta}_{\alpha p}(e_1)\} \cup \hat{\Delta}_{\alpha p}(e_2)), & \text{if } E_\alpha(e_1)=1\\ \sum\{ee_2 \mid e\in \hat{\Delta}_{\alpha p}(e_1)\}, & \text{if } E_\alpha(e_1)=0\end{cases}\\ &= \sum \hat{\Delta}_{\alpha p}(e_1e_2),\end{aligned}$$

$$\begin{aligned}D_{\alpha p}(e_1^*) &= D_{\alpha p}(e_1)e_1^* \approx \left(\sum \hat{\Delta}_{\alpha p}(e_1)\right) e_1^*\\ &\approx \sum\{ee_1^* \mid e\in \hat{\Delta}_{\alpha p}(e_1)\} = \sum \hat{\Delta}_{\alpha p}(e_1^*).\end{aligned}$$

Now we show the result for arbitrary $x \in (\mathsf{At}\cdot\Sigma)^*$ by induction on the length of x. The case $x=\varepsilon$ is trivial, and the case $x=\alpha p$ is given by the previous argument. Finally, for $x\neq\varepsilon$ and $y\neq\varepsilon$,

$$\begin{aligned}\hat{D}_{xy}(e) &= \hat{D}_y(\hat{D}_x(e))\\ &\approx \hat{D}_y\left(\sum \hat{\Delta}_x(e)\right) && \text{by Lemma 6.1}\\ &= \sum\{\hat{D}_y(d) \mid d\in \hat{\Delta}_x(e)\}\\ &\approx \sum\left\{\sum \hat{\Delta}_y(d) \,\middle|\, \in \hat{\Delta}_x(e)\right\} \approx \sum \bigcup \{\hat{\Delta}_y(d) \mid d\in \hat{\Delta}_x(e)\}\\ &= \sum(\hat{\Delta}_y \bullet \hat{\Delta}_x)(e) = \sum \hat{\Delta}_{xy}(e).\end{aligned}$$

□

Theorem 15.6.5 *The automaton* $\mathsf{Exp}_e/{\approx}$ *has at most* $2^{|e|+1}$ *accessible states.*

Proof The accessible states of $\mathsf{Exp}_e/{\approx}$ are $\{\hat{D}_x(e)/{\approx} \mid x\in(\mathsf{At}\cdot\Sigma)^*\}$, where $d/{\approx}$ is the congruence class of d modulo $\approx$. The stated bound follows from Lemmas 15.6.2, 15.6.3, and 15.6.4. □

15.6.3 Brzozowski Meets Kleene

It is possible to obtain $\mathsf{Exp}_e/{\approx}$ by a Kleene construction to obtain a nondeterministic AGS N_e with finitely many states, then apply the construction of Sect. 15.3.4 to obtain a deterministic automaton M_e with at most $2^{|e|+1}$ states. The automaton $\mathsf{Exp}_e/{\approx}$ is a homomorphic image of M_e. A version of Kleene's theorem for KAT terms and automata on guarded strings has been described previously in Kozen (2003), but the current treatment parallels more closely Brzozowski's original treatment for ordinary regular expressions (Brzozowski 1964) and aligns with the general coalgebraic structure of Bonsangue et al. (2007, 2009).

Define the nondeterministic automaton

$$N_e \stackrel{\text{def}}{=} (Q, \Delta, \mathsf{start}, \mathsf{accept}),$$

where the set of states Q is the disjoint union $\mathrm{cl}(e) + (\mathsf{At} \times \mathrm{cl}(e))$, the transition function Δ is that defined in Sect. 15.6.2, and the start and accept states are

$$\mathsf{start} \stackrel{\text{def}}{=} \{e\} \qquad\qquad \mathsf{accept} \stackrel{\text{def}}{=} \{(\alpha, d) \mid E_\alpha(d) = 1\}.$$

That Δ_α maps $\mathrm{cl}(e)$ to $2^{\mathsf{At}\times\mathrm{cl}(e)}$ is immediate from the definition of Δ_α, and that Δ_p maps $\mathsf{At} \times \mathrm{cl}(e)$ to $2^{\mathrm{cl}(e)}$ is guaranteed by (15.4) and Lemma 15.6.3.

Now let

$$M_e \stackrel{\text{def}}{=} (2^{\mathrm{cl}(e)}, \delta, \varepsilon, \mathsf{start})$$

be the deterministic automaton obtained from N_e by the subset construction as described in Sect. 15.3.4. The start state of M_e is $\{e\}$, and δ and ε are given by

$$\delta_{\alpha p}(A) = \bigcup_{d\in A} \hat{\Delta}_{\alpha p}(d) \qquad\qquad \varepsilon_\alpha(A) = \begin{cases} 1, & \text{if } \exists d \in A\ E_\alpha(d) = 1, \\ 0, & \text{otherwise.} \end{cases}$$

Note that the accessible states are all of the form $A \subseteq \mathrm{cl}(e)$, thus by Lemma 15.6.2, M_e has at most $2^{|e|+1}$ accessible states.

Theorem 15.6.6 *For $A \subseteq \mathsf{Exp}$, the map $A \mapsto (\sum A)/{\approx}$ is a* KCT*-morphism. Ignoring inaccessible states, the quotient automaton $\mathsf{Exp}_e/{\approx}$ is the image of M_e under this map.*

Proof We must show that the function $A \mapsto \sum A$ maps the start state of M_e to the start state of Exp_e, and that this function is a bisimulation modulo $\approx$. For δ,

$$\begin{aligned}\sum \delta_{\alpha p}(A) &= \sum \bigcup \{\hat{\Delta}_{\alpha p}(d) \mid d \in A\} \\ &\approx \sum \left\{\sum \hat{\Delta}_{\alpha p}(d) \middle| d \in A\right\} \\ &\approx \sum \{D_{\alpha p}(d) \mid d \in A\} && \text{by Lemma 6.4} \\ &= D_{\alpha p}\left(\sum A\right),\end{aligned}$$

therefore

$$\left(\sum \delta_{\alpha p}(A)\right) / \approx = \left(D_{\alpha p}\left(\sum A\right)\right) / \approx \; = \; D_{\alpha p}\left(\left(\sum A\right) / \approx\right).$$

For ε,

$$\varepsilon_\alpha(A) = \begin{cases} 1, & \text{if } \exists d \in A \; E_\alpha(d) = 1 \\ 0, & \text{otherwise} \end{cases} = E_\alpha\left(\sum A\right) = E_\alpha\left(\left(\sum A\right) / \approx\right).$$

The map also preserves start states:

$$\{e\} \mapsto \left(\sum \{e\}\right) / \approx \; = \; e/\approx.$$

Thus the map $A \mapsto (\sum A)/\approx$ is a KCT-morphism mapping M_e to $\mathsf{Exp}_e/\approx$. □

15.6.4 Automatic Proof Generation in PSPACE

The results of Sects. 15.6.2 and 15.6.3 give rise to a nondeterministic linear-space algorithm for deciding the equivalence of two given KAT terms. By Savitch's theorem (Savitch 1970), there is a deterministic quadratic-space algorithm. The deterministic algorithm can be used to create bisimulation proofs of equivalence or inequivalence automatically.

To obtain the linear space bound, we first show that each element of $\mathrm{cl}(e)$ corresponds to an occurrence of a subterm of e. This lets us use the occurrences of subterms of e as representatives for the elements of $\mathrm{cl}(e)$. To define the correspondence, we view terms as labeled trees; that is, as partial functions

$$e : \omega^* \to \Sigma \cup T \cup \{+, \cdot, ^*, \bar{}\,, 0, 1\}$$

with domain of definition $\mathrm{dom}\, e \subseteq \omega^*$ such that

- $\mathrm{dom}\, e$ is finite, nonempty, and prefix-closed;
- if $\sigma \in \mathrm{dom}\, e$ and $e(\sigma)$ is of arity n, then $\sigma i \in \mathrm{dom}\, e$ iff $i < n$. The arities of elements of Σ and T are 0 and those of $+, \cdot, ^*, \bar{}\,, 0, 1$ are 2, 2, 1, 1, 0, 0, respectively.

An occurrence of a subterm of e is identified by its position $\sigma \in \mathrm{dom}\, e$. The subterm at position σ is $\lambda\tau.e(\sigma\tau)$, and its domain is $\{\tau \mid \sigma\tau \in \mathrm{dom}\, e\}$.

Define a partial function $R : \omega^* \times \mathsf{Exp} \to \mathsf{Exp}$ inductively by

$$R(0\sigma, e_1 + e_2) \stackrel{\text{def}}{=} R(\sigma, e_1) \qquad R(0\sigma, e_1 e_2) \stackrel{\text{def}}{=} R(\sigma, e_1) \cdot e_2$$

$$R(1\sigma, e_1 + e_2) \stackrel{\text{def}}{=} R(\sigma, e_2) \qquad R(1\sigma, e_1 e_2) \stackrel{\text{def}}{=} R(\sigma, e_2)$$

$$R(0\sigma, e^*) \stackrel{\text{def}}{=} R(\sigma, e) \cdot e^* \qquad R(\varepsilon, e) \stackrel{\text{def}}{=} e.$$

One can show by induction that $R(\sigma, e)$ is defined iff $\sigma \in \operatorname{dom} e$, and that a term is in $\operatorname{cl}(e)$ iff it is either 1 or $R(\sigma, e)$ for some $\sigma \in \operatorname{dom} e$.

Now we show how to construct coinductive equivalence and inequivalence proofs for two given terms e_1 and e_2. Construct the two nondeterministic AGS N_{e_1} and N_{e_2} as described in Sect. 15.6.3, representing the states by $\operatorname{dom} e_1$ and $\operatorname{dom} e_2$, respectively (assume without loss of generality that $1 = R(\sigma, e_1) = R(\tau, e_2)$ for some σ and τ). If we like, we can also reduce terms modulo $\approx$, so that if $R(\sigma, e_1) \approx R(\tau, e_1)$, we only need one of σ, τ.

Place pebbles on the start states of the two automata. Nondeterministically guess a string $y \in (\mathsf{At} \cdot \Sigma)^*$ and move the pebbles to all accessible states according to the transition functions of the two machines. Halt and declare e_1 and e_2 inequivalent if there exists $\alpha \in \mathsf{At}$ such that

$$E_\alpha\left(\sum_{\tau \in A_1} R(\tau, e_1)\right) \neq E_\alpha\left(\sum_{\tau \in A_2} R(\rho, e_2)\right),$$

where A_1 and A_2 are the sets of states of N_{e_1} and N_{e_2}, respectively, currently occupied by pebbles; we have found a guarded string $x = y\alpha$ accepted by one but not by the other, since

$$L(e_i)(x) = E_\alpha(\hat{D}_y(e_i)) = E_\alpha\left(\sum_{\tau \in A_i} R(\tau, e_i)\right)$$

for $i \in \{1, 2\}$, therefore $L(e_1)(x) \neq L(e_2)(x)$.

Once we can decide equivalence in quadratic space, we can produce a bisimulation proof of equivalence in the same amount of space. We first produce the deterministic automata M_{e_1} and M_{e_2} equivalent to N_{e_1} and N_{e_2}. The states of M_{e_1} and M_{e_2} are represented by the powersets of $\operatorname{dom} e_1$ and $\operatorname{dom} e_2$, respectively. These sets are of exponential size, but they can be generated sequentially in linear space. The transition function is the action on subsets as defined in Sect. 15.3.4, and this can also be generated in linear space.

Now we attempt to construct the maximal bisimulation between the two deterministic automata. We iterate through all pairs of states, testing equivalence of each pair as described above. If the states are equivalent, we output the pair as bisimilar. The set of pairs that are ever output is the maximal bisimulation.

In case e_1 and e_2 are not equivalent, a witness for inequivalence can also be produced in *PSPACE*. A witness for inequivalence is a guarded string x accepted by one automaton but not the other. The shortest such string can be exponentially long in the worst case, but can be produced in the same way that one would produce an exponential-length accepting computation of a nondeterministic linear-space Turing machine, by a straightforward modification of the proof of Savitch's theorem (Savitch 1970).

Acknowledgements Many thanks to Jiří Adámek, Marcello Bonsangue, Helle Hvid Hansen, Raul Leal, Jan Rutten, Mehrnoosh Sadrzadeh, Luigi Santocanale, Alexandra Silva, Yde Venema, and James Worthington. An earlier version of this work (Kozen 2008) was presented at CMCS 2008 and was supported by NSF Grant CCF-0635028. The preparation of this revised version was supported by NSF grants CCF-1535952 and CCF-1637532 and by the National Security Agency.

References

Bonsangue, M. M., Rutten, J. J. M. M., & Silva, A. M. (2007). *Regular expressions for polynomial coalgebras*. Technical Report SEN-E0703, Centrum voor Wiskunde en Informatica, Amsterdam.

Bonsangue, M. M., Rutten, J. J. M. M., & Silva, A. M. (2009). A Kleene theorem for polynomial coalgebras. In L. de Alfaro (Ed.), *Proceedings of the 12th international conference foundations of software science and computation structures (FoSSaCS 2009)* (Vol. 5504, pp. 122–136)., of lecture notes in computer science New York: Springer.

Brzozowski, J. A. (1964). Derivatives of regular expressions. *Journal of the Association for Computing Machinery*, *11*, 481–494.

Chen, H., & Pucella, R. (2003). A coalgebraic approach to Kleene algebra with tests. *Electronic Notes in Theoretical Computer Science*, *82*(1),

Cohen, E., Kozen, D., & Smith, F. (1996). *The complexity of Kleene algebra with tests*. Technical Report TR96-1598, Computer Science Department, Cornell University.

Conway, J. H. (1971). *Regular algebra and finite machines*. London: Chapman and Hall.

Kaplan, D. M. (1969). Regular expressions and the equivalence of programs. *Journal of Computer and System Sciences*, *3*, 361–386.

Kleene, S. C. (1956). Representation of events in nerve nets and finite automata. In C. E. Shannon & J. McCarthy (Eds.), *Automata studies* (pp. 3–41). Princeton, NJ: Princeton University Press.

Kozen, D. (1994). A completeness theorem for Kleene algebras and the algebra of regular events. *Computing and Information*, *110*(2), 366–390.

Kozen, D. (1997). Kleene algebra with tests. *Transactions on Programming Languages and Systems*, *19*(3), 427–443.

Kozen, D. (2000). On Hoare logic and Kleene algebra with tests. *Transactions on Computational Logic*, *1*(1), 60–76.

Kozen, D. (2003). Automata on guarded strings and applications. *Matématica Contemporânea*, *24*, 117–139.

Kozen, D. (2008). *On the coalgebraic theory of Kleene algebra with tests*. Technical Report.? http://hdl.handle.net/1813/10173, Computing and Information Science, Cornell University.

Kozen, D., & Smith, F. (1996). Kleene algebra with tests: Completeness and decidability. In D. van Dalen & M. Bezem (Eds.), *Proceedings of the 10th international workshop computer science logic (CSL'96)* (Vol. 1258, pp. 244–259)., of lecture notes in computer science Utrecht: Springer.

Rutten, J. J. M. M. (1998). Automata and coinduction (an exercise in coalgebra). *Proceedings of CONCUR'98* (Vol. 1466, pp. 193–217)., lecture notes in computer science Berlin: Springer.

Savitch, W. (1970). Relationship between nondeterministic and deterministic tape complexities. *Journal of Computer and System Sciences*, *4*(2), 177–192.

Worthington, J. (2008). Automatic proof generation in Kleene algebra. In R. Berghammer, B. Möller, & G. Struth (Eds.), *10th International conference relational methods in computer science (RelMiCS10) and 5th international conference applications of Kleene algebra (AKA5)* (Vol. 4988, pp. 382–396)., of lecturer notes in computer science Berlin: Springer.

Chapter 16
Aristotle, Boole, and Categories

Vaughan Pratt

Abstract We propose new axiomatizations of the 24 assertoric syllogisms of Aristotle's syllogistic, and the 2^{2^n} n-ary operations of Boole's algebraic logic. The former organizes the syllogisms as a 6×4 table partitioned into four connected components according to which term if any must be inhabited. We give two natural-deduction style axiomatizations, one with four axioms and four rules, the second with one axiom and six rules. The table provides immediately visualizable proofs of soundness and completeness. We give an elementary category-theoretic semantics for the axioms along with criteria for determining the term if any required to be nonempty in each syllogism. We base the latter on Lawvere's notion of an algebraic theory as a category with finite products having as models product-preserving set-valued functors. The benefit of this axiomatization is that it avoids the dilemma of whether a Boolean algebra is a numerical ring as envisaged by Boole, a logical lattice as envisaged by Peirce, Jevons, and Schroeder, an intuitionistic Heyting algebra on a middle-excluding diet as envisaged by Heyting, or any of several other candidates for the "true nature" of Boolean logic. Unlike general rings, Boolean rings have only finitely many n-ary operations, permitting a uniform locally finite axiomatization of their theory in terms of a certain associative multiplication of finite 0–1 matrices.

Keywords Syllogisms · Category-theoretic semantics · Algebraic theories

16.1 Introduction

Whereas the ancient Romans excelled at civil engineering and economics, the ancient Greeks shone in geometry and logic. Book I of Euclid's *Elements* and Aristotle's assertoric syllogisms dominated the elementary pedagogy of respectively geometry and logic from the third century BC to the eighteenth century AD.

In the middle of the seventeenth century Euler abstracted Euclid's geometry to affine geometry by omitting the notions of orthogonality and rotation-

V. Pratt (✉)
Stanford, CA, USA
e-mail: pratt@cs.stanford.edu

C. Başkent et al. (eds.), *Rohit Parikh on Logic, Language and Society*,
Outstanding Contributions to Logic 11, DOI 10.1007/978-3-319-47843-2_16

invariant length, and two centuries later Boole generalized validity of Aristotle's unconditionally valid syllogisms to zeroth-order propositional logic by inventing Boolean rings. Yet subsequent literature has continued to find the original subjects of great interest. The 2013 4th World Congress and School on Universal Logic for example featured a score of presentations involving syllogisms.

Of these, logic is closer to the interests of my multidecadal colleague, coauthor, and friend Rohit Parikh. I shall therefore focus here on logic generally, and more specifically on the contributions of Aristotle and Boole, with categories as a common if somewhat slender thread.

Starting in the 1970s, Rohit and I both worked on dynamic logic (Pratt 1976; Parikh 1978), a formalism for reasoning about behaviour that expands the language of modal logic with that of regular expressions. Dynamic logic witnessed the introduction into program verification of the possible-world semantics of modal logic (Pratt 1976), and into logic of multimodal logic with unboundedly many modalities, so it would be only logical to write about Aristotle's modal syllogistic.

However I was a number theorist before I was a logician (Pratt 1975), and it has puzzled me as to why the number of Aristotle's two-premise assertoric syllogisms should factor neatly as 6×4, the inclusion of a few odd-ball syllogisms whose middle term did not interpolate its conclusion notwithstanding. A concrete and even useful answer to this question will hopefully prove of greater interest, at least to the arithmetically inclined, than more analytic observations on Aristotle's modal syllogistic.

16.2 Aristotle's Logic

As a point of departure for our axiomatization, the following three subsections rehearse some of the basic lore of syllogisms, which has accumulated in fits and starts over the 23 centuries since Aristotle got them off to an excellent albeit controversial start. As our interest in these three subsections is more technical than historical we emphasize the lore over its lorists.

16.2.1 Syntax

An assertoric syllogism is a form of argument typified by, *no cats are birds, all wrens are birds, therefore no wrens are cats.*

In the modern language of natural deduction a ***syllogism*** is a sequent $J, N \vdash C$ consisting of three ***sentences***: a maJor premise J, a miNor premise N, and a Conclusion C.

Sentences are ***categorical***, meaning that they express a relation between two terms each of which can be construed as a category, predicate, class, set, or property, all meaning the same thing for our purposes.

The language has four relations:

(i) a ***universal*** relation XaY (X all-are Y) asserting all X are Y, or that (property) Y holds of every (member of) X;
(ii) a ***particular*** relation XiY (X intersects Y) asserting some X are Y, or that Y holds of some X;
along with their respective ***contradictories***,
(iii) XoY (X obtrudes-from Y) asserting some X are not Y, or Y does not hold of some X, or not(XaY); and
(iv) XeY (X empty-intersection-with Y) asserting no X are Y, or Y holds of no X, or not(XiY).

Set-theoretically these are the binary relations of inclusion and nonempty intersection, which are considered ***positive***, and their respective contradictories, considered ***negative***. Contradiction as an operation on syllogisms interchanges universal and particular and changes sign (the relations organized as a string *aeio* reverse to become the string *oiea*) but is not itself part of the language. Nor is any other Boolean operation, the complete absence of which is a feature of syllogistic, not a bug.

It follows from all this that a syllogism contains six occurrences of terms, two in each of the three sentences. A further requirement is that there be three terms each having two occurrences in distinct sentences.

The following naming convention uniquely identifies the syllogistic form. The conclusion is of the form S–P where S and P are ***terms*** denoting subject and predicate respectively. S and P also appear in separate premises, which are arranged so that P appears in the first or ***major*** premise and hence S in the second or ***minor*** premise.

The third term is denoted M, which appears in both premises, either on the left or right in each premise, so four possibilities, which are numbered 1–4 corresponding to the respective positions LR, RR, LL, and RL for M in the respective premises. That is, Fig. 1 is when M appears on the left in the major premise and the right in the minor, and so on. The header of Table 16.1 in Sect. 16.2.5 illustrates this in more detail (note the Gray-code order).

So in the example at the start of the section, from the conclusion we infer that S = *wrens* and P = *cats*, so M has to be *birds*. That M appears on the right in both premises (RR) reveals this syllogism to be in Fig. 2. Since P (*cats*) appears in the first or major premise there is no need to switch these premises.

When the occasion arises to transform a syllogism into an equally valid one, the result may violate this naming convention. For example we may operate on the conclusion and the major premise by contradicting and exchanging them, amounting to the rule of *modus tollens* in the context of the other premise. After any such transformation we shall automatically identify the new subject, predicate, and middle according to the naming convention, which in general could be any of the six possible permutations of the original identifications. And if this results in the othewise untouched minor premise now containing P it is promoted to major premise, i.e. the premises are switched.

Syllogisms are divided syntactically into $4^3 = 64$ ***moods*** according to which of the 4 relations are chosen for each of their 3 sentences. The wren example has the form PeM, SaM $\vdash$ SeP and so its relations are e, a, and e in that order. This mood is therefore notated EAE, to which we append the figure making it the ***form*** EAE-2.

16.2.2 Semantics

Syntactically, figures and moods are independent, whence there are $4 \times 64 = 256$ forms. Semantically however not all of them constitute valid arguments. If for example we turn the above EAE-2 wren example into an EAE-4 syllogism by replacing its minor premise by *all birds are two-legged* and its conclusion by *no two-leggeds are cats*, we obtain a syllogism with premises that, while true, do not suffice to rule out the possibility of a two-legged cat. Hence EAE-4 must be judged invalid.

Taking S to be *warm-blooded* instead of *two-legged* might have made it easier to see that EAE-4 was invalid, since the conclusion *no warm-bloodeds are cats* is clearly absurd: all cats are of course warm-blooded. The two-legged example draws attention to the sufficiency of a single individual in a counterexample to EAE-4.

The semantics of syllogisms reduces conveniently to that of first order logic via the following translations of each sentence t = P-Q to a sentence $\hat{t}$ of the (monadic) predicate calculus.

PaQ: $\forall x[\neg P(x) \vee Q(x)]$
PeQ: $\forall x[\neg P(x) \vee \neg Q(x)]$
PiQ: $\exists x[P(x) \wedge Q(x)]$
PoQ: $\exists x[P(x) \wedge \neg Q(x)]$

Call a sentence of first order logic *syllogistic* when it is either of the form $\forall x[L_1(x) \vee L_2(x)]$ where the L_i's are literals with distinct predicate symbols (thereby precluding $\forall x[P(x) \vee \neg P(x)]$), or the negation of such a sentence, i.e. of the form $\exists x[L_1(x) \wedge L_2(x)]$. Call a set of sentences, of any cardinality, syllogistic when every member is syllogistic and every pair of members shares at most one predicate symbol.

Theorem 1 *Any syllogistic set of sentences having at most one universal sentence is consistent.*

Proof Let S be such a set having u as its only universal sentence if any. Form a model of S by taking its universe E to consist of the existential sentences. For each member e of E, set to *true* at e the literals of e and, if u exists, the literal of u whose predicate symbol does not appear in e. Set the remaining values of literals to *true*. This model satisfies every sentence of S, which is therefore consistent. □

With one exception, this construction does not extend to syllogistic sets with two or more universal sentences because those sentences may contain a complementary pair of literals. The exception is when S contains no existential sentence, in which case the construction produces the empty model and all the universal sentences are vacuously satisfied regardless of their contents.[1]

The following interprets a syllogism as a 3-element syllogistic set whose third element is the negation of the translation into predicate calculus of its conclusion. The validity of the syllogism is equivalent to the unsatisfiability of that set.

Corollary 2 *A valid syllogism must contain exactly one particular among its premises and contradicted conclusion.*

For if it contains two particulars it contains only one universal, whence Theorem 1 produces a counterexample, while if it contains no particulars then the exception to the non-extendability of the theorem to multiple universals produces a counterexample.

Hence of the $4^3 = 64$ moods, only $2^3 \times 3 = 24$ can be the mood of a valid syllogism. In conjunction with any of the four figures, there are therefore $24 \times 4 = 96$ candidate forms, call these ***presyllogisms***. It follows that a valid syllogism must have at least one universal premise. Furthermore if it has a particular premise then the conclusion must also be particular, and conversely if it doesn't then the conclusion must be universal.

Theorem 3 *A presyllogism $J, N \vdash C$ is valid if and only if its translation $\hat{J} \wedge \hat{N} \wedge \neg\hat{C}$ into propositional calculus is unsatisfiable.*

Proof By the corollary any counterexample requires only one individual witnessing the one particular and any other individuals may be discarded. But in that case both $\forall x$ and $\exists x$ act as the identity operation. In the translation to predicate calculus the quantifiers may be dropped and every literal $P(x)$ simplified to a propositional literal P. □

So to decide validity of a form, verify that it is a presyllogism, translate it as above, and test its satisfiability.

The $24 \times 4 = 96$ possible forms of a valid syllogism are then those whose translation into propositional calculus is a tautology. These turn out to be the 15 in the largest of the four regions of Table 16.1 in Sect. 16.2.5 (the top four rows less AAI-4). A C program using this method to enumerate them can be seen at http://boole.stanford.edu/syllenum.c, requiring 1.33 nanoseconds on my laptop to translate and test each presyllogism (today's i7's are fast).

16.2.3 The Problem of Existential Import

It is generally agreed today that there are 24 assertoric syllogisms. What of the 9 that the procedure judged invalid?

---

[1]We take the traditional proscription in logic of the empty universe to be a pointless superstition that creates more problems than it solves.

One of them, AAI-1, is illustrated by the syllogism, *all unicorns are ungulates* (hooved), *all ungulates are mammals, therefore some unicorns are mammals*. But if unicorns don't exist this is impossible.

The truth of assertions about empty classes could go either way, or even both: Schrödinger's cats could be both dead and alive if he had none.[2] To avoid inferring the existence of unicorns, and also to reflect intuitions about natural language (at least Greek), Aristotle found it convenient to deem positive assertions about empty classes to be false. This convention justifies Aristotle's principle of ***subalternation***: SiP is subalternate to, or implied by, SaP, one edge of the traditional Square of Opposition (Parsons 1997).

But Aristotle's convention obliges the truth of not-(*all unicorns are ungulates*). Aristotle himself dealt with this by not identifying it with *some unicorns are not ungulates* but many medieval logicians found it natural to make that identification, some seemingly without noticing the inconsistency.

The long history is recounted by Parsons (1997). To cut to the chase, the modern view is that the nine extra syllogisms are ***conditionally*** valid, conditioned on the nonemptiness of one of its terms. For example the foregoing AAI-1 example is valid provided unicorns exist. This approach weakens subalternation in the Square of Opposition (Parsons 1997) to something more subtle than envisaged by Aristotle.

One such subtlety is exposed by the following principle.

Say that *Y* ***interpolates*** *X* and *Z* when (i) $Y\text{a}Z$ and either $X\text{a}Y$ or $X\text{i}Y$, or (ii) the same with *X* and *Z* interchanged. Interpolation is of interest because by transitivity we may infer in case (i) either $X\text{a}Z$ or $X\text{i}Z$ according to the relation of *X* to *Y*, and in case (ii) the same with *X* and *Z* interchanged.

Theorem 4 (Interpolation principle) *In any unconditionally valid syllogism, M′ interpolates S and P′ where M′ may be either M or its complement $\bar{M}$, and likewise for P′, independently.*

For example when the mood is EIO in any figure, the premises express SiMa$\bar{P}$ whence Si$\bar{P}$, that is, SoP. For OAO-3 we have $\bar{P}$iMaS whence $\bar{P}$iS or Si$\bar{P}$ or SoP, and so on.

This reveals interpolation as the engine of unconditional validity.

We defer the proof to the end of Sect. 16.2.5, which will be much easier to see after absorbing Table 16.1.

The subtlety is that it is not the interpolation principle that drives the following conditionally valid syllogism.

EAO-4: *No donkeys are unicorns, all unicorns are mammals, therefore some mammals are not donkeys.*

Certainly the middle term seems ideally located to be the interpolant. But the conclusion SoP, that is, Si$\bar{P}$, follows from Ma$\bar{P}$ and MaS, not by transitivity of anything but from S and $\bar{P}$ having M in common under the assumption that M is

[2]This makes more sense when phrased more precisely as a property of each *member* of the empty class; with no members no opportunity for an inconsistency ever arises.

nonempty. EAO-4 shares this subtlety with two more conditionally valid syllogisms, EAO-3, which works the same way, and AAI-3 which replaces $\bar{P}$ by P.

So how would Aristotle have justified these three? Simple: by subalternation, which permits deducing AAI-3 from the unconditionally valid AII-3, namely by "strengthening" the minor premise MiS to MaS, and similarly for the other two.

If we are to drop subalternation yet keep this little flock of three black sheep, some alternative means of support must be found for them. In the next section we offer just two: taking AAI-3 as an axiom, or permitting *modus tollens*, our rule of last resort, to be applied even to conditional syllogisms.

16.2.4 Yet Another Axiomatization of Syllogistic Deduction

The first major effort to bring Aristotle's syllogistic up to the standards of rigor of 20th century logic was carried out by Lukasiewicz in his book *Aristotle's Syllogistic* (Lukasiewicz 1957). Lukasiewicz cast syllogistic deduction in the framework of predicate calculus axiomatized by a Hilbert system, rendering entire syllogisms such as EIO-1 as (in notation we might use today) single formulas $e(M, P) \wedge i(S, M) \supset o(S, P)$.

But while technically correct, the variables S, M, P range over classes and the relations a, e, i, o as binary predicates hold between classes, making this a second-order logic. Moreover it introduces Boolean connectives into a subject that had previously only seen them in translations of individual sentences such as in the third paragraph of Sect. 16.2.2 but not in systems of syllogistic deduction such as Aristotle's. The latter are much closer to natural deduction, as the case of Gentzen's sequent calculus with one consequent, than to Hilbert systems.

Oddly enough Lukasiewicz himself had previously been an early and effective contributor to natural deduction, for which Aristotle's syllogistic would have been an ideal application, so it is strange that he chose a different framework.

Sequent calculi typically reverse the ratio of axioms to rules, having many rules but as few as one axiom, frequently $P \vdash P$. More significantly, they lend themselves to systems like Aristotle's by not requiring Boolean connectives. That such connectives are frequently encountered in sequent calculus systems, for example in rules such as $\wedge$-introduction and $\forall$-elimination, is not an intrinsic property of sequent calculi but only of the particular language being axiomatized by the particular deduction rules. The four connectives a, e, i, o of Aristotle can be axiomatized just as readily in the sequent calculus as can the three connectives $\wedge, \vee, \neg$ of Boole, in each case without assistance from other connectives.

Aristotle's first system had only two axioms, AAA-1 and EAE-1, mnemonically named *Barbara* and *Celarent*, which he viewed as self-evident and therefore not in need of proof. He derived his remaining syllogisms via a number of rules based on the Square of Opposition (Parsons 1997), including the problematic notion of subalternation. This discrepancy with Lukasiewicz's second-order Hilbert-style account was pointed out by Corcoran in a series of papers in the early 1970s, culminating in a

1974 paper (Corcoran 1974) proposing formal systems D and D2 based on Aristotle's systems, along with variants D3 and DE.

Our axiomatization, which we call D4p as a natural successor to Corcoran's D3, with "p" for preliminary as will be seen shortly, is in the same natural deduction format as those of Aristotle and Corcoran. The essential differences are in its choice and formulation of axioms and rules, its preservation only of validity and not meaning (Rules 2 and 3), and its emphasis on visualizing validity and completeness graphically to make their proofs easily seen simply by staring at Table 16.1.

As axioms we share *Barbara* with Aristotle, but in place of *Celarent* we take instead the three conditionally valid syllogisms of mood AAI, taking care to keep them logically independent. As a further departure from Aristotle, instead of taking the axioms as self-evident we shall justify them semantically in terms of certain posets lending themselves to a very elementary category-theoretic treatment.

As rules we have as usual the *converse* of the intersection sentences, provided for by rules R1 and R4, namely XeY as YeX and XiY as YiX, which do not change their meaning. But Rules R2 and R3 also allow the *obverse* of any sentence, which changes only the sign of the relation (XaY to XeY, XoY to XiY, etc.) leaving unchanged the terms and whether universal or particular. This changes the meaning by complementing a term, either P or M respectively, and is therefore only applicable to two sentences with the same right hand side, which exist in every figure save Fig. 4.

Recall that a presyllogism is a syllogistic form whose premises and contradicted conclusion contain exactly one particular. No conditionally valid syllogism is a presyllogism.

System D4p

Axioms

A1. AAA-1 (Barbara)
A2. AAI-1 (Barbari)
A3. AAI-3 (Darapti)
A4. AAI-4 (Bamalip)

Rules

R1. (*cj* or *cn*) Convert any *e* or *i* premise. (E.g. SeM → MeS.)
R2. (*ojc*) In Figs. 1 and 3, obvert the major premise and the conclusion.
R3. (*ojn*) In Fig. 2, obvert both premises.
R4. (*cc*) In Fig. 3, convert any *e* or *i* conclusion and interchange the premises.
R5. (*mtj* or *mtn*) For any presyllogism, contradict and exchange the conclusion and a premise.

Although every rule modifies the syllogism it applies to, rules R1–R3 do not modify the identities of the subject, predicate, and middle term, though R2 and R3 change the sign of both instances of a term on the right. Rule R4 converts the conclusion as P-S, making P the new subject and vice versa; the premises remain untouched other than switching the identities of S and P, requiring exchanging which is major and minor. Rule R5 can entail any of the five non-identity permutations of S, P, and M.

16.2.5 6 × 4 Tabulation of the Assertoric Syllogisms

Table 16.1 lays out the 24 valid and conditionally valid syllogisms in a form designed to facilitate visual verification of soundness and completeness, and also to make it easy to see what adjustments to the axioms are needed in response to certain adjustments of the rules.

The four columns are indexed by their respective figures, numbered in Gray-code-plus-one as 1,2,4,3 and wrapping around to form a cylinder. The connectors between adjacent syllogisms denote applications of single rules. Where no single rule applies a line is drawn to indicate blocking and the need to go round via a chain of deductions, if even possible. Two instances of Rule R1 labeled $-cn-$, convertng the minor premise in each case, wrap around to connect Figs. 1 and 3.

Rule R1 appears as $-cj-$ six times and $-cn-$ five times (the two at far left are the same occurrences as at far right) denoting the converse of respectively major and minor premises, changing figure but not mood. Rule R2 appears six times as *ojc* denoting complementation of the major premise and conclusion, changing mood but not figure. Rule R3 appears three times as *ojn* denoting complementation of both premises, also changing only mood.

It is straightforward and not even terribly tedious to verify that every possible application of Rules R1–R3 appears explicitly in the table. Only two other connections are made, one instance of Rule R4, namely *rc* in Fig. 3, and one instance of

Table 16.1 Interdeducibility from the four axioms (shown in boldface)

	1		2		4		3	
	M-P S-M		P-M S-M		P-M M-S		M-P M-S	
$-cn-$	EIO	$-cj-$	EIO	$-cn-$	EIO	$-cj-$	EIO	$-cn-$
	ojc		*ojn*				*ojc*	
$-cn-$	AII		AOO		**AAI**		AII	$-cn-$
							cc	
	EAE	$-cj-$	EAE		IAI	$-cj-$	IAI	
	ojc		*ojn*		*mtj*		*ojc*	
	AAA		AEE	$-cn-$	AEE		OAO	
	EAO	$-cj-$	EAO		EAO	$-cj-$	EAO	
	ojc		*ojn*				*ojc*	
	AAI		AEO	$-cn-$	AEO		**AAI**	

Rule R5, namely *mtj* in Fig. 4 (the *j* names the particular premise in the syllogism IAI-4 as the one to contradict), both changing mood but not figure.

These 22 horizontal and vertical connections create a graph with four connected components each containing one axiom. If for a moment we replace *cc* and *mtj* by a blocking line, corresponding to allowing only Rules R1–R3, we now have six connected components. If we add AII-1 and IAI-4 as Axioms A5 and A6 respectively, each component now contains exactly one axiom.

In the following theorems we take the criterion for completeness to be deducibility of all 24 syllogisms from the listed axioms using the listed rules.

The foregoing observations give a quick proof of the following.

Theorem 5 *Axioms A1–A6 and Rules R1–R3 completely axiomatize Aristotle's syllogisms.*

The only other instance of Rule 4, not shown, is accounted for by noting that it carries AAI-3 to itself and therefore creates no new connections. This gives another easy theorem trading off axioms for rules.

Theorem 6 *Axioms A1–A5 and Rules R1–R4 completely axiomatize Aristotle's syllogisms.*

Rule R5 is much more broadly applicable. The instance *mtj* completes the connection of the 15 presyllogisms, those forms having one particular after contradicting the conclusion, namely all but AAI-4 in the top four rows. However it is straightforward to show (and intuitively obvious) that R5 carries presyllogisms to presyllogisms, whence all further instances of its use can have no further impact on the connectivity of Table 16.1.

In this way we have given a largely visual proof of the main theorem promised in the abstract:

Theorem 7 *Axioms A1–A4 and Rules R1–R5 completely axiomatize Aristotle's syllogisms.*

Rule R4's limitation to Fig. 3 allowed for a visually simple proof of Theorem 7. It is again easy to see that dropping that restriction adds three more connections, between AII-1 and IAI-4, EAE-1 and AEE-4, and AAI-4 and AAI-1. Since the first two are within the unconditional component they change nothing. However connecting AAI-4 and AAI-1 brings the former out of solitary confinement to permit a further reduction in axioms merely by broadening the applicability of a rule, summarized thusly.

Theorem 8 *Axioms A1–A3 and Rules R1–R5 with R4 unrestricted completely axiomatize Aristotle's syllogisms.*

Although Aristotle eventually adopted Fig. 4, his initial identification of it with Fig. 1 can be understood as viewing all three of AAI-4, IAI-4, and AEE-4 as being at least morally part of Fig. 1.

Similarly lifting the restriction on Rule 5 will enable the further elimination of Axiom A3. However before tackling this it is worthwhile investigating the meaning of the axioms in order to have a clearer picture of the implications of deducing one universal conditional syllogism from another.

But first this interruption: we now have enough material for the promised

Proof (of Theorem 4) The principle is easily verified by hand for AAA-1 and AII-1 as its two canonical instances for each of XaY and XiY. All remaining unconditionally valid syllogisms are reachable from these two via Rules R1–R4. Rule R1 (cj and cn) involves no relabeling of S, M, or P while Rule R4 (rc) only interchanges S and P, whence the need for part (ii) of the definition of interpolation. Rule R2 (ojc) complements P while Rule R3 complements M. Since no rule complements S there is no need for the more general S'. □

16.2.6 Axiom Semantics

We propose to interpret the three terms of each of axioms A1–A4 as three vertices S, M, P of a graph; each sentence of the form XaY as a path from X to Y, and the conclusion XiY of the conditional axioms as a pair of paths to respectively X and Y, both starting from a fourth vertex 1 and sharing a common first edge to one of S, M, or P.

Table 16.2 exhibits the explicitly given major and minor premises of each axiom as edges labeled respectively j and n, and the tacit assumption of nonemptiness of a term as an edge from 1 to that term.

The conclusion of each axiom is then readily established by the existence of a path satisfying the conditions given above for propositions of the form XaY and XiY. In 2(a) the path n; j from S to P verifies the conclusion SaP, while in 2(b) the paths s and s; n; j from 1 to respectively S and P verify SiP, and similarly for 2(d). In the case of 2(c), `AAI-3`, even though the paths m; n and m; j to S and P diverge they both start with m which we interpret as a single individual, and hence suffice to witness SiP.

Table 16.2(e) is included to show the difficulty in meeting these conditions for `AAI-2`. In order to have two paths from from 1 to S and P that start out with the

Table 16.2 (a) `AAA-1` (b) `AAI-1` (c) `AAI-3` (d) `AAI-4` (e) `AAI-2`

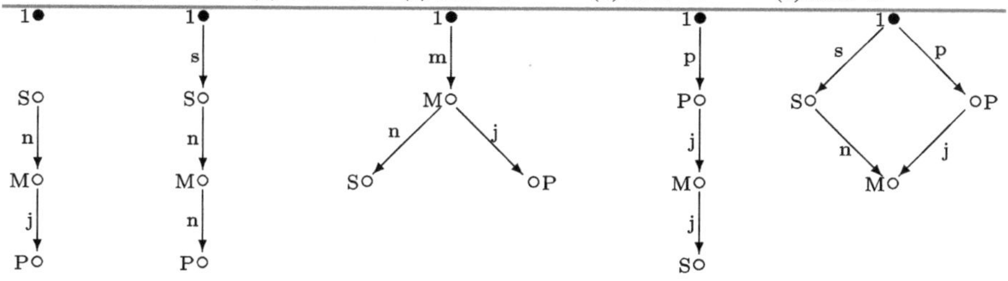

same edge to one of S, M, or P, there would need to be additional tacit assumptions besides those representable by edges from 1.

Now if by "categories" in the Introduction we had meant those of Aristotle, that would indeed be a slender thread connecting Aristotle to Boole. Rather, we had in mind that each of these five graphs freely generates a category C, namely the category whose objects are the vertices and whose morphisms are the paths. Composition of morphisms is the converse of path concatenation, that is, the concatenation $n; j$ corresponds to the composition jn and so on. Concatenation is associative, whence so is its converse. Allowing the empty path at each vertex (not shown) ensures that every object has at least an identity morphism.

The role of the black vertex in each of the five categories C in Table 16.2 is as the object representing the forgetful functor $U: C \to \mathbf{Set}$, namely via C's homfunctor. For all we know the objects of C may have much structure, including potentially billions of other members, not relevant to verification of the associated axiom's validity, which we forget by taking U to be $C(\bullet, -): C \to \mathbf{Set}$. This represents each object v of C as merely the set of morphisms (edges) from $\bullet$ to v sufficient to witness nonemptiness when needed, and each morphism $f: u \to v$ of C as the function $C(\bullet, f): C(\bullet, u) \to C(\bullet, v)$ that maps each member $a: \bullet \to u$ of $U(u)$ to the member $fa: \bullet \to v$ of $U(v)$.

In all five examples $\bullet$ has only one morphism to it from $\bullet$, namely its identity morphism $1_\bullet$, making it a singleton. In 2(a) the remaining objects could for all we care be completely empty, which they can afford to be because the problem of existential import does not arise with presyllogisms. In 2(b)–(d) they all have at least one member each, witnessed by the forgetful functor.

In 2(e) S and P have respective members s and p. We specified that C is *freely* generated, which simply means that distinct paths remain distinct as morphisms, i.e. the square does not commute and M therefore has (at least) two members, namely ns and jp. There is therefore no visible means of support in this semantics of axioms for the conclusion SiP in AAI-2.

These examples suggest the following general criterion for any syllogism with universal premises and a particular conclusion. If the diagram formed by the premises has a least member then that member is the one whose nonemptiness must be assumed. In the absence of such a member, the assumption of nonemptiness even for all the terms of the syllogism is insufficient to warrant judging that syllogism as conditionally valid, with Table 16.2(e) serving as the generic example justifying this rule.

With these insights about conditionally valid syllogisms we can now understand the meaning of applying Rule R5 to them. Formally R5 takes AAI-3 to either EAO-1 or AEO-2 depending on whether it is applied to the major or minor premise respectively. It is a nice exercise to trace the progress of the term required to be nonempty under these deductions and evaluate its meaning in the light of Table 16.2. In any event we have the following.

Theorem 9 *Axioms A1 and A2 and Rules R1–R5 with R4 and R5 unrestricted completely axiomatize Aristotle's syllogisms.*

If the goal is to minimize the number of axioms, this is clearly the way to go. But if Table 16.2 has any appeal, keeping the three conditionally valid components separate by observing the restrictions on rules R4 and R5 is the better choice, because the unrestricted use of Rule R5 does violence to the logical structure of conditional syllogisms. As the abstract makes clear we prefer the latter, making Theorem 7 the *Hauptsatz* for this section, but this is more a matter of personal taste than anything deep.

Consideration of mtj as the passage in Table 16.1 between the two halves of the unconditionally valid syllogisms also reveals great violence: whereas in IAI-4 M interpolates as PiMaS, mtj changes the interpolation radically to $\mathrm{Si}\bar{M}\mathrm{a}\bar{P}$.

If violence to logic is a concern at all, this would be a strong argument for stopping at Theorem 6 and putting up with five axioms and four rules. Or better yet, dropping the restriction on Rule R4, which was only to avoid having to insert a distracting crosslink from AAI-4 to AAI-1, thereby permitting replacing the ridiculously lonely Axiom A4 by A5, AII-1, which is much nicer as an axiom than AAI-4. We'll resume this train of thought in Sect. 16.2.8.

16.2.7 Normalization

Taking Table 16.1 as a complete enumeration of all valid syllogisms, conditional or not, validity of syllogistic forms is decidable simply by searching the table. It is nonetheless of interest to ask whether there is some less brute-force way of deciding validity, and moreover one which in the case of conditionally valid syllogisms automatically identifies the term whose nonemptiness must be assumed. We propose the following normalization procedure aimed at moving as briskly as possible from anywhere in Table 16.1 towards an axiom.

Step 1. If there exists a particular premise, apply Rule R5 to render it and the conclusion universal.

Step 2. If not already at an axiom, work through the list ojc, cj, ojn, cn of rule instances until finding one that is applicable, and apply it.

Step 3. Work backwards through the list from that point applying the previously inapplicable instances if now applicable.

The result is the normal form.

Step 1 takes a long jump towards the given syllogism's axiom if any. Step 2 measures the remaining short distance from the axiom, and Step 3 then takes that many steps to reach it if possible.

While it might seem simpler to skip Step 2 and just apply the list in the other order, unless Step 1 lands at the maximal distance from an axiom among syllogisms with universal premises, this naive method would guarantee taking a step away from the axiom instead of towards it and then getting stuck.

Theorem 10 *A syllogistic form is valid if and only if its normal form is one of the four axioms.*

Proof (Only if, completeness) By inspection of Table 16.1 the procedure carries all entries to one of the four axioms.

(If, soundness) All operations of the procedure are reversible and preserve validity in both directions. Hence the procedure cannot carry any syllogism outside the table into the table, whence any well-formed syllogism leading to an axiom must have been valid to begin with. □

The axiom arrived at then tells which term in the original syllogism if any needs to be nonempty. Note that the subject-predicate-middle identities can change in the course of the deduction; what matters is the actual term, such as *unicorn*, that arrives in that position. Thus if the unicorn started out as the predicate P and ended up as M in AAI-3, it is P in the initial syllogism that needs to be nonempty, not M.

While the procedure uses Rule R5 to leapfrog many steps, modifying it to work without it is a matter of coming up with an extension of Step 2 that can carry IAI-4 and OAO-3 etc. all the way round to AII-1. A case could be made for taking EIO as the axiom for that component, ignoring its figure, which can be reached from anywhere in the component in at most three steps.

This procedure may well be similar to at least some extant normalization procedures in the literature, but its use of Table 16.1 to make the reasoning easy to follow and its uniform method of handling the problem of existential import is to our knowledge new.

16.2.8 A Role for Subalternation

Taking Parson's history of the Square of Opposition (Parsons 1997) as definitive of the state of the art of our understanding of subalternation, all attempts to date at restoring it to the good graces of logical consistency have failed. With Table 16.1 giving a clear picture of what is at stake, we can analyze why, and see whether there is any way round the inconsistencies.

As motivation for even considering subalternation, Axioms A2–A4 of our system D4p serve to justify just the nine conditionally valid syllogisms, costing an average of a third of an axiom per syllogism. One might just as well dispense altogether with axioms and rules, along with issues of soundness and completeness, and simply list the nine along with the term each one needs to be nonempty, as typically done.

But we also struggled in D4p to connect IAI-4 and AEE-4, either via a fifth rule or giving up and settling instead for a fifth axiom. Given that the only purpose served by Rule R5 is to make one single connection, that makes it even more expensive than the conditionally valid axioms!

Subalternation to the rescue. As noted earlier it permitted deducing AAI-3 from AII-3, namely by licensing the strengthening of the minor premise from MiS to

MaS. In similar fashion we may deduce AAI-4 from IAI-4, and AAI-1 from AAA-1 (weakening the conclusion instead of strengthening a premise). This dispenses with all three conditional axioms, leaving us with just Axiom A1 provided we keep Rule R5.

Moreover we can see at a glance which term needs to be nonempty for a conditional syllogism to hold, namely the subject of the sentence modified by subalternation.

And if in Figs. 1 and 2, whose minor premise is S-M, we allow joint weakening of the minor premise and conclusion, we obtain the unconditionally sound Axiom A5, AII-1, from AAA-1, which Theorem 6 showed could replace Rule R5, with no requirement that any term be nonempty. We would then need only Axiom A1, Rules R1–R4, and subalternation in suitable form.

But as pointed out by Russell, one can expect anything to follow from an inconsistent rule, so perhaps we should not be amazed that so much can follow from this naive appeal to subalternation.

The constructive approach in such a case is to limit the offending construct to a domain of applicability small enough to avoid any inconsistency. Table 16.1 facilitates seeing where those limits should be set in the case of subalternation by explicitly listing all possible applications of Rules R1–R3, making it easy to see at a glance that none of them lead out of the region of the conditionally valid syllogisms. And it is equally easy to see that the region is closed under Rule R4 as well after noting that it is applicable only to AAI-1, AAI-4, and AAI-3 in that region, interchanging the first two and carrying the third to itself.

We are now in a position to improve our preliminary system D4p to D4. Rules R1–R4 are unchanged from D4p. In rules R5-R6, by "weaken" we mean replace *a* or *e* by *i* or *o* respectively, and the reverse for "strengthen".

System D4

Axioms

A1. AAA-1 (Barbara)

Rules

R1. Convert any *e* or *i* premise.
R2. In Figs. 1 and 3, obvert the major premise and the conclusion.
R3. In Fig. 2, obvert both premises.
R4. In Fig. 3, convert any *e* or *i* conclusion and interchange the premises.
R5. Strengthen a particular premise or weaken a universal conclusion.
R6. With premise SaM or SeM and conclusion SaP or SeP, weaken both.

The foregoing discussion, combined with the easily seen fact that Rules R5 and R6 are not applicable to any conditionally valid syllogism, makes the following immediate.

Theorem 11 *System D4 is sound and complete.*

Soundness entails consistency. The system has the additional benefit that Rule R5, as the only rule deducing a conditional from an unconditional syllogism, provides the term whose nonemptiness is required, namely the subject of the modified sentence, which Rules R1–R4 subsequently do not modify within the region of conditionally valid syllogisms.

The one axiom and six rules of System D4 is arguably a significant improvement over the four axioms and five rules of System D4p. And it is very much in the spirit of those sequent calculi that require only a single axiom, with AAA-1 as the natural counterpart in Aristotle's syllogistic of $P \vdash P$.

16.3 Boole's Logic

16.3.1 The Language Problem

In 1847 George Boole wrote a short pamphlet (Boole 1847) entitled *The Mathematical Analysis of Logic, being an essay towards a calculus of deductive reasoning.* In the preface he wrote "In the spring of the present year, my attention was directed to the question then moved between Sir W. Hamilton and Professor De Morgan; and I was induced by the interest which it inspired, to resume the almost-forgotten thread of former inquiries."

Boole was referring to an acrimonious debate that had lately arisen between University College London's Augustus De Morgan and Edinburgh University's Sir William Hamilton, 9th Baronet. The dispute regarded a matter of priority in setting Aristotle's logic on a firmer footing. It continued for nearly a decade until Hamilton's death in 1856 at age 68.

In his preface Boole asked his readers not to dismiss his algebraization of truth as inconsistent but to consider the whole without concern for its departure from conventional algebra. Taking for his language that of algebra, namely addition, subtraction, and multiplication (but not division), he lists as his "first principles" three properties of multiplication: distributivity $x(y + z) = xy + xz$ over addition, commutativity $xy = yx$, and idempotence, $x^2 = x$.

Now whereas the first two of these properties are the laws making an additive (i.e. abelian) group a commutative ring, the third makes it what today we call a Boolean ring, the basic such being the ring of integers mod 2.

This ring witnesses the consistency of Boole's logic.

Boole himself failed to recognize this, despite giving explicit definitions of both inclusive or and exclusive or. The stumbling block for both him and his critics appeared to have been their reluctance to accept $x + x = 0$ as a property of exclusive-or. Instead Jevons, Peirce, and Schroeder made the decisive move from rings to lattices by focusing on the logical operations of conjunction and disjunction as the basic operations of an algebra of logic. Peirce however believed incorrectly that every lattice was distributive, leaving to Schroeder the distinction of being the first to properly axiomatize a Boolean algebra. In modern language Schroeder defined it as a bounded distributive lattice (one with 0 and 1 as lower and upper bounds) that was *complemented*, meaning it had a unary operation $\neg$ of negation satisfying $x \vee \neg x = 1$ and $x \wedge \neg x = 0$.

The lattice-based approach to Boolean logic then developed slowly for a few more decades. But then in 1927, three-quarters of a century after Boole's ring-based proposal, Ivan Ivanovich Zhegalkin realized (Zhegalkin 1927) that there was nothing wrong with that proposal after all. Zhegalkin proposed to define a Boolean operation as what we now call variously a Zhegalkin polynomial, algebraic normal form, or Reed-Muller expansion.

Nine years later, communication between Moscow and the west having been greatly slowed by the fallout from the 1917 revolution, topologist Marshall Stone independently realized essentially the same thing (Stone 1936) in 1936. Stone had been writing up the duality he'd discovered between Boolean algebras and what we now call Stone spaces. He was in the middle of explaining Boolean algebras by analogy with rings when it dawned on him that his explanation was more than a mere analogy: a Boolean algebra actually *was* a ring! As an approach to Boolean algebras rings are an alternative to lattices. Paul Halmos's little book on Boolean algebras (Halmos 1963) adopts the ring-based definition without even mentioning the alternatives.

But there is yet another approach, via the propositional fragment of the intuitionistic logic proposed by Luitzen Egbertus ("Bertus") Jan Brouwer (Brouwer 1920) and formalized by his student Arend Heyting (Heyting 1930). Its language is that of bounded lattices: conjunction, disjunction, 0 and 1, supplemented with implication $x \to y$. Implication is defined implicitly in terms of the lattice by either $x \wedge y \leq z$ iff $y \leq x \to z$ or its equational equivalent. Negation is defined explicitly by $\neg x = x \to 0$ and hence need not be included among the basic operations and constants, which are $\wedge$, $\vee$, $\to$, and 0 (1 is definable as $0 \to 0$). These are the properties required of a Heyting algebra. A Boolean algebra can then be defined as a Heyting algebra satisfying double negation, $(x \to 0) \to 0 = x$, or equivalently the law of excluded middle, $x \vee (x \to 0) = 1$.

These insights into Boolean algebra create something of a ontological dilemma: what exactly is a Boolean algebra, and how should it be defined? In particular, on what language should it be based? A language consists not of operations but of operation *symbols*. Should we take the symbols to be those of complemented lattices, $\wedge, \vee, \neg, 0, 1$, or of rings with unit, $+, *, 0, 1$, or of intuitionistic logic, $\wedge, \vee, \to, 0, 1$, or something else again?

The following offers one approach that is indisputedly neutral, yet not as impractical as it might at first seem. The approach exploits the fact that Boolean algebras form a locally finite variety, a rarity in algebras with binary operations that we discuss in more detail later. We first described the approach in a Wikipedia article (Pratt 2006); its account here clarifies the connection with categories while fixing some technical problems as well as the problem that prior to now it was "original research", officially *verboten* on Wikipedia even though we seem to have gotten away with it.

16.3.2 The Language MaxBool

We define the language MaxBool as follows.

MaxBool is the language whose operation symbols are the finitary operations on $\{0, 1\}$.

The m input bits of an m-ary operation have 2^m possible combinations of values. The operation assigns one bit to each. Since the assignments are independent, there are 2^{2^m} possible combinations of such assignments. Hence MaxBool has that many operation symbols $f, g, \ldots$ of each finite arity $m \geq 0$.

Ordinarily the difference between syntax and semantics is that syntax is what you write while semantics is what the writing denotes in some universe. Here we are doing the opposite: interpreting the semantics of an m-ary operation in the universe $\{0, 1\}$ as an operation symbol. How could reversing the roles of syntax and semantics in this way make sense?

An m-ary operation on $\{0, 1\}$ can be understood syntactically as a string of 2^m bits,[3] namely its truth table. Every string of length 2^m constitutes an operation symbol, whose arity m is given by the logarithm to the base two of its length. Figure 16.5 gives five binary and seven ternary operation symbols as examples.

Above the line are the two binary projections or variables x and y and the three ternary such (the sense in which variables are projections will gradually become clear, in Sect. 16.3.7 if not sooner). Reading the columns from right to left it can be seen that the projections taken together are counting in binary notation (without leading zero suppression): from 0 to 3 for the binary projections and from 0 to 7 for the ternary ones. Each column of the projections uniquely identifies its horizontal position (distance from the right end) and can therefore be understood as a column index.

Below the line are some examples of truth tables: three binary and four ternary operations formed coordinatewise as Boolean combinations of operations above them. The last is the ternary majority operation, returning whichever of 0 or 1 is in the majority at its inputs.

x	1010
y	1100
$x \wedge y$	1000
$x \vee y$	1110
$x \rightarrow y$	1101

x	10101010
y	11001100
z	11110000
$x \wedge y$	10001000
$y \wedge z$	11000000
$z \wedge x$	10100000
$(x \wedge y) \vee (y \wedge z) \vee (z \wedge x)$	11101000

Fig. 16.5 Five binary and seven ternary operation symbols of MaxBool

[3]When succinctness is important symbols can be shortened by a factor of four by writing them in hexadecimal, exploited in the C program mentioned at the end of Sect. 16.2.2.

The set of all bit strings of the same length is closed under the Boolean operations applied coordinatewise and hence forms a finite Boolean algebra. For those of length 2^m, the closure under the Boolean operations of the variables $x_1, \ldots, x_m$ as represented above is the whole Boolean algebra, and constitutes the free Boolean algebra $\mathcal{F}_m$ on m generators.

A fringe benefit of taking the language to consist of all operations on $\{0, 1\}$ is that separate provision for variables becomes unnecessary because they arise as projections. The variable names x, y, z above are notationally convenient synonyms for the first three variables in the infinite supply $x_1, x_2, x_3, x_4, \ldots$. Each variable x_i for $i \geq 1$ is of arity i or more depending on context as treated in the next section.

16.3.3 MaxBool Terms

As customary in algebra, a MaxBool term is either an atom or an application. The atoms include the constants 0 and 1 and the variables $x_1, x_2, \ldots$, while the applications have the form $f(t_1, \ldots, t_n)$ where f is an n-ary operation symbol and the t_i's are MaxBool terms constituting the n ***arguments*** of f. Atoms are of height one while the height of an application is one more than that of its highest argument.

A novelty with MaxBool atoms is that every operation symbol is permitted to be an atom, subject to the constraint that all the atoms of a term have the same arity. The arity of a term is the common arity of its atoms. Since in MaxBool the variables up to x_m are representable as m-ary operation symbols, those variables are permitted in an m-ary term but no others.

An m-ary term has two personalities. One is as an ordinary term in Boolean algebra each of whose atoms is understood as an m-ary operation symbol applied to a fixed tuple $(x_1, \ldots, x_m)$ of variables, the same tuple for every atom. In that personality the variables are the real leaves of the term, having height zero (the atoms still have height one) and taking values in the Boolean algebra $2 = \{0, 1\}$. A Boolean identity is a pair of m-ary terms having the same value, 0 or 1, for all values 0 or 1 of the m variables. These form the theory T_m consisting of all Boolean identities involving at most m variables. The terms of height zero need not be mentioned in T_m because every atom is applied to the same m-tuple of them.

The other personality is as a constant term, one whose atoms play the role of constants valued in the free Boolean algebra $\mathcal{F}_m$. In this personality the height-zero variables of the other personality disappear leaving the atoms as the leaves, which serve as constants having fixed standard interpretations in $\mathcal{F}_m$. Variables are still permitted but only in the personality of constants, namely height-one m-ary operation symbols, e.g. 1010 and 1100 when $m = 2$.[4] A Boolean identity is a pair of m-ary terms evaluating to the same element of $\mathcal{F}_m$. The intent is for these to form the same theory T_m as for the other personality.

---

[4]The C program mentioned at the end of Sect. 16.2.2 represents S, P, and M as respectively 10101010, 11001100, and 11110000.

16.3.4 Reduction of Height-Two Terms in $\mathcal{F}_m$

In its second personality, an m-ary MaxBool term interpreted in $\mathcal{F}_m$ denotes an element of $\mathcal{F}_m$, namely an m-ary operation or bit string of length 2^m.

Atoms denote themselves.

Terms $f(t_1, \ldots, t_n)$ of height two are trickier. The m-ary operation denoted by such a term is the result of substituting each m-ary operation t_i for the i-th variable of the n-ary operation. In ordinary algebra this would entail substituting n m-variable polynomials for the variables of an n-variable polynomial, for example substituting $x + y$ for a and $x - y$ for b in ab, which for larger polynomials would require considerable simplification to put the result in a suitable normal form.

In MaxBool the entire process can be accomplished with a form of matrix multiplication.

Represent an n-tuple of m-ary operation symbols as an $n \times 2^m$ bit matrix whose n rows are the n operation symbols. This will be matrix B in the following. The n-ary operation f will be the matrix A with $p = 1$; the general case of arbitrary p is just as easy to describe.

Given a $p \times 2^n$ matrix A and an $n \times 2^m$ matrix B, define their ***product*** $A \circ B$ to be the $p \times 2^m$ result of replacing each column of B by the column of A indexed by the replaced column. Formally,

$$(A \circ B)_{ij} = A_{it} \quad \text{where } t = \lambda k.B_{kj} = B_{*j} \text{ (column } j \text{ of } B).$$

Figure 16.6 illustrates this product using the examples from Fig. 16.5. Take A to be all seven rows of the ternary example, and B to be the bottom three rows of the binary example, i.e. omitting the projections.

Reading B from left to right, the successive columns of B pick out columns 7, 6, 2, and 4 of A (equivalently, the columns of A whose top 3 rows match the replaced columns of B), which are then assembled in that order to form $A \circ B$.

We retained the horizontal line in A to make it block structured, mainly to show the result of multiplying the upper block of A by B, namely B itself. It should be clear that the upper block of A is in fact the 3×2^3 identity matrix for this product, yet another role for the projections taken collectively.

Fig. 16.6 The product $A \circ B$

A
10101010
11001100
11110000
10001000
11000000
10100000
11101000

$\circ$

B
1000
1110
1101

$=$

$A \circ B$
1000
1110
1101
1000
1100
1000
1100

16.3.5 An Axiomatization of Boolean Algebra

The following system, MB_m, uses the foregoing product to axiomatize T_m, the theory of Boolean algebra based on m variables. The atomic subterms of terms are m-ary operation symbols; all other operation symbols, namely those applied to n-tuples, are of arity $n \leq m$.

System MB_m.

A1. $f(t_1, \ldots, t_n) = f \circ T$ where f is an n-ary operation symbol, the t_i's are atoms, and T is the $n \times 2^m$ matrix whose i-th row is t_i.

R1. $\vdash \ t = t$.

R2. $s = t \ \vdash \ t = s$.

R3. $s = t, \ t = u \ \vdash \ s = u$.

R4. $s_1 = t_1, \ \ldots, \ s_n = t_n \ \vdash \ f(s_1, \ldots, s_n) = f(t_1, \ldots, t_n)$ where f is an n-ary operation symbol.

R1–R3 realize the properties of an equivalence relation, which R4 extends to a congruence.

Theorem 12 *MB_m is complete, meaning that it proves every equation $s = t$ of T_m.*

Proof We begin with the case when t is an atom, namely an m-ary operation. We proceed by induction on the height of s.

For height one s is an atom. But distinct m-ary atoms must denote distinct operations, whence the only s for which $s = t$ can hold must be t itself, which R1 supplies.

For height two s must equal a uniquely determined m-ary atom, which $A1$ supplies.

For height $h > 2$ with s of the form $f(s_1, \ldots, s_n)$, the induction hypothesis supplies atoms $t_1, \ldots, t_n$ such that $s_i = t_i$ has been proved for each i. Furthermore A1 supplies $f(t_1, \ldots, t_n) = a$ for some atom a, which must therefore be t. Hence by R4 and R3, $s = t$.

For general t, if $s = t$ is an identity both must reduce to the same atom a. We have shown that MB proves $s = a$ and $t = a$, whence by R2 and R3 we have $s = t$. □

The theory of Boolean algebras is then $T = \bigcup_m T_m$.

Note that this union is disjoint. This is because every term contains at least one atom, whose length determines which T_m it came from. However every member of T_m has a counterpart in T_{m+1} obtained by duplicating every atom to make it twice as long. In that sense every equation appears infinitely often, once in each T_m beyond a certain point.

As with any of the other languages for Boolean algebra and their axiomatizations, a Boolean algebra is a model of T.

16.3.6 Discussion

Propositional logic is notorious for having relatively simple theorems with long proofs. In contrast proofs in MB simply evaluate every term directly, in a number of steps linear in the number of symbols in the term, with no need for auxiliary variables. In terms of running time however there is no free lunch: MB replaces exponentially long proofs with exponentially long operation symbols, slowing down each proof step exponentially. They are not infinitely long however, which is what would happen if instead of representing say x_1 as 10 in MB_1, 1010 in MB_2 and so on we represented it once and for all as the infinite string 10101010....

This approach to axiomatization would also become infinite if used to axiomatize say commutative rings. This is because whereas there are only finitely many Boolean polynomials in the variables $x_1, \ldots, x_m$, there are infinitely many integer-coefficient polynomials in m variables. With the requirement $n \leq m$ the schema A1 expands to $2^{2^n} \times (2^{2^m})^n$ axioms summed over $0 \leq n \leq m$. For $m = 0, 1, 2, 3, \ldots$ this comes to respectively 2, 18, 4096, 4,296,016,898, ... axioms when the schema A1 is expanded out.

16.3.7 The Categorical Basis for MB

As defined by Lawvere (1963), an ***algebraic theory*** is a category having for its objects the natural numbers, such that for all n, n is the n-th power of 1. This means that for each n there exists an n-tuple $(x_1, \ldots, x_n)$ of morphisms $x_i : n \to 1$, the ***projections*** of n onto 1, such that for each n-tuple $(t_1, \ldots, t_n)$ of morphisms $t_i : m \to 1$ there exists a unique morphism $t : m \to n$ such that $x_i t = t_i$. The morphism t encodes the n-tuple of morphisms to 1 as a single morphism to n.

MaxBool adapts Lawvere's notion to the case of locally finite varieties, which permit a straightforward syntactic definition of composition. The category C_{MB} has as objects the natural numbers, as morphisms from m to n the $n \times 2^m$ bit matrices, and as composition the matrix multiplication defined above. The identity at m is the $m \times 2^m$ matrix $(x_1, \ldots, x_m)$, whose rows constitute the projections making m the m-th power of 1.

Figure 16.6 illustrates this in the case $n = 3$. The requisite 3-tuple (x_1, x_2, x_3) of projections $x_i\colon 3 \to 1$ are the first three rows of A. In the case when (t_1, t_2, t_3) consists of the 1×2^2 truth tables for respectively $x \wedge y$, $x \vee y$, and $x \to y$ at lower left of Fig. 16.5, there exists a morphism $t : 2 \to 3$ such that $x_i t = t_i$, namely the 3×2^2 matrix B, such that $x_i \circ B$ is t_i for each x_i. Furthermore any other 3×2^2 matrix must contain a row not equal to t_i for some i, whence B is the *unique* such morphism.

The category C_{MB} is easily seen to be isomorphic to the full subcategory of **Set** consisting of the finite powers of $2 = \{0, 1\}$. That m is the m-th power of 1 makes more sense when m and 1 are interpreted concretely as the power sets 2^m and 2^1. From the perspective of language theory the power 2^m can be identified with the set of bit strings of length m, with 2^0 consisting of just the empty string. Each function from 2^m to 2^n can be understood as n functions from 2^m to 2, i.e. n m-ary operations, one for each of the n bits of the result. The corresponding matrix takes these to be the n rows, each the truth table of an m-ary operation.

Two paths from m to n that are mapped by composition to the same morphism from m to n form a ***commutative diagram***. The equations of an algebraic theory are realized as its commutative diagrams to 1.

A set-valued model of an algebraic theory T is a functor from T to **Set** that preserves products. Such a functor interprets 1 as a set X, n as X^n, and n-ary operation symbols $f : n \to 1$ as functions $f: X^n \to X$. Functors preserve commutative diagrams, whence any equation holding in the theory holds of every model of the theory.

A Boolean algebra is any model of $\mathcal{C}_{MB}$ as an algebraic theory.

Algebraic theories do not assume a locally finite variety. Composition is always definable by substitution. MB is special because composition has a syntactically simple form that justifies organizing it as an axiom.

16.4 Conclusion

My main goal in this paper has been to contribute a few hopefully novel insights into two logics neither of which could be considered low-hanging fruit the way computer science was in the early 1970s. I laid out a system of deductions for Aristotle's syllogisms designed to permit evaluating their soundness and completeness visually. And I gave a locally finite axiomatization of Boolean algebras that did not need to start from any particular point of view such as that they are lattices, or rings, or Heyting algebras, etc.

I also drew attention to how such things can look when viewed through the lens of category theory. In the case of syllogisms, category theory enters at such a trivial level that the purpose served is not so much insight into syllogisms as into category theory, which is often pitched inaccessibly higher. In the case of Boolean algebra it was the other way round: it was only by contemplating Lawvere's presentation of an algebraic theory as a category closed under products that I realized how the subject could be made independent of choice of basis for the operations.

Although there have been logicians who have seen deeper into logic, particularly within the past century, to date Aristotle and Boole stand out as the inventors of two logics each of which stood the test of time longer than any other. Though Aristotle's logic dominated the field for over two millennia, it is fair to say that Boole's logic has pushed it into something of a backwater. I frequently hear calls from type theory, programming languages, category theory, and proof theory advocating the same fate for Boole's logic, but from my perspective as an exponent of both the hardware and software of digital logic it will be a while yet.

References

Boole, G. (1847). *The mathematical analysis of logic, being an essay towards a calculus of deductive reasoning*. London: Macmillan.

Brouwer, L. E. J. (1920). Intuitionische mengenlehre. *Jahresber. Dtsch. Math. Ver*, *28*, 203–208.

Corcoran, J. (1974). Aristotle's natural deduction system. In J. Corcoran (Ed.), *Ancient logic and its modern interpretations, synthese historical library* (pp. 85–131). Berlin: Springer.

Halmos, P. R. (1963). *Lectures on Boolean algebras*. Van Nostrand.

Heyting, A. (1930). Die formalen Regeln der intuitionischen Logik. In *Sitzungsberichte Die Preussischen Akademie der Wissenschaften* (pp. 42–56). Physikalische-Mathematische Klasse.

Lawvere, W. (1963). Functorial semantics of algebraic theories. *Proceedings of the National Academy of Sciences*, *50*(5), 869–873.

Lukasiewicz, J. (1957). *Aristotle's Syllogistic* (2nd ed.). Oxford: Clarendon Press.

Parikh, R. (1978). A completeness result for a propositional dynamic logic. In *Lecture notes in computer science* (Vol. 64, pp. 403–415). Berlin: Springer.

Parsons, T. (1997). The traditional square of opposition. In *Stanford encyclopedia of philosophy*. Stanford University.

Pratt, V. R. (1975). Every prime has a succinct certificate. *SIAM Journal of Computing*, *4*(3), 214–220.

Pratt, V. R. (1976). Semantical considerations on Floyd–Hoare logic. In *Proceedings of the 17th annual IEEE symposium on foundations of computer science* (pp. 109–121).

Pratt, V. R. (2006). Boolean algebras canonically defined. *Wikipedia*.

Stone, M. (1936). The theory of representations for Boolean algebras. *Transactions of the American Mathematical Society*, *40*, 37–111.

Zhegalkin, I. I. (1927). On the technique of calculating propositions in symbolic logic. *Matematicheskii Sbornik*, *43*, 9–28.

Chapter 17
Galois Theory of Algorithms

Noson S. Yanofsky

Abstract Many different programs are the implementation of the same algorithm. The collection of programs can be partitioned into different classes corresponding to the algorithms they implement. This makes the collection of algorithms a quotient of the collection of programs. Similarly, there are many different algorithms that implement the same computable function. The collection of algorithms can be partitioned into different classes corresponding to what computable function they implement. This makes the collection of computable functions into a quotient of the collection of algorithms. Algorithms are intermediate between programs and functions:

$$\text{Programs} \twoheadrightarrow \text{Algorithms} \twoheadrightarrow \text{Functions}.$$

Galois theory investigates the way that a subobject sits inside an object. We investigate how a quotient object sits inside an object. By looking at the Galois group of programs, we study the intermediate types of algorithms possible and the types of structures these algorithms can have.

Keywords Algorithms · Galois theory · Computable functions · Monoidal categories

N.S. Yanofsky (✉)
Department of Computer and Information Science,
Brooklyn College CUNY, Brooklyn, NY 11210, USA
e-mail: noson@sci.brooklyn.cuny.edu

N.S. Yanofsky
Computer Science Department, The Graduate Center CUNY,
New York, NY 10016, USA

C. Başkent et al. (eds.), *Rohit Parikh on Logic, Language and Society*,
Outstanding Contributions to Logic 11, DOI 10.1007/978-3-319-47843-2_17

17.1 Introduction

As an undergraduate at Brooklyn College in 1989, I had the good fortune to take a masters level course in theoretical computer science given by Prof. Rohit Parikh. His infectious enthusiasm and his extreme clarity turned me onto the subject. I have spent the last 25 years studying theoretical computer science with Prof. Parikh at my side. After all these years I am still amazed by how much knowledge and wisdom he has at his fingertips. His broad interests and warm encouraging way has been tremendously helpful in many ways. I am happy to call myself his student, his colleague, and his friend. I am forever grateful to him.

In this paper we continue the work in Yanofsky (2011) where we began a study of formal definitions of algorithms (knowledge of that paper is not necessary for this paper.) The previous paper generated some interest in the community: Yuri I. Manin looked at the structure of programs and algorithms from the operad/PROP point of view (Manin 2009a) (Chap. 9), see also Manin (2009b, c) where it is discussed in the context of renormalization; there is an ongoing project to extend this work from primitive recursive functions to all recursive functions in Manin and Yanofsky (2010); Ximo Diaz Boils has looked at these constructions in relations to earlier papers such as Burroni (1986), Maietti (2003), Thibault (1982); Andreas Blass, Nachum Dershowitz, and Yuri Gurevich discuss the paper in Blass et al. (2008) with reference to their definition of an algorithm.

Figure 17.1 motivates the formal definition of algorithms. On the bottom is the set of computable functions. Two examples of computable functions are given: the sorting function and the find max function. On top of the diagram is the set of programs. To each computable function on the bottom, a cone shows the corresponding set of programs that implement that function. Four such programs that implement the sorting function have been highlighted: $\texttt{mergesort}_a$, $\texttt{mergesort}_b$, $\texttt{quicksort}_x$ and $\texttt{quicksort}_y$. One can think of $\texttt{mergesort}_a$ and $\texttt{mergesort}_b$ as two implementations of the mergesort algorithm written by Ann and Bob respectively. These two programs are obviously similar but they are not the same. In the same way, $\texttt{quicksort}_x$ and $\texttt{quicksort}_y$ are two different implementations of the quicksort algorithm. These two programs are similar but not the same. We shall discuss in what sense they are "similar." Nevertheless programs that implement the mergesort algorithm are different than programs that implement the quicksort algorithm. This leads us to having algorithms as the middle level of Fig. 17.1. An algorithm is to be thought of as an equivalence class of programs that implement the same function. The mergesort algorithm is the set of all programs that implement mergesort. Similarly, the quicksort algorithm is the set of all programs that implement quicksort. The set of all programs are partitioned into equivalence classes and each equivalence class corresponds to an algorithm. This gives a surjective map from the set of programs to the set of algorithms.

One can similarly partition the set of algorithms into equivalence classes. Two algorithms are deemed equivalent if they perform the same computable function.

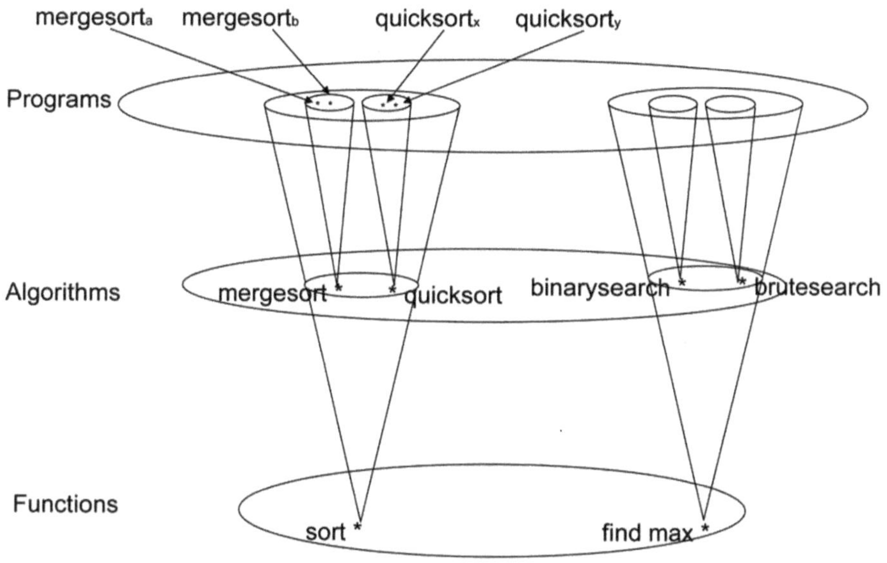

Fig. 17.1 Programs, algorithms and functions

This gives a surjective function from the set of algorithms to the set of computable functions.

This paper is employing the fact that equivalence classes of programs have more manageable structure than the original set of programs. We will find that the set of programs does not have much structure at all. In contrast, types of algorithms have better structure and the set of computable functions have a very strict structure.

The obvious question is, what are the equivalence relation that say when two programs are "similar?" In Yanofsky (2011) a single tentative answer was given to this question. Certain relations were described that seem universally agreeable. Using these equivalence relations, the set of algorithms have the structure of a category (composition) with a product (bracket) and a natural number object (a categorical way of describing recursion.) Furthermore, we showed that with these equivalence relations, the set of algorithms has universal properties. See Yanofsky (2011) for more details.

Some of the relations that describe when two programs are "similar" were:

- One program might perform $Process_1$ first and then perform an unrelated $Process_2$ after. The other program might perform the two unrelated processes in the opposite order.
- One program will perform a certain process in a loop n times and the other program will "unwind the loop" and perform it $n - 1$ times and then perform the process again outside the loop.
- One program might perform two unrelated processes in one loop, and the other program might perform each of these two processes in its own loops.

In Blass et al. (2008), the subjectivity of the question as to when two programs are considered equivalent was criticized. While writing (Yanofsky 2011), we were aware that the answer to this question is a subjective decision (hence the word "Towards" in the title), we nevertheless described the structure of algorithms in that particular case. In this paper we answer that assessment of Yanofsky (2011) by looking at the *many different sets* of equivalence relations that one can have. It is shown that with every set of equivalence relations we get a certain structure.

The main point of this paper is to explore the set of possible intermediate structures between programs and computable functions using the techniques of Galois theory. In Galois theory, intermediate fields are studied by looking at automorphism of fields. Here we study intermediate algorithmic structures by looking at automorphism of programs.

A short one paragraph review of Galois theory is in order. Given a polynomial with coefficients in a field F, we can ask if there is a solution to the polynomial in an extension field E. One examines the group of automorphisms of E that fix F, i.e., automorphisms $\phi : E \longrightarrow E$ such that for all $f \in F$ we have $\phi(f) = f$.

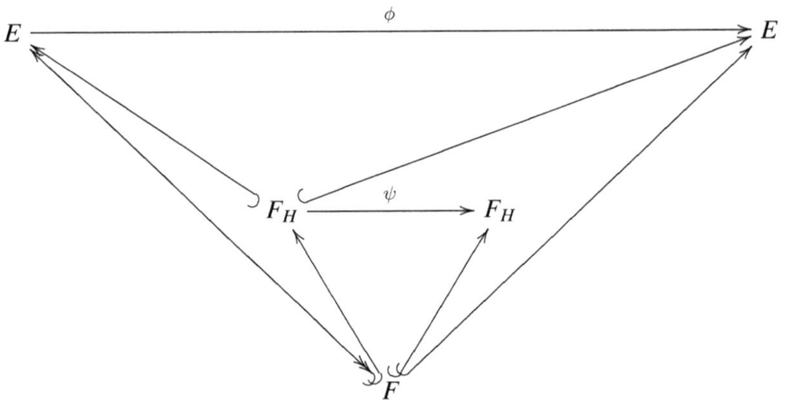

(17.1)

This group is denoted $Aut(E/F)$. For every subgroup H of $Aut(E/F)$ there is an intermediate field $F \subseteq F_H \subseteq E$. And conversely for every intermediate field $F \subseteq K \subseteq E$ there is a subgroup $H_K = Aut(E/K)$ of $Aut(E/F)$. These two maps form a Galois connection between intermediate fields and subgroups of $Aut(E/F)$. If one further restricts to normal intermediate fields and normal subgroups than there is an isomorphism of partial orders. This correspondence is the essence of the Fundamental Theorem of Galois Theory which says that the lattice of normal subgroups of $Aut(E/F)$ is isomorphic to the dual lattice of normal intermediate fields between F and E. The properties of $Aut(E/F)$ mimic the properties of the fields. The group is "solvable" if and only if the polynomial is "solvable."

In order to understand intermediate algorithmic structures we study automorphisms of programs. Consider all automorphisms of programs that respect functionality. Such automorphisms can be thought of as ways of swapping programs for other programs that perform the same function. An automorphism ϕ makes the following outer triangle commute.

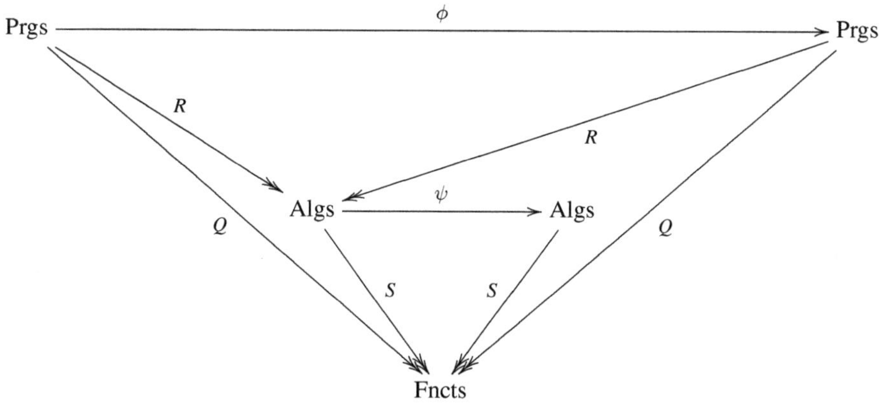

(17.2)

A subgroup of the group of all automorphisms is going to correspond to an intermediate structure. And conversely, an intermediate algorithmic structure will correspond to a subgroup. We will then consider special types of "normal" structures to get an isomorphism of partial orders. This will be the essence of the fundamental theorem of Galois theory of algorithms. This theorem formalizes the intuitive notion that two programs can be switched for one another if they are considered to implement the same algorithm. One extreme case is if you consider every program to be its own algorithm. In that case there is no swapping different programs. The other extreme case is if you consider two programs to be equivalent when they perform the same function. In that case you can swap many programs for other programs. We study all the intermediate possibilities.

Notice that the diagonal arrows in Diagram (17.2) go in the opposite direction of the arrows in Diagram (17.1) and are surjections rather than injections. In fact the proofs in this paper are similar to the ones in classical Galois theory as long as you stand on your head. We resist the urge to call this work "co-Galois theory."

All this is somewhat abstract. What type of programs are we talking about? What type of algorithmic structures are we dealing with? How will our descriptions be specified? Rather than choosing one programming language to the exclusion of others, we look at a language of descriptions of primitive recursive functions. We choose this language because of its beauty, its simplicity of presentation, and the fact that most readers are familiar with this language. The language of descriptions of primitive recursive functions has only three operations: Composition, Bracket, and Recursion. We are limiting ourselves to the set of primitive recursive functions as opposed to all computable functions for ease. By so doing, we are going to get a proper subset of all algorithms. Even though we are, for the present time, restricting ourselves, we feel that the results obtained are interesting in their own right. There is an ongoing project to extend this work to all recursive functions (Manin and Yanofsky 2010).

Another way of looking at this work is from the homotopy theory point of view. We can think of the set of programs as a graph enriched over groupoids. In detail, the 0-cells are the powers of the natural number (types), the 1-cells are the programs from a power of natural numbers to a power of natural numbers. There is a 2-cell from one program to another program if and only if they are "essentially the same". That is, the 2-cells describe the equivalence relations. By the symmetry of the equivalence relations, the 2-cells form a groupoid. (One goes on to look at the same graph structure with different enrichments. In other words, the 0-cells and the 1-cells are the same, but look at different possible isomorphisms between the 1-cells.) Now we take the quotient, or fraction category where we identify the programs at the end of the equivalences. This is the graph or category of algorithms. From this perspective we can promote the mantra:

"Algorithms are the homotopy category of programs."

Similar constructions lead us to the fact that

"Computable functions are the homotopy category of algorithms."

This is a step towards

"Semantics is the homotopy category of syntax."

Much work remains to be done.

Another way of viewing this work is about composition. Compositionality has for many decades been recognized as one of the most valuable tools of software engineering.

There are different levels of abstractions that we use when we teach computation or work in building computers, networks, and search engines. There are programs, algorithms, and functions. Not all levels of abstraction of computation admit useful structure. If we take programs to be the finest level then we may find it hard to compose programs suitably. But if we then pass to the abstract functions they compute, again we run into trouble. In between these two extremes—extreme concreteness and extreme abstractness—there can be many levels of abstraction that admit useful composition operations unavailable at either extreme.

It is our goal here to study the many different levels of algorithms and to understand the concomitant different possibilities of composition. We feel that this work can have great potential value for software engineering.

Yet another way of viewing this work is an application and a variation of some ideas from universal algebra and model theory. In the literature, there is some discussion of Galois theory for arbitrary universal algebraic structures (Cohen 1980, section II.6) and m-theoretic structures (Da Costa and Rodrigues 2007; da Costa 2005; Fleisher 1990; Pöschel 1980.) In broad philosophical terms, following the work of Galois and Klein's *erlangen* program, an object can be defined by looking at its symmetries. Primitive recursive programs are here considered as a universal algebraic structure where the generators of the structure are the initial functions while composition, bracket and recursion are the operations. This work examines the symmetries of such programs and types of structures that can be defined from those symmetries.

Section 17.2 reviews primitive recursive programs and the basic structure that they have. In Sect. 17.3 we define an algorithmic universe as the minimal structure that a set of algorithms can have. Many examples are given. The main theorems in this paper are found in Sect. 17.4 where we prove the Fundamental Theorem of Galois Theory for Algorithms. Section 17.5 looks at our work from the point of view of homotopy theory, covering spaces, and Grothendieck Galois theory. We conclude with a list of possible ways that this work might progress in the future.

17.2 Programs

Consider the structure of all descriptions of primitive recursive functions. Throughout this paper we use the words "description" and "program" interchangeably. The descriptions form a graph denoted **PRdesc**. The objects (nodes) of the graph are powers of natural numbers $\mathbb{N}^0, \mathbb{N}^1, \mathbb{N}^2, \ldots, \mathbb{N}^n, \ldots$ and the morphisms (edges) are descriptions of primitive recursive functions. In particular, there exists descriptions of initial functions: the null function (the function that only outputs a 0) $z : \mathbb{N}^0 = \mathbf{1} \longrightarrow \mathbb{N}$, the successor function $s : \mathbb{N} \longrightarrow \mathbb{N}$, and the projection functions, i.e., for all $n \in \mathbb{N}$ and for all $1 \leq i \leq n$ there are distinguished descriptions $\pi_i^n : \mathbb{N}^n \longrightarrow \mathbb{N}$.

There will be three ways of composing edges in this graph:

- Composition: For $f : \mathbb{N}^a \longrightarrow \mathbb{N}^b$ and $g : \mathbb{N}^b \longrightarrow \mathbb{N}^c$, there is a

$$(g \circ f) : \mathbb{N}^a \longrightarrow \mathbb{N}^c.$$

  Notice that this composition need not be associative. There is also no reason to assume that this composition has a unit.
- Recursion: For $f : \mathbb{N}^a \longrightarrow \mathbb{N}^b$ and $g : \mathbb{N}^{a+b+1} \longrightarrow \mathbb{N}^b$, there is a

$$(f \sharp g) : \mathbb{N}^{a+1} \longrightarrow \mathbb{N}^b.$$

  There is no reason to think that this operation satisfies any universal properties or that it respects the composition or the bracket.
- Bracket: For $f : \mathbb{N}^a \longrightarrow \mathbb{N}^b$ and $g : \mathbb{N}^a \longrightarrow \mathbb{N}^c$, there is a

$$\langle f, g \rangle : \mathbb{N}^a \longrightarrow \mathbb{N}^{b+c}.$$

  There is no reason to think that this bracket is functorial (that is respects the composition) or is in any way coherent.

At times we use trees to specify the descriptions. The leaves of the trees will have initial functions and the internal nodes will be marked with **C**, **R** or **B** for composition, recursion and bracket as follows:

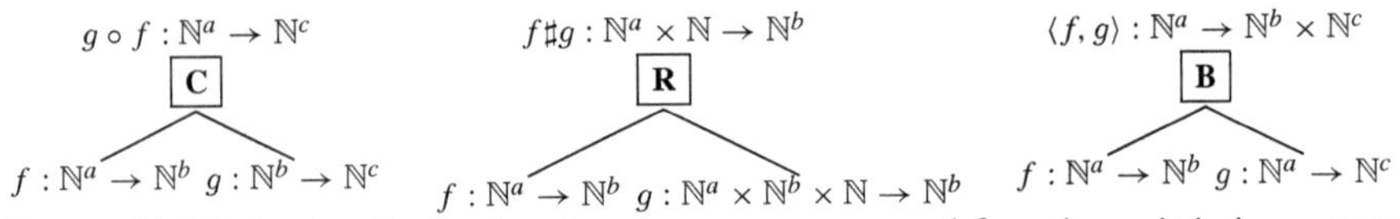

Just to highlight the distinction between programs and functions, it is important to realize that the following are all legitimate descriptions of the null function:

- $z : \mathbb{N}^0 \longrightarrow \mathbb{N}$
- $(z \circ s \circ s \circ s \circ s \circ z \circ s \circ s \circ s \circ s \circ s \circ s \circ s \circ s \circ s \circ z) : \mathbb{N}^0 \longrightarrow \mathbb{N}$
- $(z \circ (\pi_1^2 \circ \langle s, s \rangle)) \circ z : \mathbb{N}^0 \longrightarrow \mathbb{N}$
- etc.

There are, in fact, an infinite number of descriptions of the null function.

In this paper we will need "macros", that is, certain combinations of operations to get commonly used descriptions. Here are a few.

There is a need to generalize the notion of a projection. The π_i^n accepts n inputs and outputs one. A multiple projection takes n inputs and outputs m outputs. Consider $\mathbb{N}^n$ and the sequence $X = \langle x_1, x_2, \ldots, x_m \rangle$ where each x_i is in $\{1, 2, \ldots, n\}$. For every X there exists $\pi^X : \mathbb{N}^n \to \mathbb{N}^m$ as

$$\pi^X = \langle \pi_{x_1}^n, \langle \pi_{x_2}^n, \langle \ldots, \langle \pi_{x_{m-1}}^n, \pi_{x_m}^n \rangle \rangle \ldots \rangle.$$

In other words, π^X outputs the proper numbers in the order described by X. In particular

- If $I = \langle 1, 2, 3, \ldots, n \rangle$ then $\pi^I : \mathbb{N}^n \longrightarrow \mathbb{N}^n$ will be a description of the identity function.
- If $X = \langle 1, 2, 3, \ldots, n, 1, 2, 3, \ldots, n \rangle$ then $\pi^X = \triangle = \pi_{n+n}^n : \mathbb{N}^n \longrightarrow \mathbb{N}^{n+n}$ is the diagonal map.
- For $a \leq b \in \mathbb{N}$, if

  $$X = \langle b, b+1, b+2, \ldots, b+a, 1, 2, 3, \ldots, b-1 \rangle,$$

  then π^X will be the twist operator which swaps the first a elements with the second b elements. Then by abuse of notation, we shall write

  $$\pi^X = tw = \pi_{b+a}^{a+b} : \mathbb{N}^{a+b} \longrightarrow \mathbb{N}^{a+b}.$$

Whenever possible, we omit superscripts and subscripts.

Concomitant with the bracket operation is the product operation. A product of two maps is defined for a given $f : \mathbb{N}^a \to \mathbb{N}^b$ and $g : \mathbb{N}^c \to \mathbb{N}^d$ as

$$f \times g : \mathbb{N}^a \times \mathbb{N}^c \to \mathbb{N}^b \times \mathbb{N}^d.$$

The product can be defined using the bracket as

$$f \times g = \langle f \circ \pi_a^{a+c}, g \circ \pi_c^{a+c} \rangle.$$

Given the product and the diagonal, $\triangle = \pi_{a+a}^{a}$, we can define the bracket as

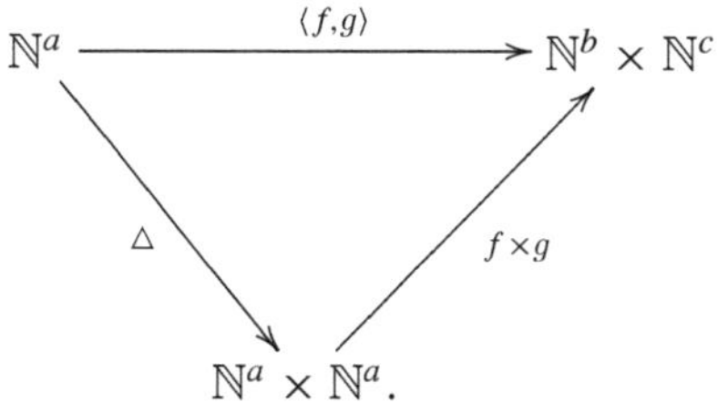

Since the product and the bracket are derivable from each other, we use them interchangeably.

That is enough about the graph of descriptions.

Related to descriptions of primitive recursive functions is the set of primitive recursive *functions*. The set of functions has a lot more structure than **PRdesc**. Rather than just being a graph, it forms a category. **PRfunc** is the category of primitive recursive functions. The objects of this category are powers of natural numbers $\mathbb{N}^0, \mathbb{N}^1, \mathbb{N}^2, \ldots, \mathbb{N}^n, \ldots$ and the morphisms are primitive recursive functions. In particular, there are specific maps $z : \mathbb{N}^0 \longrightarrow \mathbb{N}, s : \mathbb{N} \longrightarrow \mathbb{N}$ and for all $n \in \mathbb{N}$ and for all $1 \leq i \leq n$ there are projection maps $\pi_i^n : \mathbb{N}^n \longrightarrow \mathbb{N}$. Since composition of primitive recursive functions is associative and the identity functions $id = \pi_n^n : \mathbb{N}^n \longrightarrow \mathbb{N}^n$ are primitive recursive and act as units for composition, **PRfunc** is a genuine category. **PRfunc** has a categorically coherent Cartesian product $\times$. Furthermore, **PRfunc** has a strong natural number object. That is, for every $f : \mathbb{N}^a \longrightarrow \mathbb{N}^b$ and $g : \mathbb{N}^a \times \mathbb{N}^b \times \mathbb{N} \longrightarrow \mathbb{N}^b$ there exists a *unique* $h = f \sharp g : \mathbb{N}^a \times \mathbb{N} \longrightarrow \mathbb{N}^b$ that satisfies the following two commutative diagrams

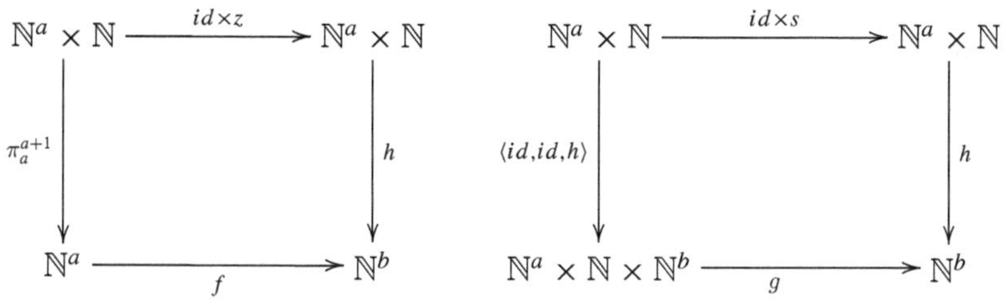

(17.3)

This category of primitive recursive functions was studied extensively by many people including (Burroni 1986; Maietti 2003; Thibault 1982; Roman 1989; Yanofsky 2011). It is known to be the initial object in the 2-category of categories, with products and strict natural number objects. Other categories in that 2-category will be primitive recursive functions with oracles. One can think of oracles as functions put on the leaves of the trees besides the initial functions.

There is a surjective graph morphism $Q : \mathbf{PRdesc} \longrightarrow \mathbf{PRfunc}$ that takes $\mathbb{N}^n$ to $\mathbb{N}^n$, i.e., is identity on objects and Q takes descriptions of primitive recursive functions in **PRdesc** to the functions they describe in **PRfunc**. Since every primitive recursive function has a—in fact infinitely many—primitive recursive description, Q is surjective on morphisms. Another way to say this is that **PRfunc** is a quotient of **PRdesc**.

Algorithms will be graphs that are "between" **PRdesc** and **PRfunc**.

17.3 Algorithms

In the last section we saw the type of structure the set of primitive recursive programs and functions form. In this section we look at the types of structures a set of algorithms can have.

Definition 1 A **primitive recursive (P.R.) algorithmic universe**, **PRalg**, is a graph whose objects are the powers of natural numbers $\mathbb{N}^0, \mathbb{N}^1, \mathbb{N}^2, \ldots, \mathbb{N}^n, \ldots$. We furthermore require that there exist graph morphisms R and S that are the identity on objects and that make the following diagram of graphs commute:

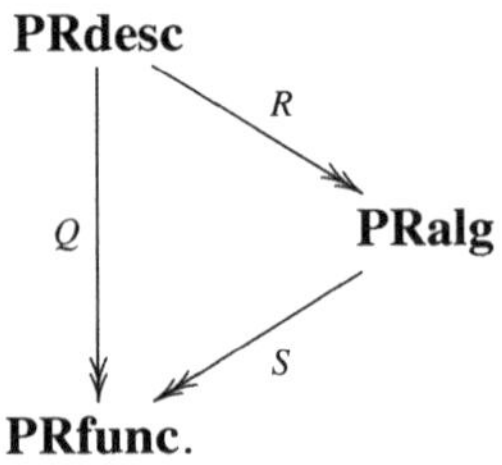

(17.4)

The image of the initial functions under R will be distinguished objects in **PRalg**: $z : \mathbb{N}^0 \longrightarrow \mathbb{N}$, $s : \mathbb{N} \longrightarrow \mathbb{N}$, and for all $n \in \mathbb{N}$ and for all $1 \leq i \leq n$ there are projection maps $\pi_i^n : \mathbb{N}^n \longrightarrow \mathbb{N}$.

In addition, a P.R. algorithmic universe *might* have the following operations: (Warning: even when they exist, these are not necessarily functors because we are not dealing with categories.)

- Composition: For $f : \mathbb{N}^a \longrightarrow \mathbb{N}^b$ and $g : \mathbb{N}^b \longrightarrow \mathbb{N}^c$, there is

$$(g \circ f) : \mathbb{N}^a \longrightarrow \mathbb{N}^c.$$

- Recursion: For $f : \mathbb{N}^a \longrightarrow \mathbb{N}^b$ and $g : \mathbb{N}^{a+b+1} \longrightarrow \mathbb{N}^b$, there is

$$(f \sharp g) : \mathbb{N}^{a+1} \longrightarrow \mathbb{N}^b$$

- Bracket: For $f : \mathbb{N}^a \longrightarrow \mathbb{N}^b$ and $g : \mathbb{N}^a \longrightarrow \mathbb{N}^c$, there is a

$$\langle f, g \rangle : \mathbb{N}^a \longrightarrow \mathbb{N}^{b+c}$$

These operations are well defined for programs but need not be well defined for equivalence classes of programs. There was never an insistence that our equivalence relations be congruences (i.e. respect the operations). We study when these operations exist at the end of the section.

Notice that although the Q graph morphism preserves the composition, bracket and recursion operators, we do not insist that R and S preserve them. We will see that this is too strict of a requirement.

Definition 2 Let **PRalg** be a P.R. algorithmic universe. A **P.R. quotient algorithmic universe** is a P.R. algorithmic universe **PRalg**$'$ and an identity on objects, surjection on edges graph map T that makes all of the following triangles commute

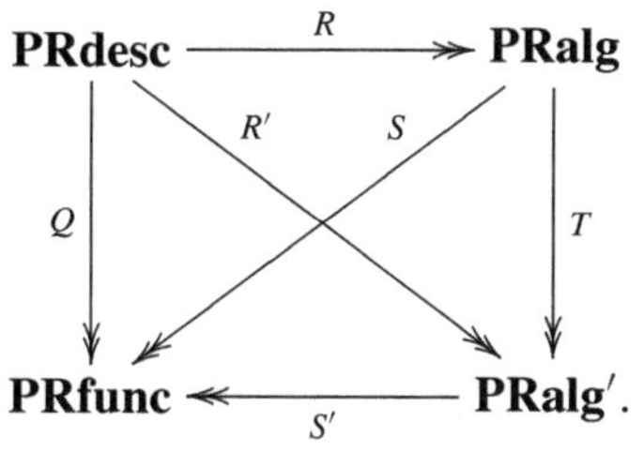

(17.5)

Examples of P.R. algorithmic universe abound:

Example **PRdesc** is the primary trivial example. In fact, all our examples will be quotients of this algorithmic universes. Here $R = id$ and $S = Q$. □

Example **PRfunc** is another trivial example of an algorithmic universe. Here $R = Q$ and $S = id$.□

Example **PRalgC** is a quotient of **PRdesc**. This is constructed by adding the following relation:

For any three composable maps f, g and h, we have

$$h \circ (g \circ f) \sim (h \circ g) \circ f. \tag{17.6}$$

In terms of trees, we say that the following trees are equivalent:

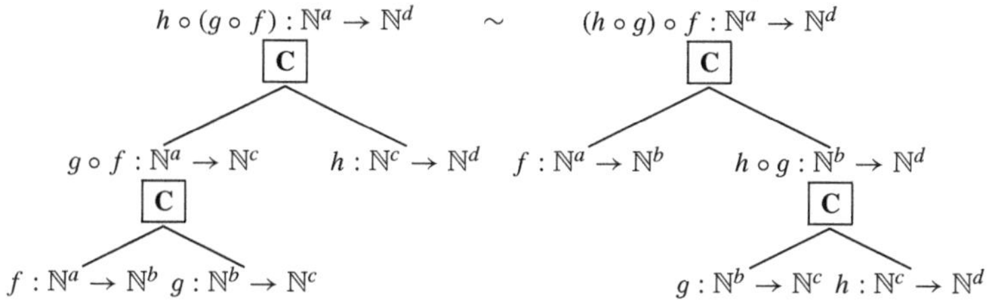

It is obvious that if there is a well-defined composition map in **PRalgC** it is associative. □

Example **PRalgI** is also a quotient of **PRdesc** that is constructed by adding in the relations that say that the projections π^Is act like identity maps. That means for any $f : \mathbb{N}^a \to \mathbb{N}^b$, we have

$$f \circ \pi^a_a \sim f \sim \pi^b_b \circ f. \tag{17.7}$$

In terms of trees:

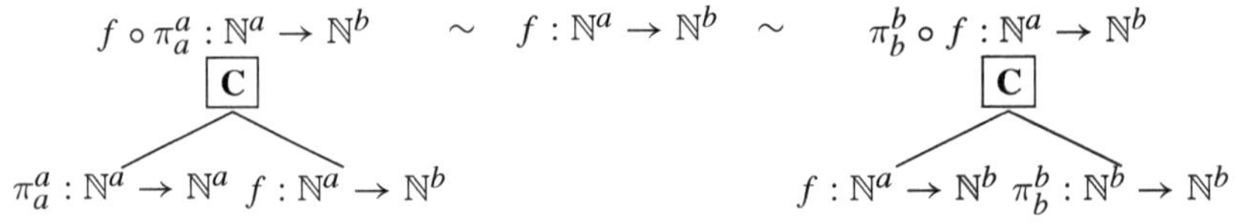

□

The composition map in **PRalgI** has a unit.

Example **PRalgCat** is **PRdesc** with both relations (17.6) and (17.7). Notice that this ensures that **PRalgCat** is more than a graph and is, in fact, a full fledged category. □

Example **PRalgCatX** is a quotient of **PRalgCat** which has a well-defined bracket/product function. We add the following relations to **PRalgCat**:

- The bracket is associative. For any three maps f, g, and h with the same domain, we have

$$\langle\langle f, g\rangle, h\rangle \sim \langle f, \langle g, h\rangle\rangle$$

In terms of trees, this amounts to

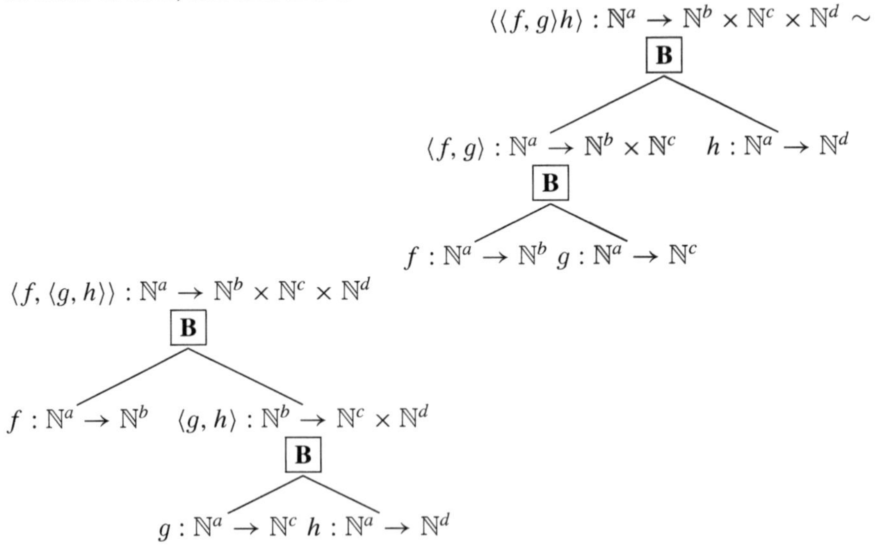

- Composition distributes over the bracket on the right. For $g : \mathbb{N}^a \to \mathbb{N}^b$, $f_1 : \mathbb{N}^b \to \mathbb{N}^c$ and $f_2 : \mathbb{N}^b \to \mathbb{N}^d$, we have

$$\langle f_1, f_2 \rangle \circ g \sim \langle f_1 \circ g, f_2 \circ g \rangle. \tag{17.8}$$

In terms of trees, this amounts to saying that these trees are equivalent:

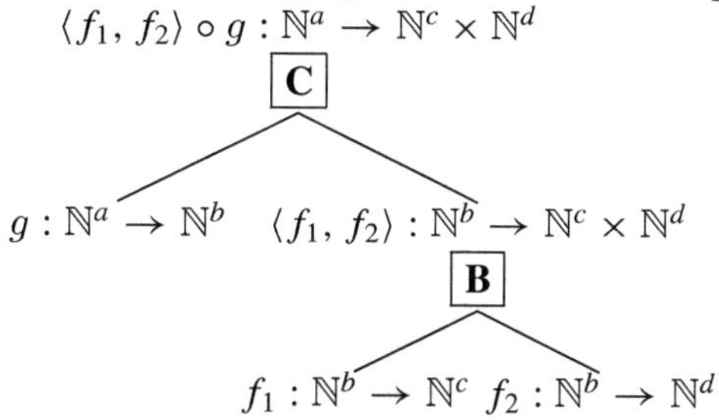

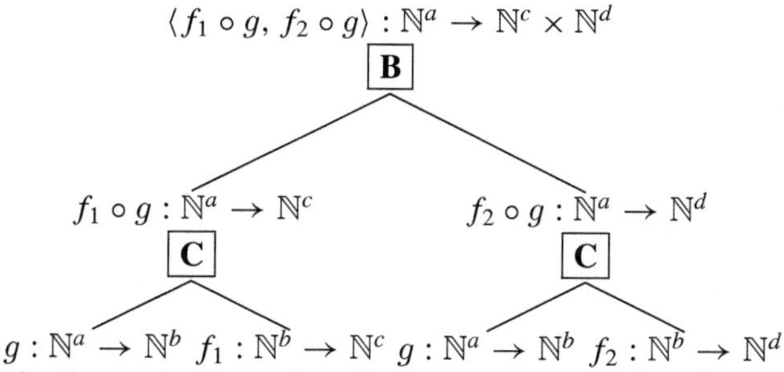

- The bracket is almost commutative. For any two maps f and g with the same domain,

$$\langle f, g \rangle \sim tw \circ \langle g, f \rangle.$$

In terms of trees, this amounts to

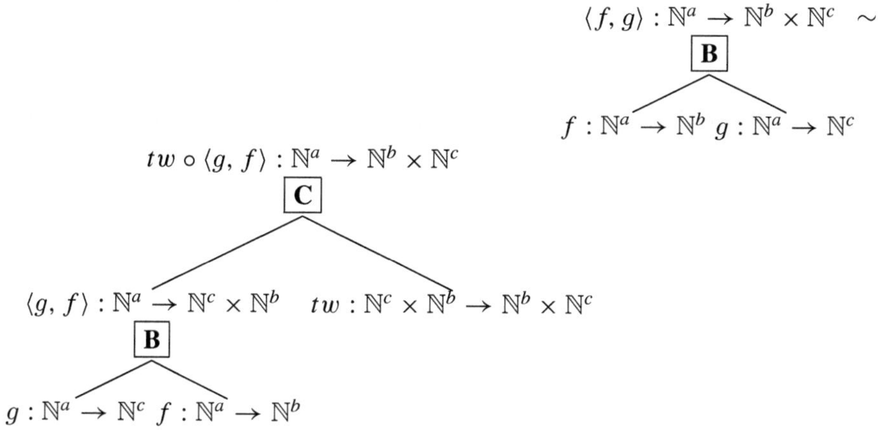

- Twist is idempotent.

$$tw_{\mathbb{N}^a,\mathbb{N}^b} \circ tw_{\mathbb{N}^a,\mathbb{N}^b} \sim id = \pi^{a+b}_{a+b} : \mathbb{N}^a \times \mathbb{N}^b \to \mathbb{N}^a \times \mathbb{N}^b.$$

- Twist is coherent. That is, the twist maps of three elements behave with respect to themselves.

$$(tw_{\mathbb{N}^b,\mathbb{N}^c} \times id) \circ (id \times tw_{\mathbb{N}^a,\mathbb{N}^c}) \circ (tw_{\mathbb{N}^a,\mathbb{N}^b} \times id) \sim (id \times tw_{\mathbb{N}^a,\mathbb{N}^b}) \circ (tw_{\mathbb{N}^a,\mathbb{N}^c} \times id) \circ (id \times tw_{\mathbb{N}^b,\mathbb{N}^c}).$$

  This is called the hexagon law or the third Reidermeister move. Given the idempotence and hexagon laws, it is a theorem that there is a unique twist map made of smaller twist maps between any two products of elements (Mac Lane 1998, Sect. XI.4). The induced product map will be coherent.

□

Example **PRalgCatN** is a category with a natural number object. It is **PRalgCat** with the following relations:

- Left square of Diagram (17.3). $(f\sharp g) \circ (id \times z) \quad \sim \quad (f \circ \pi^{a+1}_a)$.
- Right square of Diagram (17.3). $(f\sharp g) \circ (id \times s) \quad \sim \quad (f \circ \langle id, id, (f\sharp g)\rangle$.
- Natural number object and identity. If $g = \pi^{a+b+1}_b : \mathbb{N}^a \times \mathbb{N}^b \times \mathbb{N} \longrightarrow \mathbb{N}^b$ then

$$(f\sharp\pi^{a+b+1}_b) \quad \sim \quad (f \circ \pi^{a+1}_a).$$

- Natural number object and composition. This is explained in Sect. 3.5 of Yanofsky (2011).

$$g_1 \ddot{\circ} (f\sharp(g_2 \ddot{\circ} g_1)) \quad \sim \quad (g_1 \ddot{\circ} f)\sharp(g_1 \ddot{\circ} g_2).$$

□

Example **PRalgCatXN** is a category that has both a product and natural number object. It can be constructed by adding to **PRalgCat** all the relations of **PRalgCatX** and **PRalgCatN** as well as the following relations:

- Natural number object and bracket. This is explained in Sect. 3.4 of Yanofsky (2011)

$$\langle f_1, f_2\rangle\sharp(g_1 \boxtimes g_2) \quad \sim \quad \langle f_1\sharp g_1, f_2\sharp g_2\rangle.$$

□

Putting all these examples together, we have the following diagram of P.R. algorithmic universes.

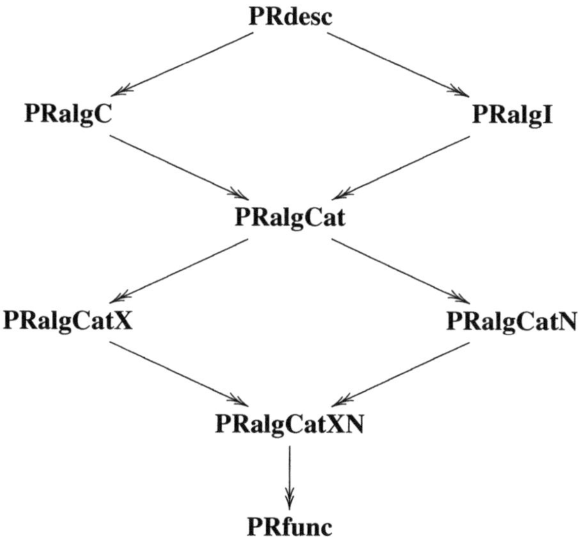

There is no reason to think that this is a complete list. One can come up with infinitely many more examples of algorithmic universes. We can take other permutations and combinations of the relations given here as well as new ones. Every appropriate equivalence relation will give a different algorithmic universe.

In Yanofsky (2011), we mentioned other relations which deal with the relationship between the operations and the initial functions. We do not mention those relations here because our central focus is the existence of well defined operations.

A word about decidability. The question is, for a given P.R. algorithmic universe determine whether or not two programs in **PRdesc** are in the same equivalence class of that algorithmic universe.

- This is very easy in the algorithmic universe **PRdesc** since every equivalence class has only one element. Two descriptions are in the same equivalence relation iff they are *exactly* the same.
- The extreme opposite is in **PRfunc**. By a theorem similar to Rice's theorem, there is no way to tell when two different programs/descriptions are the same primitive recursive function. So **PRfunc** is not decidable.
- In between **PRdesc** and **PRfunc** things get a little hairy. This is the boundary between syntax and semantics. Consider **PRalgC**, i.e., the graph with associative composition. This is decidable. All one has to do is change all the contiguous sequences of compositions to associate on the left. Do this for both descriptions and then see if the two modified programs are the same.
- One can perform a similar trick for **PRalgI**. Simply eliminate all the identities and see if the two modified programs are the same.
- For **PRalgCat** one can combine the tricks from **PRalgC** and **PRalgI** to show that it is also decidable.

- **PRalgCatX** is also decidable because of the coherence of the product. Once again, any contiguous sequences of products can be associated to the left. Also, equivalence relation (17.8) insures the naturality of the product so that products and compositions can "slide across" each other. Again, each description can be put into a canonical form and then see if the modified programs are the same.
- However we loose decidability when it comes to structures with natural number objects. See the important paper by Okada and Scott (1999). It seems that this implies that **PRalgCatN** and **PRalgCatXN** are undecidable. One can think of this as the dividing line between the decidable, syntactical structure of **PRdesc** and the undecidable, semantical structure of **PRfunc**.

17.4 Galois Theory

An automorphism ϕ of **PRdesc** is a graph isomorphism that is the identity on the vertices (i.e., $\phi(\mathbb{N}^n) = \mathbb{N}^n$). For every $a, b \in \mathbb{N}$ ϕ basically acts on the edges between $\mathbb{N}^a$ and $\mathbb{N}^b$. We are interested in automorphisms that preserve functionality. That is, automorphisms ϕ, such that for all programs p, we have that p and $\phi(p)$ perform the same function. In terms of Diagram (17.2) we demand that $Q(\phi(p)) = Q(p)$. It is not hard to see that the set of all automorphism of **PRdesc** that preserve functionality forms a group. We denote this group as $Aut(\mathbf{PRdesc}/\mathbf{PRfunc})$. One can also think of this group as the direct product, over all primitive recursive functions f, of all the permutation groups of programs realizing f. We shall look at subgroups of this group and see its relationship with intermediate fields. Let GRP denote the partial order of subgroups of $Aut(\mathbf{PRdesc}/\mathbf{PRfunc})$.

Let ALG be the partial order of intermediate algorithmic universes. One algorithmic universe, **PRalg** is greater than or equal to **PRalg**$'$ if there is a quotient algorithmic map $\mathbf{PRalg} \twoheadrightarrow \mathbf{PRalg}'$.

We shall construct a Galois connection between GRP and ALG. That is, there will be an order reversing map $\Phi : ALG \longrightarrow GRP$ and an order reversing $\Psi : GRP \longrightarrow ALG$.

In detail, for a given algorithmic universe **PRalg**, we construct the subgroup

$$\Phi(\mathbf{PRalg}) \subseteq Aut(\mathbf{PRdesc}/\mathbf{PRfunc}).$$

$\Phi(\mathbf{PRalg})$ is the set of all automorphisms of **PRdesc** that preserve that algorithmic universe, i.e., automorphisms ϕ such that for all programs p, we have p and $\phi(p)$ are in the same equivalence class in **PRalg**. That is,

$$\Phi(\mathbf{PRalg}) = \{\phi | \forall p \in \mathbf{PRdesc}, \quad [\phi(p)] = [p] \in \mathbf{PRalg}\}.$$

In terms of Diagram (17.4), this means $R(\phi(p)) = R(p)$. In order to see that $\Phi(\mathbf{PRalg})$ is a subgroup of $Aut(\mathbf{PRdesc}/\mathbf{PRfunc})$, notice that if ϕ is in $\Phi(\mathbf{PRalg})$ then we have

$$R\phi = R \quad \Rightarrow \quad SR\phi = SR \quad \Rightarrow \quad Q\phi = Q$$

which means that ϕ is in $Aut(\mathbf{PRdesc}/\mathbf{PRalg})$. In general, this subgroup fails to be normal. (Normality is about doing the order of composition of doing two functions. In general $f \circ g$ is not the same as $g \circ f$.)

If $T : \mathbf{PRalg} \twoheadrightarrow \mathbf{PRalg}'$ is a quotient algorithmic universe as in Diagram (17.5) then

$$\Phi(\mathbf{PRalg}) \subseteq \Phi(\mathbf{PRalg}') \subseteq Aut(\mathbf{PRdesc}/\mathbf{PRfunc}).$$

This is obvious if you look at $\phi \in \Phi(\mathbf{PRalg})$ then we have that

$$R\phi = R \quad \Rightarrow \quad TR\phi = TR \quad \Rightarrow \quad R'\phi = R'$$

which means that ϕ is also in $\Phi(\mathbf{PRalg}')$.

The other direction goes as follows. For $H \subseteq Aut(\mathbf{PRdesc}/\mathbf{PRfunc})$, the graph $\Psi(H)$ is a quotient of **PRdesc**. The vertices of $\Psi(H)$ are powers of natural numbers. The edges will be equivalence classes of edges from **PRdesc**. The equivalence relation $\sim_H$ is defined as

$$p \sim_H p' \quad \text{iff} \quad \text{there exists a } \phi \in H \text{ such that } \phi(p) = p' \tag{17.9}$$

The fact that $\sim_H$ is an equivalence relation follows from the fact that H is a group. In detail

- Reflexivity comes from the fact that $id \in H$.
- Symmetry comes from the fact that if $\phi \in H$ then $\phi^{-1} \in H$.
- Transitivity comes from the fact that if $\phi \in H$ and $\psi \in H$ then $\phi\psi \in H$.

If $H \subseteq H' \subseteq Aut(\mathbf{PRdesc}/\mathbf{PRfunc})$ then there will be a surjective map

$$\Psi(H) \twoheadrightarrow \Psi(H').$$

The way to see this is to realize that there are more ϕ in H' to make different programs equivalent as described in line (17.9).

Theorem 1 *The maps* $\Phi : ALG \longrightarrow GRP$ *and* $\Psi : GRP \longrightarrow ALG$ *form a Galois connection.*

Proof. We must show that for any H in GRP and any **PRalg** in ALG we have

$$H \subseteq \Phi(\mathbf{PRalg}) \text{ if and only if } \Psi(H) \twoheadrightarrow \mathbf{PRalg}.$$

This will be proven with the following sequence of implications.

$$H \subseteq \Phi(\mathbf{PRalg})$$

if and only if

$$\phi \in H \quad \Rightarrow \quad \phi \in \Phi(\mathbf{PRalg})$$

if and only if

$$\phi \in H \quad \Rightarrow \quad \forall p, [\phi(p)] = [p] \in \mathbf{PRalg}$$

if and only if

$$\forall p \quad \phi(p) \sim_H p \quad \Rightarrow \quad \phi(p) \sim_{\mathbf{PRalg}} p$$

if and only if

$$\Psi(H) \twoheadrightarrow \mathbf{PRalg}.$$

□

Every Galois connection (adjoint functor) induces an isomorphism of sub-partial orders (equivalence of categories.) Here we do not have to look at a sub-partial order of ALG for the following reason:

Theorem 2 *For any* **PRalg** *in* ALG

$$(\Psi \circ \Phi)(\mathbf{PRalg}) = \mathbf{PRalg}.$$

Proof. $\Psi(\Phi((\mathbf{PRalg})) = \sim_{\Phi((\mathbf{PRalg})}$

$$= \sim_{\Phi(\{\phi | \forall p \in \mathbf{PRdesc}, \quad [\phi(p)] = [p] \in \mathbf{PRalg}\})} = \mathbf{PRalg}. \square$$

In contrast, it is not necessarily the case that for any H in GRP, we have

$$(\Phi \circ \Psi)(H) = H.$$

We do have that

$$(\Phi \circ \Psi)(H) = \Phi(\Psi(H)) = \Phi(\sim_H) = \{\phi | \forall p \quad \phi(p) \sim_H p\} \supseteq H$$

because any ϕ in H definitely satisfies that condition. But many other ϕ might also satisfy this requirement. In general this requirement is not satisfied. H might generate a transitive action. In that case $\Phi(\Psi(H))$ will be all automorphisms.

If one wants to complete the analogy with classical Galois theory then we should consider the following definition. A subgroup H whose induced action does not extend beyond H will be important:

Definition 3 A subgroup H of $Aut(\mathbf{PRdesc}/\mathbf{PRfunc})$ is called "restricted" if $(\Phi \circ \Psi)(H) = H$.

We can sum up with the following statement:

Theorem 3 (Fundamental theorem of Galois theory) *The lattice of restricted subgroups of* $Aut(\mathbf{PRdesc}/\mathbf{PRfunc})$ *is isomorphic to the dual lattice of algorithmic universes between* **PRdesc** *and* **PRfunc**.

Notice that the algorithmic universes that we dealt with in this theorem does not necessarily have well-defined extra structure/operations. We discussed the equivalence relations of **PRdesc** and did not discuss congruences of **PRdesc**. Without the congruence, the operations of composition, bracket and recursion might not be well-defined for the equivalence classes. This is very similar to classical Galois theory where we discuss a single weak structure (fields) and discuss all intermediate objects as fields even though they might have more structure. So too, here we stick to a weak structure.

However we can go further. Our definition of algorithmic universes is not carved in stone. One can go on and define, say, a *composable* algorithmic universe. This is an algorithmic universe with a well-defined composition function. Then we can make the fundamental theorem of Galois theory for composable algorithmic universes by looking at automorphisms of **PRdesc** that preserve the composition operations. That is, automorphisms ϕ such that for all programs p and p' we have that $\phi(p \circ p') = \phi(p) \circ \phi(p')$. Such automorphisms also form a group and one can look at subgroups as we did in the main theorem. On the algorithmic universe side, we will have to look at equivalence relations that are congruences. That is, $\sim$ such that if $p \sim p'$ and $p'' \sim p'''$ then $p \circ p'' \sim p' \circ p'''$. Such an analogous theorem can easily be proved.

Similarly, one can define *recursive* algorithmic universes and *bracket* algorithmic universes. One can still go further and ask that an algorithmic universe has two well-defined operations. In that case the automorphism will have to preserve two operations. If H is a group of automorphisms, then we can denote the subgroup of automorphisms that preserve composition as H_C. Furthermore, the subgroup that preserves composition and recursion will be denoted as H_{CR}, etc. The subgroups fit into the following lattice.

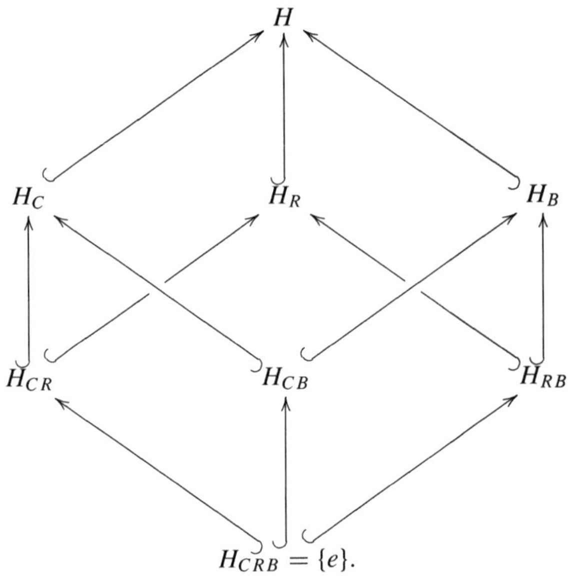

(17.10)

It is important to realize that it is uninteresting to require that the algorithmic universe have all three operations. The only automorphism that preserves all the operations is the identity automorphism on **PRdesc**. One can see this by remembering that the automorphisms preserve all the initial functions and if we ask them to preserve all the operations, then it must be the identity automorphism. This is similar to looking at an automorphism of a group that preserves the generators and the group operation. That is not a very interesting automorphism.

One can ask of the automorphisms to preserve all three operations but not preserve the initial operations. Similarly, when discussing oracle computation, one can ask the automorphisms to preserve all three operations and the initial functions, but not the oracle functions. All these suggestions open up new vistas of study.

17.5 Homotopy Theory

The setup of the structures presented here calls for an analysis from the homotopy perspective. We seem to have a covering space and are looking at automorphims of that covering space. Doing homotopy theory from this point of view makes it very easy to generalize to many other areas in mathematics and computer science. This way of doing homotopy theory is sometimes called Grothendieck's Galois theory. We gained much from Baez and Shulman (2006) and Dubuc and Sanchez de la Vega (2000).

First a short review of classical homotopy theory. Let X and C be topological spaces. $P : X \longrightarrow C$ is a covering space if P is a continuous surjective map such that for every $x \in X$, there exists an open neighborhood U of x, with $P^{-1}(U)$ (the inverse image of U under P) is a union of disjoint open sets in C, each of which is mapped homeomorphically onto U by P. (One could think of $P^{-1}(U)$ as a stack of pancakes that map onto the pancake U.) $P : X \longrightarrow C$ is a universal covering space if it is a covering space, C is connected, and the fundamental group of X is trivial. (The fundamental group is the equivalence classes of loops of X based at a point where two loops are equivalent if one can be homotopically transformed into another. It is trivial if and only if every loop in X can be homotopicly shrunk to a point.) P induces an isomorphism

$$Aut(C) \quad \cong \quad \pi_1(X) \tag{17.11}$$

where $Aut(C)$ is the set of automorphisms (homeomorphisms) of C that respect P (automorphisms $f : X \longrightarrow X$ such that $P \circ f = P$.) Such automorphisms are called "deck transformations." $\pi_1(X)$ is the fundamental group of X. This result can be extended by dropping the assumption that X is connected. We then have

$$Aut(C) \quad \cong \quad \Pi_1(X) \tag{17.12}$$

where $\Pi_1(X)$ is the fundamental groupoid of X. The fundamental groupoid of X is a category whose objects are the points of the topological space and whose reversible morphisms are equivalence classes of paths from one point to another. Another way of generalizing this result is to consider C which is not necessarily a universal, i.e. $\pi_1(C)$ is not necessarily trivial. The theorem then says that

$$Aut(C) \quad \cong \quad \pi_1(X)/P_*(\pi_1(C)). \tag{17.13}$$

That is, we look at quotients of $\pi_1(X)$ by the image of the fundamental groups of C.

Our functor Q : **PRdesc** $\longrightarrow$ **PRfunc** seems to have a feel of a covering space. The functor is bijective on objects and is surjective on morphisms. Also, for every primitive recursive function $f : \mathbb{N}^m \longrightarrow \mathbb{N}^n$ the set

$$P^{-1}(f : \mathbb{N}^m \longrightarrow \mathbb{N}^n)$$

has a countably infinite number of programs/descriptions. Furthermore the preimage of string of functions $f_1 \circ f_2 \circ \cdots \circ f_k$ is a disjoint union of a string of programs each of which maps onto the string of functions.

Our goal will be an isomorphism of the form

$$Aut(\mathbf{PRdesc}/\mathbf{PRfunc}) \quad \cong \quad \pi_1(\mathbf{PRfunc}) \tag{17.14}$$

The right side of the purported isomorphism is very easy to describe. π_1(**PRfunc**) is the group of invertible primitive recursive functions from $\mathbb{N}$ to $\mathbb{N}$. Note that because of primitive recursive isomorphisms of the form $\mathbb{N} \longrightarrow \mathbb{N}^k$ for all k (Gödel numbering functions), the elements of this group can be rather sophisticated.

One should not be perturbed by the fact that we are looking at the reversible primitive recursion functions based at $\mathbb{N}$ as opposed to $\mathbb{N}^k$ for an arbitrary k. We can also look at the fundamental group based at $\mathbb{N}^k$ and denote these two fundamental groups as $\pi_1(\mathbf{PRfunc}, \mathbb{N})$ and $\pi_1(\mathbf{PRfunc}, \mathbb{N}^k)$. By a usual trick of classical homotopy theory, these two groups are isomorphic as follows. Let $\alpha : \mathbb{N}^k \longrightarrow \mathbb{N}$ be a primitive recursion isomorphic function. Let $f : \mathbb{N} \longrightarrow \mathbb{N}$ be an element of $\pi_1(\mathbf{PRfunc}, \mathbb{N})$. We then have the isomorphism from $\pi_1(\mathbf{PRfunc}, \mathbb{N})$ to $\pi_1(\mathbf{PRfunc}, \mathbb{N}^k)$:

$$f : \mathbb{N} \longrightarrow \mathbb{N} \qquad \mapsto \qquad \alpha^{-1} \circ f \circ \alpha : \mathbb{N}^k \longrightarrow \mathbb{N} \longrightarrow \mathbb{N} \longrightarrow \mathbb{N}^k.$$

Since all these groups are isomorphic, we ignore the base point and call the group $\pi_1(\mathbf{PRfunc})$.

As far as I can find, this group has not been studied in the literature. There are many questions to ask. Are there generators to this group? What properties does this group have?(An anonymous reviewer gave a simple proof that it is not finitely generated.) What is the relationship of the groups of all recursive isomorphisms from $\mathbb{N}$ to $\mathbb{N}$ and the primitive recursive isomorphisms? What is the relationship between invertible primitive recursive functions (that is, a primitive recursive function that is an isomorphism) and all primitive recursive functions? Can every primitive recursive function be mimicked in some way (like Bennetts result about reversible computation) by a reversible/invertible primitive recursive function? In essence, the group $\pi_1(\mathbf{PRfunc})$ is the upper left of the following commutative square of monoides:

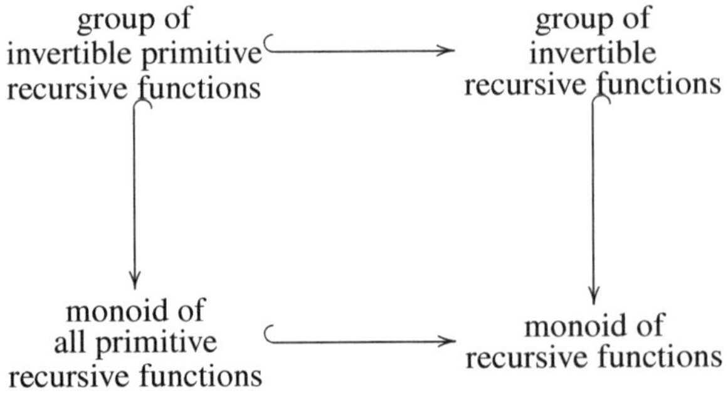

Essentially we are asking if any of these inclusion maps have some type of retract.

Unfortunately the map $Q : \mathbf{PRdesc} \longrightarrow \mathbf{PRfunc}$ fails to be a real covering map because it does not have the "unique lifting property." In topology, a covering map $P : C \longrightarrow X$ has the following lifting property: for any path $f : I \longrightarrow X$ from $f(0) = x_0$ to $f(1) = x_1$ and for any $c_0 \in C$ such that $P(c_0) = x_0$ there is a *unique* $\hat{f} : I \longrightarrow C$ such that $P(\hat{f}) = f$. In English this says that for any path in X and any starting point in C, there is a unique path in C that maps onto the path in X. In our context, such a unique lifting would mean that for every primitive recursive function made out of a sequence of functions, if you choose one program/description to start the function, then the rest of the programs/descriptions would all be forced. There is, at the moment, no reason for this to be true. The problem is that above every function, there is only a set of programs/descriptions. This set does not have any more structure.

However, all hope is not lost. Rather than look at **PRdesc** as simply a graph, look at it as graph enriched in groupoids. That is, between every two edges there is a possibility for there to be isomorphisms corresponding to whether or not the two programs are essentially the same. This is almost a bicategory or 2-groupoid but the bottom level is not a category but a graph. (I am grateful to Robert Paré for this suggestion.) If we were able to formalize this, then we would have that $P^{-1}(f : \mathbb{N}^m \longrightarrow \mathbb{N}^n)$ would be a connected groupoid and we would have some type of unique lifting property. At the moment I am not sure how to do this. Another advantage of doing this is that

$$Aut(\mathbf{PRdesc}/\mathbf{PRfunc})$$

would not be as big and as unmanageable as it is now. The automorphisms would have to respect the higher dimensional cells. Much work remains.

17.6 Future Directions

Extend to all computable functions. The first project worth doing is to extend this work to all computable functions from primitive recursive functions. One need only add in the minimization operator and look at its relations with the other operations. The study of such programs from our point is an ongoing project in Manin and Yanofsky (2010). However the Galois theory perspective is a little bit complicated because of the necessity to consider partial operations. A careful study of Rosenberg (1981) will, no doubt, be helpful.

Continuing with Galois Theory There are many other classical Galois theory theorems that need to be proved for our context. We need the Zassenhaus lemma, the Schreier refinement theorem, and culminating in the Jordan-Hölder theorem. In the context of algorithms this would be some statement about decomposing a category of algorithms regardless of the order in which the equivalence relations are given. We might also attempt a form of the Krull-Schmidt theorem.

Impossibility results. The most interesting part of Galois theory is that it shows that there are certain contingencies that are impossible or not "solvable." What would the analogue for algorithms be?

Calculate some groups. It is not interesting just knowing that there are automorphism groups. It would be nice to actually give generators and relations for some of these groups. This will also give us a firmer grip for any impossibility results.

Universal Algebra of Algorithms. In this paper we stressed looking at quotients of the structure of all programs. However there are many other aspects of the algorithms that we can look at from the universal algebraic perspective. Subalgebras: We considered all primitive recursive programs. But there are subclasses of programs that are of interest. We can for example restrict the number of recursions in our programs and get to subclasses like the Grzegorczyk's hierarchy. How does the subgroup lattice survive with this stratification? Other subclasses of primitive recursive functions such as polynomial functions and EXPTIME functions can also be studied. Superalgebras: We can also look at larger classes of algorithms. As stated above, we can consider all computable functions by simply adding in a minimization operator. Also, oracle computation can be dealt with by looking at trees of descriptions that in addition to initial functions permit arbitrary functions on their leaves. Again we ask similar questions about the structure of the lattice of automorphisms and the related lattice of intermediate algorithms. Homomorphisms: What would correspond to a homomorphism between classes of computable algorithms? Compilers. They input programs and output programs. This opens up a whole new can of worms. What does it mean for a compiler to preserve algorithms? When are two compilers similar? What properties should a compiler preserve? How are the lattices of subgroups and intermediate algorithms preserved under homomorphisms/compilers? There is obviously much work to be done.

Acknowledgements I thank Ximo Diaz Boils, Leon Ehrenpreis (of blessed memory), Thomas Holder, Roman Kossak, Florian Lengyel, Dustin Mulcahey, Robert Paré, Vaughan Pratt, Phil Scott, and Lou Thrall for helpful discussions. I am also thankful to an anonymous reviewer who was very helpful.

References

Baez, J. C., & Shulman, M. (2006). *Lectures on n-Categories and Cohomology*. http://arxiv.org/pdf/math/0608420v2.pdf.

Blass, A., Dershowitz, N., & Gurevich, Y. (2008). *When are two algorithms the same?*. http://arxiv.org/PS_cache/arxiv/pdf/0811/0811.0811v1.pdf. Downloaded Feburary 5, 2009.

Burroni, A. (1986). *Recursivite Graphique (1^e partie): Categories des fonctions recursives primitives formelles*. Cahiers De Topologie Et Geometrie Differentielle Categoriques (vol. XXVII-1)).

Cohen, P. M. (1980). *Universal algebra* (2nd ed.). Dordrecht: D. Reidel.

da Costa, N. C. A. (2005). *Remarks on abstract Galois theory*. ftp://logica.cle.unicamp.br/pub/e-prints/vol.5,n.8,2005.pdf.

Da Costa, N. C. A., & Rodrigues, A. A. M. (2007). Definability and invariance. *Studia Logica, 86*, 1–30.

Dubuc, E. J., & Sanchez de la Vega, C. (2000). *On the Galois theory of Grothendieck*. http://arxiv.org/pdf/math/0009145v1.pdf.

Fleischer, I. (1990). The abstract Galois theory: A survey. In C. H. Bergman, R. D. Maddux, & D. L. Pigozzi (Eds.), *Proceedings of the international conference on cryptology on algebraic logic and universal algebra in computer science (Sydney, Australia)* (pp. 133–137). Berlin: Springer.

Mac Lane, S. (1998). *Categories for the working mathematician* (2nd ed.). Berlin: Springer.

Maietti, M. E. (2003). Joyal's arithmetic universe via type theory. *Electronic notes in Theoretical Computer Science, 69*.

Manin, Yu I. (2009a). *A course in mathematical logic for mathematicians* (2nd ed.). Berlin: Springer.

Manin, Yu. I. (2009b). *Renormalization and computation I: Motivation and background*. http://arxiv4.library.cornell.edu/abs/0904.4921.

Manin, Yu. I. (2009c). *Renormalization and computation II: Time cut-off and the Halting problem.* http://arxiv4.library.cornell.edu/abs/0908.3430.

Manin, Yu. I., & Yanofsky, N. S. (2010). *Notes on the recursive operad.* Work in Progress.

Okada, M., & Scott, P. J. (1999). A note on rewriting theory for uniqueness of iteration. *Theory and Applications of Categories, 6*, 47–64. http://www.tac.mta.ca/tac/volumes/6/n4/n4.pdf.

Pöschel, R. (1980). *A general Galois theory for operations and relations and concrete characterization of related algebraic structures.* http://www.math.tu-dresden.de/~poeschel/poePUBLICATIONSpdf/poeREPORT80.pdf.

Roman, L. (1989). Cartesian categories with natural numbers object. *Journal of Pure and Applied Algebra, 58*, 267–278.

Rosenberg, I. G. (1004). *1981.* Springer LNM: Galois theory for partial algebras.

Thibault, M.-F. (1982). Prerecursive categories. *Journal of Pure and Applied Algebra, 24*, 79–93.

Yanofsky, N. S. (2011). Towards a definition of algorithms. *Journal of Logic Computation* (first published online May 30, 2010). doi:10.1093/logcom/exq016.

Chapter 18
Parikh's Closing Remarks

Rohit Parikh

Abstract This is a survey of my research career which began in 1960 and is still going on. During these years I have traveled from context free languages and recursion theory to social software, via non-standard analysis, Dynamic Logic, and other areas. Many people, some of them famous, have influenced me as have my students. My deep appreciation to all for the influence they have had on my work.

Keywords Context free languages · Non-standard analysis · Dynamic Logic · Epistemic Logic · Social software · Game theory

I am really grateful to all the people who have contributed to this volume. All the papers are quite good and it is good that I will get a free copy, for otherwise I would have to buy one!

The authors of this volume represent the last two thirds of my career, so let me say something also about the first third. During the first third my interests were in formal languages, recursive function theory, proof theory, and non-standard analysis. The four people who influenced me were respectively, Noam Chomsky, Hartley Rogers, Georg Kreisel and Abraham Robinson. Robinson not only introduced me to non-standard analysis but also arranged for me to come to Caltech and get a green card. Others who influenced me were W.V. Quine who taught me my first course in logic, Dreben, who taught the second course (and approved my dissertation) and Marvin Minsky, who introduced me to Chomsky. I must have been scatter brained because I was a student in the Mathematics department at Harvard, my main adviser Rogers was in Math at MIT, and Dreben who read my thesis was in the Philosophy department at Harvard. Alas, both Dreben and Rogers are gone, as are Kreisel and Quine.

Albert Meyer was the main figure who, in 1977, influenced a sharp transition. He invited me to join his group at the Lab for Computer science (at MIT) when I was teaching at Boston University. I still remember Meyer writing to me when I was

R. Parikh (✉)
City University of New York, New York, NY, USA
e-mail: RParikh@gc.cuny.edu

C. Başkent et al. (eds.), *Rohit Parikh on Logic, Language and Society*,
Outstanding Contributions to Logic 11, DOI 10.1007/978-3-319-47843-2_18

visiting Berkeley for a semester, "Rohit, you should have an NSF grant. Send me a blank piece of paper with your signature on it and I will do the rest of the work of applying." I did so, and the grant was funded, supporting me, two students at Boston University and one at MIT.

While a member of Meyer's group, I met Vaughan Pratt, David Harel, and Joseph Halpern, all of whom remain good friends. I also met Dexter Kozen who, until recently, had the largest number of joint papers with me.

So my field of interest changed from non-standard analysis to Logic of programs, founded by Floyd, Hoare, Pratt and Pnueli, and fostered by Meyer, Halpern, Harel. Ali Khan of Johns Hopkins complains that I left Non-standard analysis, but unfortunately I am a wanderer, driven by insatiable curiosity.

Meyer asked Halpern and me to referee a paper on the logic of knowledge and that fact certainly awakened my interest in this area. Over the next years I worked intensely on the logic of knowledge, and had the pleasure of meeting Johan van Benthem, working with Ramanujam and Moss and advising Georgatos and Pacuit, both of whom went on to do brilliant work. Non-CUNY students included Amy Greenwald, Laxmi Parida, Andreas Witzel, Mark Pauly, and last but not least, Horacio Arlo Costa whose death in 2011 was a great loss, both for me and for philosophical logic.

My game theoretic tendencies were strengthened by working with Bud Mishra (Greenwald and Parida were officially *his* students) and attending seminars at NYU. This resulted in some work in game theory, I met Aumann and Nash, and recently had the pleasure of having three of my students (Baskent, Greenwald and Tasdemir) present work at the 2015 conference in game theory at Stony Brook. Three other students (Johnson, Satari and Stambaugh) had presented work there previously. Another student who contributed to game theory was Pawel Krasucki who published a joint paper with me in the *Journal of Economic Theory*, and went on to publish in that journal on his own.

During this period I also met Steven Brams, Donald Saari and Brian Skyrms whose interests overlap greatly with mine. Brams and Saari are experts on elections, whereas Skyrms' main impetus seems to be in the game theory of communication.

In 1996, during a conference in Hyderabad, an afternoon was devoted to my sixtieth birthday. Alessandra Carbone, Gilbert Ndjatou and Vaughan Pratt attended from France and the US, and gave talks. I still remember my 80 year old mother insisting that we meet Alessandra at the Mumbai airport at 2 AM, for the sake of her safety.

It turned out that there was a gap in the program that afternoon and I used it to advertise *social software* which is a blend of logic of programs, analysis of algorithms and game theory. Later, elements from psychology also entered when I became aware of the work of people like Heinz Wimmer, Josef Perner, Herbert Clark and Steve Pinker.

Two side interests are philosophy of language, and vagueness which have been areas of commonality with Prashant Parikh, Robert van Rooij and Gabriel Sandu. Noson Yanofsky was once a student of mine and is now a colleague at Brooklyn College. I feel I have influenced his brilliant mind at least a little. Another person to work jointly with me was Jouko Vaananen who is simply a brilliant mathematician. Most of the technical work in our joint paper is his, but I do claim credit for starting it and for using Stalin's daughter in one of our examples! I have never written a joint paper with Mel Fitting, but he has been a friend since before I came to CUNY in 1982.

Finally, Juliet Floyd has been a great friend and support since we met at CUNY. I remember going to a talk on Wittgenstein when I noticed a beautiful young woman sitting a few rows ahead of me. Every time the speaker said something incorrect she shook her head. I said to myself, "Does CUNY really have students as brilliant as this?" But it turned out that she, like me, was a Harvard product, an assistant professor at CCNY, and had worked with both Quine and Dreben. She, along with Arthur Collins, has been my main source of learning about Wittgenstein. She is nothing if not a hard worker, and her paper in this volume shows that she has read everything I have written since 1969 and saw the common theme which inspired that body of work.

And what is that theme? She and I had a grant from CUNY to do joint work and our work was so fruitful that a magazine interviewed us about it. Reporters must be smart because the title of his article was "Downloading Wittgenstein." Both Juliet and I have seen that for all of Wittgenstein's shortcomings as a person, he really saw a wider landscape than anyone in his time and he is not just someone to be studied but someone to be mined for new ideas, ideas which were perhaps implicit in him and which we could make explicit.

One of Wittgenstein's posthumous books is dedicated to God but he says that in these times, his dedication is likely to be misunderstood. Perhaps there is no God, but perhaps there is something? And in that case, *may the force be with all of us!*

Let me finish by saying something about Social Software which is still my main interest. Social Software is interested in what makes social algorithms, and individual algorithms in a social context, work. Robinson Crusoe had no need for social software; whatever he needed, he made. But most of us rely on many others to enable us to do the simplest task. I once read as a schoolboy that making a simple needle involves a thousand people or so. We take buses, taxis, buy coffee, call up someone on the telephone etc. All these require the context which society supplies us with. But these are still individual algorithms in a social context.

But there are other algorithms which are inherently social. Building a bridge, holding an election, or invading another country are all social algorithms. We may not always know what exactly the goal was in invading some country, if indeed there was one, but even if there is no clear goal, there is a goal which is sold to the public.

Elections are an even more interesting example. Arrow, building on an example from Condorcet, proved in 1950 that there is no fair system of elections, that the very goal which we expect an election to achieve is self-contradictory. It is like Wittgenstein's East Pole. It isn't that we are unable to reach the East pole, but rather that unlike the North Pole and the South Pole, there is no such place as the East Pole.

Of course we do hold elections, and something happens which is typically better than whatever happens in a dictatorship. But if we attempt to define the *goal* of an election, satisfying some conditions of rationality, we arrive at a contradiction.

Auctions are a better defined variety of social software and Vickrey auctions have beautiful properties in that they do not punish you for being honest.

But there also needs to be a general field. Economic Design is such a field but its scope is a bit limited and it does not make use of insights from computer science, psychology or philosophy. So Social Software is needed.

It is a project which would ideally involve many kinds of talents: economists, psychologists, computer scientists and philosophers. But I think the task is urgent because if mankind is to solve its problems it needs to understand itself better.

One might close by suggesting that rather than say "the proper study of mankind is man," it is better to say, "the proper study of mankind is how society works and how it does not."[1,2]

---

[1]Thanks to Juliet Floyd, Joseph Halpern, Dexter Kozen, Carol Parikh and Vaughan Pratt for comments on the draft version of this article.

[2]An interesting fact about my recent work is that most of my recent co-authors have been women: Adriana Renero, Arancha San Ginez Ruiz, Cagil Tasdemir and Yunqi Xue. However, I do have some energetic and creative male students: Jongjin Kim, Yang Liu, Ignacio Ojea and Todd Stambaugh. I have benefitted from all eight.

Chapter 19
The Complete Bibliography of Rohit Parikh

Rohit Parikh

1. (With Garrett Birkhoff) *The Tabulation of Michell's Function*, Research report, Harvard 1958.
2. "Many one degrees of Certain Sets of Recursive Well Orderings", *Notices of the AMS* 8 (1961) 495.
3. (With J.N. Crossley) "On Isomorphisms of Recursive Well Orderings", *Journal of Symbolic Logic* (abstract) 28 (1963) 308.
4. "On Context Free Languages", *Jour. ACM* 13 (1966) 570–81.
5. "Some Generalisations of the Notion of Well Ordering", *Zeit. Math. Logik u. Grund. Math.* 12 (1966) 333–340.
6. "Some Theorems in Logic", *Math. Student* 34 (1966) 125–29.
7. "Nonuniqueness in Transfinite Progressions", *Jour. Ind. Math. Soc.* 31 (1967) 23–32.
8. "A Nonstandard Theory of Topological Groups", in *Applications of Model Theory* Ed. W.A.J. Luxemburg, Holt, Rheinhart and Winston (1969) 279–284.
9. "A Conservation Result", in *Applications of Model Theory* Ed. W.A.J. Luxemburg, Holt, Rheinhart and Winston (1969), pp. 107–108.
10. "Existence and Feasibility in Arithmetic", *Jour. Symbolic Logic* 36 (1971) 494–508.
11. "D-Structures and their Semantics", *Notices of the AMS* 19 (1972) A329.
12. (With J. Mayberry) "D-structures and *-structures", *Notices of the AMS* 19 (1972) A454.
13. (With D.H.J.de Jongh and N. Goodman) "On Regular *-structures with Classical Theories", *J. Symb. Logic* 37 (1972) 777.
14. "A Note on Rigid Substructures", *Proc. AMS* 33 (1972) 520–522.
15. (With M. Parnes) "Conditional Probability can be Defined for Arbitrary Pairs of Sets of Reals", *Advances in Math* 9 (1972) 520–522.
16. "On the Length of Proofs", *Trans. AMS* 177 (1973) 29–36.

R. Parikh (✉)
City University of New York, New York, NY, USA
e-mail: RParikh@gc.cuny.edu

C. Başkent et al. (eds.), *Rohit Parikh on Logic, Language and Society*,
Outstanding Contributions to Logic 11, DOI 10.1007/978-3-319-47843-2_19

17. "A Note on Paths through $\mathcal{O}$", *Proc. AMS* 39 (1973) 178–180.
18. (With M. Parnes) "Conditional Probabilities and Uniform Sets", *Proc. Victoria Symp. Nonstandard Analysis* Ed. Hurd and Loeb, Lecture Notes in Math #369, 180–84.
19. (Editor) *Logic Colloquium* Lecture Notes in Math #453 (1975)
20. "An $\aleph_0$-categorical Theory whose Language is Countably Infinite", *Proc. AMS*, **49 (5)**, pp. 216–18, 1975.
21. (With D.H.J. de Jongh) "Well Partial Orderings and Hierarchies", *Proc. Kon. Ned. Akad. Sci* Series A 80 (1977) 195–207.
22. "The Completeness of Propositional Dynamic Logic", *7th MFCS* (1978) LNCS #64 403–415. (LNCS=Lecture Notes in Computer Science)
23. "A Decidability Result for a Second Order Process Logic", *19th IEEE-FOCS* (1978) 177–183.
24. "Propositional Dynamic Logics of Programs: Systems, Models and Complexity", *7th ACM-POPL* (1980) 186–192.
25. "Propositional Logics of Programs: A Survey" in *Logics of Programs* Ed E. Engeler, LNCS #125 (1980) 102–144.
26. "Effectiveness", *the Philosophical Forum* XII (1980) 68–81.
27. (With D. Kozen) "An Elementary Completeness Proof for PDL" *Theor. Comp. Sci* 14 (1981) 113–118.
28. (With A. Meyer) "Definability in Dynamic Logic", *12th ACM- STOC* (1980) 1–7, also *JCSS* 23 (1981) 279–298.
29. "Models for Programs", *Proc. 1st Bangalore Conference on Foundations of Software Theory and Theoretical Computer Science (FST-TCS)* TIFR (1981) 109–121.
30. (With A. Ehrenfeucht and G. Rozenberg) "Pumping Lemmas and Regular Sets", *SIAM J. Computing* 10 (1981) 536–41.
31. (With D. Harel, D. Kozen) "Process Logic: Expressiveness, Decidability, Completeness", *JCSS* 25 (1982) 144–170.
32. "The Problem of Vague Predicates", in *Logic, Language and Method* Ed. Cohen and Wartofsky, Reidel (1982) 241–261.
33. (With A. Mahoney) "A Theory of Probabilistic Programs", in *Logic of Programs* Ed. Clarke and Kozen, LNCS #164 (1983) 396–402.
34. (with D. Kozen) "A Decision Procedure for the Propositional μ-calculus", in *Logic of Programs* (Ed Clarke and Kozen) Springer Lecture Notes in Comp. Sci. (LNCS) #164 (1983) 313–325.
35. "Propositional Logics of Programs: New Directions", *FCT-83* LNCS #158, 347–359, 1983.
36. "Propositional Game Logic", *24th IEEE-FOCS* (1983) 195–200.
37. "Some Applications of Topology to Program Semantics", *Math. Sys. Theory* 16 (1983) 111–131.
38. "Logics of Knowledge, Games and Dynamic Logic", *Foundations of Software Theory and Theoretical Computer Science* Springer LNCS #181, (1984) 202–222.

39. (With A. Chandra, J. Halpern, A. Meyer) "Equations between Regular Terms and an Application to Process Logic", *13th STOC* (1981) 384–390, also in *SIAM J. Computing* 14 (1985) 935–942.
40. "The Logic of Games and its Applications" *Annals of Discrete Math.*, 24 (1985) 111–140.
41. "Introductory Note to Gödel's paper on the Length of Proofs", in *the Collected Works of K. Gödel* Ed. by Feferman et al., Oxford 1986, pp. 394–399.
42. "Modal Logic", in *the Encyclopedia of Artificial Intelligence*, John Wiley, 1987.
43. (With R. Ramanujam) "Distributed Processes and the Logic of Knowledge", in *Logics of Programs* (Editor), Proceedings of a Conference at Brooklyn College, June 1985, Springer Lecture Notes in Computer Science #193, pp. 256–268.
44. *Logics of Programs* (Editor), Proceedings of a Conference at Brooklyn College, June 1985, Springer Lecture Notes in Computer Science #193.
45. "Levels of Knowledge in Distributed Computing", in *Proc IEEE Symposium on Logic in Computer Science,* June 1986, 314–321.
46. "Some Recent Contributions of Logic to Computer Science", in *Proc. World Computer Congress*, Dublin, Sep. 1986, 391–392.
47. "Knowledge and the Problem of Logical Omniscience" *ISMIS-87* (International Symp. on Methodology for IntelligentSystems), North Holland (1987) pp. 432–439.
48. "Some Recent Applications of Knowledge" in *FST and TCS 7*, Proceedings of a Conference in Pune, India, December 1987, LNCS #287, pp. 528–539.
49. "Decidability and Undecidabilty in Distributed Transition Systems" IBM Research report. In *A Perspective in Theoretical Computer Science*, Ed. R. Narasimhan, World Scientific Ltd. 1989 pp. 199–209.
50. "Communication, Consensus and Knowledge", (with P. Krasucki), *J. Economic Theory* **52** (1990) pp. 178–189.
51. "Recent Trends In Reasoning about Knowledge", in *Theoretical Aspects of Reasoning about Knowledge*, Morgan Kaufmann, Los Altos, California, 1990 pp. 3–10.
52. *Theoretical Aspects of Reasoning about Knowledge*, (editor) Morgan Kaufmann, Los Altos, California, 1990.
53. "Probabilistic Knowledge and Probabilistic Common Knowledge" (with Paul Krasucki and Gilbert Ndjatou), *ISMIS 90*, North Holland 1990, pp. 1–8.
54. "Dumb-founded Sets", *Bull. EATCS*, no. 43, Feb 1991, pp. 183–184.
55. "The Effect of AI on Other Fields of Research", *Proc. Phoenix Conference on Computers and Communication*, March 1991.
56. "Finite and Infinite Dialogues", in the *Proceedings of a Workshop on Logic from Computer Science*, Ed. Moschovakis, MSRI publications, Springer 1991 pp. 481–498.
57. "Monotonic and Non-monotonic Logics of Knowledge", in *Fundamenta Informatica* special issue, *Logics for Artificial Intelligence* vol XV (1991) pp. 255–274.
58. "A Test for Fuzzy Logic", *SIGACT NEWS,* **22**, 3, Summer 1991, pp. 49–50.

59. (With P. Krasucki), "Levels of knowledge in distributed computing", *Sadhana—Proc. Ind. Acad. Sci.* **17** (1992) pp. 167–191.
60. "Topological Resaoning and the Logic of Knowledge". (with Larry Moss,). in *TARK 4*, Ed. Yoram Moses, Morgan-Kaufmann publishers, March 1992, pp. 95–105.
61. "Notes of Rohit Parikh's lectures on Reasoning about Knowledge", by Anna Maria Zanaboni. (the Lectures were given in Acireale at an International School for Computer Scientists) published in Italy, summer 1993. (Cassa di Risparmio di Padova e Rovigo)
62. "Vagueness and Utility: the Semantics of Common Nouns" in *Linguistics and Philosophy* **17** 1994, 521–35.
63. "Logical omniscience", in *Logic and Computational Complexity* Ed. Leivant, Springer Lecture Notes in Computer Science no. 960, (1995) 22–29.
64. "A Logical Study of Distributed Transition Systems", with Lodaya, Ramanujam and Thiagarajan. *Information and Computation* **119** May 1995, 91–119.
65. "Knowledge based computation (Extended abstract)" in *Proceedings of AMAST-95* Montreal, July 1995, Edited by Alagar and Nivat, LNCS no. 936, 127–42.
66. "How far can we formalize language games?" in *The Foundational Debate* edited by DePauli-Scimanovich, Köhler and Stadler, Kluwer Academic (1995) pp. 89–100.
67. "Topological Reasoning and The Logic of Knowledge" (with Dabrowski and Moss) *Annals of Pure and Applied Logic* **78** (1996) 73–110
68. "Computation", *MacMillan Encyclopedia of Philosophy* - supplement (1996), pp. 86–87.
69. "Vague predicates and language games", *Theoria* Spain, vol XI, no. 27, Sep 1996, pp. 97–107.
70. "Church's theorem and the decision Problem", *the Routledge Encyclopedia of Philosophy*, 1998.
71. "Belief revision and language splitting" in *Proc. Logic, Language and Computation*, Ed. Moss, Ginzburg and de Rijke, CSLI 1999, pp. 266–278 (earlier version appeared in 1996 in the prliminary proceedings).
72. "Length and structure of proofs", in *Synthese* **114** 1998, special issue edited by J. Hintikka.
73. "Frege's puzzle and belief revision", typescript, November 1997. Presented at the World Congress of Philosophy, Boston 1998.
74. (With J. Dauben) "Mathematics in India in the 20th century" to appear in the Italian Encyclopedia of Mathematics, 1998.
75. "Some reminiscences of Kreisel", in *Kreiseliana* edited by P. Odifreddi, 1999.
76. (With Samir Chopra and Konstantinos Georgatos) "Non-Monotonic Inference on Sequenced Belief Bases", *Proceedings of the Delphi Conference in Logic*, July 1999.
77. "*D*-Structures and their semantics", in *JFAK* (Ed.s Jelle Gerbrandy, Maarten Marx, Maarten de Rijke and Yde Venema), 1999.

78. (With Horacio Arlo Costa) "Two place probabilities, beliefs and belief revision: on the foundations of iterative belief kinematics", in *Proc. 12th Amsterdam Colloquium*, December'99, edited by Paul Dekker, pp. 1–6.
79. "Propositions, propositional attitudes and belief revision" in K. Segerberg, M. Zakharyaschev, M. de Rijke, H. Wansing, editors, *Advances in Modal Logic, Volume 2*, CSLI Publications, 2001.
80. (With Laxmi Parida and Vaughan Pratt) "Sock Sorting", appeared in a volume dedicated to Johan van Benthem, University of Amsterdam, August 99, reprinted in *Logic J. of IGPL*, vol 9 (2001).
81. (With Samir Chopra) "An Inconsistency Tolerant Model for Belief Representation and Belief Revision" appeared in *Proc. IJCAI 99. Annals of Math and AI*, 2001.
82. (With Samir Chopra and Renata Wasserman) "Approximate belief revision", *Logic Journal of the IGPL*, **9(6)**, pp. 755–768, 2001.
83. "Social Software", *Synthese*, **132**, 2002, 187–211.
84. "Completeness of certain bimodal logics for subset spaces" (with Angela Weiss), *Studia Logica*, **71** (June 2002), 1–30.
85. "Language as social software" (abstract), International Congress on Logic, Methodology and Philosophy of Science (1995), page 417. Full paper in *Future Pasts: the Analytic Tradition in Twentieth Century Philosophy*, Ed. J. Floyd and S. Shieh, Oxford U. Press, 2001, 339–350.
86. 'Towards a theory of social software', in *Proceedings of DEON'02, (Deontic Logic in CS)* Imperial College, London, pages 265–277
87. (With R. Ramanujam), A Knowledge based Semantics of Messages, *J. Logic, Language and Information* **12** 2003, 453–467.
88. (With Marc Pauly), Game Logic—An Overview, in *Studia Logica*, 2003.
89. Levels of Knowledge, Games, and Group Action, *Research in Economics* **57** 2003, 267–281.
90. (With Samir Chopra and Eric Pacuit) "Knowledge-Theoretic Properties of Strategic Voting" in *Proceedings of JELIA 2004*, pp. 18–30, 2004.
91. (With Horacio Arlo Costa) "Conditional probability and defeasible inference", *Journal of Philosophical Logic*, **34** (2005) 97–119.
92. "WHAT do we know and what do WE know?", the proceedings of *Theoretical Aspects of Rationality and Knowledge*, June 2005, University of Singapore.
93. (With Jouko Vaananen), Finite information logic, *Annals of Pure and Applied Logic*, **134** (2005) 83–93.
94. (With Eric Pacuit and Eva Cogan) The logic of knowledge based obligation, *Synthese*, (2006). **149** 311–341.
95. (With Eric Pacuit) Social Interaction, Knowledge, and Social Software, in *Interactive Computation: The New Paradigm*, ed. Dina Goldin, Sott Smolka, Peter Wegner, Springer publishers, 2007, 441–462.
96. (With Melvyn Nathanson) Density of Natural Numbers and the Levy Group, in the *Journal of Number Theory*. **124** (2007) 151–158.

97. "Some Puzzles About Probability and Probabilistic Conditionals", in *Logical Foundations of Computer Science*. Eds. Sergei Artemov and Anil Nerode, pp. 449–56, Springer, 2007.
98. (With Eric Pacuit) Reasoning about communication graphs, in *Interactive Logic*, edited by van Benthem, Gabbay and Löwe, Amsterdam University Press 2007, 135–158.
99. (With Larry Moss and Chris Steinsvold) Topology and Epistemic Logic, in *Logic of Space*, edited by Johan van Benthem et al., 2007.
100. (Co-edited with Gupta and van Benthem) *Logic at the Crossroads* Allied publishers, 2007.
101. "Is there a logic of society?", in *Logic at the Crossroads* (*op cit*), 20–38.
102. (With Matt Johnson) Probabilistic Conditionals are almost Monotonic, *Review of Symbolic Logic* **1** (2008), 73–78.
103. "Sentences, belief and logical omniscience, or what does deduction tell us?", *Review of Symbolic Logic*, **1** (2008) 459–476.
104. "Knowledge, games and tales from the East", in *Logic and its Applications*, Ramanujam and Sarukkai, editors, Springer 2009, 65–76.
105. (With Jan van Eijck, Marc Pauly and Rineke Verbrugge) "Social Software and the Ills of Society", in *Discourses on Social Software* Jan van Eijck and Rineke Verbrugge (editors). Texts in Logic and Games, volume 5, pages 219–226. Amsterdam University Press, 2009.
106. "Knowledge and structure in social algorithms", in *Relations and Kleene Algebra in Computer Science*, edited by Berghammer, Jaoua and Möller, Springer lecture notes in computer science, vol. 5827 (2009) 1–12.
107. "From language games to social software", in *Reduction, Asbsraction, Analysis, proceedings of the 31st International Ludwig Wittgenstein Symposium in Kirchberg*, edited by Alexander Hieke and Hannes Leitgeb, Ontos Verlag 2009, pp. 365–376.
108. "Remarks on Knowledge, Games and Society", *Journal of Indian Council of Philosophilcal Research*, **XXVII**, Jan-March 2010, pp. 413–26.
109. (With Joseph Dauben) "Beginnings of Modern Mathematics in India", *Current Science*, **99** 3, August 2010, 1–24.
110. (With Walter Dean) "The Logic of Campaigning", in M. Banerjee and A. Seth (Eds.): *ICLA 2011*, LNAI 6521, pp. 38–49. Springer, Heidelberg (2011).
111. "Beth definability, interpolation and language splitting", *Synthese*, **179** 2, (March 2011) 211–221.
112. (With Hans van Ditmarsch and R. Ramanujam), "Logic in India: Editorial Introduction", *Journal of Philosophical Logic*, **40**, pp. 557–561, 2011.
113. (With Çağıl Taşdemir and Andreas Witzel) "Choice and Uncertainty in Games" in *Logic and Program Semantics* (Eds. Robert Constable and Alexandra Silva), Volume 7230 of the Lecture Notes in Computer Science series, pp. 244–55, Springer, 2012.
114. "States of Knowledge", *Electronic Notes in Theoretical Computer Science*, Volume 67, Pages 49–60, October 2002.

115. (With Can Başkent and Loes Olde Loohuis) "On Knowledge and Obligation", *Episteme*, **9** no. 2, pp. 171–188, 2012.
116. "On Kripke's Puzzle About Time and Thought", *Proceedings of the Indian Conference in Logic and Applications*, pp. 121–26, 2012.
117. (With Çağıl Taşdemir and Andreas Witzel) "The Power of Knowledge in Games", *International Game Theory Review*, **15.4**, 2013.
118. "Epistemic Reasoning in Life and Literature" in *David Makinson on Classical Methods for Non-Classical Problems*, Ed. Sven Ove Hansson, pp. 143–56, Springer, 2014.
119. "Thinking About Knowledge, Some Fresh Views", in *Logic and Rationality Yearbook* 2012 vol. II., Eds. Zoe Christoff et al., pp. 195–202, Amsterdam University Press, 2014.
120. (With Aranzazu San Gines) "A Pragmatistic Approach to Propositional Knowledge Based on the Successful Behavior of Belief" in *Logic and its Applications*, 6th Indian conference ICLA 2015, Eds. M. Banerjee and S. Krishna. pp. 146–57, Springer, 2015.
121. (With Çağıl Taşdemir) "The Strategy of Campaigning", in *Fields of Logic and Computation II*, LNCS vol. 9300, pp. 253–260, Springer, 2015.
122. (With Yunqi Xue) "Strategic Belief Updates through Influence in a Community", Studies in Logic, Vol. 8, No. 4, pp. 124–143, 2015.
123. "Knowledge and Action in Groups Preliminary Report", Studies in Logic, Vol. 8, No. 4, pp. 108–123, 2015.
124. (With Adriana Renero) "Justified True Belief: Plato, Gettier and Turing" in J. Floyd and A. Bokulich eds., *Philosophical Explorations of the Legacy of Alan Turing—Turing 100*, Boston Studies in the Philosophy and History of Science (Springer Verlag), 2016.
125. "Is there a Church-Turing Thesis for Social Algorithms?" in J. Floyd and A. Bokulich eds., *Philosophical Explorations of the Legacy of Alan Turing—Turing 100*, Boston Studies in the Philosophy and History of Science (Springer Verlag), 2016.

Developments in Soil Science 13

ELEMENTS OF SOIL PHYSICS

Further Titles in this Series

1. I. VALETON
BAUXITES

2. IAHR
FUNDAMENTALS OF TRANSPORT PHENOMENA
IN POROUS MEDIA

3. F.E. ALLISON
SOIL ORGANIC MATTER AND ITS ROLE IN
CROP PRODUCTION

4. R.W. SIMONSON (Editor)
NON-AGRICULTURAL APPLICATIONS OF SOIL SURVEYS

5A. G.H. BOLT and M.G.M. BRUGGENWERT (Editors)
SOIL CHEMISTRY. A. BASIC ELEMENTS

5B. G.H. BOLT (Editor)
SOIL CHEMISTRY. B. PHYSICO-CHEMICAL MODELS

6. H.E. DREGNE
SOILS OF ARID REGIONS

7. H. AUBERT and M. PINTA
TRACE ELEMENTS IN SOILS

8. M. SCHNITZER and S.U. KHAN (Editors)
SOIL ORGANIC MATTER

9. B.K.G. THENG
FORMATION AND PROPERTIES OF CLAY-POLYMER COMPLEXES

10. D. ZACHAR
SOIL EROSION

11A. L.P. WILDING, N.E. SMECK and G.F. HALL (Editors)
PEDOGENESIS AND SOIL TAXONOMY. I. CONCEPTS AND
INTERACTIONS

11B. L.P. WILDING, N.E. SMECK and G.F. HALL (Editors)
PEDOGENESIS AND SOIL TAXONOMY. II. THE SOIL ORDERS

12. E.B.A. BISDOM and J. DUCLOUX (Editors)
SUBMICROSCOPIC STUDIES OF SOILS

Developments in Soil Science 13

ELEMENTS OF SOIL PHYSICS

P. KOOREVAAR, G. MENELIK and C. DIRKSEN

Department of Soil Science and Plant Nutrition, Agricultural University of Wageningen, Wageningen, The Netherlands

ELSEVIER — Amsterdam — Oxford — New York — Tokyo 1983

ELSEVIER SCIENCE PUBLISHERS B.V.
1 Molenwerf
P.O. Box 211, 1000 AE Amsterdam, The Netherlands

Distributors for the United States and Canada:

ELSEVIER SCIENCE PUBLISHING COMPANY INC.
52, Vanderbilt Avenue
New York, N.Y. 10017

ISBN 0-444-42242-0 (Vol. 13)
ISBN 0-444-41882-7 (Series)

Printed in The Netherlands

Preface

This book is not meant to be 'just another textbook' on Soil Physics. Instead, it presents a different approach to teaching students the basic principles of Soil Physics. The main idea is that students should be forced to think through and apply immediately what they read. To achieve this goal, the text is written mostly in rather short paragraphs and is alternated with questions. Many questions are an integral part of the transfer of knowledge; these, in particular, should be answered by the student before he continues. The student can check his answers with those given at the end of each chapter.

The origin of the present book goes back to lectures notes by G. H. Bolt, which eventually resulted in two lecture syllabi written in the early sixties by G. H. Bolt, A. R. P. Janse and F. F. R. Koenigs for internal use at the Department of Soil Science and Plant Nutrition of the Agricultural University, Wageningen, The Netherlands. One syllabus was on Soil Chemistry, the other on Soil Physics; both had to be studied by the students as one unit. In the course of time, several reasons developed for revising these syllabi. First, the c.g.s. system of units needed to be replaced by the new S.I.-system. Secondly, in 1970 a new study program was introduced at the Agricultural University, for which some students needed to study only Soil Physics. Unfortunately, some subjects that were needed for Soil Physics were discussed only in the Soil Chemistry syllabus, e.g., the behaviour of clay particles and the diffuse electric double layer. Obviously, it is desirable to incorporate all subjects of Soil Physics in one syllabus. A third reason for revision was the need to introduce a consistent set of definitions, symbols, etc., based on a rigorous, lucid development from first principles, in agreement with the latest internationally accepted conventions.

The final and most elaborate, if not most important reason for revision was the need to seek ways to improve the examination results of the students. To this end, we first handed out questions at the lectures, some of which were prepared by L. Stroosnijder. Once a week the 'homework' was discussed in small groups of about 15 students and evaluated by the teacher. This system was used with about 150 students, and thus was very time-consuming for the teaching staff. For the students, the gain in examination results was marginal, in spite of their larger time investment. This led eventually to the idea to integrate the questions and answers with the subject matter in the syllabus and develop some kind of 'half-programmed' text for selfstudy. During a year of experimenting with a first draft of such a text, in which about half of the subjects were given in this new form, the students liked the new approach, obtained better results at the examinations, and even needed less time to master the material than with the old method. For the teaching staff also, the new method was far more efficient. Lectures were no longer needed and were replaced by sessions in which students could ask questions about the material studied. On the average, only about ten percent of the students made use of these sessions. From 1974 on, the new method has been used with good results.

The first author played a major role in the development and implementation of the elementary study program in Soil Physics, along the lines described above. He was assisted in this by G. Muggen of the Bureau for Development and Research of Education (B.O.O.) of the Agricultural University. This effort resulted, in 1975, in a new syllabus entitled 'Bodemnatuurkunde. Zelfinstructieve leertekst'. A. Kamphorst translated this text in English and added some material for use in a M.Sc-course on Soil Science and Water Management at the Agricultural University for students from developing countries. This English version, the experience gathered during about eight years of use with Dutch and foreign students, and many years of experience of the third author in a research environment in the USA formed the basis for the present, much expanded and revised book. G. H. Bolt wrote the epilogue.

The book is meant to be a first introduction in Soil Physics, but students who already have some knowledge in this area may also benefit from working through it. It presents a rigorous, complete and consistent treatment of the definitions and relationships for the fundamental physical processes occurring in soil. In addition, it gives detailed descriptions and solutions of most elementary problems encountered in the realm of Soil Physics. The treatment normally involves many simplifications, such as homogeneous soil, isothermal conditions, isotropy, etc. Because of its elementary level, we deemed it not necessary to include references each time a new concept was introduced. Instead, at the end of each chapter we present a list of recommended literature for further study. This literature, in turn, gives references to more specific or advanced literature, which should be adequate for starting a more thorough study of a particular subject. The teacher may set a time-table for studying the various subjects and accompany the students by means of question sessions. Students could be tested only at the end of the course, but we have had good experience with giving two or three tests at regular intervals, followed by a final examination.

The S.I.-system is used rigorously throughout the text. A few other units which are used quite often in Soil Physics and related disciplines, and are accepted – at least temporarily – alongside the S.I.-system, are also included, e.g., bar, mbar, day, hour, minute, degree Celsius, hectare. The generally used phrase 'per unit' was avoided, because it is dimensionally incorrect. For instance, bulk density, usually defined as 'mass per unit volume', is here defined as 'mass divided by volume', or 'mass per volume'. As is standard practice, symbols for 'quantities' are printed in italics, while units are printed in roman characters. Failure to recognize this may, in some instances, lead to erroneous results (see, e.g., Answer 3.24).

Finally, we wish to thank G. H. Bolt and M. G. M. Bruggenwert for their efforts which helped to bring this rather eventful undertaking to a succesful conclusion. We also want to thank Mrs. I. E. Diraouï-van Rijsinge for typing the manuscript and P. A. C. Raats and J. H. Dane for commenting on various parts of the manuscript.

Wageningen, June 1983 P. K., G. M. and C. D.

CONTENTS

List of symbols

Symbol	Description	Unit
$\mathbf{a}$	acceleration	$m\,s^{-2}$
a	length of clay platelet (chapter 1)	m
	mathematical coefficient	—
A	area	m^2
	coefficient in infiltration equation	$m\,s^{-1}$
b	width of clay platelet (chapter 1)	m
	mathematical coefficient	—
B	coefficient in infiltration equation	$m\,s^{-\frac{3}{2}}$
$\mathbf{c}$	electrostatic field strength	$N\,C^{-1}$
c	concentration in solution (chapter 1)	$mol\,m^{-3}$
	differential water capacity (chapters 4 to 7)	m^{-1}
C	differential water capacity	Pa^{-1}
C_h	differential heat capacity	$J\,m^{-3}\,K^{-1}$
d	diameter	m
D	soil water diffusivity	$m^2\,s^{-1}$
D_0	diffusion coefficient of gas in free air	$m^2\,s^{-1}$
D_g	diffusion coefficient of gas in soil	$m^2\,s^{-1}$
D_h	heat diffusivity	$m^2\,s^{-1}$
E	energy	J
f	mass fraction	—
f_h	heat flux density	$W\,m^{-2}$
f_i	flux density of gas component i	$kg\,m^{-2}\,s^{-1}$
f_w	water flux density	$kg\,m^{-2}\,s^{-1}$
f_n	net radiation flux density at the soil surface	$W\,m^{-2}$
f_a	part of the net radiation flux density used for heating of air (chapter 7)	$W\,m^{-2}$
f_e	part of the net radiation flux density used for evaporation	$W\,m^{-2}$
f_s	part of the net radiation flux density used for heating of soil	$W\,m^{-2}$
f_t	part of the net radiation flux density used for transpiration	$W\,m^{-2}$
$\mathbf{F}$	force	N
F	magnitude of force	N
$\mathbf{g}$	gravitational field strength	$N\,kg^{-1}$
g	magnitude of gravitational field strength	$N\,kg^{-1}$
G	universal gravitational constant (chapter 2)	$N\,m^{-2}\,kg^{-2}$
G_i	mass of gas component i per volume of soil	$kg\,m^{-3}$

Symbol	Description	Unit
h	pressure head of soil water	m
h_a	pneumatic head of soil water	m
h_m	matric head of soil water	m
h_o	osmotic head of soil water	m
H	hydraulic head of soil water	m
i	infiltration rate of water	$m\ s^{-1}$
I	cumulative infiltration	m
$\mathbf{k}$	unit vector in vertical direction	—
k	hydraulic conductivity	$m\ s^{-1}$
K	hydraulic conductivity	$m^2\ Pa^{-1}\ s^{-1}$
K_i	intrinsic permeability	m^2
l	length	m
m	mass	kg
M	molar mass	$kg\ mol^{-1}$
n	number of moles	mol
n_i	number of pores of size class i per area	m^{-2}
p	absolute pressure	Pa
p'	pressure relative to atmospheric pressure	Pa
p_a	pressure equivalent of pneumatic potential ($\approx$ excess gas pressure)	Pa
p_h	pressure equivalent of hydraulic potential	Pa
p_m	pressure equivalent of matric potential (= matric pressure)	Pa
p_t	tensiometer pressure	Pa
pF	$^{10}\log(-h_m/\text{cm})$	—
q	volume flux density of water	$m\ s^{-1}$
Q	volume flux of water (rate of discharge)	$m^3\ s^{-1}$
	electrical charge (chapter 2)	C
	heat content per volume of soil (chapter 4 and 7)	$J\ m^{-3}$
r	radius	m
R	radius of curvature of gas-liquid interface	m
	molar gas constant ($= 8.314\ J\ mol^{-1}\ K^{-1}$)	$J\ mol^{-1}\ K^{-1}$
s	specific surface area (chapter 1)	$m^2\ kg^{-1}$
	space coordinate (in direction of movement)	m
S	sorptivity	$m\ s^{-\frac{1}{2}}$
t	time	s
T	absolute temperature	K
u	potential extent of the double layer	m
$\mathbf{v}$	velocity	$m\ s^{-1}$

Symbol	Description	Unit
V	volume	m^3
w	wetness (gravimetric water content)	($kg\ kg^{-1}$)
x	horizontal space coordinate	m
y	horizontal space coordinate	m
z	vertical space coordinate (height)	m
	gravitational head	m
z_c	height of capillary rise	m

Greek

α	production term in continuity equation	$kg\ m^{-3}\ s^{-1}$ or $m^3\ m^{-3}\ s^{-1}$
β	dummy variable	—
γ	surface energy of solid-liquid interface	$J\ m^{-2}$
δ	thickness of clay platelet	m
η	viscosity (dynamic)	Pa s
θ	volume fraction of liquid ($= \phi_l$)	—
λ	heat conductivity (chapter 4 and 7)	$W\ m^{-1}\ K^{-1}$
	Boltzmann variable (chapter 5)	$m\ s^{-\frac{1}{2}}$
Π	osmotic pressure	Pa
ρ_α	density of phase α	$kg\ m^{-3}$
σ	surface energy of gas-liquid interface	$J\ m^{-2}$
σ_s	soil pressure	Pa
σ_g	intergranular pressure	Pa
σ_w	water pressure	Pa
τ	tortuosity factor	—
φ	angle	rad
ϕ	porosity	—
ϕ_α	volume fraction of phase α	—
χ_i	mass concentration of gas component i	$kg\ m^{-3}$
ψ	potential of soil water	$J\ kg^{-1}$
ψ_a	pneumatic potential of soil water	$J\ kg^{-1}$
ψ_g	gravitational potential of soil water	$J\ kg^{-1}$
ψ_h	hydraulic potential of soil water	$J\ kg^{-1}$
ψ_m	matric potential of soil water	$J\ kg^{-1}$
ψ_o	osmotic potential of soil water	$J\ kg^{-1}$

Symbol	Description	Unit
ψ_p	tensiometer pressure potential of soil water	$J\,kg^{-1}$
ψ_t	total potential of soil water	$J\,kg^{-1}$
ψ_α	adsorption potential of water in double layer	$J\,kg^{-1}$
ψ_δ	pressure potential of water in double layer	$J\,kg^{-1}$
ψ_ω	excess osmotic potential of water in double layer	$J\,kg^{-1}$
ω	angular frequency	s^{-1}

1 Composition and Physical properties of soils

1.1. COMPOSITION OF SOILS

1.1.1. Components and phases

Soil is the natural material that covers the land surface of the earth. It is the product of mechanical, chemical and biological weathering of parent material. Its four constituent fractions or components are

- mineral matter (both primary and secondary minerals)
- organic matter
- water solution
- soil air.

Soils are very complex in their composition and quite variable in their occurrence and properties. Their complex nature, changing behaviour upon use, and variable spatial distribution are the result of nonhomogeneous mixing and interacting of all their components.

In spite of this intricate nature, soils can be handled and studied systematically. Their components, like all other matter in nature, exist in the three states of matter, i.e. solid, liquid and gas. With some exceptions, all three states occur in soils side by side. Hence, these states are called 'phases' of the soil. Strictly speaking, the term 'phase' is not used properly here, since in chemistry it is reserved for a portion of a system having definite geometrical boundaries and uniform properties. The latter is not true here. Nevertheless, we will adhere to this terminology and discuss, successively,

- the solid phase
- the liquid phase
- the gas phase.

Question 1.1 Suggest some situations where only two of the three phases occur in soil.

Question 1.2 Is it possible for a soil to consist of only one phase?

Solid phase

The solid phase is composed of mineral and/or organic matter. The mineral portion consists of particles of various shapes, sizes and chemical composition. Minerals such as quartz, feldspar and colloidal silicates form the major part of the mineral particles. Their density (for definition, see Section *1.1.4*) varies between 2600 and 2850 kg m^{-3}. Heavier minerals, such as magnetite, garnet or hornblendes, also occur, but in small amounts. Generally, one can take 2700 kg m^{-3} as the average density of the mineral matter.

The organic matter consists of plant and animal residues. It also includes living organisms. This portion has densities of 1200 to 1500 kg m^{-3} (on the average 1400 kg m^{-3}). The presence of organic matter may highly influence the average density of the solid phase. For instance, some mineral soils with high organic matter contents have densities less than 2400 kg m^{-3}. For the same reason topsoils have lower densities than subsoils. Except for precise scientific work, 2650 kg m^{-3} can be taken as the average density of the solid phase of mineral soils. When the organic matter content exceeds 20%, soils are usually called organic rather than mineral. Their average density is usually hard to estimate and can better be measured.

Question 1.3 What would be your estimate for the density of a stone found in the soil?

The solid phase forms the matrix or 'skeleton' of the soil. The soil matrix, however, is not always as rigid as the word 'solid' suggests. Certain soils swell and shrink, i.e. the solid phase particles move with respect to one another. In other situations, part of the soil matrix may detach itself and be carried along by moving liquid. This is referred to as particle migration.

The spaces between the solid particles are called soil pores (Figure 1.1). Soil pores are cavities of different shape and size determined by the shape, size and arrangement of the solid particles. Normally the pores form a continuous space throughout the soil. Although larger pores are connected by narrow channels, there are strictly speaking no discrete pores.

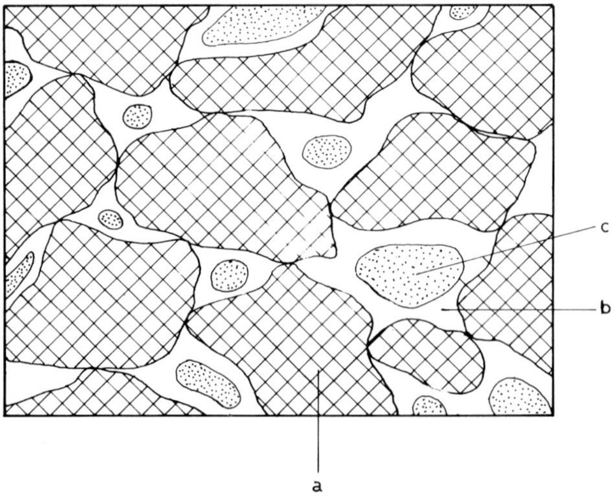

Figure 1.1 Cross-section of soil. Soil particles (a) forming soil pores, partly filled with liquid (b) and gas (c).

Within the pores, the liquid and gas phases reside and move. The portion of the pore space not occupied by the liquid phase is filled by the gas phase.

Question 1.4 Do isolated pores occur in soils?

Liquid phase

The liquid phase, also called soil solution, is an aqueous solution of various electrolytes. Ions commonly found in solution include Na^+, K^+, Ca^{2+}, Mg^{2+}, Cl^-, NO_3^- and SO_4^{2-}. Most of them are plant macro- and micro-nutrients. The liquid phase differs in composition, amount and mobility from pore to pore.

The soil solution may fill the soil pores either completely or partly. In the former case, the soil is said to be 'saturated'. In the latter 'unsaturated' case, the soil solution occurs as thin films along particle surfaces, as annular wedges around contact points of particles and as isolated bodies in narrow pore passages.

Though the density of the liquid phase depends on the composition of the electrolytes in solution, the difference from 1000 kg m^{-3} (the density of pure water at about 4°C) is of no concern except in the most precise studies. Hereafter, in order to avoid confusion, we shall use the term 'pure water' for the pure chemical constituent H_2O. The terms liquid phase, soil water, water and aqueous solution will be used interchangeably for soil solution.

Question 1.5 Will a fragment of soil, when gently immersed in water, become completely saturated with water?

Gas phase

The gas phase, also known as soil air or soil atmosphere, is composed mainly of N_2, O_2, water vapour and CO_2, with traces of other gases. The soil air may vary in composition with prevailing conditions. Presence of organic matter, activity of plant roots and soil micro-organisms, soil aeration and chemical reactions all may alter the concentration of the gas components. Soil aeration is the gas exchange between soil and atmosphere. Even in a well-aerated soil the gas phase contains a higher CO_2 concentration and has a higher relative humidity – nearly vapour saturated – than atmospheric air. In the absence of aeration chemical reduction takes place, producing gases like methane, hydrogen sulphide and nitrous oxide. Ignoring trace gases, we can say that the volume fraction of N_2 is about 0.8 (the same as in the atmosphere) and that of CO_2 and O_2 together is about 0.2.

Because of variations in the composition, the density of soil air fluctuates from 1.0 to 1.5 kg m^{-3}. For calculations, an average of 1.3 kg m^{-3} may be taken.

Question 1.6 What would be the volume fraction of CO_2, if there were no aeration (and no chemical reduction)?

Question 1.7 A surface soil in good condition for plant growth may have the following volume fractions: mineral matter 0.45, organic matter 0.05, soil solution 0.2–0.3 and soil air 0.3–0.2.

a. What is the total volume fraction of the pores in this surface soil?

b. What is the volume fraction of liquid phase required to saturate this soil?

Question 1.8 Consider 1 m^3 of the surface soil of Question 1.7.

a. What is the mass of water required to saturate it?

b. What is the mass of air in it when completely dry?

c. What is the mass of the solid phase?

d. What is the mass of the soil when it is saturated? And when dry?

e. What is the mass of the soil when the volume fraction of liquid phase is only 0.25?

Question 1.9 Convert the average densities of the solid, liquid and gas phases given in the text to g cm^3.

1.1.2 Texture and structure

The solid phase of soils consists of discrete units, called primary soil particles. These particles may vary widely in size, shape and composition. The particle-size distribution or texture of mineral soils determines to a large extent the physical and chemical behaviour. Since these distributions vary little with time, they are used to classify soils.

There are several classification systems in use, which differ only slightly in dividing the spectrum of particle sizes into ranges, called soil separates. The soil separates have been given names. Those recognized by the International Society of Soil Science (ISSS), and their equivalent diameters, are

coarse sand	2000–200 μm
fine sand	200– 20 μm
silt	20– 2 μm
clay	< 2 μm

The determination of the amounts of the various soil separates in a soil sample is called particle-size analysis. For particles larger than 50 μm this is done with sieves with circular holes or wire-mesh screens. Particle sizes smaller than 50 μm are determined by sedimentation, assuming the particles to be spherical (Section *2.1.4*). Of course, primary soil particles are not spherical. Sand and silt particles are isometric or isodiametric, while clay particles are shaped like plates, flakes, planks, or even rods. For this reason, the results of the sieving and sedimentation procedures are expressed in equivalent diameters.

The above soil separates are often divided further into subgroups. The most significant of these is the fine or colloidal clay with equivalent diameters between 1 nm and 0.2 μm. Others are the various grades of sand. Only particles with equivalent diameters smaller than 2 mm, sometimes called fine earth, are included in the particle-size distribution. Coarser soil particles are usually divided into gravel, with equivalent diameters between 2 and 75 mm, and stones which are larger than 75 mm.

Natural soils exhibit infinite combinations of mass fractions in the various soil separates. Soils are given soil textural class names according to the weight percentages of sand, silt and clay, as shown in Figure 1.2. Note that the boundary between sand and silt in this classification system is chosen at 50 μm, rather than 20 μm, All possible combinations are represented in the triangle and the soil textural classes are demarcated by heavy lines. To find the composition of the soil depicted by **X**, draw a line parallel to the sand axis to find the clay percentage, and parallel to the clay axis to find the silt percentage. The rest is sand. Soil **X** has weight percentages of 30, 30 and 40, respectively.

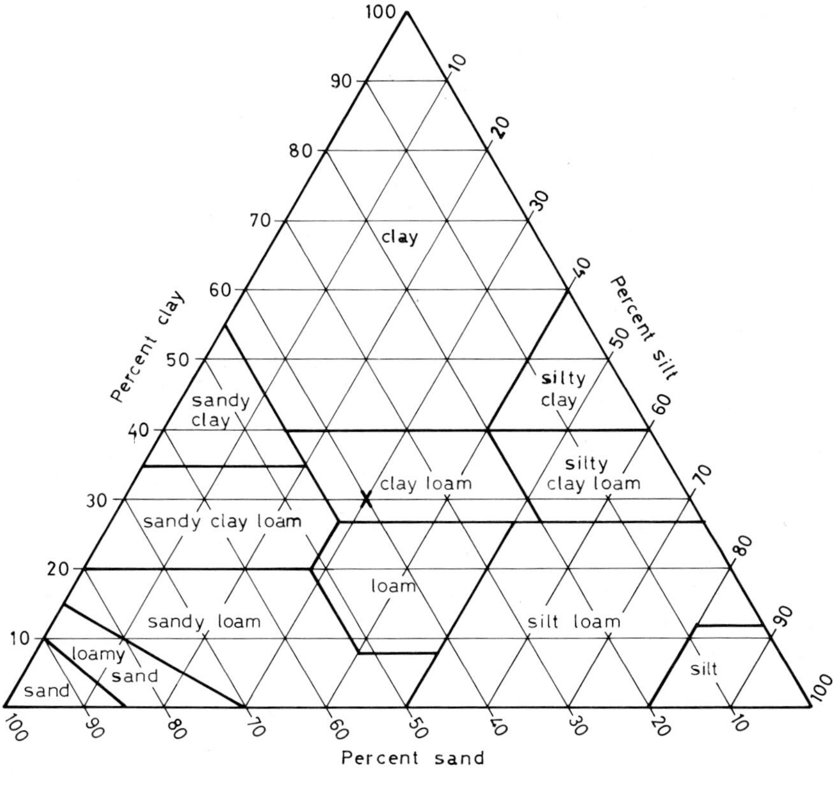

Figure 1.2 Textural soil classes determined by percentages of clay (< 2 μm), silt (2–50 μm) and sand (50–2000 μm).

Another physical property of soils is the soil structure (see also Section *1.4.4*). This is the spatial arrangement of the primary soil particles into compound particles, aggregates or peds and the stability of these units. Common names for structural types are platy, prismatic, columnar, blocky, granular and crumb. Both soil texture and soil structure directly influence the shape, size and volume fraction of soil spores. Soil structure governs characteristics of major concern for plant growth: aeration, temperature, movement of soil solution, microbial activity and root penetration.

1.1.3 Porosity

An adequate supply of soil solution and soil air, especially oxygen, to plant roots is essential for plant growth and crop production. Soil solution and soil air are stored and transported within the soil pores. Also plant roots exist in the soil pores. Thus, the volume fraction of soil pores, ϕ, commonly called porosity, is a very important property of soils. It is defined as

$$\phi = \frac{\text{total volume of pores}}{\text{volume of soil}} \qquad (1.1)$$

where volume of soil represents volume of pores plus volume of solids. The volume of soil is also referred to as bulk volume of soil.

Question 1.10 100 cm^3 of a dry soil requires 40 g of water to saturate it. Find its porosity.

Question 1.11 Assume that the soil in Question 1.10 is located on an irrigated farm. The farmer wants to saturate it down to 25 cm depth.

a. Calculate the volume of water he has to apply per cm^2 of soil surface.

b. If this were to be applied as a single application, what would be the depth of flooding?

Soils differ strongly in porosity. The following ranges have been found in cultivated surface soils:

sandy soils	0.35–0.60
clay soils	0.30–0.70
silty soils	0.30–0.60
peat soils	0.80–0.85
loamy soils	0.30–0.60

Storage, availability and transport of soil solution and soil air are not nearly as much dependent on the porosity as such, as on how the total pore space is partitioned, the so-called pore-size distribution. Soil pores differ in size and shape as a result of textural and structural arrangement (Section *1.1.2*). Based on the diameter at the narrowest point, pores may be classified as follows:

macropores	>	100 μm
mesopores	30 –	100 μm
micropores	<	30 μm

Although these limits are arbitrary, the functioning of these classes is roughly the following. Macropores conduct water only during flooding, ponding rain, etc. They soon drain after cessation of such water supply. Thus they affect aeration and drainage. Mesopores are effective in conducting water also after the macropores have become empty, such as during non-ponding rain, redistribution of water, etc. The remaining soil solution is retained or moves very slowly within the micropores. Part of this water can be taken up by plant roots.

Question 1.12 What is meant by 'drainage' here?

Sandy soils have large pores, whereas clay soils have small pores with or without larger fractures depending on shrinkage. Clay soils retain water more strongly than sandy soils (Section *3.1.5*). Generally, clay soils have a larger porosity and can retain more water than sandy soils, but this does not necessarily mean that more water is available for uptake by plant roots.

Soil porosity consists of gas-filled porosity or aeration porosity, ϕ_g, and liquid-filled porosity, ϕ_l, which is equal to the volume fraction of liquid, θ. For optimum plant growth the aeration porosity should be at least 0.10–0.12 within two to three days after irrigation or heavy rainfall.

1.1.4 Density

In Section *1.1.1* the densities of the three phases of soils were discussed without defining density. Density, also called volumic mass, is the mass of matter divided by its volume. It is expressed in kg m^{-3}. Accordingly, phase density, ρ_α, is the mass of phase α divided by the volume of phase α.

The phase density of the solid phase, also called particle density, is written as ρ_s, where s stands for solid. Those of the liquid and gas phases are written as ρ_l and ρ_g, respectively. In the same manner, the densities of the mineral and organic matter can be written as ρ_{min} and ρ_{org}, respectively. In Section *1.1.1* average densities of the solid, liquid and gas phase of soils were given. They were 2650, 1000 and 1.3 kg m^{-3}, respectively. The average solid-phase density can be calculated if the phase densities and volume fractions of all the constituents of the solid phase are known. Alternatively, it can be determined in the laboratory by measuring the volume of water displaced by a known mass of soil particles.

Question 1.13 To determine ρ_s, a surveyor took 82 g of powdered dry soil at random, put it in a glass bottle slightly larger than 100 cm^3, and added deaerated water at room temperature. After ascertaining that all the air in the soil was driven out, he filled the bottle to exactly 100 cm^3. (A bottle with which this can be done with great accuracy is called a pycnometer.) The mass of the water and soil was 151 g.

a. Calculate ρ_s.

b. Why was it necessary to drive out all the air?

The mass of soil divided by the total or bulk volume of soil is called bulk density, ${}^{b}\rho$, of the soil, where b stands for bulk. The bulk density is the volume-weighted mean density of the three phases:

$$ {}^{b}\rho = \phi_s \rho_s + \phi_l \rho_l + \phi_g \rho_g \tag{1.2} $$

where ϕ_α is the volume fraction of phase α:

$$ \phi_\alpha = \frac{\text{volume phase } \alpha}{\text{bulk volume}} \tag{1.3} $$

Question 1.14 What is the sum of the volume fractions of the three phases in any soil?

Question 1.15 Calculate the bulk density of a soil with volume fractions of solid, liquid and gas of 0.5, 0.3 and 0.2, respectively.

Question 1.16 a. Compare the contribution of each of the three phases to the bulk density in Question 1.15.

b. If only the contributions of the liquid and solid phases were taken, what would be the bulk density? Express the difference in density as a fraction of the former bulk density.

c. What do you conclude from Answers 1.16a and 1.16b?

Question 1.17 Suppose the soil in Question 1.15 were dry.

a. What would be its bulk density?

b. Compare the contribution of each of the phases to the bulk density.

c. What do you conclude from Answer 1.17b?

These answers illustrate that a distinction must be made between the wet bulk density, ${}^{b}\rho_w$, and the dry bulk density, ${}^{b}\rho_d$, of a soil sample. The former is the bulk density of the original wet sample; the latter is the bulk density of the sample after drying in an oven at 105 °C to a constant mass, based on the original volume.

Question 1.18 a. Derive an equation for dry soil analogous to Equation 1.2.

b. Express the difference between ${}^{b}\rho_w$ and ${}^{b}\rho_d$ as an equation.

Answers 1.16 and 1.17 show that the density of the gas phase is so low that the contribution of that phase to the bulk density can be neglected, regardless of the water content. Thus, the equations in Answer 1.18 can be simplified to

$$ {}^{b}\rho_w \approx \phi_s \rho_s + \phi_l \rho_l \tag{1.4} $$
$$ {}^{b}\rho_d \approx \phi_s \rho_s \tag{1.5} $$

and $^{b}\rho_w - {}^{b}\rho_d \approx \phi_l \rho_l$ (1.6)

Bulk densities can be determined by the core method, and by the clod or paraffin method. With the core method the determinations are made on undisturbed samples obtained from natural soil in a cylinder of known volume, generally 100 cm^3. With the clod method, no special effort is needed to obtain a sample of known volume. Instead, a soil clod is coated with paraffin wax and its volume determined by means of its submerged weight. This is discussed further in Chapter 2 (Answer 2.14). The core method is used whenever possible. In soils that swell and shrink, special precautions must be taken.

Question 1.19 A 100 cm^3 undisturbed soil sample has a mass of 162.5 g. After drying at 105 °C, its mass is reduced to 132.5 g.

a. Find $^{b}\rho_w$ and $^{b}\rho_d$.
b. Find ϕ_s, ϕ_l and ϕ_g.
c. Is the soil well-aerated?
d. Are the obtained values of ϕ_l, ϕ_g and ϕ_s true for every part of the soil sample?

A high dry bulk density may mean that the porosity is small and that the soil particles are closely packed or compacted. Soil compaction affects aeration, air and water permeability, infiltration rate, internal drainage, root growth and seedling emergence. Thus, soil compaction hinders crop production. Average dry bulk densities found in surface soils are

1600 kg m^{-3} for sandy soils
1100 kg m^{-3} for clay soils
1100 kg m^{-3} for loamy soils
250 kg m^{-3} for peat soils.

Dry bulk densities are needed for the conversion of gravimetric to volumetric water contents (Equation 1.7). Wet bulk densities together with dry bulk densities are used to estimate the volume fraction of air-filled pore space. Without specification, the term 'bulk density' is normally used for 'dry bulk density'.

1.1.5 Water content

It is difficult to measure soil water contents in situ. In addition, there is the problem of spatial variation. To overcome both obstacles, it is common practice to collect replicate soil samples in the field and to determine the water contents in the laboratory. The water content of a soil sample is, by convention, the amount of water removed by drying at 105 °C. This can be expressed as the mass ratio of water to solid phase, w, commonly called wetness or gravimetric water content. The water content can also be expressed as the volume fraction of water in bulk soil, θ, (Section *1.1.3*) commonly called volumetric water content (volumic water content would be a better term). The volume fraction of water is more difficult to determine than the wetness, but is more useful for calculations of water transport in soils, irrigation management, etc. We will use the term 'water content' only in a general, qualitative meaning. If specific water contents are meant, we will use volume fraction of water for θ and wetness for w.

Question 1.20 Express w in terms of ${}^{b}\rho_{w}$ and ${}^{b}\rho_{d}$.
Question 1.21 Express θ in terms of ${}^{b}\rho_{w}$ and ${}^{b}\rho_{d}$.

From Answers 1.20 and 1.21 follows immediately the often used relationship between θ and w:

$$\theta = \frac{{}^{b}\rho_{d}}{\rho_{l}} w \tag{1.7}$$

Question 1.22 For a soil sample ${}^{b}\rho_{w} = 1716 \text{ kg m}^{-3}$ and ${}^{b}\rho_{d} = 1430 \text{ kg m}^{-3}$.
Calculate:
a. the volume fraction of water, θ
b. the wetness, w
c. the gas-filled porosity, ϕ_{g}.

1.1.6 Organic matter

Organic matter consists of plant and animal residues (Section *1.1.1*). Soils of bogs, marshes and swamps are chiefly composed of organic matter. Surface layers of mineral soils have mass fractions of organic matter of only 0.01 to 0.1; those of peat soils, however, can be as high as 0.80 to 0.95.

Despite the small amounts, organic matter has a profound influence on the physical and chemical properties of mineral soils. It promotes the formation and stability of soil structure and increases water retention, infiltration and drainage. In addition, it provides a source of nitrogen, phosphorus and sulphur, which are essential plant nutrients.

Although organic matter is continuously reworked and decomposed by micro-organisms, its resistant components only break down to finer materials. These finer materials, together with proteins synthesized by soil organisms, form humus.

Humus, like the mineral component clay, is colloidal. Its capacity to hold water and nutrient ions exceeds that of clay. When it is plentiful, it keeps the soil open and loose. Clay is generally present in larger amounts and its contribution to the properties of soil usually equals or exceeds that of humus. Both humus and clay are characterized by a large specific surface area (i.e. surface area divided by mass) and surface charge.

When the surface area of any body is very large in relation to its volume or mass, the special properties of the surface molecules (higher energy and capacity to attract other molecules which approach them) influence the properties of the material. Many properties are proportional to specific surface area. If particles are so small that surface effects dominate, they are called colloids. Phenomena effected by soil colloids are dispersion, swelling, shrinkage, flocculation, cohesion and plasticity (Section 1.3 and 1.4).

Organic matter is usually specified as a separate volume fraction, ϕ_{org}. Together with ϕ_{min}, the volume fraction of mineral matter, it constitutes the volume fraction of the solid phase, ϕ_s. The mass fractions of each of these components can be obtained from the loss of weight of an oven-dry sample upon complete oxidation of the organic matter at about 900 °C.

Question 1.23 a. Calculate ${}^{b}\rho_w$, ${}^{b}\rho_d$, w, θ, ϕ_{min}, ϕ_{org}, ϕ_s, ϕ, ϕ_g and ρ_s of a certain soil sample, if

volume of core sampler	= 90 cm^3
mass of core sampler	= 120 g
mass of sampler and moist soil	= 243 g
mass of sampler and oven-dry soil	= 207 g
mass of sampler and soil after heating at 900 °C	= 167 g

b. What is the significance of ρ_s here?

A soil sample should be as representative as possible for the field or natural soil body from which it is taken. If, for instance, a stone found in a clod or core sample is deemed unrepresentative of the field, a correction should be made for the volume and mass of the stone to obtain the bulk density and other parameters of the remaining part of the sample.

Question 1.24 Correct the results of Question 1.23 for a stone found in the core sample. The stone has a mass of 13 g and a density of 2800 kg m^{-3}.

1.2 SPECIFIC SURFACE AREA

Many of the soil properties are attributable to surface phenomena that take place at the interface between the liquid and the solid phase. These properties can be correlated – they are even approximately proportional – to the specific surface area of the solid phase, s, which is defined as the surface area of particles divided by their mass.

For a homogeneous granular material, the specific surface area can be calculated from the density, size and shape of the particles. If the soil is heterogeneous, the contribution of each component must be evaluated separately, and the total specific surface area is then the mass-weighted mean specific surface area of the components.

There are many methods of estimating the specific surface area of soils. It can be derived from the particle geometry, from adsorption characteristics, heat of immersion, etc. The following is an introduction to the geometrical method.

1.2.1 Regularly shaped particles

Consider a cube of side l. Its total surface area is $6l^2$, its volume l^3, and its mass $\rho_s l^3$. Therefore, its specific surface area is

$$s = \frac{6l^2}{\rho_s l^3} = \frac{6}{\rho_s l} \tag{1.8a}$$

Thus, s is inversely proportional to l.

Question 1.25 Calculate s for cubical particles with a density of 2650 kg m^{-3} and a side of 2 cm, 200 μm and 2 μm, respectively.

For a sphere of diameter d, the surface area and volume are πd^2 and $\frac{1}{6}\pi d^3$, respectively. Its mass is $\frac{1}{6}\rho_s \pi d^3$. Hence,

$$s = \frac{\pi d^2}{\frac{1}{6}\rho_s \pi d^3} = \frac{6}{\rho_s d} \tag{1.8b}$$

This result is identical to that of Equation 1.8a, except that l is replaced by d.

Consider a regular polyhedron with n polyhedral planes and $\frac{d}{2}$ the normal from the centre of the polyhedron to each of the polyhedral planes. Subdivide the regular polyhedron into n polyhedrons, each polyhedron with apex at the centre of the regular polyhedron and with one of the regular polyhedral planes as a base. Now, each polyhedron has a height of $\frac{d}{2}$ and a volume of $\frac{1}{3} \times A \times \frac{d}{2}$, where A is the area of each polyhedral plane. Therefore, the volume of the regular polyhedron is $n \times \frac{1}{3} \times A \times \frac{d}{2} = \frac{1}{6}nAd$, its surface area is nA and its mass is $\rho_s \frac{nAd}{6}$. Hence,

$$s = \frac{nA}{\rho_s \frac{nAd}{6}} = \frac{6}{\rho_s d} \tag{1.8c}$$

Equation 1.8a to 1.8c show for all these regularly shaped particles the same relationship between the specific surface area and a characteristic length. The smaller the particles, the larger is the specific surface area.

1.2.2 Plate-shaped particles

So far, we have derived equations for specific surface area of particles characterized by equal edges, e.g., cubes or regular polyhedrons. Such particles do not occur in nature, except in sand which has more or less spheroid particles.

Consider a cube of side l, sliced parallel to one of its sides into n plates, each of thickness $\delta = l/n$. The original cube has a surface area of $6l^2$; each time a slice is cut off, the total surface area increases with $2l^2$. Thus, after cutting $n-1$ times to obtain n plates, the total surface area becomes $6l^2 + 2l^2(n-1)$.

Question 1.26 a. Using this result, derive an equation for s in n, δ and ρ_s.
b. Give an approximation of s, for n is large.

Consider a plate-shaped, rectangular particle with δ as thickness and a and b as length and width. Its volume is $ab\delta$, its mass $\rho_s ab\delta$, and its total surface area $2(ab + \delta a + \delta b)$. Thus, the specific surface area is

$$s = \frac{2(ab + \delta a + \delta b)}{\rho_s \delta ab} = \frac{2}{\rho_s \delta}\left(1 + \frac{\delta}{b} + \frac{\delta}{a}\right) \qquad (1.9a)$$

or

$$s = \frac{2}{\rho_s}\left(\frac{1}{\delta} + \frac{1}{b} + \frac{1}{a}\right) \qquad (1.9b)$$

If $\delta = a = b = l$, Equation 1.9a reduces to Equation 1.8a. If a and b are kept constant, and δ is reduced, s increases (Equation 1.9b). If $\delta \ll a$ and $\delta \ll b$, which is true for thin plates, Equation 1.9a reduces to

$$s \approx \frac{2}{\rho_s \delta} \qquad (1.9c)$$

This is the same result as obtained in Answer 1.26. If $\delta \ll a$, but $\delta = \mathrm{b}$, i.e. a rod-shaped particle, Equation 1.9a reduces to

$$s \approx \frac{4}{\rho_s \delta} \qquad (1.9d)$$

Equations 1.9 do not apply to objects with rough or bulging shapes.

Question 1.27 a. Which equations are appropriate for estimating specific surface area of sand, silt and clay particles?
b. Calculate the maximum specific surface area of sand and silt.
c. What do you expect about the specific surface area of clay?

Question 1.28 What general characteristic does one observe in Equations 1.8a to 1.9d?

Question 1.29 a. Calculate the specific surface area of plate-shaped clay particles with $\delta = 1$ nm, $a = b = 20$ nm and $\rho_s = 2750$ kg m^{-3}.
b. Give the result also in km^2 kg^{-1} and ha kg^{-1}.

c. Find the number of particles of the size specified in Question 1.29a that would make a total mass of 1 kg.

Question 1.30 a. Calculate $\frac{\delta}{a}$ for the data in Question 1.29.

b. Does s differ significantly from that in Answer 1.29a if one uses Equation 1.9c instead of Equation 1.9b?

Question 1.31 Assuming $a = b$, calculate the percent error in s when Equation 1.9c is used instead of Equation 1.9a for the following values of $\frac{\delta}{a}$: 0.05, 0.02, 0.01, 0.005, 0.002, 0.001 and 0.0005.

Question 1.32 If $\delta = b = 1$ nm and $a = 20$ nm (a rod-shaped particle), is s significantly different if calculated with Equations 1.9c and 1.9d instead of with Equation 1.9a?

Question 1.33 If clay particles would have circular bases of radius r, what would the use of Equation 1.9c imply?

1.2.3 Soils

The specific surface area of a soil of which the particle-size distribution is known, can be calculated by the summation equation

$$s = \sum_i f_i s_i \tag{1.10}$$

where f_i is the mass fraction and s_i is the specific surface area for particle size group i.

Question 1.34 Calculate the specific surface area of a soil that has the following particle-size distribution:

80% fine sand (average d = 0.1 mm)
10% silt (average d = 10 μm)
10% illite clay (average $a = b$ = 200 nm and δ = 5 nm).

The density of the solid phase for sand and silt is 2660 kg m^{-3}, and for illite clay 2750 kg m^{-3}.

Question 1.35 a. What is the contribution of each of the textural separates to the specific surface area of the soil in Question 1.34?

b. What is the textural class name of this soil?

c. To which component can most of the physical and chemical properties of such a soil be attributed?

In summary, Equations 1.8a and 1.8b show that a sphere and a cube have the same specific surface area when the diameter and the edge length are equal. If one side of a cube is reduced while the base remains the same, Equation 1.9c shows that the specific surface area increases. In other words, the thinner the particle, the larger the specific surface area. Clay consists of very thin, plate-shaped particles and thus has a much larger specific surface area than sand or silt, which consist of isodiametric particles of relatively large size. The larger the specific surface area, the larger the surface activity or colloidal activity.

1.3 INTERACTION BETWEEN SOLID AND LIQUID PHASES

Soil particles range in size from slightly larger than molecular size to stones (Section *1.1.2*). The larger particles form the framework or skeleton of the soil, while the smaller ones fill the space around the contact points and cover the surfaces of the larger particles. The liquid phase in the soil pores is in physical contact with the solid phase. At the contact surface, forces of electrical nature cause phenomena such as swelling, shrinkage, aggregation, flocculation and dispersion, which in turn influence transport of air and soil solution. They are typical of soils containing clay and humus (Section *1.1.6*).

1.3.1. Clay minerals

The clay fraction (Section *1.1.2*) consists, by definition, of solid particles less than 2 µm in equivalent diameter. It differs from the coarser sand and silt fractions in that it is composed mainly of minerals which are newly constituted from the products of weathering, which are not encountered in unweathered rocks. The so-called clay minerals constitute most, but not necessarily all, of the subfraction finer than 0.2 µm, known as colloidal clay. Clay minerals are complex in nature. Only a few essentials will be mentioned here.

Clay minerals are built up of silica and alumina sheets and are classified into types 2:1 and 1:1. The former consist of two sheets of silica with one sheet of alumina between them; the latter are formed from one sheet of silica and one sheet of alumina. The structural elements of the sheets are the silicon tetrahedron and aluminum octahedron. The silicon tetrahedron is made up of one silicon atom surrounded by four oxygen atoms; the aluminum octahedron consists of an aluminum atom surrounded by six hydroxyl groups or oxygen atoms. The tetrahydral and octahydral sheets have similar dimensions and are held together with covalent bonds, forming clay mineral layers.

Examples of clay minerals are kaolinite, various micas, illite, chlorite, vermiculite, montmorillonite and a group of allophanes. The most significant of these is montmorillonite; it is characterized by a great degree of swelling and shrinkage (Section *1.3.4*).

The specific surface area of clay minerals, which governs many soil properties, varies from one mineral to another. It is 700–800 $m^2\ g^{-1}$ for montmorillonite, 300–500 $m^2\ g^{-1}$ for vermiculite, 100–300 $m^2\ g^{-1}$ for mica and 5–100 $m^2\ g^{-1}$ for kaolinite.

The thickness of clay mineral layers varies from mineral to mineral. It is about 1 nm for montmorillonite, illite and mica, 1.4 nm for vermiculite, and 0.7 nm for kaolinite.

The planar extensions (i.e. length and width) of clay mineral layers varies from 100 nm to 2 μm. The ratio thickness/planar extension varies between about 0.5×10^{-3} and 10×10^{-3}, which means that Equation 1.9c may be applied for clay minerals.

Question 1.36 What are the symbols in Section *1.2.2* corresponding to the planar extensions and thickness of clay mineral layers?

Question 1.37 Can you guess why clay particles have different specific surface areas? (Assume that they have equal planar extensions.)

Question 1.38 a. Which clay mineral has the specific surface area calculated in Answer 1.30a?

b. For constant planar extensions as in Question 1.29a, calculate the specific surface area if δ is 1, 2, 3, 4, 5 or 10 nm.

1.3.2 Surface charge of clay minerals

During the formation of clay minerals the tetrahedral and octahedral sheets do not meet as already finished sheets. Rather, they grow together, unit by unit. During growth, Si, Al and Mg are hardly ever present in the ideal ratios required for the clay mineral. Therefore, if Al^{3+}-ions are present in excess, they may occupy Si^{4+}-sites in the crystal lattice. Similarly, Mg^{2+}-ions may occupy sites of Al^{3+}-ions if there is a shortage of Al^{3+}-ions. This process leads to an excess negative charge in the crystal lattice, which manifests itself as specific surface charge density.

The replacement of one cation by another of similar size (e.g. Si^{4+} by Al^{3+} and Al^{3+} by Mg^{2+}) is known as isomorphous substitution. It is named isomorphous because these substitutions hardly affect the shape of the crystal lattice. Only cations that fit in the lattice without making it unstable, can be substituted. These substitutions take place only while clay minerals are being formed. Once a clay particle is formed, extraction of the substituting cation is hardly possible without destroying the lattice.

In some clay minerals there is no isomorphous substitution. Then the clay particles remain electrically neutral, except that there is another source of charge. This is a result of incomplete charge compensation of the ions at crystal lattice edges. Below pH = 7, the edges of clay particles have a positive electrical charge.

1.3.3 Adsorption and dissociation of counterions

Since overall electrical neutrality is always maintained in nature, the net negative charge of a clay platelet is compensated by an equivalent positive charge of cations, such as Na^+, K^+, H^+, Mg^{2+} and Ca^{2+}, which are either adsorbed at the platelet surface or swarm in the surrounding solution. These cations counteract the negative electrical charge of the clay platelet towards its surroundings, and are, therefore, also called counterions.

Question 1.39 What is the main binding force between adsorbed cations and a clay platelet?

Question 1.40 Which influence does a clay platelet have on anions?

Question 1.41 What is the difference in electrostatic attraction by a clay platelet on Na^{+}-ions and Ca^{2+}-ions?

Question 1.42 How does the electrostatic attractive force change, qualitatively, with distance from a clay platelet?

Question 1.43 What is the electrical status of a clay platelet dried out from a soil solution?

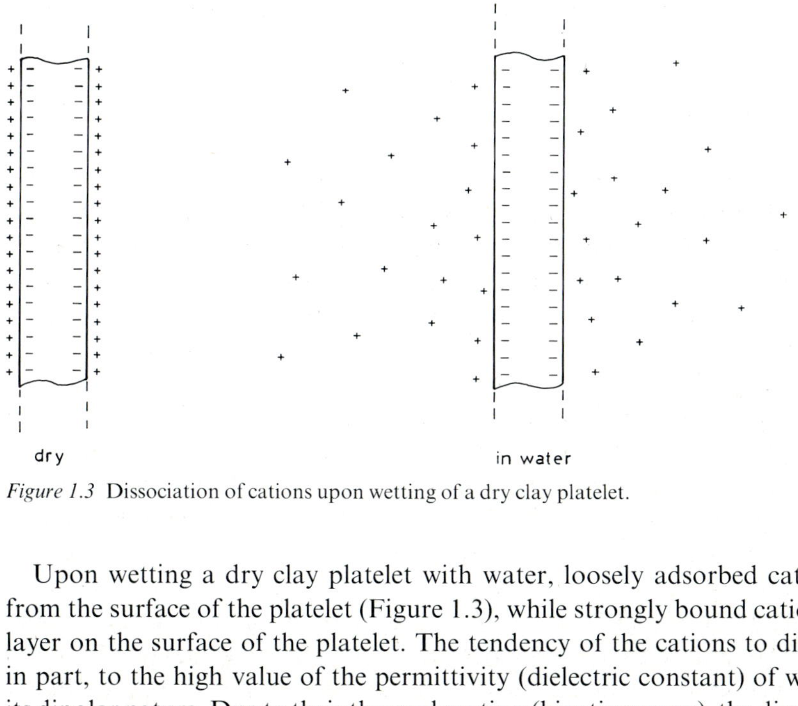

Figure 1.3 Dissociation of cations upon wetting of a dry clay platelet.

Upon wetting a dry clay platelet with water, loosely adsorbed cations dissociate from the surface of the platelet (Figure 1.3), while strongly bound cations form a thin layer on the surface of the platelet. The tendency of the cations to dissociate is due, in part, to the high value of the permittivity (dielectric constant) of water caused by its dipolar nature. Due to their thermal motion (kinetic energy), the dissociated cations have a strong tendency to diffuse away from the clay particle and attain the same concentration everywhere. This represents the condition of maximum entropy of the system (Figure 1.4b). However, the cations remain electrically attracted to the negatively charged clay particle. This tends to drive the system towards the condition of lowest energy (Figure 1.4a). The net result of these opposed influences is the condition of minimum free energy, in which the concentration of cations is very high near the surface of the clay platelet, but decreases with distance from the platelet, at first fast, and then gradually (Figure 1.4c). The portion of the soil solution where the cationic concentration is uniform, is called the bulk solution.

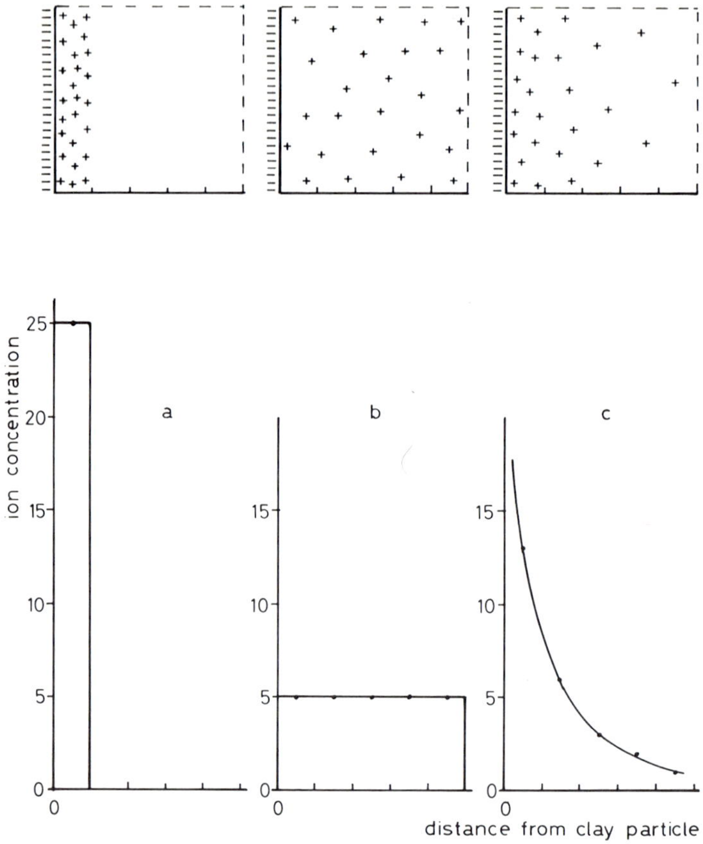

Figure 1.4 Concentration of counterions in the double layer. a: condition of minimum energy; b: condition of maximum entropy; c: condition of minimum free energy.

The charged surface of a clay platelet with its swarm of counterions is called a diffuse electrical double layer. The term double layer refers to the spatial separation of the negatively charged clay platelet and the positively charged counterions. The latter can be divided into two layers: a fixed layer, also called Stern layer, and a diffuse layer. The fixed layer consists of cations strongly adsorbed at the surface of the clay platelet, whereas the diffuse layer is the adjacent zone of more loosely bound cations. The thickness of the fixed layer is little more than the diameter of a single hydrated cation.

Question 1.44 What causes part of a double layer to be diffuse?

The extent of the double layer, u, is defined as the distance from the clay surface at which the cation concentration reaches a uniform value or a minimum. It is also the distance over which the electrical influence of the day platelet on its surroundings vanishes.

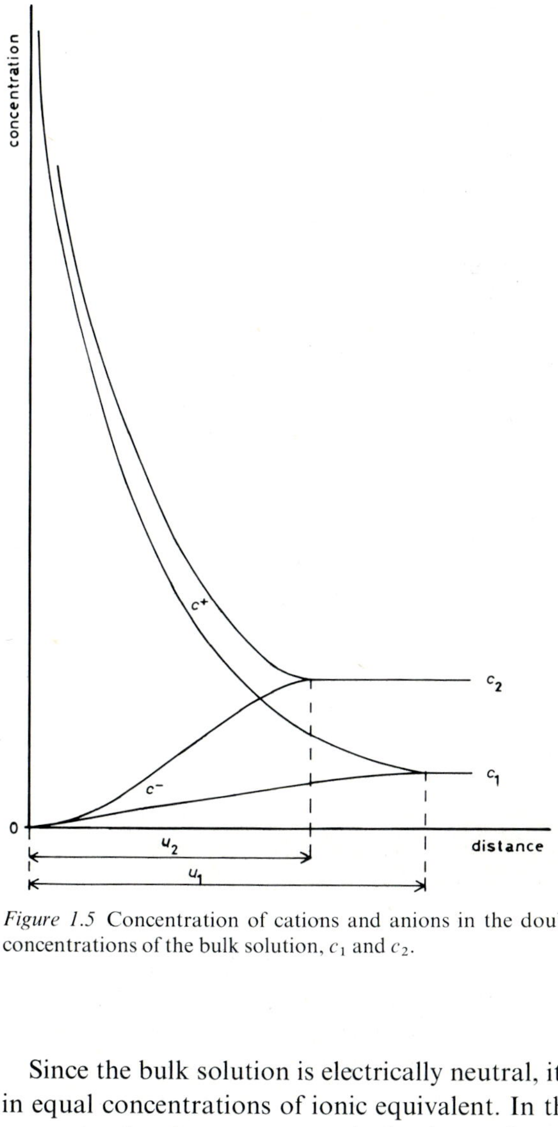

Figure 1.5 Concentration of cations and anions in the double layer of a clay particle for two different concentrations of the bulk solution, c_1 and c_2.

Since the bulk solution is electrically neutral, it contains positive and negative ions in equal concentrations of ionic equivalent. In the double layer, however, the anion concentration decreases towards the clay surface (Figure 1.5). The difference between the total positive electrical charge of the cations and the total negative charge of the anions in the double layer is equal to the negative charge of the clay platelet.

Question 1.45 Why does the anion concentration increase with distance from a clay platelet?

Question 1.46 What will be the influence of an increase in salt concentration in the bulk solution on the extent of a double layer?

Question 1.47 Suppose the cations in the diffuse electric double layer of Figure 1.5 are Na^+-ions and are replaced by an equal ionic equivalent of Ca^{2+}-ions. (Take only the lower concentration of the bulk solution.)
Describe, qualitatively, how the equilibrium distribution of ionic equivalents in the diffuse layer will change. Use a sketch.

1.3.4 Swelling and shrinkage

At low soil water contents, the soil solution around clay particles is present as thin films. The double layer is then prevented from attaining its full (potential) extent. Such a double layer is referred to as a truncated double layer. Figure 1.6 shows schematically an extended and a truncated diffuse double layer.

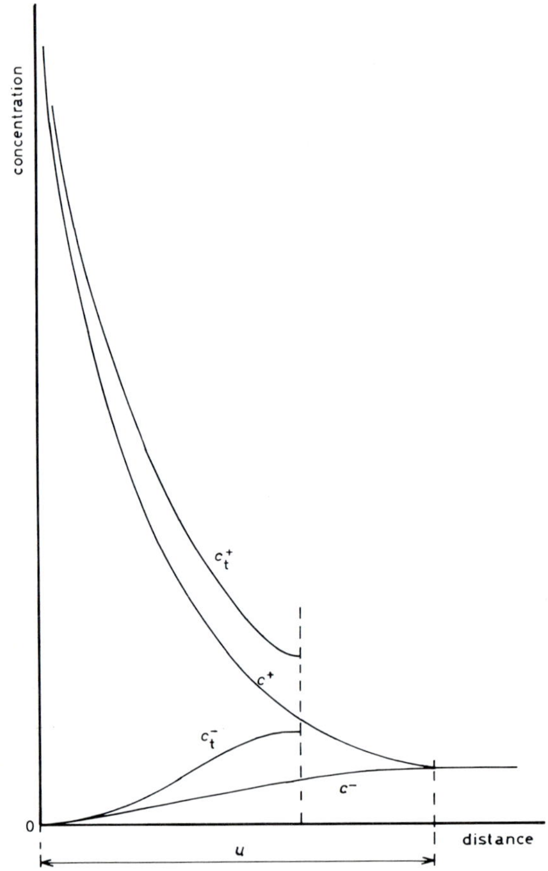

Figure 1.6 Concentration distribution in a fully extended (c^+, c^-) and in a truncated double layer (c_t^+, c_t^-).

A truncated double layer has a higher cation concentration than an extended one. Thus it exhibits the capacity to absorb water by osmosis. If the particles can move unhindered, they are pushed apart until repulsion becomes negligible. If the particles are not free to move, the repulsion force expresses itself as a swelling pressure.

The amount of swelling depends upon the potential extent of the double layer. This potential extent depends on the clay type, the types of counterions and the salt concentration of the bulk solution. For instance, montmorillonite exhibits more swelling than other clay minerals under the same conditions. Upon withdrawal of water, e.g. due to root uptake, evaporation, etc., the extent of the double layer decreases and the soil shrinks.

Swelling involves separation of the smallest, primary soil particles. This often occurs at the expense of the larger pores and cracks, without significant increase in the total porosity. As a result of swelling the permeability of a clay soil for water can be reduced by many orders of magnitude (Section *5.2.1*). Thus swelling hampers the growth of plants because it impedes the movement of soil solution and soil air. Shrinkage, on the other hand, forms cracks between soil aggregates, which increase the permeability for rain and irrigation water of clay soils which otherwise would be impermeable.

Question 1.48 Fill out 'large' or 'small' ('high' of 'low'):

A soil will swell more strongly if

a. its specific surface area is ... L
b. its potential extent, u, of the double layer is ... L
c. its actual extent of the double layer is ...
d. its clay content is ...
e. its dry bulk density is ...
f. its content of monovalent ions is ...
g. its concentration of salt in the soil solution is ...

Question 1.49 A non-saline alkali soil ($u = 20$ nm) contains 33.3% illitic clay ($s = 90$ $m^2\ g^{-1}$), while ${}^{b}\rho_w = 1700$ kg m^{-3} and ${}^{b}\rho_d = 1300$ kg m^{-3}.

a. Calculate the specific surface area of the soil.
b. Calculate the volume fraction of water.
c. Calculate the average thickness of the water films around the particles.
d. How will this soil respond to rain?

1.4 INTERACTIONS BETWEEN SOLID PHASE PARTICLES

1.4.1 Plate condensation

If the distance between adjoining clay platelets is strongly reduced by removal of water, the counterions in the double layers will be forced into the central plane between them. The positive charges of the counterions and the negative charges of the clay surfaces may then approach each other close enough to exert a very large mutual

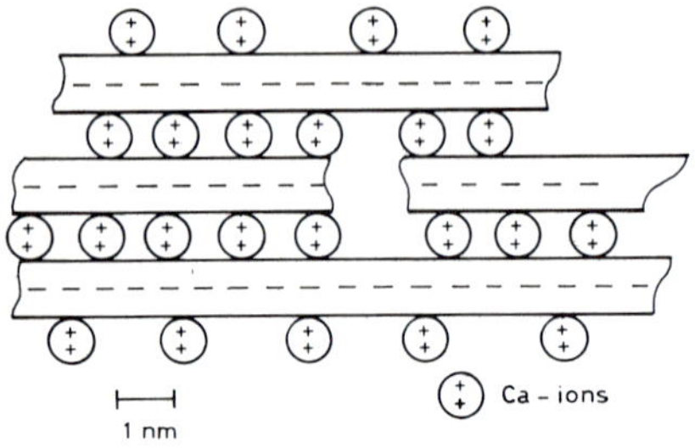

Figure 1.7 Plate-condensation of Ca-clay.

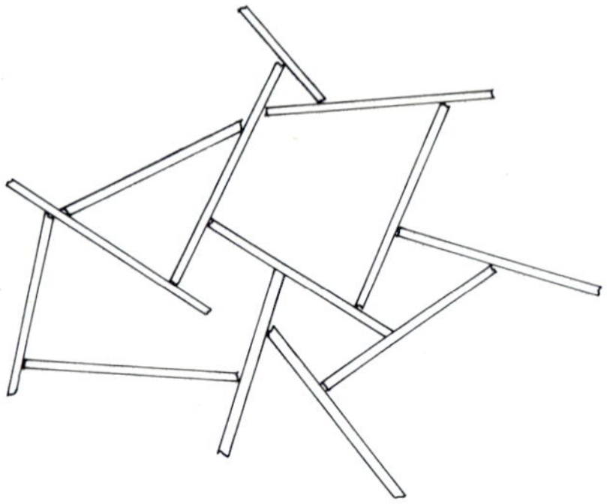

Figure 1.8 'Card-house' structure of floccules.

attraction. In clays with a high percentage of Ca^{2+}-ions (or other polyvalent ions) this attraction may be strong enough to lead to a stable condition called plate condensation (Figure 1.7), which prevents reabsorption of water. In Na-clays the attraction is, generally, not strong enough to prevent swelling. This is partly due to the higher osmotic pressure.

Question 1.50 Why is the osmotic pressure in the double layer higher for Na-clay than for Ca-clay?

Question 1.51 How is the stability of plate condensation influenced by the salt concentration of the soil solution?

Question 1.52 Can you suggest some sources of salts in soil?

1.4.2 Flocculation

The edges of clay platelets have a positive electrical charge at pH below 7. If the extent of the double layer is small, the positively charged edges may approach the negatively charged surfaces of the platelets sufficiently to form weakly bonded floccules. This process is called flocculation. Floccules are loose combinations of clay particles, similar to 'card-house' arrangements (Figure 1.8). Since the stability of these arrangements decreases with increasing extent of the double layer, a Ca-clay will flocculate at a lower salt concentration of the soil solution than a Na-clay.

Clays suspensions probably initially flocculate in card-house arrangements. With slow drying, the floccules tend to orient and stick together in the same manner as with plate-condensation.

Peptisation, dispersion or defloccualtion is the reverse process of flocculation, namely the separation into primary particles. This can be achieved either chemically or mechanically.

When a flocculated clay is dried, it forms a sediment with extensive cracks. It also tends to be crumbly. When this sediment is rewetted, the crumbs as well as the sediment as a whole are stable. When a dispersed clay is dried, it forms a hard crust. If the wetting and drying cycle is repeated, hard, large clods are formed. When the sediment is rewetted, it forms a sticky mud without structure. This is typical of Na-clays, but not of Ca-clays.

Question 1.53 Explain briefly how plate condensation and card-house type flocculation take place in a clay suspension.

Question 1.54 a. How can one deflocculate a flocculated clay?
b. Which laboratory determination already mentioned requires this treatment?

1.4.3 Cementing agents

The primary soil particles within an aggregate may be bound together by certain substances called cementing agents. The main cementing agents are organic matter, silicate clays, lime and sesquioxides.

Humus, which is colloidal organic matter like clay, adsorbs cations. If humus contains a high proportion of Ca^{2+}- and other divalent cations, its long polymer chains can form bonds with each other and with the mineral components of the solid phase. It also binds clay domains to quartz, which is the primary mineral component of silt and sand. In this way, a stable 'clay-humus complex' is formed, resulting in aggregates. In soils rich in Na^{+}- (alkali soils) or H^{+}-ions (acid soils), the bonds are unstable and the humus dissolves.

Silicate clays, too, may cement particles together, but their binding effect is much smaller than that of humus. The degree of cementing and the kind of aggregates formed, vary with the electrical charge, its distribution on the clay particles and the type of clay involved.

Other cementing agents for sand and silt are lime (mainly $CaCO_3$) and sesquioxides (Al- and Fe-oxides). Lime, when it precipitates around contact points between soil particles, acts as a cement. The binding effect of iron oxide is doubtful, but aluminum oxide is probably effective. Finally, soil organisms also may keep soil particles together by their activity and sometimes by their byproducts.

Question 1.55 How can one destroy the binding action of the different cementing agents for particle-size analysis?

1.4.4 Soil structure and structural stability

The spatial arrangement or clustering of primary soil particles into secondary units, called aggregates or peds, is known as soil structure (Section *1.1.2*). Soil structure influences water transport, soil temperature, air transport and mechanical impedance of soil to seedling emergence and root penetration. It is nearly impossible to measure soil structure directly. The spatial arrangement and orientation of soil particles and soil pores as seen in thin sections, probably represents soil structure the closest. Soil structure is often, somewhat arbitrarily, described by the sizes and shapes of aggregates. It is also characterized by porosity and pore-size distribution (Section *1.1.3*). These are an indication of the soil's ability to retain water, to allow water to infiltrate and to make water available to crops. One can also measure parameters such as soil air permeability, infiltration rate, bulk density, aeration porosity and penetration resistance. Though indirectly, these measurements all render information about soil structure.

Question 1.56 Why is soil structure important for plant growth?
Question 1.57 Why is structure only partly defined by porosity and pore-size distribution (volume fractions of pore size classes)?

Processes such as plate condensation and flocculation of clays and humus enhance structure formation, but are by themselves not sufficient. Structure and its stability depend upon the processes of formation, as well as the presence of adequate binding agents. If sufficient amounts of cementing agents are present, even sandy soils can form stable fabrics with many large and small pores produced by soil fauna, roots or frost.

Structural stability of aggregates is as relevant as structure itself. It is the resistance of soil structure to mechanical and physico-chemical destructive forces. This resistance is determined by the attractive forces between the soil constituents under various conditions. For instance, lack of stability of soil structure may seal the soil surface and reduce water infiltration, hamper aeration and impede emergence of seedlings. The process by which dry soil aggregates disintegrate upon wetting is called slaking. This may be aggravated by air locked inside aggregates, while a dry soil is being flooded or infiltrated. The pressure of occluded air may then build up such that the aggregates explode. When the disintegrated aggregates dry, a soil crust may be formed.

Other physico-chemical forces include swelling, shrinkage, flocculation and dispersion. Frost action may also improve soil structure. Growing ice lenses displace soil particles and form large pores. The water being accumulated in the ice lenses causes decreasing water contents elsewhere, often leading to plate condensation, flocculation, etc.

Examples of mechanical, external forces are tillage, movement of farm machines, treading by animals and raindrop splash. Tillage loosens the soil during seedbed preparation. Farm machines compact the soil as they move over the field for various purposes. Animals compact the soil by treading or puddle the soil when it is wet. Raindrops disperse the surface soil by impact; splashed particles are detached and displaced or carried away.

Question 1.58 a. Do you know a tillage procedure where the surface soil is dispersed on purpose?

b. Why is this done?

1.5 SUGGESTED LITERATURE

BAVER, L. D., W. H. GARDNER and W. R. GARDNER, 1972. Soil Physics (fourth edition). Wiley, 489 p. Chapters 1 and 5.

BEERS, W. F. J. VAN, 1980. Soils and Soil Properties. Chapter 1 in: Drainage Principles and Applications. Publication 16, Vol. 1, Int. Inst. for Land Reclamation and Improvement (ILRI), Wageningen, The Netherlands, 2nd edition.

BOLT, G. H. and M. G. M. BRUGGENWERT (eds), 1978. Soil Chemistry A. Basic Elements. Elsevier Scientific Publishing Co. 281 p. Chapters 1 and 3.

HILLEL, D., 1980. Fundamentals of Soil Physics. Academic Press, 413 p. Chapters 2, 4, 5 and 6.

MARSHALL, T. J. and J. W. HOLMES, 1979. Soil Physics. Cambridge University Press, 345 p. Chapters 1 and 8.

TAYLOR, S. A. and G. L. ASHCROFT, 1972. Physical Edaphology. Freeman and Co, 533 p. Chapter 11.

YONG, R. N. and B. P. WARKENTIN, 1975. Soil properties and behaviour. Elsevier Scientific Publishing Co., 449 p. Chapters 1, 2 and 3.

Answer 1.1 The only combinations of two phases that occur together in soil are:

a. the solid and liquid phase;

b. the solid and gas phase.

The former occurs when the soil pores are completely filled with the liquid phase. The latter occurs when the soil is absolutely dry.

Answer 1.2 The only possibility for a soil to consist of one phase would be solid rock, and this is not considered to be a soil.

Answer 1.3 A good estimate for the density of a stone is 2700 kg m^{-3}, unless there are a great deal of clearly recognizable heavy minerals.

Answer 1.4 Isolated pores may occur in soil. Since they are not part of the continuous pore space, they do not contribute to the transport of the liquid and gas phases. In cultivated agricultural soils, they hardly exist.

Answer 1.5 In general, the fragment of soil will not become completely saturated. There usually are pores from which air escapes only partially, or not at all, during wetting. Even though the soil is often called saturated, this is strictly speaking not correct, as long as such pores exist. The air in such pores is known as entrapped air or occluded air. This occurs, in practice, during infiltration of water, and to a lesser extent when the ground water is rising.

Answer 1.6 If there is no aeration, all the O_2 would be consumed and replaced by CO_2 produced by biochemical reactions. Thus, the volume fraction of CO_2 would be up to 20%.

Answer 1.7 a. Since the solid phase, i.e. organic and mineral matter, occupies 50% by volume, the rest (50%) is pore space.

b. When the soil is saturated, the volume of the liquid equals that of the pore space. Hence, it is 50%.

Answer 1.8 a. When saturated, 50% by volume of the soil is filled with water. This amounts to 0.5 m^3 of 500 kg water.

b. When completely dry, there is 0.5 m^3 air in the soil. For an average density of 1.3 kg m^{-3}, this is 0.65 kg air.

c. The mineral matter occupies 45% of the volume and the organic matter 5%, or 0.45 m^3 and 0.05 m^3, respectively. Since the densities are approximately 2700 kg m^{-3} and 1400 kg m^{-3}, the mass contributions are 1215 kg and 70 kg, respectively. Thus the mass of the solid phase is 1285 kg.

d. When the soil is saturated, its mass is 1785 kg (Answers 1.8a and 1.8c). When it is dry, its mass is 1285.65 kg (Answers 1.8b and 1.8c).

e. The mass of the liquid phase is 0.25×1000 kg $= 250$ kg. The mass of the air is 0.25×1.3 kg $= 0.325$ kg. Thus, the total mass is $(1285 + 250 + 0.325)$ kg $=$ 1535.325 kg.

Of course, this can be rounded off to 1535 kg (see Answer 1.16).

Answer 1.9 The average density of the solid phase is:

$$2650 \text{ kg m}^{-3} = \frac{2650 \times 10^3 \text{ g}}{10^6 \text{ cm}^3} = 2.65 \text{ g cm}^{-3}$$

The average density of the liquid phase is:

$$1000 \text{ kg m}^{-3} = \frac{1000 \times 10^3 \text{ g}}{10^6 \text{ cm}^3} = 1.0 \text{ g cm}^{-3}$$

The average density of the gas phase is:

$$1.3 \text{ kg m}^{-3} = \frac{1.3 \times 10^3 \text{ g}}{10^6 \text{ cm}^3} = 0.0013 \text{ g cm}^{-3}$$

Answer 1.10 If the density of the water is 1.0 g cm^{-3}, 40 cm^3 is needed to completely fill the pore space. Hence, per definition, the porosity is 40 cm^3/100 cm^3 = 0.40.

Answer 1.11 a. The volume of soil to be saturated under every cm^2 of soil surface is 25 cm^3. Since the porosity is 0.40, the water in this volume is 0.40 × 25 cm^3 = 10 cm^3. Therefore, he has to apply 10 cm^3 per cm^2 surface area.

b. The depth of flooding would be 10 cm^3/1 cm^2 = 10 cm.

Answer 1.12 By 'drainage' is meant the loss of excess water from the root zone to the lower regions of the soil profile, under the influence of gravity. Lack of drainage impedes the normal functioning of roots because they cannot obtain enough oxygen. Excess drainage causes plants to suffer from lack of water.

Answer 1.13 a. Volume of pycnometer = 100 cm^3

mass of solids and water = 151 g

mass of soil solids = 82 g

mass of water only = (151–82) g = 69 g

$$\text{volume of water} = \frac{69 \text{ g}}{1 \text{ g cm}^{-3}} = 69 \text{ cm}^3$$

volume of soil solids = (100–69) cm^3 = 31 cm^3

$$\rho_s = \frac{\text{mass of solids}}{\text{volume of solids}} = \frac{82 \text{ g}}{31 \text{ cm}^3} \approx 2.65 \text{ g cm}^{-3} = 2650 \text{ kg m}^{-3}$$

b. If all the air in the soil is not driven out, the mass of solids and water in the pycnometer will be too small. As a result, the calculated volume of the solid is larger than what it actually is. This results in a value for ρ_s which is too small.

Answer 1.14 A volume fraction is defined as the ratio of the volume of a phase and the total or bulk volume of soil. Since the sum of the volumes of the soil phases is equal to the total volume of the soil, the sum of the volume fractions equals unity, i.e. $\phi_s + \phi_l + \phi_g = 1$.

Answer 1.15 $^b\rho = \phi_s\rho_s + \phi_l\rho_l + \phi_g\rho_g = (0.5 \times 2650 + 0.3 \times 1000 + 0.2 \times 1.3)$ kg m^{-3} = 1625.26 kg m^{-3}.

Answer 1.16 a. The contributions of the three phases are 1325 kg, 300 kg and 0.26 kg for the solid, liquid and gas phases, respectively. Thus the contributions of the liquid and solid phases are each more than 1000 times that of the gas phase.

b. It would be 1625.0 kg m^{-3}.

The deviation is $\frac{(1625.26 - 1625.0)\ \text{kg m}^{-3}}{1625.26\ \text{kg m}^{-3}} \approx 1.6 \times 10^{-4}$

c. The contribution of the gas phase is negligible compared to the contributions of the other two phases. Therefore, it can safely be neglected.

Answer 1.17 a. Since the soil is dry, the volume of the liquid phase is taken over by the gas phase. Therefore, the gas phase occupies 50% of the bulk volume.

$^{b}\rho = \phi_s \rho_s + \phi_g \rho_g = (0.5 \times 2650 + 0.5 \times 1.3)\ \text{kg m}^{-3} = 1325.65\ \text{kg m}^{-3}$

b. The solid phase contributes 1325 kg m^{-3}, while the gas phase only 0.65 kg m^{-3}.

Thus, the contribution of the solid phase is at least 2000 times that of the gas phase.

c. Despite the fact that the gas phase now occupies half of the volume of the soil, its contribution is still negligible compared to the contribution of the solid phase.

Answer 1.18 a. Since the soil is dry, we have only gas and solid masses. Thus

$^{b}\rho_d = \phi_s \rho_s + \phi \rho_g$

b. $^{b}\rho_w = \phi_s \rho_s + \phi_l \rho_l + \phi_g \rho_g$

$^{b}\rho_d = \phi_s \rho_s + \phi \rho_g$

$^{b}\rho_w - {}^{b}\rho_d = \phi_l \rho_l - \rho_g (\phi - \phi_g)$

Answer 1.19 a. $^{b}\rho_w = \frac{162.5\ \text{g}}{100\ \text{cm}^3} = 1625\ \text{kg m}^{-3}$

$^{b}\rho_d = \frac{132.5\ \text{g}}{100\ \text{cm}^3} = 1325\ \text{kg m}^{-3}$

b. The mass of the solid phase is 132.5 g. Since $\rho_s = 2650\ \text{kg m}^{-3}$, this mass occupies $\frac{0.1325\ \text{kg}}{2650\ \text{kg m}^{-3}} = 50\ \text{cm}^3$. Thus $\phi_s = \frac{50}{100} = 0.50$.

The difference in mass before and after drying is 30 g. Since $\rho_l = 1000\ \text{kg m}^{-3} = 1\ \text{g cm}^{-3}$, this water occupied 30 cm^3.

Thus $\phi_l = \frac{30}{100} = 0.30$.

$\phi_s + \phi_l + \phi_g = 1$. Since $\phi_s = 0.50$ and $\phi_l = 0.30$, $\phi_g = 0.20$.

c. The soil is well-aerated, because ϕ_g is more than 0.12 (Section *1.1.3*).

d. The obtained values are only average values for the whole sample.

Answer 1.20 For an undisturbed soil sample of known volume:

$$w = \frac{\text{mass liquid}}{\text{mass solid}} = \frac{\phi_l \rho_l \times \text{volume of soil}}{\phi_s \rho_s \times \text{volume of soil}} = \frac{\phi_l \rho_l}{\phi_s \rho_s}$$

Substitution of Equation 1.6 in the numerator and Equation 1.5 in the denominator yields:

$$w = \frac{\phi_l \rho_l}{\phi_s \rho_s} \approx \frac{^{b}\rho_w - {}^{b}\rho_d}{^{b}\rho_d}$$

Answer 1.21 Since $\phi_l = \theta$, it follows from Equation 1.6:

$$\theta \approx \frac{{}^{\mathrm{b}}\rho_{\mathrm{w}} - {}^{\mathrm{b}}\rho_{\mathrm{d}}}{\rho_{\mathrm{l}}}$$

Answer 1.22 a. $\theta = \phi_{\mathrm{l}} \approx \frac{{}^{\mathrm{b}}\rho_{\mathrm{w}} - {}^{\mathrm{b}}\rho_{\mathrm{d}}}{\rho_{\mathrm{l}}} = 0.286$

b. $w = \frac{{}^{\mathrm{b}}\rho_{\mathrm{w}} - {}^{\mathrm{b}}\rho_{\mathrm{d}}}{{}^{\mathrm{b}}\rho_{\mathrm{d}}} = 0.200$

c. $\phi_{\mathrm{g}} = 1 - (\phi_{\mathrm{s}} + \phi_{\mathrm{l}}) \approx 1 - (\frac{{}^{\mathrm{b}}\rho_{\mathrm{d}}}{\rho_{\mathrm{s}}} + \phi_{\mathrm{l}}) = 1 - (\frac{1430}{2650} + 0.286) \approx 0.174$

Answer 1.23 a. Moist soil = (243 – 120) g = 123 g

Ovendry soil = (207 – 120) g = 87 g

Oxidized soil = (167 – 120) g = 47 g

$${}^{\mathrm{b}}\rho_{\mathrm{w}} = \frac{123\ \mathrm{g}}{90\ \mathrm{cm}^3} = 1.367\ \mathrm{g\ cm^{-3}} = 1367\ \mathrm{kg\ m^{-3}}$$

$${}^{\mathrm{b}}\rho_{\mathrm{d}} = \frac{87\ \mathrm{g}}{90\ \mathrm{cm}^3} = 0.967\ \mathrm{g\ cm^{-3}} = 967\ \mathrm{kg\ m^{-3}}$$

$$w = \frac{(123 - 87)\ \mathrm{g}}{87\ \mathrm{g}} = 0.414$$

$$\text{or } w \approx \frac{{}^{\mathrm{b}}\rho_{\mathrm{w}} - {}^{\mathrm{b}}\rho_{\mathrm{d}}}{{}^{\mathrm{b}}\rho_{\mathrm{d}}} = \frac{(1367 - 967)\ \mathrm{kg\ m^{-3}}}{967\ \mathrm{kg\ m^{-3}}} = 0.414$$

$$\theta = \frac{(123 - 87)\ \mathrm{g}/1\ \mathrm{g\ cm^{-3}}}{90\ \mathrm{cm^{-3}}} = 0.400$$

$$\text{or } \theta \approx \frac{{}^{\mathrm{b}}\rho_{\mathrm{w}} - {}^{\mathrm{b}}\rho_{\mathrm{d}}}{\rho_{\mathrm{l}}} = \frac{(1367 - 967)\ \mathrm{kg\ m^{-3}}}{1000\ \mathrm{kg\ m^{-3}}} = 0.400$$

$$\text{or } \theta \approx w\frac{{}^{\mathrm{b}}\rho_{\mathrm{d}}}{\rho_{\mathrm{l}}} = 0.414 \times \frac{967}{1000} = 0.400$$

$$\phi_{\mathrm{min}} = \frac{47\ \mathrm{g}/2.7\ \mathrm{g\ cm^{-3}}}{90\ \mathrm{cm}^3} \approx 0.193$$

$$\phi_{\mathrm{org}} = \frac{(87 - 47)\ \mathrm{g}/1.4\ \mathrm{g\ cm^{-3}}}{90\ \mathrm{cm}^3} \approx 0.317$$

$\phi_{\mathrm{s}} = \phi_{\mathrm{min}} + \phi_{\mathrm{org}} = 0.193 + 0.317 = 0.510$

$\phi = 1 - \phi_{\mathrm{s}} = 0.490$

$\phi_{\mathrm{g}} = \phi - \phi_{\mathrm{l}} = 0.490 - 0.400 = 0.090$

$$\rho_{\mathrm{s}} \approx \frac{{}^{\mathrm{b}}\rho_{\mathrm{d}}}{\phi_{\mathrm{s}}} = \frac{967\ \mathrm{kg\ m^{-3}}}{0.510} = 1896\ \mathrm{kg\ m^{-3}}$$

b. ρ_{s} is the average density of the solid phase. The low value indicates that the soil is chiefly made up of organic matter, with a relatively small amount of mineral matter. This illustrates that one has to be careful with taking $\rho_{\mathrm{s}} = 2650\ \mathrm{kg\ m^{-3}}$. This value is valid only for a mineral soil with a low organic matter content.

Answer 1.24 Moist soil = (123 – 13) g = 110 g

Ovendry soil = (87 – 13) g = 74 g

Oxidized soil = (47 – 13) g = 34 g

$$\text{Volume of stone} = \frac{13\ \text{g}}{2.8\ \text{g cm}^{-3}} = 4.643\ \text{cm}^3$$

$$\text{Volume of soil} = (90 - 4.643)\ \text{cm}^3 = 85.357\ \text{cm}^3$$

$$^{b}\rho_w = \frac{110\ \text{g}}{85.357\ \text{cm}^{-3}} = 1.289\ \text{g cm}^{-3} = 1289\ \text{kg m}^{-3}$$

$$^{b}\rho_d = \frac{74\ \text{g}}{85.357\ \text{cm}^3} = 0.867\ \text{g cm}^{-3} = 867\ \text{kg m}^{-3}$$

$$w = \frac{(110 - 74)\ \text{g}}{74\ \text{g}} = 0.487$$

$$\text{or } w \approx \frac{^{b}\rho_w - {^{b}\rho_d}}{^{b}\rho_d} = \frac{1289 - 867}{867} = 0.487$$

$$\theta = \frac{(110 - 74)\ \text{g}/1\ \text{g cm}^{-3}}{85.357\ \text{cm}^3} = 0.422$$

$$\text{or } \theta \approx \frac{^{b}\rho_w - {^{b}\rho_d}}{\rho_l} = \frac{1289 - 867}{1000} = 0.422$$

$$\text{or } \theta \approx w\frac{^{b}\rho_d}{\rho_l} = 0.487 \times \frac{867}{1000} = 0.422$$

$$\phi_{min} = \frac{34\ \text{g}/2.7\ \text{g cm}^{-3}}{85.357\ \text{cm}^{-3}} \approx 0.148$$

$$\phi_{org} = \frac{(74 - 34)\ \text{g}/1.4\ \text{g cm}^{-3}}{85.357\ \text{cm}^3} \approx 0.335$$

$$\phi_s = \phi_{min} + \phi_{org} = 0.148 + 0.335 = 0.483$$

$$\phi = 1 - \phi_s = 0.517$$

$$\phi_g = \phi - \phi_l = 0.095$$

$$\rho_s \approx \frac{^{b}\rho_d}{\phi_s} = \frac{867\ \text{kg m}^{-3}}{0.482} = 1799\ \text{kg m}^{-3}$$

Answer 1.25 For $l = 2$ cm,

$$s = \frac{6}{2650\ \text{kg m}^{-3} \times 2 \times 10^{-2}\ \text{m}} = 0.113\ \text{m}^2\ \text{kg}^{-1}$$

For $l = 200$ μm,

$$s = \frac{6}{2650\ \text{kg m}^{-3} \times 2 \times 10^{-4}\ \text{m}} = 11.3\ \text{m}^2\ \text{kg}^{-1}$$

For $l = 2$ μm,

$$s = \frac{6}{2650\ \text{kg m}^{-3} \times 2 \times 10^{-6}\ \text{m}} = 1132\ \text{m}^2\ \text{kg}^{-1}$$

Answer 1.26 a. The specific surface area is

$$s = \frac{6l^2 + 2l^2(n-1)}{\rho_s l^3} = \frac{4 + 2n}{\rho_s l} = \frac{4 + 2n}{\rho_s n\delta}$$

b. The result of Answer 1.26a can be written as

$$s = \frac{4 + 2n}{\rho_s n\delta} = \frac{2 + 4/n}{\rho_s \delta}$$

For large n this becomes: $s \approx \dfrac{2}{\rho_s \delta}$.

Answer 1.27 a. Equations 1.8 can be used for sand and silt and Equations 1.9 for clay. (For clay particles δ is negligible in comparison to a and/or b; Section *1.3.1*).

b. Since sand has sizes that range from 2 mm to 20 μm, its specific surface area ranges between

$$s = \frac{6}{2650 \times 2 \times 10^{-3}}\ \mathrm{m^2\ kg^{-1}} = 1.13\ \mathrm{m^2\ kg^{-1}}\ \text{and}$$

$$s = \frac{6}{2650 \times 2 \times 10^{-5}}\ \mathrm{m^2\ kg^{-1}} = 113\ \mathrm{m^2\ kg^{-1}}.$$

Hence, the maximum specific surface area for sand is about 113 $\mathrm{m^2\ kg^{-1}}$.

Silt particles range from 20 μm to 2 μm. Thus, similarly, the maximum specific surface area for silt is about 1130 $\mathrm{m^2\ kg^{-1}}$.

c. Since clay particles are smaller than those of silt, the specific surface area of clay is even larger than that of silt.

Answer 1.28 The general characteristic is: the smaller the particles, the larger the specific surface area.

Answer 1.29 a. According to Equation 1.9b:

$$s = \frac{2}{\rho_s}\left(\frac{1}{\delta} + \frac{1}{a} + \frac{1}{b}\right) = \frac{2}{2750}\left(\frac{1}{10^{-9}} + \frac{1}{20 \times 10^{-9}} + \frac{1}{20 \times 10^{-9}}\right)\ \mathrm{m^2\ kg^{-1}}$$

$= 8.0 \times 10^5\ \mathrm{m^2\ kg^{-1}}$.

b. $8.0 \times 10^5\ \mathrm{m^2\ kg^{-1}}$ equals $0.8\ \mathrm{km^2\ kg^{-1}}$

$8.0 \times 10^5\ \mathrm{m^2\ kg^{-1}}$ equals 80 ha $\mathrm{kg^{-1}}$.

c. The volume of one clay particle is

$10^{-9} \times 20 \times 10^{-9} \times 20 \times 10^{-9}\ \mathrm{m^3} = 4 \times 10^{-25}\ \mathrm{m^3}$.

The mass of a single clay particle is then

$4 \times 10^{-25}\ \mathrm{m^3} \times 2750\ \mathrm{kg\ m^{-3}} = 1.1 \times 10^{-21}$ kg.

Thus, the number of particles that compose 1 kg of this clay is about 10^{21}.

Answer 1.30 a. $\dfrac{\delta}{a} = \dfrac{1\ \mathrm{nm}}{20\ \mathrm{nm}} = 0.05$

b. In Answer 1.29a, Equation 1.9b yielded $s = 8 \times 10^5\ \mathrm{m^2\ kg^{-1}}$. This equation involves no approximation.

Equation 1.9c gives:

$$s \approx \frac{2}{\rho_s \delta} = \frac{2}{2750 \times 10^{-9}}\ \mathrm{m^2\ kg^{-1}} \approx 7.3 \times 10^5\ \mathrm{m^2\ kg^{-1}}$$

Thus, for these plate-shaped particles, Equation 1.9c introduces about 9% error.

Answer 1.31 According to Equation 1.9a, $s_1 = \dfrac{2}{\rho_s \delta}\left(1 + \dfrac{2\delta}{a}\right)$.

An approximation of s, according to Equation 1.9c, is $s_2 = \dfrac{2}{\rho_s \delta}$.

Thus, the percent error in s_2 is $\frac{s_1 - s_2}{s_1} \times 100\% = \frac{2\delta/a}{1 + 2\delta/a} \times 100\%$.

δ/a	0.05	0.02	0.01	0.005	0.002	0.001	0.0005
percent error	9.1	3.8	2.0	1.0	0.4	0.2	0.1

Answer 1.32 According to Equation 1.9a,

$$s = \frac{2 \times 2.05}{2750 \times 10^{-9}} \text{ m}^2 \text{ kg}^{-1} = 1.49 \times 10^6 \text{ m}^2 \text{ kg}^{-1}.$$

Equation 1.9c gives $s = \frac{2}{2750 \times 10^{-9}} \text{ m}^2 \text{ kg}^{-1} = 0.73 \times 10^6 \text{ m}^2 \text{ kg}^{-1}$.

Equation 1.9d gives $s = \frac{4}{2750 \times 10^{-9}} \text{ m}^2 \text{ kg}^{-1} = 1.46 \times 10^6 \text{ m}^2 \text{ kg}^{-1}$.

Equation 1.9c introduces a significant error of about 50%, but Equation 1.9d, which was derived specifically for rod-shaped particles, introduces only a 2% error. The large deviation introduced by Equation 1.9c is due to the fact that it was derived for plate-shaped particles ($\delta \ll a$, $\delta \ll b$). For these rod-shaped particles ($\delta = b$), Equation 1.9c cannot be used.

Answer 1.33 If clay particles have circular bases, then

$$s = \frac{2\pi r^2 + 2\pi r\delta}{\rho_s \pi r^2 \delta} = \frac{2}{\rho_s \delta} + \frac{2}{\rho_s r} = \frac{2}{\rho_s}\left(\frac{1}{\delta} + \frac{1}{r}\right)$$

Equation 1.9c can be derived from this equation by neglecting $\frac{2}{\rho_s r}$.

This implies that the contribution of the side area, $2\pi r\delta$, to s is neglected in Equation 1.9c.

Answer 1.34 For the sand and silt fraction, we can use Equation 1.8a.

Sand: $s \approx \frac{6}{\rho_s d} = \frac{6}{2660 \times 10^{-4}} \text{ m}^2 \text{ kg}^{-1} \approx 22.6 \text{ m}^2 \text{ kg}^{-1}$

Silt: $s \approx \frac{6}{2660 \times 10^{-5}} \text{ m}^2 \text{ kg}^{-1} \approx 226 \text{ m}^2 \text{ kg}^{-1}$

For the clay fraction, we can use Equation 1.9a.

$$s \approx \frac{2}{\rho_s \delta}\left(1 + 2\frac{\delta}{a}\right) = \frac{2(1 + 2 \times 0.025)}{2750 \times 5 \times 10^{-9}} \text{ m}^2 \text{ kg}^{-1} \approx 152700 \text{ m}^2 \text{ kg}^{-1}$$

The specific surface area of the soil is then

$$s = \sum_i f_i s_i = (0.8 \times 22.6 + 0.1 \times 226 + 0.1 \times 152700) \text{ m}^2 \text{ kg}^{-1}$$

$$= (18.1 + 22.6 + 15270) \text{ m}^2 \text{ kg}^{-1} \approx 15310 \text{ m}^2 \text{ kg}^{-1}.$$

Answer 1.35 a. The sand contributes $\frac{18.1}{15310} \approx 0.12\%$

the silt contributes $\frac{22.6}{15310} \approx 0.15\%$

the clay contributes $\frac{15270}{15310} \approx 99.7\%$

b. The soil is classified as loamy sand. This can be found in the triangle with textural class names (Figure 1.2).

c. Since the 10% illite clay contributes 99.7% of the specific surface area of the soil, the chemical and physical properties of the soil are attributable to the clay.

Answer 1.36 The symbols are a and b for the planar extensions, and δ for the thickness.

Answer 1.37 Equation 1.9c shows that the greater δ, the smaller s. This indicates that stacking of clay mineral layers within crystals, thereby forming thicker particles, is the most likely reason for different specific surface areas.

Answer 1.38 a. In Answer 1.30a, $s = 8 \times 10^5\ \mathrm{m^2\ kg^{-1}} = 800\ \mathrm{m^2\ g^{-1}}$. This is a very high value typical of montmorillonite.

b. With $a = b = 20$ nm and $\rho_s = 2750\ \mathrm{kg\ m^{-3}}$, Equation 1.9a gives:

δ (nm)	1	2	3	4	5	10
s ($10^6\ \mathrm{m^2\ kg^{-1}}$)	0.80	0.44	0.32	0.26	0.22	0.15

Answer 1.39 The main binding force is an electrostatic force. The cations are electrically attracted but are not chemically bound.

Answer 1.40 Since anions have a negative charge, they are repelled by the negative charge of a clay platelet.

Answer 1.41 Ca^{2+}-ions are attracted twice as strongly by a clay surface as Na^{+}-ions, because their electrical charge is twice as large.

Answer 1.42 The electrostatic attractive force decreases with distance from a clay platelet (in a rather complicated way).

Answer 1.43 All counterions belonging to a clay platelet dried out from a soil are adsorbed on its surface. Thus, the unit is electrically neutral.

Answer 1.44 The diffuse part of the double layer is a result of the electrostatic attraction of the clay platelet on the counterions and the tendency of the counterions to diffuse away from the surface of the platelet, where the concentration is relatively high.

Answer 1.45 Anions are repelled electrostatically by a negatively charged clay particle. The repulsive force decreases with distance. Hence, the anion concentration will decrease with distance from a clay particle up to the extent of its double layer. increase

Answer 1.46 An increase in the salt concentration of the bulk solution suppresses the tendency of counterions to diffuse away from a clay platelet. The extent of its double layer will therefore decrease (Figure 1.5).

Answer 1.47 Because Ca^{2+}-ions are attracted twice as strongly as Na^{+}-ions, they will on the average be closer to the clay particle than the Na^{+}-ions. Hence, the charge concentration of Ca^{2+}-ions is higher near the surface of the clay particle than the original charge concentration of Na^{+}-ions. The area between the cation and anion concentration curves represents the total electrical charge per surface area of the excess cations in the diffuse layer (see Figure below). Since this is equal to the surface charge density of the clay particle, which is constant, the area between the cation and anion curves must be the same for the Na^+-ions and the Ca^{2+}-ions. Since near the clay surface the charge concentration of Ca^{2+}-ions is higher than that of Na^+-ions, it must be lower farther away. The distance over which the concentration of the bulk solution is attained is, therefore, also shorter for Ca^{2+}-ions. Thus, the extent of the double layer decreases when Na^+-ions are replaced by Ca^{2+}-ions. In general, the thickness of the diffuse double layer will decrease as the average valency of the counterions is increased.

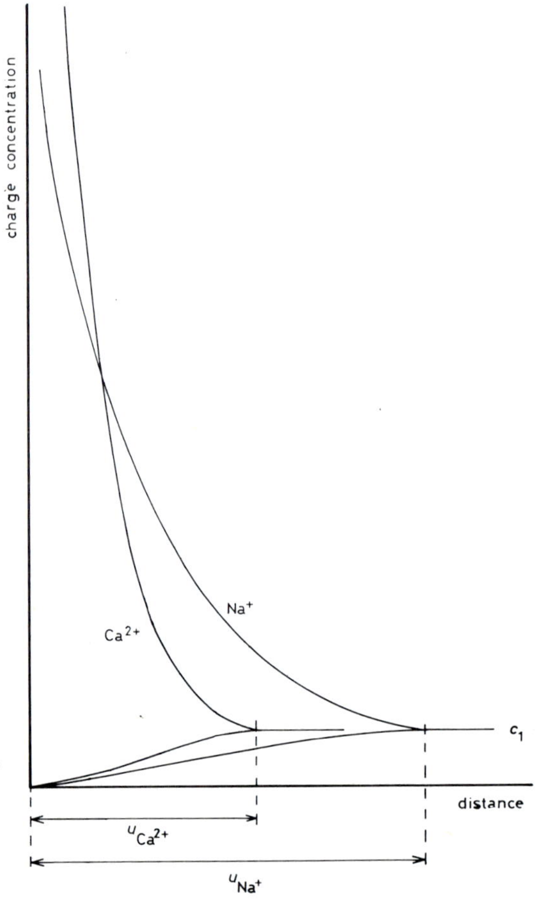

Answer 1.48 a. Large
b. Large
c. Small
d. High
e. High

f. High

g. Low

Answer 1.49 a. The contribution of the sand and silt fractions to s is negligible (Answer 1.34). Thus, the specific surface area is:

$$s \approx \frac{33.3}{100} \times 9 \times 10^4\ \mathrm{m^2\ kg^{-1}} = 3 \times 10^4\ \mathrm{m^2\ kg^{-1}}$$

b. $$\theta = \phi_l \approx \frac{{}^{b}\rho_w - {}^{b}\rho_d}{\rho_l} = \frac{1700 - 1300}{1000} = 0.4$$

c. A volume fraction of water of 0.4 amounts to $\dfrac{0.4}{1300\ \mathrm{kg\ m^{-3}}} = 3.08 \times 10^{-4}\ \mathrm{m^3}$ water per kg soil. Since per kg soil the surface area is $3 \times 10^4\ \mathrm{m^2}$ and the water content $3.08 \times 10^{-4}\ \mathrm{m^3}$, the average thickness of the water films is

$$\frac{3.08 \times 10^{-4}\ \mathrm{m^3\ kg^{-1}}}{3 \times 10^4\ \mathrm{m^2\ kg^{-1}}} \approx 10\ \mathrm{nm}.$$

d. Since the thickness of the water films is less than the potential extent of the double layer ($u = 20$ nm), the soil will swell when it starts raining.

Answer 1.50 There are twice as many Na^+-ions than Ca^{2+}-ions needed to compensate the surface charge of the clay platelet. So in a truncated double layer of the same volume, the osmotic pressure in the double layer is about twice as high for a Na-clay as for a Ca-clay, because the osmotic pressure is proportional to the number of particles in solution (Law of Van 't Hoff).

Answer 1.51 The high salt concentration reduces the swelling pressure and thus increases the stability of plate condensation.

Answer 1.52 Irrigation water containing salts, fertilizer application, rain with dissolved salts, weathering of soil minerals, decaying organic matter, etc. are some of the sources of salts in soil.

Answer 1.53 When the double layers of two clay particles in suspension begin to mesh, they start repulsing each other, because their surface charges are no longer completely shielded from each other. If their kinetic energy of approach can overcome this repulsive force to the point where the London-Van der Waals forces between clay faces take effect, plate condensation will take place. If it overcomes the repulsive force during an approach of a positive edge and a negative face, then 'card-house' type of flocculation will occur.

Answer 1.54 a. Deflocculation is enhanced by saturating the clay with Na^+-ions, by lowering the salt concentration, by increasing the pH, by shaking and stirring, etc.

b. This treatment is required in particle-size analysis.

Answer 1.55 H_2O_2 is used for the destruction of humus, while HCl helps the dissolution of lime and sesquioxides.

Answer 1.56 Movement and storage of soil water and gas take place in the soil pores. Hence, the porosity and the size of the pores determine the supply of water, dissolved nutrients and oxygen to plant roots.

Answer 1.57 Soil structure of a heterogeneous soil is characterized, in addition to porosity and pore-size distribution, by the spatial distribution of these factors. The shapes of the pores and their distribution and arrangement in the soil matrix determine the transport possibilities of liquid and gaseous components. For instance, a thin dense surface layer overlying an otherwise homogeneous porous soil profile can block virtually all vertical transport of water and gases.

Answer 1.58 a. Surface soil of ricefields is puddled before sowing or transplanting.
b. It is done to restrict the downward percolation of water in the flooded field.

2 Equilibrium in force fields and theory of potentials

In Chapter 1 it was mentioned that several forces act on soil water due to the nature of the solid particles. Because these forces operate simultaneously and have different magnitudes and directions, it is necessary to know how to add and multiply vectors in order to describe and predict water transport in soils. As in mechanics, the forces acting in soils obey the laws of vector operation. In this chapter we shall first discuss the basic vector operations and the associated conditions for equilibrium. In the second part we shall introduce the concept of potential and define various forms and components.

2.1 EQUILIBRIUM CONDITIONS IN FORCE FIELDS

2.1.1 Forces and equilibria

Intuitively, a force can be defined as an influence that tends to produce a change in the motion of objects. However, a force can also be balanced by another force acting on the object, without change in motion. Fundamentally, a force is defined by Newton's second law of motion

$$\boldsymbol{F} = m\boldsymbol{a} \tag{2.1a}$$

where $\boldsymbol{F}$ is force in N (kg m s^{-2}), m is mass in kg and $\boldsymbol{a}$ is acceleration in m s^{-2}. Force and acceleration are vectors (printed in bold face) which have the same direction, whereas mass is scalar. Vectors are quantities that require specification of direction as well as magnitude for complete characterization. Scalars are quantities that are completely characterized by their magnitude only.

When more than one force is acting on an object at the same time, the forces may be added in the proper way. Scalars can be added just by adding their magnitudes; vectors must be added according to direction and magnitude. This can be done by placing them tail to head in any order. The vector sum or resultant vector is then equal to the single vector leading from the tail of the first vector to the head of the last vector. The component of the resultant force in a certain direction, for instance the vertical direction, is equal to the sum of the components of the individual vectors in that direction.

Question 2.1 Determine the resultant vector, $\boldsymbol{R}$, of the vectors with their tail in the origin and their heads at the x, y-coordinates $\boldsymbol{A}(2.2)$, $\boldsymbol{B}(-3,1)$ and $\boldsymbol{C}(3,-4)$, respectively.

The sum of all forces acting on a body simultaneously, is the vector sum or resultant force, $\Sigma\, \boldsymbol{F}$. Equation 2.1a is also true for the resultant force. That is:

$$\Sigma\, \boldsymbol{F} = m\,\boldsymbol{a} \tag{2.1b}$$

Therefore, if the vector sum of all forces equals zero, the acceleration is zero, and the body is said to be in mechanical equilibrium:

$$\Sigma\, \boldsymbol{F} = 0 \rightleftharpoons \text{mechanical equilibrium} \tag{2.2}$$

The double arrow indicates that Equation 2.2 can be read from left to right and from right to left. When Equation 2.2 is valid, there is no acceleration, but the body can still be in uniform motion.

Question 2.2 Describe Newton's second law of motion in words.
Question 2.3 Define mechanical equilibrium in words.
Question 2.4 A body is placed on a table, where it stays indefinitely. What is its equilibrium status?

The forces in Equation 2.2 can be divided into two types: static forces, $\boldsymbol{F}^{\mathrm{s}}$, and dynamic or drag forces, $\boldsymbol{F}^{\mathrm{d}}$, with $\Sigma\, \boldsymbol{F} = \Sigma\, \boldsymbol{F}^{\mathrm{s}} + \Sigma\, \boldsymbol{F}^{\mathrm{d}}$. Static forces are present irrespective of movement. Drag forces may arise whenever static forces cause a body to move. Thus, they are reaction forces, associated with movement. When mechanical equilibrium and motion occur simultaneously, Equation 2.2 becomes $\Sigma\, \boldsymbol{F} = \Sigma\, \boldsymbol{F}^{\mathrm{s}} + \Sigma\, \boldsymbol{F}^{\mathrm{d}} = 0$. This implies $\Sigma\, \boldsymbol{F}^{\mathrm{s}} = -\Sigma\, \boldsymbol{F}^{\mathrm{d}}$. In the absence of motion there are no drag forces, i.e. $\Sigma\, \boldsymbol{F}^{\mathrm{d}} = 0$. In such a situation $\Sigma\, \boldsymbol{F}$ equals $\Sigma\, \boldsymbol{F}^{\mathrm{s}}$, and Equation 2.2 is reduced to $\Sigma\, \boldsymbol{F} = \Sigma\, \boldsymbol{F}^{\mathrm{s}} = 0$. This situation, called static equilibrium, is described by

$$\Sigma\, \boldsymbol{F} = \Sigma\, \boldsymbol{F}^{\mathrm{s}} = \Sigma\, \boldsymbol{F}^{\mathrm{d}} = 0 \rightleftharpoons \text{static equilibrium} \tag{2.3}$$

Thus, static equilibrium is a special case of mechanical equilibrium.

Question 2.5 Can you imagine a situation (not on earth) to which Equation 2.3 would not apply if read from left to right?
Question 2.6 In Answer 2.4 the sum of the forces is zero ($\Sigma\, \boldsymbol{F} = 0$)
a. What are the forces to be summed?
b. Is there a drag force?
c. How would you redefine its equilibrium status?

Static forces acting on objects are either body forces, $\boldsymbol{F}^{\mathrm{s}}_{\mathrm{b}}$, or contact forces, $\boldsymbol{F}^{\mathrm{s}}_{\mathrm{c}}$. Contact forces are local reaction forces resulting from contact with other objects. An example of a body force is the gravitational force. Body forces are associated with force fields, which are discussed in the next section. Static forces are the sum of body and contact forces, i.e.

$$\Sigma\, \boldsymbol{F}^{\mathrm{s}} = \Sigma\, \boldsymbol{F}^{\mathrm{s}}_{\mathrm{b}} + \Sigma\, \boldsymbol{F}^{\mathrm{s}}_{\mathrm{c}} \tag{2.4}$$

At static equilibrium, $\Sigma\, \boldsymbol{F}^{\mathrm{s}} = 0$. Then the sum of the contact forces is equal, but oppositely directed, to the sum of the body forces, i.e.

$$\Sigma \boldsymbol{F}_{c}^{s} = -\Sigma \boldsymbol{F}_{b}^{s} \tag{2.5}$$

2.1.2 *Force fields*

A force field is a concept which facilitates the computation of forces, especially when they differ with position. The concept cannot be defined and described here exhaustively. It will only be illustrated by a few examples. The first is the earth's gravitational attraction. The gravitational force field is described by

$$\boldsymbol{F} = m\boldsymbol{g} \tag{2.6}$$

$\boldsymbol{F}$ is the gravitational force, or weight (N), of a body with mass m (kg), and $\boldsymbol{g}$ is the gravitational field strength, which is the gravitational force divided by mass (N kg^{-1}). The gravitational force acts on a body, irrespective whether the body is static or in motion and whether it is in contact with other bodies or not. It is, therefore, also called a body force.

An object falling freely (in a vacuum) in the earth's gravitational field obtains an acceleration described by Equation 2.1a. Experimentally, one can determine that the magnitude of $\boldsymbol{a}$ is 9.81 m s^{-2}. One can show that then also in Equation 2.6, the magnitude of $\boldsymbol{g}$ is 9.81 N kg^{-1}. In this book the magnitude of $\boldsymbol{g}$ will, for convenience, be rounded off to g = 10.0 N kg^{-1}.

Question 2.7 a. Define weight in words.
b. What is the weight of a body with a mass of 5 kg?

Question 2.8 Newton's law of gravitation can be written as

$$F = \frac{G m_1 m_2}{r^2} \tag{2.7}$$

where m_1 and m_2 are the masses of two bodies, r is the distance between their centres, G is the universal gravitational constant and F is the magnitude of the attractive force between the masses. By equating Newton's law of gravitation to Equation 2.6, derive an expression for the magnitude of $\boldsymbol{g}$ (g). Take $m = m_1$, as the mass of a body and m_2 as the mass of the earth.

A distinction can be made between homogeneous and nonhomogeneous force fields. A homogeneous force field is a force field in which the magnitude and direction of the force are independent of position, whereas the magnitude and direction of a nonhomogeneous force field vary with position.

Question 2.9 a. Is the gravitational field strength homogeneous? Explain.
b. Is the centrifugal field in a centrifuge homogeneous? Explain.

The field strength of electrostatical force fields can be defined as the force divided by electrical charge. This definition is similar to Equation 2.6, except that mass must be replaced by electrical charge, Q. Equation 2.6 becomes

$$\boldsymbol{F} = Q\boldsymbol{c} \tag{2.8}$$

where $\boldsymbol{c}$ is the electrostatic field strength. This field strength can be translated into a force divided by mass if the specific charge (i.e. charge divided by mass) of the material is known. This subject matter is pertinent to the charges and forces within diffuse double layers of clay particles (Section *1.3.3*).

Question 2.10 A body has a mass of 0.1 g and an electrical charge of 10^{-7} C (coulomb). It experiences a force due to the presence of other charges. The electric field strength at the location of the body is 10^3 N C^{-1}.

a. Express its acceleration as a function of mass, charge and field strength.
b. Compute the acceleration if the body is free to move.
c. Express the force and field strength analogously to Answer 2.8.

2.1.3 Forces and pressures in static equilibrium

Contact forces act not only between different bodies, but also within a body, where they cause material stresses. This can be explained by considering a concrete pillar at static equilibrium in the gravitational force field (Figure 2.1). Each mass m in this pillar is subject to a body force $m\boldsymbol{g}$. But, being at static equilibrium, the body force acting on each mass must be counterbalanced by a net contact force, $-m\boldsymbol{g}$. (The minus sign indicates that the contact force is directed oppositely to the field strength $\boldsymbol{g}$.) This can be shown by dividing the pillar into a number of horizontal, thin slices, each with a mass m. The body force of the top slice exerts a contact force $m\boldsymbol{g}$ on the second slice. Since the pillar, and thus the uppermost slice, is at static equilibrium, the second slice reacts with a contact force $-m\boldsymbol{g}$ on the top slice. The second slice exerts a contact force $2m\boldsymbol{g}$ on the third slice, because the total mass resting on the third slice is that of the upper two slices, i.e. $2m$. But then the third slice reacts with a contact force on the second slice equal to $-2m\boldsymbol{g}$. Thus the net contact force on the second slice is $m\boldsymbol{g} + (-2m\boldsymbol{g}) = -m\boldsymbol{g}$, i.e. $m\boldsymbol{g}$ from above and $-2m\boldsymbol{g}$ from below (Figure 2.1b). Similarly, the third slice is subject to a contact force of $-3m\boldsymbol{g}$ from the fourth slice and the net contact force on the third slice is $2m\boldsymbol{g} + (-3m\boldsymbol{g}) = -m\boldsymbol{g}$. And so forth.

The last paragraph indicates that the contact force in any horizontal plane is the product of $-m\boldsymbol{g}$ and the number of overlying slices, n. Thus, $\boldsymbol{F}_c^s = -nm\boldsymbol{g}$, where nm is the total mass of all the overlying material above the horizontal plane. When this contact force is divided by the area of the horizontal plane, it is called normal material stress, and is expressed in Pa. By 'normal' is meant perpendicular to the horizontal plane. This distinction is necessary, since solid bodies may also exhibit frictional tan-

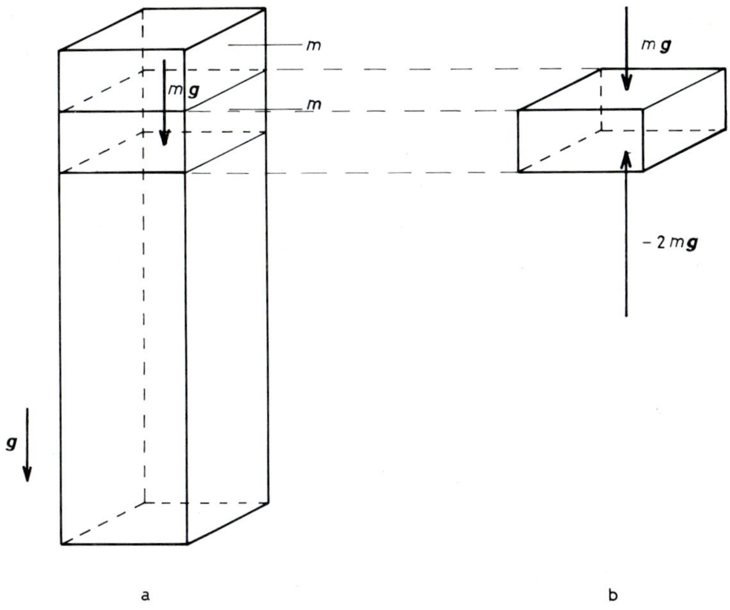

Figure 2.1 a. Concrete pillar divided in horizontal slices, each of mass m.
b. Contact forces on the second slice.

gential stresses in other directions, which will not be discussed here. We conclude that, in bodies supported from below, the material stress increases in absolute value in the direction of the gravitational field.

In fluids – substances without rigidity, such as water and air – the material stress acts with equal magnitude in all directions. In that case, the stress can be measured as a pressure, p (Pa), and is, unlike force and acceleration, a scalar quantity.

Question 2.11 Why is p scalar?

Consider now a thin layer in a column of water at rest (Figure 2.2), with horizontal upper and lower square planes with edges l and vertical planes with height Δz. As usual, a coordinate system is chosen with x and y in horizontal directions and z, positive upwards, in the vertical direction. Suppose the average pressure on the vertical plane at the left-hand side is p_1. Then the magnitude of the horizontal force on this plane is $F_1 = l \times \Delta z \times p_1$. Similarly, the magnitude of the horizontal force on the plane at the right is $F_2 = l \times \Delta z \times p_2$. Since the layer is at rest, and $\boldsymbol{F}_1$ and $\boldsymbol{F}_2$ are acting in exactly opposite directions, the magnitudes of $\boldsymbol{F}_1$ and $\boldsymbol{F}_2$ must be the same. Thus, $p_1 = p_2$. By letting Δz approach zero, it follows that the pressure at every point in a continuous horizontal layer of water at rest is everywhere the same. From this it follows easily that the horizontal forces on any body completely surrounded by a liquid at rest, always counterbalance each other. Therefore, there is no net horizontal force

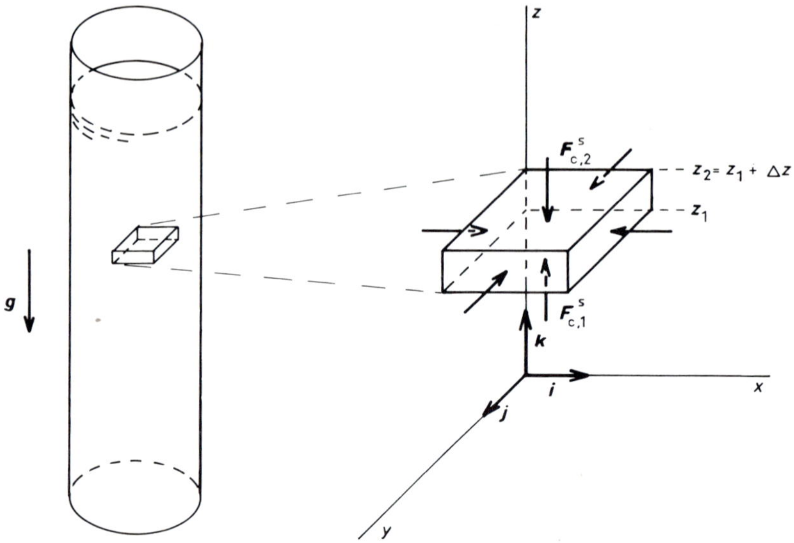

Figure 2.2 Contact forces on a thin layer of water in a column at static equilibrium.

acting on the vertical planes of the water layer in Figure 2.2. Only vertical forces need to be taken into account.

The vertical upward contact force on the lower plane of the layer in Figure 2.2 is

$$\boldsymbol{F}^{s}_{c,1} = p_1\, l^2\, \boldsymbol{k} = p_1 \frac{V}{\Delta z}\, \boldsymbol{k} \tag{2.9}$$

where p_1 is the pressure at that plane, $\boldsymbol{k}$ is the unit vector in the z-direction and V is the volume of the layer.

The contact force on the upper layer, directed downward, is similarly

$$\boldsymbol{F}^{s}_{c,2} = -p_2\, l^2\, \boldsymbol{k} = -p_2 \frac{V}{\Delta z}\, \boldsymbol{k} \tag{2.10}$$

The net vertical contact force on the layer, $\Sigma\, \boldsymbol{F}^{s}_{c}$, is then

$$\Sigma\, \boldsymbol{F}^{s}_{c} = -\frac{\Delta p}{\Delta z} V \boldsymbol{k} \tag{2.11}$$

where $\Delta p = p_2 - p_1$.

As Δz approaches zero, the net vertical contact force divided by volume becomes

$$\frac{\Sigma\, \boldsymbol{F}^{s}_{c}}{V} = -\,\boldsymbol{k} \lim_{\Delta z \to 0} \frac{\Delta p}{\Delta z} = -\frac{\mathrm{d}p}{\mathrm{d}z}\, \boldsymbol{k} \tag{2.12}$$

Now, using $\rho_l = \frac{m}{V}$ for the density of the water and Equations 2.5 and 2.6, we obtain

$$-\frac{1}{\rho_l}\frac{dp}{dz}\boldsymbol{k} = \frac{\Sigma \boldsymbol{F}_c^s}{m} = -\frac{\boldsymbol{F}_b^s}{m} = -\boldsymbol{g} \qquad (2.13)$$

Since $\boldsymbol{g} = -g\boldsymbol{k}$, Equation 2.13 can be reduced to

$$\frac{dp}{dz} = -\rho_l g \qquad (2.14)$$

Assuming ρ_l is not dependent on p, i.e. water is incompressible, integration of Equation 2.14 yields

$$p = -\rho_l g z + C \qquad (2.15)$$

where C is an integration constant. If $z = 0$ is taken at the water surface, where the pressure is the atmospheric pressure, p_{atm}, then $C = p_{atm}$ and thus:

$$p = -\rho_l g z + p_{atm} \qquad (2.16)$$

Thus, in a liquid at rest, the pressure increases linearly with depth. The difference in pressure at two heights is equal to the magnitude of the weight of the liquid column stretching between those two heights, divided by the cross-sectional area of the column.

Question 2.12 A cube is submerged in a container of water, with two faces parallel to the water surface.

a. Derive general equations for the vertical net contact force, the body force and the total static force on the cube.
b. Calculate $\Sigma \boldsymbol{F}^s$ if the cube has a mass of 30 g and a volume of 27 cm^3.
c. What is the effect of $\Sigma \boldsymbol{F}^s$? (Discuss also $\boldsymbol{F}^d$.)

The result obtained in Answer 2.12a for a cube, is valid for a body of any irregular shape. This is known as Archmides' law: A fluid acts on a body immersed in it with a net contact force that is directed vertically upward and equal in magnitude to the weight of the fluid displaced by the body. The upward force is called buoyant force. This principle is used in soil physics for the determination of the bulk density of irregularly shaped soil samples.

Question 2.13 A clod with a mass of 8.99 g, perfectly sealed with paraffin wax, is lowered into a container of water on a string attached to a torsion balance. At equilibrium, the balance reads 2.79 g. Assume that the masses of the wax coating and the string are negligible.

a. Calculate the weight of the clod.
b. Calculate the body force, the net vertical static force and the net contact force on the clod by the water.
c. What is the bulk density of the clod?

d. Why was it necessary to seal the clod perfectly with paraffin wax? How can you check that it was perfectly sealed?

In soils, water is not only subject to the gravitational field, but also to forces of adsorption by the solid phase (Sections 1.3 and *3.1.1*). There, the pressure in the water increases in the direction of the adsorbing surface, similar to the pressure in the gravitational field.

Another force field sometimes used in soil physics, is the centrifugal force field in a centrifuge.

Question 2.14 A liquid in a centrifuge tube rotates horizontally with a constant angular frequency ω about a vertical axis (centrifuge centre, Figure 2.3). Its centripetal acceleration is $\boldsymbol{a} = -\omega^2 r \boldsymbol{i}$ and r_0 is the distance from the vertical axis to the nearest liquid surface in the tube. ($\boldsymbol{i}$ is the unit vector in the direction of r.)

a. If $p = p_{atm}$ ($\approx 10^5$ Pa) at $r = r_0$, derive p at any distance r. (Neglect gravity.)

b. If the centrifuge is turning at 6000 revolutions per minute, $\rho_l = 1000$ kg m^{-3} and $r_0 = 0.2$ m, calculate the pressure at $r = 0.3$ m.

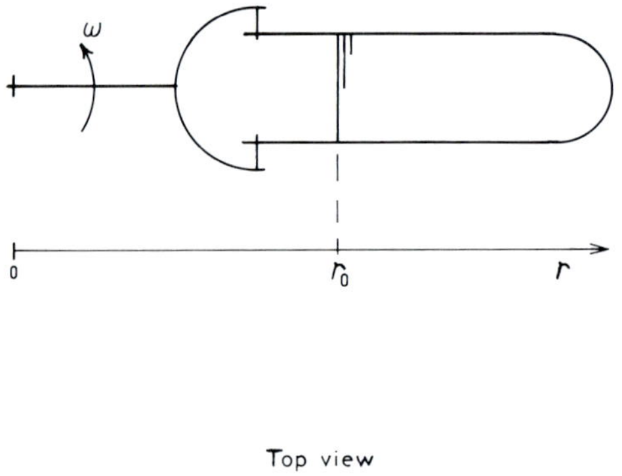

Figure 2.3 Liquid in centrifuge tube rotating at angular frequency ω.

2.1.4 Sedimentation velocity as mechanical equilibrium

The concept of mechanical equilibrium in the gravitational force field is used in particle-size analysis by sedimentation (Section *1.1.2*). When a soil particle is suspended in water, it first accelerates due to the influence of gravity. As the velocity increases, the drag force exerted by the water on the particle due to the viscosity of the water also increases. The net static force (weight minus buoyant force), however, remains constant. When the drag force has become equal to the net static force, the resultant force is zero, and thus also the acceleration. Then the particle continues to move at constant velocity (Figure 2.4).

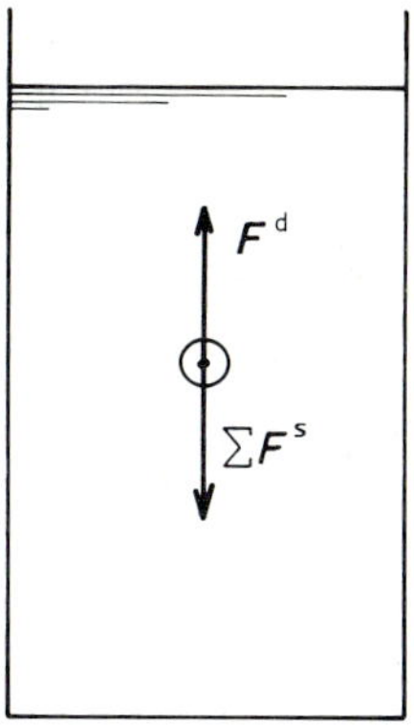

Figure 2.4 Net static force, $\Sigma \boldsymbol{F}^{s}$, and drag force, $\boldsymbol{F}^{d}$, on a small solid particle moving through a viscous fluid under the influence of gravity.

When a solid, spherical particle of radius r and density ρ_s moves through a viscous liquid, the net static force acting downward is (Answer 2.12)

$$\Sigma \boldsymbol{F}^{s} = \frac{4}{3}\pi r^3 (\rho_s - \rho_l)\, \boldsymbol{g} \tag{2.17}$$

where ρ_l is the density of the liquid. The drag force on a spherical particle under laminar conditions was shown by Stokes to be

$$\boldsymbol{F}^{d} = -6\pi\eta r\, \boldsymbol{v} \tag{2.18}$$

where η is the viscosity of the liquid, $\boldsymbol{v}$ is the terminal velocity of the particle, and the minus sign indicates that $\boldsymbol{F}^{d}$ is directed oppositely to $\boldsymbol{v}$ (and of course also to $\boldsymbol{F}^{s}$). Laminar conditions will persist as long as the Reynolds number

$$Re = \frac{2rv\rho_l}{\eta} \tag{2.19}$$

remains smaller than about 0.1, which is true for particles $< 50\ \mu m$. The sum of forces is zero when the sedimentation velocity has reached the value for which

$$\Sigma \boldsymbol{F}^{s} + \boldsymbol{F}^{d} = \frac{4}{3}\pi r^3 \boldsymbol{g} (\rho_s - \rho_l) - 6\pi\eta r \boldsymbol{v} = 0 \tag{2.20}$$

The terminal velocity, also known as settling or sedimentation velocity, is then

$$\boldsymbol{v} = \frac{2\boldsymbol{g} r^2 (\rho_s - \rho_l)}{9\eta} \tag{2.21}$$

Since the particles attain a terminal velocity almost instantly, the radius of particles that travel a distance s during a time interval t is then:

$$r = \left(\frac{9\eta s}{2g(\rho_s - \rho_l)t}\right)^{1/2} \tag{2.22}$$

This sedimentation equation is based on a number of simplifying assumptions (see Answer 2.15e). When the equation is used for particle-size analysis, the obtained radius r is not the actual radius but only an equivalent radius of the particle. For nonspherical particles, such as clays, r is the radius of a hypothetical spherical particle with the same theoretical settling velocity.

Question 2.15 A soil sample was taken from a field for particle-size analysis by sedimentation. It was dried and by various means reduced to its primary soil particles. The largest grain sizes were 3 mm and the total mass of solid phase was 65 g. A 1250-cm^3 sedimentation cylinder filled with water at 20 °C ($\eta = 10^{-3}$ Pa s), and a 20-cm^3 pipette were used for estimation of grain-size distribution.

a. Outline briefly the steps for estimating the textural components of the soil sample.
b. Which sieve sizes are required?
c. The mass of grains larger than 50 μm and 2 mm were 15 g and 10 g, respectively. The rest of the grains are dispersed in the cylinder. Give the time of sampling after the commencement of sedimentation for a few grain sizes, if samples are taken at 15 cm depth.
d. What is the textural classification of the samples taken at $t \approx 6.82$ min and $t \approx 682$ min?
e. Is the silt content of the sample taken at $t \approx 6.82$ min the same as at $t = 0$?
f. If the mass of solid taken at times about 6.82 min and 682 min are, respectively, 0.3 and 0.1 g in 20 cm^3, find the mass fraction of the different textural fractions in the original soil sample.
g. What is the textural classification of the original soil sample?
h. Justify the use of Stokes' law in the sedimentation method.

2.2 POTENTIALS IN STATIC EQUILIBRIUM

In Section 2.1, we have described equilibrium conditions, their underlying principles and their significance. For a few cases we calculated the resultant force by summing all the component forces according to the rules of vector addition. Unfortunately, the magnitudes and directions of forces are often not easily identified, making vector addition difficult. Therefore, the criterion for static equilibrium stated in Section *2.1.1* ($\Sigma \boldsymbol{F} = 0$ and $\Sigma \boldsymbol{F}^s = 0$), correct as it may be, is not very practical. Much easier to handle is the concept of potential. Since potential is a scalar quantity, it involves only algebraic addition of its components. The potential of soil water which we shall discuss is the energy associated with position. The kinetic energy of soil water, which is the energy associated with its motion, can be neglected, since soil water always moves very slowly.

2.2.1 Potential theory

Consider a mass m subjected to a uniform force field (Figure 2.5). The distance

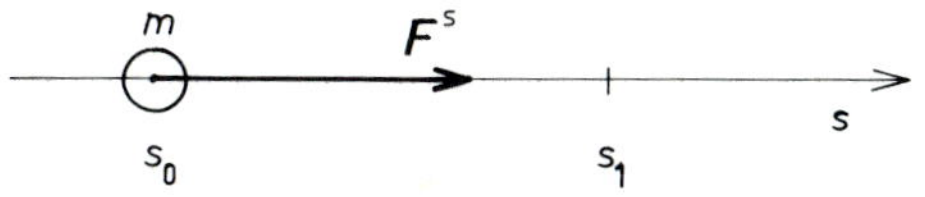

Figure 2.5 A mass m moving in the direction of a force field.

s increases (is positive) in the direction of $\boldsymbol{F}^s$, thus $s_1 > s_0$ and $\Delta s = s_1 - s_0 > 0$. When the mass m moves from s_0 to s_1 under the influence of force $\boldsymbol{F}^s$, the system expends an amount of useful work or energy equal to $F^s \Delta s$. To move the mass m back from s_1 to s_0, an external force, acting in the opposite direction must be applied over the distance s_1 to s_0. If the translocation is to occur isothermally and reversibly, this force must be only an infinitesimally small amount larger than $\boldsymbol{F}^s$. In this way the energy of the system is restored to its original value. The energy lost or gained in these transitions between s_0 and s_1 is the difference in potential energy of the mass m in the force field between the two positions. When this potential energy difference is divided by the mass m, one obtains, per definition, the potential difference, $\Delta\psi$:

$$\Delta\psi = -\frac{F^s}{m}\Delta s \tag{2.23}$$

Potential is potential energy divided by mass and is expressed in J kg^{-1}. The minus sign in Equation 2.23 is required, since the potential decreases as the mass moves in the direction of the force. Thus, the force is in the direction of decreasing potential.

Equation 2.23 is valid when F^s is constant over the distance Δs. This is generally not true for soil water. If Δs is confined to a very small (infinitesimal) distance ds, over which F^s can be assumed constant, then Equation 2.23 can be written as:

$$\frac{F^s}{m} = -\lim_{\Delta s \to 0} \frac{\Delta\psi}{\Delta s} = -\frac{d\psi}{ds} \tag{2.24}$$

Or, again in the form of Equation 2.23:

$$d\psi = -\frac{F^s}{m} ds \tag{2.25}$$

By choosing a reference position s_0 where $\psi = 0$, the potential for any point s' can be obtained by integrating Equation 2.25:

$$\psi = -\int_{s_0}^{s'} \frac{F^s}{m} \mathrm{d}s \tag{2.26}$$

In words, the potential at a point in a force field is the amount of useful work or energy required to transfer a body from a reference point s_0 to the point under consideration s', divided by that body's mass. It is not possible to assign an absolute value to the potential at point s' without defining the potential at the reference point s_0. Fortunately, in soil physics, we are never concerned with absolute potentials, but only with differences in potential between points. Thus we can assign any arbitrary value (often zero) to the potential at s_0.

Question 2.16 What influence has ψ_0 at s_0 on the numerical value of ψ at s?

In Figure 2.5, $\boldsymbol{F}^s$ and s have the same direction, but this is not generally true. If the two directions make an angle φ, $\boldsymbol{F}^s$ can be regarded as the sum of two components: one in the direction of s with magnitude $F^s \cos\varphi$ and one perpendicular to s with magnitude $F^s \sin\varphi$ (Figure 2.6). The amount of useful work expended by the component in the direction of s is $F^s \mathrm{d}s \cos\varphi$, while that of the component of $\boldsymbol{F}^s$ perpendicular to s is zero, because this component cannot bring about a translocation in the direction of s. Thus, the total amount of useful work expended by $\boldsymbol{F}^s$ equals $F^s \mathrm{d}s \cos\varphi$.

The product $F^s \mathrm{d}s \cos\varphi$, which is a scalar quantity, is known as the scalar or dot product of the vectors $\boldsymbol{F}^s$ and $\mathrm{d}\boldsymbol{s}$:

$$\boldsymbol{F}^s \cdot \mathrm{d}\boldsymbol{s} = F^s \mathrm{d}s \cos\varphi \tag{2.27}$$

Thus, for any combination of directions of $\boldsymbol{F}^s$ and $\mathrm{d}\boldsymbol{s}$, Equation 2.26 becomes:

$$\psi = -\int_{s_0}^{s'} \frac{\boldsymbol{F}^s}{m} \cdot \mathrm{d}\boldsymbol{s} = -\int_{s_0}^{s'} \frac{F^s}{m} \cos\varphi \, \mathrm{d}s \tag{2.28}$$

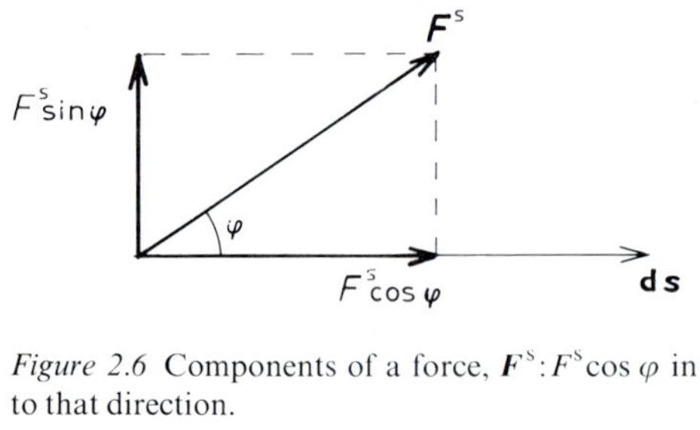

Figure 2.6 Components of a force, $\boldsymbol{F}^s$: $F^s \cos\varphi$ in the direction of movement and $F^s \sin\varphi$ perpendicular to that direction.

If more force fields are present in the same system, each force field contributes its own potential, i.e. the partial potential ψ_i due to the ith force field. The total potential ψ_t is then the algebraic sum of the partial potentials of the different force fields:

$$\psi_t = \sum_i \psi_i = \sum_i - \int \frac{\boldsymbol{F}_i^s}{m} \cdot d\boldsymbol{s} = - \int \frac{\Sigma_i \boldsymbol{F}_i^s}{m} \cdot d\boldsymbol{s}$$

At static equilibrium, where $\Sigma \boldsymbol{F}_i^s = 0$, one finds:

$$\psi_t = - \int \frac{\Sigma \boldsymbol{F}_i^s}{m} \cdot d\boldsymbol{s} = - \int 0 \, d\boldsymbol{s} = C \qquad (2.30)$$

where C is a constant of integration.

In a system at static equilibrium, the total potential is constant, i.e. invariant with position in the system. This conclusion is true for all components in the soil.

2.2.2 Hydrostatic equilibrium in the gravitational field. Hydraulic potential

'Hydrostatics' is the study that deals with potentials, pressures, etc. of liquids at hydrostatic equilibrium.

Figure 2.7 shows a cylinder filled with water. The water is subject to the gravitational field and is in hydrostatic equilibrium. Therefore, according to Equation 2.30, ψ_t is constant and depends only on the choice of the reference points for the various partial potentials. In the gravitational field, the water has a gravitational potential, ψ_g. Intro-

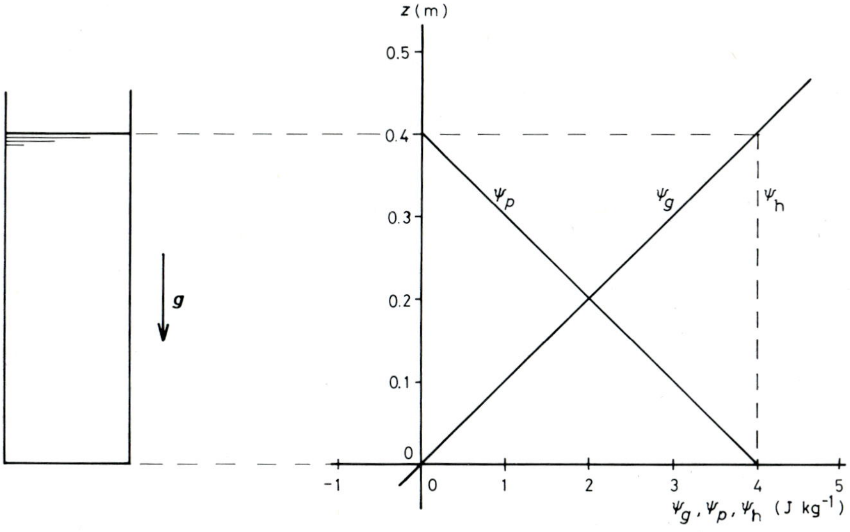

Figure 2.7 Potential diagram of a static column of water in the gravitational field.

duction of ψ_g and F_b^s for ψ and F^s in Equation 2.28, respectively, yields:

$$\psi_g = -\int F_b^s \cdot ds = -\int g \cdot ds = -\int g \cos\varphi \, dz = gz + C \tag{2.31}$$

where z refers to the height of the point under consideration and φ is the angle between the direction of z and g. Since z increases in the direction opposite to the gravitational field, $\cos\varphi = -1$. Usually, the reference for the gravitational potential is chosen such that $\psi_g = 0$ at $z = 0$. Then the integration constant in Equation 2.31 vanishes, and

$$\psi_g = gz \tag{2.32}$$

Equation 2.32 shows that ψ_g increases linearly with height z. Accordingly, Figure 2.7 shows a linearly increasing ψ_g with height, with $z = 0$ chosen at the bottom of the cylinder.

Question 2.17 a. Compute ψ_g at $z = 0$, 0.1, 0.4, 0.5 and -0.1 m, when $z = 0$ at the bottom of the cylinder.
b. Interpret the meaning of the positive and negative values of ψ_g.

At static equilibrium, the gravitational potential, ψ_g, is compensated by the pressure potential, ψ_p, in the same way that a body force is compensated by a contact force. The compensation of one by the other keeps their sum constant.

The general equation for ψ_p can be obtained from Equation 2.28 by replacing ψ by ψ_p and F^s by ΣF_c^s. Thus,

$$\psi_p = -\int \frac{\Sigma F_c^s}{m} \cdot ds = -\int \frac{\Sigma F_c^s}{m} \cdot dz \tag{2.33}$$

By substituting $-\frac{1}{\rho_l}\frac{dp}{ds} k$ for $\frac{\Sigma F_c^s}{m}$ (Equation 2.13) and $k\,dz$ for dz and using $k \cdot k = 1$, we obtain:

$$\psi_p = -\int -\frac{1}{\rho_l}\frac{dp}{dz} dz = \int \frac{1}{\rho_l} dp \tag{2.34}$$

If ρ_l is independent of p, i.e. ρ_l is constant, ψ_p becomes:

$$\psi_p = \frac{1}{\rho_l} p + C \tag{2.35}$$

where C is a constant.

Equation 2.35 is valid for water, because water is only slightly compressible and thus ρ_l can be regarded as independent of p.

The reference for ψ_p is, by convention, selected at atmospheric pressure. Thus, $\psi_p = 0$ if $p = p_{atm}$. For instance, in Figure 2.7 the reference for ψ_p is at the water surface, since there the pressure is atmospheric. Now, after substituting p_{atm} for p and zero for ψ_p into Equation 2.35, the integration constant becomes $-\frac{1}{\rho_l}p_{atm}$. This yields

$$\psi_p = \frac{1}{\rho_l}p - \frac{1}{\rho_l}p_{atm} = \frac{1}{\rho_l}(p - p_{atm}) = \frac{1}{\rho_l}p' \quad (2.36)$$

where $p' = p - p_{atm}$ (2.37)

In hydrostatics, as well as in hydraulic flow of the soil solution, the gravitational potential and the pressure potential, generally, are the only two components of the total potential that need to be taken into account. The sum of the pressure potential and the gravitational potential is called the hydraulic potential, ψ_h.

$$\psi_h = \psi_p + \psi_g \quad (2.38)$$

In Figure 2.7, both ψ_p and p' are zero at the water surface, where p equals p_{atm}. At the same height, $z = 0.4$ m, the gravitational potential ψ_g is
$\psi_g = gz = 10 \text{ N kg}^{-1} \times 0.4 \text{ m} = 4 \text{ J kg}^{-1}$
Therefore, at $z = 0.4$ m the hydraulic potential, ψ_h, according to Equation 2.38, is
$\psi_h = \psi_p + \psi_g = 0 + 4 \text{ J kg}^{-1} = 4 \text{ J kg}^{-1}$
According to Equations 2.30 and 2.38, ψ_h is uniform at static equilibrium, i.e. invariant with position. Hence, $\psi_h = 4 \text{ J kg}^{-1}$ throughout the cylinder. Now, ψ_p can be calculated with $\psi_p = \psi_h - \psi_g$. The result of this calculation is shown in Figure 2.7. This illustrates that at hydrostatic equilibrium, the ψ_p and ψ_g-lines have equal but opposite slopes.

Question 2.18 Explain why ψ_p has to be positive below the water surface.
Question 2.19 If the reference point for ψ_g were at half the height of the water column in Figure 2.7:
a. What would be ψ_h?
b. Would there be a change in the plot of ψ_p?

2.2.3 *Potential on volume basis. Pressure equivalent*

It is sometimes easier to work with potentials on volume basis (J m^{-3}) than on mass basis (J kg^{-1}). Potential on volume basis is obtained from potential on mass basis by multiplying by density ρ (i.e. it becomes $\rho\psi$). It has the dimension (units) of pressure: $\text{kg m}^{-3} \times \text{J kg}^{-1} = \text{J m}^{-3} = \text{N m}^{-2} = \text{Pa}$. Particularly for water, the density of which is almost constant, it is convenient to express potentials as pressure equivalents. For convenience, we will assume a constant density of the soil solution of 1000 kg m^{-3}.

On the basis of volume, the gravitational potential becomes

$$\rho_l\psi_g = \rho_l gz \quad (2.39)$$

and the pressure potential becomes

$$\rho_1 \psi_p = \rho_1 \frac{1}{\rho_1} p' = p' \tag{2.40}$$

Hence, the pressure equivalent of the hydraulic potential, p_{h}, is

$$p_{\mathrm{h}} = \rho_1 \psi_{\mathrm{h}} = \rho_1 g z + p' \tag{2.41}$$

Pressure equivalent should be expressed in Pa, but, traditionally, the units bar (= 10^5 Pa) and mbar are frequently used in soil physics and related disciplines.

Question 2.20 Convert the potential scale in Figure 2.7 to a Pa scale and a mbar scale.

Question 2.21 What is the water pressure at the bottom of the cylinder in Figure 2.7?

Question 2.22 The barometric pressure at two subsequent days differs by 1 cm mercury. How does this influence the hydraulic potential of the water?

2.2.4 *Potential on weight basis. Head equivalent*

Just as the potential on mass basis was converted to a potential on volume basis (Section 2.2.3), it can also be converted to a potential related to the weight of the soil solution. Since weight = mass × g (Equation 2.6), to go from J kg^{-1} to J N^{-1} requires dividing by g. The convenience of this form of the potential is immediately apparent from the gravitational potential

$$\frac{\psi_g}{g} = \frac{gz}{g} = z \tag{2.42}$$

Thus, the gravitational potential on weight basis has the dimension of height and is called gravitational head. Similarly, the head equivalent of the pressure potential is the pressure head, h,

$$\frac{\psi_p}{g} = h \tag{2.43}$$

The head equivalent of the hydraulic potential is then the hydraulic head, H,

$$\frac{\psi_{\mathrm{h}}}{g} = H = h + z = \frac{\psi_p}{g} + \frac{\psi_g}{g} \tag{2.44}$$

The convenience of using head equivalents will become obvious in Chapters 3 and 5.

Question 2.23 Convert the potential scale in Figure 2.7 to a head scale.

2.2.5 Equilibrium of a compressible medium in the gravitational field

In Answer 2.20 the water pressure increased proportionally with depth, because ρ_l was uniform (incompressible medium). If density is dependent on pressure, the relationship between pressure and height is more complex. Such is the case, for instance, in the atmosphere, because air is compressible. If the air in the atmosphere behaves as an ideal gas and temperature is constant with height, the relationship between air pressure and height can be calculated for the equilibrium state. Then the total potential, $\psi_t = \psi_g + \psi_p$, is constant (Equations 2.30 and 2.38).
Hence (Equations 2.32 and 2.34),

$$\psi_t = gz + \int \frac{1}{\rho_a} dp = \text{constant} \tag{2.45}$$

where ρ_a is the density of the air.

The relationship between ρ_a and p can be derived from the ideal gas law

$$pV = nRT \tag{2.46}$$

where p is pressure, V is volume, n is amount of gas (mol), R is molar gas constant and T is temperature (K).

$$\text{Also } \rho_a = \frac{nM}{V} = \frac{pM}{RT} \tag{2.47}$$

where M is molar mass of air.
Substitution of ρ_a in Equation 2.45 gives:

$$\psi_t = gz + \int \frac{RT}{pM} dp = C_1 \tag{2.48}$$

or

$$gz + \frac{RT}{M} \ln p = C_2 \tag{2.49}$$

or

$$\ln p = -\frac{M}{RT} gz + C_3 \tag{2.50}$$

or

$$p = C \exp\left(-\frac{M}{RT} gz\right) \tag{2.51}$$

where C_1, C_2, C_3 and C are constants.

Taking $p = p_0$ at $z = 0$ (sea level) leads to

$$p = p_0 \exp\left(-\frac{M}{RT} gz\right) \tag{2.52}$$

With M (air) ≈ 28.8 g mol^{-1}, $g \approx 10$ N kg^{-1}, $R \approx 8.3$ J mol^{-1} K^{-1} and $T \approx 290$ K, Equation 2.52 reduces to

$$p = p_0 \exp(-1.2 \times 10^{-4} \mathrm{m}^{-1} \times z) \tag{2.53}$$

At a height of 1 km, for instance, this gives

$$p = p_0 \exp(-0.12) \approx 0.89\, p_0$$

Question 2.24 If the pressure at sea level is 1000 mbar,

a. what is the pressure at 1, 2 and 3 km height?
b. Plot pressure against height above sea level.
c. What do you conclude?

Density generally increases with pressure. Therefore, pressure increases more than proportionally in the direction of the force field. Although this is to some extent true for water, in soil physics water is always assumed to be incompressible and thus uniform in density.

2.3 SUGGESTED LITERATURE

BAVER, L. D., W. H. GARDNER and W. R. GARDNER, 1972. Soil Physics (fourth edition). Wiley, 489 p. Chapter 1.

BOLT, G. H. et al. 1976. Final Report Second Committee on Terminology in Soil Physics. ISSS Bulletin No 49, p. 26–36.

CHILDS, E. C., 1969. An introduction to the physical basis of soil water phenomena. Wiley, 493 p. Chapter 5.

FEYNMAN, R. P., R. B. LEIGHTON and M. SANDS, 1966. The Feynman Lectures on Physics. Vol. 1. Addison-Wesley Publ. Co. Chapters 11, 12, 13 and 14.

Answer 2.1 Sum of x-components: $2 - 3 + 3 = 2$
Sum of y-components: $2 + 1 - 4 = -1$
Thus, the x,y-components of the resultant vector, $\boldsymbol{R}$, are $(2, -1)$. The diagram below shows that the same result is obtained by placing the vectors tail to head in the order $\boldsymbol{A}$, $\boldsymbol{B}$ and $\boldsymbol{C}$. Any other order gives the same result. (Check this.)

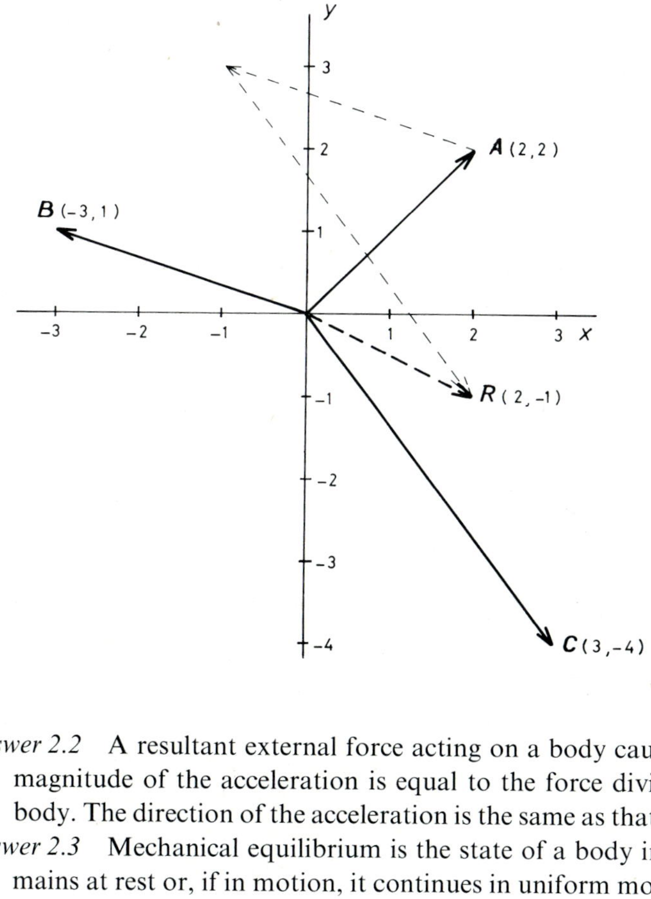

Answer 2.2 A resultant external force acting on a body causes it to accelerate. The magnitude of the acceleration is equal to the force divided by the mass of the body. The direction of the acceleration is the same as that of the resultant force.

Answer 2.3 Mechanical equilibrium is the state of a body in which, if at rest, it remains at rest or, if in motion, it continues in uniform motion. No resultant force acts on that body.

Answer 2.4 The body is at rest; there is no acceleration and no resultant force. Thus, the body is (at least) in mechanical equilibrium.

Answer 2.5 Equation 2.3 does not apply to an object moving with constant velocity in outer space. Since it is moving in a virtual vacuum, the resistance to its motion is negligible, or $\Sigma \boldsymbol{F}^{\mathrm{d}} \approx 0$. Since it is moving with constant velocity, there can be no net static force acting on it, i.e. $\Sigma \boldsymbol{F}^{\mathrm{s}} = 0$. However, because it is moving, it is not at static equilibrium. Thus Equation 2.3 would not apply to this situation.

Answer 2.6 a. The forces to be summed are those acting on the body. There are two: the one by gravity (weight) and the one exerted by the table (reaction to the weight).

b. Since the body is at rest, there are no drag forces acting on it: $\Sigma \boldsymbol{F}^{\mathrm{d}} = 0$.

c. Since $\Sigma \boldsymbol{F} = \Sigma \boldsymbol{F}^{\mathrm{s}} + \Sigma \boldsymbol{F}^{\mathrm{d}} = 0$, it follows from Answer 2.6b that $\Sigma \boldsymbol{F}^{\mathrm{s}} = 0$. Thus, the body is, according to Equation 2.3, also at static equilibrium.

Answer 2.7 a. Weight is the gravitational force exerted by the earth on a body.

b. The magnitude of the weight is $F = mg = 5\ \mathrm{kg} \times 10\ \mathrm{N\ kg^{-1}} = 50\ \mathrm{N}$. Its weight is 50 N.

Answer 2.8 By equating the two equations, we find for the magnitude of $\boldsymbol{g}$:

$$g = \frac{Gm_2}{r^2}$$

Answer 2.9 a. Answer 2.8 shows that g depends only on r, because G and m_2 are constants. Thus g decreases with height above the earth surface. Because the radius of the earth is not constant and for several other reasons, even the gravitational field strength is nonhomogeneous if viewed on a large scale. However, on a small scale (e.g. the depth of a soil profile) it is, for all practical purposes, homogeneous.

b. The magnitude of the centrifugal force exerted on a body in a centrifuge is $F = m\omega^2 r$. $\boldsymbol{F}$ is at all times directed away from the centre of rotation, ω is the angular frequency, r is the distance from the centre of rotation and m is the mass of the body. Thus, the centrifugal force increases with distance from the centre of rotation and changes direction constantly. Therefore, the centrifugal field is nonhomogeneous.

Answer 2.10 a. Since $\boldsymbol{F} = m\boldsymbol{a} = Q\boldsymbol{c}$, $\boldsymbol{a} = \dfrac{Q}{m}\boldsymbol{c}$

b. Since $Q = 10^{-7}\ \mathrm{C}$, $m = 0.1\ \mathrm{g}$ and $c = 10^3\ \mathrm{N\ C^{-1}}$,

$$a = \frac{10^{-7}\ \mathrm{C}}{10^{-4}\ \mathrm{kg}} \times \frac{10^3\ \mathrm{N}}{\mathrm{C}} = 1\ \mathrm{N\ kg^{-1}} = 1\ \mathrm{m\ s^{-2}}$$

c. Analogously to Newton's law of gravitation,

$$F = \frac{kQ_1Q_2}{r^2}$$

where Q_1 and Q_2 are two charges, F is the mutual force of attraction between them, k is constant and r is the distance separating the charges. This equation is known as Coulomb's law. The magnitude of the field strength $\boldsymbol{c}$ is now: $c = \dfrac{kQ}{r^2}$ where Q is the charge of the body due to which the electric field is present.

Answer 2.11 The pressure in fluids is scalar, because it has only a magnitude and not a particular direction.

Answer 2.12 a. Combination of Equations 2.12 and 2.14, using $g\boldsymbol{k} = -\boldsymbol{g}$, gives:

$$\Sigma \boldsymbol{F}^{\mathrm{s}}_{\mathrm{c}} = \rho_1 g V \boldsymbol{k} = -\rho_1 V \boldsymbol{g}$$

The body force is $\boldsymbol{F}_{\mathrm{b}}^{\mathrm{s}} = m\boldsymbol{g} = \rho_{\mathrm{c}} V \boldsymbol{g}$
where ρ_{c} is the density of the cube.
Thus, $\Sigma \boldsymbol{F}^{\mathrm{s}} = (\rho_{\mathrm{c}} V - \rho_{\mathrm{l}} V)\boldsymbol{g} = (\rho_{\mathrm{c}} - \rho_{\mathrm{l}}) V \boldsymbol{g}$

b. $\Sigma \boldsymbol{F}^{\mathrm{s}} = (30\ \mathrm{g} - 27\ \mathrm{cm}^3 \times 1\ \mathrm{g\ cm}^{-3}) \times (-10\ \mathrm{N\ kg}^{-1}) = -0.03\ \mathrm{N}$ (downward).

c. Since $\Sigma \boldsymbol{F}^{\mathrm{s}}$ is negative, the cube will accelerate downwards. $\boldsymbol{F}^{\mathrm{d}}$ will increase with speed (magnitude of velocity) until $\boldsymbol{F}^{\mathrm{d}} = -\Sigma \boldsymbol{F}^{\mathrm{s}}$. From that moment on, the cube will sink at constant velocity until it collides with the bottom of the container (See Section *2.1.4*).

Answer 2.13 a. The weight of the clod is $\boldsymbol{F} = m\boldsymbol{g} = 8.99\ \mathrm{g} \times (-10^{-2})\ \mathrm{N\ g}^{-1} = -8.99 \times 10^{-2}\ \mathrm{N}$.

b. $\boldsymbol{F}_{\mathrm{b}}^{\mathrm{s}} = m\boldsymbol{g} = -8.99 \times 10^{-2}\ \mathrm{N}$. The clod is in static equilibrium, thus $\Sigma \boldsymbol{F}^{\mathrm{s}} = 0$ and $\Sigma \boldsymbol{F}_{\mathrm{c}}^{\mathrm{s}} = 8.99 \times 10^{-2}\ \mathrm{N}$. Part of this force, $2.79\ \mathrm{g} \times 10^{-2}\ \mathrm{N\ g}^{-1} = 2.79 \times 10^{-2}$ N, is provided by the balance. Thus, the net contact force by the water (upthrust) equals $8.99 \times 10^{-2}\ \mathrm{N} - 2.79 \times 10^{-2}\ \mathrm{N} = 6.2 \times 10^{-2}\ \mathrm{N}$.

c. The net contact force by the water is 6.2×10^{-2} N (Answer 2.13b) and this force equals $-\rho_{\mathrm{l}} V \boldsymbol{g}$ (Answer 2.12a).

Therefore, $V = \dfrac{6.2 \times 10^{-2}\ \mathrm{N}}{-\rho_{\mathrm{l}} g} = \dfrac{6.2 \times 10^{-2}\ \mathrm{N}}{-1\ \mathrm{g\ cm}^{-3} \times (-10^{-2})\ \mathrm{N\ g}^{-1}} = 6.2\ \mathrm{cm}^3$. The mass of the clod is 8.99 g. Thus, its bulk density is $\dfrac{8.99\ \mathrm{g}}{6.2\ \mathrm{cm}^3} = 1.45\ \mathrm{g\ cm}^{-3}$ or $1450\ \mathrm{kg\ m}^{-3}$.

d. The clod needed to be perfectly sealed to prevent water from entering. If water enters the clod, the balance reads higher and one would calculate a higher density than the actual density of the clod. Perfect sealing can be checked by reweighing the clod after it is taken out of the water and comparing the weight with that before it was immersed in water.

Answer 2.14 a. For an observer who is part of the rotating system, a body of liquid is in static equilibrium, because it does not move in the direction of r.

The centrifugal force field exerts a body force $\dfrac{\boldsymbol{F}_{\mathrm{b}}^{\mathrm{s}}}{m} = \omega^2 r \boldsymbol{i}$, directed away from the axis of rotation. This body force is compensated by an equal, but oppositely directed net contact force exerted by the surrounding liquid.

Thus, analogously to Equation 2.13:

$$\frac{\Sigma \boldsymbol{F}_{\mathrm{c}}^{\mathrm{s}}}{m} = -\frac{1}{\rho_{\mathrm{l}}} \frac{\mathrm{d}p}{\mathrm{d}r} \boldsymbol{i} = -\omega^2 r \boldsymbol{i}$$

For an observer outside the system, there is no static equilibrium. The velocity of the liquid changes in direction as a result of the centripetal acceleration $-\omega^2 r \boldsymbol{i}$ (towards the axis of rotation). This acceleration results from the net contact force $\dfrac{\Sigma \boldsymbol{F}_{\mathrm{c}}^{\mathrm{s}}}{m}$, which leads to the same conclusion as above.

Therefore, $dp = \rho_l \omega^2 r\, dr$ or, after integration, $p = \frac{1}{2} \rho_l \omega^2 r^2 + C$, where C is an integration constant.

Because $p = p_{atm}$ at $r = r_0$, $C = p_{atm} - \frac{1}{2} \rho_l \omega^2 r_0^2$

Thus, $p = \frac{1}{2} \rho_l \omega^2 (r^2 - r_0^2) + p_{atm}$

b. Since 1 revolution equals 2π rad, $\omega = \dfrac{6000 \times 2\pi}{60 \text{ s}} = 200\pi \text{ s}^{-1}$ (radians are dimensionless).

Therefore, $p = \frac{1}{2} \rho_l \omega^2 (r^2 - r_0^2) + p_{atm}$

$$= \frac{1}{2} \times 1000 \text{ kg m}^{-3} \times (200\pi)^2 \text{ s}^{-2} \times (0.09 - 0.04) \text{ m}^2 + p_{atm}$$

$$\approx (10^7 + 10^5) \text{ Pa} = 1.01 \times 10^7 \text{ Pa}.$$

Answer 2.15 a. The procedure is as follows:

1. Sieve the grains.
2. Disperse the finer grains in water in such a way that a homogeneous suspension is formed.
3. Sample the suspension at a given depth at various times.
4. Dry the samples and determine the mass fraction of each textural fraction by weighing.

b. 2 mm and 50 μm sieves are required. A 2 mm sieve is needed to separate grains larger than 2 mm from the smaller ones. Grains between 2 and 3 mm are classified as gravel. These are not included in the textural analysis of soils or fine earth (Section 1.1.2). Grains larger than 50 μm can easily be separated by sieves.

c. Equation 2.22 can be written as

$$t = \frac{9\eta s}{2gr^2(\rho_s - \rho_l)}$$

To find the sand content according to the ISSS classification, sample for $r = 10$ μm at 15 cm depth at time:

$$t = \frac{9 \times 10^{-3} \text{ Pa s} \times 0.15 \text{ m}}{2 \times 10 \text{ N kg}^{-1} \times (10\ \mu\text{m})^2 \times 1.65 \times 10^3 \text{ kg m}^{-3}} \approx 409 \text{ s} \approx 6.82 \text{ min}$$

To find the silt content, sample for $r = 1$ μm at:

$$t = \frac{9 \times 10^{-3} \text{ Pa s} \times 0.15 \text{ m}}{2 \times 10 \text{ N kg}^{-1} \times (1\ \mu\text{m})^2 \times 1.65 \times 10^3 \text{ kg m}^{-3}} \approx 40909 \text{ s} \approx 682 \text{ min} =$$

11 h 22 min

d. At $t \approx 6.82$ min, all particles larger than 20 μm are beyond 15 cm depth; therefore, the sample contains only clay and silt. At $t \approx 682$ min, all silt particles are also beyond 15 cm depth, thus only clay remains in the sample.

e. Yes. At $t = 0$, all particle sizes are assumed to be uniformly distributed throughout the cylinder. It is also assumed that all particles attain their terminal velocity instantly or within a negligible distance and that they all settle independently of each other. Then at any depth, the concentration of particles of a particular size remains at the original value, until it drops instantly to zero when the particles of that size that started settling at the water surface, pass that depth. Thus a sample taken at $t \approx 6.82$ min from a layer of infinitesimal thickness at 15 cm depth contains all the particles smaller than $r = 10$ μm in their original concentration, and none of the particles larger than $r = 10$ μm.

f. The clay content in the sedimentation cylinder, as well as in the original soil sample, is: $\dfrac{0.1\ \text{g}}{20\ \text{cm}^3} \times 1250\ \text{cm}^3 = 6.25$ g.

Clay and silt content in the original soil sample is:

$$\frac{0.3\ \text{g}}{20\ \text{cm}^3} \times 1250\ \text{cm}^3 = 18.75\ \text{g}.$$

Thus, the silt content is: 18.75 g − 6.25 g = 12.50 g.

The mass of grains less than 2 mm is 65 g − 10 g = 55 g (Question 2.15c).

So the sand content is 55 g − 18.75 g = 36.25 g.

Of this, (15 − 10) g = 5 g is between 2 mm and 50 μm (Question 2.15c) and (36.25 − 5) g = 31.25 g is between 50 and 20 μm.

The mass fractions are then:

sand $\dfrac{36.25\ \text{g}}{55\ \text{g}} \approx 66\%$

silt $\dfrac{12.50\ \text{g}}{55\ \text{g}} \approx 23\%$

clay $\dfrac{6.25\ \text{g}}{55\ \text{g}} \approx 11\%$

g. The textural classification of the soil sample is sandy loam (Figure 1.2).

h. Stokes' law is valid for $Re < 0.1$.

Combination of $Re = \dfrac{2rv\rho_l}{\eta}$ and $v = \dfrac{2gr^2(\rho_s - \rho_l)}{9\eta}$ gives

$$r^3 = \frac{9\,Re\,\eta^2}{4g\rho_l(\rho_s - \rho_l)}$$

$$\text{At } 20\,^\circ\text{C, } r^3 = \frac{9 \times 0.1 \times 10^{-6}\ (\text{Pa s})^2}{4 \times 10\ \text{N kg}^{-1} \times 10^3\ \text{kg m}^{-3} \times 1.65 \times 10^3\ \text{kg m}^{-3}}$$

$$\approx 1.36 \times 10^{-14}\ \text{m}^3$$

$$r \approx 2.39 \times 10^{-5}\ \text{m} \approx 24\ \mu\text{m}$$

Thus the maximum particle diameter for which Stokes' law can be used is about 50 μm. This is the same as the maximum grain size used in the sedimentation method. Therefore, it is justified to use Stokes' law.

Answer 2.16 Carrying out the integration of Equation 2.25, assuming $\frac{F^{\rm s}}{m}$ independent of s, yields

$$\psi = -\frac{F^{\rm s}}{m}s + C$$

where C is a constant of integration.

Substitution of ψ_0 for ψ at $s = s_0$ yields

$C = \psi_0 + \frac{F^{\rm s}}{m}s_0$, and thus

$$\psi = -\frac{F^{\rm s}}{m}(s - s_0) + \psi_0$$

This result shows that the numerical value of ψ at s depends on the value of ψ_0 at s_0. The difference in ψ between two points is not affected by the choice of ψ_0 at s_0.

Answer 2.17 a. From Equation 2.32, $\psi_g = gz = 10\ \mathrm{N\ kg^{-1}} \times z$. For $z = 0, 0.1, 0.4, 0.5$ and -0.1 m, $\psi_g = 0, 1, 4, 5$ and $-1\ \mathrm{J\ kg^{-1}}$, respectively.

b. A positive value of ψ_g means that the point is located above the selected reference level for ψ_g, here the bottom of the cylinder. A negative value means that the point is located below the reference level.

Answer 2.18 According to Equation 2.36, $\psi_p = \frac{1}{\rho_{\rm l}}(p - p_{\rm atm})$. Below the water surface p is greater than $p_{\rm atm}$ because of the overlying water (remember $\boldsymbol{F}^{\rm s}_{\rm b}$). Therefore, ψ_p is positive.

Answer 2.19 a. Now, at the surface, $z = 0.2$ m and $\psi_p = 0$, so $\psi_g = gz = 10\ \mathrm{N\ kg^{-1}} \times 0.2\ \mathrm{m} = 2\ \mathrm{J\ kg^{-1}}$. Therefore, $\psi_{\rm h} = \psi_p + \psi_g = 2\ \mathrm{J\ kg^{-1}}$.

b. The plot of ψ_p does not change because the reference point for ψ_p remains at the water surface.

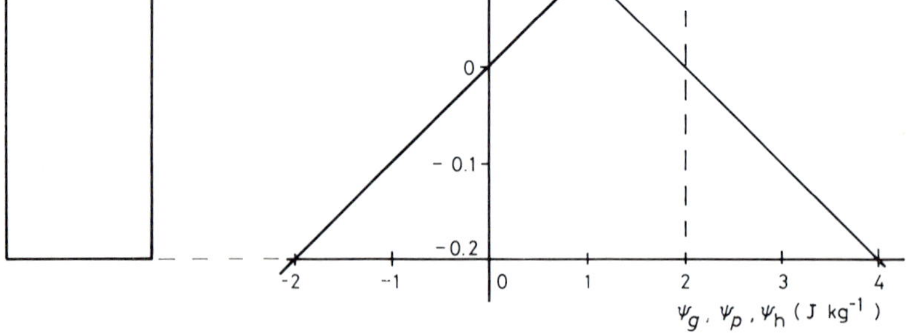

Answer 2.20 This amounts to changing potential on mass basis (J kg^{-1}) into potential on volume basis (J m^{-3}). If the density of water is assumed constant at 1000 kg m^{-3}, then the pressure equivalent of $\psi = 1$ J kg^{-1} is $\rho_l\psi = 10^3$ kg $m^{-3} \times 1$ J $kg^{-1} = 10^3$ J $m^{-3} = 1$ kPa $= 10$ mbar. Thus the potential scale can be changed into a pressure equivalent scale by substituting 1 kPa or 10 mbar for every J kg^{-1}.

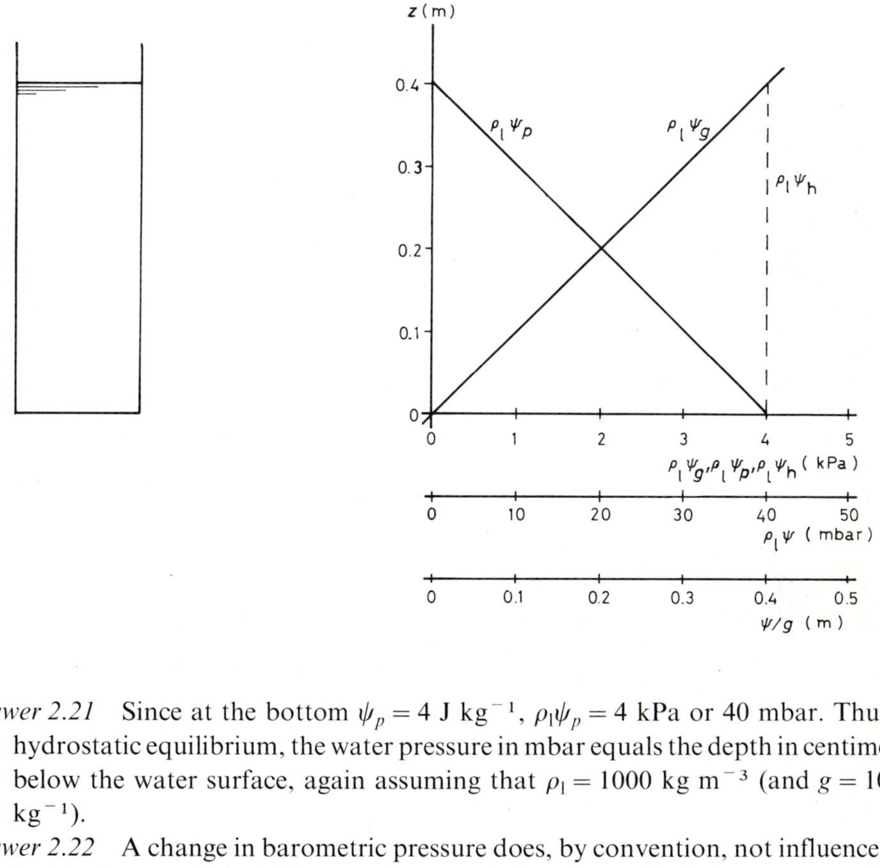

Answer 2.21 Since at the bottom $\psi_p = 4$ J kg^{-1}, $\rho_l\psi_p = 4$ kPa or 40 mbar. Thus at hydrostatic equilibrium, the water pressure in mbar equals the depth in centimeter below the water surface, again assuming that $\rho_l = 1000$ kg m^{-3} (and $g = 10$ N kg^{-1}).

Answer 2.22 A change in barometric pressure does, by convention, not influence the hydraulic potential of the water, because ψ_p is expressed relative to atmospheric pressure. At the water surface, ψ_p remains zero, irrespective of the barometric pressure.

Answer 2.23 Converting a potential scale into a head scale amounts to changing potential on mass basis (J kg^{-1}) into potential on weight basis (J N^{-1}). Therefore, the potentials on mass basis have to be divided by g. If $\psi = 1$ J kg^{-1}, then $\psi/g = 0.1$ J $N^{-1} = 0.1$ m. So every J kg^{-1} on the potential scale is substituted by 0.1 m. This scale is also indicated in the figure of Answer 2.20.

Answer 2.24 a. $p = p_0 \exp -(1.2 \times 10^{-4}\ \text{m}^{-1} \times z) = 1000\ \text{mbar} \times \exp -(1.2 \times 10^{-4}\ \text{m}^{-1} \times z)$

At $z = 1$ km, $p = 887$ mbar; at $z = 2$ km, $p = 787$ mbar, and at $z = 3$ km, $p = 698$ mbar.

b.

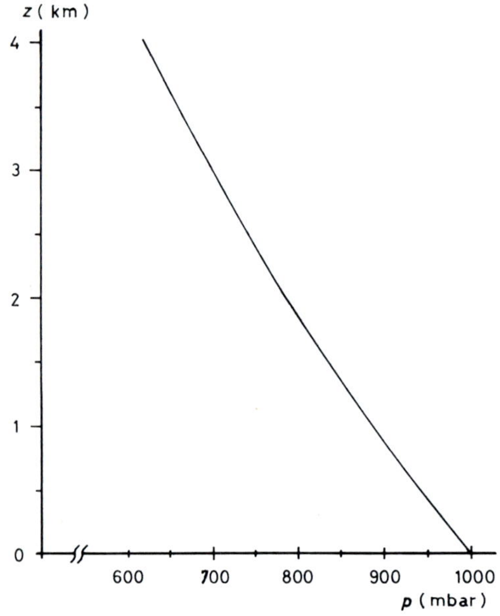

c. In an isothermal atmosphere, the pressure decreases exponentially with height.

3 Static equilibria in soils

3.1 HYDROSTATIC EQUILIBRIUM

3.1.1 Binding of water by soils

As described in Chapter 1, soils consist of a solid skeleton (matrix) with pores in between. The pores have different sizes, shapes and spatial distributions and provide the space for storage and transport of soil water and gas.

Storage or retention of water by soils is a result of attractive forces between the solid and liquid phases. These 'matric' forces enable the soil to hold water against forces or processes such as gravity, evaporation, uptake by plant roots, etc. There are three mechanisms for binding of water to the solid matrix:

- direct adhesion of water molecules to solid surfaces by London-van der Waals forces
- capillary binding of water
- osmotic binding of water in double layers.

Adhesion is the attraction of dissimilar substances for each other; cohesion is the mutual attraction of particles of the same substance.

Water molecules are attracted to solid surfaces by various types of London-van der Waals forces. These are strong, but very short-range forces; they diminish with about the sixth power of the distance. Thus only a very thin water layer is adsorbed in this way around soil particles. Since these adhesive forces are so strong that the water cannot be extracted by plant roots, this form of water retention in itself is insignificant for storage, transport and plant growth. However, the adhesive forces, together with the cohesive forces between water molecules, form the basis for capillary binding of soil water. This is the most important mechanism of binding of water in coarse soils. In clay soils the osmotic binding of water in diffuse electric double layers may exceed the capillary binding.

The capillary and osmotic binding of water in soils, the components of the soil water potential associated with them, and standard equipment for measuring some of these components are discussed in detail in the remainder of this chapter. This discussion is started with an analysis of the binding of water in a glass capillary to provide a sound physical basis for understanding capillary binding of water in soils.

3.1.2 Binding of water in a capillary

The adhesive and cohesive forces together are responsible for the binding of water in capillaries.

The adhesion or repulsion of water by solid surfaces can be characterized with the potential energy of water molecules near the solid surface, divided by area of solid-water interface, γ (J m^{-2}). Taking as reference, as usual, zero potential for bulk water, a positive value of γ means that the energy of water molecules near solid surfaces is higher than the energy of molecules in bulk water. Water will not adhere to such surfaces. They are called hydrophobic. Surfaces with negative γ values will spontaneously attract water, if available, because near such surfaces water molecules have a lower energy than in bulk water. Such surfaces are called hydrophilic. The water is said to 'wet' such surfaces. This is the normal situation for soil particles, glass, etc.

A similar characterization can be made for the cohesive forces between water molecules. Inside bulk water a water molecule experiences the same cohesive attraction in all directions. Thus, the net force on that molecule is zero. In contrast, water molecules at an air-water interface are attracted only towards the interior, because there are no water molecules on the air side of the interface. Thus the energy content of water molecules at the interface is higher than that of molecules inside bulk water. This is characterized by the surface tension, σ, which is the interface potential energy divided by interface area (J m^{-2}). Again, taking the potential of water molecules in bulk as zero, σ is positive. The result is that water in contact with air will always attempt to reduce its surface area. The smallest surface area for a given mass of water is that of a sphere. This explains the formation of spherical droplets of rain, sprays, etc. Any other shape would represent a larger surface area and thus a higher energy content. For the same reason the meniscus of water in a capillary assumes the spherical shape which is compatible with the angle of contact, φ (Figure 3.1a).

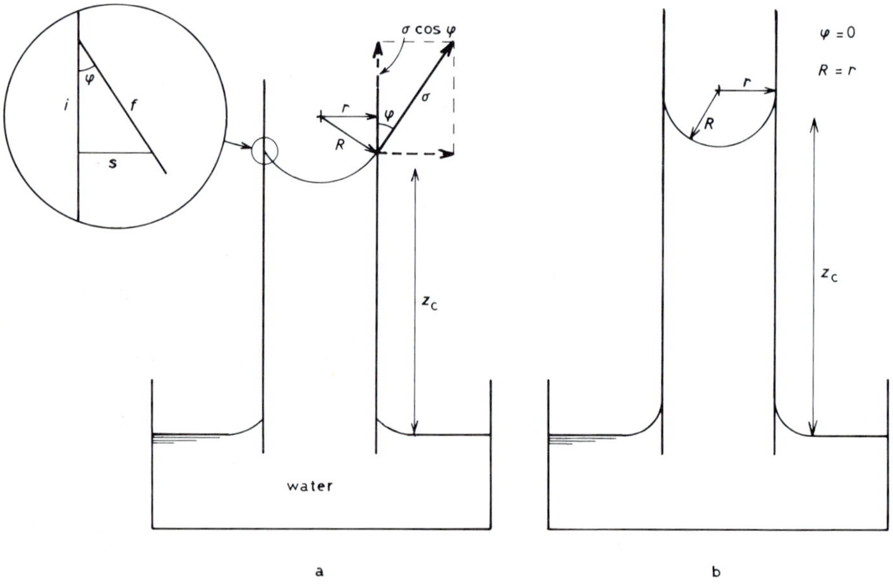

Figure 3.1 a. Capillary rise in a capillary with wetting angle φ.
b. Same, with wetting angle $\varphi = 0$ (complete wetting).

When a solid, water and air phase come together at one point, such as in a capillary, the shape of the interface, as characterized by the angle of contact, φ, depends on whether γ is positive, negative or zero, and on its magnitude with respect to σ. This can be shown using the enlargement of the situation around the point of contact shown in Figure 3.1a. The angle between the meniscus and the solid surface at that point is φ. Consider the water in the triangular volume of unit width perpendicular to the paper, for which the maximum distance to the solid surface is s. The distance s is arbitrary, except that it should be large compared to the distance over which the adhesive and cohesive forces are active, and small compared to the radius of curvature of the meniscus. The area of solid-water interface associated with the triangular volume is i, that of the water-air interface is f. The total energy associated with both types of interfaces is then:

$$E = \sigma f + \gamma i = \sigma s/\sin\varphi + \gamma s/\tan\varphi \tag{3.1}$$

The angle φ will adjust itself, until E is at a minimum. For this situation

$$\frac{\mathrm{d}E}{\mathrm{d}\varphi} = -\frac{\sigma s\cos\varphi}{\sin^2\varphi} - \frac{\gamma s}{\sin^2\varphi} = 0 \tag{3.2}$$

$$\text{or } \cos\varphi = -\frac{\gamma}{\sigma} \tag{3.3}$$

Thus, $\varphi = 180°$, when $\gamma = \sigma$
$\varphi = 90°$, when $\gamma = 0$
$\varphi = 0°$, when $\gamma = -\sigma$

Since $\varphi = 180°$ and $\varphi = 0$ are the physical limits, it should be clear that $\varphi = 180°$ for $\gamma \geqslant \sigma$ and $\varphi = 0$ for $\gamma \leqslant -\sigma$.
The latter case is normal for soil-water and glass-water systems. It is called complete wetting. An example of nonwetting is mercury on clean glass, for which $\varphi = 140°$. Clean steel has no attraction, nor repulsion for water, and thus $\gamma = 0$ and $\varphi = 90°$.

To illustrate the principle of capillary binding of water in soils, consider a clean glass capillary being inserted in a container with water, such as in Figure 3.1a. Since $\gamma < -\sigma$, the angle of contact φ is zero. Since $\gamma < 0$, the inside and outside surfaces of the capillary attract a water film. The large air-water interfacial area of these films represent a relatively high energy ($\sigma > 0$) and thus the system would move spontaneously to a state of lower potential energy if the air-water interfacial area could be reduced. On the outside surface of the capillary there is no inherent physical possibility to reduce the air-water interfacial area. But inside the capillary the interfacial area would be reduced if the capillary would fill with water. The reduction in air-water interfacial energy can be used to draw water up into the capillary against gravity. Above and below the water level in the capillary, the walls of the capillary are covered with a water film. Thus, after the initial wetting of the capillary walls, the energy of the solid-water interfaces remains the same independent of the height of capillary rise inside the capillary. Thus, water will continue to enter the capillary until the reduction in the air-water interfacial energy brought about by an additional incremental volume of water has become equal to the increase in gravitational energy of that incremental volume in the capillary. Since $\varphi = 0$, the meniscus will assume a spherical shape with radius of curvature, R, equal to the radius of the capillary, r.

Now suppose the water in the capillary has reached its final height of capillary rise, z_c (Figure 3.1b). A further incremental rise of Δz would decrease the air-water interfacial area by $2\pi r \Delta z$ (m^2) and thus the interfacial energy by $2\pi r \Delta z \sigma$ (J). The water that would fill the thin layer Δz would have to come from the container. This would mean an increase in potential energy in the gravitational force field of $\pi r^2 \Delta z \rho_l g z_c$ (J). At the point of maximum capillary rise these two amounts of energy are the same, thus:

$$2\pi r \sigma \Delta z = \pi r^2 \rho_l g z_c \Delta z \tag{3.4}$$

$$\text{or } z_c = \frac{2\sigma}{\rho_l g r} \tag{3.5}$$

As the synonym 'surface tension' suggests, the air-water interfacial energy divided by area, σ, manifests itself as a force divided by length ($J\ m^{-2} = N\ m^{-1}$). By considering σ in this way, the influence of the angle of contact, φ, on height of capillary rise is easily derived. In Figure 3.1a the surface tension force acts over the entire circumference, $2\pi r$, and is everywhere directed tangentially to the meniscus and thus at an angle φ with the wall. The vertical component of the resultant force is then $2\pi r \sigma \cos \varphi$, directed upward. The horizontal component of the resultant force is zero, because the horizontal components, $\sigma \sin \varphi$, along the circumference cancel each other. The resultant upward force is balanced by the downward force on the liquid in the capillary tube which equals $\pi r^2 z_c \rho_l g$. Thus

$$2\pi r \sigma \cos \varphi = \pi r^2 z_c \rho_l g \tag{3.6}$$

from which:

$$z_c = \frac{2\sigma \cos \varphi}{\rho_l g r} \tag{3.7}$$

From Figure 3.1a it is clear that the radius of curvature, R, of the meniscus in a capillary with radius r and contact angle φ is:

$$R = r/\cos \varphi \tag{3.8}$$

Question 3.1 Calculate the height of capillary rise of water in a clean glass tube of 70 μm inner diameter. The surface tension of water $\sigma = 0.07\ N\ m^{-1}$.

The water in the capillary tube of Figure 3.1 is at static equilibrium. Thus its potential components can be calculated in the same way as was done in connection with

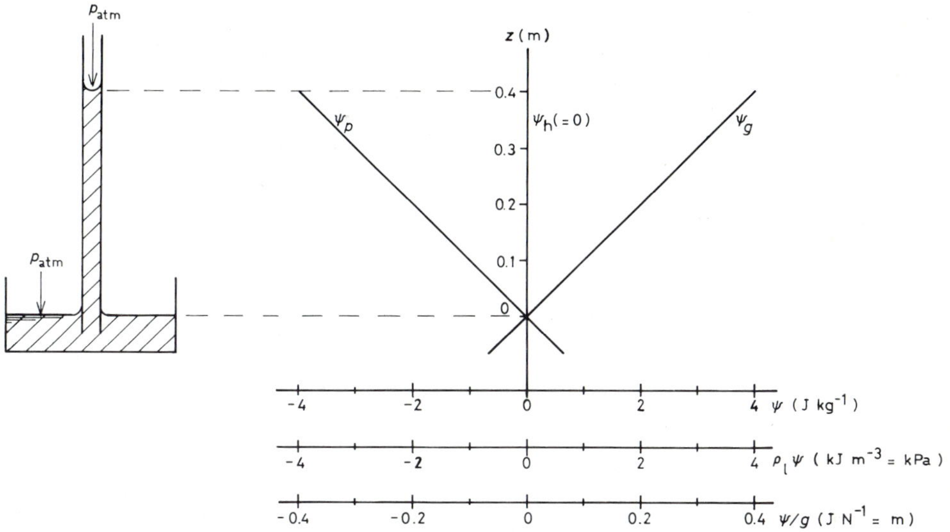

Figure 3.2 Potential diagram of water in a capillary.

Figure 2.6. Figure 3.2 presents a diagram of the potentials on mass basis, as well as on volume and weight basis. As in Figure 2.6, the pressure potential in the big container increases with depth from zero at the water surface to positive values below the surface and compensates the decreasing gravitational potential. Likewise, the pressure potential of the water in the capillary tube decreases with height to compensate the increasing gravitational potential. Thus, the pressure of the water in the capillary tube is lower than atmospheric pressure, creating a pressure difference between both sides of the curved meniscus.

Question 3.2 What is in Figure 3.2 the pressure potential of the water at the air-water interface in the capillary tube? Give your answer in J kg^{-1}, Pa and m.

Question 3.3 Find an expression for the pressure potential on volume basis (Pa) of the water at the meniscus, based on Equation 3.7.

The pressure potential in the air above the curved meniscus in the capillary equals zero. Therefore, Answer 3.3 gives also the pressure potential difference across a curved interface. Since, according to Equation 3.8, $r = R \cos \varphi$, a general equation for the pressure difference across a curved interface with radius of curvature R is

$$\Delta p = 2\sigma / R \tag{3.9}$$

where the highest pressure is at the concave side. Thus the pressure difference across a curved interface is inversely proportional to the radius of curvature. This is why reference points for pressure potentials are not chosen at curved menisci but at flat water surfaces under atmospheric pressure, free from any influence of solid surfaces, etc.

Question 3.4 Derive Equation 3.9 for a free spherical droplet of water.

Consider now the capillary tube of Figure 3.2 first being immersed in a water container and then raised slowly (Figure 3.3). Except for the gradual decrease of the radius of curvature of the meniscus, the capillary will remain filled with water until its top reaches a height above the water level in the container which is equal to the maximum height of capillary rise, z_c, in this case, $z_c = 0.4$ m. If the capillary is raised still further, air will enter from above, such that the water column remains 0.4 m high. This is why the pressure equivalent, $-\rho_l g z_c$, of the maximum height of capillary rise in a capillary is called air-entry value. In the case of positive gas pressure, it is also called bubbling pressure. Since $-z_c$, and its pressure equivalent $-\rho_l g z_c$, is also the pressure head of the water at the curved air-water interface, the concept can be extended to capillaries, pores, etc., in general. Thus, if the pressure potential of the water in a capillary, pore, etc. tends to become lower than the air-entry value, because of gravity or any other reason, the adhesive and cohesive forces are no longer able to hold onto the water and the capillary will empty to the extent necessary to keep the pressure potential of the water at the air-water interface at the air-entry value.

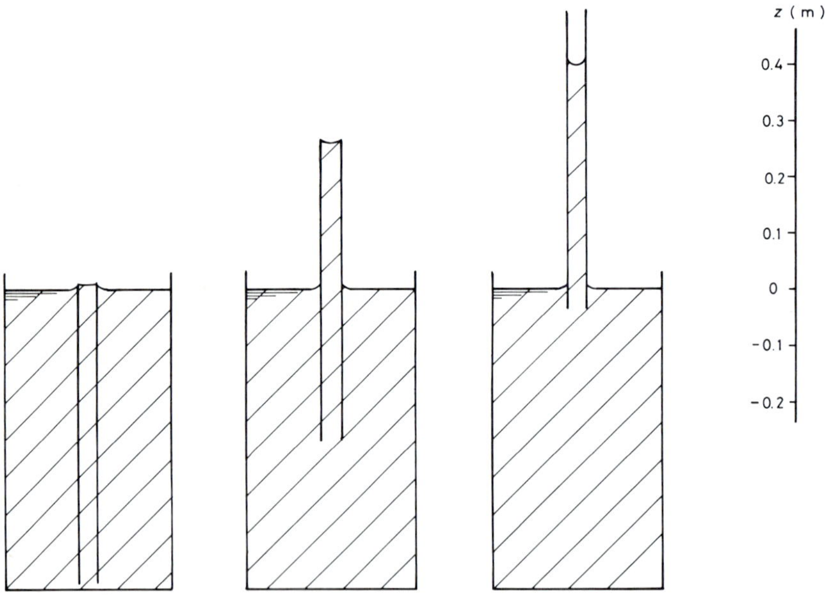

Figure 3.3 Illustration of air-entry value.

Question 3.5 A cylindrical capillary has a diameter of 20 μm. What is the air-entry value of this capillary in case of complete wetting?

Capillary rise of water takes place also in noncylindrical capillaries. The equivalent diameter of such a capillary is defined as the diameter of the cylindrical capillary that has the same height of capillary rise (or the same air-entry value).

Question 3.6 A noncylindrical capillary with complete wetting has an air-entry value of –14 kPa. What is its height of capillary rise and the equivalent diameter of the capillary?

A glass capillary is a first, crude approximation or model of a soil pore. With respect to the capillary binding of water, soils can be considered to consist of an assembly of irregularly shaped capillaries. Therefore, the results obtained in this section can, qualitatively, be applied to the capillary binding of water in soils. This will be done in the following sections.

3.1.3 Hydraulic potential of soil water at static equilibrium

A cylinder with an overflow at 0.4 m above the bottom is equipped with a noncapillary glass gauge (Figure 3.4). The cylinder and the gauge are filled with water up to

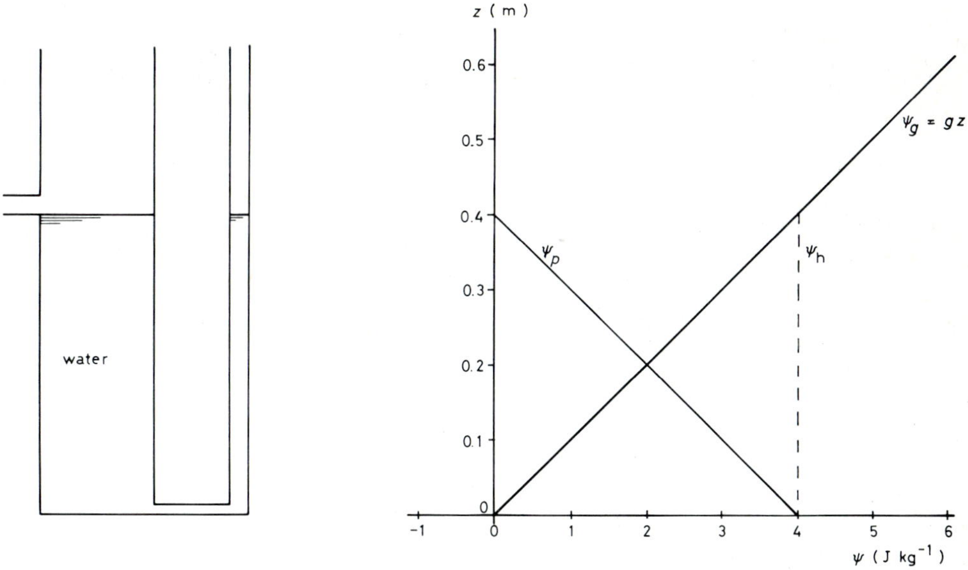

Figure 3.4 Potential diagram of water in container with noncapillary glass gauge and overflow.

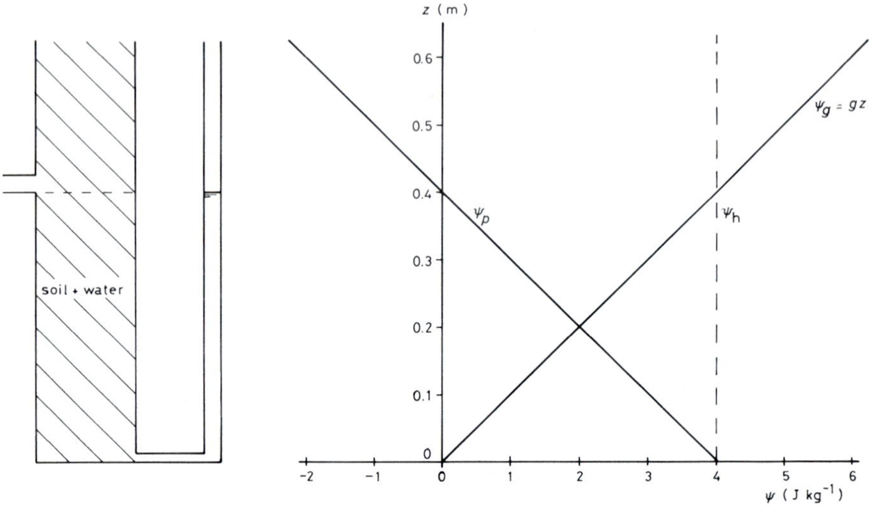

Figure 3.5 Potential diagram of water in container filled with soil, with noncapillary glass gauge and overflow.

the level of the overflow. If the reference point for the gravitational potential is chosen at the bottom of the cylinder, the potential diagram is exactly the same as that in Figure 2.7.

The empty cylinder is filled to the rim with dry soil and water is added slowly through the gauge until it starts discharging from the overflow. The water surface in the gauge will then again establish itself at the level of the overflow (Figure 3.5). This means that below $z = 0.4$ m the potential diagram is the same as in Figure 3.4. The presence of soil decreases the volume fraction of water but leaves the pressure potential unchanged.

Question 3.7 Why is the pressure potential below $z = 0.4$ m still the same as in Figure 3.4?

The soil above the overflow will attract water, due to capillary rise in the soil pores, osmotic binding, etc., until static equilibrium is attained.

Question 3.8 What is the condition for static equilibrium of the water above the overflow?

Since the hydraulic potential is everywhere the same and the gravitational potential increases linearly with height, the pressure potential must decrease linearly with height. Above the overflow, then, ψ_p is negative, because at the overflow, where the water surface is flat under atmospheric pressure, ψ_p is zero. This negative pressure potential is caused by the capillary action of the soil pores (similar to Figure 3.2) and other matric forces (Section *3.1.1*). The level where the pressure potential in a soil column or profile is zero, is called phreatic level. In field situations it is also called the groundwater table.

Question 3.9 A hole is made in the cylinder of Figure 3.5 at 0.1 m above the bottom.

a. What happens to the phreatic level?

b. What is the new hydraulic potential, ψ_h?

Draw the potential diagram for the new equilibrium situation that will be established after some time. (N.B.: do not change the reference level of the gravitational potential!)

Question 3.10 Two cylinders, filled with soil and placed at different levels, are connected with a tube. The tube connection is closed by means of a stopcock. The phreatic level in each cylinder is at the level of its overflow (Figure 3.6).

a. Plot in one diagram the water potentials in the cylinders against height. Choose the reference level of the gravitational potential at the phreatic level of the lowest cylinder.

b. Describe what happens when the stopcock is opened until a new static equilibrium is attained.

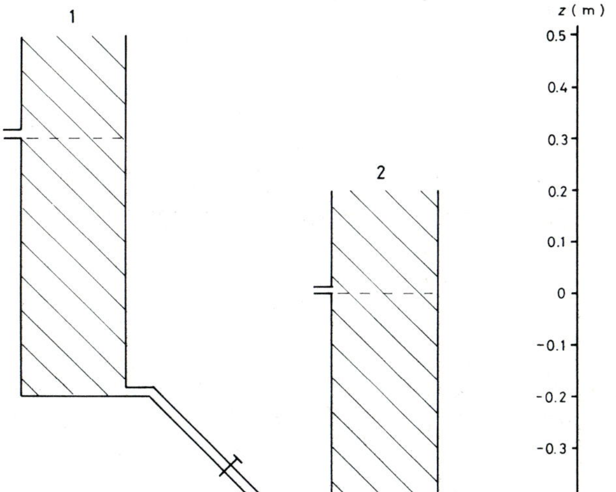

Figure 3.6 Two cylinders filled with soil and water, connected by a tube with stopcock, at different phreatic levels.

Below the phreatic level or groundwater table, water has a hydrostatic pressure which is larger than atmospheric pressure. Therefore, it will displace any air present at atmospheric pressure. However, some 'blocked' air may remain present at higher than atmospheric pressure, because it could not escape during wetting of the soil. The volume of this blocked air is usually neglected and the soil is still called saturated. However, the blocked air may amount to 10% of the pore volume, especially when the soil is saturated quickly from above.

The soil will also be saturated above a groundwater table in the zone where the pressure potential is higher than the air-entry value of the largest capillaries. This saturated zone above the groundwater table is called the full-capillary zone. Above the full-capillary zone, progressively smaller pores with lower air-entry values are empty. Thus, in a homogeneous soil, the water content decreases with height above the full-capillary zone.

Question 3.11 Which are the extreme values of the volume fraction of liquid, θ, above the groundwater table?

The presence of a full-capillary zone makes that the groundwater table is not the 'visible' boundary between saturated and unsaturated soil zones. A groundwater table can be made visible only by augering a vertical hole in which a free water surface can be formed.

Question 3.12 So far in this section only potentials on mass basis were used. Make a potential diagram for the soil column of Figure 3.5 using potentials on weight basis ('heads').

3.1.4 The piezometer and tensiometer

A piezometer is a tube of a few cm inner diameter, open at both ends, which is installed in a soil profile. If the lower end is below the groundwater table, a piezometer is partially filled with water. By determining the height of the water level in a piezometer it is possible to calculate the (positive) pressure potential of the soil water at the lower end of the tube. The diameter of piezometers is chosen large enough that capillary rise and resistance to water flow are negligible. As a result, any variation in hydraulic potential that may arise inside the piezometer, is instantaneously equalized. Thus, even if the pressure potential at the lower end is changing rapidly, the water inside a piezometer goes through a series of static equilibria and at any moment it can be assumed that the hydraulic head is uniform and equal to the hydraulic head of the soil water at the open lower end. There exchange of water takes place such that the pressure is always locally uniform. The static hydraulic head in piezometers can be determined by measuring the depth of the water level, since at the flat air-water interface the pressure potential is zero.

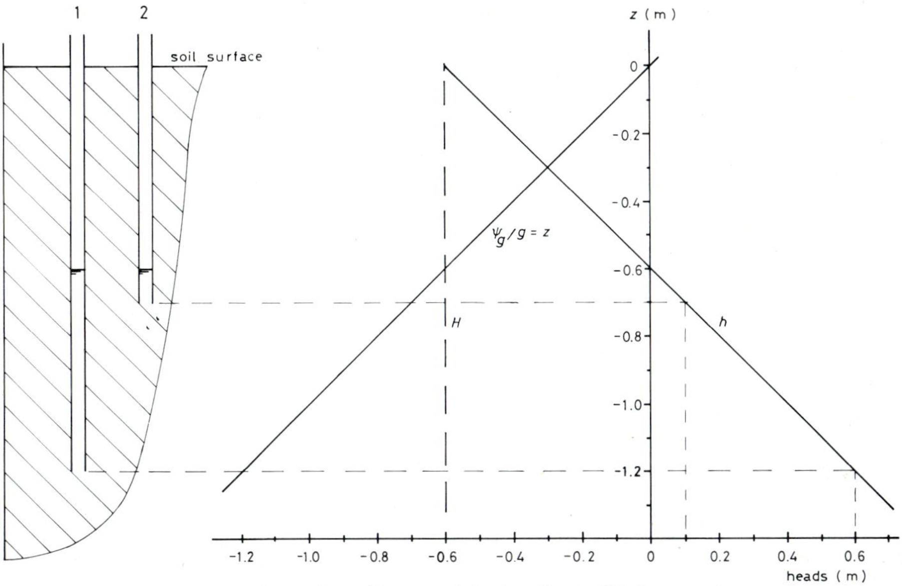

Figure 3.7 Two piezometers in a soil profile at static hydraulic equilibrium.

Figure 3.7 shows two piezometers in a soil profile in which the water is at static equilibrium. In the potential diagram the reference point $z = 0$ is taken at the soil surface. In both piezometers, at the water level, $H = h + z = 0 - 0.6$ m $= -0.6$ m. Thus at both lower ends H is also -0.6 m, which indicates that indeed the soil water is at static equilibrium.

From the water level in piezometer 1 at $z = -0.6$ m, it follows that at its lower end, where $z = -1.2$ m, $h = H - z = (-0.6 + 1.2)$ m $= 0.6$ m. Similarly, for piezometer 2, the pressure head at its lower end, where $z = -0.7$ m, is $h = 0.1$ m. With these results the potential diagram can be drawn.

Question 3.13 At what level is the groundwater table in Figure 3.7?

Since the water in the soil in Figure 3.7 is at static equilibrium, the potential diagram for the soil water is also valid for the water in each of the piezometers. In general, at static equilibrium of the soil water, the water level in piezometers is at the level of the groundwater table, independent of the depths of the lower ends. These results are somewhat trivial for this case of static equilibrium of the soil water, but this is no longer true when the water in the soil is flowing. Although water transport in soils is discussed extensively in Chapter 5, the functioning of piezometers in such a case is illustrated in the next question.

Question 3.14 Suppose that in Figure 3.7 the liquid level is in piezometer 1 at $z = -0.65$ m, and in piezometer 2 at $z = -0.55$ m.

a. Draw a potential diagram, assuming that the hydraulic head is a linear function of height.
b. At what height is the groundwater table?
c. Is this an equilibrium situation?
d. In which direction will the water flow?

Piezometers cannot be used to measure negative pressure heads because any water in the tubes will be absorbed by the soil. Negative pressure heads can be measured with so-called tensiometers. A tensiometer consists of a liquid-filled unglazed porous ceramic cup connected to a pressure measuring device, such as a vacuum gauge, via a liquid-filled tube (Figure 3.8). If the ceramic cup is embedded in soil, the soil solution can flow into or out of the tensiometer through the very small pores in the ceramic cup. Analogously to the situation discussed for piezometers, this flow continues until the pressure potential of the liquid in the cup has become equal to the pressure potential of the soil water around the cup. Thus the pressure potential of soil water can be measured with a tensiometer. It is, therefore, also called tensiometer pressure potential, and the pressure potential on volume basis is also called tensiometer pressure, p_t.

The vacuum gauge does not indicate the pressure in the cup when there is a difference in height between the two, such as in Figure 3.8. The liquid in the tube between the

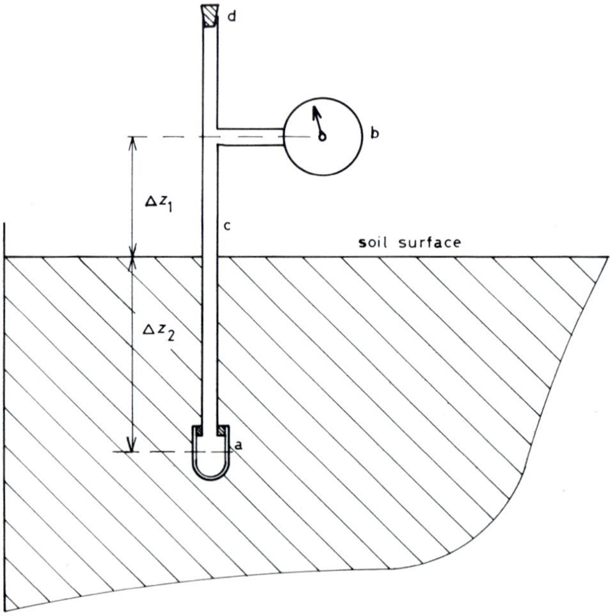

Figure 3.8 Tensiometer with vacuum gauge. a. porous cup; b. vacuum gauge; c. connecting tube; d. removable rubber stopper for deaeration.

cup and the vacuum gauge is at static equilibrium and thus the pressure in this liquid column increases linearly with depth. Therefore, in general, the tensiometer pressure in the cup is:

$$p_t = p_{gauge} + \rho_l g(\Delta z_1 + \Delta z_2) \tag{3.10}$$

where p_{gauge} is the pressure relative to atmospheric pressure of the liquid in the vacuum gauge (which is usually, but not necessarily the gauge reading), Δz_1 is the height of the gauge above the soil surface (which can also be negative!), Δz_2 is the depth of the tensiometer cup below the soil surface and ρ_l is the density of the liquid in the connecting tube. The pressure head of the liquid in the tensiometer cup is then:

$$h = h_{gauge} + \Delta z_1 + \Delta z_2 \tag{3.11}$$

where h_{gauge} is the pressure head of the liquid in the gauge.

As in piezometers, the liquid inside the tensiometer is always at static equilibrium. Thus, with $z = 0$ at the soil surface and using Equation 3.11:

$$H_{tens} = h_{gauge} + \Delta z_1 = h - \Delta z_2 = \text{uniform} \tag{3.12}$$

where H_{tens} is the hydraulic head of the liquid in the tensiometer.

While there is no resistance to flow in piezometers, so that they are always instantaneously at equilibrium with the soil water at the lower, open end, this is not necessarily true for tensiometers. The porous cup usually presents considerable resistance to flow and the water pressure inside may adjust only slowly to changes in the potential of the soil water at the cup. If the two are at equilibrium with each other, then the hydraulic head of the soil water at the cup is

$$H = h_{gauge} + \Delta z_1 \tag{3.13}$$

From this it follows immediately that for two tensiometers at equilibrium with the soil water, which have the gauge at the same height above the soil surface (Δz_1), the difference in H of the soil water at the cups is equal to the difference in gauge readings, independent of the depths of the cups.

If the hydraulic potential in the soil is not uniform, the tensiometer can only indicate an average hydraulic potential of the soil water around the cup. If the hydraulic potential varies mostly with height (Answer 3.14), which is usually the case, the vertical dimension of the cup should be kept small and Δz_2 should be related to the average depth of the permeable part of the cup.

Question 3.15 The cups of tensiometers 1 and 2 are at a depth of 0.6 m and 0.8 m, respectively. The gauges are 0.2 m above the soil surface. The gauge of tensiometer 1 indicates $h_{gauge} = -0.9$ m.

a. Draw the potential diagram, assuming that the water in the soil is at static equilibrium.

b. Calculate the gauge reading of tensiometer 2 and comment on the result.

Question 3.16 At another moment the tensiometers of Question 3.15 indicate $h_{\text{gauge}} = -0.9$ m for tensiometer 1 and $h_{\text{gauge}} = -1.0$ m for tensiometer 2.

a. Draw the potential diagram for this new situation, assuming H is linear with z.

b. What is the heigth of the groundwater table?

c. How can one easily find the difference in hydraulic potential of the soil water at the two cups?

Instead of vacuum gauges, mercury manometers can be used to measure the pressure head of the liquid in the tensiometer (Figure 3.9). The pressure in the cup relative

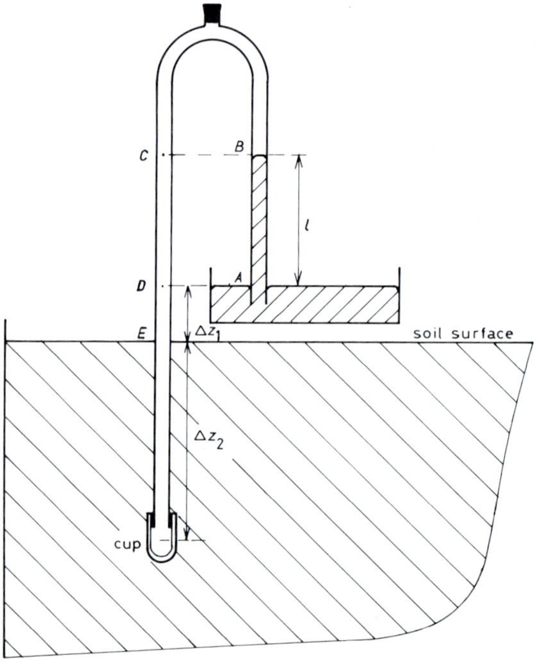

Figure 3.9 Tensiometer with mercury manometer.

to atmospheric pressure (= tensiometer pressure, p_t) can be calculated by starting at the flat air-mercury interface in the reservoir where the pressure potential is zero, and moving through the manometer to the cup while applying two rules for the pressure in a uniform liquid at rest:

a. the pressure in a horizontal plane is uniform

b. the pressure changes with height according to $\Delta p = -\rho_l g \Delta z$.

Thus,

$$p_t = 0 - \rho_{Hg} g l + \rho_l g l + \rho_l g \Delta z_1 + \rho_l g \Delta z_2 \qquad (3.14)$$
$$\quad A \quad B,C \quad D \quad E \quad \text{cup}$$

The letters under the terms in Equation 3.14 refer to the positions indicated in Figure 3.9, l is the length of the mercury column above the reservoir, Δz_1 the height of the mercury level in the reservoir above the soil surface and Δz_2 the depth of the tensiometer cup below the soil surface. Dividing Equation 3.14 by $\rho_l g$ gives the pressure head in the cup, and thus of the soil water in contact with the cup:

$$h = -\frac{\rho_{Hg} - \rho_l}{\rho_l} l + \Delta z_1 + \Delta z_2 \qquad (3.15)$$

Using $\rho_{Hg} = 13600$ kg m^{-3} and $\rho_l = 1000$ kg m^{-3} for the density of mercury and liquid in the cup, respectively, and assuming that the variation in the mercury level in the reservoir is negligible, Equation 3.15 reduces to

$$h = -12.6 l + \Delta z_1 + \Delta z_2 \qquad (3.16)$$

What was said in connection with Equation 3.12 for tensiometers with pressure gauges, is also true for tensiometers with mercury manometers if h_{gauge} is replaced by $-12.6\, l$. Therefore, subject to the same restriction ($z = 0$ at the soil surface),

$$H = -12.6 l + \Delta z_1 \qquad (3.17)$$

Thus, for manometers in the same reservoir, the difference in lengths of mercury columns, multiplied by -12.6, is equal to the difference in hydraulic head of the soil water at the cups, again irrespective of the depths of the cups.

Question 3.17 Two tensiometers are installed in a soil profile and connected to the same mercury reservoir. The cups of tensiometers 1 and 2 are at depths 0.4 and 0.8 m, respectively. The mercury level in the reservoir is at 0.1 m above the soil surface. The length of the mercury column in tensiometer 1 is 7.5 cm, that of tensiometer 2 is 9.0 cm.

a. Draw the potential diagram assuming H is linear with z.
b. Calculate the height of the groundwater table.
c. What do you expect about the mercury levels in the manometer tubes when the water in the soil is at static equilibrium?

Nowadays so-called pressure transducers are frequently used instead of pressure gauges or mercury manometers. A pressure transducer is a device that converts pressures to voltages linearly. With these devices tensiometer pressures can be measured very accurately and they can also be registered automatically. Pressure transducers require very small displacements of their sensing element to register changing pressures. This displacement must be effected by minute volumes of liquid flowing into or out of the tensiometer cup. In contrast, mercury manometers need exchange of considerable volumes of water between the soil and the tensiometer. Thus it takes considerable time for a tensiometer with a mercury manometer to equilibrate with the pressure potential of the soil water. Tensiometers with a pressure transducer often equilibrate almost instantaneously. A disadvantage of the use of pressure transducers with tensiometers is that the readings are usually strongly influenced by changing temperatures, mainly due to thermal expansion and contraction of the liquid.

The lowest soil water pressure potential that, theoretically, can be measured with tensiometers is limited by the lowest possible absolute pressure in the instrument, which is the vapour pressure of water (e.g., 2.3 kPa at 20 °C). In practice, however, already at absolute pressures below ≈ 10 kPa (which is −90 kPa or −9.0 m relative to an atmospheric pressure of 100 kPa or 10.0 m) problems arise from dissolution of gases from the water.

The measuring range of tensiometers is further limited by the size of the largest pores in the cup. If the pressure in the cup falls below the air-entry value of the largest pores, air will enter the tensiometer and all the water may be absorbed by the soil.

Question 3.18 What is the equivalent diameter of the largest pore in a tensiometer cup that can be used to measure pressure heads of −9.0 m (at complete wetting)?

Since air is compressible and has a large thermal expansion coefficient, isolated air bubbles inside tensiometers make them very sluggish in following changes in soil water potential and make them sensitive to temperature changes. If air is abundant in the system, it can make accurate measurements altogether impossible.

The problems mentioned above can be nearly eliminated by filling tensiometers with deaerated water and selecting ceramic cups with pores smaller than 2 μm. However, air can still diffuse through the walls of the tubing connecting the cup with the pressure measuring device (often nylon tubing) and through the water in the pores of the cup walls. The former can be eliminated by using impermeable tubing, such as copper, but the latter can only be reduced by using thick ceramic walls of low porosity. Therefore, tensiometers must be flushed periodically with deaerated water to drive accumulated air out of the system.

3.1.5 Water retention (pF curve)

In dealing with problems of flow of water in unsaturated soil, there is need for a relationship between the volume fraction of water and an appropriate component of the water potential. Recalling the discussion in Section *3.1.3*, the pressure potential would appear to be suitable for that purpose, but there is one problem associated with it. This problem is illustrated in Figure 3.10. A soil column with a tensiometer is placed in a container, in which the air pressure can be made to differ from atmospheric. The vacuum gauge on the tensiometer is located outside the container, 1.0 m above the groundwater table in the column.

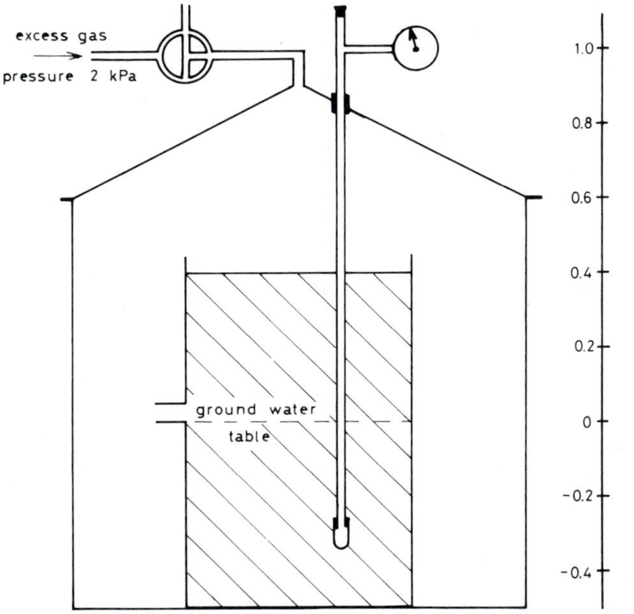

Figure 3.10 Apparatus for illustrating the influence of nonatmospheric pressure of the soil gas phase on tensiometer readings.

Question 3.19 What is the pressure head at the vacuum gauge when the soil column is under atmospheric pressure?

Question 3.20 The gas pressure inside the container is raised to 2 kPa above atmospheric pressure by turning the stopcock.

a. What is the change in the pressure head at the gauge?

b. What do you conclude about the pressure potential of the soil water in the column?

Question 3.21 Has the volume fraction of water changed at any position in the soil profile?

Answers 3.20 and 3.21 illustrate that there is no unique relationship between the volume fraction of water and the tensiometer pressure potential of soil water, if the soil air pressure changes with respect to atmospheric pressure. This occurs occasionally in laboratory and field conditions when the soil air becomes isolated from the atmosphere by a zone with a low permeability for air, such as a water-saturated zone. In such situations, the nonuniqueness of the relationship makes it unsuitable for solving problems of water flow in unsaturated soil. This problem can be circumvented by dividing the tensiometer pressure potential into two components: the matric potential, ψ_m, resulting exclusively from the soil matrix, and the pneumatic potential, ψ_a, due to the soil gas pressure in excess to atmospheric pressure.

Thus $\psi_p = \psi_m + \psi_a$ (3.18)

The pressure equivalents or potentials on volume basis are:

$$p_t = p_m + p_a \tag{3.19}$$

where p_m is the matric pressure and p_a is the excess gas pressure. The head equivalents or potentials on weight basis are:

$$h = \frac{p_m}{\rho_l g} + \frac{p_a}{\rho_l g} = h_m + h_a \tag{3.20}$$

where h_m is the matric head and h_a is the pneumatic head.

These equations show that under normal atmospheric conditions ($p_a = 0$), the matric potential is equal to the tensiometer pressure potential.

The relationship between the volume fraction of water and the matric potential is not dependent on the pressure in the gas phase. It is characteristic for different types of soil and is, therefore, often called soil water characteristic. A more descriptive name for it is soil water retentivity curve or function. Figure 3.11 shows the water retentivity function of a coarse sandy soil with grains of more or less equal size.

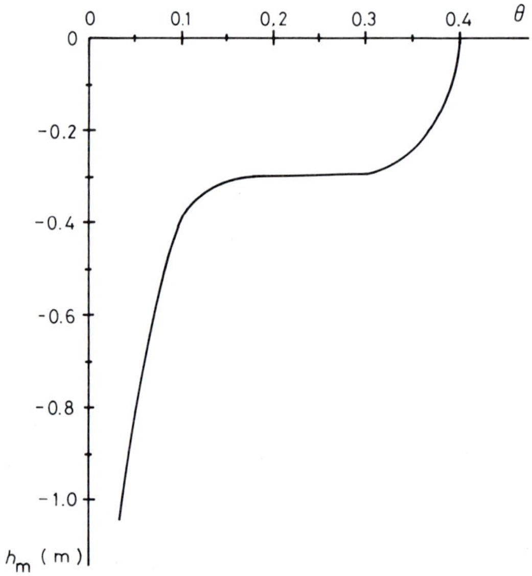

Figure 3.11 Water retentivity curve of a coarse sandy soil.

Question 3.22 What is the porosity of this soil?

If water is extracted from the soil of Figure 3.11, starting from saturation, h_m drops at first quite sharply from zero to about −0.3 m. Thereafter, much water can be extracted without a substantial change of h_m. At these high h_m values the binding of water takes place almost exclusively by capillary action. Therefore, the above soil water characteristic shows that the soil contains a large number of pores with the same equivalent diameter.

Question 3.23 What is this equivalent diameter with complete wetting?

As the water content is decreased further, the capillary binding of the remaining water becomes stronger, because ever smaller capillaries are emptied, leaving only progressively smaller radii of curvature and lower values of the matric head. The last amounts of soil water are bound very strongly. This water occurs only in the smallest parts of the pores around the contact points between grains and as thin films around soil particles bound by adsorptive forces.

In soils that contain a wider range of grain sizes one finds a more gradual release of water when the matric head is decreased from $h_m = 0$ to $h_m \approx -10^4$ m. In such cases it is difficult to plot the whole soil water characteristic in one graph, unless h_m is plotted on a logarithmic scale. The value of $^{10}\log(-h_m/\text{cm})$, where $h_m < 0$, is called the *pF* value and the graph showing the relationship between the *pF* value and θ is known as the *pF* curve of a soil.

Question 3.24 a. Give the *pF* values corresponding with the following h_m values: $h_m = -160, -10, -1, -0.1, -0.01$ and 0 m.
b. Do negative *pF* values exist?
c. What is the *pF* value at the phreatic level?
d. What is the *pF* value at 10 cm below the groundwater table (at static equilibrium)?

Question 3.25 Draw the *pF* curve corresponding to Figure 3.11 and compare the two curves.

The shape of *pF* curves is determined by the contributions of different matric forces to the matric potential. Capillary binding of water can take place only at *pF* values of less than about 3 ($h_m < -10$ m). This is due to the fact that negative absolute pressures do not exist.

Question 3.26 Why are negative absolute pressures not possible?

Matric heads smaller than $h_m = -10$ m are caused exclusively by the adhesive binding of water on solid surfaces and by the osmotic binding of water in double layers (Section *3.1.9*). Figure 3.12 presents the *pF* curves of a number of Dutch soils. The curves reflect that sandy soils have mainly capillary binding and that clay soils are

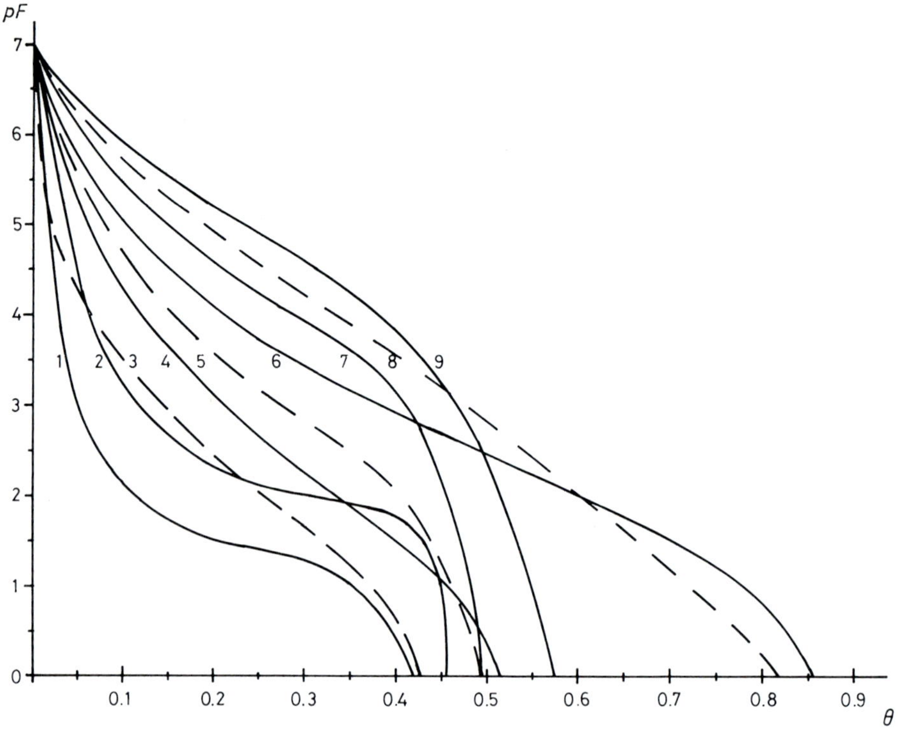

Figure 3.12 *pF* curves of some Dutch soils: 1. Dune sand; 2. loamy sand; 3. calcareous fine sandy loam; 4. calcareous loam; 5. silt loam derived from loess; 6. young oligotrophous peat soil; 7. marine clay; 8. eutrophous peat soil; 9. river basin clay. (Data: ICW, Wageningen.)

dominated by adhesive and osmotic binding.

Question 3.27 How can you see this from the *pF* curves?

Question 3.28 a. What can you say about loamy and peaty soils in this context?

b. What is the most significant difference between these two soil types that can be observed in the *pF* curves?

The soil water characteristic produced by progressive extraction of soil water from saturation (desorption) differs from that obtained by gradual addition of water to air-dry soil (absorption). This phenomenon is called hysteresis. Hysteresis can be caused by a number of factors. It is to a large extent due to the irregular and composite sizes and shapes of soil pores. In addition, the contact angle (Section *3.1.2*) may be lower during wetting than during drying and cause hysteresis. Also during wetting usually some air is entrapped which keeps the water content lower than it could be. Finally, clays can change in water content while remaining saturated. Different arrangements of the clay particles during drying and wetting can cause hysteresis. Question and Answer 3.29 explain the principle of hysteresis as a result of varying pore geometry.

Question 3.29 Figure 3.13a shows a 'pore' that has the shape of a cylindrical bottle with an upper and lower neck. The diameter and height of the open-ended bottle necks are a and $2a$, respectively. The belly has a diameter $2a$ and a height $4a$. The pore is inserted into a cylinder with diameter $2\frac{1}{2}a$ and height $8a$ (Figure 3.13b). The space between the walls of the pore and the cylinder is filled with solid material of zero porosity. This total configuration can be used as a model for an undisturbed soil sample.

a. What is its porosity?

b. The pore is completely filled with water. What is the volume fraction of water of the 'soil sample' at saturation?

c. Water is extracted from the soil sample by applying a gradually decreasing pressure at the lower end of the pore. Draw the water characteristic (desorption curve) of the soil sample if given that $\sigma = 7 \times 10^{-2}$ N m^{-1}, $a = 14$ µm and that there is complete wetting. (The pore is small enough that differences in the gravitational potential can be ignored.)

d. What happens when the pressure is gradually increased again to zero? Draw the absorption curve in the same graph.

e. Is it correct to draw the curves with sharp angles?

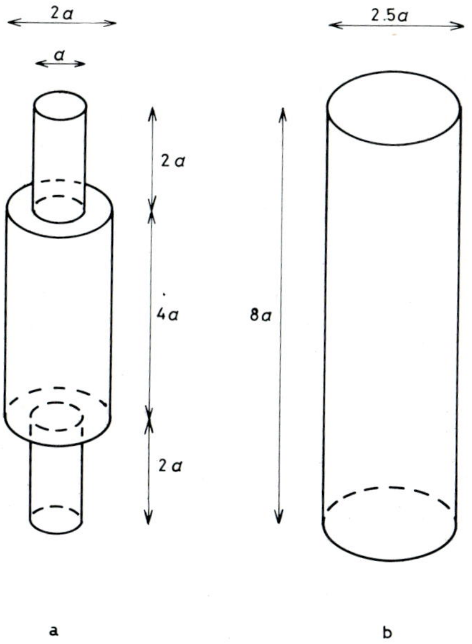

Figure 3.13 Elements of soil pore model used to explain the principle of hysteresis.

Answer 3.29 shows that the water content at a certain matric potential is lower during absorption than during desorption. In general, after any reversal the soil is wetter during drying than during wetting. The same principle applies to real soils, but the curves are smoother, as shown in Figure 3.14.

Question 3.30 Can you explain this?

Figure 3.14 shows so-called primary scanning curves inside the envelope formed by the characteristic, main, or boundary absorption and desorption curves of a soil. The latter are obtained when the soil is dried from saturation to air-dryness, or wetted from air-dryness to saturation, without reversal of the process somewhere in between these two states. Anywhere on a boundary curve one could reverse the process from wetting to drying, or vice versa, and create a primary scanning curve. The relationship between matric potential and volume fraction of water then depends on the value at which the reversal is made. Reversal on a primary scanning curve produces a secondary scanning curve, etc. Thus the recent history of wetting and drying must be known to be able to obtain the water content for a particular matric potential, or vice versa, from a given set of scanning curves of a soil. The *pF* curves in Figure 3.12 are characteristic desorption curves.

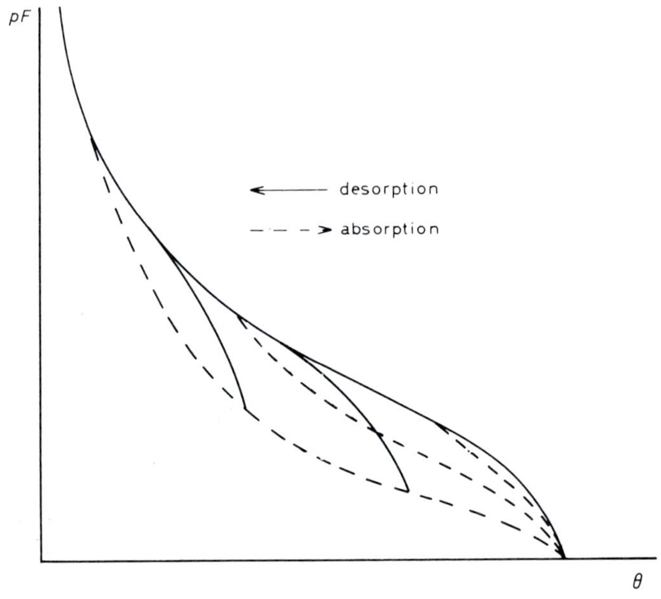

Figure 3.14 Absorption and desorption characteristics with primary scanning curves.

Absorption of water by plant roots becomes more difficult as the matric potential decreases. Yield reductions may result from still relatively high matric heads. This depends much on atmospheric conditions, type of crop, growth stage of crop, etc. However, most plants are unable to take up any more water when the matric head has decreased to $h = -160$ m or $pF \approx 4.2$. This value is, therefore, called the permanent wilting point of a soil. Water held at very high matric potentials is also not available to plants if it percolates through the root zone to the subsoil under the influence of gravity. The water content in a soil profile when deep percolation has nearly stopped after the soil was thoroughly wetted by heavy rain or irrigation, is called the field capacity. The matric potential at field capacity varies with the circumstances. When the groundwater table is stable at a relatively shallow depth, field capacity corresponds with hydrostatic equilibrium with the groundwater table. Then the water content profile is identical to an inverted soil water retentivity curve, such as in Figure 3.11.

Soils with deep groundwater tables at field capacity generally show matric heads of about -3.0 m. The corresponding pF value is about 2.5.

The difference between the water contents at field capacity and wilting point is called the available soil water for plants. One should use this concept with caution. On the one hand, this amount is often augmented with water supplied by capillary rise from a groundwater table, and precipitation and irrigation water during the growing season. On the other hand, yield reductions result well before soils reach $pF = 4.2$, while plants are seldom able to extract water uniformly throughout the root zone.

Question 3.31 Calculate for soils 1, 3, 6, 7 and 9 in Figure 3.12 the water available for plants between $pF = 2$ and $pF = 4.2$. Give your answer in volume fraction, θ, and in equivalent mm (rainfall) in a root zone of 0.6 m.

Question 3.32 A rain shower of 8 mm falls on soil 4 of Figure 3.12, which is at wilting point.

a. Calculate the penetration depth of the rainwater if the soil is wetted up to field capacity ($pF = 2$).

b. How many cubic metres of water are required per hectare to wet 0.3 m of this soil to field capacity?

Question 3.33 a. Make for soils 1 and 7 in Figure 3.12 a diagram of θ versus depth for static equilibrium with a groundwater table at 1.0 m depth.

b. Do the same for a soil that consists of soil 1 in the upper 0.5 m and soil 7 in the subsoil, with the groundwater table at a depth of 1.0 m.

c. Do the same for a soil that has 0.5 m of soil 7 overlying a subsoil of soil 1, again with the groundwater table at 1.0 m depth.

d. Comment on the results.

3.1.6 *Equipment for determining soil water characteristics*

Normally, the soil water characteristics determined in the laboratory are main desorption curves. These are obtained by subjecting a wet soil sample to a constant h_m value until hydraulic equilibrium is reached, after which the soil water content is determined.

For relatively high values of h_m a water-saturated soil sample is placed on a fine-porous plate. Thereafter the matric head in the pores of the suction plate is decreased by means of a hanging water column (Figure 3.15). After some time, depending on the head value required and the kind of soil, hydraulic equilibrium is established. Then:

$$H = h + z = \text{uniform} \tag{3.21}$$

At the overflow, $h = 0$ and $H = z_{\text{overflow}}$

$$\text{Thus: } h_{\text{sample}} = H - z_{\text{sample}} = z_{\text{overflow}} - z_{\text{sample}} \tag{3.22}$$

If $z = 0$ is taken at the soil sample as in Figure 3.15, then

$$h_{\text{sample}} = z_{\text{overflow}} \tag{3.23}$$

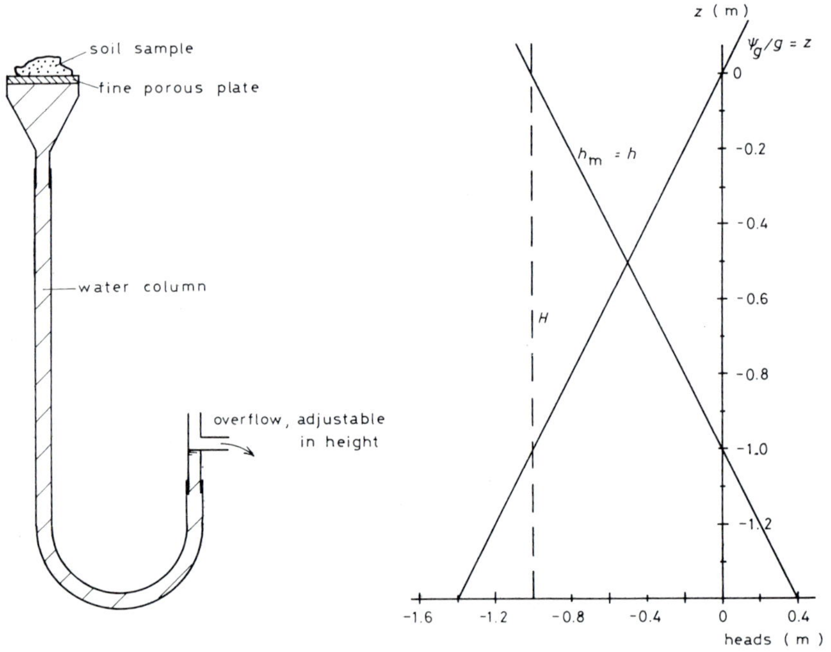

Figure 3.15 Apparatus for determination of soil water characteristics at high matric heads.

Because the system is under atmospheric pressure, $p_a = 0$ and $h_m = h$ (Equations 3.19 and 3.20). Thus the pressure head of the sample is also its matric head.

Question 3.34 a. What is the value of the matric head for the soil sample in Figure 3.15?
b. What must be done to determine the water content at $pF = 1$?

Of course, the pores of the porous plate must be small enough to prevent that the air-entry value is reached. The minimum head for which this apparatus can be used is about –8 m. The lower values between about –5 m and –8 m are usually obtained with a partial vacuum rather than a hanging water column because of problems with dissolution of gas.

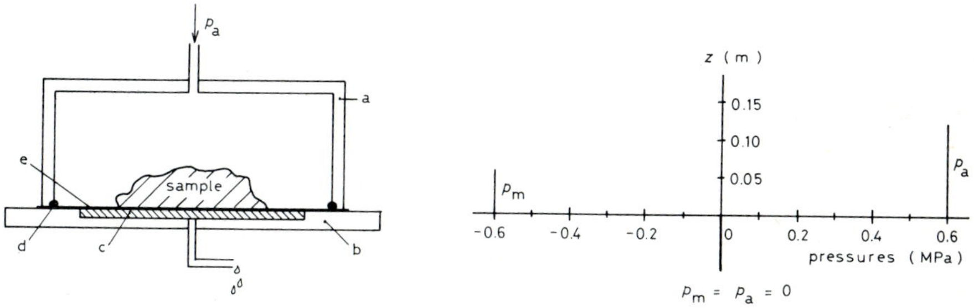

Figure 3.16 Pressure membrane apparatus for measuring desorption characteristic at low matric heads: a. steel cylinder; b. steel base plate; c. porous plate; d. O-ring; e. membrane.

For lower head values so-called pressure plates or pressure membranes are used. In such apparatus moist soil samples are placed on fine-porous plates or membranes and subjected to external gas pressures. Below the plate or membrane the water is discharged at atmospheric pressure (Figure 3.16). Thus, for this 'free' water, $h = 0$ and $H = z_{discharge}$. At hydraulic equilibrium,

$$H_{sample} = H_{discharge} = z_{discharge} \tag{3.24}$$

Equations 2.44 and 3.20 for the water in the soil sample give:

$$H_{sample} = h_m + h_a + z_{sample} \tag{3.25}$$

Disregarding the small difference in height between the soil sample and the discharging water, combination of Equations 3.24 and 3.25 gives immediately:

$$h_m = -h_a \tag{3.26}$$

or when multiplied by $\rho_l g$:

$$p_m = -p_a \tag{3.27}$$

The pressure membrane apparatus is used for head values between ≈ -5 and -200 m.

Question 3.35 What is the external gas pressure required for the determination of the water content at the permanent wilting point ($pF = 4.2$)?

3.1.7 Potentials of the pure component water. Vapour equilibrium

In general, dissolved salts can move freely with water through soil. Salts also pass porous materials such as tensiometer cups and porous plates and membranes used for the determination of soil water characteristics. Therefore, osmotic effects did not need to be considered up to this point and only potentials of the soil solution were discussed. There are circumstances, however, under which only the pure component water can move from one place to another. This is the case when there are semi-permeable membranes in the system, such as in plant roots. Another example is a gas barrier, because water can be transmitted via the vapour phase, but solutes remain behind in the liquid phase. Then osmotic effects must be taken into account.

The total potential of the pure component water is not only influenced by gravity, matric forces and excess gas pressure, but also by the presence of dissolved salts. To account for this effect, the so-called osmotic potential, ψ_o, is introduced. Pure water is taken as reference for the osmotic potential, i.e. $\psi_o = 0$ for pure water. The total potential of the pure component water is then:

$$\psi_t = \psi_p + \psi_g + \psi_o \tag{3.28}$$

While the gravitational potential, $\psi_g = gz$, is the same for the pure component water as for the soil solution as a whole, the tensiometer pressure potential of the pure component is slightly different from the tensiometer pressure potential of the soil liquid, $\psi_p = \frac{1}{\rho_l} p_t$. This is due to the small difference between the density of the soil solution, ρ_l, and that of pure water, ρ_w. In practice, however, this difference can be neglected.

The osmotic potential can be illustrated by Figure 3.17. A container, divided in two parts by a semi-permeable membrane, holds a salt solution at one side of the membrane and pure water at the other side. Both sides are at atmospheric pressure. Because the concentration of water molecules is higher in the pure water than in the solution, more water molecules will move in the direction of the solution than in the other direction. As a result, the surface of the soil solution will rise slowly and that of the pure water will drop. One can prevent this flow of water across the membrane and maintain static equilibrium by exerting on the salt solution a gas pressure in excess of atmospheric pressure. At the surface of the solution the total potential of the pure water is then:

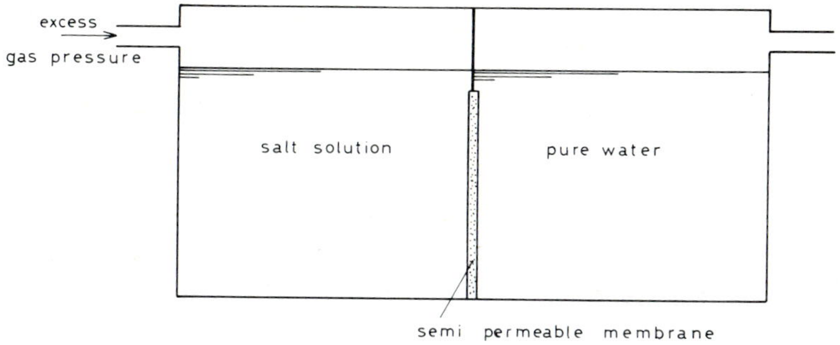

Figure 3.17 Apparatus for illustrating the concept of osmotic pressure.

$$\psi_t = \frac{1}{\rho_w} p_a + \psi_g + \psi_o \tag{3.29}$$

where p_a is the gas pressure in excess of atmospheric pressure. At the surface of the pure water, the total potential is:

$$\psi_t = \psi_p + \psi_g + \psi_o = 0 + \psi_g + 0 = \psi_g \tag{3.30}$$

At static equilibrium, with the liquid surfaces at the same level, ψ_t and ψ_g are equal at both sides. Thus, from Equation 3.29 and 3.30:

$$\frac{1}{\rho_w} p_a + \psi_o = 0 \tag{3.31}$$

The excess gas pressure needed to maintain equilibrium is equal to the so-called osmotic pressure, Π, of the solution. Then the osmotic potential can be written as:

$$\psi_o = -\frac{1}{\rho_w} \Pi \tag{3.32}$$

The total potential of the component water in a soil solution is then, according to Equations 3.28 and 3.32:

$$\psi_t = \frac{1}{\rho_w} p_t + \psi_g - \frac{1}{\rho_w} \Pi = gz + \frac{1}{\rho_w}(p_t - \Pi) \tag{3.33}$$

Water is present in soil not only in the liquid phase, but also in the gas phase, i.e. as water vapour. At static equilibrium the total potentials in both phases are equal. Water vapour consists of pure water, and thus $\psi_o = 0$. Also, it is not influenced by matric forces, and thus $\psi_m = 0$. Therefore, its total potential is determined only by its vapour pressure, p, and by its position in the gravitational field (Section *2.2.5*):

$$\psi_t\,(\text{vapour}) = \int \frac{1}{\rho_{vap}} \mathrm{d}p + gz \tag{3.34}$$

where ρ_{vap} is the density of the water vapour. Substitution of ρ_{vap} with an equation similar to Equation 2.47 yields for the total potential of the water vapour:

$$\psi_t\,(\text{vapour}) = \int \frac{RT}{pM}\,\mathrm{d}p + gz = \frac{RT}{M}\ln p + gz + C \tag{3.35}$$

where M is the molar mass of water and C is an integration constant.

Per definition, the saturated vapour pressure, p_0, is the vapour pressure at a free surface of pure water under atmospheric pressure. The total potential of the liquid water at this water surface, according to Equation 3.33, is:

$$\psi_t\,(\text{liquid}) = \psi_g = gz \tag{3.36}$$

Question 3.36 Why is this so?

At the gas-liquid interface, gz is the same for the liquid water and the water vapour. Thus:

$$\psi_t\,(\text{liquid}) = gz = \psi_t\,(\text{vapour}) = \frac{RT}{M}\ln p_0 + C + gz \tag{3.37}$$

It follows:

$$C = -\frac{RT}{M}\ln p_0 \tag{3.38}$$

Substitution of Equation 3.38 into Equation 3.35 yields the total potential of water vapour with a vapour pressure p:

$$\psi_t\,(\text{vapour}) = \frac{RT}{M}\ln\frac{p}{p_0} + gz \tag{3.39}$$

At static equilibrium, the total potential of the component water in the soil solution is equal to the total potential of the water vapour in the soil air. Thus, at the gas-liquid interface of the soil water is, according to Equations 3.33 and 3.39:

$$\frac{1}{\rho_w}(p_t - \Pi) = \frac{RT}{M}\ln\frac{p}{p_0} \tag{3.40}$$

With $\rho_w \approx 1000$ kg m^{-3}, $RT \approx 2.4 \times 10^3$ J mol^{-1} at 15°C and $M \approx 18 \times 10^{-3}$ kg mol^{-1}, this reduces to:

$$\ln\frac{p}{p_0} = 7.5 \times 10^{-9}\,\text{Pa}^{-1}(p_t - \Pi) \tag{3.41}$$

This equation indicates that the relative vapour pressure p/p_0 (also called relative humidity) at the gas-liquid interface is determined by the tensiometer pressure and the osmotic pressure of the soil solution. If the soil air is at atmospheric pressure, Equation

3.41 can be written as:

$$\ln \frac{p}{p_0} = 7.5 \times 10^{-9}\,\text{Pa}^{-1}\,(p_m - \Pi) \qquad (3.42)$$

or in the head equivalent form, using $\rho_l g \approx 10^4\ \text{N m}^{-3}$,

$$\ln \frac{p}{p_0} = 7.5 \times 10^{-5}\,\text{m}^{-1}\,(h_m + h_o) \qquad (3.43)$$

in which $h_o = -\dfrac{\Pi}{\rho_l g}$ is the osmotic pressure head of the soil solution.

Question 3.37 A small soil sample and a cup containing a saturated NaCl-solution are placed side by side in a closed, air-tight jar. The osmotic pressure of the NaCl-solution is 300 bar.

a. What is the relative vapour pressure in the jar?

b. What is the *pF* value of the soil sample if the salt concentration of the soil water is negligible?

Question 3.38 The relative humidity in a laboratory is maintained at 50 %.

a. What is the *pF* value of 'air-dry' soil samples for which the osmotic potential can be neglected?

b. Derive a general relationship between *pF* and relative humidity, under the same restriction.

c. What is the head value registered by a tensiometer kept in this laboratory?

Question 3.39 Make a graph showing the relative vapour pressure as a function of *pF*. Plot the *pF* values in the range between 0 and 7 on the horizontal scale of the graph.

Question 3.40 A colomn of fine-sandy soil, with an air-entry value of –1.2 m, has a groundwater table at 1.0 m below the surface. The water in the soil column has a salt concentration of 50 mol NaCl per m^3. The osmotic pressure of the water can be calculated by $\Pi = RTc$, where c is the ionic concentration.

An air-tight container is placed over the soil column. The air is under atmospheric pressure and contains water vapour in equilibrium with the soil-water system.

a. A soil sample A (Figure 3.18) is placed in direct contact with the surface of the soil column. Calculate h_m of the water in sample A at equilibrium.

b. A second sample B is placed on the soil column in a small beaker preventing direct contact. What is the *pF* value in sample B, if the salt concentration in sample B is negligible?

c. Calculate the relative vapour pressure above the soil column.

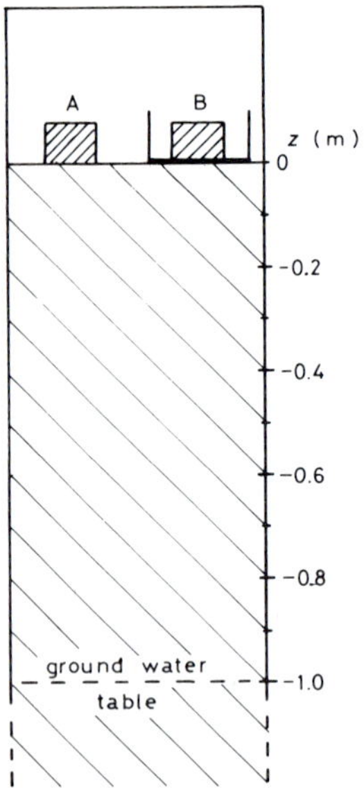

Figure 3.18 Soil column with groundwater table under air-tight container. Soil sample A is in direct contact with soil column, soil sample B is not.

3.1.8 Criteria for equilibrium of the component water

So far, two different cases of equilibrium of soil water have been discussed. These can be described as follows:

- If dissolved salts or solutes can move freely through the soil, which is the case in sandy soils, there will be equilibrium of the soil solution if the hydraulic potential $\psi_h = \psi_p + \psi_g$ is uniform, i.e. invariant with position. If the concentration of solutes varies with position, they will diffuse in the direction of decreasing concentration until the concentration is everywhere the same. Any increase in volume occupied by the solutes associated with this diffusion will be compensated by the movement of an equal volume of the component water in the opposite direction. Thus there will be total equilibrium only if ψ_h is uniform and ψ_o is uniform.
- If the system contains semi-permeable barriers that allow movement of the component water but not of solutes, there is equilibrium only if the total potential $\psi_t = \psi_p + \psi_g + \psi_o$ is uniform. This situation is found where the transport has to take place through the gas phase (Question 3.37), e.g. in dry soils where the water is stored in isolated small pores and/or in double layers.

A third situation can be distinguished in soils with barriers that obstruct the transport of solutes without completely preventing it. This can be illustrated by Figure

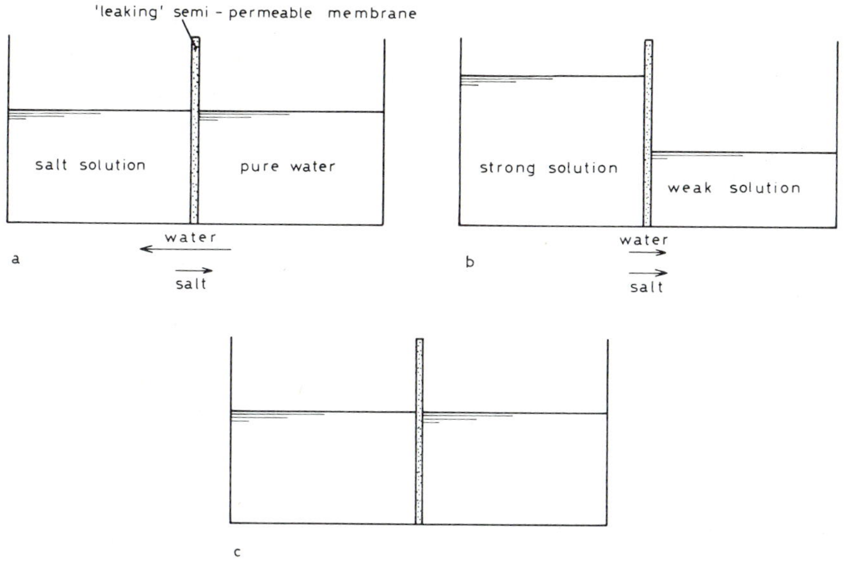

Figure 3.19 Salt solution and pure water separated by a 'leaking' semi-permeable membrane.
a. Shortly after installation.
b. Some time after direction of flow of water reverses.
c. Equilibrium with equal concentrations at both sides.

3.19, where a container is divided in two parts by an imperfectly semi-permeable membrane. Initially, there is a salt solution at one side of the membrane and pure water at the other side, with the liquid surfaces at the same height. This will result in net diffusion of water in the direction of the solution and leakage of very small quantities of solutes through the membrane in the direction of the pure water. The liquid level of the salt solution will rise at first, but after some time a situation will be reached where the total potential of the component water will be equal at both sides. Then ψ_t = uniform, while the difference in hydraulic potential at both sides balances the difference in osmotic potential. But, because the solute concentration is still different at both sides, the leakage of solutes through the membrane will continue. Final equilibrium will be reached when the solute concentration and the liquid levels have both become equal at both sides of the membrane. Thus, total equilibrium will exist only if ψ_h = uniform and ψ_o = uniform.

This situation is found in moist clay soils, where the soil water is present in thin films on the clay particles. Salt transport is then difficult because the anions are repelled from the double layer by the negative electrostatic charge of the clay particles.

Summarizing these three situations one can state that generally there is hydraulic equilibrium (no bulk transport) if ψ_h is uniform. Exceptions to this are found in all nearly dry soils and in moist clay soils. In these situations there will be bulk transport if ψ_o is not uniform, even if ψ_h is uniform.

3.1.9 *Osmotic binding of water in the double layer*

Water molecules in the double layer are subject to several kinds of forces. Close to the surface of a clay particle water molecules are attracted by London-van der Waals forces. These short range forces decrease very rapidly with distance from the clay surface and are not dependent on the composition and extent of the double layer.

Due to their dipolar nature, water molecules orient themselves in the electrostatic field resulting from the negative charge of clay particles. Because the field strength decreases with distance from the clay particle, the water molecules experience a net static electrical force in the direction of the clay surface. This force decreases gradually with distance from the surface and is zero at the end of the double layer (either truncated or fully extended).

Question 3.41 Why is the electrostatic field strength, resulting from the charge of a clay particle, zero at the end of the double layer?

The effect of the two kinds of forces mentioned above can be represented by an 'adsorption' potential, ψ_α. The reference $\psi_\alpha = 0$ is chosen at a point where these forces are zero, i.e. outside the double layer. Within the double layer, ψ_α decreases towards the surface of the clay particle to a very low (negative) value (Figure 3.20).

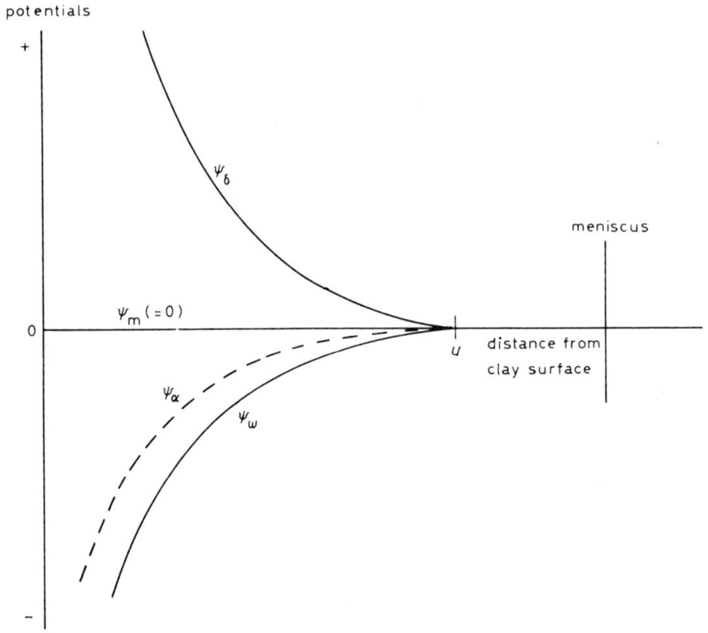

Figure 3.20 Water potentials in a fully extended double layer near a flat gas-liquid interface.

A third kind of force acting on water molecules in the double layer results from the increasing ion concentration towards the solid surface. This results in an osmotic potential in the double layer, $\psi_{o,dl}$, which decreases from the outside soil solution towards the clay particle. Similar to the situation of water in the gravitational field, where the decreasing gravitational potential is compensated by an increasing pressure potential, the decrease of ψ_α and $\psi_{o,dl}$ is compensated by an increase of a pressure potential, ψ_δ. This potential will be taken relative to the pressure of the soil air. The total potential of the component water in the double layer can then be written as:

$$\psi_t = \psi_\alpha + \psi_{o,dl} + \psi_\delta + \psi_a + \psi_g \tag{3.44}$$

The total potential for water outside the double layer is given by Equation 3.28:

$$\psi_t = \psi_p + \psi_g + \psi_o \tag{3.28}$$

Combination of Equations 3.18, 3.28 and 3.44 and neglecting very small differences of ψ_g yields:

$$\psi_m + \psi_o = \psi_\alpha + \psi_{o,dl} + \psi_\delta \tag{3.45}$$

Defining $\psi_\omega = \psi_{o,dl} - \psi_o$ as the osmotic potential due to the ions in the double layer in excess to those in the soil solution, Equation 3.45 can be written as:

$$\psi_m = \psi_\alpha + \psi_\omega + \psi_\delta \tag{3.46}$$

Figure 3.20 shows the various potentials in a fully extended double layer adjacent to the soil solution with which it is in equilibrium. If the soil solution has a flat meniscus with the gas phase (at the same height of the double layer), then the matric potential of the soil solution is zero. (A tensiometer in situ would read $\psi_m = 0$.) Both ψ_α and ψ_ω are also zero in the solution outside the double layer and thus ψ_δ there is also zero. In fact, ψ_δ is the partial potential of the water associated with the actual pressure in the water relative to the pressure of the gas phase. Near the clay surface ψ_δ, and thus the actual pressure in the water, increases to very high values (probably 10 à 100 bar). If the meniscus is curved, ψ_δ at the meniscus may have either a negative or positive value depending on whether the meniscus is concave or convex. In both cases is at the gas-water interface

$$\psi_m = \psi_\delta \tag{3.47}$$

Figure 3.21 shows the various potentials when the double layer is truncated (see Figure 1.7). The value of ψ_δ at the gas-liquid interface again depends on the radius of curvature of that interface. Figure 3.21 is drawn for a situation where ψ_δ has a negative value at the interface. ψ_ω now has a finite negative value at the interface, because the ionic concentration is larger than it would be if the double layer were fully extended, in equilibrium with the soil solution. As stated earlier, at the interface ψ_α is zero. Thus, at the gas-water interface of a truncated double layer:

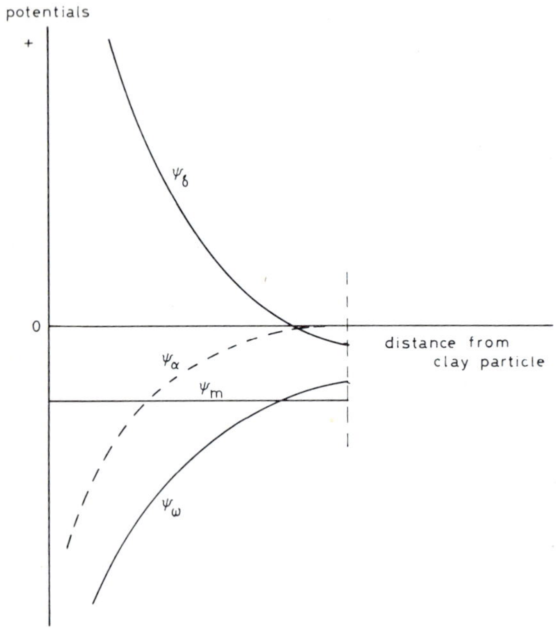

Figure 3.21 Water potentials in a truncated double layer.

$$\psi_m = \psi_\delta + \psi_\omega \tag{3.48}$$

Equations 3.47 and 3.48 describe the behaviour of double layer systems with respect to their environment. They show that the matric potential is equal to the potential due to the actual liquid pressure at the gas-liquid interface relative to the pressure of the gas phase, plus the osmotic potential due to the excess ions at the same interface, if the double layer is truncated. It follows that for soils with truncated double layers – that is, for almost every soil except pure sand – Equation 2.41 must be rewritten as:

$$p_h = p_t + \rho_l g z \tag{3.49}$$

or, if the soil gas phase has atmospheric pressure:

$$p_h = p_m + \rho_l g z \tag{3.50}$$

Equation 3.48 explains why matric pressures of less than –1 bar (matric heads less than –10 m) can occur. The matric pressure is the pressure *equivalent* of the matric potential and is not the actual pressure (relative to the pressure in the gas phase) in the liquid. The actual absolute pressure will always be larger than zero (Question 3.26), or the pressure relative to atmospheric pressure always higher than ≈ –1 bar. Matric pressures less than –1 bar are caused mainly by the osmotic potential, ψ_ω, and in very dry soils also by a negative value of ψ_α.

3.2 STATIC EQUILIBRIUM OF THE SOIL SYSTEM

3.2.1 Soil pressure

The condition of static equilibrium in the gravitational force field was so far applied to the soil water only. In a similar way, this condition can be applied to the soil system as a whole, if the various (partial) potentials of the soil system are defined properly. By 'soil system' we mean the soil matrix and all substances (mass) in the pore space associated with the soil matrix.

At static equilibrium the total potential of the soil system, ψ_t, has to be uniform, which means that the gradient of the gravitational potential, ψ_g, has to be compensated by an equal but opposite gradient of the pressure potential, ψ_p. Thus, similar to Equation 2.38, but now for the soil system:

$$\psi_t = \psi_g + \psi_p = \text{uniform} \quad (3.51)$$

If the water in the cylinder of Figure 2.7 is replaced by soil, the potential diagram for the soil system is the same as it was for water, if the same reference points are chosen for ψ_g and ψ_p. This is so, because these potentials are defined on mass basis. If, however, potentials on volume basis (pressure equivalents) are used, the diagram for the soil system will be different from that for water. This is due to the difference between the density of water, ρ_l, and the bulk density of the soil, ${}^b\rho$. Then, static equilibrium of the soil system is described by:

$$ {}^b\rho\psi_t = {}^b\rho\psi_g + {}^b\rho\psi_p = \text{uniform} \quad (3.52)$$

In soil mechanics, the pressure potential on volume basis is called soil pressure and is indicated with the symbol σ_s:

$$\sigma_s = {}^b\rho\psi_p \quad (3.53)$$

Since soil is usually rigid, σ_s is hardly a real pressure. It is better to define σ_s as the 'normal' component (i.e. the component perpendicular to a certain plane) of the material stress at a certain location. The term soil pressure will here be used for the vertical force exerted by the weight of the overlying material, divided by surface area of a horizontal plane.

Question 3.42 What is the soil pressure at static equilibrium at a depth of 1 m in a dry sandy soil with a dry bulk density of 1500 kg m^{-3}? Draw a diagram of the soil potentials in pressure units.

3.2.2 Intergranular pressure

The intergranular pressure, σ_g, is defined as the total contact force between the grains divided by area. In a dry soil, the vertical intergranular pressure equals the soil pressure, because at any depth the weight of the overlying soil material is borne solely by the contact forces between the grains at that depth. Hence:

$$\sigma_s = \sigma_g \tag{3.54}$$

The intergranular pressure causes friction between the grains, which in turn determines the resistance against motion of grains relative to each other. For instance, the friction determines the resistance of the soil against ploughing, its 'bearing capacity' for buildings and its resistance against compaction by heavy farm machinery.

Question 3.43 What is σ_g at 1 m depth for the soil of Question 3.42?

In a water-saturated soil, the weight of the overlying soil is borne not only by the contact force between the grains, but also by the contact force in the water:

$$\sigma_s = \sigma_g + \sigma_w \tag{3.55}$$

Similar to σ_s, σ_w is called water pressure, and it is defined as $\sigma_w = \rho_l \psi_p$. Then, according to Equation 2.37,

$$\sigma_w = p' \tag{3.56}$$

Thus σ_w is the actual pressure of the water relative to atmospheric pressure.

Question 3.44 a. Make a diagram of σ_s and σ_w versus depth for the soil of Question 3.42, if the groundwater table is at the soil surface and $\phi = 0.40$.
b. What is now the value of σ_g at a depth of 1 m?
c. Is the bearing capacity of the wet soil higher or lower than the bearing capacity of the dry soil?

Question 3.45 Calculate σ_g at a depth of 1 m when the groundwater table is at 0.10 m above the surface of the sandy soil (submerged condition). What is now σ_g at the soil surface?

In a rigid soil not saturated with water, σ_w is negative. This negative pressure is supported by the soil particles, which are pressed together with a force in excess of the weight of the overlying soil. The water pressure is active only in the water-filled pore space, while σ_w, σ_s and σ_g are defined as if they act over the total cross-sectional area. For an unsaturated, rigid soil Equation 3.55 must be changed to:

$$\sigma_s = \sigma_g + \frac{\theta}{\phi}\sigma_w \tag{3.57}$$

The coefficient θ/ϕ in front of σ_w varies from zero at air-dryness to one at saturation. This is as it should be, according to Equations 3.54 and 3.55. In nonrigid or swelling soils the situation is more complex and beyond the scope of this text.

Question 3.46 What is σ_g at the surface of the sandy soil if the groundwater table is at a depth of 0.20 m and the air-entry value of the soil equals –0.30 m?

Answer 3.46 shows that if the groundwater table is lowered from the soil surface, the intergranular pressure at the surface will increase. This continues until the air-entry value is reached. After that σ_g will decrease again, due to the decrease of θ/ϕ, according to:

$$\sigma_{g(\text{surface})} = -\frac{\theta}{\phi}\sigma_w \tag{3.58}$$

This general rule can be confirmed at the beach. Beyond the waterline it is difficult to use a bicycle, because $\sigma_s = 0$ at the surface. The same applies to the dry sand higher up. In between those areas is a zone where the sand is saturated with water at a negative σ_w value. There the intergranular pressure at the surface is positive and cycling or walking is most easy.

3.3 SUGGESTED LITERATURE

BAVER, L. D., W. H. GARDNER and W. R. GARDNER, 1972. Soil Physics (fourth edition). Wiley, 489 p. Chapter 8.

CHILDS, E. C., 1969. An introduction to the physical basis of soil water phenomena. Wiley, 493 p. Chapters 7 and 8.

HILLEL, D., 1980. Fundamentals of Soil Physics. Academic Press, 413 p. Chapters 3 and 7.

MARSHALL, T. J. and J. W. HOLMES, 1979. Soil Physics. Cambridge University Press, 345 p. Chapters 2 and 3.

TAYLOR, S. A., and G. L. ASHCROFT, 1972. Physical Edaphology. Freeman and Co, 533 p. Chapters 9 and 10.

STAKMAN, W. P., 1980. Measuring Soil Moisture. Chapter 3 in: Drainage Principles and Applications. Publication 16, Vol. 3, Int. Inst. for Land Reclamation and Improvement (ILRI), Wageningen, The Netherlands, 2nd edition.

YONG, R. N., and B. P. WARKENTIN, 1975. Soil properties and behaviour. Elsevier Scientific Publishing Co., 449 p. Chapter 4.

Answer 3.1 The contact angle of water on clean glass is zero, and thus cos $\varphi = 1$. The radius of the capillary tube is 35×10^{-6} m. Using Equation 3.7:

$$z_c = \frac{2 \times 0.07 \text{ N m}^{-1} \times 1}{1000 \text{ kg m}^{-3} \times 10 \text{ N kg}^{-1} \times 35 \times 10^{-6} \text{ m}} = 0.40 \text{ m}$$

Answer 3.2 The pressure potential of the water at the meniscus can be read from the diagram. It is -4 J kg^{-1}, -4000 Pa or -0.4 m.

Answer 3.3 If the reference point for z is taken at the flat water surface, then everywhere $\psi_h = 0$ and $\psi_p = -\psi_g$. So, the pressure potential on volume basis at the meniscus is:

$$\rho_l \psi_p = -\rho_l \psi_g = -\rho_l g z_c = -\frac{2\sigma \cos\varphi}{r}$$

Answer 3.4 Consider the forces acting on the left-hand hemisphere of the spherical droplet in the figure below. If the pressure inside the shere is p_i, then the total horizontal force on the vertical cross-section, directed towards the left, is $F_i = \pi r^2 p_i$.

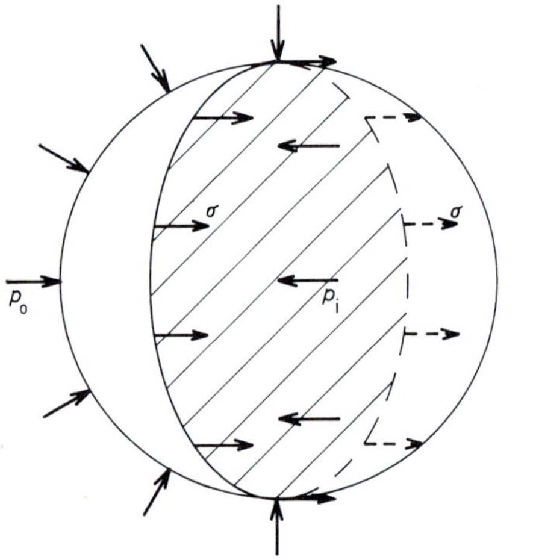

Similarly, the net horizontal force on the hemispherical outside surface as a result of the outside pressure p_o, directed towards the right, is $F_o = \pi r^2 p_o$.
The vertical components of the outside forces cancel each other. The surface tension in the spherical surface acts over the circumference of the cross-sectional area in a horizontal direction. This total force is $F_\sigma = 2\pi r \sigma$, exerted by the right-hand hemisphere on the left-hand one and, therefore, directed towards the right.
At static equilibrium, $\Sigma \boldsymbol{F} = 0$. Thus
$\pi r^2 p_o + 2\pi r \sigma = \pi r^2 p_i$

From this follows immediately: $p_i - p_o = \dfrac{2\sigma}{r}$

Thus the inside pressure is higher than the outside. By using the sign convention for the radius of concave (positive) and convex (negative) surfaces, the proper sign of the pressure difference is automatically obtained.

Answer 3.5 At complete wetting, the height of capillary rise for a capillary of 20 μm diameter is:

$$z_c = \frac{2\sigma \cos\varphi}{\rho_l g r} = \frac{2 \times 0.07 \times 1}{1000 \times 10 \times 10^{-5}} \text{ m} = 1.4 \text{ m}$$

Thus the air-entry value is $-z_c = -1.4$ m or $-\rho_l g z_c = -14$ kPa.

Answer 3.6 The air-entry value of the noncylindrical capillary is the same as that of the cylindrical capillary of Question 3.5. Therefore, its height of capillary rise is 1.4 m and its equivalent diameter is 20 μm.

Answer 3.7 At static equilibrium the gradient of the gravitational potential must be compensated by an equal but opposite gradient of the pressure potential. The reference level of the pressure potential is again at the overflow, since there the water surface is flat and in contact with atmospheric pressure. So, the pressure potential of the water below $z = 0.4$ m remains unchanged.

Answer 3.8 At static equilibrium the hydraulic potential is everywhere the same. Thus the hydraulic potential above the overflow is equal to that below the overflow.

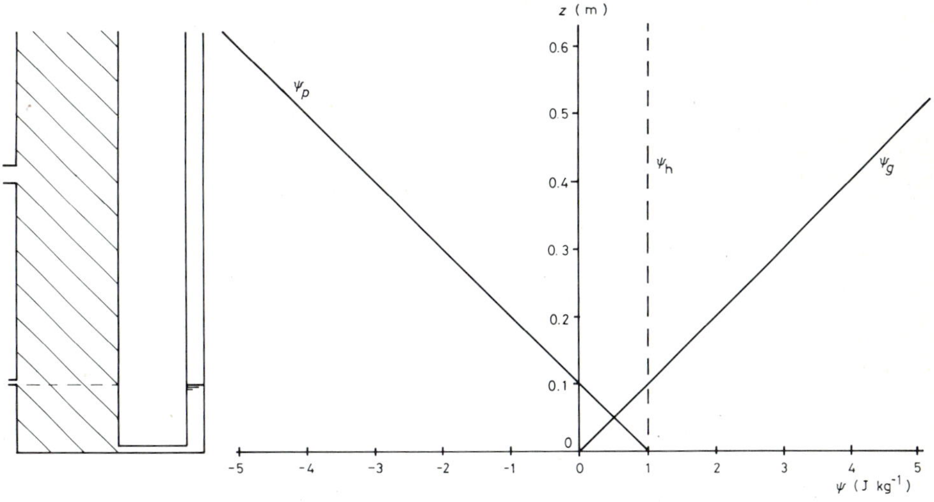

Answer 3.9 a. After the hole is made, the pressure there drops to atmospheric. Thus the phreatic level drops 0.30 m to the level of the hole.

b. The drop in phreatic level causes a decrease of ψ_p, and hence a decrease of ψ_h, of 3 J kg^{-1}. Thus, the new hydraulic potential of the water is everywhere 1 J kg^{-1}.

Answer 3.10 a. With the stopcock closed the water in the two cylinders has different hydraulic potentials. The hydraulic potential in cylinder 2, $\psi_h(2)$, is zero, because the gravitational and pressure potential have the same reference level, namely the phreatic level. In cylinder 1, $\psi_h(1) = 3\ \mathrm{J\ kg^{-1}}$, because its phreatic level is located at $z = 0.3$ m.

b. When the stopcock is opened water will flow from cylinder 1 to cylinder 2, because the hydraulic potential in the former is higher than in the latter. This water will be discharged through the overflow of cylinder 2. When the phreatic level in cylinder 1 has dropped to the phreatic level in cylinder 2, a new static equilibrium will be attained. In this equilibrium situation the hydraulic potential in both cylinders is the same, and equal to the hydraulic potential in cylinder 2 before the stopcock was opened.

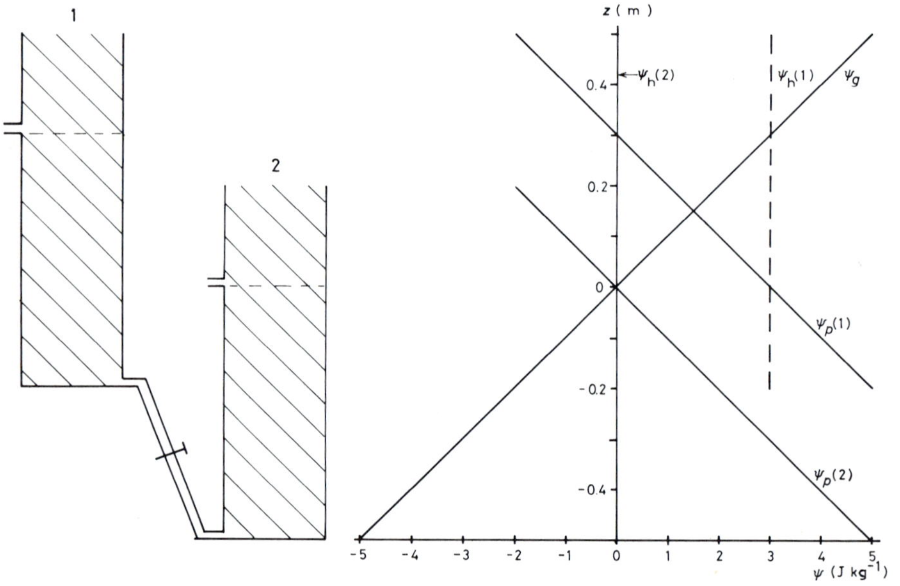

Answer 3.11 $0 \leqslant \theta \leqslant \phi$, which is the range from completely dry to saturation.

Answer 3.12 Conversion of potentials on mass basis to potentials on weight basis requires division by g (Section *2.2.4*). The result is given in the diagram below.

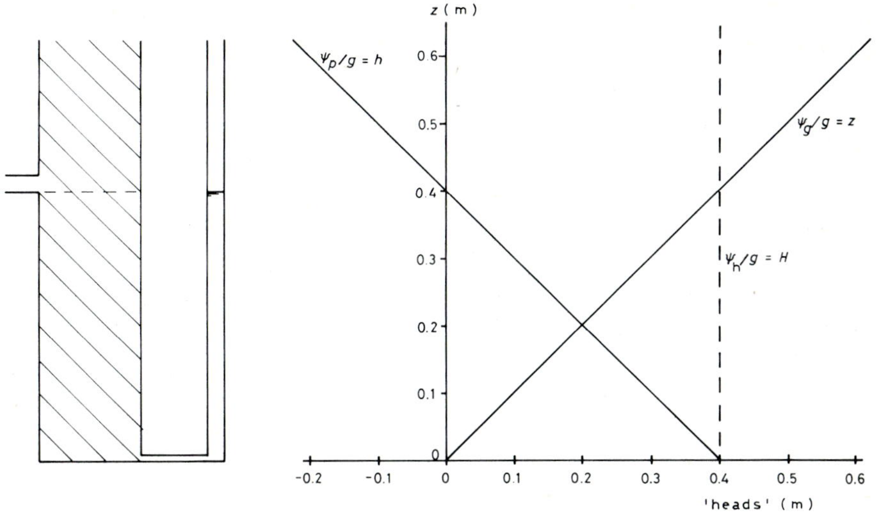

Answer 3.13 The diagram shows that the pressure head is zero at $z = -0.6$ m. So, the groundwater table is at $z = -0.6$ m.

Answer 3.14 a. The hydraulic head of the water in piezometer 1 is constant and equal to that at the air-water interface; thus $H = h + z = 0 - 0.65$ m $= -0.65$ m. This is also the valuc of H in the soil at the depth of the lower end, thus $H = -0.65$ m at $z = -1.20$ m. Similarly, from piezometer 2 it follows that $H = -0.55$ m at $z = -0.70$ m. Since it is given that H is a linear function of z, H is determined, as indicated in the potential diagram. The h line can be obtained from $h = H - z$.

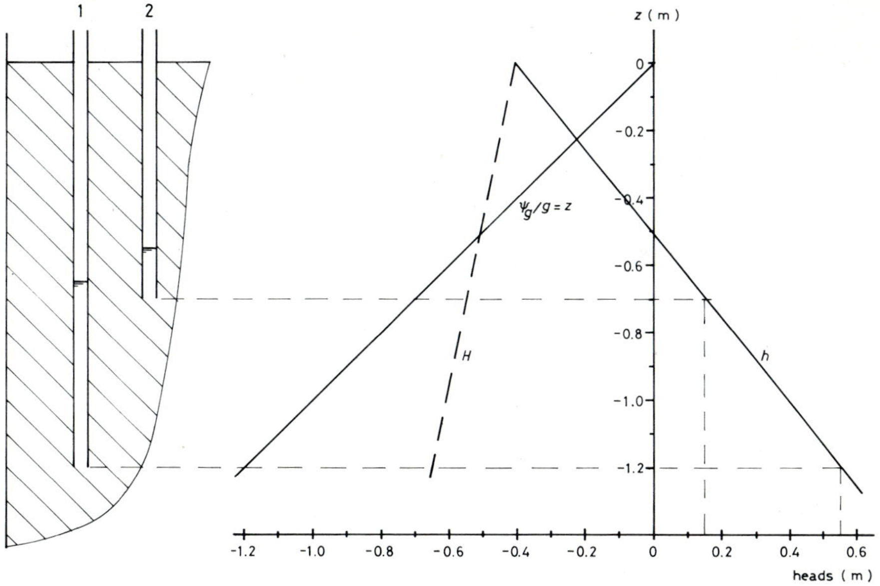

b. The groundwater table is at the depth where $h = 0$. The diagram shows this at $z = -0.51$ m.

c. This is not an equilibrium situation since H is not constant with height.

d. Water flows in the direction of decreasing hydraulic potential. Thus, in this case, the water flows downwards.

Answer 3.15 a. For the soil water at tensiometer cup 1 is, according to Equation 3.11:

$h = h_{gauge} + \Delta z_1 + \Delta z_2 = (-0.9 + 0.2 + 0.6)$ m $= -0.1$ m and $H = h + z = (-0.1 - 0.6)$ m $= -0.7$ m. The potential diagram at static equilibrium can now be drawn.

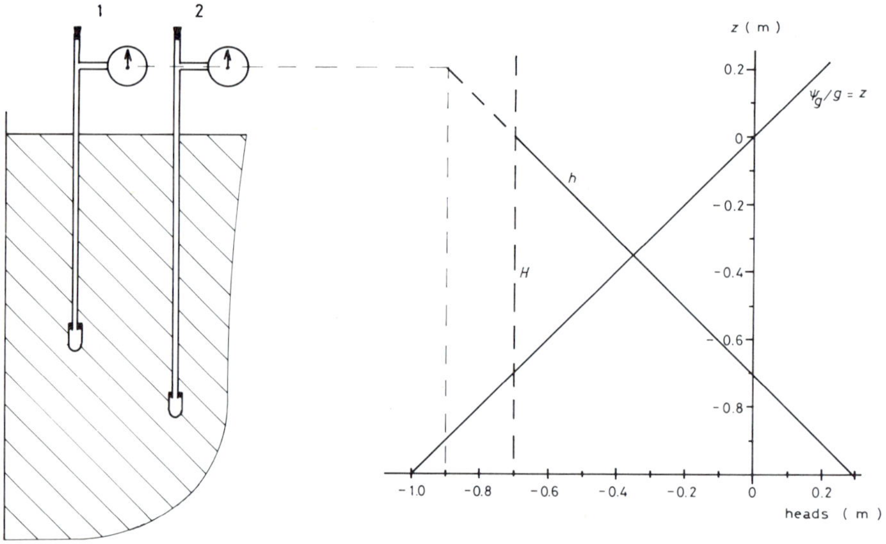

b. For tensiometer cup 2: $h = H - z = -0.7$ m $- (-0.8$ m$) = 0.1$ m. Thus the cup is below the groundwater table. While negative pressure heads cannot be measured with piezometers, it is of course possible to measure positive pressure heads with tensiometers.

For tensiometer 2, $h_{gauge} = h - \Delta z_1 - \Delta z_2 = (0.1 - 0.8 - 0.2)$ m $= -0.9$ m. This is the same value as for tensiometer 1. Thus, like the piezometers in Figure 3.7, at static equilibrium the tensiometer gauges give the same readings if they are at the same height, irrespective of the depths of the cups. Note that at static equilibrium the pressure heads in the soil and in both tensiometer tubes have the same value at the same height. The hydraulic head, of course, has everywhere the same value.

Answer 3.16 a. The gauge reading of tensiometer 1 is the same as in Answer 3.15 and thus $H = -0.7$ m and $h = -0.1$ m at $z = -0.6$ m. For tensiometer cup 2 at $z = -0.8$ m, $h = (-1.0 + 1.0)$ m $= 0$ and $H = 0 - 0.8$ m $= -0.8$ m. Assuming H is linear with z, the diagram can now be drawn.

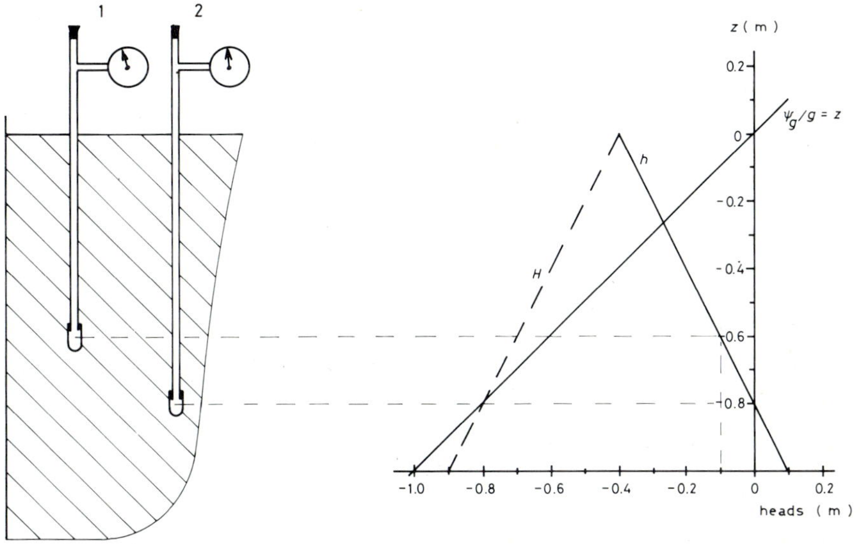

b. The groundwater table is at the cup of tensiometer 2, where $h = 0$.
c. If the water in the tensiometers is at static equilibrium with the soil water in contact with the cups, then, according to Equation 3.13, the difference in hydraulic head of the soil water at the cups is equal to the difference in the readings of the gauges (0.1 m), because the gauges are located at the same height above the soil surface.

Answer 3.17 a. Applying Equation 3.16 to tensiometer 1 gives:
$h = (-12.6 \times 0.075 + 0.1 + 0.4)$ m $= -0.445$ m.
Similarly for tensiometer 2:
$h = (-12.6 \times 0.09 + 0.1 + 0.8)$ m $= -0.234$ m.
Using $H = h + z$, and assuming H is linear with z, the diagram can be drawn.
b. The h line can be represented by $h = az + b$
at cup 1: $-0.445 = -0.4a + b$
at cup 2: $-0.234 = -0.8a + b$.
Subtraction of these two equations gives: $-0.211 = 0.4a$ or $a = -0.5275$.
From the first equation: $b = -0.445 - 0.4 \times 0.5275 = -0.656$.

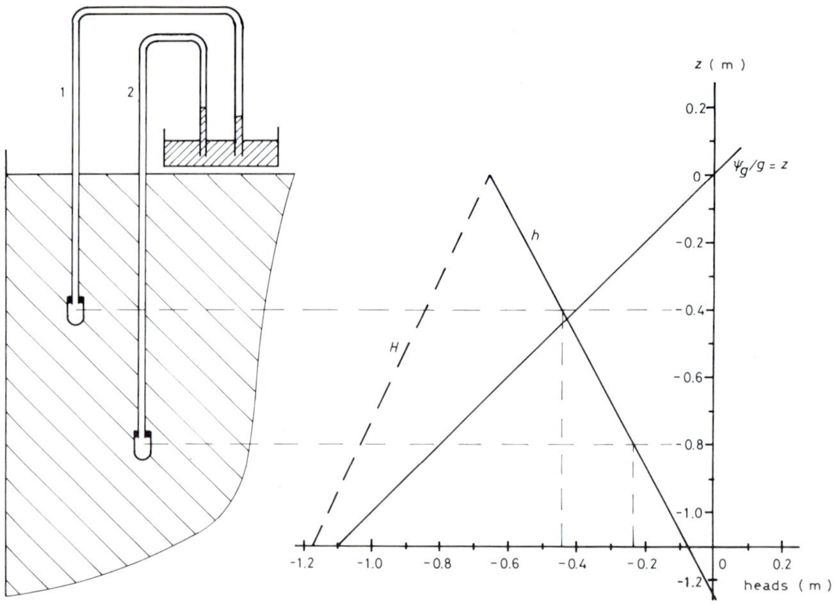

At the groundwater table $h = 0$, and thus $z = -b/a$. Thus the groundwater table is at $z = -1.244$ m.

c. When the water in the soil is at static equilibrium, H is uniform with position. Then, according to Equation 3.17, the mercury levels in the manometers will be at the same height.

Answer 3.18 According to Equations 3.8 and 3.9, the maximum pressure difference that can be maintained across a curved air-water interface in a capillary with radius r is:

$$\Delta p_{max} = \frac{2\sigma}{r}$$

Thus the maximum equivalent pore diameter that can occur in a tensiometer cup for measuring pressure heads of -9.0 m is:

$$d_{max} = 2r_{max} = \frac{4\sigma}{\Delta p_{max}} = \frac{-4\sigma}{\rho_1 g h_{min}} = \frac{4 \times 0.07 \text{ N m}^{-1}}{1000 \text{ kg m}^{-3} \times 10 \text{ N kg}^{-1} \times 9.0 \text{ m}} = 3.1 \text{ μm}$$

Answer 3.19 The water in the soil and that in the tensiometer are both at static equilibrium and are also in equilibrium with each other. Thus the pressure head in the soil and in the tensiometer tube is zero at the level of the groundwater table. The pressure head at the gauge is then -1.0 m, because the gauge is at 1.0 m above the groundwater table.

Answer 3.20 a. The increase in gas pressure of 2 kPa inside the container increases the pressure on the soil water everywhere with the same amount. Thus the pressure head of the soil water increases with

$$\frac{2\text{ kPa}}{\rho_1 g} = \frac{2000\text{ Pa}}{10^3\text{ kg m}^{-3} \times 10\text{ N kg}^{-1}} = 0.2\text{ m}$$

and the gauge will now indicate $-1.0\text{ m} + 0.2\text{ m} = -0.8\text{ m}$.

b. Since the pressure head of the soil water is read, per definition, relative to atmospheric pressure, it has changed 0.2 m.

Answer 3.21 According to Equation 3.9, the curvatures of the air-water interfaces in the soil, and with them the volume fraction of water, are determined by the pressure difference across the interface and not by the absolute values. Since the pressure in the water and in the air changed everywhere by the same amount, the soil water content will not change.

Answer 3.22 The porosity of this soil is 40%, because at $h_m = 0$ the soil is saturated with water.

Answer 3.23 Pores that are emptied at $h_m = -0.3$ m have, according to Answers 3.5 and 3.6, an equivalent diameter:

$$2r = \frac{4\sigma}{\rho_1 g |h_m|} = \frac{4 \times 7 \times 10^{-2}}{10^3 \times 10 \times 0.3}\text{ m} \approx 93\ \mu\text{m}$$

Answer 3.24 a. $pF = {}^{10}\log(-h_m/\text{cm}) = {}^{10}\log(160\text{ m/cm}) = {}^{10}\log 16000 \approx 4.2$

Similarly, for h_m =

h_m	pF
−10 m,	3
−1 m,	2
−0.1 m,	1
−0.01 m,	0
0	$-\infty$

b. Negative pF values do exist for $-1.0\text{ cm} < h_m < 0$.

c. At the phreatic level $pF = \log 0 = -\infty$.

d. At 10 cm below the groundwater table there is no negative pressure head and hence no pF value.

Answer 3.25

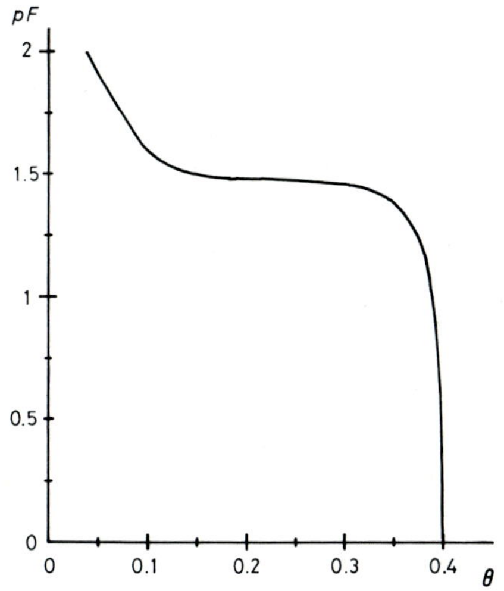

Note that the scale on the vertical axis has been compressed in the lower matric potential range and stretched in the higher matric potential range.

Answer 3.26 If the absolute pressure is zero, there is vacuum. So negative absolute pressures do not exist.

Answer 3.27 Sandy soils (curve 1 and 2) release most of their water below $pF = 3$, where mainly capillary binding is active. Clay soils release much of their water at higher pF values.

Answer 3.28 a. Loamy and peaty soils exhibit both capillary and osmotic binding of water.

b. The most significant difference between these two soil types is that the total amount of water that can be stored, and hence the total porosity, is much larger in peat soils than in loam soils.

Answer 3.29 a. Volume of pore $= 2 \times \frac{1}{4}\pi a^2 \times 2a + \pi a^2 \times 4a = \pi a^3 + 4\pi a^3 = 5\pi a^3$

Volume of sample $= \dfrac{25}{16}\pi a^2 \times 8a = 12.5\pi a^3$

Thus, $\phi = \dfrac{5\pi a^3}{12.5\pi a^3} = 0.4$

b. $\theta = \phi = 0.4$ at $h_m = 0$.

c. The opening of the upper bottle-neck will be able to 'hold' the meniscus against the applied negative pressure as long as h_m is larger than the air-entry value, or

$$h_m > \frac{-2\sigma}{\rho_l g r} = \frac{-2 \times 7 \times 10^{-2}}{1000 \times 10 \times 7 \times 10^{-6}}\ \text{m} = -2.0\ \text{m}$$

Thus the pore remains full while h_m is being decreased from 0 to -2.0 m, except for a slight decrease of the water content due to a gradual change of the shape of the meniscus from flat to concave hemispherical. At $h_m = -2.0$ m the whole pore will empty at once. The corresponding desorption curve is drawn below.

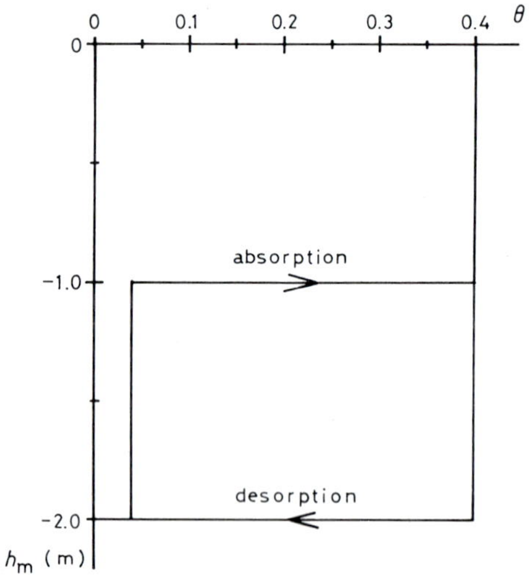

d. The lower bottle-neck will refill with water as soon as h_m becomes larger than its air-entry value, i.e. at $h_m = -2.0$ m.

The belly will refill at $h_m = \frac{-2\sigma}{\rho_l gr} = \frac{-2 \times 7 \times 10^{-2}}{1000 \times 10 \times 14 \times 10^{-6}}$ m $= -1.0$ m.

As soon as the belly is filled, the upper bottle-neck will also fill immediately, because its air-entry value is lower than -1.0 m. When the whole pore is filled, $\theta = 0.4$. When only the lower bottle-neck is filled,

$$\theta = \frac{\frac{1}{4}\pi a^2 \times 2a}{12.5\pi a^3} = \frac{1}{25} = 0.04$$

The corresponding absorption curve is drawn in the same figure.

e. As indicated under c, the meniscus at the top will change gradually from flat to a hemisphere. At maximum curvature the decrease in water content, associated with this curvature, is

$$\frac{1}{2} \times \frac{4}{3} \times \pi \times (\tfrac{1}{2}a)^3 = \frac{1}{12}\pi a^3$$

This is 1/60-th of the total pore volume of $5\pi a^3$. Thus, the vertical branch of the desorption curve should deflect gradually from $\theta = 0.4$ to $\theta = 0.393$ at $h_m = -2.0$ m. This is barely more than the accuracy with which the figure can be drawn. The same holds true, of course, for the absorption curve.

Answer 3.30 The smoother absorption and desorption curves are due to the fact that real soils have pores of many different sizes and shapes, which change gradually into each other.

Answer 3.31 The volume fraction of water available in soil 1 for plant use is 0.09. Integrated over a root zone of 0.6 m this is equivalent to a rainfall of 0.09×600 mm $= 54$ mm. The equivalents for the various soils are tabulated in the last column.

Soil	θ at $pF = 2$	θ at $pF = 4.2$	$\Delta\theta$	equivalent mm rainfall
1	0.11	0.02	0.09	54
3	0.31	0.09	0.22	132
6	0.60	0.18	0.42	252
7	0.47	0.26	0.21	126
9	0.50	0.36	0.14	84

Answer 3.32 a. Figure 3.12 gives for soil 4:

Wilting point, $pF = 4.2$, $\theta \approx 0.10$

Field capacity, $pF = 2.0$, $\theta \approx 0.35$

Thus, to wet this soil from wilting point to field capacity requires filling a volume fraction of about $0.35 - 0.10 = 0.25$ with water. The penetration depth of the rainshower is then 8 mm/0.25 = 32 mm.

b. 0.3 m topsoil is wetted to field capacity by a rainshower of 0.25×300 mm $= 75$ mm. Per hectare this is a volume of 10^4 m$^2 \times 75 \times 10^{-3}$ m $= 750$ m^3.

Answer 3.33 a. At static equilibrium, $-h_m$ is equal to the height above the groundwater table. From the definition $pF = {}^{10}\log(-h_m/\text{cm})$ it follows:
height above groundwater table $= -h_m = 10^{pF}$ cm.

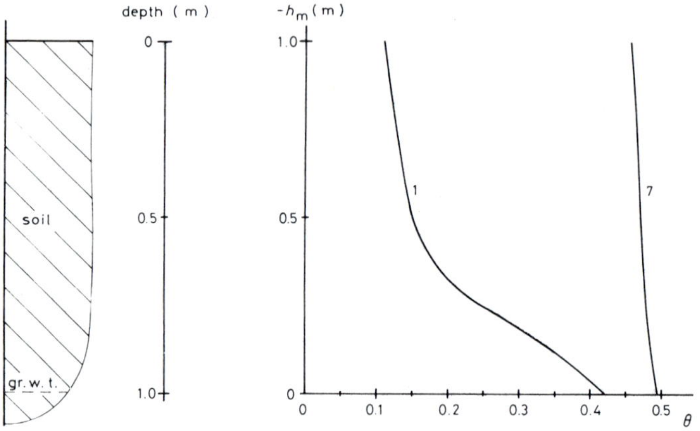

b/c.

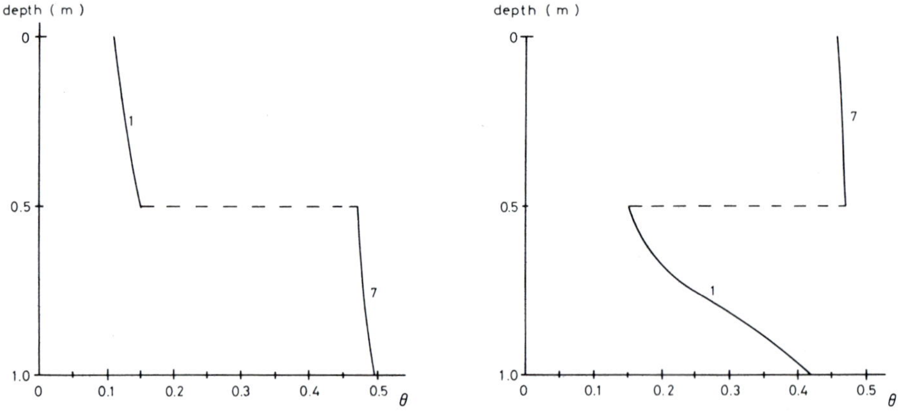

d. Even though there is static equilibrium and soil 7 has lower values of h_m than soil 1, soil 7 is much wetter than soil 1. This result illustrates that in an inhomogeneous soil the water content may be discontinuous. The matric potential, however, is always continuous.

Answer 3.34 a. The matric head of the soil sample varies from -1.00 m at the bottom to -1.05 at the top of the sample. This difference usually causes a negligible difference in water content of the top and bottom. When the reference for z is chosen half way the height of the soil sample, the average water content should give a satisfactory result.

b. $pF = 1.0$ when $h_m = -10$ cm. So the overflow should be kept at 10 cm below the soil sample. With the set-up of Figure 3.15 it is physically impossible to do this for the whole sample. At this high value of h_m the difference between the top and bottom of the soil sample is no longer negligible. The thickness of the soil sample should then be limited to about 1 cm.

Answer 3.35 At $pF = 4.2$, the matric head is $-10^{4.2}$ cm ≈ -158.5 m. Using Equations 3.20 and 3.26, one finds for the air pressure required:

$p_a = -\rho_l g h_m \approx 1.6 \times 10^6$ Pa.

Answer 3.36 In pure water $\Pi = 0$; at the 'free' water surface under atmospheric pressure $p_t = 0$. Thus, according to Equation 3.23, $\psi_t = gz$.

Answer 3.37 a. For the conditions stated, Equation 3.40 is valid, and at 15 °C thus also Equations 3.41 to 3.43. At the free surface of the NaCl-solution, $p_m = 0$. Hence, according to Equation 3.42,

$$\frac{p}{p_0} = \exp(7.5 \times 10^{-9}\ \text{Pa}^{-1} \times (-3 \times 10^7\ \text{Pa})) = \exp(-0.225) \approx 0.80 \text{ or } 80\%.$$

b. If the salt concentration of the soil water is negligible, the osmotic pressure head h_o of the soil solution is negligible and the total potential of the soil water is composed only of the matric potential and the gravitational potential.

Thus the matric potential of the soil water is equal to the osmotic potential of the solution, or -300 bar. Therefore:

$$h_m = -\frac{300 \times 10^5\ \text{Pa}}{\rho_l g} = -3000 \text{ m and}$$

$$pF = {}^{10}\log\left(\frac{-h_m}{\text{cm}}\right) = {}^{10}\log\left(\frac{3000\ \text{m}}{\text{cm}}\right) \approx 5.48$$

Answer 3.38 a. When the relative humidity is 50%, then according to Equation 3.43:

$$h_m + h_o = \frac{\ln 0.50}{7.5 \times 10^{-5}\ \text{m}^{-1}} = -9.24 \times 10^3 \text{ m}$$

If the salt in the soil sample is negligible, then $h_o = 0$ and

$$pF = {}^{10}\log\left(\frac{-h_m}{\text{cm}}\right) = {}^{10}\log\left(\frac{9.24 \times 10^3\ \text{m}}{\text{cm}}\right) = 5.97$$

b. $$pF = {}^{10}\log\left(\frac{-h_m}{\text{cm}}\right) = {}^{10}\log\left(\frac{-100\, h_m}{\text{m}}\right) =$$

$$ {}^{10}\log\left(\frac{-100 \ln(p/p_0)}{7.5 \times 10^{-5}}\right) \approx 6.12 + {}^{10}\log(-\ln(p/p_0))$$

c. A tensiometer can not register pF values above 3. The water in the tensiometer will evaporate from the porous cup until the whole tensiometer is filled with air only.

Answer 3.39 Using Equation 3.43 the following table is easily composed:

pF	h_m (m)	$\ln (p/p_0)$	p/p_0
0	$-$ 0.01	-7.5×10^{-7}	1.0000
1	$-$ 0.10	-7.5×10^{-6}	1.0000
2	$-$ 1.0	-7.5×10^{-5}	0.9999
3	-10	-7.5×10^{-4}	0.9993
4	-10^2	-7.5×10^{-3}	0.993
5	-10^3	-7.5×10^{-2}	0.928
5.5	-3.2×10^3	-0.237	0.789
6	-10^4	-0.750	0.472
6.5	-3.2×10^4	-2.37	0.093
7	-10^5	-7.50	0.0006

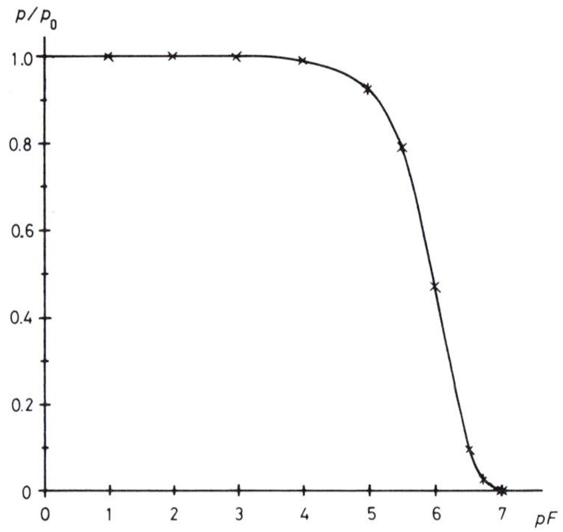

Answer 3.40 a. At equilibrium the salt concentration will be the same in sample A and in the soil column, because the salt can diffuse freely.
The condition for equilibrium is, therefore:
$\psi_h =$ uniform, or $H =$ uniform.
Since $p_a = 0$, $h = h_m$ and $H = h_m + z =$ uniform.
Thus at the same height, h_m is the same in the sample as in the soil column. Hence, $h_m = -1.0$ m, since sample A is located 1.0 m above the groundwater table (neglecting the thickness of the sample).
b. In the case of sample B, salt diffusion is not possible but water can diffuse between the sample and the column through the vapour phase. It is through the vapour phase that equilibrium is attained. The condition for equilibrium is, therefore:
ψ_t (sample) $= \psi_t$ (vapour) $= \psi_t$ (column surface)
Neglecting the salt concentration in sample B and the small differences in height:
ψ_m (sample) $= \psi_m + \psi_o$ (column surface), or

h_m (sample) $= h_m + h_o$ (column surface)
h_m (column surface) $= -1.0$ m
The NaCl is fully dissociated, and thus
$c = 2 \times 50 \text{ mol m}^{-3} = 100 \text{ mol m}^{-3}$.

Then $h_o = -\dfrac{\Pi}{\rho_l g} = -\dfrac{RTc}{\rho_l g} = -\dfrac{2.4 \times 10^3 \times 100}{10^3 \times 10} \text{ m} = -24 \text{ m}$

Hence, h_m (sample) $= -25$ m and $pF = {}^{10}\log\left(\dfrac{-h_m}{\text{cm}}\right) = {}^{10}\log 2500 \approx 3.4$.

c. Using Equation 3.43, $\ln(p/p_0) = 7.5 \times 10^{-5} \times (-25) \approx -18.8 \times 10^{-4}$.
Hence $p/p_0 \approx 0.998$.

Answer 3.41 Irrespective of the extension of the double layer, the charge of a clay particle is always fully compensated by counterions. So, outside the double layer there is no electrostatic field.

Answer 3.42 The reference point for the gravitational potential is taken at the soil surface, and thus everywhere ${}^b\rho_d\psi_t = 0$.
Then, according to Equations 3.52 and 3.53:
$\sigma_s = {}^b\rho_d\psi_p = -{}^b\rho_d\psi_g = -{}^b\rho_d g z$
With this the soil potentials are determined as indicated in the diagram.

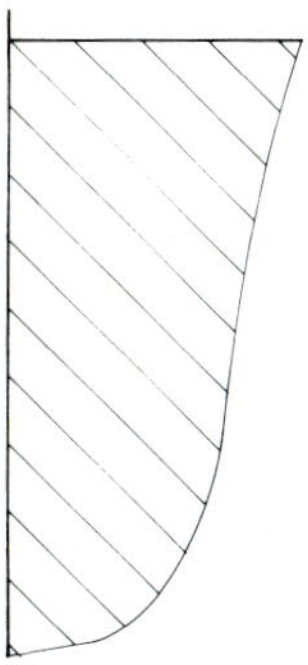

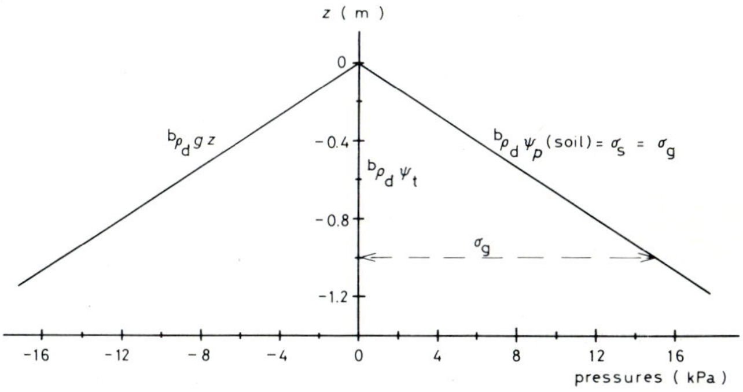

At a depth of 1 m:
$\sigma_s = -{}^b\rho_d g z = -1500 \times 10 \times (-1) \text{ N m}^{-2} = 15 \text{ kPa}$.

Answer 3.43 In a dry soil $\sigma_g = \sigma_s$. Thus at 1 m depth σ_g is also 15 kPa.

Answer 3.44 a. ${}^b\rho_w = (1500 + 400) \text{ kg m}^{-3} = 1900 \text{ kg m}^{-3}$.
Then, similar to Answer 3.42, at 1 m depth:
$\sigma_s = -{}^b\rho_w g z = -1900 \times 10 \times (-1) \text{ Pa} = 19 \text{ kPa}$.
Also at 1 m depth, $\sigma_w = -\rho_l g z = -1000 \times 10 \times (-1) \text{ Pa} = 10 \text{ kPa}$.
At the soil surface $\sigma_s = \sigma_w = 0$. With this information the diagram can be drawn.

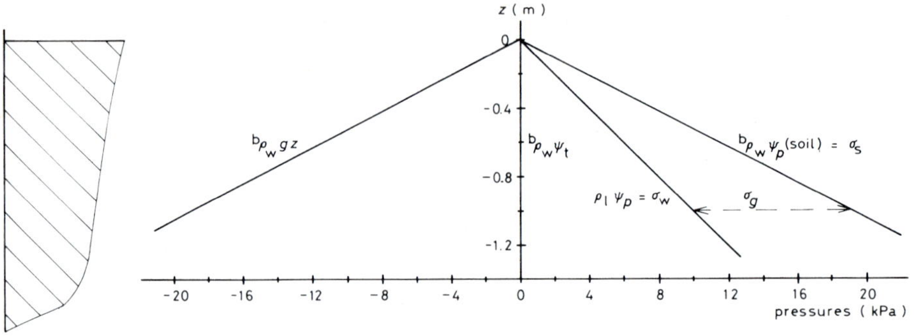

b. According to Equation 3.55, at 1 m depth, $\sigma_g = \sigma_s - \sigma_w = (19 - 10)$ kPa = 9 kPa.

c. Compared with Answer 3.43, the intergranular pressure at 1 m depth has decreased from 15 kPa to 9 kPa. Thus the bearing capacity of the wet soil is lower than that of the dry soil.

Answer 3.45 If there is a 0.10 m layer of water on the soil surface, both σ_s and σ_w increase by 1 kPa. Hence, σ_g remains everywhere the same, as in Answer 3.44. At 1 m depth it is still 9 kPa and the soil surface it is zero. This is shown in the diagram.

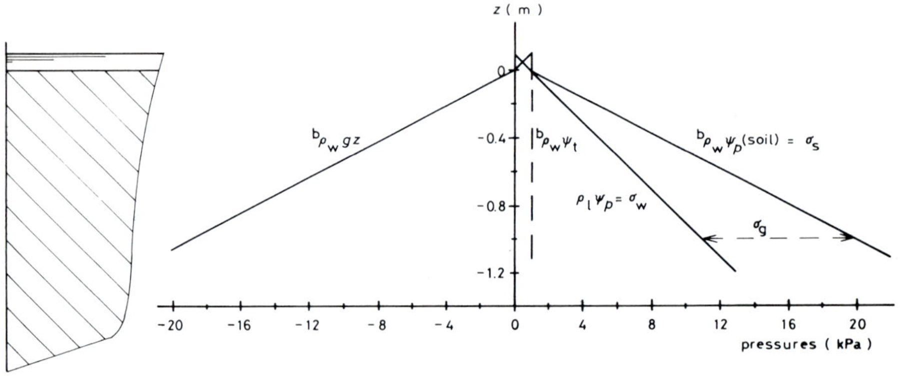

Thus, at static equilibrium, the intergranular pressure is zero at the soil surface, both in a completely dry soil and in a submerged soil.

Answer 3.46 At the soil surface, $\sigma_s = 0$ and thus, according to Equation 3.57, $\sigma_g = -\dfrac{\theta}{\phi}\sigma_w$.

Also at the soil surface, the soil is still saturated, so that $\theta/\phi = 1$ and $\sigma_w = -2$ kPa. Hence at the soil surface, $\sigma_g = 2$ kPa.

4 General concepts of transport processes in soil

So far we have considered soils, and the substances and phenomena occurring within them, as being at static equilibrium. This simplification is seldom, if ever, valid. In nature, conditions change continually due to influences from within, as well as from outside the soil. Disturbances of equilibrium conditions tend to be dissipated until new equilibrium is attained. This occurs by so-called transport processes which may involve water, heat, gases, solutes, etc. The prime concern of soil physics is to describe, measure, predict and possibly control these transport processes.

All transport processes have a number of aspects in common. To emphasize this basic similarity we introduce in this chapter concepts which can be used to describe most transport processes occurring in soils. We do this using flow of water as an example, since this forms a logical extension of the subject matter presented in the preceding chapters. We then apply these concepts also to the diffusion of gases and the conduction of heat in soils, respectively. Each of the three transport processes is treated more extensively in subsequent chapters.

4.1 FLOW OF WATER

4.1.1 Movement and accumulation

Flow of water in soils involves two distinct aspects:

- movement, which is the change of position of water with respect to a chosen frame of reference, usually the soil matrix
- accumulation, which is the increase or decrease in time of the amount of water at a given position. Accumulation normally occurs as a result of flow under influence of changing external conditions. It can also be due to exchange with other phases, local production or consumption (negative accumulation), etc.

Question 4.1 Give examples of the latter type of accumulation of water in soils.

4.1.2 Driving force. Hydraulic potential gradient

When a system is at static equilibrium there is, by definition, no transport of matter taking place in that system. Transport occurs when equilibrium conditions are disturbed. A condition for static equilibrium was given in Chapter 2:

$$\Sigma \boldsymbol{F}^{s} = 0 \rightleftharpoons \text{static equilibrium} \tag{2.3}$$

It follows that $\Sigma \boldsymbol{F}^{s}$ can be considered as the driving force for a transport process ensuing from a disturbance from static equilibrium:

$$\text{driving force} = \Sigma \boldsymbol{F}^{s} \tag{4.1}$$

Conditions for static equilibrium can also be expressed in terms of the total potential ψ_t (Equation 2.30).

For the flow of water in soil, in general, only the hydraulic potential ψ_h needs to be considered. As defined in Equation 2.38 the hydraulic potential is the sum of the pressure potential, ψ_p, and the gravitational potential, ψ_g. If Equation 2.24 is applied to $\Sigma \boldsymbol{F}^s$, the corresponding potential in the case of flow of water in soil is ψ_h, and thus

$$\frac{\Sigma \boldsymbol{F}^{s}}{m} = -\frac{\partial \psi_h}{\partial s} \tag{4.2}$$

Equations 4.1 and 4.2 indicate that the driving force divided by mass for flow of water in soil is in magnitude equal to the gradient of the hydraulic potential. The minus sign indicates that the driving force, and the ensuing flow of water, are directed towards decreasing values of the hydraulic potential.

In Chapter 2 it was shown that hydraulic potentials can be defined not only as energy divided by mass (ψ_h in J kg^{-1}), but also as energy divided by volume (p_h in J m^{-3} = Pa) or as energy divided by weight (H in J N^{-1} = m). Similarly, hydraulic potential gradients can be expressed per mass, per volume or per weight.

Question 4.2 What is the unit of hydraulic potential gradient expressed per mass, per volume and per weight?

From Answer 4.2 it follows that, similar to Equation 4.2, for flow of water in soil is also true:

$$\text{driving force divided by volume} = -\frac{\partial p_h}{\partial s} \tag{4.3}$$

$$\text{driving force divided by weight} = -\frac{\partial H}{\partial s} \tag{4.4}$$

4.1.3 Flux density

The rate of movement of water in soil is generally expressed as a flux density. The flux density, f_w, of water moving in the direction s is the amount of water passing through a plane perpendicular to the direction s during a time interval, divided by the area of that plane and the magnitude of the time interval. In soil physics the term 'flux' is used often when 'flux density' is meant. Use of the term 'flux' should be restricted to volume of water divided by time.

The amount of water, like potentials, can be expressed as a mass, a volume or a weight. The flux density of water flowing in soil is usually expressed as a volume.

The symbol commonly used for the volume flux density is q.

Question 4.3 Define the flux density on volume basis for the flow of water in soil and state its dimension (unit).

4.1.4 Flux density equation.

The relationship between the driving force and the ensuing flux density depends on the transport mechanism and the conditions under which it takes place. We will call this relationship the flux density equation. Flow of water in soils is dominated by the large friction force per volume of water between the soil and the water. This determines, as will be shown next, the form of the flux density equation.

When a driving force, $\Sigma \boldsymbol{F}^{\mathrm{s}}$, starts acting on a mass of water at rest in a soil, the water will start moving with an acceleration according to Newton's second law of motion (Equation 2.1b). Because of this motion the water will experience drag or friction forces as discussed in Section *2.1.1*.

In Section *2.1.4* the drag force on a spherical particle falling through a liquid at rest was, according to Stokes' law, proportional to the sedimentation velocity. For Stokes' law to be valid the conditions around the falling spherical particle must remain laminar. This means that the velocity must remain so small that turbulence does not occur. Similarly, for small flow velocities of water in soils the drag forces between the water molecules and the solid surfaces will be proportional to the flow velocity. The area of contact between water and soil particles is inherently very large. This means that drag forces divided by mass of water increase very fast with flow velocity. On the other hand, driving forces acting on water in soil are usually not very large. As a result, the drag force acting on a mass of water as it increases with velocity, will soon reach a magnitude equal to that of the driving force. Since these forces act in opposite directions the net force at that instance will become zero. The water will no longer be accelerated but maintain its attained velocity until the balance of forces is disturbed. For all normal flow of water in soil the average 'terminal' velocity, $\overline{v}$, is much smaller than the value at which the drag force would no longer be proportional to the velocity. Thus, when $\Sigma \boldsymbol{F}^{\mathrm{d}} = -\Sigma \boldsymbol{F}^{\mathrm{s}}$,

$$\overline{v} = \text{constant and proportional to } \frac{-\Sigma \boldsymbol{F}^{\mathrm{d}}}{m} = \frac{\Sigma \boldsymbol{F}^{\mathrm{s}}}{m} \tag{4.5}$$

We can derive similar relationships when the forces are expressed per volume or per weight. Using equations 4.3 and 4.4, and the fact that the flux density, q, is proportional to $\overline{v}$, we obtain the flux density equations, also called Darcy's law:

$$q = -K\frac{\partial p_{\mathrm{h}}}{\partial s} = -K\left(\frac{\partial p_{\mathrm{t}}}{\partial s} + \rho_{\mathrm{l}} g \frac{\partial z}{\partial s}\right) \tag{4.6}$$

$$q = -k\frac{\partial H}{\partial s} = -k\left(\frac{\partial h}{\partial s} + \frac{\partial z}{\partial s}\right) \tag{4.7}$$

where K and k are proportionality coefficients.

The last parts of Equations 4.6 and 4.7 are obtained by substituting Equations 3.49 and 2.44, respectively. From the derivation of the latter equations it is obvious that the first terms represent the contribution of the tensiometer pressure potential gradient to the flux density and the second terms that of the gravitational potential gradient. From this it follows that for horizontal flow the second terms vanish, since then the gravitational potential does not change. The flux density equation in terms of ψ_h is omitted, since it is seldom used.

4.1.5 Transport coefficient. Hydraulic conductivity

The proportionality coefficients K and k in Equations 4.6 and 4.7 are called hydraulic conductivity, because they are a measure of the ability of the soil to conduct the flow of water. They are special cases of what generally is called the transport coefficient of a medium for a particular transport process. Its dimension depends on the choices of the dimension of the other variables in the flux density equation. The relationship between K and k follows immediately from the second terms on the right hand side of Equations 4.6 and 4.7, since each represents the contribution of the gravitational potential to the flux density:

$$k = \rho_l g K \tag{4.8}$$

Question 4.4 What are the units of K and k?

In some transport processes, e.g. heat conduction in a solid material (Section 4.3), the transport coefficient is nearly independent of the potential of the transported entity. The hydraulic conductivity, however, varies strongly with water content and thus with the matric potential.

Question 4.5 Why does the hydraulic conductivity of a soil vary with the matric potential?

The hydraulic conductivity also differs greatly between soils. The reason is that the friction forces acting on the flowing water depend not so much on the total amount of water in the soil, but on the geometry of the liquid-filled pore space. This, in turn, depends on the pore-size distribution of the soil, and thus on the soil texture and soil structure. Finally, the hydraulic conductivity depends also on the friction between water molecules and thus on the viscosity of the water. Although the viscosity varies significantly with temperature, this factor is usually neglected.

Because the hydraulic conductivity varies over many orders of magnitude between air-dryness and saturation, it is very difficult to solve practical soil water flow problems. This will be discussed further in Chapter 5.

4.1.6 Continuity equation

Transport processes always obey the laws of conservation of matter and energy. Mathematical formulations of transport processes must reflect these laws. Soil water flow problems are usually considered to be isothermal and the law of conservation of energy can then be omitted. The law of conservation of matter is expressed in the so-called continuity equation. For one-dimensional flow of water in soil the continuity equation can be written as:

$$\frac{\partial \theta}{\partial t} = -\frac{\partial q}{\partial s} \qquad (4.9)$$

where θ is the volume fraction of water (Section *1.1.5*).

Question 4.6 Derive Equation 4.9, using Figure 4.1.

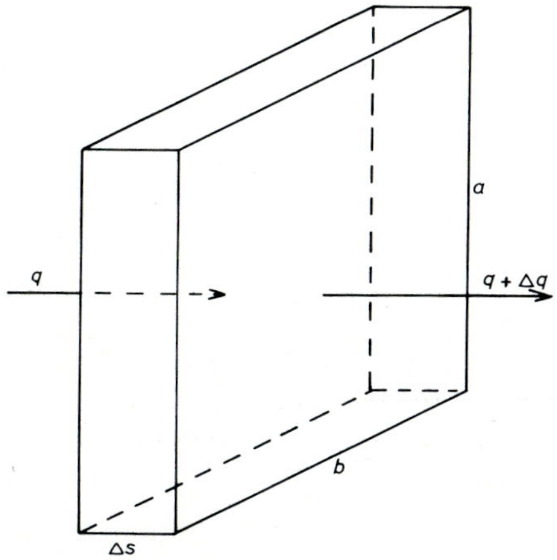

Figure 4.1 Volume element for deriving continuity equation.

Question 4.7 If the flux density at s is q_1 and at $s + \Delta s$ is q_2, fill in either <, = or > in the following statements:

a. if $q_2 > q_1$, then $\frac{\partial q}{\partial s} \ldots 0$ and $\frac{\partial \theta}{\partial t} \ldots 0$

b. if $q_2 = q_1$, then $\frac{\partial q}{\partial s} \ldots 0$ and $\frac{\partial \theta}{\partial t} \ldots 0$

c. if $q_2 < q_1$, then $\frac{\partial q}{\partial s} \ldots 0$ and $\frac{\partial \theta}{\partial t} \ldots 0$

In some transport situations, positive and negative accumulation of a component are due not only to changes in flux density in the direction of transport, but also to local additions to or extractions from outside the system (Answer 4.1). For the flow of water in soils the proper continuity equation is then

$$\frac{\partial \theta}{\partial t} = -\frac{\partial q}{\partial s} + \alpha \qquad (4.10)$$

where α is the volume of water added or extracted divided by volume of soil and by time. Often the production term, α, is small relative to the other terms in the continuity equation and can be ignored. This will be done in the rest of this chapter.

Question 4.8 Give a unit analysis of Equation 4.10.

4.1.7 *Differential capacity*

The flux density equation and the continuity equation must both be satisfied in solving a particular water flow problem. These equations involve the four variables θ, p_t (or h), t and s, while s may be composed of up to three coordinates. Obtaining solutions of water flow problems can often be made easier by eliminating either θ or p_t (or h) by introducing a differential capacity. A familiar example of differential capacity is the electrical capacity of a conductor, which is the ratio of a change in its electrical charge and the resulting change in its electrical potential. Similarly, the differential capacity of soils for water can be defined as

$$C = \mathrm{d}\theta/\mathrm{d}p_m \qquad (4.11)$$

or $$c = \mathrm{d}\theta/\mathrm{d}h_m \qquad (4.12)$$

depending on the form in which the matric potential is used. One cannot use the hydraulic potential for this purpose, because it includes the gravitational potential. The tensiometer pressure potential cannot be used either, because it includes the pneumatic potential. Both potentials have no direct influence on the volume fraction of water (Section *3.1.5*). However, normally the gas phase in the soil is at atmospheric pressure, so that $p_m = p_t$ and $h_m = h$.

Question 4.9 Which relationship can be used to determine the differential water capacity of a soil?

Question 4.10 a. Is the differential water capacity dependent on the volume fraction of water?

b. What value has the differential water capacity of a soil at saturation?

Question 4.11 How can c be determined from the slope of a pF curve?

4.1.8 *Diffusivity*

Transport problems can often be solved more easily by introducing a diffusivity. This is the ratio of the transport coefficient and the differential capacity. The soil water diffusivity is defined as

$$D = \frac{K}{C} = \frac{K}{\dfrac{d\theta}{dp_m}} = K\frac{dp_m}{d\theta} \tag{4.13}$$

$$\text{or}\quad D = \frac{k}{c} = \frac{k}{\dfrac{d\theta}{dh_m}} = k\frac{dh_m}{d\theta} \tag{4.14}$$

Since $p_m = \rho_l g h_m$ and $K = k/\rho_l g$ (Equation 4.8) the diffusivities in Equations 4.13 and 4.14 are identical.

Question 4.12 Is D a function of the volume fraction of water?
Question 4.13 What is the value of D in a region where the soil is saturated?

If the pressure in the soil gas phase is atmospheric, so that $p_t = p_m$ and $h = h_m$, the flux density equations 4.6 and 4.7 can be rewritten, respectively, as follows:

$$q = -K\left(\frac{dp_m}{d\theta}\frac{\partial\theta}{\partial s} + \rho_l g\frac{\partial z}{\partial s}\right) = -D\frac{\partial\theta}{\partial s} - \rho_l g K\frac{\partial z}{\partial s} \tag{4.15}$$

$$q = -k\left(\frac{dh_m}{d\theta}\frac{\partial\theta}{\partial s} + \frac{\partial z}{\partial s}\right) = -D\frac{\partial\theta}{\partial s} - k\frac{\partial z}{\partial s} \tag{4.16}$$

These equations show why the diffusivity is especially useful for horizontal flow problems. Then the second terms vanish and the flux density is proportional to the gradient of the volume fraction of water.

Question 4.14 What does this suggest about the transport mechanism of the flow of water in soils? Is this correct?
Question 4.15 Which flux density equation would you use in an inhomogeneous soil?
Question 4.16 Which effect does hysteresis have on the validity of these flux density equations?

4.1.9 *General flow equations*

To solve a particular water flow problem, the flux density equation best suited for the particular problem can be combined with the continuity equation to obtain a general flow equation. For example, substitution of flux density Equation 4.6 or 4.7 into continuity Equation 4.9 yields, respectively,

$$\frac{\partial\theta}{\partial t}=\frac{\partial}{\partial s}\left(K\frac{\partial p_\mathrm{t}}{\partial s}\right)+\rho_\mathrm{l}g\frac{\partial}{\partial s}\left(K\frac{\partial z}{\partial s}\right)\tag{4.17}$$

$$\frac{\partial\theta}{\partial t}=\frac{\partial}{\partial s}\left(k\frac{\partial h}{\partial s}\right)+\frac{\partial}{\partial s}\left(k\frac{\partial z}{\partial s}\right)\tag{4.18}$$

If the soil air is at atmospheric pressure ($p_\mathrm{t}=p_\mathrm{m}$ and $h=h_\mathrm{m}$) the left hand sides can be written as

$$\frac{\partial\theta}{\partial t}=\frac{\mathrm{d}\theta}{\mathrm{d}p_\mathrm{m}}\frac{\partial p_\mathrm{m}}{\partial t}=\frac{\mathrm{d}\theta}{\mathrm{d}h}\frac{\partial h_\mathrm{m}}{\partial t}=C\frac{\partial p_\mathrm{t}}{\partial t}=c\frac{\partial h}{\partial t}\tag{4.19}$$

For vertically upward flow, for which $\mathrm{d}s=\mathrm{d}z$, Equations 4.17 and 4.18 become, respectively,

$$C\frac{\partial p_\mathrm{t}}{\partial t}=\frac{\partial}{\partial z}\left(K\frac{\partial p_\mathrm{t}}{\partial z}\right)+\rho_\mathrm{l}g\frac{\partial K}{\partial z}\tag{4.20}$$

$$c\frac{\partial h}{\partial t}=\frac{\partial}{\partial z}\left(k\frac{\partial h}{\partial z}\right)+\frac{\partial k}{\partial z}\tag{4.21}$$

Question 4.17 Show that these equations remain the same for vertically downward flow.

For horizontal flow, $\partial z/\partial s=0$, and thus the second terms on the right hand side of Equations 4.17 and 4.18 vanish. The first terms represent the contribution of the tensiometer pressure potential gradient, as was already discussed in connection with Equations 4.6 and 4.7.

One can also substitute the diffusivity-type flux density Equations 4.15 or 4.16 into continuity Equation 4.9 and obtain, respectively,

$$\frac{\partial\theta}{\partial t}=\frac{\partial}{\partial s}\left(D\frac{\partial\theta}{\partial s}\right)+\rho_\mathrm{l}g\frac{\partial}{\partial s}\left(K\frac{\partial z}{\partial s}\right)\tag{4.22}$$

$$\frac{\partial\theta}{\partial t}=\frac{\partial}{\partial s}\left(D\frac{\partial\theta}{\partial s}\right)+\frac{\partial}{\partial s}\left(k\frac{\partial z}{\partial s}\right)\tag{4.23}$$

Again, for horizontal flow the second terms vanish, after which the two equations are identical. The reason for this is of course that, by using the diffusivity, the pressure potential has been eliminated in favour of the volume fraction of water, θ. A substitution similar to Equation 4.19 is not needed, since the number of variables is already reduced to three.

4.1.10 *Initial and boundary conditions*

The general flow equations discussed in the preceding section are mathematical representations which are valid for most water flow processes in soils. However, they do not yet specify a particular flow problem. The mathematical description of a particular flow problem must include specifications on boundary conditions and, in the case of time-variant problems, also initial conditions. These are usually values of water content, pressure potential or flux density at certain positions and at certain times.

Question 4.18 Give examples of initial and boundary conditions for water flow problems in soil.

The solution of a particular flow problem differs yet from soil to soil, depending on the soil hydraulic properties. Thus, a particular soil water flow problem can only be described completely when also the soil water characteristic (Section *3.1.6*) and the soil hydraulic conductivity or diffusivity are specified as functions of water content or matric potential. Particular soil water flow problems are discussed in Chapter 5.

4.2 GAS DIFFUSION

In soils, the gas pressure generally is constant and equal to atmospheric pressure. Hence, gas components (e.g. O_2, CO_2) move only due to gradients of their partial pressures. Such transport is known as gas diffusion.

The driving force for gas diffusion of a component *i* can be represented by the partial pressure gradient $\partial p_i/\partial s$, where p_i is the pressure due to gaseous component *i*. One can also express the driving force as a gradient of the mass concentration, χ_i, which is the mass of component *i* divided by volume of gas (kg m^{-3}). χ_i is proportional to the partial pressure at low total gas pressures. Experimentally, it is easier to measure χ_i than p_i.

Question 4.19 a. What law describes the proportionality between partial pressure and mass concentration?

b. Derive from this law the relationship between p_i and χ_i.

Using mass concentration gradient as driving force, the flux density equation for gas diffusion is

$$f_i = -D_g \frac{\partial \chi_i}{\partial s} \tag{4.24}$$

Equation 4.24 is called Fick's law, and the transport coefficient D_g is the diffusion coefficient for the particular gaseous component in the particular soil. If the flux density, f_i, is expressed in kg m^{-2} s^{-1}, the unit of D_g is m^{-2} s^{-1} (check this).

The continuity equation is then

$$\frac{\partial G_i}{\partial t} = -\frac{\partial f_i}{\partial s} (+\alpha) \tag{4.25}$$

where G_i is the amount of component i divided by volume of soil, in kg m^{-3}.

The differential capacity of the soil for component i is:

$$\frac{dG_i}{d\chi_i} = \phi_g \tag{4.26}$$

where ϕ_g is the air-filled porosity.

Question 4.20 Derive Equation 4.26.

The general transport equation for gas diffusion in soils is obtained by combining Equations 4.24, 4.25 and 4.26:

$$\frac{\partial \chi_i}{\partial t} = \frac{1}{\phi_g} \frac{\partial}{\partial s} \left(D_g \frac{\partial \chi_i}{\partial s} \right) \tag{4.27}$$

4.3 HEAT CONDUCTION

Heat has no mass and thus the formulation of the driving force for heat conduction cannot be the same as that in Equations 4.2, 4.3 and 4.4. The similarity with material transport can be maintained by choosing as driving force the temperature gradient, $\partial T/\partial s$ (K m^{-1}). As unit of heat flux density, f_h, can be taken J m^{-2} s^{-1} (= W m^{-2}). The flux density equation for heat conduction in soils is then

$$f_h = -\lambda \frac{\partial T}{\partial s} \tag{4.28}$$

The transport coefficient λ is called the heat conductivity and has the unit J m^{-1} s^{-1} K^{-1} (= W m^{-1} K^{-1}). Equation 4.28 is called Fourier's law.

Question 4.21 Check the unit of heat conductivity.

The continuity equation for heat flow is:

$$\frac{\partial Q}{\partial t} = -\frac{\partial f_h}{\partial s} (+\alpha) \tag{4.29}$$

where Q is the heat content divided by volume of the soil in J m^{-3} (check this). Combination of Equations 4.28 and 4.29 and introduction of the differential heat capacity $C_h = dQ/dT$ lead to the general heat conduction equation:

$$\frac{\partial T}{\partial t} = \frac{1}{C_h} \frac{\partial}{\partial s} \left(\lambda \frac{\partial T}{\partial s} \right) \tag{4.30}$$

where C_h, normally called volumic heat capacity, is in J m^{-3} K^{-1}.

Temperatures can be measured easily, as well as λ and C_h, since they are nearly independent of temperature. The latter makes heat conduction problems much easier to solve than water flow problems.

If λ is independent of s, and thus also of T (homogeneous soils with uniform water content), this equation can be simplified to

$$\frac{\partial T}{\partial t} = \frac{\lambda}{C_h}\frac{\partial^2 T}{\partial s^2} = D_h \frac{\partial^2 T}{\partial s^2} \tag{4.31}$$

The heat diffusivity D_h is a measure for the rate at which an imposed temperature gradient in the soil is dissipated by conduction.

Question 4.22 Derive the unit of D_h.

The similarities and differences between the transport processes discussed in this chapter, are summarized in Tables 4.1 and 4.2.

4.4 SUGGESTED LITERATURE

BIRD, R. B., W. E. STEWART and E. N. LIGHTFOOT, 1960. Transport Phenomena. Wiley, 780 p.
CARSLAW, H. S. and J. C. JAEGER, 1959. Conduction of Heat in Solids. Oxford Univ. Press, 510 p.
CRANK, J., 1964. The Mathematics of Diffusion. Oxford Univ. Press, 347 p.

Table 4.1 Summary of equations for water flow, gas diffusion and heat conduction (without production).

	water flow	gas diffusion	heat conduction
a. flux density equation	$q=-K\frac{\partial p_h}{\partial s}=-k\frac{\partial H}{\partial s}$	$f_i=-D_g\frac{\partial \chi_i}{\partial s}$	$f_h=-\lambda\frac{\partial T}{\partial s}$
b. continuity equation	$\frac{\partial\theta}{\partial t}=-\frac{\partial q}{\partial s}$	$\frac{\partial G_i}{\partial t}=-\frac{\partial f_i}{\partial s}$	$\frac{\partial Q}{\partial t}=-\frac{\partial f_h}{\partial s}$
c. general transport equation	$\frac{\partial\theta}{\partial t}=\frac{\partial}{\partial s}\left(K\frac{\partial p_h}{\partial s}\right)=\frac{\partial}{\partial s}\left(k\frac{\partial H}{\partial s}\right)$	$\frac{\partial G_i}{\partial t}=\frac{\partial}{\partial s}\left(D_g\frac{\partial \chi_i}{\partial s}\right)$	$\frac{\partial Q}{\partial t}=\frac{\partial}{\partial s}\left(\lambda\frac{\partial T}{\partial s}\right)$
d. equivalent of c	$C\frac{\partial p_m}{\partial t}=\frac{\partial}{\partial s}\left(K\frac{\partial p_m}{\partial s}\right)$ $c\frac{\partial h_m}{\partial t}=\frac{\partial}{\partial s}\left(k\frac{\partial H}{\partial s}\right)$	$\frac{\partial \chi_i}{\partial t}=\frac{1}{\phi_g}\frac{\partial}{\partial s}\left(D_g\frac{\partial \chi_i}{\partial s}\right)$	$\frac{\partial T}{\partial t}=\frac{1}{C_h}\frac{\partial}{\partial s}\left(\lambda\frac{\partial T}{\partial s}\right)$
e. diffusivity form of c	$\frac{\partial\theta}{\partial t}=\frac{\partial}{\partial s}\left(D\frac{\partial\theta}{\partial s}\right)$ (horizontal only)	same as above	$\frac{\partial T}{\partial t}=\frac{\partial}{\partial s}\left(D_h\frac{\partial T}{\partial s}\right)$
f. e for constant D	not applicable	$\frac{\partial \chi_i}{\partial t}=\frac{D_g}{\phi_g}\frac{\partial^2 \chi_i}{\partial s^2}$	$\frac{\partial T}{\partial t}=D_h\frac{\partial^2 T}{\partial s^2}$

Table 4.2 Symbols and units for parameters used in equations of Table 4.1

parameter	water flow		gas diffusion	heat conduction
flux density	$f_w=q$ $m^3\,m^{-2}\,s^{-1}=m\,s^{-1}$		f_i $kg\,m^{-2}\,s^{-1}$	f_h $J\,m^{-2}\,s^{-1}=W\,m^{-2}$
potential	$p_h=p_t+\rho_l gz$ Pa	$H=h+z$ m	χ_i $kg\,m^{-3}$	T K
transport coefficient	K $m^2\,Pa^{-1}\,s^{-1}$	k $m\,s^{-1}$	D_g $m^2\,s^{-1}$	λ $J\,m^{-1}\,s^{-1}\,K^{-1}$
content	θ $m^3\,m^{-3}$ (dimensionless)		G_i $kg\,m^{-3}$	Q $J\,m^{-3}$
differential capacity	$C=\frac{d\theta}{dp_m}$ Pa^{-1}	$c=\frac{d\theta}{dh_m}$ m^{-1}	$\phi_g=\frac{dG_i}{d\chi_i}$ $m^3\,m^{-3}$ (dimensionless)	$C_h=\frac{dQ}{dT}$ $J\,m^{-3}\,K^{-1}$
diffusivity	$D=\frac{K}{C}$ $m^2\,s^{-1}$	$D=\frac{k}{c}$ $m^2\,s^{-1}$	$\frac{D_g}{\phi_g}$ $m^2\,s^{-1}$	$D_h=\frac{\lambda}{C_h}$ $m^2\,s^{-1}$

Answer 4.1 Exchanges of liquid water in soil with other phases can occur as a result of freezing, thawing, evaporation or condensation. An example of local consumption is the uptake of water by plant roots.

Answer 4.2 The unit of hydraulic potential gradient per mass is J kg^{-1} m^{-1} = N kg^{-1}, i.e. force divided by mass.

The unit of hydraulic potential gradient per volume is J m^{-3} m^{-1} = N m^{-3}, i.e. force divided by volume.

The unit of hydraulic potential gradient per weight is J N^{-1} m^{-1} = N N^{-1}, i.e. force divided by weight. It can also be considered as m m^{-1}, that is as hydraulic head gradient. Both forms are ultimately dimensionless.

Answer 4.3 The flux density of water on volume basis, q, is the volume of water flowing through a specified cross-section of soil perpendicular to the direction of flow during a certain time interval, divided by the area of the cross-section and the magnitude of the time interval. Thus $[q] = m^3\ m^{-2}\ s^{-1} = m\ s^{-1}$, which is the dimension of a velocity. However, q is only an equivalent velocity.

Answer 4.4 $$[K] = \left[\frac{q}{\partial p_h/\partial s}\right] = \frac{m\ s^{-1}}{J\ m^{-3}\ m^{-1}} = \frac{m\ s^{-1}}{Pa\ m^{-1}} = m^2\ Pa^{-1}\ s^{-1}$$

$$[k] = \left[\frac{q}{\partial h/\partial s}\right] = \frac{m\ s^{-1}}{m\ m^{-1}} = m\ s^{-1}$$

Thus k has the dimension of a velocity.

Answer 4.5 As the matric potential decreases, water will be present only in ever smaller pores and in ever thinner films on the soil particles. Thus, the area of contact between the solid phase and the water, and thus the friction forces, per volume of water will increase. This means that the 'terminal' velocity that will be reached with a certain driving force, decreases as the matric potential decreases. Thus the ability of the soil to conduct water flow, the hydraulic conductivity, decreases with the matric potential. While the hydraulic conductivity is a function of the matric potential, it is in the first place a function of the volume fraction of water.

Answer 4.6 The volume of water entering the volume element during time interval Δt is: $qab\Delta t$.

The volume of water leaving the volume element during time interval Δt is: $(q + \Delta q)ab\Delta t$.

The volume of water accumulated in the volume element during Δt then equals $qab\Delta t - (q + \Delta q)ab\Delta t = -ab\Delta q\Delta t$.

This volume of water has accumulated in a volume $ab\Delta s$. Hence:

$$\Delta\theta = \frac{-ab\Delta q\Delta t}{ab\Delta s} \text{ or } \frac{\Delta\theta}{\Delta t} = -\frac{\Delta q}{\Delta s}$$

For infinitesimal time and distance, this finite difference equation yields the continuity equation in differential form:

$$\frac{\partial \theta}{\partial t} = -\frac{\partial q}{\partial s}.$$

Answer 4.7 a. if $q_2 > q_1$, then $\frac{\partial q}{\partial s} > 0$ and $\frac{\partial \theta}{\partial t} < 0$

b. if $q_2 = q_1$, then $\frac{\partial q}{\partial s} = 0$ and $\frac{\partial \theta}{\partial t} = 0$

c. if $q_2 < q_1$, then $\frac{\partial q}{\partial s} < 0$ and $\frac{\partial \theta}{\partial t} > 0$

Answer 4.8 The unit of $\partial\theta/\partial t$ is $m^3\, m^{-3}\, s^{-1} = s^{-1}$
The unit of $\partial q/\partial s$ is $m\, s^{-1}\, m^{-1} = s^{-1}$
The unit of α is $m^3\, m^{-3}\, s^{-1} = s^{-1}$
Thus, the units are all the same, as they should be.

Answer 4.9 The differential water capacity of a soil can be derived from the relationship between the volume fraction of liquid and the matric potential, known as the soil water characteristic. It is the slope (cotangent of inclination) of a graph of this relationship (Figure 3.11).

Answer 4.10 a. The slope of the graph varies with the volume fraction of water. Thus the differential water capacity depends on the volume fraction of water.
b. At saturation, a small change in matric potential does not cause any change of the volume fraction of water. This means that the differential water capacity is zero.

Answer 4.11 From the pF curve one can read $d\theta/d(pF)$, while we have to find $c = d\theta/dh_m$. The relationship between the two differentials can be found using:
$pF = {}^{10}\log(-h_m/\text{cm})$, from which $h_m = -10^{pF}$ cm, and
$d\,10^a = d[e^{\ln 10}]^a = d\,e^{a \ln 10} = \ln 10\, e^{a \ln 10} da = 10^a \ln 10\, da$

$$\text{Thus, } c = \frac{d\theta}{dh_m} = \frac{d\theta}{-1\text{ cm } d\,10^{pF}} = \frac{d\theta}{-1\text{ cm} \times 10^{pF} \ln 10\, d(pF)} = \frac{1}{h_m \ln 10} \frac{d\theta}{d(pF)} = \frac{0.4343}{h_m} \frac{d\theta}{d(pF)}$$

Answer 4.12 Both K and C are a function of the volume fraction of water, and thus D is also. D usually varies less with volume fraction of water than K.

Answer 4.13 Since, at saturation, the differential water capacity is zero (Answer 4.10b), the diffusivity is infinitely large.

Answer 4.14 These flux density equations suggest that the gradient of the volume fraction of water is the driving force for the flow of water in soils. This would make it similar to the diffusion of salts in solution, for which the concentration gradient can be taken as driving force. From this similarity stems the name diffusivity. This representation suggests that the water moves through the soil by the process of diffusion (see also Section 4.2). This is not correct. Water moves under influence of hydraulic potential gradients. Only when the volume fraction of water is a continuous, singlevalued function of the pressure potential can the chain rule for differentials be used to arrive at equations 4.15 and 4.16.

Answer 4.15 In an inhomogeneous soil, such as at a boundary between two layers, the volume fraction of water is discontinuous (Answer 3.33), making $dp_m/d\theta$ and $dh_m/d\theta$ undefined. Then also D is undefined and Equations 4.15 and 4.16 cannot be used. The water potential is never discontinuous, so Equations 4.6 and 4.7 can always be used.

Answer 4.16 When hysteresis occurs, the soil water characteristic is no longer a singlevalued function. This again makes the diffusivity undefined. Thus, Equations 4.15 and 4.16 are valid only in monotonically drying or wetting, homogeneous soils in which the gas pressure is atmospheric.

Answer 4.17 For downward flow, s is positive downward. In the second term of Equations 4.17 and 4.18, z stands for the gravitational potential. Therefore, z must be taken positive upward and $ds = -dz$. In these equations ∂s occurs twice in each term. Upon substitution of $\partial s = -\partial z$, the two minus signs cancel and Equations 4.20 and 4.21 result again.

Answer 4.18 Possible forms of initial and boundary conditions for water flow problems in soil are:

- water content distribution at $t = 0$
- hydraulic potential distribution at $t = 0$
- rainfall intensities as function of time
- rate of evaporation from the soil surface
- zero flux at a specified depth.

Answer 4.19 a. The proportionality is described by the ideal gas law (Equation 2.46).
b. For a gas component i in a gas mixture of total volume V, the relationship between the mass concentration χ_i and the partial pressure p_i is, according to this law,

$$\chi_i = \frac{m_i}{V} = \frac{nM_i}{V} = \frac{M_i}{RT} p_i$$

Answer 4.20 Let A_i be the amount of gas i in a soil with volume V_s, of which the gas-filled volume is V_g. Then

$$G_i = \frac{A_i}{V_s} \text{ and } \chi_i = \frac{A_i}{V_g}$$

Therefore, $G_i = \dfrac{V_g}{V_s}\chi_i = \phi_g \chi_i$

or in differential form:

$$\partial G_i = \phi_g \partial \chi_i \text{ or } \frac{\partial G_i}{\partial \chi_i} = \phi_g$$

Answer 4.21 $[\lambda] = \left[\dfrac{-f_h}{\partial T/\partial s}\right] = \dfrac{\mathrm{J\,m^{-2}\,s^{-1}}}{\mathrm{K\,m^{-1}}} = \mathrm{J\,m^{-1}\,s^{-1}\,K^{-1}}$

Answer 4.22 $\left[\dfrac{\partial T}{\partial t}\right] = \mathrm{K\,s^{-1}}$

$$\left[\frac{\partial^2 T}{\partial s^2}\right] = \mathrm{K\,m^{-2}}$$

So $[D_h] = \dfrac{\mathrm{K\,s^{-1}}}{\mathrm{K\,m^{-2}}} = \mathrm{m^2\,s^{-1}}$

5 Flow of water in soil

5.1 INTRODUCTION

Water is present in every soil profile, but the amount varies with time and place as a result of supply and demand by its environment. If the supply of water at the soil surface (e.g. rainfall, irrigation, dew, flood) during a certain period exceeds the extraction of water from a given soil profile (e.g. evaporation from soil surface, water uptake by plant roots), the excess supply during that period is stored in the soil profile or discharged to lower regions (drainage). The supply at the soil surface can also be smaller than the sum of transpiration and evaporation. In that case, the topsoil will dry out. Upward transport of water from the subsoil (capillary rise) may also take place.

In Chapters 3 and 4 we found:

- at static hydraulic equilibrium:
  p_h (or H) = uniform and $q = 0$
- without static equilibrium,

$$\frac{\partial \theta}{\partial t} = -\frac{\partial q}{\partial s} \text{ and } q = -K\frac{\partial p_h}{\partial s} = -k\frac{\partial H}{\partial s}$$

In this chapter the subject matter of Chapters 3 and 4 is developed further for vertical flow in soil profiles. Except in one instance (Section *5.2.4*), it is assumed that the soil is homogeneous and isotropic. Three different types of flow will be considered:

1. steady flow in saturated soil:

$$\frac{\partial \theta}{\partial t} = 0 \text{ and } \theta = \phi$$

   q, K (or k) and θ are constant in time and position.

2. steady flow in unsaturated soil:

$$\frac{\partial \theta}{\partial t} = 0 \text{ and } \theta < \phi$$

   q is constant in time and position, but K (or k) and θ only in time.

3. nonsteady flow in unsaturated soil:

$$\frac{\partial \theta}{\partial t} \neq 0 \text{ and } \theta < \phi$$

   q, K (or k) and θ vary with time and position.

Nonsteady flow in saturated soil, in which θ is constant but q varies in time, occurs in drainage processes. This type of flow is usually treated separately and will not be considered here.

5.2 FLOW OF WATER IN SATURATED SOIL

5.2.1 Saturated hydraulic conductivity

A major division in types of flow of water in soil is that between saturated flow and unsaturated flow. The reason is that the hydraulic conductivity of saturated soils is constant in time and, in a uniform soil, also in position. This simplifies transport equations such as Equation 4.18 and makes their solution relatively easy. Flow processes for which the hydraulic conductivity is not constant, are much more complex.

In Chapter 4 two hydraulic conductivities were distinguished. One, K, is associated with the hydraulic potential, $p_h = p_t + \rho_l g z$, expressed as pressure equivalent, and has the unit $m^2\ Pa^{-1}\ s^{-1}$. The other, k, is associated with the hydraulic potential, $H = h + z$, expressed as head equivalent, and has the unit $m\ s^{-1}$. The relationship between the two hydraulic conductivities is:

$$k = \rho_l g K \approx 10^4\ Pa\ m^{-1} \times K \tag{5.1}$$

In Chapter 4 it was stated that the hydraulic conductivity of porous media is determined by the internal friction in the water, as characterized by the viscosity, η, and by the friction between the pore walls and the water. To gain a better understanding of the latter, a porous medium can, as a first approximation, be considered as a bundle of capillary tubes.

According to Poiseuille's law (for derivation, see Appendix 1) the rate of discharge, Q, of a capillary tube with radius r is

$$Q = -\frac{\pi r^4}{8\eta}\frac{\partial p_h}{\partial s} \tag{5.2}$$

Suppose a soil column contains cylindrical, parallel pores of one size only, and the number of pores divided by the cross-sectional area perpendicular to the flow is n. Then the flux density through the column is

$$q = nQ \tag{5.3}$$

If the pores, instead of being all the same, are of various size classes, the total flux density is the sum of the flux densities of each of the pore size classes. If n_i represents the number of pores in the ith pore size class with radius r_i divided by total cross-sectional area, and Q_i is the discharge rate per pore in that group, then the total flux density through the soil column is:

$$q = \Sigma n_i Q_i = -\frac{1}{8\eta}\frac{\partial p_h}{\partial s}\Sigma \pi n_i r_i^4 \qquad (5.4)$$

Equation 5.4 is valid if all capillary pores run parallel to the flux direction. In soils, however, they never do. Therefore, a tortuosity factor, τ, must be introduced to account for the lengthened distance over which ∂p_h acts and which the flow must traverse ($\tau \approx 2$). The flux density is then

$$q = -\frac{1}{8\eta}\frac{\partial p_h}{\tau \partial s}\Sigma \pi n_i r_i^4 \qquad (5.5)$$

Since $\pi n_i r_i^2$ is the contribution, $(\Delta\theta)_i$, of the pores with radius r_i to the total volume fraction of water, θ, the flux density is also

$$q = -\frac{1}{8\eta}\frac{\partial p_h}{\tau \partial s}\Sigma (\Delta\theta)_i r_i^2 \qquad (5.6)$$

A comparison of Equation 5.6 with Equation 4.6 yields immediately:

$$K = \frac{1}{8\eta\tau}\Sigma (\Delta\theta)_i r_i^2 \qquad (5.7)$$

Thus the hydraulic conductivity of a porous medium can be estimated from the size distribution of the water-filled pores.

Equation 5.7 provides an explanation for the sharp decrease of the hydraulic conductivity with drying and swelling. Since K is a function of the number of water-filled pores and proportional to the square of the pore radius, its value is determined mainly by the number and size of the largest water-filled pores. During drying, the large pores are the first ones to empty and K is determined only by the smaller pores that remain full (Section *5.3.1*). During swelling of clay soils the total porosity increases, but the smaller pores expand at the expense of the larger ones, so that K decreases sharply.

In sandy soils water is largely present in capillary pores. This makes it possible to estimate the pore-size distribution from the soil water characteristic. From the contribution of the different pore-size classes to θ, a rough estimate of K can be made.

Question 5.1 Figure 5.1 is the soil water characteristic of a sandy soil. Estimate the saturated hydraulic conductivity on the basis of the three pore-size classes shown, using the following information:
$\sigma = 7 \times 10^{-2}$ N m^{-1}, $\eta = 10^{-3}$ Pa s and $\tau = 2$.

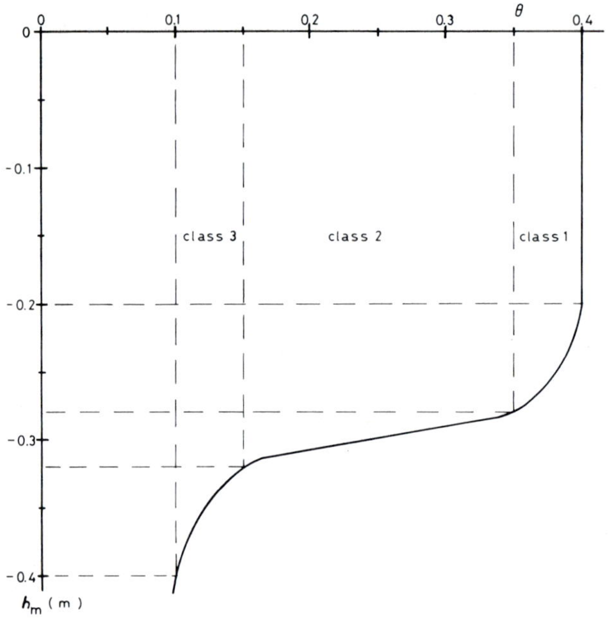

Figure 5.1 Soil water characteristic of sandy soil divided in three pore size classes.

The procedure used in Answer 5.1 gives only a very rough estimate of K, even for a sandy soil. In the literature one can find various procedures to calculate the hydraulic conductivity from the pore-size distribution, but they all lack accuracy.

Saturated hydraulic conductivities vary strongly between soils, due to their different pore-size distributions. Hydraulic conductivities in excess of 1 m d^{-1} are more than adequate for most agricultural purposes, whereas hydraulic conductivities less than 0.01 m d^{-1} may cause problems. Actual hydraulic conductivities can only be judged in relation with prevalent climatic conditions and the use of the soil.

5.2.2 *Intrinsic permeability*

Equation 5.7 contains the viscosity as the only parameter related to the fluid being transported. This suggests that

$$K_\mathrm{i} = \eta K = \frac{1}{8\tau} \Sigma(\Delta\theta)_i r_i^2 \tag{5.8}$$

is a parameter which depends only on the geometry of the porous medium.

Question 5.2 Check the units of Equation 5.8.

The dimension of m^2 for K_i indicates that it is indeed only a function of the geometric arrangement of the pores and that it is an intrinsic property of a stable porous medium. It is, therefore, called intrinsic permeability.

Equation 5.8 still contains the parameter θ, which is related to water as the permeating fluid. Its presence in Equation 5.8 was obscured in Answer 5.2 because it is dimensionless. The intrinsic permeability of a stable porous medium is the same for gas flow as for water flow and thus a more consistent formulation for K_i is:

$$K_i = \frac{1}{8\tau} \Sigma(\Delta\phi)_i r_i^2 \tag{5.9}$$

where $(\Delta\phi)_i$ is the contribution of all the pores with radius r_i to the porosity, ϕ.

Question 5.3 a. What is the intrinsic permeability of the soil in Question 5.1?
b. What is the intrinsic permeability of highly conducting and poorly conducting soils?

In slaking and swelling soils the pore size distribution varies with wetness. The ratio K_{ia}/K_{iw} can then be used as an index for the degree of slaking or swelling, where K_{ia} and K_{iw} are the intrinsic permeabilities calculated from a measurement with air and water, respectively.

5.2.3 Steady downward flow in homogeneous soil

The flux density equation for the flow of water in saturated soil, with the hydraulic potential expressed on weight basis, was given in Chapter 4:

$$q = -k\frac{\partial H}{\partial s} = -k\left(\frac{\partial h}{\partial s} + \frac{\partial z}{\partial s}\right) \tag{4.7}$$

This relationship is called Darcy's law, after the Frenchman who found this relationship experimentally in 1856 for the flow of water in saturated sand.

In a homogeneous, water-saturated soil k is constant with position and in time. For steady one-dimensional flow, as in the soil column in Figure 5.2, q is also constant in time and with height. From Equation 4.7 it follows then that $\partial H/\partial s$ is also constant with height, or that H varies linearly with height.

The lower end of the soil column in Figure 5.2 is supported by a screen, such that water can flow out freely under atmospheric pressure. This means that at that height, $h = 0$. If the reference level for the gravitational potential is chosen at the same height, then there also $z = 0$ and thus, $H = h + z = 0$.

At the free water surface above the soil column, $z = 0.9$ m and $h = 0$, thus $H = h + z = 0.9$ m. The water in the layer from $z = 0.8$ m to $z = 0.9$ m can be assumed hydrostatic. This means that in this layer H is constant. Thus at the soil surface, $H = 0.9$ m, $z = 0.8$ m and thus $h = 0.1$ m. Since within the soil H, and also h, vary linearly with height, their values are now determined as indicated in Figure 5.2.

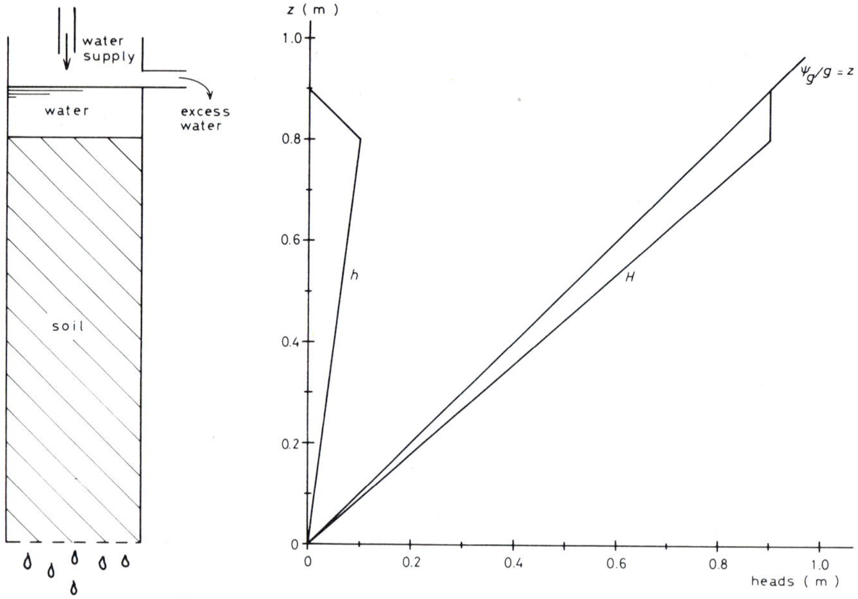

Figure 5.2 Steady saturated flow by inundation.

The system described above and depicted in Figure 5.2 is used to measure the saturated hydraulic conductivity of packed soil columns, as well as undisturbed soil cores taken from the field to the laboratory.

Question 5.4 Calculate the saturated hydraulic conductivity, k, of the soil in Figure 5.2 if the rate of discharge of water at the bottom of the column is 1.62 cm^3 per minute and the internal cross-sectional area of the cylinder is 20 cm^2.

Question 5.5 a. What happens if the cylinder in Figure 5.2 is perforated all around at a height $z = 0.5$ m?

b. What will be the value of h at that height soon after the cylinder is perforated?

c. Do you expect a continuous discharge of water from the perforations? Explain your answer.

d. Make a diagram of the potentials and calculate the flux densities in the upper and lower halves of the column. How much water (cm^3) will be discharged from the perforations in one hour?

5.2.4 Steady downward flow in layered soil

For saturated flow of water in a soil profile or soil column with layers of different hydraulic conductivity, the flux density is the same in all the layers.

Question 5.6 Why is this so?

Because the flux density is everywhere the same, the gradient of H is inversely proportional to the values of k in the different layers. From this H and h can be calculated.

This will be illustrated for the soil column in Figure 5.2 after the bottom 0.2 m is replaced by a layer of which k is one sixth of that in the overlying soil (Figure

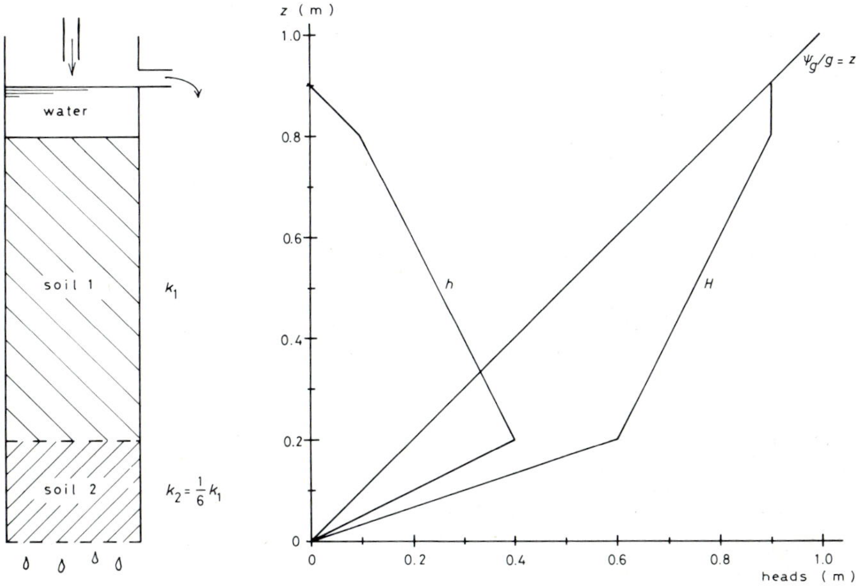

Figure 5.3 Steady saturated flow through two soil layers with different k values.

5.3). With the subscripts referring to soil 1 and soil 2:

$$q_1 = -k_1 (\mathrm{d}H/\mathrm{d}z)_1 = -k_2 (\mathrm{d}H/\mathrm{d}z)_2 = q_2 \text{ or} \tag{5.10}$$

$$k_1 \frac{\Delta H_1}{0.6 \text{ m}} = k_2 \frac{\Delta H_2}{0.2 \text{ m}} \tag{5.11}$$

ΔH is the difference of H between the upper and lower boundaries of a layer. With $k_1 = 6k_2$, one then finds $\Delta H_1 = \frac{1}{2}\Delta H_2$. Because $H = 0.9$ m at the upper surface of the soil column and $H = 0$ at the lower surface, $\Delta H_1 + \Delta H_2 = 0.9$ m. It easily follows that $\Delta H_1 = 0.30$ m and $\Delta H_2 = 0.60$ m. The gradients of H in the two layers can now be calculated and the values of h can be found from the equation $h = H - z$.

Question 5.7 Check the potential diagram of Figure 5.3 and calculate the flux density when the hydraulic conductivity of soil 1 is that in Answer 5.4.

A situation similar to that in Figure 5.3 may occur in the field. Excessive implement traffic may compact the soil, while the top layer is kept loose by ploughing. In this way less permeable layers may develop, which hinder, either temporarily or permanently, the downward flow of soil water. When the water stagnates at the interface between the layers and develops a positive hydrostatic pressure, as in Figure 5.3, this causes a so-called perched water table. Since this is detrimental to normal root growth, deep tillage and subsoiling are often applied to break up the compacted layers.

Question 5.8 Assume that the situation in Figure 5.3 represents a flooded rice field. Calculate in mm and m^3 per hectare the daily supply required to maintain a constant water depth at the surface.

Question 5.9 Suppose an effort is made to intercept the water lost to the subsoil by installing closely spaced tile drains at the boundary between layers 1 and 2. (The water discharged by the drainage system could be pumped to a neighbouring field for reuse.)

a. Draw a new potential diagram.

b. Calculate the discharge through layers 1 and 2.

c. Is the drainage system effective in reducing the 'percolation losses' through layer 2?

Quite a different situation arises when the less permeable layer is on top. This may be caused by loss of soil structure due to puddling or crust formation due to the impact of raindrops. It may also be due to the treading by animals or compaction by heavy machinery.

Question 5.10 The soil profile of a rice field consists of an upper layer of 0.10 m and a lower layer of 0.70 m. Below these layers is coarse gravel. The saturated hydraulic conductivity of the lower layer is 3.5 times that of the puddled layer. The water level in the field is maintained at 0.10 m above the soil surface.

a. Make a schematic drawing of the system and a potential diagram, assuming that the air-entry value is not exceeded.

b. Calculate the flux density to the gravel layer if the lower layer is the same soil as used in Figure 5.2.

c. Why is the assumption 'the air-entry value is not exceeded' necessary?

Figure 5.3 and Answer 5.10 indicate that in stratified soils the flow is mostly influenced by the layer with the lowest hydraulic conductivity.

5.2.5 Steady upward flow

Upward flow of water may occur in a soil profile due to transpiration by vegetation and/or evaporation from the soil surface. If the water is supplied by groundwater, this upward flow will lower the groundwater table, unless the groundwater is replenished by water from elsewhere. For instance, replenishment of groundwater may occur by seepage from higher surroundings, as found in many 'polders' in the Netherlands. Another typical example is seepage from irrigation canals or irrigated fields to the groundwater in neighbouring fields. If the seepage balances the evapotranspiration, the groundwater level remains stable. In the laboratory such a situation can be simulated by a system as depicted in Figure 5.4.

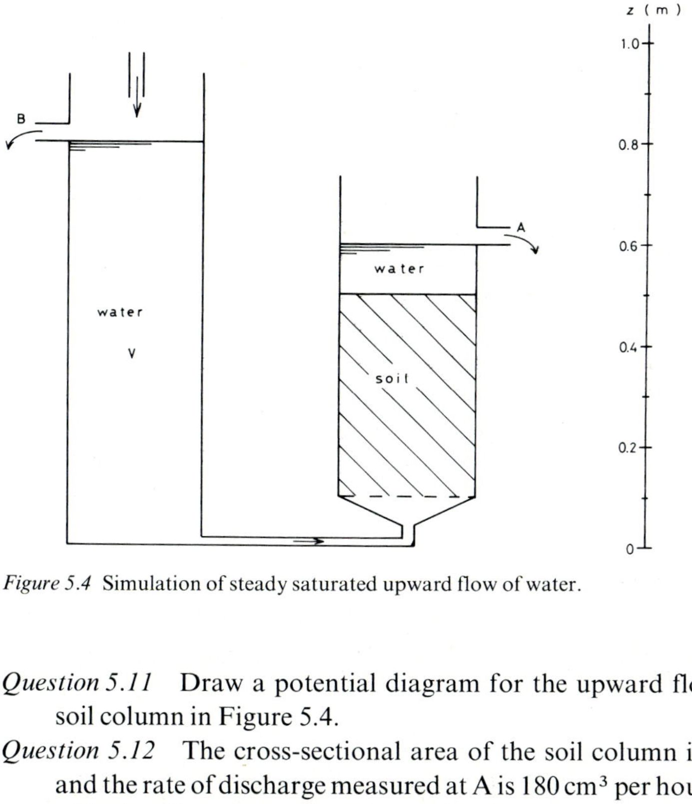

Figure 5.4 Simulation of steady saturated upward flow of water.

Question 5.11 Draw a potential diagram for the upward flow in the homogeneous soil column in Figure 5.4.

Question 5.12 The cross-sectional area of the soil column in Figure 5.4 is 100 cm^2 and the rate of discharge measured at A is 180 cm^3 per hour. Calculate the saturated hydraulic conductivity of the soil column.

5.2.6 *Intergranular pressure*

In Chapter 3 the relationships between the intergranular pressure, σ_g, the soil pressure, σ_s, and the water pressure, σ_w, were discussed for static equilibrium of the liquid phase. These relationships are illustrated once again in Figure 5.5, which shows the pressure diagram of a sand column with a layer of water on top, in a cylinder closed at the bottom (${}^b\rho_d = 1300$ kg m^{-3} and $\rho_s = 2600$ kg m^{-3}).

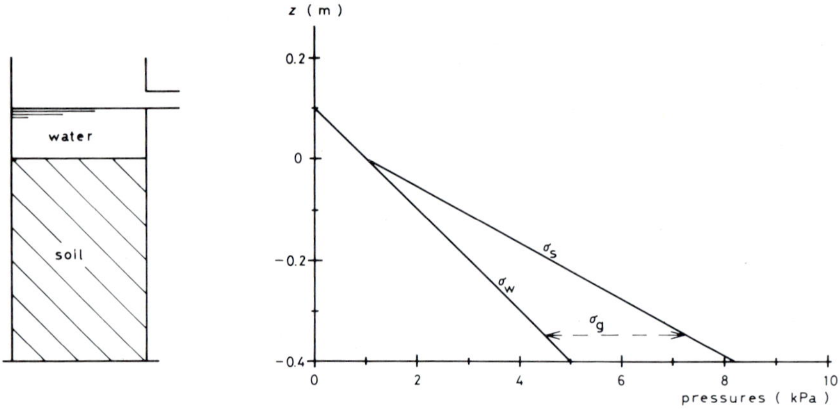

Figure 5.5 Pressures in a soil column at static equilibrium.

Question 5.13 a. Verify the pressure diagram in Figure 5.5.
b. What is the value of σ_g at the bottom?

If the bottom of the cylinder in Figure 5.5 is replaced by a screen and a steady downward saturated water flow is maintained, the pressure distribution will change.

Question 5.14 a. Draw a new pressure diagram for this situation.
b. Determine the intergranular pressure at the bottom of the soil column.

Question 5.15 Suppose the flow conditions in the soil column of Question 5.14 are changed to those in Figure 5.4.
a. Draw a diagram for σ_s and σ_w and determine the intergranular pressure at the bottom of the column.
b. Calculate the intergranular pressure in the soil column if the overflow B is raised to $z = 0.92$ m.

A comparison of Answers 5.14 and 5.15 indicates that the intergranular pressure decreases due to upward water transport. As a result, the bearing capacity of the soil, which is wholly dependent on the friction forces between the grains, is reduced. In extreme cases, i.e. where a strong upward water transport occurs, the intergranular pressure may completely disappear, resulting in quicksand. Such a situation of strong seepage may occur for instance next to dikes in a polder or in deep construction pits.

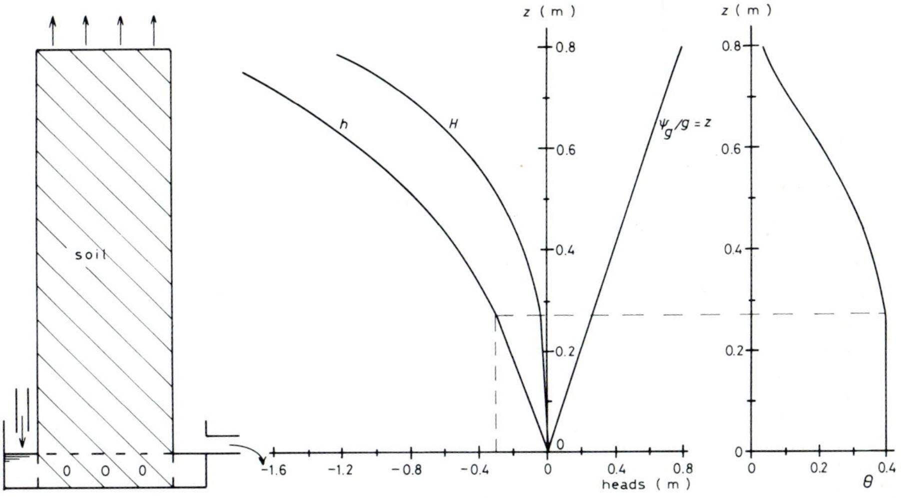

Figure 5.8 Steady upward flow of water, constant evaporation from a stable groundwater table.

For the flow to be directed upwards, H must decrease with height, making $\mathrm{d}H/\mathrm{d}z$ negative and q positive. Thus h, and with it θ, must decrease even more with height than in Answer 5.19. Since k decreases fast with θ (Figure 5.6), it decreases fast with height. To maintain the same flux density at all heights, a certain rate of decrease of k with height must be compensated by the same rate of increase with height of the absolute value of $\mathrm{d}H/\mathrm{d}z$. Thus the H curve in the potential diagram must be increasingly flatter towards the soil surface, as indicated in Figure 5.8.

The above situation is often found in the field during periods of drought, when a rather sharp transition from a dry upper layer to a moist lower zone is developed due to the sharp decrease of k with decreasing water content. The formation of such a sharp transition is the result of a self-accelerating mechanism, which can be explained as follows. When evaporation starts, θ as well as h and H decrease near the surface. But, because k decreases too, the new H gradient is too small to maintain the upward flow needed to satisfy the evaporative demand. Hence, θ decreases still further, causing a further decrease in H, but again not sufficient to offset the accompanying decrease in k, etc. This process continues until the surface layer has become so dry that practically all flow of water ceases. The formation of this dry surface layer is of practical importance because it protects the soil against large evaporation losses. This so-called soil mulch is the principle upon which dry-land farming is practised.

The exact values of H and h in the steady situation represented in Figure 5.8 are difficult to calculate, and then only when the $k(\theta)$ and $h_{\mathrm{m}}(\theta)$ functions are known accurately. The gradients in the lower, water-saturated part of the profile can be calculated easily, provided the saturated k, the evaporation rate and the air-entry value are known.

Question 5.21 Is the relationship of H and h with z curvilinear in the water-saturated part of the profile in Figure 5.8?

Question 5.22 Estimate the evaporation rate in Figure 5.8, assuming a steady state situation, a saturated k value of 0.05 m d^{-1} and $h = -0.22$ m at $z = 0.20$ m.

Question 5.23 a. Determine the height of the saturated zone for the situation in Question 5.22.

b. Does this height remain constant if the evaporation rate changes?

5.3.2.2 Steady downward flow

Figure 5.9 shows the potential diagram for a field situation similar to Figure 5.2. The water at the surface is applied by sprinklers and the groundwater table is stabilized by tile drains. The sprinkling rate is such that, without runoff, the layer of water on the soil surface remains constant.

Question 5.24 Calculate the flux density in Figure 5.9 if $k = 10$ cm d^{-1} and the water layer is 10 cm deep.

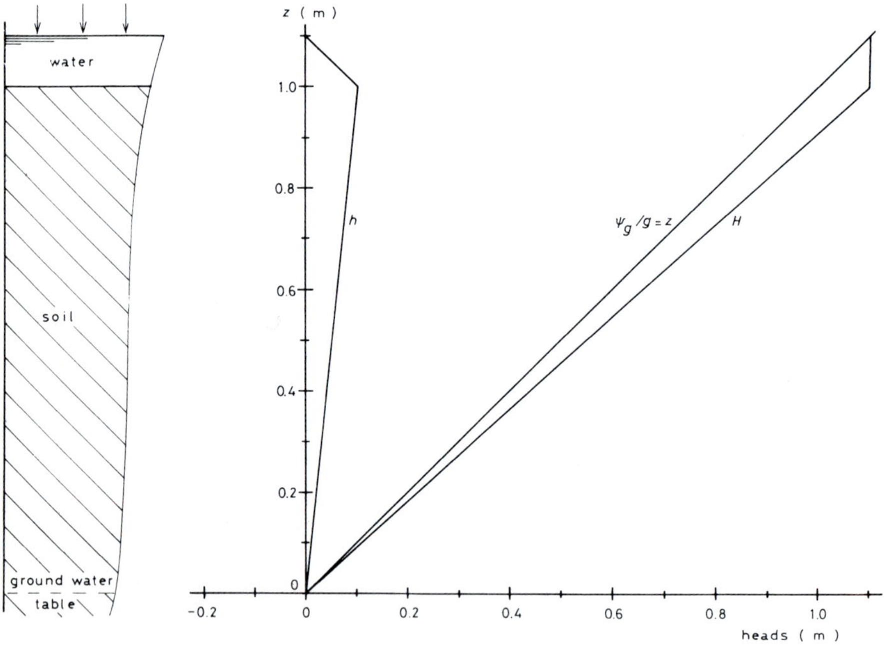

Figure 5.9 Steady downward flow of water at a high sprinkling rate (flooded soil).

If the sprinkling rate is decreased to another constant value, a new steady state situation arises, where the downward flux density still equals the sprinkling rate. This new situation is accompanied by a decrease in the H gradient, i.e. both H and h at the soil surface decrease. If the sprinkling rate falls below a certain critical value, the air-entry value of the soil will be exceeded and an unsaturated zone will develop in the top of the soil profile. Since k decreases with θ in the unsaturated zone and q is constant, the H gradient must increase with height, as shown in Figure 5.10. Consequently, the h gradient must also increase with height, which means that the h curve becomes steeper towards the soil surface.

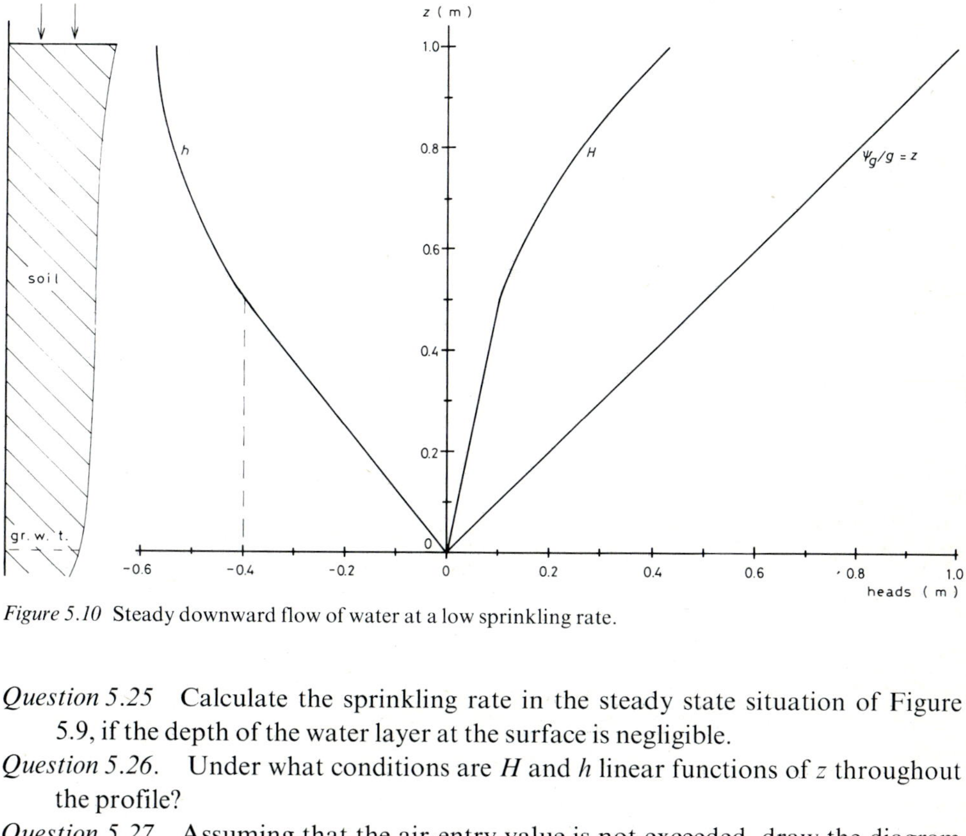

Figure 5.10 Steady downward flow of water at a low sprinkling rate.

Question 5.25 Calculate the sprinkling rate in the steady state situation of Figure 5.9, if the depth of the water layer at the surface is negligible.

Question 5.26. Under what conditions are H and h linear functions of z throughout the profile?

Question 5.27 Assuming that the air-entry value is not exceeded, draw the diagram for the steady state situation with a sprinkling rate of 8 cm d^{-1}.

Question 5.28 Calculate the height of the saturated zone above the groundwater table in the steady state situation if the sprinkling rate is 2 cm d^{-1} and the air-entry value is -0.40 m.

Question 5.29 What is the highest possible H gradient that can be attained in a steady state downward flow as shown in Figure 5.10?

Question 5.30 a. Draw as logical as possible the potential diagram for steady downward flow with a sprinkling rate of 1 cm d^{-1}.
b. Why is the phrase 'as logical as possible' needed in a?

Figure 5.11 presents a summary of the h profiles for the various cases of steady vertical flow of water discussed in this Section *5.3.2*. The numbers next to the curves are the values of the flux density. Upward flux densities ($q > 0$) represent evaporation and downward flux densities ($q < 0$) represent drainage. Zero flux density ($q = 0$) is the special case of hydrostatic equilibrium.

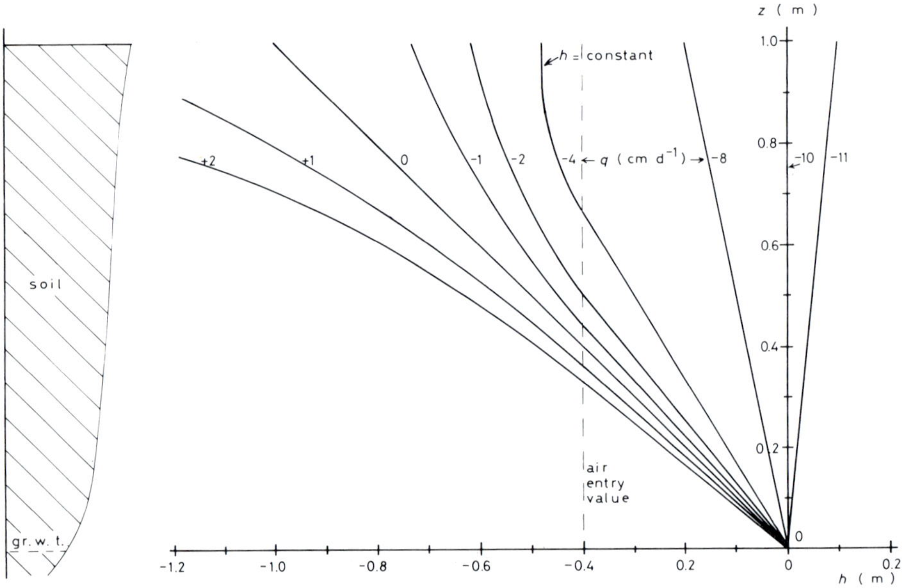

Figure 5.11 Summary of h profiles for steady vertical flow of water with a stable groundwater table.

5.3.3 Nonsteady flow of water in unsaturated soil

5.3.3.1 Analysis of H profiles

In the field, water transport seldom is steady, because precipitation, evapotranspiration, irrigation, drainage, etc. change continually. Transport processes which change with time, are called nonsteady. One can obtain insight into nonsteady water flow processes in the field, qualitatively, by obtaining tensiometer data at different times and depths in the soil profile. With tensiometers one obtains information on the pressure head h, as well as the hydraulic head $H = h + z$ (Section *3.1.4*). From the pressure head at different times one can derive whether the volume fraction of water is increasing or decreasing. From the gradient of H one can deduce in which direction the water is moving, and possibly predict values of θ and directions of flow at future dates.

Question 5.31 Which is the direction of flow if $\partial H/\partial z$ is positive, negative or zero?

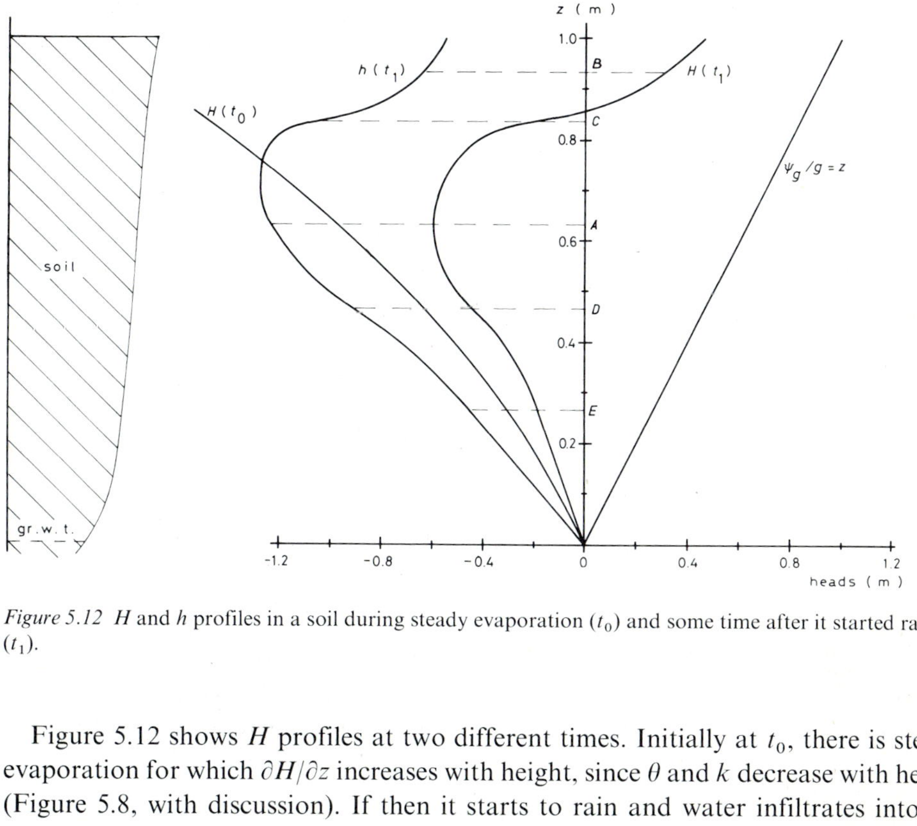

Figure 5.12 H and h profiles in a soil during steady evaporation (t_0) and some time after it started raining (t_1).

Figure 5.12 shows H profiles at two different times. Initially at t_0, there is steady evaporation for which $\partial H/\partial z$ increases with height, since θ and k decrease with height (Figure 5.8, with discussion). If then it starts to rain and water infiltrates into the soil profile, the H profile may after some time have changed to that indicated by t_1. From this profile one can tell in which direction the water is moving, and whether $\partial\theta/\partial t > 0$ or $\partial\theta/\partial t < 0$.

Question 5.32 Is there a water layer on top of the soil at time t_1?
Question 5.33 Indicate the depths where there is upward, downward or no movement of water at time t_1.
Question 5.34 By using the continuity equation explain, whether at height A in Figure 5.12, θ decreases, increases or remains constant after t_1.

For height A it is easy to tell whether θ will increase or not, just by looking at the H profile (Answer 5.34). For other heights it is not so easy to predict the changes in θ. Take for instance heights B and C in Figure 5.12. At both heights the flux density is downward.

Since C is the inflection point of the H profile above height A, the H gradient at B is smaller than at C. However, θ is greater at B than at C, as can be interpreted from the h profile. This means that k is greater at B than at C. Therefore, without knowing the exact numerical values of k and $\partial H/\partial z$, it cannot be concluded at which height the greater flux density occurs. This reasoning is true for every height above C. Thus, it is not possible to conclude whether above C, θ will increase or decrease, or even remain constant.

The same problem arises in the interval between heights D and E, where D is the inflection point of the H profile below A, and E is the height where the H profile becomes linear (check this).

In contrast, between A and D, the absolute value of $\partial H/\partial z$ as well as k decrease with height, i.e. $\partial q/\partial z < 0$. Hence, $\partial\theta/\partial t$ is positive at every height in this interval. A similar reasoning can be given for the interval between A and C. Also in this region $\partial q/\partial z$ is negative (check this). Therefore, $\partial\theta/\partial t$ will be positive for every height between C and D.

Question 5.35 Suppose the weather becomes foggy for an extended period of time directly after the rain has stopped.

a. What will be the ultimate shape of the H profile?
b. Compare θ at different heights in this final stage with those at time t_1.

By the preceding reasoning one can acquire only a qualitative understanding of the water flow processes in soil. Since in unsaturated soil k varies with θ, it is not possible to calculate q as a function of height directly from H profiles as shown in Figure 5.12. For this it is necessary to know, in addition, $k(\theta)$ and $\theta(h_m)$ functions such as in Figure 5.6. The procedure that then should be followed to estimate $\partial\theta/\partial t$ is outlined in Figure 5.13. In this figure, the subscript z indicates height, s stands for a small increment in height and $\bar{\theta}$ is the average volume fraction of water.

Question 5.36 Describe in words the steps followed in Figure 5.13 in order to estimate $\partial\theta/\partial t$ at a height z in a soil profile.

The procedure outlined in Figure 5.13 makes use of average values of θ, and not of hydraulic conductivity or pressure head. This amounts to assuming that θ changes linearly with height. This assumption is usually not far from reality and introduces small errors in $\partial\theta/\partial t$. An alternative approach would be to average the k values for each of the pertinent θ values. This amounts to assuming that k changes linearly with θ, which obviously introduces much more serious errors in $\partial\theta/\partial t$ because of the strong nonlinear $k(\theta)$ function (Figure 5.6). An intermediate approach would be to use average h values.

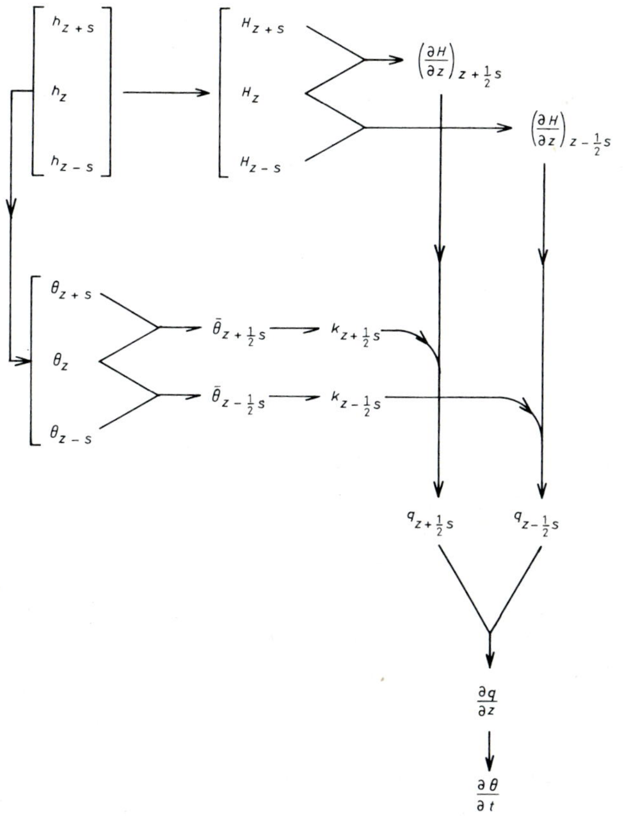

Figure 5.13 Calculation scheme for analysing a θ profile.

Question 5.37 In a medium fine sandy soil, for which the $\theta(h_m)$ and $k(\theta)$ functions are given in Figure 5.6, three tensiometers are installed at depths of 0.30, 0.40 and 0.50 m below the surface. At a certain moment the tensiometers indicate h values of -2.00, -0.80 and -0.40 m, respectively. Estimate $\partial\theta/\partial t$ at a depth of 0.40 m using the scheme given in Figure 5.13. Hysteresis may be neglected.

Question 5.38 a. Also estimate $\partial\theta/\partial t$ at a depth of 0.40 m using average h values. Do the same using average k values.

b. Compare the values for k and $\partial\theta/\partial t$ obtained by the three calculation schemes.

Question 5.39 In a loamy soil profile nine tensiometers are installed at depths of 0, 0.10, 0.20, ... and 0.80 m below the soil surface. Estimate $\partial\theta/\partial t$ at depths 0.10, ... and 0.70 m for the moment that the tensiometers indicate h values of 0, -0.30, -0.80, -1.10, -1.60, -1.00, -0.50, $+0.20$ and 0 m, respectively. Use the $\theta(h_m)$ and $k(\theta)$ functions in Figure 5.6. Hysteresis may be neglected.

5.3.3.2 General description of infiltration

A typical example of nonsteady unsaturated flow of water is the infiltration of water into soils. When water is applied at the soil surface (e.g. under flood irrigation or inundation), it enters the soil profile and changes the water content distribution with depth. After irrigation has continued for some time, the following zones can usually be distinguished in the water content profile (Figure 5.14):

- saturated zone
- transition zone
- transmission zone
- wetting zone
- wetting front.

The saturated zone is a thin zone at the soil surface. The transition zone is a zone of decreasing water content between the saturated zone and the nearly saturated transmission zone. Its lower end may reach from a few millimeters to a few centimeters below the surface. These upper two zones are not always clearly distinguishable, especially in the laboratory. They are caused by structural changes at the soil surface and by entrapped air.

The transmission zone is the conveyance zone for the infiltrating water. While all the other zones remain nearly constant in thickness, this zone continues to elongate as long as water is supplied at the soil surface. Its water content, though slightly changing with depth, is rather constant and close to saturation.

The wetting zone is the normally thin zone where the water content changes from its initial value to the value of the transmission zone.

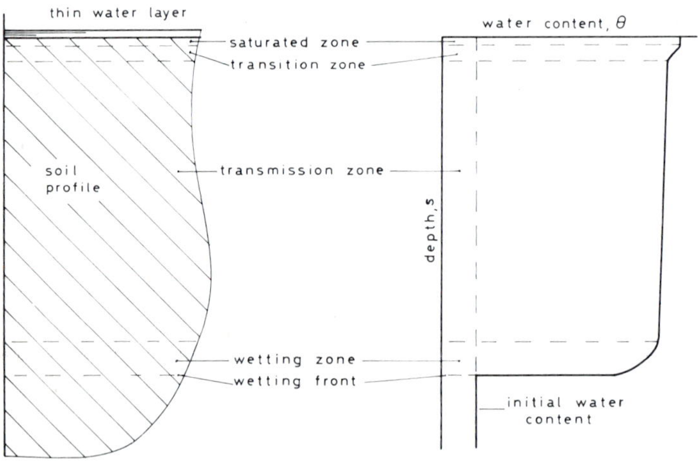

Figure 5.14 Zones of water content profile some time after start of flood irrigation.

The wetting front is the visible limit of water penetration, where the gradient of the pressure head is very large.

If the water, initially present in the soil profile, is assumed at rest, the cumulative infiltration, I, which is the volume of water that has infiltrated into the soil divided by the surface area, is

$$I = \int_{s=0}^{\infty} (\theta - \theta_i)\, ds \tag{5.14}$$

where s is distance in the direction of flow (here depth in the soil profile), θ is the volume fraction of water at distance s and θ_i is the initial uniform value of θ. The cumulative infiltration is, of course, an increasing function of time.

Question 5.40 Derive Equation 5.14.

The driving forces for the water entering the soil are the gradient of the pressure head between the wetting front and the soil surface, and gravity (Equation 4.7). While the latter is constant, the gradient of the pressure head decreases with time, because of the advancing wetting front. As a result, the flux density through the soil surface, also called the infiltration rate, i, decreases monotonically with time and approaches asymptotically a constant value, as gravity becomes the main driving force (Figure 5.15).

The infiltration rate may be expressed as:

$$i = \frac{dI}{dt} \tag{5.15}$$

Question 5.41 Check Equation 5.15.

Integration of Equation 5.15 gives the cumulative infiltration as a function of time:

$$I = \int_{0}^{t} i\,dt \tag{5.16}$$

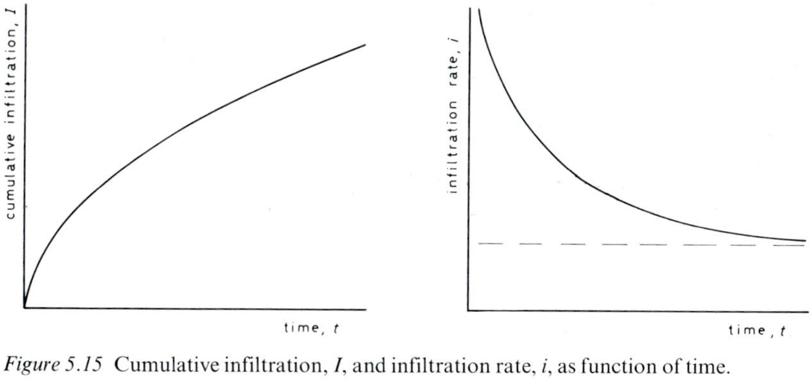

Figure 5.15 Cumulative infiltration, I, and infiltration rate, i, as function of time.

Question 5.42 Derive Equation 5.16 from Equation 5.15.

The cumulative infiltration as a function of time may be measured in the field, but such an experiment does not give any information about the water content distribution or the depth of the wetting front. To obtain the volume fraction of water as a function of depth and time, $\theta(s,t)$, a general flow equation such as

$$\frac{\partial \theta}{\partial t} = \frac{\partial}{\partial s}\left(D\frac{\partial \theta}{\partial s}\right) + \frac{\partial}{\partial s}\left(k\frac{\partial z}{\partial s}\right) \tag{4.23}$$

must be solved, using appropriate initial and boundary conditions.

Equation 4.23 is a nonlinear partial differential equation of order two. It is nonlinear, because both D and k are functions of θ (Figure 5.6). Due to this nonlinearity general analytical solutions of Equation 4.23 are not available.

Water flow problems can be solved numerically with computers, using calculation schemes of the type shown in Figure 5.13 and Answers 5.37, 5.38 and 5.39. Unfortunately, the large computing times required for obtaining solutions for the practical time spans often make it prohibitively expensive. One must then resort to simplifications which make the problem solvable, while retaining the essential characteristics of the flow process. Examples of this will be discussed in the remainder of this chapter.

5.3.3.3 Horizontal infiltration. Boltzmann transformation

For the special case of infiltration from a thin layer of water into a homogeneous soil with uniform initial volume fraction of water, the initial and boundary conditions are:

$$\begin{aligned} &\theta = \theta_i \text{ for } s > 0 \text{ and } t = 0 \\ &\theta = \theta_i \text{ for } s \rightarrow \infty \text{ and } t \geqslant 0 \\ &\theta = \theta_0 \text{ for } s = 0 \text{ and } t \geqslant 0 \end{aligned} \tag{5.17}$$

where θ_0 is the constant (saturated) volume fraction of water at the soil surface ($s = 0$).

In the early stages of vertical infiltration the term $k\,\partial z/\partial s$ in Equation 4.23 is much smaller than the term $D\,\partial\theta/\partial s$ and thus the driving force due to gravity can be neglected. In the case of infiltration into a horizontal soil column in the laboratory, often called absorption, the term $k\,\partial z/\partial s$ is zero at all times, because $\partial z/\partial s$ is zero. For both cases, Equation 4.23 reduces to:

$$\frac{\partial \theta}{\partial t} = \frac{\partial}{\partial s}\left(D\frac{\partial \theta}{\partial s}\right) \tag{5.18}$$

Equation 5.18 may be transformed into an ordinary differential equation by introducing the variable λ defined as:

$$\lambda = st^{-1/2} \tag{5.19}$$

This is known as the Boltzmann transformation. Equation 5.18 then becomes (Appendix 2):

$$-\frac{\lambda}{2}\frac{d\theta}{d\lambda} = \frac{d}{d\lambda}\left(D\frac{d\theta}{d\lambda}\right) \tag{5.20}$$

The boundary conditions 5.17 can also be transformed:

$$\begin{aligned} \theta &= \theta_i \text{ for } \lambda \to \infty \\ \theta &= \theta_0 \text{ for } \lambda = 0 \end{aligned} \tag{5.21}$$

At this point the absorption problem is expressed entirely in terms of θ and λ. Therefore, one may expect that the solution will be a unique function $\theta(\lambda)$, which for each soil is determined by its particular $D(\theta)$ function and the values of θ_i and θ_0.

Equations 5.20 and 5.21 can be solved analytically if D is assumed constant. This solution is given in detail in Appendix 2. The result is:

$$\theta = \theta_i + (\theta_0 - \theta_i)\,\text{erfc}\left(\frac{\lambda}{2\sqrt{D}}\right) \tag{5.22}$$

where erfc stands for complementary error function. A plot of this function is given in Figure 5.16.

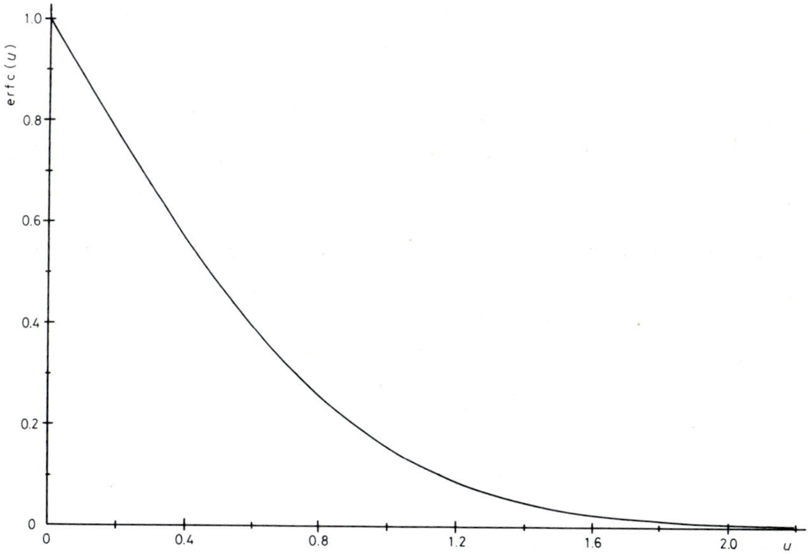

Figure 5.16 The complementary error function erfc (u) for $u \geqslant 0$.

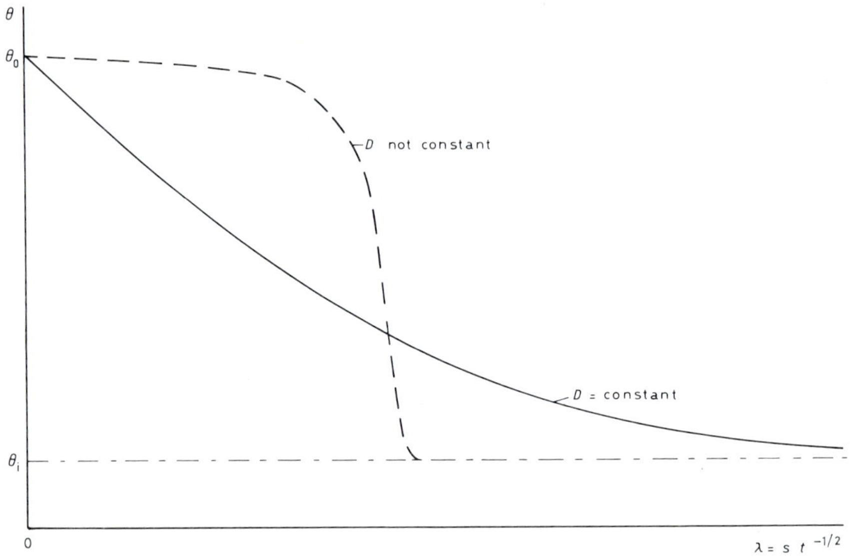

Figure 5.17 Solutions of Equation 5.18 for D = constant and for D of a soil (D increasing with increasing θ).

For a chosen set of values of θ_i, θ_0 and D, Equation 5.22 is plotted in Figure 5.17. When this graph is compared with Figure 5.14, it is obvious that this solution, a gradually decreasing volume fraction of water with distance, is far from reality. The reason for this discrepancy is that D is not constant over the range of θ, but varies with a factor of up to about 10^5.

Equations 5.20 and 5.21 cannot be solved analytically for variable D. However, various techniques can be found in the literature for solving Equations 5.20 and 5.21 by numerical methods. Discussion of these techniques goes beyond the scope of this book. The result of such calculations, for the same values of θ_i and θ_0 but a realistic $D(\theta)$ function, is also plotted in Figure 5.17. This figure shows, as expected, a unique relationship between θ and λ. This implies that $s/\sqrt{t}\,(=\lambda)$ is constant for a given θ.

Question 5.43 Which relationship do you expect between s and t for the visible wetting front and for any other value of θ between θ_i and θ_0?

Combination of Equations 5.14 and 5.19 gives:

$$I = \int_{s=0}^{\infty} (\theta - \theta_i)\mathrm{d}s = \int_{\lambda=0}^{\infty} (\theta - \theta_i)\mathrm{d}\lambda t^{1/2} = t^{1/2} \int_{\lambda=0}^{\infty} (\theta - \theta_i)\mathrm{d}\lambda \qquad (5.23)$$

Introduction of the parameter

$$S = \int_{\lambda=0}^{\infty} (\theta - \theta_i) \mathrm{d}\lambda \tag{5.24}$$

leads to:

$$I = St^{1/2} \tag{5.25}$$

and, consequently,

$$i = \frac{\mathrm{d}I}{\mathrm{d}t} = \tfrac{1}{2}St^{-1/2} = \frac{I}{2t} \tag{5.26}$$

S is called the sorptivity and is a measure for the capacity of a soil to absorb water. It is the cumulative infiltration during the first unit of time (Equation 5.25).

Question 5.44 Explain how S can be found from the $\theta(\lambda)$ curve (for variable D) given in Figure 5.17.

Question 5.45 Show that S can also be given by: $S = \int_{\theta_i}^{\theta_0} \lambda \mathrm{d}\theta$

Question 5.46 Would you expect S to be constant for a given soil?

For initially dry soils ($\theta_i = 0$) the sorptivity varies from about 5×10^{-5} m s$^{-1/2}$ (≈ 0.04 cm min$^{-1/2}$) for a heavy clay soil to about 2×10^{-3} m s$^{-1/2}$ (≈ 1.5 cm min$^{-1/2}$) for a coarse sand.

Question 5.47 a. Calculate the cumulative horizontal infiltration in an initially dry, heavy clay soil during 24 hours.
b. Do the same for a coarse sand.

5.3.3.4 Vertical infiltration

So far, the infiltration process was described only for cases in which the gravitational potential can be neglected. During vertical infiltration, the influence of gravity becomes more important as time progresses. Then Equation 4.23, together with the initial and boundary conditions of Equation 5.17, must be solved. A number of different techniques for solving Equation 4.23 are available. A solution obtained by one of these techniques is

$$s(\theta, t) = a_1 t^{1/2} + a_2 t + a_3 t^{3/2} + a_4 t^2 + \ldots \tag{5.27}$$

where a_1, a_2, a_3, a_4, ... are all functions of θ. By means of Equation 5.27, the depth of a given value of θ at time t can be evaluated when a_1, a_2, etc. are known. These coefficients can be evaluated by numerical methods using the $D(\theta)$ and $k(\theta)$ functions.

The cumulative infiltration obtained by the above technique is given by:

$$I = St^{1/2} + At + Bt^{3/2} + Ct^2 + \ldots \tag{5.28}$$

The coefficients of Equation 5.28 are evaluated in a similar way as those of Equation 5.27.

The infiltration rate can be derived from Equation 5.28 by differentiating with respect to t (Equation 5.15):

$$i = \tfrac{1}{2}St^{-1/2} + A + \tfrac{3}{2}Bt^{1/2} + 2Ct + \ldots \tag{5.29}$$

Equations 5.28 and 5.29 usually are truncated after the first two terms, because for not too large t these series converge rapidly. In that case Equations 5.28 and 5.29 become, respectively,

$$I = St^{1/2} + At \tag{5.30}$$

and

$$i = \tfrac{1}{2}St^{-1/2} + A \tag{5.31}$$

S is again the sorptivity defined in Equation 5.24. It is the dominant parameter in the early stage of infiltration. As time progresses, the first term becomes negligible and the importance of A, which represents the main part of the gravitational influence, increases.

Question 5.48 a. Make a graph of the cumulative infiltration, I, and the infiltration rate, i, as a function of time for vertical infiltration during 4 hours in a fine sandy loam for which:

$S = 7.0 \times 10^{-4}$ m s$^{-1/2}$
$A = 1.0 \times 10^{-6}$ m s^{-1}
$B = 1.0 \times 10^{-9}$ m s$^{-3/2}$

Use Equations 5.28 and 5.29 (first three terms).

b. Give in the same figure the results when Equations 5.30 and 5.31 are used.

c. Compare the results.

5.3.3.5 Green and Ampt model

A simple but useful model for the infiltration process was initially (in 1911) proposed by Green and Ampt. The advantage of this model is that it leads to equations which can be solved analytically, both for horizontal and vertical infiltration. It is based on the following assumptions (Figure 5.18):

- throughout the wetted zone, the volume fraction of water, θ_t, is uniform and constant with time
- the change of θ_i to θ_t at the wetting front takes place in a layer of negligible thickness
- the pressure head at the wetting front, h_f, has a constant value, independent of the position of the wetting front, s_f.

These assumptions are quite realistic for infiltration into coarse-textured soils with low initial water content, since wetting fronts are, generally, very sharp under those

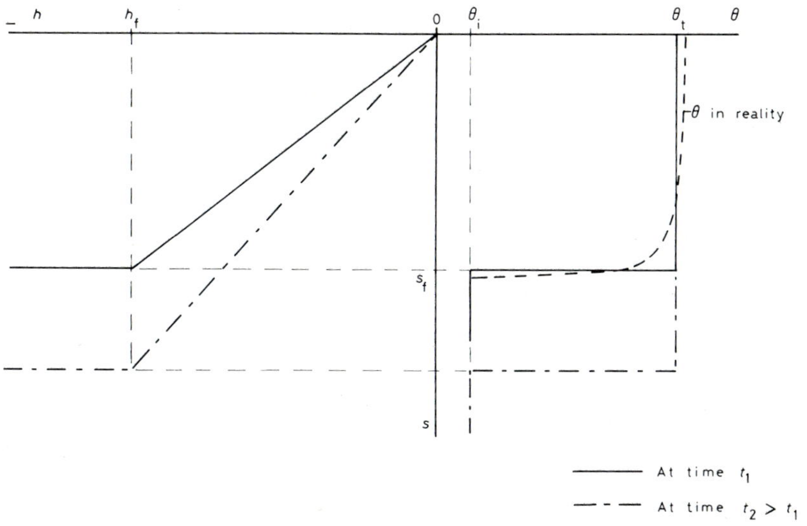

Figure 5.18 Green and Ampt model of infiltration. θ and h profiles at two different times.

conditions.

From the first assumption it follows that throughout the transmission zone the hydraulic conductivity, k_t, has a constant value. Besides, the flux density is the same everywhere in the transmission zone.

Question 5.49 Explain how these conclusions can be made.

Assuming negligible thickness of the water layer on the soil surface, $h = 0$ at $s = 0$, and using the second and third assumption, Darcy's law for this case can be written as:

$$i = q = -k\frac{\partial H}{\partial s} = -k\left(\frac{\partial h}{\partial s} + \frac{\partial z}{\partial s}\right) = -k_t\left(\frac{h_f}{s_f} - 1\right) \tag{5.32}$$

Question 5.50 Explain why one obtains the correct flux density, even though in the above equation h_f/s_f is used instead of $\partial h/\partial s$.

Question 5.51 Is the flux density constant with time?

The cumulative infiltration is:

$$I = \int_{s=0}^{\infty} (\theta_t - \theta_i)\mathrm{d}s = (\theta_t - \theta_i)\, s_f \tag{5.33}$$

and thus:

$$i = \frac{dI}{dt} = (\theta_t - \theta_i)\frac{ds_f}{dt} \tag{5.34}$$

Equations 5.32 and 5.34 yield:

$$k_t\left(1 - \frac{h_f}{s_f}\right) = (\theta_t - \theta_i)\frac{ds_f}{dt} \tag{5.35}$$

When gravity can be neglected (early stage of infiltration or horizontal infiltration), the term $\partial z/\partial s$ in Equation 5.32 vanishes and Equation 5.35 becomes:

$$-k_t\frac{h_f}{s_f} = (\theta_t - \theta_i)\frac{ds_f}{dt} \tag{5.36}$$

Rearranging and integrating Equation 5.36 obtains

$$-k_t h_f t = \tfrac{1}{2}(\theta_t - \theta_i)s_f^2 + C \tag{5.37}$$

Since $s_f = 0$ at $t = 0$, the integration constant C is zero, and thus Equation 5.37 can be rearranged to

$$s_f = \left(\frac{-2k_t h_f}{\theta_t - \theta_i}\right)^{1/2} t^{1/2} \tag{5.38}$$

Because k_t, θ_t, θ_i and h_f remain constant during the flow process, Equation 5.38 can be read as $s_f/\sqrt{t} = \text{constant}$, i.e. the depth of the wetting front is proportional with the square root of time (verify that this is the same conclusion as in Question and Answer 5.43).

Combination of Equations 5.33 and 5.38 leads to

$$I = [-2k_t h_f(\theta_t - \theta_i)]^{1/2} t^{1/2} = St^{1/2} \tag{5.39}$$

where the sorptivity is now

$$S = [-2k_t h_f(\theta_t - \theta_i)]^{1/2} \tag{5.40}$$

Question 5.52 a. How can there be a minus sign within the brackets of Equation 5.40, when the square root of a negative number is undefined?

b. Check that S remains constant during the infiltration process.

Question 5.53 Two horizontal columns of the same soil, with initial volume fractions of water θ_1 and θ_2, respectively, are infiltrated with water. Assume θ_t and h_f independent of θ_i. Denote parameters pertaining to sample 1 by subscript 1 and to sample 2 by subscript 2. Using the Green and Ampt model, derive in terms of θ_1, θ_2 and θ_t the ratios of:

a. the distances of the wetting front at time t
b. the cumulative infiltrations at time t
c. the times required for the wetting front to reach s
d. the sorptivities

e. the infiltration rates at time t.

Question 5.54 Paraphrase the relationships between s_1 and s_2, I_1 and I_2, etc. in Question 5.53 if $\theta_1 < \theta_2$.

When gravity cannot be neglected, Equation 5.35 can be written as:

$$\frac{k_t}{\theta_t - \theta_i}\,\mathrm{d}t = \frac{s_f}{s_f - h_f}\,\mathrm{d}s_f \qquad (5.41)$$

Integration, using $\frac{s_f}{s_f - h_f} = 1 + \frac{h_f}{s_f - h_f}$, yields:

$$\frac{k_t}{\theta_t - \theta_i}\,t = s_f + h_f \ln(s_f - h_f) + C \qquad (5.42)$$

The integration constant C can be found from the condition $s_f = 0$ for $t = 0$:

$$C = -h_f \ln(-h_f) \qquad (5.43)$$

The final result is then:

$$t = \frac{\theta_t - \theta_i}{k_t}\left[s_f + h_f \ln\left(\frac{s_f - h_f}{-h_f}\right)\right] \qquad (5.44)$$

The physical parameters θ_t, θ_i, k_t and h_f for a given soil must be found experimentally. The values for θ_t and k_t are both not far from the values at saturation. h_f cannot be measured directly, but can be derived from Equation 5.40 by measuring S, k_t and $\theta_t - \theta_i$. Values for h_f vary from about -0.05 m to -0.8 m for different soils. Once these parameters are known, the time needed for the wetting front to reach a certain depth can be calculated directly with Equation 5.44. To find s_f for a certain time t with this equation is a little more difficult, since s_f cannot be expressed explicitly as a function of t. However, it is possible to graph t versus s_f and read s_f for various t from the graph.

Once s_f is known, the cumulative infiltration and the infiltration rate can be found using Equations 5.32 and 5.33.

Question 5.55 For a fine sandy loam $k_t = 1.38 \times 10^{-6}$ m s^{-1}, $\theta_i = 0.1$, $\theta_t = 0.5$ and $h_f = -0.4$ m.

a. Calculate s_f, I and i for horizontal infiltration in this soil at $t = 30$ min.

b. Do the same for $t = 60$ min.

c. Compare the results of the calculations for $t = 30$ min and $t = 60$ min.

Question 5.56 a. Find s_f at $t = 60$ min for vertical infiltration into the soil of Question 5.55, using Equation 5.44. Use the following procedure:

- calculate t using the value found for s_f in Answer 5.55b
- calculate t also for a somewhat larger value of s_f (larger, because gravity is included now)

- find a better guess for s_f by linear interpolation
- calculate again t for the value of s_f just found
- if necessary, repeat these steps.

b. Calculate also I and i using the value found for s_f in a.

c. Compare the results of these calculations with those for horizontal infiltration (Answer 5.55b).

For not too large t, an approximation for s_f can be derived mathematically from Equation 5.44. The logarithmic part of Equation 5.44 can be written as:

$$\ln\left(\frac{s_f - h_f}{-h_f}\right) = \ln\left(1 - \frac{s_f}{h_f}\right) = \ln(1+y) \tag{5.45}$$

where $y = -s_f/h_f$. For $y^2 < 1$, ln $(1 + y)$ can be written in series form, which leads to the solution (for details see Appendix 3):

$$s_f = f_1 t^{1/2} + f_2 t + f_3 t^{3/2} + \ldots \tag{5.46}$$

$$\text{where } f_1 = \left(\frac{-2k_t h_f}{\theta_t - \theta_i}\right)^{1/2} \tag{5.47}$$

$$f_2 = \frac{2k_t}{3(\theta_t - \theta_i)} \tag{5.48}$$

$$f_3 = \left[-\left(\frac{k_t}{\theta_t - \theta_i}\right)^3 \cdot \frac{1}{18\,h_f}\right]^{1/2} \tag{5.49}$$

Using Equation 5.33, Equation 5.46 can be transformed into an equation identical to Equation 5.28:

$$I = St^{1/2} + At + Bt^{3/2} + \ldots \tag{5.50}$$

where $S = [-2k_t h_f(\theta_t - \theta_i)]^{1/2}$, as in Equation 5.40, and

$$A = \frac{2}{3}k_t \tag{5.51}$$

$$B = \left(\frac{-k_t^3}{18h_f(\theta_t - \theta_i)}\right)^{1/2} \tag{5.52}$$

Thus the first term in Equation 5.50 is identical to the result for horizontal infiltration (Equation 5.39) and represents the contribution of the pressure head gradient. The following terms represent the contribution of gravity. Of course, the infiltration rate associated with Equation 5.50 is, in form, identical to Equation 5.29.

It should be stressed that these results of the Green and Ampt model are only in form identical to the general solution represented by Equations 5.28 and 5.29. The coefficients of Equation 5.50 have different values and are evaluated in a quite different way. The results can only be used for small times ($y^2 < 1$) and for soils which exhibit sharp wetting fronts.

Question 5.57 Check that for the soil given in Question 5.55, the approximation given above can be used as long as $t < 9$ h 53 min.

Question 5.58 a. Calculate s_f, I and i for vertical infiltration in the soil given in Question 5.55 for $t = 60$ min, using Equations 5.46 and 5.50.

b. Compare the results with those of Answer 5.56.

Question 5.59 a. Calculate s_f for $t = 9$ h 53 min, using Equation 5.46.

b. What is the correct value for s_f when Equation 5.44 is used?

c. Draw a conclusion about the error made.

5.3.3.6 Redistribution

When infiltration of water into a dry soil has ceased, it will be followed by redistribution until the hydraulic potential is uniform throughout the profile. Water present above the wetting front at the time of cessation of infiltration (Figure 5.19 at $t = t_1$) will start to drain due to gravity and the soil below that wetting front will also become wet ($t = t_2$). After some time, the soil below the original wetting front may also dry after it was firstly wetted (see shaded area in Figure 5.19). It should be clear that redistribution is a very complex process, in which hysteresis (Section *3.1.5*) plays an important role. Because of hysteresis, the drying upper soil will, at equilibrium, be wetter than the wetting soil below. The hydraulic potential gradients diminish in time

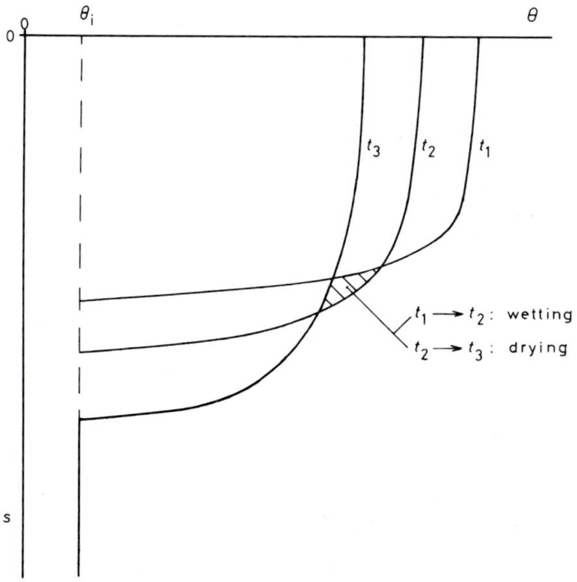

Figure 5.19 Water content profiles at end of infiltration (t_1) and at two later points in time as a result of redistribution.

and may become very small, although equilibrium is hardly ever reached. Thus, hysteresis retards and diminishes redistribution. True equilibrium is only possible when a groundwater table is present (see Figure 5.7). Without such a groundwater table, gravity prevents the water from ever reaching an equilibrium state. Thus, the concept of field capacity (Section *3.1.5*) has no firm theoretical basis, but has practical significance because after one or two days, flux densities usually have become negligible.

5.4 SUGGESTED LITERATURE

BAVER, L. D., W. H. GARDNER and W. R. GARDNER, 1972. Soil physics (fourth edition). Wiley, 489 p. Chapters 9 and 10.

BRUCE, R. R., K. W. FLACH and H. M. TAYLOR (eds.), 1973. Field Soil Water Regime. SSSA Special Publication no. 5, 212 p. Chapter 5.

CHILDS, E. C., 1969. An introduction to the physical basis of soil water phenomena. Wiley, 493 p. Chapters 12 and 13.

GROENEVELT, P. H. and J. W. KIJNE, 1980. Physics of Soil Moisture. Chapter 5 in: Drainage Principles and Applications. Publication 16, Vol. 1, Int. Inst. for Land Reclamation and Improvement (ILRI), Wageningen, The Netherlands, 2nd edition.

HILLEL, D., 1980. Fundamentals of Soil Physics. Academic Press, 413 p. Chapters 8 and 9.

HILLEL, D., 1980. Applications of Soil Physics. Academic Press, 385 p. Chapters 2 and 3.

KIRKHAM, D. and W. L. POWERS, 1972. Advanced Soil Physics. Wiley, 534 p. Chapters 6 and 7.

KLUTE, A., 1969. Movement of Water in Unsaturated Soils. In: The Progress of Hydrology, Vol. 2 p. 821–888. Dept. of Civil Engineering, Urbana I11. USA.

MARSHALL, T. J. and J. W. HOLMES, 1979. Soil Physics. Cambridge University Press, 345 p. Chapters 4 and 5.

NIELSEN, D. R., R. D. JACKSON and J. W. CARY (eds.), 1972. Soil Water. ASA, Madison, Wis. 175 p. Chapter 4.

PHILIP, J. R., 1969. Theory of Infiltration. Adv. Hydroscience, 5. Academic Press, p. 215–296.

TAYLOR, S. A., and G. L. ASHCROFT, 1972. Physical Edaphology. Freeman and Co, 533 p. Chapter 8.

YONG, R. N. and B. P. WARKENTIN, 1975. Soil properties and behaviour. Elsevier Scientific Publishing Co., 449 p.

Answer 5.1 From Figure 5.1 it is obvious that there is no change in water content from $h_m = 0$ to $h_m = -0.20$ m. Thus the air-entry value of the largest pores is -0.20 m. The pores in class 1 have an average air-entry value $h_m = -0.24$ m.
The corresponding equivalent pore radius is:

$$r = \frac{2\sigma}{-\rho_l g h_m} = \frac{2 \times 7 \times 10^{-2} \text{ N m}^{-1}}{10^4 \times 0.24 \text{ N m}^{-2}} \approx 58 \text{ µm}$$

Thus, in Equation 5.7:
$\Delta\theta_1 = 0.05$ and $r_1 \approx 58$ µm.
The same procedure for classes 2 and 3 yields:
$\Delta\theta_2 = 0.20$ and $r_2 \approx 47$ µm ($h_m = -0.30$ m)
$\Delta\theta_3 = 0.05$ and $r_3 \approx 39$ µm ($h_m = -0.36$ m)
A rough estimate of the saturated hydraulic conductivity is then

$$K = \frac{1}{8 \times 10^{-3} \times 2}(0.05 \times 58^2 + 0.02 \times 47^2 + 0.05 \times 39^2) \times 10^{-12} \text{ m}^2 \text{ Pa}^{-1} \text{ s}^{-1}$$
$$= 4.3 \times 10^{-8} \text{ m}^2 \text{ Pa}^{-1} \text{ s}^{-1}.$$

And, therefore, $k = 4.3 \times 10^{-4}$ m s^{-1}.
One could also first take the average value of θ of each class, read off the corresponding value of h_m, and calculate the value of r for this value of h_m. This yields an only slightly different value of k.
In general, the step sizes in h_m should be kept small, because h_m is directly (not linearly) related to the pore size. The step sizes in θ have no direct influence on the accuracy of the calculations, since they only reflect the number of pores within a certain size range.

Answer 5.2 The unit of $K_i = \eta K$ is Pa s × m^2 Pa^{-1} s^{-1} = m^2.
The right hand side of Equation 5.8 also has m^2 as unit, because both τ and $(\Delta\theta)_i$ are dimensionless and the unit of r_i^2 is m^2.

Answer 5.3 a. In Answer 5.1, $K = 4.3 \times 10^{-8}$ m^2 Pa^{-1} s^{-1}.
With $\eta = 10^{-3}$ Pa s, $K_i = 10^{-3} \times 4.3 \times 10^{-8}$ m^2 $= 4.3 \times 10^{-11}$ m^2.
b. A highly conducting soil has $k \approx 1$ m d$^{-1} \approx 1.2 \times 10^{-5}$ m s^{-1} and $K = k/\rho_l g \approx 1.2 \times 10^{-9}$ m^2 Pa^{-1} s^{-1}.
With $\eta = 10^{-3}$ Pa s, $K_i \approx 1.2 \times 10^{-12}$ m^2.
A poorly conducting soil has $k \approx 0.01$ m d^{-1}, i.e. 0.01 times the value of a highly conducting soil. Thus, for a poorly conducting soil, $K_i \approx 1.2 \times 10^{-14}$ m^2.

Answer 5.4 The flux density is downwards, so it is negative. The rate of discharge is $\dfrac{1.62 \text{ cm}^3}{1 \text{ min}} = \dfrac{1.62 \times 10^{-6} \text{ m}^3}{60 \text{ s}}$.
The cross-sectional area is 20 cm^2 $= 20 \times 10^{-4}$ m^2.

$$\text{Hence, } q = -\frac{\text{rate of discharge}}{\text{area}} = -\frac{1.62 \times 10^{-6} \text{ m}^3}{60 \times 20 \times 10^{-4} \text{ s m}^2} = -1.35 \times 10^{-5} \text{ m s}^{-1}.$$

$$\frac{dH}{dz} = \frac{0.9\text{ m}}{0.8\text{ m}} = 1.125\text{ m m}^{-1}$$

Then, according to Darcy's law,

$$k = \frac{-q}{dH/dz} = \frac{1.35 \times 10^{-5}\text{ m s}^{-1}}{1.125} = 1.2 \times 10^{-5}\text{ m s}^{-1}$$

Answer 5.5 a. If the cylinder is perforated, water will squirt out because the water pressure at $z = 0.5$ m is greater than atmospheric pressure.

b. h will become zero.

c. Soon after the perforations are made, h is zero both at the top and the bottom of the lower part of the column. This means, that for the lower part $\Delta H = \Delta z$ and $dH/dz = 1$. In the upper part of the column, $dH/dz > 1$. Since both parts are saturated and have the same value of k, the flux density through the upper part will be larger than through the lower part. Thus the lower part will remain saturated, the excess flux density will discharge continuously from the perforations and h will become zero everywhere in the bottom part.

d. From Answer 5.2 and the reasoning above, the diagram below should be self-explanatory. The flux density in the upper part is:

$$q = -k\frac{dH}{dz} = -1.2 \times 10^{-5} \times \frac{0.4}{0.3}\text{ m s}^{-1} = -1.6 \times 10^{-5}\text{ m s}^{-1}$$

In the lower part, $dH/dz = 1$ and thus

$$q = -k = -1.2 \times 10^{-5}\text{ m s}^{-1}$$

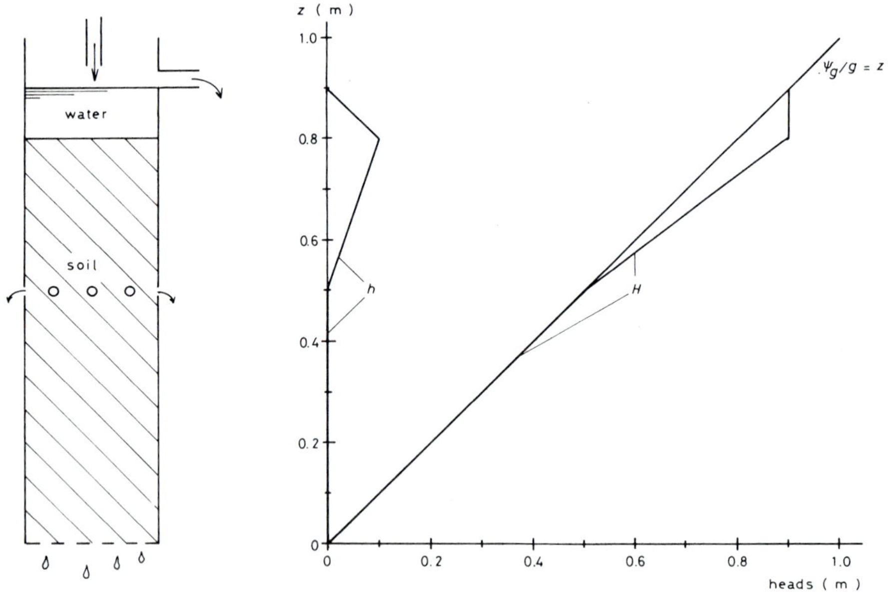

The difference between these two flux densities is 0.4×10^{-5} m s^{-1}.

In a soil column with a cross-section of 20 cm^2, this yields a rate of discharge of $0.4 \times 10^{-5} \times 20 \times 10^{-4}$ m^3 s^{-1} or 28.8 cm^3 per hour.

Answer 5.6 The whole column is and remains saturated. Thus, $\partial\theta/\partial t = 0$. Then, according to the continuity equation without production term:

$$\frac{\partial \theta}{\partial t} = -\frac{\partial q}{\partial s} = 0$$

This means that the flux density is constant with height. The hydraulic conductivity and its variation with depth do not enter into the continuity equation and thus have no effect on this result.

Answer 5.7 At $z = 0$, $H = 0$. Since $\Delta H_2 = 0.6$ m, $H = 0.6$ m at $z = 0.2$ m.

With this the diagram can be drawn.

The flux density in soil 1 is:

$$q = -k\,(\mathrm{d}H/\mathrm{d}z) = -1.2 \times 10^{-5} \times \frac{0.30}{0.60}\ \mathrm{m\ s^{-1}} = -6.0 \times 10^{-6}\ \mathrm{m\ s^{-1}}.$$

Answer 5.8 The flux density is -6.0×10^{-6} m s^{-1} $= -6.0 \times 10^{-6} \times 3600 \times 24$ m d^{-1} ≈ 0.518 m d^{-1}. Therefore, the daily supply should be 518 mm or 5180 m^3 per hectare.

Answer 5.9 a. The drainage system provides atmospheric pressure; hence, $h = 0$ at $z = 0.2$ m. The potential diagram will now be as follows:

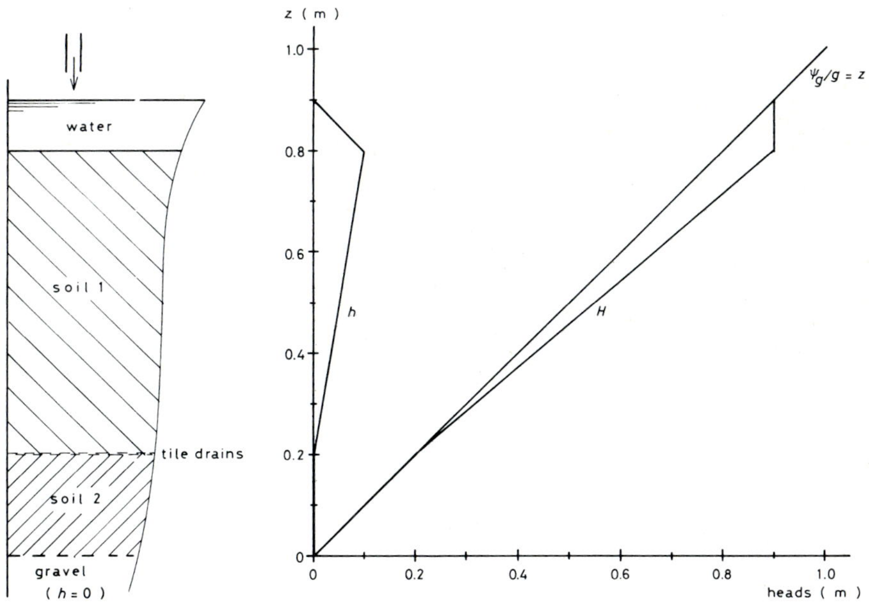

b. From the diagram and the hydraulic conductivities, it follows:

$$q_1 = -1.2 \times 10^{-5} \times \frac{0.7}{0.6} \text{ m s}^{-1} = -1.4 \times 10^{-5} \text{ m s}^{-1}$$

$$q_2 = -0.2 \times 10^{-5} \times \frac{0.2}{0.2} \text{ m s}^{-1} = -2.0 \times 10^{-6} \text{ m s}^{-1}.$$

c. The absolute value of the flux density through the second layer has been reduced from 6.0×10^{-6} m s^{-1} to 2.0×10^{-6} m s^{-1}.
Consequently, the percolation losses have been reduced from 518 mm d^{-1} to 173 mm d^{-1}. Though the latter figure is still very high, the drainage system is effective for recycling purposes.

Answer 5.10 a. Application of Equation 5.10 to this problem gives:

$$\frac{\Delta H_1}{\Delta H_2} = \frac{k_2\,\Delta z_1}{k_1\,\Delta z_2} = 3.5 \times \frac{0.10}{0.70} = 0.5$$

At $z = 0.80$ m, $H = h + z = (0.10 + 0.80)$ m $= 0.90$ m
At $z = 0$, $H = 0$
Thus $\Delta H_1 + \Delta H_2 = \Delta H_1 + 2\Delta H_1 = 0.90$ m
Then $\Delta H_1 = 0.30$ m and $\Delta H_2 = 0.60$ m
With this the potential diagram is determined.

b. The flux density in layer 2 is:

$$q = -k_2 \frac{\Delta H_2}{\Delta z_2} = -1.2 \times 10^{-5} \times \frac{0.60}{0.70} \text{ m s}^{-1} = -1.03 \times 10^{-5} \text{ m s}^{-1}$$

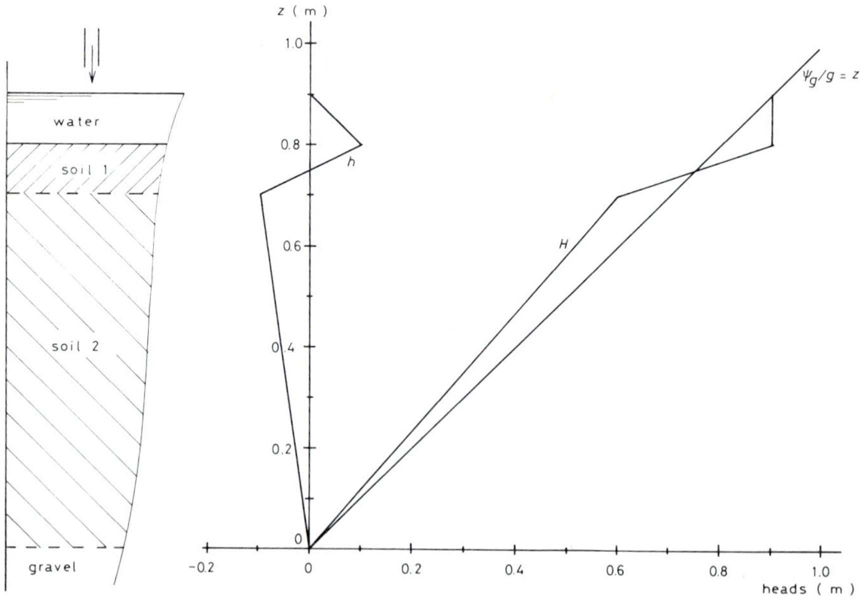

c. The potential diagram shows that h is negative in nearly the entire profile. If the air-entry value would be exceeded, the soil layers would no longer be saturated and their hydraulic conductivity would decrease. This would change the flux density and the potential gradients and make the problem undetermined, unless the change in hydraulic conductivity with matric head is specified. Even then, the problem becomes far more complex.

Answer 5.11 At the bottom of the vessel V, $h = 0.80$ m. If the reference level for z is chosen at the same place, $H = 0.80$ m at $z = 0$. The resistance of the connection pipe between the vessel and the bottom of the soil column can be assumed so small, that the hydraulic head loss there is negligible and thus $H = 0.80$ m at $z = 0.10$ m, at the bottom of the soil column. At $z = 0.60$ m, $h = 0$ and thus $H = 0.60$ m. As before, H is constant in the water layer above the soil column. Thus, at $z = 0.50$ m, $H = 0.60$ m and $h = 0.10$ m. With this the potential diagram is determined.

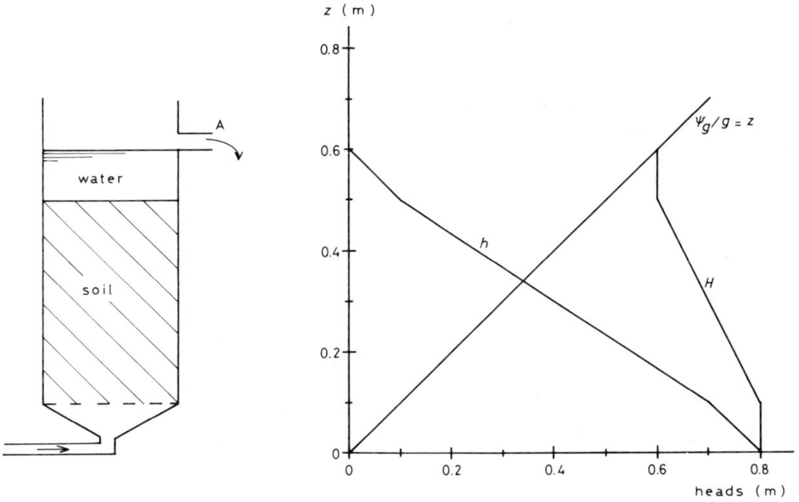

Answer 5.12 The flux density is:

$$q = \frac{180 \times 10^{-6}\ \mathrm{m}^3}{3600\ \mathrm{s} \times 100 \times 10^{-4}\ \mathrm{m}^2} = 5 \times 10^{-6}\ \mathrm{m\ s}^{-1}$$

Across the soil column:

$$\frac{\Delta H}{\Delta z} = \frac{-0.20\ \mathrm{m}}{0.40\ \mathrm{m}} = -0.5\ \mathrm{m\ m}^{-1}$$

$$\text{Thus } k = \frac{-q}{\Delta H/\Delta z} = \frac{5 \times 10^{-6}\ \mathrm{m\ s}^{-1}}{0.5} = 1.0 \times 10^{-5}\ \mathrm{m\ s}^{-1}$$

Answer 5.13 a. Let $z = 0$ at the top of the soil column. Then the groundwater table is at $z = 0.10$ m.

$$\phi \approx 1 - \frac{{}^{b}\rho_d}{\rho_s} = 1 - \frac{1300\ \mathrm{kg\ m^{-3}}}{2600\ \mathrm{kg\ m^{-3}}} = 0.50$$

$${}^{b}\rho_w \approx {}^{b}\rho_d + \phi \times 1000\ \mathrm{kg\ m^{-3}} = 1800\ \mathrm{kg\ m^{-3}}$$

For calculating the soil pressure, two different values for ${}^{b}\rho_w$ have to be used here:

${}^{b}\rho_w = 1000\ \mathrm{kg\ m^{-3}}$ for $0 < z < 0.10$ m

${}^{b}\rho_w = 1800\ \mathrm{kg\ m^{-3}}$ for $-0.40\ \mathrm{m} < z < 0$

At $z =$ 0.10 m, σ_s and σ_w are zero. At $z = 0$, σ_s and σ_w are the same and are $1000\ \mathrm{kg\ m^{-3}} \times 10\ \mathrm{N\ kg^{-1}} \times 0.10\ \mathrm{m} = 1.0$ kPa. In the soil, σ_w increases with 10 kPa per m depth and σ_s increases with $1800\ \mathrm{kg\ m^{-3}} \times 10\ \mathrm{N\ kg^{-1}} \times 1\ \mathrm{m} = 18$ kPa per m depth.

b. At the bottom of the column, $\sigma_w = (1.0 + 0.40 \times 10)$ kPa $= 5.0$ kPa, $\sigma_s = (1.0 + 0.40 \times 18)$ kPa $= 8.2$ kPa and $\sigma_g = \sigma_s - \sigma_w = 3.2$ kPa.

Answer 5.14 a. The whole soil, being static and saturated, has the same value for σ_s as in Answer 5.13, but σ_w changes. At $z = 0$, σ_w is still 1.0 kPa, but at the bottom of the column, σ_w is zero, because the water is under atmospheric pressure at $z = -0.40$ m. The new pressure diagram is then as follows:

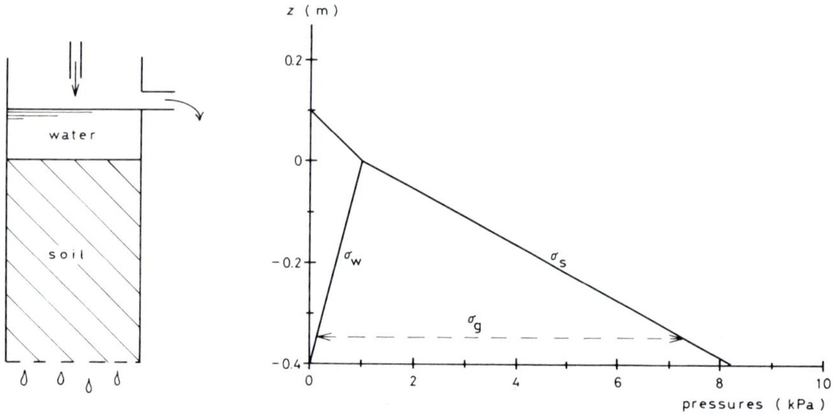

b. At $z = -0.40$ m, $\sigma_s = 8.2$ kPa and $\sigma_w = 0$. Thus, at that height, $\sigma_g = \sigma_s - \sigma_w = 8.2$ kPa. The decrease in σ_w is compensated by the increase in σ_g. In general, downward water transport in a soil profile causes an increase of the intergranular pressure relative to the static equilibrium situation.

Answer 5.15 a. σ_s remains the same as in Answer 5.14, but the pressure in the water, p', has now the same values as in Answer 5.11. Because in a sand column, $\sigma_w = p'$ (Equation 3.56), σ_w at the bottom of the column has increased to 7.0 kPa. Hence, at the bottom of the column, $\sigma_g = (8.2 - 7.0)$ kPa $= 1.2$ kPa.

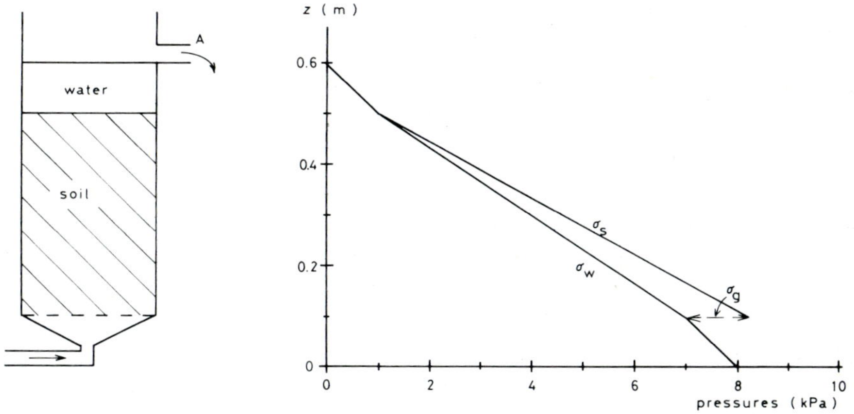

b. If the overflow B is raised from $z = 0.8$ m to $z = 0.92$ m, σ_w at the bottom of the column increases to 8.2 kPa, which equals the value of σ_s. In the whole column then, we find $\sigma_s = \sigma_w$ or $\sigma_g = 0$.

Answer 5.16 At the same water content the water-filled pores of loam are smaller than those of sand. Thus the contact area per volume, and with it the drag force per volume, between the solid phase and the water is the largest for loam. This results in a higher resistance to flow, or a lower hydraulic conductivity.

Answer 5.17 First, use the soil water characteristic to find: at $h_m = -1.0$ m, $\theta \approx 0.425$. Then use the $k(\theta)$ curve to find: at $\theta \approx 0.425$, $k = 6 \times 10^{-3}$ m d^{-1}.

Answer 5.18 For a given value of h_m, the water content may be different due to hysteresis. Since k is in the first place a function of θ, the hydraulic conductivity will be different. Other causes could be entrapped air or structural decay of the soil.

Answer 5.19 The following diagram should be self-explanatory.

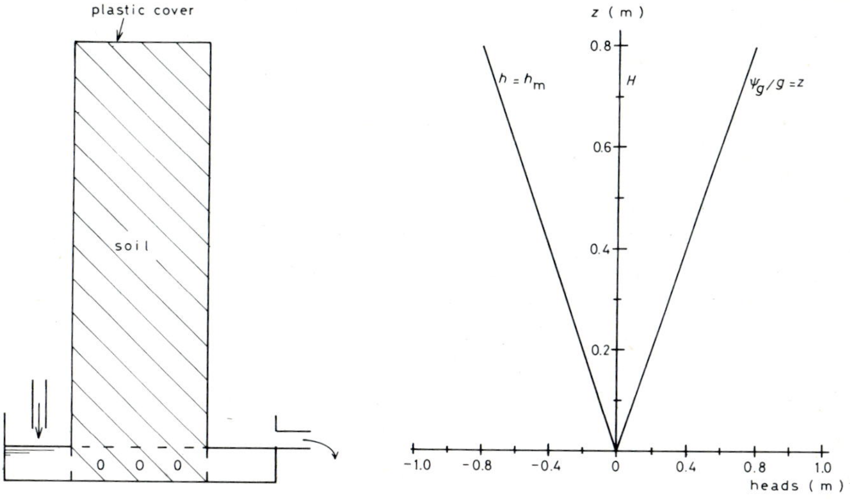

Answer 5.20 The soil is saturated below $z = 0.3$ m. Since at $z = 0.3$ m, $h_m = -0.30$ m, the air-entry value is -0.30 m.

Answer 5.21 In the water-saturated part of the profile, k is constant with height. In steady saturated water transport, the flux density is also constant with height. This implies that dH/dz is constant with height or that H changes linearly with height, and not curvilinearly. The same is true for h.

Answer 5.22 Since q is constant with height, the evaporation rate at the surface is the same as the flux density in the saturated zone. If $h = -0.22$ m at $z = 0.20$ m, then $H = -0.02$ m at that height, and thus:

$$q = -k\frac{dH}{dz} = -0.05 \text{ m d}^{-1}\frac{-0.02}{0.20} = 5 \text{ mm d}^{-1}$$

Answer 5.23 a. In the saturated zone is

$$\frac{dH}{dz} = \frac{q}{-k} = \frac{0.005 \text{ m d}^{-1}}{-0.05 \text{ m d}^{-1}} = -0.1 \text{ m m}^{-1}$$

Then $\frac{dh}{dz} = \frac{dH}{dz} - \frac{dz}{dz} = (-0.1 - 1.0) \text{ m m}^{-1} = -1.1 \text{ m m}^{-1}$

or $h = -1.1\, z + C$

Since $h = 0$ at $z = 0$, $C = 0$.

At the top of the saturated zone, h is equal to the air-entry value, or $h = -0.30$ m (Answer 5.20). Thus the height of the saturated zone is

$$z = \frac{h}{-1.1} = \frac{-0.30}{-1.1} \text{ m} \approx 0.27 \text{ m}$$

b. No. The larger the evaporation rate, the larger the absolute value of the hydraulic potential gradient, and thus also the h gradient, the smaller the height over which the air-entry value will be reached, and thus the smaller the saturated zone.

Answer 5.24 $q = -k\frac{dH}{dz} = -0.10 \text{ m d}^{-1} \times \frac{1.1}{1.0} = -0.11 \text{ m d}^{-1}$

Answer 5.25 In this situation, $h = 0$ at the soil surface and in the whole profile, because $h = 0$ also at $z = 0$ (Answer 5.5d). This means that $H = z$ and $dH/dz = 1$ m m^{-1}.

The sprinkling rate then equals the value of the hydraulic conductivity, 10 cm d^{-1}.

Answer 5.26 H and h are linear functions of z only when dH/dz = constant, i.e. when k is constant throughout the profile. Because the soil is saturated at the groundwater table, it should be saturated throughout the profile.

Answer 5.27 A flux density of 8 cm d^{-1} corresponds to $\frac{dH}{dz} = -\frac{q}{k} = 0.8 \text{ m m}^{-1}$.

Hence, at the soil surface, $H = 0.8$ m and $h = -0.2$ m.

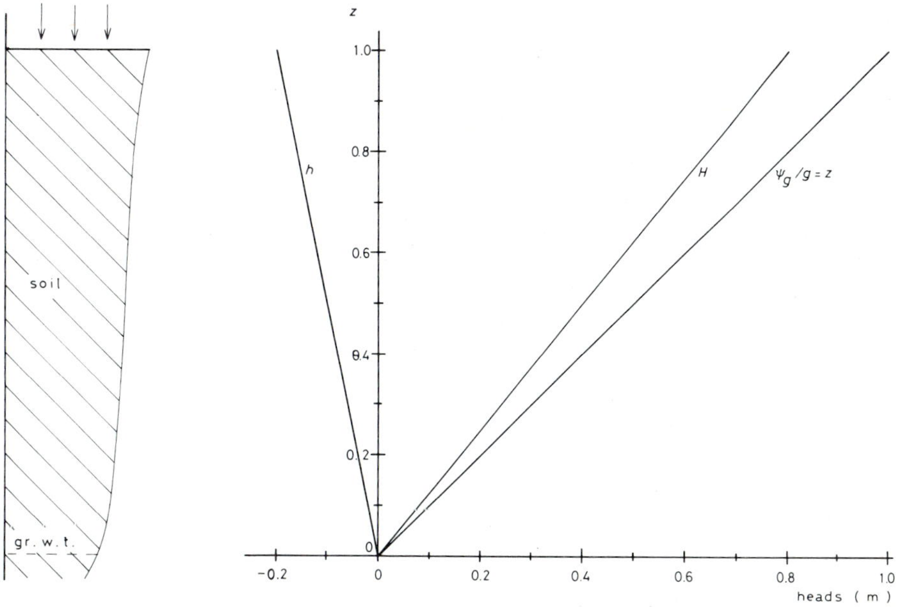

Answer 5.28 The soil will be saturated up to $z = 0.5$ m. See Answer 5.23a for the procedure of solving this problem.

Answer 5.29 In the unsaturated zone, where h is decreasing with height, k is also decreasing with height. Therefore, $\mathrm{d}H/\mathrm{d}z$ must increase with height to satisfy the condition for steady state: $q = -k\,\mathrm{d}H/\mathrm{d}z = \text{constant}$.

Consequently, $\mathrm{d}h/\mathrm{d}z$ must also increase with height, i.e. become less negative. Now suppose $\mathrm{d}h/\mathrm{d}z$ would become positive at a certain height. Then both h and k would increase with height. To keep the flux density constant, $\mathrm{d}H/\mathrm{d}z$ would have to decrease with height, and thus also $\mathrm{d}h/\mathrm{d}z$. This is in conflict with the supposition made above. Thus, we can conclude that the highest possible value of $\mathrm{d}h/\mathrm{d}z$ is zero and that of $\mathrm{d}H/\mathrm{d}z$ is unity. In this situation h has reached the constant value at which k is equal to the absolute value of the imposed flux density. The water then flows through the soil under the gravitational potential gradient of $1\ \mathrm{m\,m^{-1}}$.

Answer 5.30 a. From $q = -1.0\ \mathrm{cm\,d^{-1}}$ and $k = 10\ \mathrm{cm\,d^{-1}}$ it follows:

$$\frac{\mathrm{d}H}{\mathrm{d}z} = -\frac{q}{k} = 0.1\ \mathrm{m\,m^{-1}}\ \text{and}$$

$$\frac{\mathrm{d}h}{\mathrm{d}z} = \frac{\mathrm{d}H}{\mathrm{d}z} - \frac{\mathrm{d}z}{\mathrm{d}z} = (0.1 - 1)\ \mathrm{m\,m^{-1}} = -0.9\ \mathrm{m\,m^{-1}}$$

The air-entry value is $h = -0.40$ m (Question 5.28). This value is reached at $z = 0.44$ m (for calculation procedure, see Answer 5.23a). Thus the soil is saturated below $z = 0.44$ m. Above $z = 0.44$ m, the gradient $\mathrm{d}H/\mathrm{d}z$ increases with height,

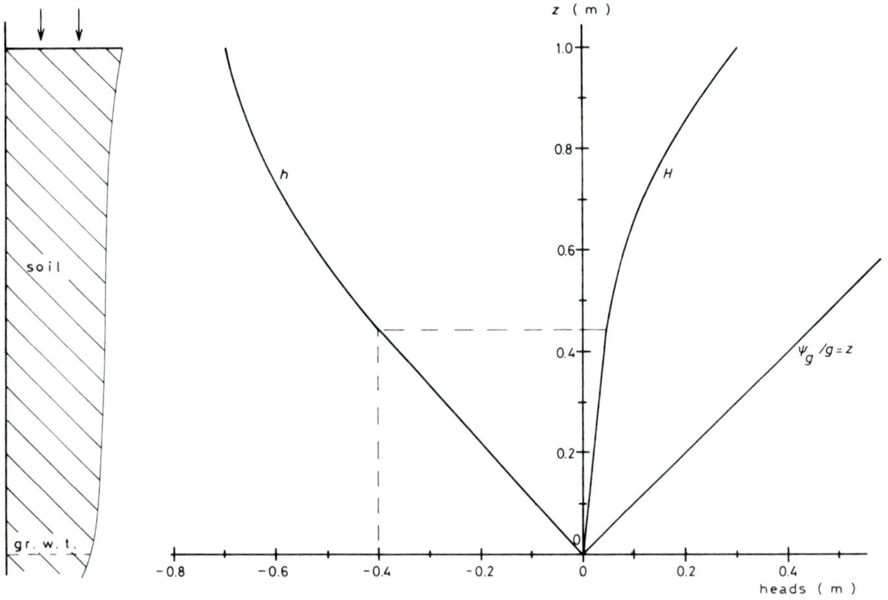

but cannot exceed the value for which $\mathrm{d}h/\mathrm{d}z = 0$. With these considerations the potential diagram can be drawn.

b. Above $z = 0.44$ m, the soil is unsaturated and θ as well as k vary. Since k is not given in the unsaturated zone, we only can guess the most probable shapes of the H and h profiles. Even if k were known, calculation of these profiles is complex because of the nonlinear nature of the problem.

Answer 5.31 Since z is defined positive upward, a positive flux density is directed upward and a negative flux density downward. Also $q = -k\ \partial H/\partial z$. Thus when $\partial H/\partial z$ is positive, then q is negative, and thus directed downward. Similarly, when $\partial H/\partial z$ is negative, the flux density is directed upward. There is no flow when $\partial H/\partial z = 0$.

Answer 5.32 At the soil surface, $h < 0$. Thus, there can not be free water on the soil surface.

Answer 5.33 Below height A there is still upward movement, because $\partial H/\partial z < 0$. Above height A, the transport is now downward. At height A there is no movement, because $\partial H/\partial z = 0$.

Answer 5.34 At some height z_1 above A the flux density q_1 is negative, because there $\partial H/\partial z$ is positive. At some height z_2 below A the flux density is positive, because $\partial H/\partial z$ is negative. Therefore, at height A,

$$\frac{\partial q}{\partial z} = \lim_{(z_1 \to A \text{ and } z_2 \to A)} \left(\frac{q_1 - q_2}{z_1 - z_2}\right) < 0,$$

because $(q_1 - q_2) < 0$ and $(z_1 - z_2) > 0$.

Thus, at height A, $\frac{\partial \theta}{\partial t} = -\frac{\partial q}{\partial z} > 0$

This means that θ increases with time. Even without the above proof, it is easy to see that the water content at A must increase, because water is moving towards A both from below and from above.

Answer 5.35 a. When there is fog, there is no evaporation. But, downward and upward water flow within the soil profile will continue until finally static equilibrium is attained. Since at static equilibrium $\partial H/\partial z = 0$, H is constant and zero throughout the soil profile, because $h = 0$ at $z = 0$.

b. At equilibrium, h is linear with height and equal to $-z$.

A little above C, $H = 0$ at time t_1, and thus at that height, h at this final stage will be the same as h at time t_1. Therefore, the water content also will remain the same at that height. Above this point, h and θ will decrease compared to their values at t_1. Below this point, both h and θ will increase. At point E and below, the soil is saturated at time t_1. Although there h will increase with time, because the flux will vanish, the water content cannot increase any further.

Answer 5.36 By means of tensiometers, measure h at three different heights, i.e. at z, $z + s$ and $z - s$. Add z to each h to obtain H. Estimate $\partial H/\partial z$ at the heights $z + \frac{1}{2}s$ and $z - \frac{1}{2}s$, assuming that H changes linearly with height in the intervals from z to $z + s$ and from $z - s$ to z. By means of a predetermined soil water characteristic, determine θ that corresponds to h at z, $z + s$ and $z - s$ (assuming $h = h_m$). Assuming that θ changes linearly with height, find θ at the heights $z + \frac{1}{2}s$ and $z - \frac{1}{2}s$. By means of a predetermined $k(\theta)$ function, find the k values at heights $z + \frac{1}{2}s$ and $z - \frac{1}{2}s$. Now,

$$q_{z+\frac{1}{2}s} = -k_{z+\frac{1}{2}s}\left(\frac{\partial H}{\partial z}\right)_{z+\frac{1}{2}s} \text{ and } q_{z-\frac{1}{2}s} = -k_{z-\frac{1}{2}s}\left(\frac{\partial H}{\partial z}\right)_{z-\frac{1}{2}s}$$

Finally, a good estimate of $\left(\frac{\partial \theta}{\partial t}\right)_z$ is

$$\left(\frac{\partial \theta}{\partial t}\right)_z = -\frac{q_{z+\frac{1}{2}s} - q_{z-\frac{1}{2}s}}{(z+\frac{1}{2}s) - (z-\frac{1}{2}s)}$$

Answer 5.37 By reading the appropriate values from Figure 5.6 the following scheme can be set up:

depth	z	h	H	θ	$\bar{\theta}$	k	$\frac{\partial H}{\partial z}$	q	$\frac{\partial \theta}{\partial t}$
m	m	m	m			mm d^{-1}	m m^{-1}	mm d^{-1}	d^{-1}
0.30	0.20	−2.00	−1.80	0.090					
					0.14	0.19	−11	2.1	
0.40	0.10	−0.80	−0.70	0.190					0.105
					0.24	4.2	−3	12.6	
0.50	0	−0.40	−0.40	0.295					

Answer 5.38 a. Using average h values the calculation scheme is:

depth	z	h	H	$\bar{h}$	θ	k	$\frac{\partial H}{\partial z}$	q	$\frac{\partial \theta}{\partial t}$
m	m	m	m	m		mm d^{-1}	m m^{-1}	mm d^{-1}	d^{-1}
0.30	0.20	−2.00	−1.80						
				−1.40	0.11	0.08	−11	0.88	
0.40	0.10	−0.80	−0.70						0.117
				−0.60	0.24	4.2	−3	12.6	
0.50	0	−0.40	−0.40						

Using average k values yields:

depth	z	h	H	θ	k	$\bar{k}$	$\frac{\partial H}{\partial z}$	q	$\frac{\partial \theta}{\partial t}$
m	m	m	m		mm d^{-1}	mm d^{-1}	m m^{-1}	mm d^{-1}	d^{-1}
0.30	0.20	−2.00	−1.80	0.090	0.03				
						0.39	−11	4.29	
0.40	0.10	−0.80	−0.70	0.190	0.75				0.57
						20.4	−3	61.2	
0.50	0	−0.40	−0.40	0.295	40				

b. The results in Answer 5.37 show that the θ values derived from the measured h values vary nearly linearly with depth. We will, therefore, consider these results as standard against which the other two calculation schemes can be evaluated.
A comparison of the results of the first two schemes shows that using average values of h leads to an underestimation of k for the drier soil. The error in the final value of $\partial\theta/\partial t$ is about 10%. The third scheme shows that using average k values leads to a value of $\partial\theta/\partial t$ which is almost six times too large. This illustrates the extreme nonlinear character of unsaturated water transport.

Answer 5.39 The calculation scheme is as follows:

depth	z	h	H	θ	$\bar{\theta}$	k	$\frac{\partial H}{\partial z}$	q	$\frac{\partial \theta}{\partial t}$
m	m	m	m			mm d^{-1}	m m^{-1}	mm d^{-1}	d^{-1}
0	0.80	0	0.80	0.50					
					0.49	50	4.0	−200	
0.10	0.70	−0.30	0.40	0.48					1.10
					0.46	15	6.0	−90	
0.20	0.60	−0.80	−0.20	0.44					0.68
					0.42	5.5	4.0	−22	
0.30	0.50	−1.10	−0.60	0.40					0.10
					0.36	2.0	6.0	−12	
0.40	0.40	−1.60	−1.20	0.32					0.24
					0.37	2.4	−5.0	12	
0.50	0.30	−1.00	−0.70	0.42					0.24
					0.445	9.0	−4.0	36	
0.60	0.20	−0.50	−0.30	0.47					0.04
					0.475	20	−2.0	40	
0.70	0.10	−0.20	−0.10	0.48					0.10
					0.49	50	−1.0	50	
0.80	0	0	0	0.50					

Answer 5.40 A column of soil with surface area A can be divided into n slices, each with thickness Δs. The volume of water absorbed by each slice is then $A(\theta-\theta_i)\Delta s$, where θ is the average volume fraction of water of the slice.

The volume of water absorbed by the whole column then equals
$V = A \sum_n (\theta - \theta_i)\Delta s$

For $\Delta s \to 0$ this can be written as the integral

$$V = A \int_{s=0}^{\infty} (\theta-\theta_i)\,ds$$

The volume divided by the surface area is then

$$I = \frac{V}{A} = \int_{s=0}^{\infty} (\theta-\theta_i)\,ds$$

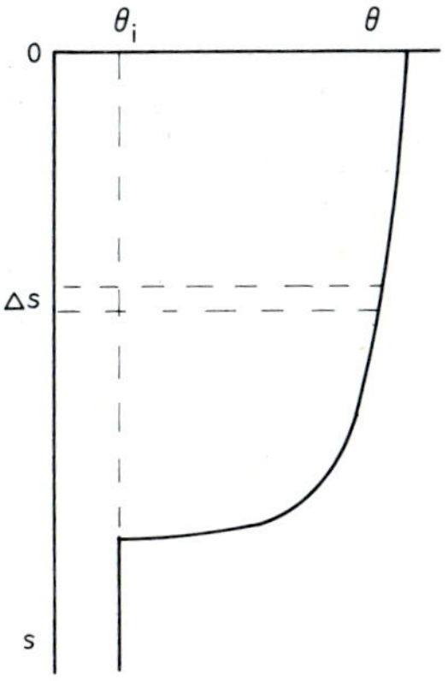

(Whereas θ_i has been assumed constant it could also vary with s, without changing Equation 5.14. The integral would only get another value.)

Answer 5.41 Suppose at time t, the cumulative infiltration is I and at time $t + \Delta t$, $I + \Delta I$. The mean flux density during the time interval Δt is then the increase of the cumulative infiltration ΔI, divided by Δt. i.e. $\Delta I/\Delta t$.

The flux density (infiltration rate) at time t can be found by letting Δt approach zero:

$$i = \lim_{\Delta t \to 0} \left(\frac{\Delta I}{\Delta t}\right) = \frac{dI}{dt}$$

Answer 5.42 Equation 5.15 may be written as: $dI = i dt$.

Integration with boundary condition $I = 0$ for $t = 0$ gives:

$$I = \int_0^t i dt$$

Answer 5.43 Since the ratio $s/\sqrt{t}$ is constant for a given value of θ, the distance from the wetting surface to the visible wetting front (or any other value of θ) increases proportional with the square root of time.

Answer 5.44 Since $S = \int_{\lambda=0}^{\infty} (\theta - \theta_i)d\lambda$, S can be found by evaluating the area between the θ axis, the θ_i line and the $\theta(\lambda)$ curve.

Answer 5.45 Using the product rule $d(\lambda\theta) = \lambda d\theta + \theta d\lambda$ and the boundary conditions $\theta = \theta_i$ for $\lambda \to \infty$ and $\theta = \theta_0$ for $\lambda = 0$,

$$\int_{\lambda=0}^{\infty} (\theta - \theta_i)d\lambda = \int_{\lambda=0}^{\infty} d(\theta - \theta_i)\lambda - \int_{\theta_0}^{\theta_i} \lambda d\theta =$$

$$(\theta - \theta_i)\lambda \Big|_{\lambda=0}^{\infty} - \int_{\theta_0}^{\theta_i} \lambda d\theta = 0 - 0 - \int_{\theta_0}^{\theta_i} \lambda d\theta = \int_{\theta_i}^{\theta_0} \lambda d\theta$$

Also, from Figure 5.17 it is obvious that $\int_{\theta_i}^{\theta} \lambda d\theta$ and $\int_0^{\infty} (\theta - \theta_i)d\lambda$ represent the same area.

Answer 5.46 S is not constant for a given soil. The $\theta(\lambda)$ curve is the solution of Equation 5.16 subject to conditions 5.17. Therefore, S depends on θ_i, θ_0 and on the $D(\theta)$ function of the soil. While θ_0 (saturation) and the $D(\theta)$ function normally do not vary for a given soil, S will still depend on θ_i.

Answer 5.47 a. Using $I = S\sqrt{t}$, we find for heavy clay:

$I = 5 \times 10^{-5}$ m s$^{-1/2} \times \sqrt{24 \times 3600 \text{ s}} \approx 1.47 \times 10^{-2}$ m or 1.47 cm

b. For coarse sand:

$I = 2 \times 10^{-3}$ m s$^{-1/2} \times \sqrt{24 \times 3600 \text{ s}} \approx 0.59$ m or 59 cm

Answer 5.48 a and b:

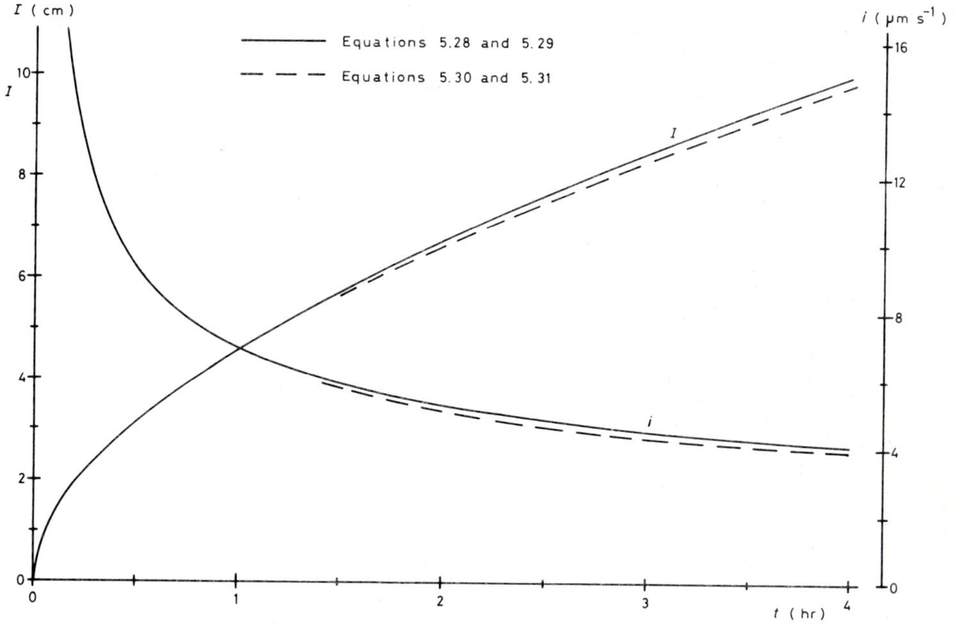

c. Truncating Equations 5.28 and 5.29 after the second term introduces an error which increases with time. In this case, the error is still small after 4 hours of infiltration.

Answer 5.49 Since k is only a function of θ, and θ is constant, k must be constant. According to the continuity equation: if θ is constant with time ($\partial\theta/\partial t = 0$), then $\partial q/\partial s$ is zero, i.e. q is independent of s.

Answer 5.50 Since in the wetted zone q and k_t are independent of s, $\partial h/\partial s$ also has the same value throughout this zone, i.e. h is a linear function of s. Thus h decreases uniformly from zero at the soil surface to h_f at the wetting front and $\partial h/\partial s$ may be replaced by h_f/s_f.

Answer 5.51 While k_t and h_f are constant, s_f increases because the wetting front advances. Thus, $q = -k_t\left(\frac{h_f}{s_f} - 1\right)$ decreases with time.

Answer 5.52 a. The quantity between brackets is positive because h_f is negative.
b. Since k_t, θ_t, θ_i and h_f are constant, $S = [-2k_t h_f(\theta_t - \theta_i)]^{1/2}$ is constant.

Answer 5.53 a. For sample 1, $s_1 = \left(\frac{-2k_t h_f}{\theta_t - \theta_1}\right)^{1/2} t^{1/2}$

For sample 2, $s_2 = \left(\frac{-2k_t h_f}{\theta_t - \theta_2}\right)^{1/2} t^{1/2}$

Thus, $\frac{s_1}{s_2} = \left(\frac{\theta_t - \theta_2}{\theta_t - \theta_1}\right)^{1/2}$

b. $I_1 = (\theta_t - \theta_1)s_1$ and $I_2 = (\theta_t - \theta_1)s_2$

Therefore, $\frac{I_1}{I_2} = \frac{\theta_t - \theta_1}{\theta_t - \theta_2}\left(\frac{\theta_t - \theta_2}{\theta_t - \theta_1}\right)^{1/2} = \left(\frac{\theta_t - \theta_1}{\theta_t - \theta_2}\right)^{1/2}$

c. $t_1 = \frac{\theta_t - \theta_1}{-2k_t h_f} s^2$ and $t_2 = \frac{\theta_t - \theta_2}{-2k_t h_f} s^2$

Thus, $\frac{t_1}{t_2} = \frac{\theta_t - \theta_1}{\theta_t - \theta_2}$

d. $S_1 = [-2k_t h_f(\theta_t - \theta_1)]^{1/2}$, $S_2 = [-2k_t h_f(\theta_t - \theta_2)]^{1/2}$

Thus, $\frac{S_1}{S_2} = \left(\frac{\theta_t - \theta_1}{\theta_t - \theta_2}\right)^{1/2}$

e. $i_1 = \frac{S_1}{2\sqrt{t}}$ and $i_2 = \frac{S_2}{2\sqrt{t}}$

Therefore, $\frac{i_1}{i_2} = \frac{S_1}{S_2} = \left(\frac{\theta_t - \theta_1}{\theta_t - \theta_2}\right)^{1/2}$

Answer 5.54 When $\theta_1 < \theta_2$, $s_1 < s_2$ for a given t

$I_1 > I_2$ for a given t

$t_1 > t_2$ for a given s

$S_1 > S_2$ independent of s and t

$i_1 > i_2$ for a given t

Answer 5.55 For these calculations we use Equations 5.38, 5.39 and 5.26.

a. At $t = 30$ min:

$s_f = \left(\frac{-2k_t h_f}{\theta_t - \theta_i}\right)^{1/2} t^{1/2} =$

$\left(\frac{-2 \times 1.38 \times 10^{-6}\ \text{m s}^{-1} \times (-0.4\ \text{m})}{0.5 - 0.1}\right)^{1/2} \times (1800\ \text{s})^{1/2} \approx$

$1.66 \times 10^{-3}\ \text{m s}^{-1/2} \times 42.43\ \text{s}^{1/2} \approx 0.07048$ m or 7.05 cm

$I = [-2k_t h_f(\theta_t - \theta_i)]^{1/2} t^{1/2} =$

$6.645 \times 10^{-4}\ \text{m s}^{-1/2} \times 42.43\ \text{s}^{1/2} \approx 2.819$ cm

$i = \frac{I}{2t} = \frac{2.819 \times 10^{-2}\ \text{m}}{2 \times 1800\ \text{s}} \approx 7.832 \times 10^{-6}\ \text{m s}^{-1}$

(The results are not rounded further because they are used again in Answer 5.58.)

b. For $t = 60$ min, similar calculations yield

$s_f \approx 9.968$ cm

$I \approx 3.987$ cm and

$i \approx 5.538 \times 10^{-6}\ \text{m s}^{-1}$

c. Both from the equations used and the results calculated, it follows that s_f and I are proportional with the square root of time, and i is inversely proportional with the square root of time.

Answer 5.56 a. Using $s_f = 9.97$ cm in Equation 5.44 gives $t = 3096$ s. Taking $s_f = 11$

cm gives $t = 3716$ s. By linear interpolation a better guess for s_f is

$$\frac{3600 - 3096}{3716 - 3096} \times (11 - 9.97)\text{ cm} + 9.97\text{ cm} = 10.81\text{ cm}$$

For $s_f = 10.81$ cm, $t \approx 3598$ s, which is accurate enough.

Thus, $s_f = 10.81$ cm.

b. Using Equations 5.32 and 5.33 we find:

$I = (\theta_t - \theta_i)\, s_f \approx 4.32$ cm

$$i = k_t \left(1 - \frac{h_f}{s_f}\right) = 1.38 \times 10^{-6}\text{ m s}^{-1} \times \left(1 - \frac{-0.4}{0.1081}\right) \approx 6.49 \times 10^{-6}\text{ m s}^{-1}$$

d. Due to the influence of gravity, the values of s_f, I and i at $t = 60$ min for vertical infiltration are all larger than those for horizontal infiltration.

Answer 5.57 The series approximation is valid only for $y^2 = \left(\frac{-s_f}{h_f}\right)^2 < 1$, or for $s_f < |h_f|$. Substituting $s_f = -h_f = 0.4$ m in Equation 5.44, gives: $t \approx 35577$ s or 9 h 52 min and 57 s. Thus the approximation can be used as long as $t < 9$ h 53 min. Beyond that time $s_f > |h_f|$.

Answer 5.58 a. The first terms of Equations 5.46, 5.50 and 5.29 are equal to the right hand side of Equations 5.38, 5.39 and 5.26, respectively. These are already calculated in Answer 5.55. Therefore, we need to calculate only the second and third terms of Equations 5.46, 5.50 and 5.29.

At $t = 30$ min:

$s_f \approx (7.048 + 0.414 + 0.018)$ cm ≈ 7.48 cm

$I \approx (2.819 + 0.166 + 0.007)$ cm ≈ 2.99 cm

$i \approx (7.832 + 0.920 + 0.061) \times 10^{-6}$ m s^{-1} $\approx 8.81 \times 10^{-6}$ m s^{-1}

At $t = 60$ min:

$s_f \approx (9.968 + 0.828 + 0.052)$ cm ≈ 10.85 cm

$I \approx (3.987 + 0.331 + 0.021)$ cm ≈ 4.34 cm

$i \approx (5.538 + 0.920 + 0.086) \times 10^{-6}$ m s^{-1} $\approx 6.54 \times 10^{-6}$ m s^{-1}

b. The differences with the results found in Answer 5.56 are negligible (less than 1%).

Answer 5.59 a. Using Equation 5.46, we find for s_f at $t = 35580$ s:

$s_f \approx (0.3134 + 0.0818 + 0.0160)$ m ≈ 41.12 cm.

b. The correct value is, of course, 40 cm (Answer 5.57).

c. Even at this limit for the time of infiltration, for which the approximation can be used, the error is less than 3%, and thus more than acceptable for any practical problem.

6 Gas transport in soil

6.1 AMOUNT AND COMPOSITION OF SOIL AIR

The gas phase of soil is characterized by amount or volume and composition. The volume fraction of gas generally decreases with depth in a soil profile due to decreasing porosity as a result of compaction, biological activity, etc. and due to increasing water content. As discussed in Chapter 1, the volume fractions of gas and water are each others complement; together they fill the available pore volume. The critical volume fraction of gas for plant roots depends on the kind of plant and many other factors. A volume fraction of gas of 5–15% is generally considered adequate.

The composition of the soil air also varies with depth and time. Due to free diffusion of the atmospheric components through the soil surface, the composition of the soil air near the surface is almost the same as in the atmosphere, i.e. 80% N_2, with O_2 and CO_2 together forming the remaining 20%. With depth, the concentration of CO_2 increases and that of O_2 decreases because of biological activity in the soil (respiration of plant roots, micro-organisms and soil fauna).

6.2 EQUILIBRIUM CONDITIONS FOR SOIL AIR

Under equilibrium and isothermal conditions, the pressure of air in the gravitational field can be expressed as (Section *2.2.5*):

$$p = p_0 \exp\left(-\frac{M}{RT} gz\right) \qquad (2.52)$$

If $z = 1$ m and $T = 290$ K, Equation 2.52 gives $p = 0.99988\ p_0$. Thus, air pressure changes by 0.012% over a height of 1 m. If this small gradient (0.12 mbar m^{-1}) is ignored, air pressure can be regarded as independent of height (i.e. p is uniform). This assumption applies not only to the total gas pressure, but also to the partial pressures of the different gaseous components. Therefore, soil air and the atmosphere at the soil surface are in equilibrium with each other only if the total gas pressure *and* the composition of the soil air are the same as in the atmosphere.

6.3 BULK FLOW OF GAS

Bulk flow of gas in soil results from a gradient of total gas pressure. This can be caused, for instance, by variations in barometric pressure of the atmosphere, displacement of soil air by infiltrating rain or irrigation water, rise of the groundwater table or temperature changes. We can ignore compression of soil air, because of the very small gas-pressure differences encountered in soils. The bulk flow of soil air can then be expressed analogously to the transport of soil water:

$$f_a = -K_a \frac{\partial p}{\partial s} \tag{6.1}$$

where f_a is the air flux density in $m^3\ m^{-2}\ s^{-1}$, or $m\ s^{-1}$, and $\partial p/\partial s$ is the air pressure gradient in $Pa\ m^{-1}$. Thus, the air conductivity, K_a, has the same unit as the water conductivity ($m^2\ Pa^{-1}\ s^{-1}$).

The air conductivity depends on the volume fraction, size, continuity and geometry of the gas-filled pores. It is, therefore, highly dependent on the volume fraction of water. The air conductivity of a completely dry soil is about 50 times the water conductivity of the same soil when completely saturated, provided the porosity and pore geometry of the soil are the same in both conditions. Of course, for most soils this is not true. The factor 50 is the ratio of the viscosities of water and air ($\eta_w \approx 1.0$ mPa s, $\eta_a \approx 20$ μPa s).

The intrinsic permeability of a soil for air, K_{ia}, can be calculated by an equation similar to that used for water (Section *5.2.2*):

$$K_{ia} = \eta_a K_a \tag{6.2}$$

The value of K_{ia} depends only on the geometry of the soil pores.

Question 6.1 Give examples of changing porosity and pore geometry of soils upon wetting.

Question 6.2 An undisturbed soil sample is first dried completely to measure the air conductivity and then saturated completely to measure the saturated water conductivity. The values found for K_a and K are 5.6×10^{-8} and 0.11×10^{-8} $m^2\ Pa^{-1}\ s^{-1}$, respectively.

a. Calculate K_{ia} and K_{iw}.
b. What do you conclude about this soil?

Saturated air conductivity can be measured in the laboratory with the apparatus depicted in Figure 6.1. By keeping the counterweight slightly less than the weight of the mobile cylinder, the air in the gas cylinder has a slightly higher pressure than the atmosphere. Thus, air will flow through the soil sample as soon as the stopcock, d, is opened. The saturated air conductivity can then be calculated from the air flux density and the pressure difference between the air in the cylinder and the atmosphere.

Question 6.3 In Figure 6.1:

cross-sectional area of cylinder b	= 500	cm^2
pressure difference indicated by manometer	= 5	mbar
cross-sectional area of soil sample	= 10	cm^2
length of soil sample	= 10	cm
velocity of cylinder b	= 0.005	$cm\ s^{-1}$

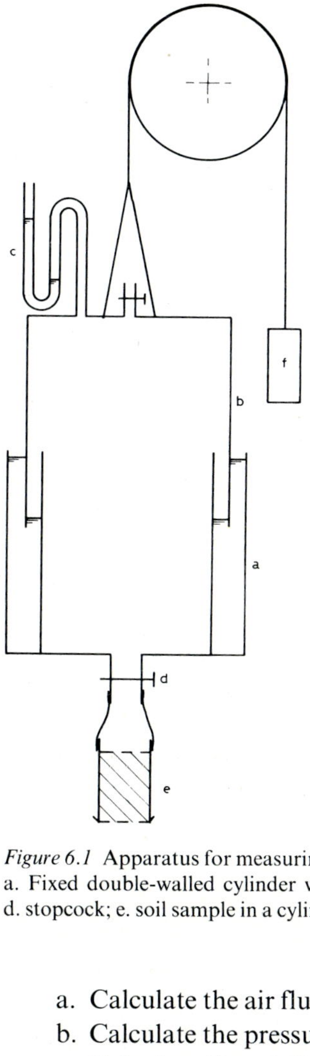

Figure 6.1 Apparatus for measuring air conductivity of a soil sample.
a. Fixed double-walled cylinder with water between the walls; b. mobile cylinder; c. water manometer; d. stopcock; e. soil sample in a cylinder; f. counterweight.

a. Calculate the air flux density through the soil sample.
b. Calculate the pressure gradient in the soil sample.
c. Calculate K_a and K_{ia}.

6.4 GAS DIFFUSION

The composition of soil air changes due to consumption of O_2 and production of CO_2 as a result of respiration of soil organisms and plant roots (Section 6.1). The volume fraction of CO_2 in soil air is more than the 0.03% in the atmosphere. In general, the volume fractions of CO_2 and O_2 together are nearly constant at about 20%. Root growth and other activities of most plants are hindered if the volume fraction of CO_2 is more than about 5% (O_2 fraction < 15%), while the upper limit for plants is about

10% CO_2. Supply of O_2 and removal of CO_2 by exchange with the atmosphere is called soil aeration.

Since bulk flow of soil air contributes very little to the aeration of the root zone (Section 6.3), exchange of O_2 and CO_2 with the atmosphere takes place mainly by diffusion. Diffusion can be described by Fick's law (Section 4.2):

$$f_i = -D_g \frac{\partial \chi_i}{\partial s} \tag{4.24}$$

where D_g is the diffusion coefficient for gases in the particular soil. Gas diffusion in soil due to a partial pressure gradient of one component is always accompanied by counter-current diffusion of other components. The molecules diffusing in opposite directions hinder each other to such a high degree that the friction of the pore walls is negligible compared to the internal friction in the gas. The influence of pore sizes is, therefore, so small that it is possible to estimate the value of D_g from the diffusion coefficient, D_0, of the gas mixture in free air. The value of D_0 only needs to be corrected for the following effects:

- Only a fraction of the cross-sectional area normal to the direction of the flux is available for diffusion. It equals the volume fraction of gas, ϕ_g.
- Diffusion through the pores is not in a straight line. The increase of the effective diffusion distance due to the tortuosity of the pore space and the blocking by water of connections between pores is generally estimated at a factor of about 2, even though this tortuosity factor, τ, will be larger in a wet soil than in a dry soil.

Taking into account the above effects yields the following estimate of D_g:

$$D_g \approx \tfrac{1}{2}\phi_g D_0 \tag{6.3}$$

The flux density equation then becomes:

$$f_i = -\tfrac{1}{2}\phi_g D_0 \frac{\partial \chi_i}{\partial s} \tag{6.4}$$

Question 6.4 Explain why counter-current diffusion takes place during diffusion of gases.

6.5 CO_2 DIFFUSION IN SOIL PROFILES

The CO_2 concentration in the gas phase of a soil generally increases with depth. This is one of the reasons why root development and activity is limited to the topsoil. Under average conditions in the root zone of natural soils, the production rate of CO_2, α_{CO_2}, varies between 0.2 and 0.6 mg m^{-3}s^{-1}.

If the production rate and the diffusion coefficient of CO_2 for the soil are known, the CO_2 concentration as a function of depth can be calculated for relatively simple conditions, e.g. when the CO_2 concentration does not change with time (steady state) and the production rate is constant throughout the root zone and zero below the root zone. At steady state, the gradient of the CO_2 concentration at a certain depth must be such that all CO_2 produced below that depth is removed by upward diffusion. Thus, the CO_2 flux density is the highest at the soil surface. It is somewhat easier to solve the problem if a height scale is used, rather than a depth scale, and if $z = 0$ is taken at the bottom of the root zone. The diffusion and continuity equations are then, respectively (Section 4.2):

$$f_{CO_2} = -D_g \frac{\partial \chi_{CO_2}}{\partial z} \tag{6.5}$$

and
$$\frac{\partial G_{CO_2}}{\partial t} = -\frac{\partial f_{CO_2}}{\partial z} + \alpha_{CO_2} \tag{6.6}$$

Equation 6.6 can be simplified because, at steady state, $\partial G/\partial t = 0$ and so

$$\frac{\mathrm{d}f_{CO_2}}{\mathrm{d}z} = \alpha_{CO_2} \text{ or } \mathrm{d}f_{CO_2} = \alpha_{CO_2}\mathrm{d}z \tag{6.7}$$

Note that normal differentials can be used because, at steady state, f and χ are no longer time-dependent, but only a function of height. If α_{CO_2} is assumed constant throughout the root zone, integration of Equation 6.7 yields:

$$f_{CO_2} = \alpha_{CO_2} z + C_1 \tag{6.8}$$

where the integration constant, C_1, is zero because $f_{CO_2} = 0$ at $z = 0$.

Question 6.5 Why is $f_{CO_2} = 0$ at $z = 0$?

Substitution of $\alpha_{CO_2} z$ for f_{CO_2} in Equation 6.5 yields:

$$\alpha_{CO_2} z = -D_g \frac{\partial \chi_{CO_2}}{\partial z} \text{ or } \alpha_{CO_2} z \mathrm{d}z = -D_g \mathrm{d}\chi_{CO_2} \tag{6.9}$$

If D_g is uniform, integration yields:

$$\alpha_{CO_2} z^2 = -2D_g \chi_{CO_2} + C_2 \tag{6.10}$$

The integration constant, C_2, can be found from the condition that $\chi_{CO_2} = 0$ at the soil surface where $z = z_s$ (because CO_2 in the atmosphere is negligible):

$$C_2 = z_s^2 \alpha_{CO_2}$$

So,
$$\chi_{CO_2} = \frac{\alpha_{CO_2}}{2D_g}(z_s^2 - z^2) \tag{6.11}$$

Thus, the concentration of CO_2 is a quadratic function of height.

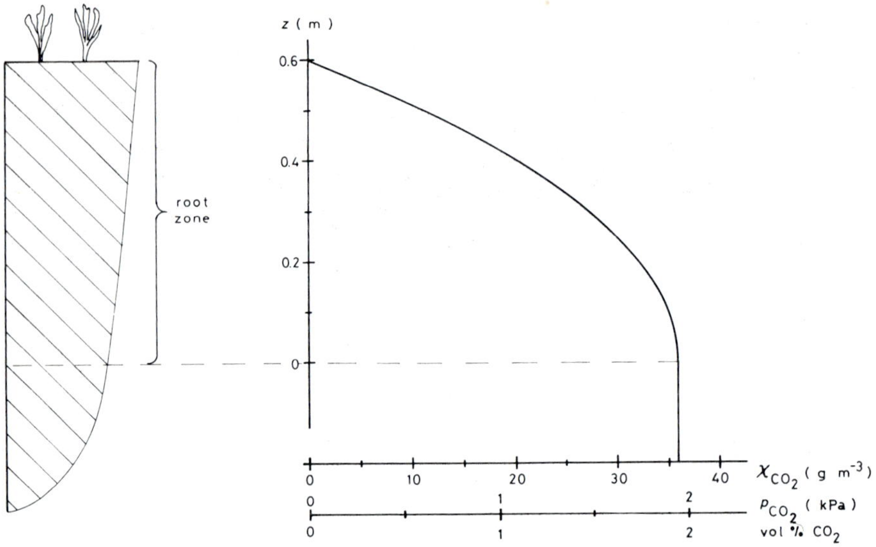

Figure 6.2 Concentration of CO_2 in soil with constant and uniform production of CO_2 and diffusion coefficient.

For a soil with a production rate of CO_2 of 0.6 mg m^{-3} s^{-1}, a 0.6 m root zone and $\phi_g = 0.3$, $D_0 = 2 \times 10^{-5}$ m^2 s^{-1} and $\tau = 2$, Equation 6.11 becomes:

$$\chi_{CO_2} = 0.1 \text{ kg m}^{-5} [(0.6 \text{ m})^2 - z^2] \qquad (6.12)$$

At the lower end of the root zone (where $z = 0$) and below the root zone, the concentration of CO_2 is uniform at 36 g m^{-3}. Figure 6.2 shows the concentration of CO_2 as a function of height as calculated with Equation 6.12.

Question 6.6 Explain why the mass concentration of CO_2 is uniform below the root zone.

Question 6.7 Express Equation 6.11 in terms of depth, s, if $s = 0$ at the soil surface.

Mass concentrations of CO_2 can be converted to partial pressures by the ideal gas law (Answer 4.19):

$$p_{CO_2} = \frac{RT}{M_{CO_2}} \chi_{CO_2} \qquad (6.13)$$

Using $RT = 2.4$ kJ mol^{-1} and $M_{CO_2} = 44$ g mol^{-1}, the partial pressure of CO_2 at the bottom of the root zone is about 1960 Pa. This corresponds with a volume fraction of CO_2 of only about 2%, even though the maximum rate of CO_2 production was assumed. Therefore, damaging concentrations of CO_2 can be expected only in soils with extremely low gas-filled porosities, due to compaction, extreme wetness, etc.

Question 6.8 A soil profile has a root zone of 0.7 m. At steady state, the oxygen consumption is 0.4 mg m^{-3} s^{-1} uniform throughout the root zone. Consumption of O_2 below the root zone is negligible. Calculate the partial pressure of O_2 as a function of height with the following information: $\phi_g = 0.32$, $D_0 = 2 \times 10^{-5}$ m^2 s^{-1}, $RT = 2.4$ kJ mol^{-1}, $M_{O_2} = 32$ g mol^{-1} and volume fraction of O_2 in the atmosphere = 20% (partial pressure = 20 kPa).

The above given solution to estimate the CO_2 concentration as a function of height is based on the simplification that ϕ_g and α_{CO_2} are constant throughout the root zone. Besides, it is limited to steady state. In reality, volume fraction of gas is a function of total porosity and volume fraction of liquid, which usually vary with depth. Also root activity is a function of depth. When the production term and the diffusion coefficient can be expressed as relatively simple functions of height, solutions can sometimes be found analytically for steady state conditions. If not, or if they are time-dependent, one has to use numerical methods (and if available, a digital computer) to find a solution.

6.6 SUGGESTED LITERATURE

BAVER, L. D., W. H. GARDNER and W. R. GARDNER, 1972. Soil Physics (fourth edition). Wiley, 489 p. Chapter 6.

HILLEL, D., 1980. Fundamentals of Soil Physics. Academic Press, 413 p. Chapter 11.

KIRKHAM, D. and W. L. POWERS, 1972. Advanced Soil Physics. Wiley, 534 p. Chapter 9.

TAYLOR, S. A. and G. L. ASHCROFT, 1972. Physical Edaphology. Freeman and Co., 533 p. Chapter 12.

Answer 6.1 The porosity of clay soils changes upon wetting due to swelling, when large pores either disappear or decrease in size. Another example is the slaking of fine-sandy and silty soils, when the porosity and the number of large pores decrease.

Answer 6.2 a. $K_{ia} = \eta_a K_a = 20 \times 10^{-6}$ Pa s $\times\ 5.6 \times 10^{-8}$ m^2 Pa^{-1} s$^{-1} \approx$ 1.1×10^{-12} m^2

$K_{iw} = \eta_w K = 1.0 \times 10^{-3}$ Pa s $\times\ 0.11 \times 10^{-8}$ m^2 Pa^{-1} s$^{-1} \approx 1.1 \times 10^{-12}$ m^2

b. The soil does not swell or slake upon wetting.

Answer 6.3 a. Velocity of cylinder b $= 5 \times 10^{-5}$ m s^{-1} and cross-sectional area of cylinder b $= 0.05$ m^2. Thus, flow rate of air $= 5 \times 10^{-5} \times 0.05$ m^3 s^{-1} $=$ 2.5×10^{-6} m^3 s^{-1}.

Cross-sectional area of soil sample $= 10^{-3}$ m^2. Thus, flux density of air,

$$f_a = \frac{2.5 \times 10^{-6}\ \text{m}^3\ \text{s}^{-1}}{10^{-3}\ \text{m}^2} = 2.5 \times 10^{-3}\ \text{m s}^{-1}.$$

b. If s is taken positive downwards (in the direction of the flux), the pressure gradient $\dfrac{dp}{ds} = \dfrac{-5\ \text{mbar}}{0.1\ \text{m}} = \dfrac{-500\ \text{Pa}}{0.1\ \text{m}} = -5\ \text{kPa m}^{-1}$.

c. With Equation 6.1,

$$K_a = -\frac{2.5 \times 10^{-3}\ \text{m s}^{-1}}{-5 \times 10^3\ \text{Pa m}^{-1}} = 5 \times 10^{-7}\ \text{m}^2\ \text{Pa}^{-1}\ \text{s}^{-1}$$

and with Equation 6.2,

$K_{ia} = 5 \times 10^{-7}$ m^2 Pa^{-1} s$^{-1} \times 20 \times 10^{-6}$ Pa s $= 10^{-11}$ m^2.

Answer 6.4 If the pressure in a gas mixture is to remain constant, the increase in partial pressure of one component must be compensated by an equal decrease in partial pressure of other components. Thus, the partial pressure gradient of the one component is compensated by reverse gradients of other components. These gradients cause the other gas components to move (diffuse) in an opposite direction. That is counter-current diffusion.

Answer 6.5 There is no CO_2 production below $z = 0$. Hence, no CO_2 has to be removed from below $z = 0$. A downward flux also is unlikely, because there usually are layers in the subsoil which block gas diffusion, for instance a groundwater table or rocks. So, below the root zone, $f_{CO_2} = 0$.

Answer 6.6 Below the root zone, there is no flux: $f_{CO_2} = 0$. Therefore, according to Equation 6.5, $\partial \chi_{CO_2}/\partial z = 0$, which means χ_{CO_2} is constant with depth.

Answer 6.7 The continuity equation for the steady state with a production term is:

$$\frac{df_{CO_2}}{ds} = \alpha_{CO_2}$$

For constant α_{CO_2}, integration yields

$f_{CO_2} = \alpha_{CO_2} s + C_1$

Since $f_{CO_2} = 0$ at $s = 0.6$ m, it follows:

$C_1 = -(0.6\text{ m})\,\alpha_{CO_2}$

and $f_{CO_2} = (s - 0.6\text{ m})\,\alpha_{CO_2}$

Substitution in the flux density equation yields

$$(s - 0.6\text{ m})\,\alpha_{CO_2} = -D_g \frac{d\chi_{CO_2}}{ds}$$

and after rearranging and integrating with D_g = uniform:

$$\frac{s^2}{2} - (0.6\text{ m})\,s \quad \alpha_{CO_2} = -D_g \chi_{CO_2} + C_2$$

Since $\chi_{CO_2} = 0$ at $s = 0$, $C_2 = 0$. Therefore,

$$\chi_{CO_2} = \frac{\alpha_{CO_2}}{2D_g}[(1.2\text{ m})\,s - s^2]$$

Hence, for $\phi_g = 0.3$, $\alpha_{CO_2} = 0.6$ mg m^{-3} s^{-1}, $\tau = 2$ and $D_0 = 2 \times 10^{-5}$ m^2 s^{-1}, $\chi_{CO_2} = 0.1$ kg m^{-5} $[(1.2\text{ m})\,s - s^2]$

This equation can also be derived from Equation 6.12 by substituting 0.6 m $- s$ for z, since $s = -z + 0.6$ m. (Check this.)

Answer 6.8 The solution is analogous to that for CO_2, except that the flux is in the opposite direction and the production is negative (consumption). Again $z = 0$ is chosen at the bottom of the root zone, where $f_{O_2} = 0$.

$$\frac{\partial G_{O_2}}{\partial t} = -\frac{\partial f_{O_2}}{\partial z} + \alpha_{O_2} = 0 \text{ and thus } \frac{df_{O_2}}{dz} = \alpha_{O_2}$$

Integration yields $f_{O_2} = \alpha_{O_2} z + C_1$, where $C_1 = 0$ because at $z = 0$, $f_{O_2} = 0$.

Now, the diffusion equation becomes:

$$\alpha_{O_2} z = -D_g \frac{d\chi_{O_2}}{dz} \quad \text{or} \quad \alpha_{O_2} z\,dz = -D_g d\chi_{O_2}$$

After integration: $\alpha_{O_2} z^2 = -2D_g \chi_{O_2} + C_2$

Substituting $\frac{M_{O_2}}{RT} p_{O_2}$ for χ_{O_2} (analogous to Equation 6.13) yields

$$\alpha_{O_2} z^2 = -\frac{2D_g M_{O_2}}{RT} p_{O_2} + C_2$$

C_2 can be found from the condition $p_{O_2} = 20$ kPa at $z = 0.7$ m:

$$C_2 = \frac{2D_g M_{O_2}}{RT} \times 2 \times 10^4\text{ Pa} + (0.7\text{ m})^2\,\alpha_{O_2}$$

$$\text{Hence, } [z^2 - (0.7\text{ m})^2]\,\alpha_{O_2} = \frac{2D_g M_{O_2}}{RT}[2 \times 10^4\text{ Pa} - p_{O_2}]$$

from which

$$p_{O_2} = -\frac{\alpha_{O_2}}{2\,D_g}\frac{RT}{M_{O_2}}[z^2 - (0.7\text{ m})^2] + 2 \times 10^4\text{ Pa}$$

For $\alpha_{O_2} = -0.4 \text{ mg m}^{-3} \text{ s}^{-1}$, $D_0 = 2 \times 10^{-5} \text{ m}^2 \text{ s}^{-1}$, $\phi_g = 0.32$,
$RT = 2.4 \text{ kJ mol}^{-1}$ and $M_{O_2} = 32 \text{ g mol}^{-1}$:
$p_{O_2} = 4.69 \text{ kPa m}^{-2} [z^2 - 0.49 \text{ m}^2] + 20 \text{ kPa}$, or
$p_{O_2} = 17.7 \text{ kPa} + 4.69 \text{ kPa m}^{-2} \times z^2$
This result is shown in the figure below. The partial pressure of O_2 is a quadratic function of height, but decreases with depth. Below the root zone, the partial pressure is about 17.7 kPa, which corresponds with a volume fraction of O_2 in the soil air of about 17.7%.

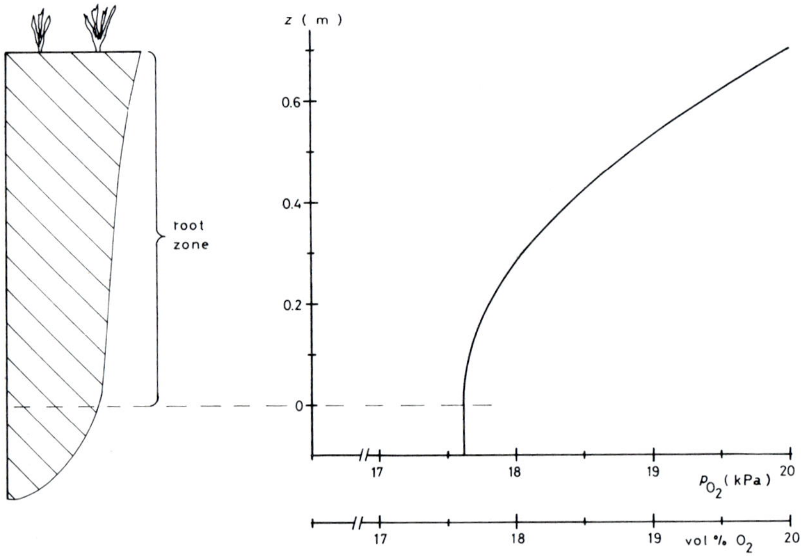

7 Heat transport in soil

7.1 INTRODUCTION

An important aspect of the behaviour and use of soils is their thermal regime. This regime is characterized to a large extent by the temperature. Many processes occurring in soils are strongly influenced by temperature. This is true especially for biological processes, such as germination of seeds, plant growth, root development and activity, microbial activity, etc. Physical and chemical processes influenced by temperature are frost heaving, weathering, decomposition of organic matter, etc.

Soil temperatures are determined by the transport processes of heat within the soil and by exchange of heat between the soil and the atmosphere. There are three basically different processes whereby heat can be transported: conduction, convection and radiation. Conduction of heat occurs by transmission of thermal energy of motion from one microscopic particle to another. Transport of heat by a fluid in motion is called convection. Heat convection with accompanying phase changes can increase heat transfer tremendously. This is especially true for water, which has very high values of latent heat of condensation/evaporation and freezing/melting. Radiation is transfer of thermal energy from a body to its surroundings by electromagnetic waves. Thus, in contrast to conduction and convection, radiation can occur through a vacuum.

Transport of heat within soils can occur by conduction and by convection, with or without latent heat transport. Heat conduction is governed by the thermal soil properties, volumic heat capacity and heat conductivity. The thermal exchange processes at the soil surface are dominated by the meteorological conditions and occur by radiation, conduction and convection, with or without phase changes. The thermal soil properties are strongly dependent on water content. A complete description of the thermal regime of soils is very complex and falls outside the scope of this text. Only the main aspects will be presented.

7.2 THERMAL SOIL PROPERTIES

7.2.1 Volumic heat capacity

The change of heat content of soil divided by volume and by change of temperature is called volumic heat capacity, C_h. It is expressed in J m^{-3} K^{-1} (Section 4.3). The volumic heat capacity of a soil can be obtained by summing the contributions of the different soil components:

$$C_h = \sum_i C_{h,i}\, \phi_i = \sum_i \rho_i\, c_i\, \phi_i \qquad (7.1)$$

where $C_{h,i}$ and c_i are the volumic and specific heat capacities, and ϕ_i and ρ_i are the volume fraction and density of component i. The volumic heat capacities of various soil components are presented in Table 7.1.

Table 7.1 Thermal properties of soil components at 10 °C

	c kJ kg^{-1} K^{-1}	C_h MJ m^{-3} K^{-1}	λ J m^{-1} s^{-1} K^{-1} (W m^{-1} K^{-1})
quartz	0.76	2.0	8.8
clay minerals	0.73	2.0	2.9
organic matter	1.8	2.5	0.25
water	4.2	4.2	0.57
ice (0 °C)	2.1	1.9	2.18
air (saturated with water vapour)	1.0	0.0013	0.025

Question 7.1 a. Derive an expression for the volumic heat capacity of a mineral soil (without organic matter) as a function of ϕ and ϕ_l, using Equation 7.1.
b. What can you conclude about the contribution of the soil air to the volumic heat capacity of the soil?

Answer 7.1 shows that the heat capacity of a soil is a linear function of the water content θ ($= \phi_l$) (Figure 7.1).

7.2.2. *Heat conductivity*

The heat conductivity, λ, of a soil is defined as the heat flux density by conduction through the soil divided by the temperature gradient. It is expressed in J m^{-1} s^{-1} K^{-1} (Section 4.3). The heat conductivity has no simple relationship with the heat conductivities of the individual soil components, because the conduction of heat takes place through all kinds of sequences of the conducting materials, in series and in parallel. The value of λ depends highly on the way in which the best conducting mineral particles are interconnected by the less conducting water phase and are separated by the poorly conducting gas phase.

Figure 7.1 shows the heat conductivity of a soil as a function of volume fraction of liquid. At very low water contents, λ is generally smaller than 0.5 J m^{-1} s^{-1} K^{-1}. The heat transport then takes place mainly through the narrow points of contact between the soil particles. The contribution of the soil air is very small due to the very low heat conductivity of air. A small increase in water content of a dry soil causes only a modest increase of λ, because this water forms thin films around the soil particles. Further increases in water content cause a sharp increase of the heat conductivity, because water has a much higher heat conductivity than air and this water collects

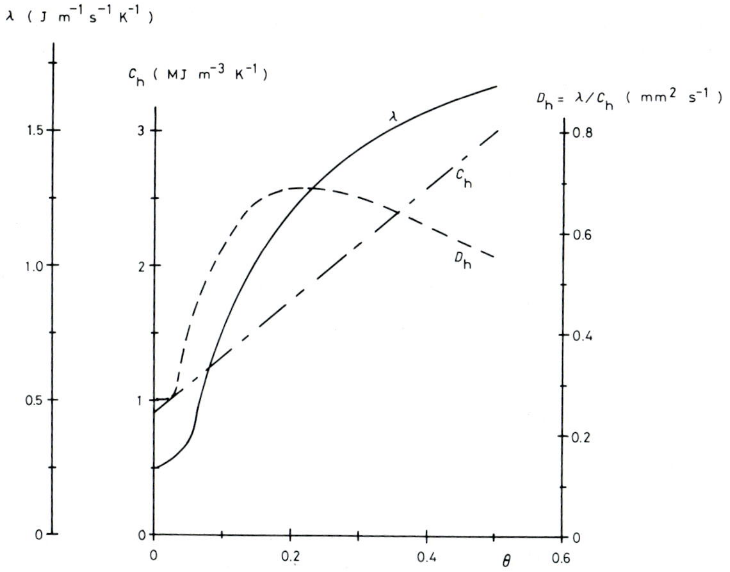

Figure 7.1 Thermal soil properties as a function of volume fraction of liquid.

around the contact points between the soil particles. Water around contact points forms very effective 'bridges' for conduction of heat. Upon further increase of the water content, the value of λ increases ever more gradually, because the conducting cross-sectional area of the water 'bridges' increases ever more slowly. The maximum value of λ is reached at water saturation. For mineral soils this value is generally between 1.5 and 2.0 $\mathrm{J\,m^{-1}\,s^{-1}\,K^{-1}}$.

7.3 HEAT CONDUCTION IN DRY SOIL

In Section 4.3, one-dimensional conduction of heat was described with the equation

$$\frac{\partial T}{\partial t} = \frac{\lambda}{C_h}\frac{\partial^2 T}{\partial s^2} \tag{4.31}$$

This equation is valid if λ is independent of position, which is usually not true. The following sections contain solutions of this equation for three simple situations, all involving dry soil for which Equation 4.31 is valid.

7.3.1 Steady heat conduction

With steady heat conduction the temperature, per definition, does not change in time, i.e. $\partial T/\partial t = 0$. Since λ and C_h are constant and non-zero for dry soil, Equation

4.31 reduces to:

$$\frac{d}{ds}\left(\frac{dT}{ds}\right) = 0 \qquad (7.2)$$

Integrating twice yields:

$$T = C_1 s + C_2 \qquad (7.3)$$

The integration constants, C_1 and C_2, can be determined if the temperature is known for two values of s.

Question 7.2 In a steady situation the temperature in a homogeneous dry soil profile is 30 °C at the soil surface ($z = 0$) and 15 °C at a depth of 1 m ($z = -1$ m).
a. Express the temperature as a function of height.
b. Calculate the temperature gradient.

7.3.2 Cyclic variation of surface temperature

In nature, soil temperatures fluctuate due to more or less cyclic variations of the surface temperature, $T_{0,t}$. These variations are the diurnal (day-night) and the annual (summer-winter) cycles. Both can be approximated by a sinus function (Figure 7.2). The mathematical derivation of the solution of Equation 4.31 for these cyclic boundary conditions will not be given here. The solution for a soil with a constant 'average' temperature, T_{av}, is:

$$T_{s,t} = T_{av} + A_0 \exp(-s/d) \sin(\omega t - s/d) \qquad (7.4)$$

where $T_{s,t}$ is the temperature at depth s and time t, A_0 is the amplitude of the sinusoidal temperature variation at the soil surface, ω is the angular frequency of the temperature variation (wave), t is time and d is the damping depth. The angular frequency is

$$\omega = \frac{2\pi}{t_c} \qquad (7.5)$$

where t_c is the time needed to complete one cycle of the wave.

Question 7.3 Make a graph of the temperature at the soil surface, $T_{0,t}$, as function of time if $T_{av} = 20$ °C, $A_0 = 10$ °C and $\omega = 2\pi/24$ radians per hour ($t_c = 24$ h).

The sinusoidal temperature fluctuation at the surface penetrates into the soil by heat conduction. However, due to the finite heat diffusivity of the soil, the amplitude of the heat wave decreases with soil depth (Figure 7.2). In Equation 7.4, this phenomenon is accounted for by the introduction of the so-called damping depth, d, which is the depth where the amplitude of the temperature fluctuation has decreased to $A_0/e \approx 0.37 \times A_0$.

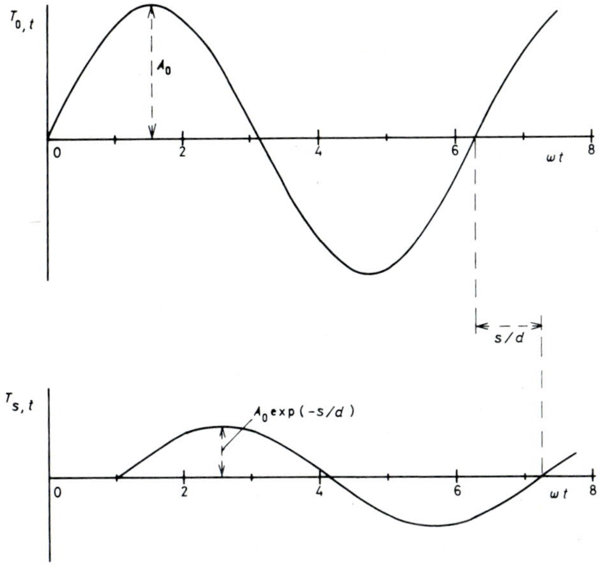

Figure 7.2 Cyclic variation of temperature at the soil surface and at depth *s*.

Question 7.4 Check this!

The damping depth depends on the thermal properties of the soil and on the angular frequency of the temperature variation according to

$$d = \sqrt{\frac{2\lambda}{\omega C_h}} = \sqrt{\frac{2D_h}{\omega}} \tag{7.6}$$

The finite heat diffusivity of the soil also causes a phase shift between the heat wave at the soil surface and at a certain depth *s* (Figure 7.2). This is represented in Equation 7.4 by the variable $-s/d$ in the sinus function ($-s/d$ is the phase angle).

Question 7.5 A soil has the following thermal properties:
$\lambda = 0.6\ \mathrm{J\ m^{-1}\ s^{-1}\ K^{-1}}$
$C_h = 1.6\ \mathrm{MJ\ m^{-3}\ K^{-1}}$
Calculate the damping depth for a diurnal and for an annual temperature fluctuation.

Question 7.6 a. Calculate the amplitude of the daily temperature fluctuation at 0.3 m depth in the soil above, if the amplitude at the soil surface is 10 K.
b. Make a graph of the temperature wave and compare it with the graph in Answer 7.3 (take $T_{av} = 20\,°\mathrm{C}$).

Question 7.7 What is the phase difference between the temperature fluctuations at the soil surface and at a depth $s = d$?

7.3.3 *Step-increase of surface temperature*

If a soil profile has a uniform initial temperature ($T = T_i$ for $s \geqslant 0$ and $t = 0$) and the surface temperature is suddenly increased and then maintained at that higher temperature ($T = T_0$, for $s = 0$ and $t > 0$), heat will penetrate into the soil from the surface. This so-called step-increase hardly occurs in nature; it is artificial. However, with the boundary conditions just stated, Equation 4.31 can be solved analytically. The solution is similar to that for infiltration of water in soil (Section 5.3.3.3 and Appendix 2):

$$\frac{T - T_i}{T_0 - T_i} = \text{erfc}\left(\frac{s}{2\sqrt{(\lambda/C_h)t}}\right) \tag{7.7}$$

where $\frac{T - T_i}{T_0 - T_i}$ is the temperature change ($T - T_i$) at time t and distance s expressed as a fraction of the temperature change at the surface ($T_0 - T_i$). Since T can vary only between T_i and T_0, it can easily be checked that this fraction can only vary between 0 and 1. It is, therefore, often called dimensionless temperature. The value of erfc (u) for any positive value of u can be found from Figure 5.16. Consistent with the above conclusion, erfc (u) varies between 0 and 1.

Question 7.8 A soil with the thermal properties stated in Question 7.5 has a constant initial temperature. At $t = 0$ the temperature is raised by 10 K. Calculate the increase of the temperature at 0.10 m depth after 5 h and 40 min.

Question 7.9 At what time will the same temperature increase occur at 0.20 m depth and at 0.30 m depth?
Note: Determine this in a simple way, without complicated calculations.

Question 7.10 What is, obviously, the relationship between the depth s and the time t at which a certain increase of temperature is found?

7.4 THE HEAT BALANCE OF THE SOIL SURFACE

As mentioned in Section 7.3, the temperature at the soil surface varies periodically. During daylight, most of the incoming direct and indirect short-wave radiation is absorbed by the soil surface, the remainder is reflected back into the atmosphere. During darkness as well as daylight, the soil surface loses energy by emitting long-wave heat radiation. The net radiation flux density, f_n (J m^{-2} s^{-1} = W m^{-2}), is the difference between the unreflected incoming short-wave radiation and the outgoing long-wave heat radiation. Normally, during the day $f_n > 0$, and at night $f_n < 0$. The value of f_n can be measured directly or estimated from meteorological data.

The net radiation flux density is used for:

- evaporation of water at the soil surface (f_e)
- transpiration of water by the vegetative cover (f_t)

- heating of the soil (f_s)
- heating of the atmospheric air by conduction at the soil surface (f_a).

The heat balance of the soil can thus be described by:

$$f_n = f_e + f_t + f_s + f_a \tag{7.8}$$

The distribution of f_n over f_e, f_t, f_s and f_a is important, because f_e and f_t determine the amount of water that can be evaporated, whereas f_s determines the soil temperature and, thus, the growth of crops.

Generally, it is not easy to calculate the distribution of f_n over the four terms in Equation 7.8, because the four processes are strongly interrelated. An estimate of this distribution is more easily made for the following two special situations:

- A rather dry, bare topsoil. Here, transpiration is absent and evaporation from the soil surface is negligible. Thus Equation 7.8 can be simplified to $f_n = f_s + f_a$. The distribution of f_n over f_s and f_a can then be estimated on the basis of calculations of heat conduction in dry soil and turbulent heat transport in the atmosphere. Such calculations yield, on average, a ratio of 1:1 between f_s and f_a. The exact value of this ratio depends on the thermal properties of the soil and the wind velocity.
- A rather wet soil. Here, evaporation of water consumes a significant part of the net radiation flux density. The soil acts mainly as a heat reservoir, absorbing heat during the day and discharging it again at night. As a result, the soil temperature changes little over a complete diurnal cycle, making f_s negligible over that interval. The net radiation flux density during one daily cycle thus will be distributed over f_e, f_t and f_a. The relative magnitudes of f_e, f_t and f_a can be estimated theoretically, but this will not be discussed here further.

Question 7.11 The net radiation flux density at a bare, dry soil surface ($C_h = 0.8$ MJ m^{-3} K^{-1}) is 20 W m^{-2}. Assuming $f_s/f_a = 1$, calculate:

a. the thickness of the surface layer in which the temperature is raised, on average, 1 K per hour;

b. the thickness of the layer of air ($C_h = 1.3$ kJ m^{-3} K^{-1}) above the soil surface that can be heated, on average, 1 K per hour.

7.5 THERMAL REGIME OF UPPER SOIL LAYERS

The temperature fluctuations in the soil profile resulting from the cyclic temperature variations at the soil surface were discussed in Section *7.3.2*. The influence of λ and C_h on these fluctuations was also described. This insight, together with the heat balance concept, can be used to characterize the thermal regime of the upper soil layers.

In dry soil, evaporation consumes very little energy from the net radiation, leaving nearly all energy available for heating the soil. At the same time, the heat penetrates into the soil only slowly, due to the low values of the heat diffusivity and damping depth (Figure 7.1). This combination of circumstances causes a large accumulation of heat in the surface layer with a relatively low heat capacity, resulting in a rapid rise of the soil surface temperature during daytime. At night, the surface of a dry soil cools down more rapidly than a wet soil surface. Thus, a dry soil is subject to much more stronger temperature variations in the upper layer than a wet soil. Due to the heat exchange between the soil and the atmosphere, the air layers above dry soil also will have large temperature fluctuations. This explains the large differences between day- and night-temperatures in deserts.

In temperate climates, dry soils will be warmer than wet soils during day-time. In early spring, this offers more favourable conditions for the germination of seeds and early development of crops. On the other hand, dry surfaces will cool down more strongly during the night and have a higher incidence of night frost. Thus, for a given climate, differences in thermal regime of the surface layers are determined to a large extent by differences in water content.

Question 7.12 Explain why much incoming radiation is required to heat up the surface layer of a wet soil.

7.6 COMBINED HEAT AND WATER VAPOUR TRANSPORT

7.6.1 Isothermal water vapour diffusion

Water vapour is one of the components of the gas phase of the soil. Like the other components, it can be transported both by bulk flow and by diffusion. Since bulk flow of soil air is generally negligible (Section 6.3), the principal transport mechanism of water vapour in soil is diffusion. The water vapour flux density by diffusion, f_v, according to Fick's law is (Section 6.4):

$$f_v = -D_g \frac{\partial \chi_v}{\partial s} \qquad (4.24)$$

The diffusion coefficient is the corrected form of the molecular diffusion coefficient of water vapour in free air:

$$D_g \approx \tfrac{1}{2}\phi_g D_0 \qquad (6.3)$$

The concentration gradient can be replaced by the vapour pressure gradient according to the ideal gas law (Answer 4.19).

In Chapter 3, it was shown that the relative humidity of the soil air is 99.3% at $pF = 4$ and approximately 93% at $pF = 5$. This means that differences in vapour pressure between, for instance, a very dry topsoil and a moist subsoil are very small, provided the temperature is constant throughout the profile. Therefore, isothermal water vapour diffusion can generally be neglected.

7.6.2 Non-isothermal water vapour diffusion

The saturated vapour pressure depends on temperature, as shown in Table 7.2. Since the soil atmosphere is normally saturated with water vapour, a temperature gradient in the soil will induce a considerable vapour pressure gradient.

Table 7.2 Saturated vapour pressure of water

T(°C)	0	5	10	15	20	25	30	35	40	45
p_0 (kPa)	0.61	0.87	1.23	1.70	2.33	3.16	4.23	5.61	7.36	9.56

Question 7.13 Does water vapour diffusion take place in the direction of higher or lower temperatures?

Question 7.14 What is the vapour pressure gradient induced by a temperature gradient of 1 K cm^{-1} at 15°C?

Question 7.15 What is the concentration gradient of the water vapour for the pressure gradient of Answer 7.14? ($RT \approx 2.4$ kJ mol^{-1}).

Question 7.16 Calculate the water vapour flux density resulting from the concentration gradient of Answer 7.15, if $D_0 \approx 2 \times 10^{-5}$ m^2 s^{-1} and $\phi_g = 0.30$.

Question 7.17 Convert the water vapour flux density in Answer 7.16 to an equivalent liquid water flux density.

Upon partially drying to, e.g., $pF = 4$, a poorly conducting soil may attain a hydraulic conductivity as low as 10^{-6} m d^{-1}. Such drying, generally, coincides with large hydraulic head gradients of the order of 10^2 to 10^3 m m^{-1}. Under those conditions, the liquid flux density still will be of the order of 0.1 to 1 mm d^{-1}. This liquid transport is about 5 to 50 times as high as the maximum equivalent water flux density calculated in Answer 7.17. Under wetter conditions, the vapour flux density will be even smaller relative to the liquid flux density. Thus, the contribution of the vapour flux density to the total water transport in soils is generally negligible. Moreover, temperature gradients that cause vapour diffusion tend to be dissipated by that diffusion because of the large latent heat transport associated with it.

Question 7.18 Explain this last statement.

7.6.3 Evaporation

The upward transport of water vapour away from the soil surface into the atmosphere is called evaporation. Except in a very thin laminar boundary layer, this transport is turbulent. Air turbulance is much more effective in transporting water vapour than diffusion. The energy required at the soil surface to sustain evaporation is provided mainly by the net solar radiation. This is the limiting factor for evaporation from a bare, wet soil.

In a bare, dry soil, the limiting factor for evaporation is the supply of water to the soil surface. As shown in Section 5.3.2.1, water in the soil is conserved by the formation of a very dry topsoil with a low water conductivity. In Section *7.6.2*, it was shown that the combined liquid and vapour transport of water in a topsoil at $pF = 4$ may produce a total flux density of no more than approximately 1 mm d^{-1}.

7.6.4 Evapotranspiration

When the soil is covered by vegetation, the situation is quite different. Water can then be transported from anywhere in the root zone, via the roots, stems and leaves into the atmosphere, the so-called soil-plant-atmosphere-continuum (SPAC). The loss of water from soil via plants is called transpiration. Transpiration is mostly a passive process. While leaf stomates are open for the exchange of CO_2 and O_2 for the process of photosynthesis, water vapour is lost to the atmosphere. The deficit is replenished by water uptake by the roots. The driving force for the transpiration stream is the large potential difference between the water in the soil and in the atmosphere. Again, except for a thin boundary layer, the transport of water vapour from the leaves into the atmosphere is turbulent. The stomates close when the roots cannot take up enough water to satisfy the evaporative demand. Then, with the transpiration, photosynthesis is also interrupted, which will result in yield reductions.

Under a full vegetative cover, evaporation from the soil can, generally, be neglected. When the soil is partially covered with vegetation, water is lost by both evaporation and transpiration, which usually are lumped together as evapotranspiration.

Evapotranspiration is a very complex process, which is intimately connected with the discipline of micrometeorology. The shorter the time span over which one wants to know the amount of evapotranspiration, the more complex the problem becomes. Evapotranspiration can be much higher than the evaporation from a bare soil, up to 10 mm d^{-1}. The determining factors are, in general:

- the turbulent transport of water vapour in the atmosphere, as determined by the wind velocity
- the supply of energy for evaporation from the leaves and the soil surface, mainly provided by solar radiation
- the availability of water in the soil and the flow to individual roots
- the type of crop and its growth stage.

7.7 SUGGESTED LITERATURE

BAVER, L. D., W. H. GARDNER and W. R. GARDNER, 1972. Soil Physics (fourth edition). Wiley, 489 p. Chapter 7.

BRUCE, R. R., K. W. FLACH and H. M. TAYLOR (eds.), 1973. Field Soil Water Regime. SSSA Special Publication no. 5, 212 p. Chapter 3.

Hanks, R. J. and G. L. Ashcroft, 1980. Applied Soil Physics. Springer-Verlag, 159 p. Chapters 4 and 5.

Hillel, D., 1980. Fundamentals of Soil Physics. Academic Press, 413 p. Chapter 12.

Hillel, D., 1980. Applications of Soil Physics. Academic Press, 385 p. Chapters 5, 6 and 7.

Kirkham, D. and W. L. Powers, 1972. Advanced Soil Physics. Wiley, 534 p. Chapter 10.

Marshall, T. J. and J. W. Holmes, 1979. Soil Physics. Cambridge University Press, 345 p. Chapters 11 and 12.

Taylor, S. A. and G. L. Ashcroft, 1972. Physical Edaphology. Freeman and Co., 533 p. Chapter 13.

Answer 7.1 a. The volumic heat capacity is composed of the following contributions

minerals:	$(1-\phi)\times 2.0$	$\mathrm{MJ\,m^{-3}\,K^{-1}}$
water:	$\phi_l \times 4.2$	$\mathrm{MJ\,m^{-3}\,K^{-1}}$
air:	$(\phi-\phi_l)\times 0.0013$	$\mathrm{MJ\,m^{-3}\,K^{-1}}$
total ≈	$2.0\,(1-\phi)+4.2\,\phi_l$	$\mathrm{MJ\,m^{-3}\,K^{-1}}$

b. The contribution of soil air to the volumic heat capacity of a soil is negligible due to the low volumic heat capacity of air.

Answer 7.2 a. To express the temperature as a function of height, substitute z for s in Equation 7.3:

$T = C_1 z + C_2$

Since $T = 30\,^\circ\mathrm{C} = 303\ \mathrm{K}$ at $z = 0$, $C_2 = 303\ \mathrm{K}$ and $T = C_1 z + 303\ \mathrm{K}$.

Substitution of $T = 15\,^\circ\mathrm{C} = 288\ \mathrm{K}$ and $z = -1.0\ \mathrm{m}$ gives:

$288\ \mathrm{K} = -1\ \mathrm{m} \times C_1 + 303\ \mathrm{K}$ or $C_1 = 15\ \mathrm{K\,m^{-1}}$.

Thus $T = 15\ \mathrm{K\,m^{-1}} \times z + 303\ \mathrm{K}$.

b. The temperature gradient is:

$$\frac{\mathrm{d}T}{\mathrm{d}z} = C_1 = 15\ \mathrm{K\,m^{-1}}$$

Answer 7.3 For the soil surface, where $s = 0$, Equation 7.4 reduces to:

$$T_{0,t} = T_{av} + A_0 \sin \omega t = 20\,^\circ\mathrm{C} + 10\,^\circ\mathrm{C} \sin\left(\frac{\pi t}{12\ \mathrm{h}}\right)$$

$$t = 0 \rightarrow T_{0,0} = (20 + 10 \sin 0)\,^\circ\mathrm{C} = 20\,^\circ\mathrm{C}$$

$$t = 6\ \mathrm{h} \rightarrow T_{0,6} = (20 + 10 \sin \tfrac{\pi}{2})\,^\circ\mathrm{C} = 30\,^\circ\mathrm{C}$$

$$t = 12\ \mathrm{h} \rightarrow T_{0,12} = (20 + 10 \sin \pi)\,^\circ\mathrm{C} = 20\,^\circ\mathrm{C}$$

$$t = 18\ \mathrm{h} \rightarrow T_{0,18} = (20 + 10 \sin \tfrac{3\pi}{2})\,^\circ\mathrm{C} = 10\,^\circ\mathrm{C}$$

$$t = 24\ \mathrm{h} \rightarrow T_{0,24} = (20 + 10 \sin 2\pi)\,^\circ\mathrm{C} = 20\,^\circ\mathrm{C}$$

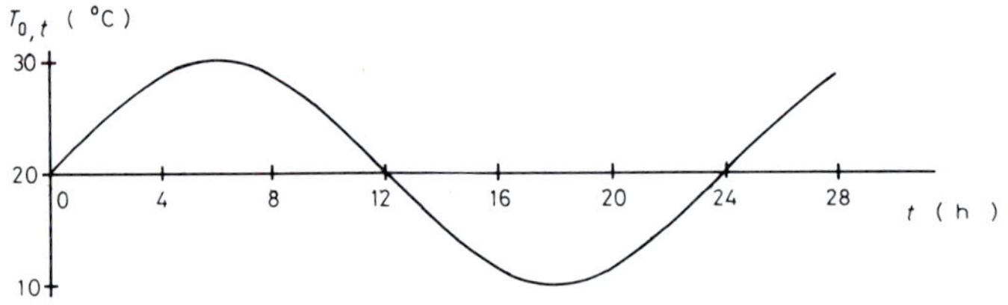

Answer 7.4 At depth $s = d$, the equation is:

$T_{d,t} = T_{av} + A_0 \exp(-1) \sin(\omega t - 1)$

Thus the amplitude of the temperature fluctuation at $s = d$ is A_0/e.

Answer 7.5 For the daily fluctuation, $\omega = \dfrac{2\pi}{24 \times 3600}\ \mathrm{s}^{-1}$.

$$\text{Thus, } d = \sqrt{\frac{2 \times 0.6 \times 24 \times 3600}{2\pi \times 1.6 \times 10^6}\ \mathrm{m}^2} = 0.10\ \mathrm{m}$$

$$\text{For the annual wave, } \omega = \frac{2\pi}{365 \times 24 \times 3600}\ \mathrm{s}^{-1}$$

$$\text{Thus, } d = \sqrt{\frac{2 \times 0.6 \times 365 \times 24 \times 3600}{2\pi \times 1.6 \times 10^6}\ \mathrm{m}^2} = 1.94\ \mathrm{m}$$

As Equations 7.5 and 7.6 indicate, d is proportional to $\sqrt{t_c}$.

Answer 7.6 a. At 0.3 m depth the amplitude equals:

$A_0 \exp(-s/d) = 10\ \mathrm{K} \times \exp(-0.3/0.1) \approx 0.50\ \mathrm{K}$

b. The phase shift with the soil surface equals:

$$-s/d = \frac{-0.3}{0.1} = -3\ \mathrm{rad} \approx -172^\circ$$

A comparision with Answer 7.3 shows that the temperature at 0.3 m depth attains its minimum value when the surface temperature is near its maximum value.

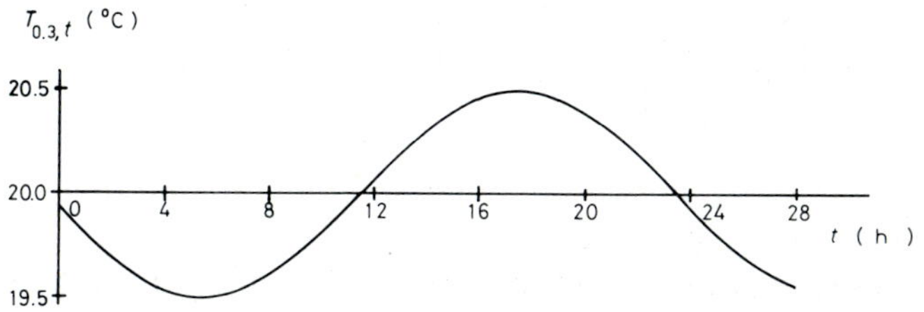

Answer 7.7 According to the equation in Answer 7.4 the phase difference is -1 radian.

Answer 7.8 Substitution in Equation 7.7 of $T_0 - T_i = 10$ K, $s = 0.1$ m, $\lambda = 0.6\ \mathrm{J\ m^{-1}\ s^{-1}\ K^{-1}}$, $C_h = 1.6\ \mathrm{MJ\ m^{-3}\ K^{-1}}$ and $t = 5$ h 40 min $= 20400$ s gives:

$$\frac{T - T_i}{10\ \mathrm{K}} = \mathrm{erfc}\ \frac{0.1}{2\sqrt{\dfrac{0.6 \times 20400}{1.6 \times 10^6}}} = \mathrm{erfc}\ (0.57)$$

With the aid of Figure 5.16 we find:

$$\frac{T - T_i}{10\ \mathrm{K}} = \mathrm{erfc}\ (0.57) \approx 0.4 \text{ and } T - T_i \approx 4\ \mathrm{K}$$

Answer 7.9 The same increase of the dimensionless temperature means that $\dfrac{T - T_i}{T_0 - T_i}$

again equals erfc (0.57) (Answer 7.8), and thus

$$\frac{s}{2\sqrt{(\lambda/C_h)t}} = 0.57$$

If $s = 0.20$ m, which is twice the value of s in Question 7.8, then also $\sqrt{t}$ must be twice as large and t must be four times as large. Hence, t = 22 h 40 min. Similarly, for $s = 0.30$ m, t must be nine times as large, or 51 h.

In both cases t can also be computed by substituting the appropriate values for s, λ and C_h.

Answer 7.10 Obviously, for a certain increase of temperature, we find that $s/\sqrt{t}$ is constant. The reason is, that a certain value of $\frac{T - T_i}{T_0 - T_i}$ yields only one particular value of $\frac{s}{2\sqrt{(\lambda/C_h)t}}$, because erfc ($u$) is a singlevalued function. If λ and C_h are constant, then $s/\sqrt{t}$ must be constant.

Answer 7.11 At a bare, dry soil surface, $f_e = f_t = 0$. Therefore, $f_n = f_s + f_a = 20$ W m^{-2}. Assuming $f_s/f_a = 1$, $f_s = f_a = 10$ W m^{-2} = 10 J m^{-2} s^{-1}.

a. Recalling the definition of differential heat capacity (Table 4.2), this radiation flux density heats the soil according to

$$\frac{36000 \text{ J m}^{-2}\text{ h}^{-1}}{0.8 \times 10^6 \text{ J m}^{-3}\text{ K}^{-1}} = 4.5 \times 10^{-2} \text{ m K h}^{-1}$$

Thus, the soil layer heated 1 K in one hour is 4.5×10^{-2} m or 4.5 cm.

b. Similarly, for air:

$$\frac{36000 \text{ J m}^{-2}\text{ h}^{-1}}{1.3 \times 10^3 \text{ J m}^{-3}\text{ K}^{-1}} = 27.7 \text{ m K h}^{-1}$$

Thus, the air layer heated 1 K in one hour is 27.7 m thick.

Answer 7.12 The surface layer of a wet soil requires much incoming radiation to heat up, because

a. much of the incoming radiation is used for evaporation

b. the remaining heat is distributed over a large depth, due to the large damping depth

c. the heat remaining in the topsoil causes a relatively small increase of temperature, due to the high value of the heat capacity.

Answer 7.13 Water vapour diffusion occurs from higher to lower vapour pressures, thus from higher to lower temperatures.

Answer 7.14 At 15°C, a temperature difference of 1 K induces a water vapour pressure difference of approximately 0.1 kPa (Table 7.2). Thus a temperature gradient of 1 K cm^{-1} produces a water vapour pressure gradient of about 10 kPa m^{-1}.

Answer 7.15 Since $\chi_v = \frac{M_v}{RT} p_v$, a pressure gradient of 10 kPa m^{-1} corresponds with a concentration gradient

$$\frac{\partial \chi_v}{\partial s} = \frac{M_v}{RT}\frac{\partial p_v}{\partial s} \approx \frac{18 \text{ g mol}^{-1} \times 10^4 \text{ Pa m}^{-1}}{2.4 \times 10^3 \text{ J mol}^{-1}} = 7.5 \times 10^{-2} \text{ kg m}^{-4}$$

Answer 7.16 The absolute value of the water vapour flux density is:

$$|f_v| \approx \tfrac{1}{2}\phi_g D_0 \frac{\partial \chi_v}{\partial s} = \tfrac{1}{2} \times 0.3 \times 2 \times 10^{-5} \times 7.5 \times 10^{-2} \text{ kg m}^{-2} \text{ s}^{-1} = 2.25 \times 10^{-7}$$

$\text{kg m}^{-2} \text{ s}^{-1}$

Answer 7.17 Since $\rho_w \approx 10^3$ kg m^{-3}, the equivalent liquid water flux density is 2.25×10^{-10} m s^{-1} or about 0.02 mm d^{-1}.

Answer 7.18 The latent heat of evaporation of water is high. Therefore, much heat is consumed where water vapour is formed and produced where it condenses. The net effect of this is an often large heat flux density in the direction of the vapour flux, as follows:

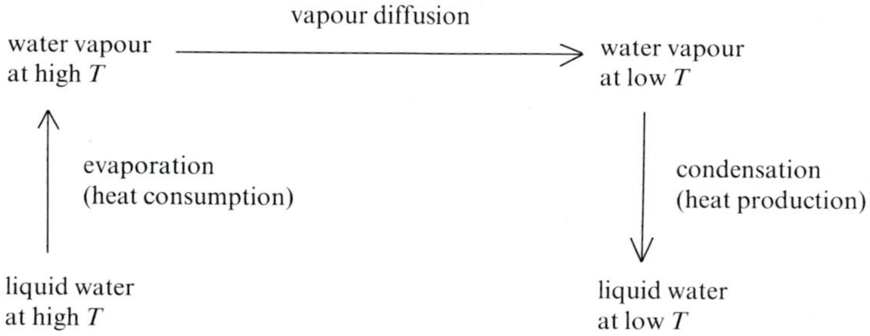

One might say that the effective heat conductivity has increased. This heat transport associated with the water vapour diffusion will eventually wipe out the temperature gradient, unless it is maintained by outside influences.

8 Epilogue

G. H. Bolt

The preceding text was intended to convey an understanding of only the most basic aspects of physical processes occurring in soil. This epilogue appraises the additional understanding needed for analysing and managing the functions of soil in the real world.

Several such functions come to mind, ranging from the classic 'habitat for plants', accidental storage and/or passing station for unwanted spills, assigned storage reservoir for water or energy, to that of foundation of structures and other engineering functions. Of these functions, the first one mentioned requires the broadest insight, because it involves the support of life. It is traditionally one of the major areas of interest of Soil Physics. Accordingly, most of the following considerations are directed towards that function. The other functions usually pose simpler, though not always simple, problems. The role of soil as storage/passing station for water-carried solutes involves the retardive and dispersive soil properties. These phenomena, also referred to as miscible displacement, are often treated under the heading of Soil Chemistry (see, e.g., Bolt and Bruggenwert (1), chapter 7). Under conditions of flooding the effect of cracks in soil may well be decisive for the break-through of surface-added pollutants into the groundwater. Storage of heat in soil and heat production around power cables and gas mains with the ensuing transport of water away from the heat source, may be considered as soil engineering problems. A useful extension on this subject, beyond the introductory material in Chapter 7, can be found in Kirkham and Powers (5), chapter 10. Freezing phenomena represent another form of coupled heat and water transport in soils. A review of past and present work on this subject is given by Miller (Hillel (3), chapter 11). A general introduction to Soil Physics from the engineering point of view is given by Yong and Warkentin (8).

Focussing now on the ecological significance of soil with respect to plant growth and crop production, a starting point can be that, with a few exceptions, plant roots require an adequate supply of water and oxygen. These two components are complementary: maximum presence of the one implies absence of the other. Therefore, plant roots can function only in the so-called 'unsaturated zone', making transport and retention of water and diffusion of gases of prime importance. Water enters soil mostly from the surface with a climate- or human-controlled intermittent pattern. The water retentivity and conductivity functions (see Figure 3.12 and 5.6, respectively) of a soil determine which fraction of the water will percolate beyond the root zone, and at which rate this will occur. The remaining water is available for evaporation from the soil surface and uptake by plant roots. Transpiration is mostly a passive process. While leaf stomates are open for the exchange of CO_2 and O_2 for the process of photosynthesis, water vapour is lost to the atmosphere. When the roots cannot take up enough water to replenish the plant and satisfy the evaporative demand, the stomates close and photosynthesis is interrupted. As a result, there is a relationship between the amount of water transpired and the amount of dry matter produced. Plant roots must respire to sustain their physiological functions, and thus they consume O_2 and produce CO_2. These gases must be exchanged with the atmosphere through the same pore system that must ensure a satisfactory water supply.

The above simplistic sketch clearly indicates the intimate connections between crop productivity and physical soil properties, within constraints of climatic conditions, human activity, etc. It also hints at the difficulties to be encountered when attempting to switch from *understanding* an observed crop growth pattern for a given set of conditions, to *predicting* such a pattern. Predictive use of soil physical knowledge is much more complicated than explanatory use. As is often the case in nature, the whole crop production system is one of checks and balances, a rather well-buffered system. Before the computer era, solutions of the second-order, nonlinear, partial differential flow equations (e.g., Equations 4.18 and 4.23) were possible only for simple boundary conditions (see, e.g., Philip (6)). These solutions are inadequate for predicting the distribution of water content in time and depth within the unsaturated zone following a 'normal' pattern of intermittent water supply.

When computers became available in the sixties, the possibilities for predictive applications of soil physics appeared to increase tremendously. In the early years, computers were used primarily to find numerical solutions to the partial differential flow equations for complex boundary conditions and multilayered columns (for a rather complete survey of such methods, see Remson, Hornberger and Molz (7)). When the ever increasing capacity and speed of computers made it possible to solve more and more complex problems, other 'limiting' factors became apparent. For instance, computation of water infiltration in a layered soil column requires complete information on the soil hydraulic functions for all layers. Such information is hardly ever available with any degree of completeness. The moment one admits alternating wetting and drying periods – mandatory for simulating crop growth under field conditions – one must include complications due to hysteresis, which requires 'families' of water retention curves for each layer (see Figure 3.14). The corresponding $k(\theta)$ functions, to a first approximation, can be assumed nonhysteretic. Not so much the difficulties of coping with such families of curves when performing predictive calculations, but the lack of information on the precise shape of such families for each layer in the soil profile prevents detailed studies of this nature. Swelling and shrinkage of clay soils present even more vexing problems. In short, calculation facilities very soon exceed the possibilities of obtaining experimental information on relevant parameters with a detail commensurate with the complexity of the calculation models.

Towards the end of the seventies, two new developments took place. The first one was the introduction of dynamic simulation models for more complete systems. In the context of the present discussion on soil as a physical habitat for plants, for instance, one would attempt to model the entire growth process of a crop during a season using simplified assumptions (see, e.g., Feddes, Kowalik and Zaradny (2)), rather than try to model precisely the transport of water in a single soil column. In such simulation models, water supply at the soil surface, climatic conditions, distributed water uptake by plant roots and other pertinent factors are simulated, according to assumed interactive relationships, to predict the rate of crop growth. These relationships should be derived from the best available theoretical and experimental information on the various processes. In the earlier stages of simulation-modeling, quite extreme simplifications of many simultaneously considered processes were needed. Increased sophistication of computer hardware and software now makes it possible to cope with a high degree of complexity of subprocesses. Quite generally, the predictive value of computer programs now being developed for simulating crop production, involving soil physical processes, depends critically on the insight of the model builder to separate important aspects of these physical processes from subordinate ones. Even if the hard to obtain, detailed information on soil physical parameters is available, clear insight in the basic processes is still indispensable.

The other development that took place in soil physics, en route towards predicting field behaviour, was the increasing awareness of the so-called spatial variability of soil properties and the consorted efforts to cope with it (for a review by Warrick and Nielsen, see Hillel (3), chapter 13). Two aspects of spatial variability of soil properties require special attention. The first aspect is: how should one sample, within the realm of practicability, for determining a particular soil property? This problem concerns the statistics of so-called regionalised variables or geostatistics. The second aspect of spatial variability is the need for acceptable averaging methods to predict the physical behaviour of three-dimensional field units rather than that of one-dimensional soil columns. Transport processes in soil involve 'series' as well as 'parallel' conductive pathways, which require different methods of averaging. Perhaps a clear example of this is the obvious dominance, for vertical transport of water through soil, of the very small pores in a *horizontal* clay pan, in contrast to that of a single big 'pore' provided by a *vertical* crack. In general, the field-average *effect* of a certain soil physical process is not likely to be similar to that of a homogeneous soil column with field-average *values* of the material properties of concern. Here again, soil physicists who must decide on sampling procedures and/or select parameter values suitable for modeling soil physical processes on a field scale, need a clear insight in the basic features of these processes.

Recently, remote sensing with multispectral radiation, active and passive radar, or radiation temperature appears able to provide relatively cheap, bird's-eye view information on the spatial variability of some physical soil properties (Holz (4)). Admitting that still much remains to be done on the interpretation of such data, combination of dynamic simulation, geostatistical analysis and remote sensing may well lead, eventually, to the much needed predictive insight in soil physical processes on a field scale.

LITERATURE

BOLT, G. H. and M. G. M. BRUGGENWERT (eds.), 1978. Soil Chemistry A. Basic Elements. Elsevier Scientific Publishing Co., 281 p.

FEDDES, R. A., P. J. KOWALIK and H. ZARADNY, 1978. Simulation of field water use and crop yield. Pudoc, Wageningen, 188 p.

HILLEL, D., 1980. Applications of Soil Physics. Academic Press, 385 p.

HOLZ, R. K. (ed.). The Surveillant Science, Remote Sensing of the Environment. Houghton Mifflin Co. 390 p.

KIRKHAM, D. and W. L. POWERS, 1972. Advanced Soil Physics. Wiley, 534 p.

PHILIP, J. R., 1969. Theory of Infiltration. Adv. Hydroscience, 5. Academic Press, p. 215–296.

REMSON, I., G. M. HORNBERGER and F. J. MOLZ, 1971. Numerical methods in subsurface hydrology. Wiley, 389 p.

YONG, R. N. and B. P. WARKENTIN, 1975. Soil properties and behaviour. Elsevier Scientific Publishing Co., 449 p.

Appendix 1

Laminar flow of liquid in a cylindrical tube. Poiseuille's law.

The rate of discharge, Q, of a liquid in a cylindrical capillary tube with fixed radius, R (Figure A1.2), depends on the driving force acting on the liquid and on the internal friction forces between liquid molecules, which are characterized by the viscosity of the liquid, η.

The driving force per volume is the gradient of the hydraulic potential, $-\mathrm{d}p_\mathrm{h}/\mathrm{d}x$. This driving force will be considered constant from here on.

The viscosity can be demonstrated with the aid of a layer of liquid between two parallel plates of a solid material (Figure A1.1).

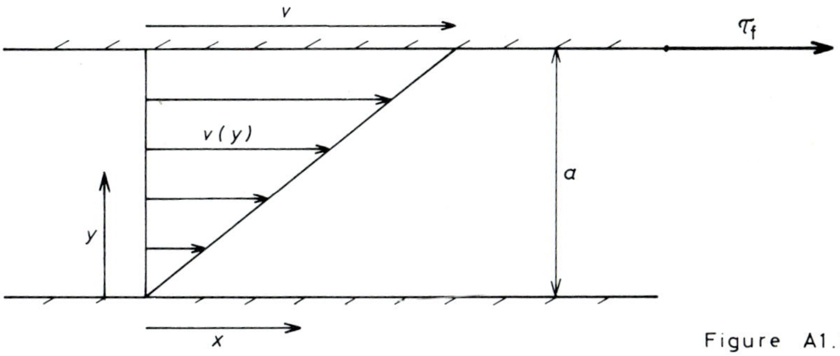

Figure A1.1

If the upper plate is moved relative to the lower plate with a constant velocity, v, the velocity of the liquid will be zero at the boundary with the lower plate and equal to v at the boundary with the upper plate, due to adhesion. In between the plates the velocity $v(y)$ increases linearly with the distance y to the lower plate.

The force divided by area of the upper plate that must be applied to maintain the velocity, v, is equal and opposite to the internal friction force per area, τ_f, in the liquid. This force appears to be inversely proportional to the distance, a, between the plates. This means that τ_f is proportional to the gradient of the velocity $\frac{\mathrm{d}v}{\mathrm{d}y}$. The proportionality factor is the viscosity, η:

$$\tau_\mathrm{f} = \eta \frac{\mathrm{d}v}{\mathrm{d}y} \tag{A1.1}$$

The rate of discharge can be calculated when the velocity distribution in the tube is known as a function of r.

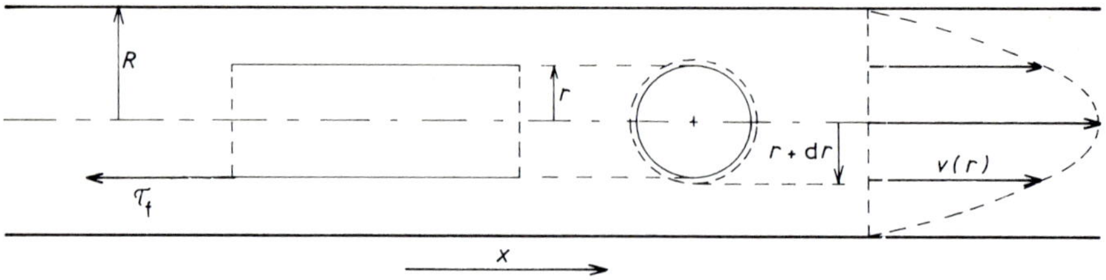

Figure A1.2

For laminar flow the velocity of the liquid is zero at the wall of the tube and has its maximum value in the centre of the tube. Intermediate velocities depend only on the radial distance from the centre, r, and thus are constant in concentric cylindrical surfaces around the centre.

There is a friction force between adjoining concentric liquid layers with different velocities. Consider the liquid cylinder with radius r and length l. The friction force on the cylindrical liquid body per volume is then

$$F^{\mathrm{d}} = \tau_{\mathrm{f}} \frac{2\pi r l}{\pi r^2 l} = \tau_{\mathrm{f}} \frac{2}{r} \tag{A1.2}$$

At mechanical equilibrium (steady laminar flow) the sum of the forces acting on the cylindrical body equals zero:

$$-\frac{\mathrm{d}p_{\mathrm{h}}}{\mathrm{d}x} + \tau_{\mathrm{f}} \frac{2}{r} = 0 \tag{A1.3}$$

Substitution of $\eta \frac{\mathrm{d}v}{\mathrm{d}r}$ for τ_{f} yields:

$$\eta \frac{2}{r} \frac{\mathrm{d}v}{\mathrm{d}r} = \frac{\mathrm{d}p_{\mathrm{h}}}{\mathrm{d}x} \tag{A1.4}$$

$$\text{or } \mathrm{d}v = \frac{1}{2\eta} \frac{\mathrm{d}p_{\mathrm{h}}}{\mathrm{d}x} r\mathrm{d}r \tag{A1.5}$$

Integration yields:

$$v = \frac{1}{4\eta} \frac{\mathrm{d}p_{\mathrm{h}}}{\mathrm{d}x} (r^2 + C) \tag{A1.6}$$

where C is an integration constant. From the boundary condition $v = 0$ for $r = R$ follows: $C = -R^2$.

Thus the velocity of the liquid at radial distance r is:

$$v_r = \frac{1}{4\eta} \frac{\mathrm{d}p_{\mathrm{h}}}{\mathrm{d}x} (r^2 - R^2) \tag{A1.7}$$

The velocity of the liquid in an infinitely thin cylindrical layer between r and $r + \mathrm{d}r$ is then also v_r. The surface area of the cross-section of this cylindrical layer is $2\pi r \mathrm{d}r$.

The discharge rate of the cylindrical layer is then:

$$\mathrm{d}Q = 2v_r \pi r \mathrm{d}r = \frac{\pi}{2\eta} \frac{\mathrm{d}p_\mathrm{h}}{\mathrm{d}x} (r^2 - R^2)\, r\mathrm{d}r \tag{A1.8}$$

and the total discharge rate of the whole tube is

$$Q = \iint_0^R \mathrm{d}Q = \frac{\pi}{2\eta} \frac{\mathrm{d}p_\mathrm{h}}{\mathrm{d}x} \iint_0^R (r^3 - R^2 r)\, \mathrm{d}r \tag{A1.9}$$

$$\text{or } Q = -\frac{\pi R^4}{8\eta} \frac{\mathrm{d}p_\mathrm{h}}{\mathrm{d}x} \tag{A1.10}$$

Equation A1.10 is known as Poiseuille's law.

Appendix 2

Boltzmann substitution.

We seek the solution of the partial differential Equation 5.18

$$\frac{\partial \theta}{\partial t} = \frac{\partial}{\partial s}\left(D\frac{\partial \theta}{\partial s}\right) \tag{5.18}$$

with initial and boundary conditions

$$\begin{aligned} &\theta = \theta_i \ \text{ for } s > 0 \text{ and } t = 0 \\ &\theta = \theta_i \ \text{ for } s \to \infty \text{ and } t \geqslant 0 \\ &\theta = \theta_0 \ \text{ for } s = 0 \text{ and } t \geqslant 0 \end{aligned} \tag{5.17}$$

The partial differential equation can be transformed to an ordinary differential equation by means of the so-called Boltzmann substitution. This substitution consists of the introduction of a new independent variable λ defined as:

$$\lambda = st^{-1/2} \tag{5.19}$$

The substitution leads to a solution of the equation, since the boundary conditions can also be expressed unambiguously in the variable λ:

$$\begin{aligned} &\theta = \theta_i \ \text{ for } \lambda \to \infty \\ &\theta = \theta_0 \ \text{ for } \lambda = 0 \end{aligned} \tag{5.21}$$

The transformation of Equation 5.18 can be done as follows. From Equation 5.19 it follows that:

$$\frac{\partial \lambda}{\partial s} = t^{-1/2} \text{ and } \frac{\partial \lambda}{\partial t} = -\tfrac{1}{2}st^{-3/2} = -\tfrac{1}{2}\lambda t^{-1}$$

Hence, using the chain rule, the left-hand side of Equation 5.18 can be written as

$$\frac{\partial \theta}{\partial t} = \frac{\mathrm{d}\theta}{\mathrm{d}\lambda}\frac{\partial \lambda}{\partial t} = -\tfrac{1}{2}\lambda t^{-1}\frac{\mathrm{d}\theta}{\mathrm{d}\lambda} \tag{A2.1}$$

Again, using the chain rule,

$$D\frac{\partial \theta}{\partial s} = D\frac{\mathrm{d}\theta}{\mathrm{d}\lambda}\frac{\partial \lambda}{\partial s} = t^{-1/2}D\frac{\mathrm{d}\theta}{\mathrm{d}\lambda}$$

and thus the right-hand side of Equation 5.18 can be written as

$$\frac{\partial}{\partial s}\left(D\frac{\mathrm{d}\theta}{\mathrm{d}s}\right) = \frac{\partial}{\partial s}\left(t^{-1/2}D\frac{\mathrm{d}\theta}{\mathrm{d}\lambda}\right) = t^{-1/2}\frac{\partial}{\partial s}\left(D\frac{\mathrm{d}\theta}{\mathrm{d}\lambda}\right) =$$

$$t^{-1/2}\frac{\partial\lambda}{\partial s}\frac{\mathrm{d}}{\mathrm{d}\lambda}\left(D\frac{\mathrm{d}\theta}{\mathrm{d}\lambda}\right)=t^{-1}\frac{\mathrm{d}}{\mathrm{d}\lambda}\left(D\frac{\mathrm{d}\theta}{\mathrm{d}\lambda}\right) \quad \text{(A2.2)}$$

Combination of Equations 5.18, A2.1 and A2.2 gives

$$-\frac{\lambda}{2}\frac{\mathrm{d}\theta}{\mathrm{d}\lambda}=\frac{\mathrm{d}}{\mathrm{d}\lambda}\left(D\frac{\mathrm{d}\theta}{\mathrm{d}\lambda}\right) \quad \text{(5.20)}$$

Equation 5.20 can be rearranged to

$$-\frac{\lambda}{2}\mathrm{d}\lambda=\frac{\mathrm{d}\lambda}{\mathrm{d}\theta}\mathrm{d}\left(D\frac{\mathrm{d}\theta}{\mathrm{d}\lambda}\right)$$

If D is a constant:

$$-\frac{\lambda}{2}\mathrm{d}\lambda=D\frac{\mathrm{d}\lambda}{\mathrm{d}\theta}\mathrm{d}\left(\frac{\mathrm{d}\theta}{\mathrm{d}\lambda}\right)=\frac{D}{\mathrm{d}\theta/\mathrm{d}\lambda}\mathrm{d}\left(\frac{\mathrm{d}\theta}{\mathrm{d}\lambda}\right)$$

Integration yields

$$-\frac{\lambda^2}{4}=D\ln\left(\frac{\mathrm{d}\theta}{\mathrm{d}\lambda}\right)+C$$

$$\text{or } -\frac{\lambda^2}{4D}=\ln\left(C_1\frac{\mathrm{d}\theta}{\mathrm{d}\lambda}\right)$$

$$\text{or } \exp(-\lambda^2/4D)=C_1\frac{\mathrm{d}\theta}{\mathrm{d}\lambda} \quad \text{(A2.3)}$$

where C and C_1 are constants.
Rearranging Equation A2.3 gives:

$$\mathrm{d}\theta=\frac{\exp(-\lambda^2/4D)}{C_1}\mathrm{d}\lambda=\frac{2\sqrt{D}\exp(-(\lambda/2\sqrt{D})^2)}{C_1}\mathrm{d}(\lambda/2\sqrt{D})$$

Substitution of $(\lambda/2\sqrt{D})=u$ gives:

$$\mathrm{d}\theta=\frac{2\sqrt{D}}{C_1}\exp(-u^2)\,\mathrm{d}u$$

or after integration

$$\theta=\frac{2\sqrt{D}}{C_1}\int\exp(-u^2)\,\mathrm{d}u \quad \text{(A2.4)}$$

According to the initial and boundary conditions, λ (and thus also u) varies between 0 and ∞, while θ varies from θ_i to θ_0. Thus Equation A2.4 can be written as

$$\theta(u) = \frac{2\sqrt{D}}{C_1} \int_0^u \exp(-\beta^2)\,d\beta + C_2 \tag{A2.5}$$

where C_2 is an integration constant and β is a dummy variable for u.

The initial condition $\theta = \theta_i$ for $\lambda \to \infty$, and thus $u \to \infty$, gives

$$\theta_i = \frac{2\sqrt{D}}{C_1} \int_0^\infty \exp(-u^2)\,du + C_2$$

and, using $\int_0^\infty \exp(-u^2)\,du = \frac{\sqrt{\pi}}{2}$, it follows:

$$C_2 = \theta_i - \frac{\sqrt{\pi D}}{C_1}$$

and thus Equation A2.5 becomes:

$$\theta(u) = \theta_i + \frac{2\sqrt{D}}{C_1} \int_0^u \exp(-\beta^2)\,d\beta - \frac{\sqrt{\pi D}}{C_1} \tag{A2.6}$$

The boundary condition $\theta = \theta_0$ for $\lambda = 0$, and thus $u = 0$, gives:

$$\theta_0 = \theta_i - \frac{\sqrt{\pi D}}{C_1} \text{ and thus } C_1 = \frac{-\sqrt{\pi D}}{\theta_0 - \theta_i}$$

Substitution of C_1 in Equation A2.6 gives

$$\theta(u) = \theta_i + (\theta_0 - \theta_i)\left(1 - \frac{2}{\sqrt{\pi}} \int_0^u \exp(-\beta^2)\,d\beta\right) \tag{A2.7}$$

The function $\frac{2}{\sqrt{\pi}} \int_0^u \exp(-\beta^2)\,d\beta$ is known as the error function erf (u), while $1 - \text{erf}(u)$ is the complementary error function, erfc (u).

Using this function and substituting $\lambda/2\sqrt{D}$ for u, one finally finds θ as a function of λ:

$$\theta(\lambda) = \theta_i + (\theta_0 - \theta_i)\,\text{erfc}\,(\lambda/2\sqrt{D}) \tag{A2.8}$$

or, by substituting $s/\sqrt{t}$ for λ, as a function of s and t:

$$\theta(s,t) = \theta_i + (\theta_0 - \theta_i)\,\text{erfc}\,(s/2\sqrt{Dt}) \tag{A2.9}$$

Appendix 3

Green and Ampt model. Vertical infiltration.

This appendix shows the derivation of Equation 5.46 in detail.
The logarithmic term of Equation 5.44, can be written as

$$\ln\left(\frac{s_f - h_f}{-h_f}\right) = \ln\left(1 - \frac{s_f}{h_f}\right) = \ln(1+y) \tag{5.45}$$

where $y = -\frac{s_f}{h_f}$. For $y^2 < 1$, $\ln(1+y)$ can be expanded as

$$\ln(1+y) = y - \frac{y^2}{2} + \frac{y^3}{3} - \frac{y^4}{4} + \dots \tag{A3.1}$$

Thus, Equation 5.44 becomes

$$t = \frac{\theta_t - \theta_i}{k_t}\left(s_f - s_f - \frac{s_f^2}{2h_f} - \frac{s_f^3}{3h_f^2} - \frac{s_f^4}{4h_f^3} - \dots\right) =$$

$$-\frac{\theta_t - \theta_i}{2k_t h_f} s_f^2 \left(1 + \frac{2s_f}{3h_f} + \frac{s_f^2}{2h_f^2} + \dots\right) \tag{A.3.2}$$

To find s_f as a function of t we first take the square root of Equation A3.2:

$$t^{1/2} = \left(-\frac{\theta_t - \theta_i}{2k_t h_f}\right)^{1/2} s_f \left(1 + \frac{2s_f}{3h_f} + \frac{s_f^2}{2h_f^2} + \dots\right)^{1/2} \tag{A3.3}$$

In general, if $P = a + bx + cx^2 + dx^3 + \dots$, then

$$P^{1/2} = a^{1/2}\left[1 + \frac{b}{2a}x + \left(\frac{c}{2a} - \frac{b^2}{8a^2}\right)x^2 + \dots\right] \tag{A3.4}$$

Using this in Equation A3.3 and rearranging terms leads to:

$$\left(\frac{-2k_t h_f}{\theta_t - \theta_i} t\right)^{1/2} = s_f + \frac{1}{3h_f} s_f^2 + \frac{7}{36\,h_f^2} s_f^3 + \dots \tag{A3.5}$$

Again, in general, if $Q = ax + bx^2 + cx^3 + \dots$, where $a \neq 0$, then

$$x = AQ + BQ^2 + CQ^3 + \dots \tag{A3.6}$$

where $A = \frac{1}{a}$, $B = -\frac{b}{a^3}$, $C = \frac{1}{a^5}(2b^2 - ac)$, etc.

This can be used to reverse Equation A3.5. The result will give s_f as a function of t in series form:

$$s_f = f_1 t^{1/2} + f_2 t + f_3 t^{3/2} + \dots \tag{5.46}$$

$$\text{where } f_1 = \left(\frac{-2k_t h_f}{\theta_t - \theta_i} \right)^{1/2} \tag{5.47}$$

$$f_2 = \frac{2k_t}{3(\theta_t - \theta_i)} \tag{5.48}$$

$$f_3 = \left(\frac{-k_t^3}{18(\theta_t - \theta_i)^3 h_f} \right)^{1/2} \tag{5.49}$$

This series is convergent for t smaller than a certain value, which depends on the value of the different coefficients f_1, f_2, f_3, etc.

Subject index

D

E

F

G

N

O

P

U

V

W